Linear Systems and Signals

The Oxford Series in Electrical and Computer Engineering

Adel S. Sedra, Series Editor

Linear Systems
and
Signals

B. P. Lathi

California State University, Sacramento

New York   Oxford
OXFORD UNIVERSITY PRESS

Oxford University Press

Oxford   New York
Athens Auckland Bangkok Bogotá Buenos Aires Cape Town
Chennai Dar es Salaam Delhi Florence Hong Kong Istanbul Karachi
Kolkata Kuala Lumpur Madrid Melbourne Mexico City Mumbai Nairobi
Paris São Paulo Shanghai Singapore Taipei Tokyo Toronto Warsaw

and associated companies in
Berlin   Ibadan

Published by Oxford University Press, Inc.
198 Madison Avenue, New York, New York, 10016
http://www.oup-usa.org

Oxford is a registered trademark of Oxford University Press

Library of Congress Cataloging-in-Publication Data available upon request

ISBN 0-19-515129-1

Printing number: 9 8 7 6 5 4

Printed in the United States of America
on acid-free paper

Contents

1 Introduction to Systems

PART II: TIME-DOMAIN ANALYSIS OF LTI SYSTEMS

2 Time-Domain Analysis: Continuous-Time Systems

PART IV: SIGNAL ANALYSIS

6 Continuous-Time Signal Analysis: The Fourier Series — 415

Preface

In this book I have attempted a comprehensive treatment of signals and linear systems at an introductory level. The book emphasizes the physical appreciation of concepts rather than mere mathematical manipulation of symbols. There is a tendency to treat engineering subjects, such as this, as a branch of applied mathematics. Such an approach ignores the physical meaning behind various derivations and deprives a student of intuitive understanding of the subject. Here I have used mathematics not so much to prove axiomatic theory as to enhance physical and intuitive understanding. Wherever possible, theoretical results are interpreted heuristically and are supported by carefully chosen examples and analogies.

Organization

For convenience, the book is divided into five parts:

1. Introduction.
2. Time-domain analysis of linear time-invariant (LTI) systems.
3. Frequency-domain (transform) analysis of LTI systems.
4. Signal Analysis.
5. State-space analysis of LTI systems.

The organization of the book permits great deal of flexibility in teaching the continuous-time and discrete-time concepts. The natural sequence of chapters is meant to integrate continuous-time and discrete-time analysis. It is also possible to use a sequential approach in which all the continuous-time analysis is covered first, followed by discrete-time analysis.

Notable Features

The notable features of the book include the following:

1. Intuitive and heuristic understanding of the concepts and physical meaning of mathematical results are emphasized throughout the book. Such an approach leads to deeper appreciation and easier comprehension of the concepts. Most reviewers of the book have noted the unusual clarity of presentation. As one reviewer put it, "I believe the strongest point of this book is the author's ability to describe very difficult material in a very clear and simple way."
2. Many students are handicapped by an inadequate background in basic material such as complex numbers, sinusoids, sketching and mathematically describing functions, Cramer's rule, partial fraction expansion, and matrix algebra. I have

added a chapter that addresses these basic and pervasive topics in electrical engineering. Response by student has been unanimously enthusiastic.

3. There are over 200 worked examples along with exercises (with answers) for students to test their understanding. There is also a large number of selected problems of varying difficulty at the end of each chapter. Many problems are provided with hints to steer a student in the proper direction.

4. For those who like to get students involved with computers, computer solutions of several examples are provided using Matlab, which is becoming a standard software package in an electrical engineering curriculum. The problem set also contains several computer problems. Use of computer examples or problems, however, is not essential for the use of this book.

5. The discrete-time and continuous-time systems may be treated in sequence, or they may be integrated by using a parallel approach.

6. The summary at the end of each chapter proves helpful to students in summing up essential developments in the chapter.

7. There are several historical notes to enhance student's interest in the subject. These facts introduce students to historical background that influenced the development of electrical engineering.

Suggestions for Using This Book

The book can be readily tailored for a variety of courses of 30 to 60 lecture hours or more. Most of the material in the first eight chapters can be covered in about 43 hours (including tests). To treat continuous and discrete-time systems by using an integrated (or parallel) approach, the appropriate sequence of chapters is 1, 2, 3, 4, 5, 6, 7, and 8. For a sequential approach, where the continuous-time analysis is followed by discrete-time analysis, the proper chapter sequence is 1, 2, 4, 3, 5, 6, 7, and 8.

Logically, the Fourier transform should precede the Laplace transform. I tried such an order in my previous books. In this book, however, I have reversed this order because experience shows that "Laplace before Fourier" goes better for most of the students. The order can be reversed, however, if the instructor so desires.

A Typical Course in Signals and Systems

A typical course in signals and systems may be divided into three roughly equal parts:

1. Time-domain analysis of LTI systems (Chapters 1, 2, 3).
2. Frequency-domain (Transform) analysis of LTI systems (Chapters 4, 5).
3. Signal analysis [Chapters 6, 7, 8, and 9 (optional)].

Each of the three parts can be covered in roughly 14 hours (including tests). I have used this sequence successfully, where the first eight chapters were covered with the following omissions:

Chapter 1: Sections 1.3-2, 1.3-3, 1.4, 1.5, and 1.6.

Chapter 2: Section 2.8. The results of Sec. 2.6-1 were stated but not proved. Students were assigned Sec. 2.7 for self study.

Chapter 3: Section 3.7. The results of section 3.6-1 were stated but not proved.

Chapter 4: Section 4.8.

Chapter 5: Sections 5.6 and 5.7.
Chapter 6: Section 6.3.
Chapter 7: Nothing omitted.
Chapter 8: Section 8.3.

Testing

Because the continuous-time and the discrete-time concepts are so similar, there is an understandable tendency on the part of students to confuse them. To minimize this difficulty, I find it helpful to test students more often. One test is given on each of these chapters: 2, 3, 4, 5, 6, and one test is given on Chapters 7 and 8 (six tests in all). The test for Chapter 6 (Fourier series) is shorter and is weighed at only 50% of the tests on the other tests. The homework is assigned the weight of one full test. Giving six tests may appear too demanding a task for an instructor. However, because of similarity of the continuous-time and discrete-time concepts, it is possible to make tests shorter (quiz) and to cover all the topics in pairs of chapters. For example, the time-domain analysis concepts in Chapters 2 (continuous-time systems) and 3 (discrete-time systems) are almost the same. Hence half the topics may be tested in Chapter 2, and the remaining half in Chapter 3. The case with Chapters 4 and 5 is similar. Thus, the six tests need not make more demands on the instructor than three or four tests given normally.

Credits

The photographs of Gauss (p. 5), Laplace (p. 263), Heaviside (p. 263), Fourier (p. 436), Michelson (p. 462) have been reprinted courtesy of the Smithsonian Institution. The photographs of Cardano (p. 5) and Gibbs (p. 462) have been reprinted courtesy of the Library of Congress.

The MATLAB examples and problems were prepared by Prof. Rory Cooper. In the second printing, Prof. O. P. Mandhana of University of Kentucky (Lexington) thoroughly revised or modified these programs partly by fixing the bugs, or adding some extra parts, or by a better usage of MATLAB syntax or programming. Some of the examples have been changed completely, and some examples are implemented using an alternate (simpler) method.

Acknowledgments

Several individuals have helped me in the preparation of this book. I am grateful for the helpful suggestions of the reviewers Professors. K.S. Arun (University of Illinois, Urbana), J.B. Cruz (University of California, Riverside), Kevin Donahue (Kentucky), Gary Ford (University of California, Davis), K.S.P. Kumar (Minnesota), Amy Reibman (Princeton), James Thorpe (Cornell), H. Valenzuela (Penn State), R. Yantorno (Temple). Prof. D. Arvind made many important and useful suggestions to improve the pedagogy of the book. Special thanks go to Paul Hinz, Professor Anne-Louise Radimsky, and Professor Isaac Ghansah (computer assistance), Professor Armand Seri (computer plots), and Joseph Coniglio (cartoons) for their generous help. The beautiful illustrations were prepared by Robert Griswold, James Roberts, Benjamin Park, and Mike Alie. The cover and graphic design is by Mike Alie. Typesetting is by Qin He. Other individuals who helped in a variety of ways include Tej Pandey, Shafique Rahman, Bharat Singh, Ernie Claussen,

Alok Rathie, Phong Vu Dao, Yuzon Hao, Janet Zhou, Arnel Guanlao, Dhananjay Rao, Vijay Pande, and E. Wanguhu. Finally I would like to mention the enormous but invisible sacrifices of my family, Rajani, Anjali, and Pandit in this endeavor.

B. P. Lathi

Matlab

Throughout this book, examples have been provided to familiarize the reader with computer tools for systems design and analysis using the powerful and versatile software package MATLAB. Much of the time and cost associated with the analysis and design of systems can be reduced by using computer software packages for simulation. Many corporations will no longer support the development systems without prior computer simulation, and numerical results which suggest a design will work. The examples and problems in this book will assist the reader in learning the value of computer packages for systems design and simulation.

MATLAB is the software package used throughout this book. MATLAB is a powerful package developed to perform matrix manipulations for system designers. MATLAB is easily expandable, and uses its own high level language. This makes developing sophisticated systems easier. In addition, MATLAB has been carefully written to yield numerically stable results to produce reliable simulations.

All the computer examples in this book are verified to work with the student edition of the MATLAB when used according to the instructions given in its manual. Make sure that \MATLAB\BIN is added in the DOS search path. MATLAB can be invoked by executing the command MATLAB. The MATLAB banner will appear after a moment with the prompt '>>'. MATLAB has a useful on-line help. To get help on a specific command, type HELP COMMAND NAME and then press the ENTER key. DIARY FILE is a command to record all the important keyboard inputs to a file and the resulting output of your MATLAB session to be written on the named file. MATLAB can be used interactively, or by writing functions (subroutines) often called M files because of the .M extension used for these files. Once familiar with the basics of MATLAB, it is easy to learn how to write functions and to use MATLAB's existing functions.

Rory Cooper
O. P. Mandhana

I

INTRODUCTION

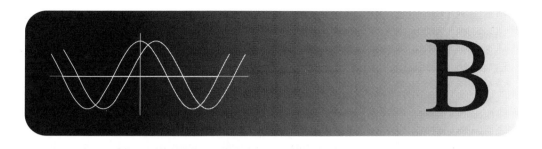

Background

The topics discussed in this chapter are not totally new to students taking this course. You have already studied many of these topics in earlier courses or are expected to know them from your previous training. Even so, this background material deserves a review because it is so pervasive in the area of signals and systems. Investing a little time in such a review will pay big dividends later. Furthermore, this material is useful not only for this course but also for several courses that follow. It will also be helpful as reference material in your future professional career.

B.1 COMPLEX NUMBERS

Complex numbers are an extension of ordinary numbers and are an integral part of the modern number system. Complex numbers, particularly **imaginary numbers**, sometimes seem mysterious and unreal. This feeling of unreality comes from their unfamiliarity and novelty rather than their supposed nonexistence! It was a blunder in mathematics to call these numbers "imaginary" because the term immediately prejudices perception. Had these numbers been called by some other name, they would have become demystified long ago, just as irrational numbers or negative numbers were. Many futile attempts were made to ascribe some physical meaning to imaginary numbers. This effort was needless. In mathematics we assign symbols and operations any meaning as long as there is an internal consistency. A healthier approach would have been to define a symbol i (with any name but "imaginary"), which has a property $i^2 = -1$. The history of mathematics is full of entities which were unfamiliar and held in abhorrence until familiarity made them acceptable. This fact will become clear from the following historical note.

B.1-1 A Historical Note

Among early people the number system consisted only of natural numbers (positive integers) that were needed to count the number of children, cattle, and quivers of arrows. They had no use for fractions. Whoever heard of two-and-one-half children or three-and-one-fourth cows!

With the advent of agriculture, people needed to measure continuously varying quantities, such as the length of a field, the weight of a quantity of butter, and so on. The number system therefore was extended to include fractions. The ancient Egyptians and Babylonians knew how to handle fractions, but **Pythagoras** discovered that some numbers (like the diagonal of a unit square) could not be expressed as a whole number or a fraction. Pythagoras, a number mystic who regarded numbers as the essence and principle of all things in the universe, was so appalled at his discovery that he swore his followers to secrecy and provided a death penalty for divulging this secret.[1] These numbers, however, were included in the number system by the time of Descartes, and they are now known as **irrational numbers**.

Until recently **negative numbers** were not a part of the number system. The idea of negative numbers must have appeared absurd to early man. However, the medieval Hindus had had a clear understanding of the significance of positive and negative numbers.[2,3] They were also the first to recognize the existence of absolute negative quantities.[4] The works of **Bhaskar** (1114-1185) on arithmetic (*Līlāvatī*) and algebra (*Bījaganit*) not only use the decimal system but also give rules for dealing with negative quantities. Bhaskar recognized that positive numbers have two square roots.[5] Much later in Europe, the banking system that arose in Florence and Venice during the late Renaissance (fifteenth century) is credited with developing a crude form of negative numbers. The seemingly absurd subtraction of 7 from 5 seemed reasonable when bankers began to allow their clients to draw seven gold ducats while their deposit stood at five. All that was necessary for this purpose was to write the difference, 2, on the debit side of a ledger.[6]

Thus the number system was once again broadened (generalized) to include negative numbers. The acceptance of negative numbers made it possible to solve equations such as $x + 5 = 0$, which had no solution before. Yet for equations such as $x^2 + 1 = 0$, leading to $x^2 = -1$, the solution could not be found in the real number system. It was therefore necessary to define a completely new kind of number with its square equal to -1. During the time of Descartes and Newton, imaginary (or complex) numbers came to be accepted as part of the number system, but they were still regarded as algebraic fiction. The Swiss mathematician **Leonhard Euler** introduced the notation i (for **imaginary**) around 1777 to represent $\sqrt{-1}$. (Electrical engineers use the notation j instead of i to avoid confusion with the notation i often used for electrical current.) Thus

$$j^2 = -1$$

and

$$\sqrt{-1} = \pm j$$

This allows us to determine the square root of any negative number. For example,

$$\sqrt{-4} = \sqrt{4} \times \sqrt{-1} = \pm 2j$$

When we include imaginary numbers in the number system, the resulting numbers are called **complex numbers**.

Origins of Complex Numbers

Ironically (and contrary to popular belief), it was not the solution of a quadratic equation, such as $x^2 + 1 = 0$, but a cubic equation with real roots that made

Gerolamo Cardano (left) and Karl Friedrich Gauss (right).

imaginary numbers plausible and acceptable to early mathematicians. They could dismiss $\sqrt{-1}$ as pure nonsense when it appeared as a solution to $x^2 + 1 = 0$ because this equation has no real solution. But in 1545, **Gerolamo Cardano** of Milan published *Ars Magna* (*The Great Art*), the most important algebraic work of the Renaissance. In this book he gave a method of solving a general cubic equation in which a root of a negative number appeared in an intermediate step. According to his method, the solution to a third-order equation†

$$x^3 + ax + b = 0$$

is given by

$$x = \sqrt[3]{-\frac{b}{2} + \sqrt{\frac{b^2}{4} + \frac{a^3}{27}}} + \sqrt[3]{-\frac{b}{2} - \sqrt{\frac{b^2}{4} + \frac{a^3}{27}}}$$

For example, to find a solution of $x^3 + 6x - 20 = 0$, we substitute $a = 6$, $b = -20$ in the above equation to obtain

$$x = \sqrt[3]{10 + \sqrt{108}} + \sqrt[3]{10 - \sqrt{108}} = \sqrt[3]{20.392} - \sqrt[3]{0.392} = 2$$

We can readily verify that 2 is indeed a solution of $x^3 + 6x - 20 = 0$. But when Cardano tried to solve the equation $x^3 - 15x - 4 = 0$ by this formula, his solution

---

†This equation is known as the *depressed cubic* equation. A general cubic equation
$$y^3 + py^2 + qy + r = 0$$
can always be reduced to a depressed cubic form by substituting $y = x - \frac{p}{3}$. Therefore any general cubic equation can be solved if we know the solution to the depressed cubic. The depressed cubic was independently solved, first by **Scipione del Ferro** (1465-1526) and then by **Niccolo Fontana** (1499-1557). The latter is better known in the history of mathematics as **Tartaglia** ("Stammerer"). Cardano learned the secret of the depressed cubic solution from Tartaglia. He then showed that by using the substitution $y = x - \frac{p}{3}$, a general cubic is reduced to a depressed cubic.

was

$$x = \sqrt[3]{2 + \sqrt{-121}} + \sqrt[3]{2 - \sqrt{-121}}$$

What was Cardano to make of this equation in the year 1545? In those days negative numbers were themselves suspect, and a square root of a negative number was doubly preposterous! Today we know that

$$(2 \pm j)^3 = 2 \pm j11 = 2 \pm \sqrt{-121}$$

Therefore Cardano's formula gives

$$x = (2 + j) + (2 - j) = 4$$

We can readily verify that $x = 4$ is indeed a solution of $x^3 - 15x - 4 = 0$. Cardano tried to explain halfheartedly the presence of $\sqrt{-121}$ but ultimately dismissed the whole enterprise as being "as subtle as it is useless." A generation later, however, **Raphael Bombelli** (1526-1573), after examining Cardano's results, proposed acceptance of imaginary numbers as a necessary vehicle that would transport the mathematician from the *real* cubic equation to its *real* solution. In other words, while we begin and end with real numbers, we seem compelled to move into an unfamiliar world of imaginaries to complete our journey. To mathematicians of the day, this seemed incredibly strange.[7] Yet they could not dismiss the idea of imaginary numbers so easily because it yielded the real solution of an equation. It took two more centuries for the full importance of complex numbers to become evident in the works of Euler, Gauss, and Cauchy. Still, Bombelli deserves credit for recognizing that such numbers have a role to play in algebra.[7]

In 1799, the German mathematician **Karl Friedrich Gauss**, at the ripe age of 22, proved the fundamental theorem of algebra, namely that every algebraic equation in one unknown has a root in the form of a complex number. He showed that every equation of the nth order has exactly n solutions (roots), no more and no less. Gauss was also one of the first to give a coherent account of complex numbers and to interpret them as points in a complex plane. It is he who introduced the term *complex numbers* and paved the way for general and systematic use of complex numbers. The number system was once again broadened or generalized to include imaginary numbers. Ordinary (or real) numbers became a special case of generalized (or complex) numbers.

The utility of complex numbers can be understood readily by an analogy with two neighboring countries X and Y as shown in Fig. B.1. If we want to travel from City a to City b (both in Country X), the shortest route is through Country Y, although the journey begins and ends in Country X. We may, if we desire, perform this journey by an alternate route that lies exclusively in X, but this alternate route is longer. In mathematics we have a similar situation with real numbers (Country X) and complex numbers (Country Y). All real-world problems must start with real numbers, and all the final results must also be in real numbers. But the derivation of results is considerably simplified by using complex numbers as an intermediary. It is also possible to solve all real-world problems by an alternate method, using real numbers exclusively, but this would increase the work needlessly.

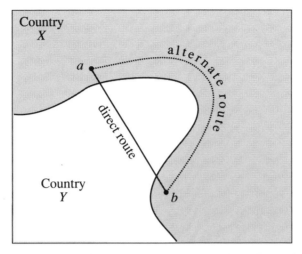

Fig. B.1 Use of complex numbers can reduce the work.

B.1-2 Algebra of Complex Numbers

A complex number (a, b) or $a + jb$ can be represented graphically by a point whose Cartesian coordinates are (a, b) in a complex plane (Fig. B.2). Let us denote this complex number by z so that

$$z = a + jb \tag{B.1}$$

The numbers a and b (the abscissa and the ordinate) of z are the **real part** and the **imaginary part**, respectively, of z. They are also expressed as

$$\text{Re } z = a$$

$$\text{Im } z = b$$

Note that in this plane all real numbers lie on the horizontal axis, and all imaginary numbers lie on the vertical axis.

Complex numbers may also be expressed in terms of polar coordinates. If (r, θ) are the polar coordinates of a point $z = a + jb$ (see Fig. B.2), then

$$a = r \cos \theta$$

$$b = r \sin \theta$$

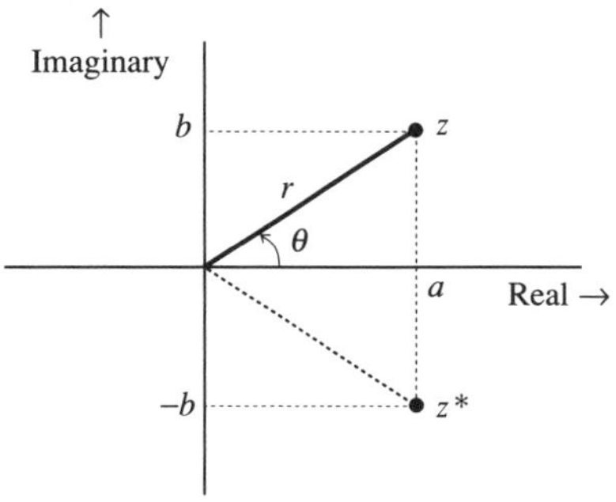

Fig. B.2 Representation of a number in the complex plane.

and

$$z = a + jb = r\cos\theta + jr\sin\theta$$
$$= r(\cos\theta + j\sin\theta) \tag{B.2}$$

The **Euler formula** states that

$$e^{j\theta} = \cos\theta + j\sin\theta$$

To prove Euler's formula, we expand $e^{j\theta}$, $\cos\theta$, and $\sin\theta$ using a Maclaurin series:

$$e^{j\theta} = 1 + j\theta + \frac{(j\theta)^2}{2!} + \frac{(j\theta)^3}{3!} + \frac{(j\theta)^4}{4!} + \frac{(j\theta)^5}{5!} + \frac{(j\theta)^6}{6!} + \cdots$$

$$= 1 + j\theta - \frac{\theta^2}{2!} - j\frac{\theta^3}{3!} + \frac{\theta^4}{4!} + j\frac{\theta^5}{5!} - \frac{\theta^6}{6!} - \cdots$$

$$\cos\theta = 1 - \frac{\theta^2}{2!} + \frac{\theta^4}{4!} - \frac{\theta^6}{6!} + \frac{\theta^8}{8!}\cdots$$

$$\sin\theta = \theta - \frac{\theta^3}{3!} + \frac{\theta^5}{5!} - \frac{\theta^7}{7!} + \cdots$$

From these results, we conclude that

$$e^{j\theta} = \cos\theta + j\sin\theta \tag{B.3}$$

Substituting (B.3) in (B.2) yields

$$z = a + jb$$
$$= re^{j\theta} \tag{B.4}$$

Thus a complex number can be expressed in Cartesian form $a + jb$ or polar form $re^{j\theta}$ with

$$a = r\cos\theta, \qquad\qquad b = r\sin\theta \tag{B.5}$$

and

$$r = \sqrt{a^2 + b^2}, \qquad\qquad \theta = \tan^{-1}\left(\frac{b}{a}\right) \tag{B.6}$$

Observe that r is the distance of the point z from the origin. For this reason, r is also called the **magnitude** (or **absolute value**) of z and is denoted by $|z|$. Similarly θ is called the angle of z and is denoted by $\angle z$. Therefore

$$|z| = r, \qquad\qquad \angle z = \theta$$

and

$$z = |z|e^{j\angle z} \tag{B.7}$$

Also

$$\frac{1}{z} = \frac{1}{re^{j\theta}} = \frac{1}{r}e^{-j\theta} = \frac{1}{|z|}e^{-j\angle z} \tag{B.8}$$

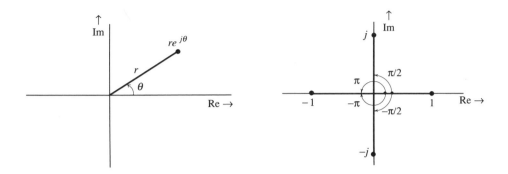

Fig. B.3 Understanding some useful identities in terms of $re^{j\theta}$.

Conjugate of a Complex Number

We define z^*, the **conjugate** of $z = a + jb$, as

$$z^* = a - jb = re^{-j\theta} \tag{B.9a}$$

$$= |z|e^{-j\angle z} \tag{B.9b}$$

The graphical representation of a number z and its conjugate z^* is shown in Fig. B.2. Observe that z^* is a mirror image of z about the horizontal axis. **To find the conjugate of any number, we need only to replace j by $-j$ in that number** (which is the same as changing the sign of its angle).

The sum of a complex number and its conjugate is a real number equal to twice the real part of the number:

$$z + z^* = (a + jb) + (a - jb) = 2a = 2\operatorname{Re} z \tag{B.10a}$$

The product of a complex number z and its conjugate is a real number $|z|^2$, the square of the magnitude of the number:

$$zz^* = (a + jb)(a - jb) = a^2 + b^2 = |z|^2 \tag{B.10b}$$

Understanding Some Useful Identities

In a complex plane, $re^{j\theta}$ represents a point at a distance r from the origin and at an angle θ with the horizontal axis as shown in Fig. B.3a. For example, the number -1 is at a unit distance from the origin and has an angle π or $-\pi$ (in fact, any odd multiple of $\pm\pi$), as seen from Fig. B.3b. Therefore,

$$1e^{\pm j\pi} = -1$$

In fact,

$$e^{\pm jn\pi} = -1 \qquad n \text{ odd integer} \tag{B.11}$$

The number 1, on the other hand, is also at a unit distance from the origin, but has an angle 2π (in fact, $\pm 2n\pi$ for any integral value of n). Therefore,

$$e^{\pm j2n\pi} = 1 \qquad n \text{ integer} \tag{B.12}$$

The number j is at unit distance from the origin and its angle is $\pi/2$ (see Fig. B.3b). Therefore

$$e^{j\pi/2} = j$$

Similarly,

$$e^{-j\pi/2} = -j$$

Thus

$$e^{\pm j\pi/2} = \pm j \qquad \text{(B.13a)}$$

In fact,

$$e^{\pm jn\pi/2} = \pm j \qquad n = 1, 5, 9, 13, \cdots \qquad \text{(B.13b)}$$

and

$$e^{\pm jn\pi/2} = \mp j \qquad n = 3, 7, 11, 15, \cdots \qquad \text{(B.13c)}$$

These results are summarized in Table B.1.

TABLE B.1

r	θ	$re^{j\theta}$	
1	0	$e^{j0} = 1$	
1	$\pm\pi$	$e^{\pm j\pi} = -1$	
1	$\pm n\pi$	$e^{\pm jn\pi} = -1$	n odd integer
1	$\pm 2\pi$	$e^{\pm j2\pi} = 1$	
1	$\pm 2n\pi$	$e^{\pm j2n\pi} = 1$	n integer
1	$\pm\pi/2$	$e^{\pm j\pi/2} = \pm j$	
1	$\pm n\pi/2$	$e^{\pm jn\pi/2} = \pm j$	$n = 1, 5, 9, 13, \ldots$
1	$\pm n\pi/2$	$e^{\pm jn\pi/2} = \mp j$	$n = 3, 7, 11, 15, \ldots$

This discussion shows the usefulness of the graphic picture of $re^{j\theta}$. This picture is also helpful in several other applications. For example, to determine the limit of $e^{(\alpha+j\omega)t}$ as $t \to \infty$, we note that

$$e^{(\alpha+j\omega)t} = e^{\alpha t} e^{j\omega t}$$

Now the magnitude of $e^{j\omega t}$ is unity regardless of the value of ω or t because $e^{j\omega t} = re^{j\theta}$ with $r = 1$. Therefore, $e^{\alpha t}$ determines the behavior of $e^{(\alpha+j\omega)t}$ as $t \to \infty$ and

$$\lim_{t\to\infty} e^{(\alpha+j\omega)t} = \lim_{t\to\infty} e^{\alpha t} e^{j\omega t} = \begin{cases} 0 & \alpha < 0 \\ \infty & \alpha > 0 \end{cases} \qquad \text{(B.14)}$$

In future discussions you will find it very useful to remember $re^{j\theta}$ as a number at a distance r from the origin and at an angle θ with the horizontal axis of the complex plane.

A Warning about Using Electronic Calculators in Computing Angles

From Cartesian form $a + jb$ we can readily compute polar form $re^{j\theta}$ [see Eq. (B.6)]. Electronic calculators provide ready conversion of rectangular into polar and vice versa. However, if a calculator is used to compute an angle of a complex number using an inverse trigonometric function $\theta = \tan^{-1}(b/a)$, proper attention must be paid to the quadrant in which the number is located. For instance, θ corresponding to the number $-2 - j3$ is $\tan^{-1}(\frac{-3}{-2})$. This is not the same as $\tan^{-1}(\frac{3}{2})$. The former is $-123.7°$, whereas the latter is $56.3°$. An electronic calculator cannot make this distinction and can give a correct answer only for angles in the first and fourth quadrant. It will read $\tan^{-1}(\frac{-3}{-2})$ as $\tan^{-1}(\frac{3}{2})$, which is clearly wrong. In computing inverse trigonometric functions if the angle is in the second or third quadrant, the answer of the calculator is off by $180°$. The correct answer is obtained by adding or subtracting $180°$ to the value found with the calculator (either adding or subtracting yields the correct answer). For this reason it is advisable to draw the point in the complex plane and determine the quadrant in which the point lies. This issue will be clarified by the following examples.

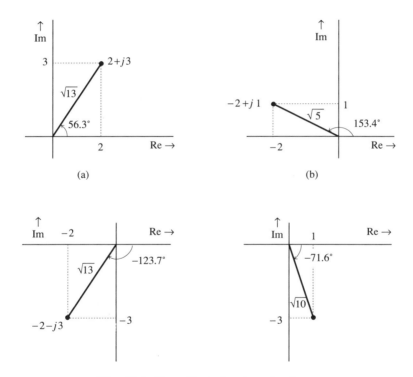

Fig. B.4 From Cartesian to polar form.

■ **Example B.1**
   Express the following numbers in polar form:
   (a) $2 + j3$ **(b)** $-2 + j1$ **(c)** $-2 - j3$ **(d)** $1 - j3$
   (a)
$$|z| = \sqrt{2^2 + 3^2} = \sqrt{13} \qquad \angle z = \tan^{-1}\left(\frac{3}{2}\right) = 56.3°$$
In this case the number is in the first quadrant, and a calculator will give the correct value of $56.3°$. Therefore (see Fig. B.4a)
$$2 + j3 = \sqrt{13}\, e^{j56.3°}$$

(b)
$$|z| = \sqrt{(-2)^2 + 1^2} = \sqrt{5} \qquad \angle z = \tan^{-1}\left(\frac{1}{-2}\right) = 153.4°$$

In this case the angle is in the second quadrant (see Fig. B.4b), and therefore the answer found with the calculator $(\tan^{-1}(\frac{1}{-2}) = -26.6°)$ is off by 180°. The correct answer is $(-26.6 \pm 180)° = 153.4°$ or $-206.6°$. Both values are correct because they represent the same angle. As a matter of convenience we choose an angle whose numerical value is less than 180°, which in this case is 153.4°. Therefore

$$-2 + j1 = \sqrt{5}e^{j153.4°}$$

(c)
$$|z| = \sqrt{(-2)^2 + (-3)^2} = \sqrt{13} \qquad \angle z = \tan^{-1}(\frac{-3}{-2}) = -123.7°$$

In this case the angle is in the third quadrant (see Fig. B.4c), and therefore the answer found with the calculator $(\tan^{-1}(\frac{-3}{-2}) = 56.3°)$ is off by 180°. The correct answer is $(56.3 \pm 180)° = 236.3°$ or $-123.7°$. As a matter of convenience we choose the latter and (see Fig. B.4c)

$$-2 - j3 = \sqrt{13}e^{-j123.7°}$$

(d)
$$|z| = \sqrt{1^2 + (-3)^2} = \sqrt{10} \qquad \angle z = \tan^{-1}(\frac{-3}{1}) = -71.6°$$

In this case the angle is in the fourth quadrant (see Fig. B.4d), and therefore the answer found with the calculator $(\tan^{-1}(\frac{-3}{1}) = -71.6°)$ is correct (see Fig. B.4d).

$$1 - j3 = \sqrt{10}e^{-j71.6°} \quad \blacksquare$$

⊙ **Computer Example CB.1**
Express the following numbers in polar form: **(a)** $2 + j3$ **(b)** $1 - j3$

(a)
```
z=2+i*3
magz=abs(z)                    % magz = |z|
argz_in_radian=angle(z)        % ∠z (in radians)
argz_in_deg=angle(z)*(180/pi)  % ∠z (in degrees)
disp('strike any key for part (b)')
pause
clc
```

(b)
```
z=1-3*i
magz=abs(z)
argz_in_radian=angle(z)
argz_in_deg=angle(z)*(180/pi)   ⊙
```

■ **Example B.2**
Represent the following numbers in the complex plane and express them in Cartesian form: **(a)** $2e^{j\pi/3}$ **(b)** $4e^{-j3\pi/4}$ **(c)** $2e^{j\pi/2}$ **(d)** $3e^{-j3\pi}$ **(e)** $2e^{j4\pi}$ **(f)** $2e^{-j4\pi}$.

(a) $2e^{j\pi/3} = 2\left(\cos\frac{\pi}{3} + j\sin\frac{\pi}{3}\right) = 1 + j\sqrt{3}$ (see Fig. B.5a)

(b) $4e^{-j3\pi/4} = 4\left(\cos\frac{3\pi}{4} - j\sin\frac{3\pi}{4}\right) = -2\sqrt{2} - j2\sqrt{2}$ (see Fig. B.5b)

(c) $2e^{j\pi/2} = 2\left(\cos\frac{\pi}{2} + j\sin\frac{\pi}{2}\right) = 2(0 + j1) = j2$ (see Fig. B.5c)

(d) $3e^{-j3\pi} = 3(\cos 3\pi - j\sin 3\pi) = 3(-1 + j0) = -3$ (see Fig. B.5d)

(e) $2e^{j4\pi} = 2(\cos 4\pi + j\sin 4\pi) = 2(1 + j0) = 2$ (see Fig. B.5e)

(f) $2e^{-j4\pi} = 2(\cos 4\pi - j\sin 4\pi) = 2(1 - j0) = 2$ (see Fig. B.5f) ■

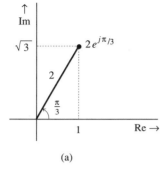

(a)

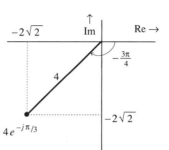

(b)

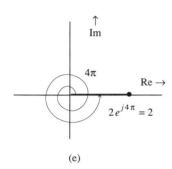

(c)

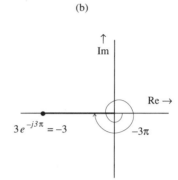

(d)

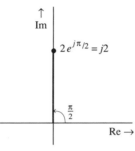

(e)

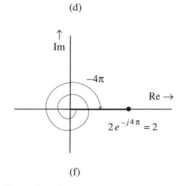

(f)

Fig. B.5 From polar to Cartesian form.

⊙ **Computer Example CB.2**

Represent the following numbers in the complex plane and express them in Cartesian form: **(a)** $2e^{-j\frac{\pi}{3}}$ **(b)** $4e^{-j\frac{3\pi}{4}}$

(a)
```
z=2*exp(-i*pi/3)
```
(b)
```
z=4*exp(-i*3*pi/4)
```
⊙

Arithmetical Operations, Powers, and Roots of Complex Numbers

To perform addition and subtraction, complex numbers should be expressed in Cartesian form. Thus, if

$$z_1 = 3 + j4 = 5e^{j53.1°}$$

$$z_2 = 2 + j3 = \sqrt{13}e^{j56.3°}$$

then

$$z_1 + z_2 = (3 + j4) + (2 + j3) = 5 + j7$$

If z_1 and z_2 are given in polar form, we would need to convert them into Cartesian form for the purpose of adding (or subtracting). Multiplication and division, however, can be carried out in either Cartesian or polar form, although the latter proves to be much more convenient. This is because if z_1 and z_2 are expressed in polar form as

$$z_1 = r_1 e^{j\theta_1} \quad \text{and} \quad z_2 = r_2 e^{j\theta_2}$$

then

$$z_1 z_2 = \left(r_1 e^{j\theta_1}\right)\left(r_2 e^{j\theta_2}\right) = r_1 r_2 e^{j(\theta_1 + \theta_2)} \tag{B.15a}$$

and

$$\frac{z_1}{z_2} = \frac{r_1 e^{j\theta_1}}{r_2 e^{j\theta_2}} = \frac{r_1}{r_2} e^{j(\theta_1 - \theta_2)} \tag{B.15b}$$

Moreover,

$$z^n = \left(re^{j\theta}\right)^n = r^n e^{jn\theta} \tag{B.15c}$$

and

$$z^{1/n} = \left(re^{j\theta}\right)^{1/n} = r^{1/n} e^{j\theta/n} \tag{B.15d}$$

This shows that the operations of multiplication, division, powers, and roots can be carried out with remarkable ease when the numbers are in polar form.

■ **Example B.3**

Determine $z_1 z_2$ and z_1/z_2 for the numbers

$$z_1 = 3 + j4 = 5e^{j53.1°}$$

$$z_2 = 2 + j3 = \sqrt{13}e^{j56.3°}$$

We shall solve this problem in both polar and Cartesian forms.

Multiplication: Cartesian Form

$$z_1 z_2 = (3 + j4)(2 + j3) = (6 - 12) + j(8 + 9) = -6 + j17$$

Multiplication: Polar Form

$$z_1 z_2 = \left(5e^{j53.1°}\right)\left(\sqrt{13}e^{j56.3°}\right) = 5\sqrt{13}e^{j109.4°}$$

Division: Cartesian Form

$$\frac{z_1}{z_2} = \frac{3 + j4}{2 + j3}$$

In order to eliminate the complex number in the denominator, we multiply both the numerator and the denominator of the right-hand side by $2 - j3$, the denominator's conjugate. This yields

$$\frac{z_1}{z_2} = \frac{(3+j4)(2-j3)}{(2+j3)(2-j3)} = \frac{18-j1}{2^2+3^2} = \frac{18-j1}{13} = \frac{18}{13} - j\frac{1}{13}$$

Division: Polar Form

$$\frac{z_1}{z_2} = \frac{5e^{j53.1°}}{\sqrt{13}e^{j56.3°}} = \frac{5}{\sqrt{13}}e^{j(53.1°-56.3°)} = \frac{5}{\sqrt{13}}e^{-j3.2°} \quad \blacksquare$$

It is clear from this example that multiplication and division are easier to accom plish in polar form than in Cartesian form.

⊙ **Computer Example CB.3**

Determine $z_1 z_2$ and z_1/z_2 if $z_1 = 3 + j4$ and $z_1 = 2 + j3$

Multiplication and Division: Cartesian Form

```
z1=3+4*i;
z2=2+3*i;
disp('multiplication and division using cartesian form of z1 and z2')
z1z2=z1*z2
z1divz2=z1/z2
disp('multiplication and division using polar Form')
z1z2 = (abs(z1)*exp(i*angle(z1))*abs(z2)*exp(i*angle(z2)))
z1divz2 = (abs(z1)*exp(i*angle(z1)))/(abs(z2)*exp(i*angle(z2)))   ⊙
```

■ **Example B.4**

For $z_1 = 2e^{j\pi/4}$ and $z_2 = 8e^{j\pi/3}$, find **(a)** $2z_1 - z_2$ **(b)** $\frac{1}{z_1}$ **(c)** $\frac{z_1}{z_2^2}$ **(d)** $\sqrt[3]{z_2}$

(a) Since subtraction cannot be performed directly in polar form, we convert z_1 and z_2 to Cartesian form:

$$z_1 = 2e^{j\pi/4} = 2\left(\cos\tfrac{\pi}{4} + j\sin\tfrac{\pi}{4}\right) = \sqrt{2} + j\sqrt{2}$$

$$z_2 = 8e^{j\pi/3} = 8\left(\cos\tfrac{\pi}{3} + j\sin\tfrac{\pi}{3}\right) = 4 + j4\sqrt{3}$$

Therefore

$$2z_1 - z_2 = 2(\sqrt{2} + j\sqrt{2}) - (4 + j4\sqrt{3})$$
$$= (2\sqrt{2} - 4) + j(2\sqrt{2} - 4\sqrt{3})$$
$$= -1.17 - j4.1$$

(b)

$$\frac{1}{z_1} = \frac{1}{2e^{j\pi/4}} = \frac{1}{2}e^{-j\pi/4}$$

(c)

$$\frac{z_1}{z_2^2} = \frac{2e^{j\pi/4}}{(8e^{j\pi/3})^2} = \frac{2e^{j\pi/4}}{64e^{j2\pi/3}} = \frac{1}{32}e^{j(\frac{\pi}{4}-\frac{2\pi}{3})} = \frac{1}{32}e^{-j\frac{5\pi}{12}}$$

(d)

$$\sqrt[3]{z_2} = z_2^{1/3} = \left(8e^{j\pi/3}\right)^{\frac{1}{3}} = 8^{\frac{1}{3}}\left(e^{j\pi/3}\right)^{1/3} = 2e^{j\pi/9} \quad \blacksquare$$

■ **Example B.5**

Consider $F(\omega)$, a complex function of a real variable ω:

$$F(\omega) = \frac{2 + j\omega}{3 + j4\omega} \tag{B.16a}$$

(a) Express $F(\omega)$ in Cartesian form, and find its real and imaginary parts. (b) Express $F(\omega)$ in polar form, and find its magnitude $|F(\omega)|$ and angle $\angle F(\omega)$.

(a) To obtain the real and imaginary parts of $F(\omega)$, we must eliminate imaginary terms in the denominator of $F(\omega)$. This is readily done by multiplying both the numerator and denominator of $F(\omega)$ by $3 - j4\omega$, the conjugate of the denominator $3 + j4\omega$ so that

$$F(\omega) = \frac{(2 + j\omega)(3 - j4\omega)}{(3 + j4\omega)(3 - j4\omega)} = \frac{(6 + 4\omega^2) - j5\omega}{9 + 16\omega^2} = \frac{6 + 4\omega^2}{9 + 16\omega^2} - j\frac{5\omega}{9 + \omega^2} \tag{B.16b}$$

This is the Cartesian form of $F(\omega)$. Clearly the real and imaginary parts $F_r(\omega)$ and $F_i(\omega)$ are given by

$$F_r(\omega) = \frac{6 + 4\omega^2}{9 + 16\omega^2}, \qquad F_i(\omega) = \frac{-5\omega}{9 + 16\omega^2}$$

(b)

$$F(\omega) = \frac{2 + j\omega}{3 + j4\omega} = \frac{\sqrt{4 + \omega^2}\, e^{j\tan^{-1}\left(\frac{\omega}{2}\right)}}{\sqrt{9 + 16\omega^2}\, e^{j\tan^{-1}\left(\frac{4\omega}{3}\right)}}$$

$$= \sqrt{\frac{4 + \omega^2}{9 + 16\omega^2}}\, e^{j\left[\tan^{-1}\left(\frac{\omega}{2}\right) - \tan^{-1}\left(\frac{4\omega}{3}\right)\right]} \tag{B.16c}$$

This is the polar representation of $F(\omega)$. Observe that

$$|F(\omega)| = \sqrt{\frac{4 + \omega^2}{9 + 16\omega^2}}, \qquad \angle F(\omega) = \tan^{-1}\left(\frac{\omega}{2}\right) - \tan^{-1}\left(\frac{4\omega}{3}\right) \tag{B.17}$$

■

B.2 SINUSOIDS

Consider the sinusoid

$$f(t) = C\cos\left(2\pi\mathcal{F}_0 t + \theta\right) \tag{B.18}$$

We know that

$$\cos\varphi = \cos(\varphi + 2n\pi) \qquad n = 0, \pm1, \pm2, \pm3, \cdots$$

Therefore, $\cos\varphi$ repeats itself for every change of 2π in the angle $\angle\varphi$. For the sinusoid in Eq. (B.18), the angle $2\pi\mathcal{F}_0 t + \theta$ changes by 2π when t changes by

$1/\mathcal{F}_0$. Clearly this sinusoid repeats every $1/\mathcal{F}_0$ seconds. As a result, there are $\mathcal{F}_0$ repetitions per second. This is the **frequency** of the sinusoid, and the repetition interval T_0 given by

$$T_0 = \frac{1}{\mathcal{F}_0} \tag{B.19}$$

is the **period**. For the sinusoid in Eq. (B.18), C is the **amplitude**, $\mathcal{F}_0$ is the **frequency** (in **Hertz**), and θ is the phase. Let us consider two special cases of this sinusoid when $\theta = 0$ and $\theta = -\pi/2$ as follows:

(a) $f(t) = C \cos 2\pi \mathcal{F}_0 t$ ($\theta = 0$)

(b) $f(t) = C \cos \left(2\pi \mathcal{F}_0 t - \frac{\pi}{2} \right) = C \sin 2\pi \mathcal{F}_0 t$ ($\theta = -\pi/2$)

The angle or phase can be expressed in units of degrees or radians. Although the radian is the proper unit, in this book we shall often use the degree unit because students generally have a better feel for the relative magnitudes of angles when expressed in degrees rather than in radians. For example, we relate better to angle $24°$ than to 0.419 radians. Remember, however, when in doubt, to use the radian unit and, above all, be consistent. In other words, in a given problem or an expression do not mix the two units.

It is convenient to use the variable ω_0 (*radian frequency*) to express $2\pi \mathcal{F}_0$:

$$\omega_0 = 2\pi \mathcal{F}_0 \tag{B.20}$$

With this notation, the sinusoid in Eq. (B.18) can be expressed as

$$f(t) = C \cos (\omega_0 t + \theta)$$

in which the period T_0 is given by [see Eqs. (B.19) and (B.20)]

$$T_0 = \frac{1}{\omega_0/2\pi} = \frac{2\pi}{\omega_0} \tag{B.21a}$$

and

$$\omega_0 = \frac{2\pi}{T_0} \tag{B.21b}$$

In future discussions, we shall often refer to ω_0 as the frequency of the signal $\cos(\omega_0 t + \theta)$, but it should be clearly understood that the frequency of this sinusoid is $\mathcal{F}_0$ Hz ($\mathcal{F}_0 = \omega_0/2\pi$), and ω_0 is actually the **radian frequency**.

The signals $C \cos \omega_0 t$ and $C \sin \omega_0 t$ are shown in Figs. B.6a and B.6b respectively. A general sinusoid $C \cos(\omega_0 t + \theta)$ can be readily sketched by shifting the signal $C \cos \omega_0 t$ in Fig. B.6a by the appropriate amount. Consider, for example,

$$f(t) = C \cos (\omega_0 t - 60°)$$

This signal can be obtained by shifting (delaying) the signal $C \cos \omega_0 t$ (Fig. B.6a) to the right by a phase (angle) of $60°$. We know that a sinusoid undergoes a $360°$ change of phase (or angle) in one cycle. A quarter-cycle segment corresponds to a $90°$ change of angle. Therefore, an angle of $60°$ corresponds to two-thirds of a

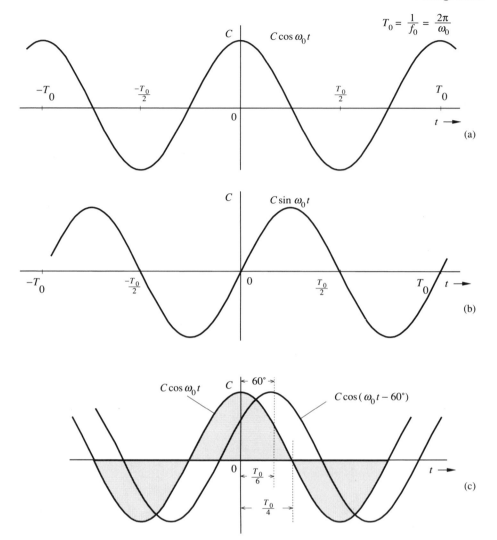

Fig. B.6 Sketching a sinusoid.

quarter-cycle segment. We therefore shift (delay) the signal in Fig. B.6a by two-thirds of a quarter-cycle segment to obtain $C\cos(\omega_0 t - 60°)$, as shown in Fig. B.6c.

Observe that if we delay $C\cos\omega_0 t$ in Fig. B.6a by a quarter-cycle (angle of 90° or $\pi/2$ radians), we obtain the signal $C\sin\omega_0 t$ shown in Fig. B.6b. This verifies the well-known trigonometric identity

$$C\cos\left(\omega_0 t - \tfrac{\pi}{2}\right) = C\sin\omega_0 t \tag{B.22a}$$

Alternatively, if we advance $C\sin\omega_0 t$ by a quarter-cycle, we obtain $C\cos\omega_0 t$. Therefore

$$C\sin\left(\omega_0 t + \tfrac{\pi}{2}\right) = C\cos\omega_0 t \tag{B.22b}$$

This observation means $\sin\omega_0 t$ lags $\cos\omega_0 t$ by 90° ($\pi/2$ radians), or $\cos\omega_0 t$ leads $\sin\omega_0 t$ by 90°.

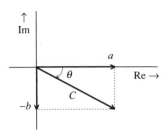

Fig. B.7 Phasor addition of sinusoids.

B.2-1   Addition of Sinusoids

Two sinusoids having the same frequency but different phases add to form a single sinusoid of the same frequency. This fact is readily seen from the well-known trigonometric identity:

$$C \cos{(\omega_0 t + \theta)} = C \cos\theta \, \cos\omega_0 t - C \sin\theta \, \sin\omega_0 t$$

$$= a \cos\omega_0 t + b \sin\omega_0 t \tag{B.23a}$$

in which

$$a = C \cos\theta, \qquad\qquad b = -C \sin\theta$$

Therefore

$$C = \sqrt{a^2 + b^2} \tag{B.23b}$$

$$\theta = \tan^{-1}\left(\frac{-b}{a}\right) \tag{B.23c}$$

Caution:

Do not confuse θ in Eq. (B.23c) with the angle θ of complex number $a + jb$, which is $\tan^{-1}(\frac{b}{a})$ as seen from Eq. (B.6).

To summarize,

$$a \cos\omega_0 t + b \sin\omega_0 t = C \cos{(\omega_0 t + \theta)}$$

in which C and θ are given by Eqs. (B.23b) and (B.23c) respectively.

The process of adding two sinusoids with the same frequency can be clarified by using **phasors** to represent sinusoids. We represent the sinusoid $C \cos(\omega_0 t + \theta)$ by a phasor of length C, and at an angle θ with the horizontal axis. Clearly the sinusoid $a \cos\omega_0 t$ is represented by a horizontal phasor of length a ($\theta = 0$), while $b \sin\omega_0 t = b \cos(\omega_0 t - \frac{\pi}{2})$ is represented by a vertical phasor of length b at an angle $-\pi/2$ with the horizontal (Fig. B.7). Adding these two phasors results in a phasor of length C at an angle θ, as shown in Fig. B.7. From this figure, we verify the values of C and θ found in Eqs. (B.23b) and (B.23c) respectively.

Proper care should be exercised in computing θ. Recall that $\tan^{-1}(\frac{-b}{a}) \neq$ $\tan^{-1}(\frac{b}{-a})$. Similarly, $\tan^{-1}(\frac{-b}{-a}) \neq \tan^{-1}(\frac{b}{a})$. Electronic calculators cannot make this distinction. When calculating such an angle, it is advisable to note the quadrant where the angle lies and not to rely exclusively on an electronic calculator. The following examples clarify this point.

■ **Example B.6**

In the following cases, express $f(t)$ as a single sinusoid.

(a) $f(t) = \cos \omega_0 t - \sqrt{3} \sin \omega_0 t$

(b) $f(t) = -3 \cos \omega_0 t + 4 \sin \omega_0 t$

(c) $f(t) = \cos \omega_0 t + \sin \omega_0 t$

(d) $f(t) = -2 \cos \omega_0 t - 3 \sin \omega_0 t$

(a) In this case, $a = 1, b = -\sqrt{3}$ and from Eqs. (B.23)

$$C = \sqrt{1^2 + (\sqrt{3})^2} = 2$$

$$\theta = \tan^{-1}\left(\tfrac{\sqrt{3}}{1}\right) = 60°$$

Therefore

$$f(t) = 2 \cos(\omega_0 t + 60°)$$

We can verify this result by drawing phasors corresponding to the two sinusoids. The sinusoid $\cos \omega_0 t$ is represented by a phasor of unit length at a zero angle with the horizontal. The phasor $\sin \omega_0 t$ is represented by a unit phasor at an angle of $-90°$ with the horizontal. Therefore, $-\sqrt{3} \sin \omega_0 t$ is represented by a phasor of length $\sqrt{3}$ at $90°$ with the horizontal, as shown in Fig. B.8a. The two phasors added yield a phasor of length 2 at $60°$ with the horizontal (also shown in Fig. B.8a). Therefore

$$f(t) = 2 \cos(\omega_0 t + 60°)$$

Observe that a phase shift of $\pm\pi$ amounts to multiplication by -1. Therefore, $f(t)$ can also be expressed alternatively as

$$f(t) = -2 \cos(\omega_0 t + 60° \pm 180°)$$
$$= -2 \cos(\omega_0 t - 120°)$$
$$= -2 \cos(\omega_0 t + 240°)$$

In practice, an expression with an angle whose numerical value is less than $180°$ is preferred.

(b) In this case, $a = -3, b = 4$ and from Eqs. (B.23)

$$C = \sqrt{(-3)^2 + 4^2} = 5$$

$$\theta = \tan^{-1}\left(\tfrac{-4}{-3}\right) = -126.9°$$

Observe that

$$\tan^{-1}\left(\tfrac{-4}{-3}\right) \neq \tan^{-1}\left(\tfrac{4}{3}\right) = 53.1°$$

Therefore

$$f(t) = 5 \cos(\omega_0 t - 126.9°)$$

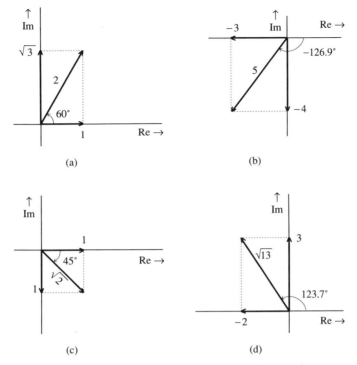

Fig. B.8 Phasor addition of sinusoids in Example B.6.

This result is readily verified from the phasor diagram in Fig. B.8b.

(c) In this case, $a = b = 1$ and from Eqs. (B.23)

$$C = \sqrt{1^2 + 1^2} = \sqrt{2}$$
$$\theta = \tan^{-1}\left(\frac{-1}{1}\right) = -45°$$

Therefore

$$f(t) = \sqrt{2}\cos\left(\omega_0 t - 45°\right)$$

This result is verified from the phasor diagram in Fig. B.8c.

(d) In this case, $a = -2, b = -3$ and from Eqs. (B.23)

$$C = \sqrt{(-2)^2 + (-3)^2} = \sqrt{13}$$
$$\theta = \tan^{-1}\left(\frac{3}{-2}\right) = 123.7°$$

Observe that

$$\tan^{-1}\left(\frac{3}{-2}\right) \neq \tan^{-1}\left(\frac{-3}{2}\right) = -56.3°$$

Therefore

$$f(t) = \sqrt{13}\cos\left(\omega_0 t + 123.7°\right)$$

This result is verified from the phasor diagram in Fig. B.8d. ■

⊙ **Computer Example CB.4**

Express $f(t) = -3\cos \omega_0 t + 4\sin \omega_0 t$ as a single sinusoid.

```
a=-3; b=4;
C=norm([a b])
disp('angle in radian is')
theta = atan((-b)/a)
disp('angle in degrees is')
theta = atan((-b)/a)*(180/pi)-180   % Keep track of the quadrant.
disp('Hence f(t) = -3 cosw0*t+4 sinw0*t is 5 cos(w0*t-126.9)')   ⊙
```

We can also perform the reverse operation of expressing

$$f(t) = C\cos(\omega_0 t + \theta)$$

in terms of $\cos \omega_0 t$, and $\sin \omega_0 t$ using the trigonometric identity

$$C\cos(\omega_0 t + \theta) = C\cos\theta \cos\omega_0 t - C\sin\theta \sin\omega_0 t$$

For example,

$$10\cos(\omega_0 t - 60°) = 5\cos\omega_0 t + 5\sqrt{3}\sin\omega_0 t$$

B.2-2 Sinusoids in Terms of Exponentials: Euler's Formula

Sinusoids can be expressed in terms of exponentials using Euler's formula [see Eq. (B.3)]:

$$\cos\varphi = \frac{1}{2}\left(e^{j\varphi} + e^{-j\varphi}\right) \tag{B.24a}$$

$$\sin\varphi = \frac{1}{2j}\left(e^{j\varphi} - e^{-j\varphi}\right) \tag{B.24b}$$

Inversion of these equations yields

$$e^{j\varphi} = \cos\varphi + j\sin\varphi \tag{B.25a}$$

$$e^{-j\varphi} = \cos\varphi - j\sin\varphi \tag{B.25b}$$

From Eq. (B.24a), it follows that

$$C\cos(\omega t + \theta) = \frac{C}{2}\left[e^{j(\omega t + \theta)} + e^{-j(\omega t + \theta)}\right]$$

$$= \left(\frac{C}{2}e^{j\theta}\right)e^{j\omega t} + \left(\frac{C}{2}e^{-j\theta}\right)e^{-j\omega t}$$

$$= De^{j\omega t} + D^* e^{-j\omega t} \tag{B.26}$$

in which

$$D = \frac{C}{2}e^{j\theta} \quad \text{and} \quad D^* = \frac{C}{2}e^{-j\theta} \tag{B.27}$$

It is clear that a sinusoid of frequency ω can be expressed as a sum of two exponentials as shown in Eq. (B.26).

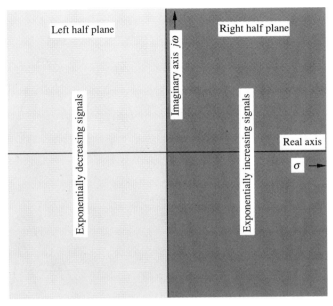

Fig. B.9 Complex frequency plane.

B.2-3 Generalized Sinusoids: Complex Frequency

Conventional sinusoids are oscillating signals with constant amplitudes. We can generalize the notion of a sinusoid to include sinusoids with exponentially varying amplitudes. This generalization is conveniently achieved by using the concept of **complex frequency**. The conventional sinusoid $\cos(\omega t + \theta)$ with a radian frequency ω can be expressed as a sum of exponentials $e^{j\omega t}$ and $e^{-j\omega t}$ as shown in Eq. (B.26). Observe that $j\omega$ and $-j\omega$, the indices of these exponentials, are imaginary and conjugates. Generalization of these indices from imaginary to complex numbers results in a sinusoid with exponentially varying amplitude. Multiplication of both sides of Eq. (B.26) with $e^{\sigma t}$ yields

$$
\begin{aligned}
Ce^{\sigma t}\cos(\omega t + \theta) &= De^{(\sigma + j\omega)t} + D^* e^{(\sigma - j\omega)t}\\
&= De^{st} + D^* e^{s^* t}
\end{aligned}
\tag{B.28a}
$$

in which

$$
s = \sigma + j\omega \qquad \text{and} \qquad s^* = \sigma - j\omega
\tag{B.28b}
$$

This result shows that an exponentially varying sinusoid $Ce^{\sigma t}\cos(\omega t + \theta)$ can be expressed as a sum of exponentials e^{st} and $e^{s^* t}$. The conventional sinusoid (with constant amplitude) is a special case in which $s = j\omega$ ($\sigma = 0$). We call s the **complex frequency** of e^{st}. The complex frequency s can be conveniently represented on a **complex frequency plane** (s plane) as shown in Fig. B.9. The horizontal axis is the real axis (σ axis), and the vertical axis is the imaginary axis ($j\omega$ axis). The imaginary part (ω) of s, the *radian* frequency, indicates the frequency of oscillation of e^{st}; the real part σ (the *neper* frequency) gives information about the rate of increase or decrease of the amplitude of e^{st}. For signals whose complex frequencies lie on the real axis (σ-axis where $\omega = 0$), the frequency of oscillation

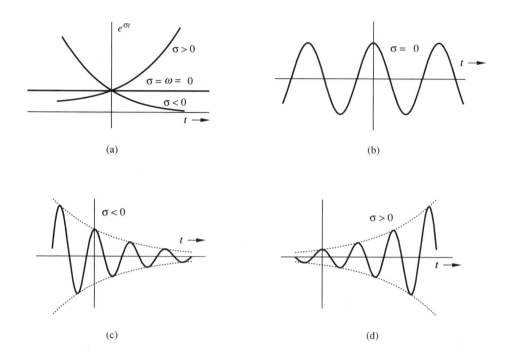

Fig. B.10 Sinusoids of complex frequency $\sigma + j\omega$.

is zero. Consequently these signals are monotonically increasing or decreasing exponentials (Fig. B.10a). For signals whose frequencies lie on the imaginary axis ($j\omega$-axis where $\sigma = 0$), $e^{\sigma t} = 1$. Therefore, these signals are conventional sinusoids with constant amplitude (Fig. B.10b). The case $s = 0$ ($\sigma = \omega = 0$) corresponds to a constant (dc) signal because $e^{0t} = 1$. For the signals shown in Figs. B.10c and B.10d, both σ and ω are nonzero; the frequency s is complex and does not lie on either axis. The signal in Fig. B.10c decays exponentially. Therefore, σ is negative, and s lies to the left of the imaginary axis. On the other hand, the signal in Fig. B.10d grows exponentially. Therefore, σ is positive, and s lies to the right of the imaginary axis. Thus the s-plane (Fig. B.9) can be differentiated into two parts: the left half-plane (LHP) corresponding to exponentially decaying signals and the right half-plane (RHP) corresponding to exponentially growing signals. The imaginary axis separates the two regions and corresponds to signals of constant amplitude.

An exponentially growing sinusoid $e^{2t}\cos{(5t + \theta)}$, for example, can be expressed as a sum of exponentials $e^{(2+j5)t}$ and $e^{(2-j5)t}$ with complex frequencies $2 + j5$ and $2 - j5$ respectively, which lie in the RHP. An exponentially decaying sinusoid $e^{-2t}\cos{(5t + \theta)}$ can be expressed as a sum of exponentials $e^{(-2+j5)t}$ and $e^{(-2-j5)t}$ with complex frequencies $-2 + j5$ and $-2 - j5$, respectively, which lie in the LHP. A constant amplitude sinusoid $\cos{(5t + \theta)}$ can be expressed as a sum of exponentials e^{j5t} and e^{-j5t} with complex frequencies $\pm j5$, which lie on the imaginary axis. Observe that the monotonic exponentials $e^{\pm 2t}$ are also generalized sinusoids with complex frequencies ± 2.

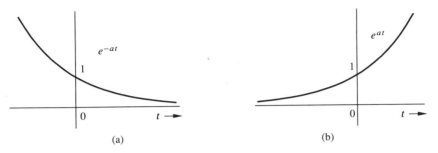

(a)   (b)

Fig. B.11 Monotonic exponentials.

B.3   SKETCHING SIGNALS

In this section we discuss sketching a few useful signals, starting with exponentials.

B.3-1   Monotonic Exponentials

The signal e^{-at} decays monotonically, and the signal e^{at} grows monotonically with t (assuming $a > 0$) as shown in Fig. B.11. In much of our discussion, the signals begin at $t = 0$. In order to accommodate such signals, we define the unit step function $u(t)$ shown in Fig. B.12a as

$$u(t) = \begin{cases} 1 & t \geq 0 \\ 0 & t < 0 \end{cases} \tag{B.29}$$

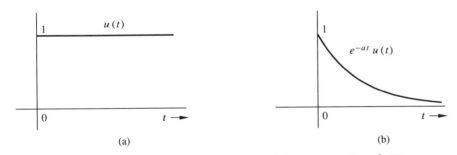

(a)   (b)

Fig. B.12 (a) Unit step function $u(t)$ (b) exponential $e^{-at}u(t)$.

If we want a signal to start at $t = 0$ (so that it has a value of zero for $t < 0$), we only need to multiply the signal with $u(t)$. Figure B.12b shows the signal $e^{-at}u(t)$.

The signal $e^{-at}u(t)$ has a unit value at $t = 0$. At $t = 1/a$, the value drops to $1/e$ (about 37% of its initial value) as shown in Fig. B.13a. This time interval over which the exponential reduces by a factor e (that is, drops to about 37% of its value) is known as the **time constant** of the exponential. Therefore, the time constant of e^{-at} is $1/a$. Observe that the exponential is reduced to 37% of its initial value over any time interval of duration $1/a$. This can be shown by considering any set of instants t_1 and t_2 separated by one time constant so that

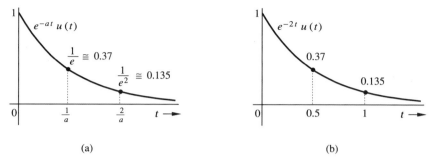

Fig. B.13 (a) Sketching $e^{-at}u(t)$ (b) sketching $e^{-2t}u(t)$.

$$t_2 - t_1 = \tfrac{1}{a}$$

Now the ratio of e^{-at_2} to e^{-at_1} is given by

$$\frac{e^{-at_2}}{e^{-at_1}} = e^{-a(t_2 - t_1)} = \tfrac{1}{e} \approx 0.37$$

We can use this fact to sketch an exponential quickly. For example, consider

$$f(t) = e^{-2t}u(t)$$

The time constant in this case is $1/2$. The value of $f(t)$ at $t = 0$ is 1. At $t = 1/2$ (one time constant) it is $1/e$ (about 0.37). The value of $f(t)$ continues to drop further by the factor $1/e$ (37%) over the next half-second interval (one time constant). Thus $f(t)$ at $t = 1$ is $(1/e)^2$. Continuing in this manner, we see that $f(t) = (1/e)^3$ at $t = 3/2$ and so on. A knowledge of the values of $f(t)$ at $t = 0$, 0.5, 1, and 1.5 allows us to sketch the desired signal† as shown in Fig. B.13b. For a monotonically growing exponential e^{at}, the waveform increases by a factor e over each interval of $1/a$ seconds.

B.3-2   The Exponentially Varying Sinusoid

We shall now discuss sketching an exponentially varying sinusoid

$$f(t) = Ae^{-at}\cos(\omega_0 t + \theta)$$

Let us consider the following specific example:

$$f(t) = 4e^{-2t}\cos(6t - 60°)\,u(t)$$

We shall sketch $4e^{-2t}u(t)$ and $\cos(6t - 60°)$ separately and then multiply them.

(i)   Sketching $4e^{-2t}\mathbf{u(t)}$

This monotonically decaying exponential has a time constant $1/2$ second and an initial value of 4 at $t = 0$. Therefore, its values at $t = 0.5$, 1, 1.5, and 2 are $4/e$,

---

†If we wish to refine the sketch further, we could consider intervals of half the time constant over which the signal decays by a factor $1/\sqrt{e}$. Thus, at $t = 0.25$, $f(t) = 1/\sqrt{e}$, and at $t = 0.75$, $f(t) = 1/e\sqrt{e}$, etc.

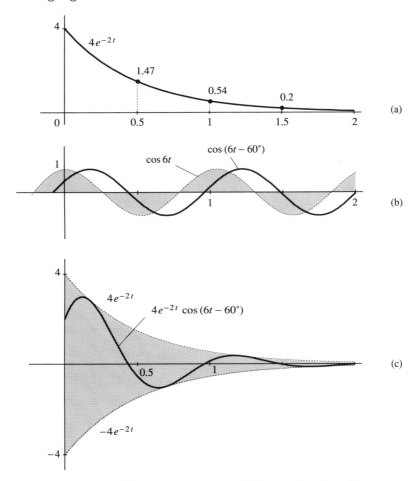

Fig. B.14 Sketching an exponentially varying sinusoid

$4/e^2$, $4/e^3$, and $4/e^4$, or about 1.47, 0.54, 0.2, and 0.07 respectively. Using these values as a guide, we sketch $4e^{-2t}u(t)$ as shown in Fig. B.14a.

(ii)   Sketching $\cos(6t - 60°)$

The procedure for sketching $\cos(6t - 60°)$ is discussed in Sec. B.2 (Fig. B.6c). Here the period $T_0 = 2\pi/6 \approx 1$, and there is a phase delay of 60°, or two-thirds of a quarter-cycle, which is equivalent to about a $(60/360)(1) \approx 1/6$ second delay (see Fig. B.14b).

(iii)   Sketching $4e^{-2t}\cos(6t - 60°)\,u(t)$

We now multiply the waveforms in **(i)** and **(ii)**. The multiplication amounts to forcing the amplitude of the sinusoid $\cos(6t - 60°)$ to decrease exponentially with time constant of 0.5. The initial amplitude (at $t = 0$) is 4, decreasing to $4/e$ (=1.47) at $t = 0.5$, to $1.47/e$ (=0.54) at $t = 1$, and so on. This is shown in Fig. B.14c. Note that at the instants where $\cos(6t - 60°)$ has a value of unity (peak amplitude),

$$4e^{-2t}\cos(6t - 60°) = 4e^{-2t}$$

Therefore, $4e^{-2t}\cos(6t-60°)$ touches $4e^{-2t}$ at those instants where the sinusoid $\cos(6t-60°)$ is at its positive peaks. Clearly $4e^{-2t}$ is an envelope for positive amplitudes of $4e^{-2t}\cos(6t-60°)$. Similarly, at those instants where the sinusoid $\cos(6t-60°)$ has a value of -1 (negative peak amplitude),

$$4e^{-2t}\cos(6t-60°) = -4e^{-2t}$$

and $4e^{-2t}\cos(6t-60°)$ touches $-4e^{-2t}$ at its negative peaks. Therefore, $-4e^{-2t}$ is an envelope for negative amplitudes of $4e^{-2t}\cos(6t-60°)$. Thus, to sketch $4e^{-2t}\cos(6t-60°)$, we first draw the envelopes $4e^{-2t}$ and $-4e^{-2t}$ (the mirror image of $4e^{-2t}$ about the horizontal axis), and then sketch the sinusoid $\cos(6t-60°)$, with these envelopes acting as constraints on the sinusoid's amplitude (see Fig. B.14c).

In general, $Ke^{-at}\cos(\omega_0 t+\theta)$ can be sketched in this manner, with Ke^{-at} and $-Ke^{-at}$ constraining the amplitude of $\cos(\omega_0 t+\theta)$.

⊙ **Example CB.5**
Plot (i) $4e^{-2t}u(t)$ (ii) $4e^{-2t}\cos(6t-60°)u(t)$.

(i) $4e^{-2t}u(t)$:

```
t1=0:.1:2;t=t';
z=4*exp(-2*t1);
plot(t1,z),grid
clear
```

(ii) $4e^{-2t}\cos(6t-60°)u(t)$:

```
t2=0:.01:4;t2=t2';
z1=4*exp(-2*t2); z2 = cos(6*t2-pi/3);
z=z1.*z2;
plot(t2,z),grid    ⊙
```

B.3-3 The Amplitude-Modulated Sinusoid

The above procedure can be generalized for sketching the amplitude-modulated sinusoid

$$f(t) = m(t)\cos(\omega_0 t+\theta)$$

Observe that

$$f(t) = \begin{cases} m(t) & \cos(\omega_0 t+\theta)=1 \\ -m(t) & \cos(\omega_0 t+\theta)=-1 \end{cases} \tag{B.30}$$

Thus, $f(t)$ touches $m(t)$ at those instants where the sinusoid $\cos(\omega_0 t+\theta)$ is at its positive peaks, while touching $-m(t)$ at those instants where the sinusoid $\cos(\omega_0 t+\theta)$ is at its negative peaks. Therefore, $m(t)$ and $-m(t)$ act as envelopes for $m(t)\cos(\omega_0 t+\theta)$. To sketch $m(t)\cos(\omega_0 t+\theta)$, we need to draw $m(t)$ and $-m(t)$ (the mirror image of $m(t)$ about the horizontal axis) and fill in $\cos(\omega_0 t+\theta)$ with these envelopes bounding the sinusoid's amplitude. Figure B.15 shows an example of $m(t)$ and the corresponding sketch of $m(t)\cos(\omega_0 t+\theta)$.† The signal $m(t)\cos(\omega_0 t+\theta)$ is

---

†We are not paying attention to the fine details of the phase. For example, when $m(t)$ changes its sign from positive to negative (or vice versa), the phase of the sinusoid changes by π instantaneously. Here, we are more interested in the envelope (general shape) of $m(t)\cos(\omega_0 t+\theta)$, and we therefore ignore the details of its phase.

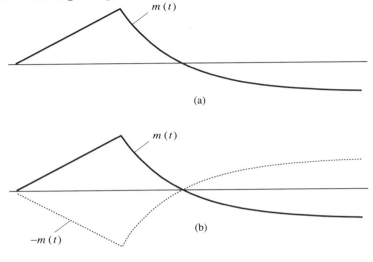

(a)

(b)

$-m(t)$

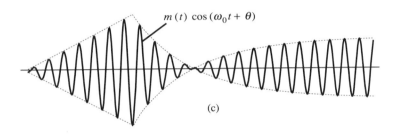

(c)

Fig. B.15 Sketching an amplitude modulated signal.

known as an **amplitude-modulated** signal because the amplitude of the sinusoid $\cos(\omega_0 t + \theta)$ is varied (modulated) in proportion to $m(t)$.

B.4   SOME USEFUL SIGNAL OPERATIONS

We discuss here three useful signal operations: shifting, scaling, and inversion. Since the independent variable in our signal description is time, these operations are discussed as time shifting, time scaling, and time inversion (or folding). However, this discussion is valid for functions having independent variables other than time (e.g., frequency or distance).

B.4-1 Time Shifting

Consider a signal $f(t)$ (Fig. B.16a) and the same signal delayed by T seconds (Fig. B.16b), which we shall denote as $\phi(t)$. Whatever happens in $f(t)$ [Fig. B.16a] at some instant t also happens in $\phi(t)$ (Fig. B.16b) T seconds later at the instant $t + T$. Therefore

$$\phi(t + T) = f(t)$$

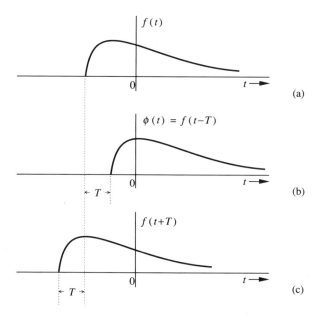

Fig. B.16 Time-shifting a signal.

and

$$\phi(t) = f(t - T) \tag{B.31}$$

Therefore, to time-shift a signal by T, we replace t with $t - T$. Thus $f(t - T)$ represents $f(t)$ time-shifted by T seconds. If T is positive, the shift is to the right (delay). If T is negative, the shift is to the left (advance). Thus, $f(t - 2)$ is $f(t)$ delayed (right-shifted) by 2 seconds, and $f(t + 2)$ is $f(t)$ advanced (left-shifted) by 2 seconds.

B.4-2 Time Scaling

The compression or expansion of a signal in time is known as **time scaling**. Consider the signal $f(t)$ of Fig. B.17a. The signal $\phi(t)$ in Fig. B.17b is $f(t)$ compressed in time by a factor of 2. Therefore, whatever happens in $f(t)$ at some instant t also happens to $\phi(t)$ at the instant $t/2$ so that

$$\phi\left(\tfrac{t}{2}\right) = f(t)$$

and

$$\phi(t) = f(2t) \tag{B.32}$$

Observe that because $f(t) = 0$ at $t = T_1$ and T_2, the same thing must happen in $\phi(t)$ at half these values. Therefore, $\phi(t) = 0$ at $t = T_1/2$ and $T_2/2$, as shown in Fig. B.17b. If $f(t)$ were recorded on a tape and played back at twice the normal recording speed, we would obtain $f(2t)$. In general, if $f(t)$ is compressed in time by a factor a $(a > 1)$, the resulting signal $\phi(t)$ is given by

$$\phi(t) = f(at) \tag{B.33}$$

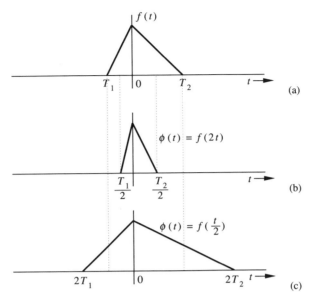

Fig. B.17 Time-scaling a signal.

Using a similar argument, we can show that $f(t)$ expanded (slowed down) in time by a factor a $(a > 1)$ is given by

$$\phi(t) = f\left(\tfrac{t}{a}\right) \tag{B.34}$$

Figure B.17c shows $f(\tfrac{t}{2})$, which is $f(t)$ expanded in time by a factor of 2.

In summary, to time-scale a signal by a factor a, we replace t with at. If $a > 1$, the scaling is compression, and if $a < 1$, the scaling is expansion.

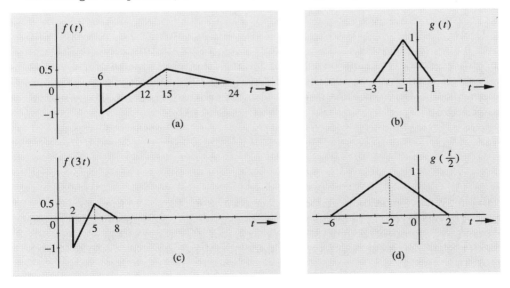

Fig. B.18 Examples of time-compression and time-expansion of signals.

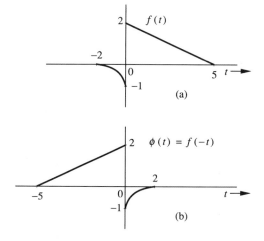

Fig. B.19 Time-inversion (reflection) of a signal.

■ **Example B.7**

Figures B.18a and B18b show the signals $f(t)$ and $g(t)$ respectively. Sketch **(a)** $f(3t)$ and **(b)** $g(t/2)$.

(a) $f(3t)$ is $f(t)$ compressed by a factor of 3. This means that the values of $f(t)$ at $t = 6$, 12, 15, and 24 occur in $f(3t)$ at the instants $t = 2$, 4, 5, and 8 respectively, as shown in Fig. B.18c.

(b) $g(t/2)$ is $g(t)$ expanded (slowed down) by a factor of 2. The values of $g(t)$ at $t = 1, -1$, and -3 occur in $g(t/2)$ at instants 2, -2, and -6 respectively, as shown in Fig. B.18d. ■

B.4-3   Time Inversion (or Folding)

Time inversion may be considered a special case of time scaling with $a = -1$ in Eq. (B.33). Consider the signal $f(t)$ in Fig. B.19a. We can view $f(t)$ as a rigid wire frame hinged at the vertical axis. To invert $f(t)$, we rotate this frame 180° about the vertical axis. This time inversion or folding (the mirror image of $f(t)$ about the vertical axis) gives us the signal $\phi(t)$ (Fig. B.19b). Observe that whatever happens in Fig. B.19a at some instant t also happens in Fig. B.19b at the instant $-t$. Therefore

$$\phi(-t) = f(t)$$

and

$$\phi(t) = f(-t) \tag{B.35}$$

Therefore, to time-invert a signal we replace t with $-t$. Thus, the time inversion of signal $f(t)$ yields $f(-t)$. Consequently, the mirror image of $f(t)$ about the vertical axis is $f(-t)$. Recall also that the mirror image of $f(t)$ about the horizontal axis is $-f(t)$. Figure B.20 shows the inversion of the unit step function $u(t)$ to obtain $u(-t)$.

■ **Example B.8**

For the signal $f(t)$ shown in Fig. B.21a, sketch $f(-t)$.

Fig. B.20 Time-inversion of a unit step $u(t)$ results in $u(-t)$.

The instants -1 and -5 in $f(t)$ are mapped into instants 1 and 5 in $f(-t)$. If $f(t) = e^{t/2}$, then $f(-t) = e^{-t/2}$. The signal $f(-t)$ is shown in Fig. B.21b. ■

B.4-4   Combined Operations

The three operations discussed above are quite simple when performed singly. Often, however, we need to combine more than one operation. For example, $f(T-t)$ involves both time inversion and time shifting. This fact is evident from the presence of the negative sign of t and the additive term T.

The most general operation involving all the three operations is $f(at-b)$. This operation requires shifting and scaling (including inversion if a is negative) and can be obtained by using either of the following two sequences of operations:

1. Time-shift $f(t)$ by b to obtain $f(t-b)$. Now time-scale the shifted signal $f(t-b)$ by a (that is, replace t with at) to obtain $f(at-b)$.

2. Time-scale $f(t)$ by a to obtain $f(at)$. Now time-shift $f(at)$ by $\frac{b}{a}$ (that is, replace t with $(t - \frac{b}{a})$ to obtain $f[a(t - \frac{b}{a})] = f(at-b)$. In either case, if a is negative, time scaling involves time inversion.

For instance, we can obtain $f(5t+10)$ from $f(t)$ by either of the following two procedures:

1 Left-shift $f(t)$ by 10 to obtain $f(t+10)$. Now time-compress the shifted signal $f(t+10)$ by factor 5 to obtain $f(5t+10)$.

2 Time-compress $f(t)$ by factor 5 to obtain $f(5t)$. Now left-shift the compressed signal $f(5t)$ by 2 to obtain $f[5(t+2)] = f(5t+10)$.

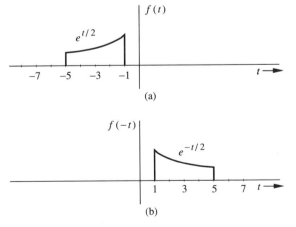

Fig. B.21 Another example of time inversion.

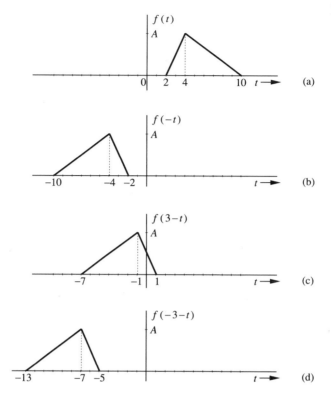

Fig. B.22 Combined operations on a signal.

■ Example B.9

For the signal $f(t)$ shown in Fig. B.22a, sketch $f(T - t)$ as a function of t for $T = 3$ and $T = -3$.

We can obtain $f(T - t)$ from $f(t)$ by either of the following two procedures:

1. Left-shift $f(t)$ by T to obtain $f(t + T)$, and then time-invert (take the mirror image about the vertical axis) of the resulting signal to obtain $f(-t + T) = f(T - t)$. Recall that time-inverting amounts to replacing t with $-t$.

2. Time-invert $f(t)$ to obtain $f(-t)$ and then right-shift the resulting signal $f(-t)$ by T to obtain $f[-(t - T)] = f(T - t)$. In short, $f(T - t)$ is $f(-t)$ right-shifted by T.

This operation, performed according to the second procedure, is shown in Fig. B.22 for $T = 3$ and -3. Figure B.22b shows $f(-t)$. The function $f(T - t)$ is obtained from $f(-t)$ by right-shifting it T seconds; Figure B.22c shows $f(T - t)$ for $T = 3$ [positive or right shift of $f(-t)$ by 3 seconds], and Fig. B.22d shows $f(T - t)$ for $T = -3$ [negative or left shift of $f(-t)$ by 3 seconds]. ■

B.5 MATHEMATICAL DESCRIPTION OF A SIGNAL FROM ITS SKETCH (GRAPHICAL DESCRIPTION)

In Sec. B.3, we discussed a procedure for sketching signals from their mathematical descriptions. The other side of this problem is obtaining a mathematical (analytical) description of a signal when it is given in its graphical form (sketch). Consider, for example, the rectangular pulse shown in Fig. B.23a. We can express such a pulse in terms of familiar step functions by observing that the pulse $f(t)$ can

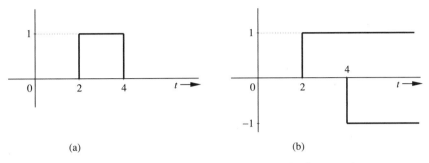

(a)   (b)

Fig. B.23 Representation of a rectangular pulse by step functions.

be expressed as the sum of the two delayed unit step functions as shown in Fig.
B.23b. The unit step function $u(t)$ delayed by T seconds is $u(t - T)$. From Fig.
B.23b, it is clear that

$$f(t) = u(t - 2) - u(t - 4) \tag{B.36}$$

The signal shown in Fig. B.24a can be conveniently handled by breaking it up
into the two components $f_1(t)$ and $f_2(t)$, shown in Figs. B.24b and B.24c respectively. Here, $f_1(t)$ can be obtained by multiplying the ramp t by the gate pulse
$u(t) - u(t - 2)$, as shown in Fig. B.24b. Therefore

$$f_1(t) = t\left[u(t) - u(t - 2)\right]$$

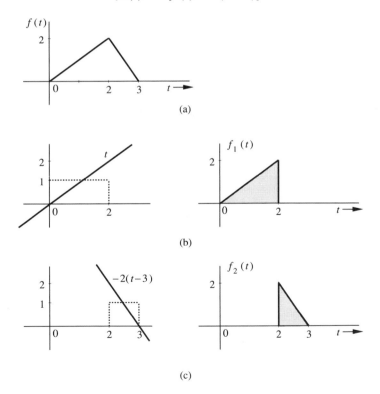

Fig. B.24 Representation of a piecewise-continuous signal.

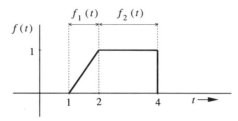

Fig. B.25 Piecewise-continuous signal for Example B.10.

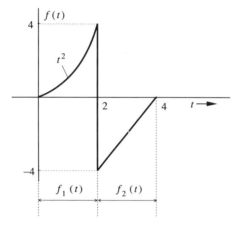

Fig. B.26 Piecewise-continuous signal for Example B.11.

The signal $f_2(t)$ can be obtained by multiplying another ramp by the gate pulse shown in Fig. B.24c. This ramp has a slope -2; hence it can be described by $-2t+c$. Now, because the ramp has a zero value at $t=3$, the constant $c=6$, and the ramp can be described by $-2(t-3)$. Also, the gate pulse in Fig. B.24c is $u(t-2)-u(t-3)$. Therefore

$$f_2(t) = -2(t-3)\left[u(t-2) - u(t-3)\right]$$

and

$$\begin{aligned}
f(t) &= f_1(t) + f_2(t) \\
&= t\left[u(t) - u(t-2)\right] - 2(t-3)\left[u(t-2) - u(t-3)\right] \\
&= tu(t) - 3(t-2)u(t-2) + 2(t-3)u(t-3)
\end{aligned} \tag{B.37}$$

■ **Example B.10**

Describe mathematically (analytically) the signal shown in Fig. B.25.

This signal can be broken up into two components, $f_1(t)$ and $f_2(t)$, as shown in Fig. B.25:

$$\begin{aligned}
f(t) &= f_1(t) + f_2(t) \\
&= \underbrace{(t-1)\left[u(t-1) - u(t-2)\right]}_{f_1(t)} + \underbrace{\left[u(t-2) - u(t-4)\right]}_{f_2(t)} \\
&= (t-1)u(t-1) - (t-2)u(t-2) - u(t-4)
\end{aligned} \tag{B.38}$$

■

■ **Example B.11**

Find the mathematical (analytical) description of the signal shown in Fig. B.26.

From Fig. B.26, we observe that

$$f(t) = f_1(t) + f_2(t) \tag{B.39}$$

$$= \underbrace{t^2\left[u(t) - u(t-2)\right]}_{f_1(t)} + \underbrace{2(t-4)\left[u(t-2) - u(t-4)\right]}_{f_2(t)} \tag{B.40a}$$

$$= t^2 u(t) - (t^2 - 2t + 8)u(t-2) - 2(t-4)u(t-4) \tag{B.40b}$$

∎

B.6   EVEN AND ODD FUNCTIONS

A function $f_e(t)$ is said to be an **even function** of t if

$$f_e(t) = f_e(-t) \tag{B.41}$$

and a function $f_o(t)$ is said to be an **odd function** of t if

$$f_o(t) = -f_o(-t) \tag{B.42}$$

An even function has the same value at the instants t and $-t$ for all values of t. Clearly, $f_e(t)$ is symmetrical about the vertical axis, as shown in Fig. B.27a. On the other hand, the value of an odd function at the instant t is the negative of its value at the instant $-t$. Therefore, $f_o(t)$ is anti-symmetrical about the vertical axis, as shown in Fig. B.27b.

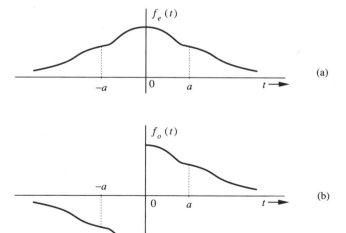

Fig. B.27 An even and an odd function of t.

B.6-1   Some Properties of Even and Odd Functions

Even and odd functions have the following property:

$$\text{even function} \times \text{ odd function} = \text{odd function}$$

$$\text{odd function} \times \text{ odd function} = \text{even function}$$

$$\text{even function} \times \text{ even function} = \text{even function}$$

The proofs of these facts are trivial and follow directly from the definition of odd and even functions [Eqs. (B.41) and (B.42)]. We leave them as exercises for the reader.

Area

Because $f_e(t)$ is symmetrical about the vertical axis, it follows from Fig. B.27a that

$$\int_{-a}^{a} f_e(t)\, dt = 2 \int_{0}^{a} f_e(t)\, dt \tag{B.43a}$$

It is also clear from Fig. B.27b that

$$\int_{-a}^{a} f_o(t)\, dt = 0 \tag{B.43b}$$

These results can also be proved formally by using the definitions in Eqs. (B.41) and (B.42). We leave them as an exercise for the reader.

B.6-2   Even and Odd Components of a Signal

Every signal $f(t)$ can be expressed as a sum of even and odd components because

$$f(t) = \underbrace{\tfrac{1}{2}\left[f(t) + f(-t)\right]}_{\text{even}} + \underbrace{\tfrac{1}{2}\left[f(t) - f(-t)\right]}_{\text{odd}} \tag{B.44}$$

From the definitions in Eqs. (B.41) and (B.42), it can be seen that the first component on the right-hand side is an even function, while the second component is odd. This is readily seen from the fact that replacing t by $-t$ in the first component yields the same function. The same maneuver in the second component yields the negative of that component.

Consider the function

$$f(t) = e^{-at}u(t)$$

Expressing this function as a sum of the even and odd components $f_e(t)$ and $f_o(t)$, we obtain

$$f(t) = f_e(t) + f_o(t)$$

in which [from Eq. (B.44)]

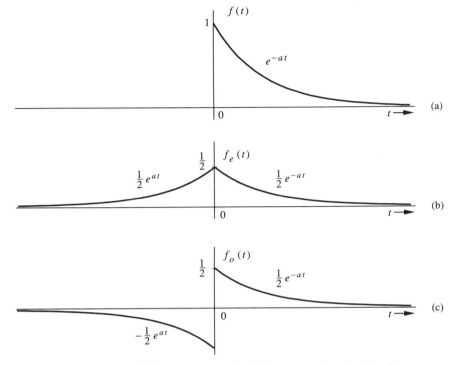

Fig. B.28 Finding an even and odd components of a signal.

$$f_e(t) = \tfrac{1}{2}\left[e^{-at}u(t) + e^{at}u(-t)\right] \tag{B.45a}$$

and

$$f_o(t) = \tfrac{1}{2}\left[e^{-at}u(t) - e^{at}u(-t)\right] \tag{B.45b}$$

The function $e^{-at}u(t)$ and its even and odd components are shown in Fig. B.28.

■ **Example B.12**
  Find the even and odd components of e^{jt}.
  From Eq. (B.44)

$$e^{jt} = f_e(t) + f_o(t)$$

in which

$$f_e(t) = \tfrac{1}{2}\left[e^{jt} + e^{-jt}\right] = \cos t$$

and

$$f_o(t) = \tfrac{1}{2}\left[e^{jt} - e^{-jt}\right] = j\sin t \quad ■$$

B.7   CRAMER'S RULE

This is a very convenient rule used in solving simultaneous linear equations. Consider a set of n linear simultaneous equations in n unknowns $x_1, x_2, \ldots, x_n$:

$$a_{11}x_1 + a_{12}x_2 + \cdots + a_{1n}x_n = y_1$$

$$a_{21}x_1 + a_{22}x_2 + \cdots + a_{2n}x_n = y_2$$

$$\cdots\cdots\cdots\cdots\cdots\cdots\cdots\cdots\cdots\cdots\cdots$$ \hfill (B.46)

$$a_{n1}x_1 + a_{n2}x_2 + \cdots + a_{nn}x_n = y_n$$

These equations can be expressed in matrix form as

$$
\begin{bmatrix}
a_{11} & a_{12} & \cdots & a_{1n} \\
a_{21} & a_{22} & \cdots & a_{2n} \\
\cdots & \cdots & \cdots & \cdots \\
a_{n1} & a_{n2} & \cdots & a_{nn}
\end{bmatrix}
\begin{bmatrix}
x_1 \\ x_2 \\ \vdots \\ x_n
\end{bmatrix}
=
\begin{bmatrix}
y_1 \\ y_2 \\ \vdots \\ y_n
\end{bmatrix}
\tag{B.47}
$$

We denote the matrix on the left-hand side formed by the elements a_{ij} as $\mathbf{A}$. The determinant of $\mathbf{A}$ is denoted by $|\mathbf{A}|$. If the determinant $|\mathbf{A}|$ is not zero, the set of equations (B.46) has a unique solution given by Cramer's formula:

$$x_k = \frac{|\mathbf{D}_k|}{|\mathbf{A}|} \qquad k = 1, 2, \ldots, n \tag{B.48}$$

where $|\mathbf{D}_k|$ is obtained by replacing the kth column of $|\mathbf{A}|$ by the column on the right-hand side of Eq. (B.47) (with elements $y_1, y_2, \ldots, y_n$).

We shall demonstrate the use of this rule with an example.

■ **Example B.13**

Using Cramer's rule, solve the following simultaneous linear equations in three unknowns:

$$2x_1 + x_2 + x_3 = 3$$

$$x_1 + 3x_2 - x_3 = 7$$

$$x_1 + x_2 + x_3 = 1$$

In matrix form these equations can be expressed as

$$
\begin{bmatrix}
2 & 1 & 1 \\
1 & 3 & -1 \\
1 & 1 & 1
\end{bmatrix}
\begin{bmatrix}
x_1 \\ x_2 \\ x_3
\end{bmatrix}
=
\begin{bmatrix}
3 \\ 7 \\ 1
\end{bmatrix}
$$

Here,

$$
|\mathbf{A}| =
\begin{vmatrix}
2 & 1 & 1 \\
1 & 3 & -1 \\
1 & 1 & 1
\end{vmatrix}
= 4
$$

Since $|\mathbf{A}| = 4 \neq 0$, a unique solution exists for x_1, x_2, and x_3. This solution is given by Cramer's rule (B.48) as follows:

$$x_1 = \frac{1}{|\mathbf{A}|} \begin{vmatrix} 3 & 1 & 1 \\ 7 & 3 & -1 \\ 1 & 1 & 1 \end{vmatrix} = \frac{8}{4} = 2$$

$$x_2 = \frac{1}{|\mathbf{A}|} \begin{vmatrix} 2 & 3 & 1 \\ 1 & 7 & -1 \\ 1 & 1 & 1 \end{vmatrix} = \frac{4}{4} = 1$$

$$x_3 = \frac{1}{|\mathbf{A}|} \begin{vmatrix} 2 & 1 & 3 \\ 1 & 3 & 7 \\ 1 & 1 & 1 \end{vmatrix} = \frac{-8}{4} = -2 \quad \blacksquare$$

⊙ **Example CB.6**
Using a Computer, solve Example B.13.
Solving on a computer using Cramer's rule:

```
A = [2 1 1;1 3 -1;1 1 1];  % Defines matrix  A.
B = [3 7 1]';
detA = det(A); % Determinant of matrix    A.
x1=det([B A(:,2:3)])/detA
x2=det([A(:,1) B A(:,3)])/detA
x3=det([A(:,1) A(:,2) B])/detA   ⊙
```

B.8 PARTIAL FRACTION EXPANSION

In the analysis of linear time-invariant systems, we encounter functions that are ratios of two polynomials in a certain variable, say x. Such functions are known as **rational functions**. A rational function $F(x)$ can be expressed as

$$F(x) = \frac{b_m x^m + b_{m-1} x^{m-1} + \cdots + b_1 x + b_0}{x^n + a_{n-1} x^{n-1} + \cdots + a_1 x + a_0} \tag{B.49}$$

$$= \frac{P(x)}{Q(x)} \tag{B.50}$$

The function $F(x)$ is **improper** if $m \geq n$ and **proper** if $m < n$. An improper function can always be separated into the sum of a polynomial in x and a proper function. Consider, for example, the function

$$F(x) = \frac{2x^3 + 9x^2 + 11x + 2}{x^2 + 4x + 3} \qquad \text{(B.51a)}$$

Because this is an improper function, we divide the numerator by the denominator until the remainder has a lower degree than the denominator:

$$
\begin{array}{r}
2x + 1 \\
\hline
x^2 + 4x + 3 \overline{\smash{\big)}\ 2x^3 + 9x^2 + 11x + 2} \\
2x^3 + 8x^2 + 6x \\
\hline
x^2 + 5x + 2 \\
x^2 + 4x + 3 \\
\hline
x - 1
\end{array}
$$

Therefore, $F(x)$ can be expressed as

$$F(x) = \frac{2x^3 + 9x^2 + 5x + 2}{x^2 + 4x + 3} = \underbrace{2x + 1}_{\text{polynomial in } x} + \underbrace{\frac{x - 1}{x^2 + 4x + 3}}_{\text{proper function}} \qquad \text{(B.51b)}$$

A proper function can be further expanded into partial fractions. The remaining discussion in this section is concerned with various ways of doing this.

B.8-1   Partial Fraction Expansion: Method of Clearing Fractions

This method consists of writing a rational function as a sum of appropriate partial fractions with unknown coefficients, which are determined by clearing fractions and equating the coefficients of similar terms on the two sides. This procedure is demonstrated by the following example.

■ **Example B.14**

Expand the following rational function $F(x)$ into partial fractions:

$$F(x) = \frac{x^3 + 3x^2 + 4x + 6}{(x + 1)(x + 2)(x + 3)^2}$$

This function can be expressed as a sum of partial fractions with denominators $(x+1)$, $(x+2)$, $(x+3)$, and $(x+3)^2$ as shown below:

$$F(x) = \frac{x^3 + 3x^2 + 4x + 6}{(x + 1)(x + 2)(x + 3)^2} = \frac{k_1}{x + 1} + \frac{k_2}{x + 2} + \frac{k_3}{x + 3} + \frac{k_4}{(x + 3)^2}$$

To determine the unknowns k_1, k_2, k_3, and k_4 we clear fractions by multiplying both sides by $(x + 1)(x + 2)(x + 3)^2$ to obtain

$$x^3 + 3x^2 + 4x + 6 = k_1(x^3 + 8x^2 + 21x + 18) + k_2(x^3 + 7x^2 + 15x + 9)$$
$$+ k_3(x^3 + 6x^2 + 11x + 6) + k_4(x^2 + 3x + 2)$$
$$= x^3(k_1 + k_2 + k_3) + x^2(8k_1 + 7k_2 + 6k_3 + k_4)$$
$$+ x(21k_1 + 15k_2 + 11k_3 + 3k_4) + (18k_1 + 9k_2 + 6k_3 + 2k_4)$$

Equating coefficients of similar terms on both sides yields

$$k_1 + k_2 + k_3 = 1$$

$$8k_1 + 7k_2 + 6k_3 + k_4 = 3$$

$$21k_1 + 15k_2 + 11k_3 + 3k_4 = 4$$

$$18k_1 + 9k_2 + 6k_3 + 2k_4 = 6$$

Solution of these four simultaneous equations yields

$$k_1 = 1, \qquad k_2 = -2, \qquad k_3 = 2, \qquad k_4 = -3$$

Therefore

$$F(x) = \frac{1}{x+1} - \frac{2}{x+2} + \frac{2}{x+3} - \frac{3}{(x+3)^2} \qquad \blacksquare$$

Although this method is straightforward and applicable to all situations, it is not necessarily the most efficient. We now discuss other methods which can reduce numerical work considerably.

B.8-2   Partial Fractions: The Heaviside "Cover-Up" Method

1.   Unrepeated Factors of $Q(x)$

We shall first consider the partial fraction expansion of $F(x) = P(x)/Q(x)$, in which all the factors of $Q(x)$ are unrepeated. Consider the proper function

$$F(x) = \frac{b_m x^m + b_{m-1} x^{m-1} + \cdots + b_1 x + b_0}{x^n + a_{n-1} x^{n-1} + \cdots + a_1 x + a_0} \qquad m < n$$

$$= \frac{P(x)}{(x - \lambda_1)(x - \lambda_2) \cdots (x - \lambda_n)} \qquad \text{(B.52a)}$$

We can show that $F(x)$ in Eq. (B.52a) can be expressed as the sum of partial fractions:

$$F(x) = \frac{k_1}{x - \lambda_1} + \frac{k_2}{x - \lambda_2} + \cdots + \frac{k_n}{x - \lambda_n} \qquad \text{(B.52b)}$$

To determine the coefficient k_1, we multiply both sides of Eq. (B.52b) by $x - \lambda_1$ and then let $x = \lambda_1$. This yields

$$(x - \lambda_1)F(x)|_{x=\lambda_1} = k_1 + \frac{k_2(x - \lambda_1)}{(x - \lambda_2)} + \frac{k_3(x - \lambda_1)}{(x - \lambda_3)} + \cdots + \frac{k_n(x - \lambda_1)}{(x - \lambda_n)}\Big|_{x=\lambda_1}$$

On the right-hand side, all the terms except k_1 vanish. Therefore

$$k_1 = (x - \lambda_1)F(x)|_{x=\lambda_1} \qquad \text{(B.53)}$$

Similarly, we can show that

$$k_r = (x - \lambda_r)F(x)|_{x=\lambda_r} \qquad r = 1, 2, \cdots, n \qquad (\text{B.54})$$

■ Example B.15

Expand the following rational function $F(x)$ into partial fractions:

$$F(x) = \frac{2x^2 + 9x - 11}{(x+1)(x-2)(x+3)} = \frac{k_1}{x+1} + \frac{k_2}{x-2} + \frac{k_3}{x+3}$$

To determine k_1, we let $x = -1$ in $(x+1)F(x)$. Note that $(x+1)F(x)$ is obtained from $F(x)$ by omitting the term $(x+1)$ from its denominator. Therefore, to compute k_1 corresponding to the factor $(x+1)$, we cover up the term $(x+1)$ in the denominator of $F(x)$ and then substitute $x = -1$ in the remaining expression (mentally conceal the term $(x+1)$ in $F(x)$ with a finger and then let $x = -1$ in the remaining expression). The procedure is explained step by step below:

$$F(x) = \frac{2x^2 + 9x - 11}{(x+1)(x-2)(x+3)}$$

Step 1: Cover up (conceal) the factor $(x+1)$ from $F(x)$:

$$\frac{2x^2 + 9x - 11}{\boxed{(x+1)}(x-2)(x+3)}$$

Step 2: Substitute $x = -1$ in the remaining expression to obtain k_1:

$$k_1 = \frac{2 - 9 - 11}{(-1-2)(-1+3)} = \frac{-18}{-6} = 3$$

Similarly, to compute k_2, we cover up the factor $(x-2)$ in $F(x)$ and let $x = 2$ in the remaining function, as shown below:

$$k_2 = \frac{2x^2 + 9x - 11}{(x+1)\boxed{(x-2)}(x+3)}\bigg|_{x=2} = \frac{8 + 18 - 11}{(2+1)(2+3)} = \frac{15}{15} = 1$$

and

$$k_3 = \frac{2x^2 + 9x - 11}{(x+1)(x-2)\boxed{(x+3)}}\bigg|_{x=-3} = \frac{18 - 27 - 11}{(-3+1)(-3-2)} = \frac{-20}{10} = -2$$

Therefore

$$F(x) = \frac{2x^2 + 9x - 11}{(x+1)(x-2)(x+3)} = \frac{3}{x+1} + \frac{1}{x-2} - \frac{2}{x+3} \quad ■$$

Complex Factors in $F(x)$

The procedure above works regardless of whether the factors of $Q(x)$ are real or complex. Consider, for example,

$$F(x) = \frac{4x^2 + 2x + 18}{(x+1)(x^2 + 4x + 13)} \qquad (\text{B.55})$$

$$= \frac{4x^2 + 2x + 18}{(x+1)(x+2-j3)(x+2+j3)}$$

$$= \frac{k_1}{x+1} + \frac{k_2}{x+2-j3} + \frac{k_3}{x+2+j3}$$

in which

$$k_1 = \left[\frac{4x^2 + 2x + 18}{(x + 1)(x^2 + 4x + 13)} \right]_{x=-1} = 2$$

Similarly,

$$k_2 = \left[\frac{4x^2 + 2x + 18}{(x + 1)(x + 2 - j3)(x + 2 + j3)} \right]_{x=-2+j3} = 1 + j2 = \sqrt{5}e^{j63.43°}$$

$$k_3 = \left[\frac{4x^2 + 2x + 18}{(x + 1)(x + 2 - j3)(x + 2 + j3)} \right]_{x=-2-j3} = 1 - j2 = \sqrt{5}e^{-j63.43°}$$

Therefore

$$F(x) = \frac{2}{x + 1} + \frac{\sqrt{5}e^{j63.43°}}{x + 2 - j3} + \frac{\sqrt{5}e^{-j63.43°}}{x + 2 + j3} \tag{B.56}$$

Note that k_2 and k_3, the coefficients corresponding to the complex conjugate factors, are also conjugates of each other. This is generally true when the coefficients of a rational function are real. Thus, for complex conjugate factors, we need to compute only one of the coefficients.

2.   Quadratic Factors

Often we are required to combine the two terms arising from complex conjugate factors into one quadratic factor. For example, $F(x)$ in Eq. (B.55) can be expressed as

$$F(x) = \frac{4x^2 + 2x + 18}{(x + 1)(x^2 + 4x + 13)} = \frac{k_1}{x + 1} + \frac{c_1 x + c_2}{x^2 + 4x + 13}$$

The coefficient k_1 is found by the Heaviside method to be 2. Therefore

$$\frac{4x^2 + 2x + 18}{(x + 1)(x^2 + 4x + 13)} = \frac{2}{x + 1} + \frac{c_1 x + c_2}{x^2 + 4x + 13} \tag{B.57}$$

The values of c_1 and c_2 are determined by clearing fractions and equating the coefficients of similar powers of x on both sides of the resulting equation. Clearing fractions on both sides of Eq. (B.57) yields

$$4x^2 + 2x + 18 = 2(x^2 + 4x + 13) + (c_1 x + c_2)(x + 1)$$
$$= (2 + c_1)x^2 + (8 + c_1 + c_2)x + (26 + c_2) \tag{B.58}$$

Equating terms of similar powers yields $c_1 = 2$, $c_2 = -8$, and

$$\frac{4x^2 + 2x + 18}{(x + 1)(x^2 + 4x + 13)} = \frac{2}{x + 1} + \frac{2x - 8}{x^2 + 4x + 13} \tag{B.59}$$

Short-Cuts

The values of c_1 and c_2 in Eq. (B.57) can also be determined by using short-cuts. After computing $k_1 = 2$ by the Heaviside method as before, we let $x = 0$ on both sides of Eq. (B.57) to eliminate c_1. This gives us

$$\frac{18}{13} = 2 + \frac{c_2}{13}$$

Therefore

$$c_2 = -8$$

To determine c_1, we multiply both sides of Eq. (B.57) by x and then let $x \to \infty$. Remember that when $x \to \infty$, only the terms of the highest power are significant. Therefore

$$4 = k_1 + c_1 = 2 + c_1$$

and

$$c_1 = 2$$

In the procedure discussed here, we let $x = 0$ to determine c_2 and then multiply both sides by x and let $x \to \infty$ to determine c_1. However, nothing is sacred about these values ($x = 0$ or $x = \infty$). We use them because they reduce the number of computations involved. We could just as well use other convenient values for x, such as $x = 1$. Consider the following example:

$$F(x) = \frac{2x^2 + 4x + 5}{x(x^2 + 2x + 5)}$$

$$= \frac{k}{x} + \frac{c_1 x + c_2}{x^2 + 2x + 5}$$

We find $k = 1$ by the Heaviside method in the usual manner. As a result,

$$\frac{2x^2 + 4x + 5}{x(x^2 + 2x + 5)} = \frac{1}{x} + \frac{c_1 x + c_2}{x^2 + 2x + 5} \tag{B.60}$$

To determine c_1 and c_2 we could try letting $x = 0$ in Eq. (B.60); however, this gives us ∞ on both sides. Therefore let us choose $x = 1$. This yields

$$F(1) = \frac{11}{8} = 1 + \frac{c_1 + c_2}{8}$$

or

$$c_1 + c_2 = 3$$

We can now choose some other value for x, such as $x = 2$, to obtain one more relationship to use in determining c_1 and c_2. In this case, however, it will be simpler to multiply both sides of Eq. (B.60) by x and then let $x \to \infty$. This yields

$$2 = 1 + c_1$$

so that

$$c_1 = 1 \quad \text{and} \quad c_2 = 2$$

Therefore

$$F(x) = \frac{1}{x} + \frac{x+2}{x^2 + 2x + 5}$$

3.   Repeated Factors in $Q(x)$

If a function $F(x)$ has a repeated factor in its denominator, it has the form

$$F(x) = \frac{P(x)}{(x - \lambda)^r (x - \alpha_1)(x - \alpha_2) \cdots (x - \alpha_j)} \tag{B.61}$$

Its partial fraction expansion is given by

$$F(x) = \frac{a_0}{(x - \lambda)^r} + \frac{a_1}{(x - \lambda)^{r-1}} + \cdots + \frac{a_{r-1}}{(x - \lambda)}$$

$$+ \frac{k_1}{x - \alpha_1} + \frac{k_2}{x - \alpha_2} + \cdots + \frac{k_j}{x - \alpha_j} \tag{B.62}$$

The coefficients k_1, k_2, ..., k_j corresponding to the unrepeated factors in this equation are found by the Heaviside method, as before [Eq. (B.54)]. To find the coefficients a_0, a_1, a_2, ..., a_{r-1}, we multiply both sides of Eq. (B.62) by $(x - \lambda)^r$. This gives us

$$(x - \lambda)^r F(x) = a_0 + a_1(x - \lambda) + a_2(x - \lambda)^2 + \cdots + a_{r-1}(x - \lambda)^{r-1}$$

$$+ k_1 \frac{(x - \lambda)^r}{x - \alpha_1} + k_2 \frac{(x - \lambda)^r}{x - \alpha_2} + \cdots + k_n \frac{(x - \lambda)^r}{x - \alpha_n} \tag{B.63}$$

If we let $x = \lambda$ on both sides of Eq. (B.63), we obtain

$$(x - \lambda)^r F(x)\big|_{x=\lambda} = a_0 \tag{B.64a}$$

Therefore, a_0 is obtained by concealing the factor $(x - \lambda)^r$ in $F(x)$ and letting $x = \lambda$ in the remaining expression (the Heaviside "cover up" method). If we take the derivative (with respect to x) of both sides of Eq. (B.63), the right-hand side is a_1+ terms containing a factor $(x - \lambda)$ in their numerators. Letting $x = \lambda$ on both sides of this equation, we obtain

$$\frac{d}{dx}[(x - \lambda)^r F(x)]\bigg|_{x=\lambda} = a_1$$

Thus a_1 is obtained by concealing the factor $(x - \lambda)^r$ in $F(x)$, taking the derivative of the remaining expression, and then letting $x = \lambda$. Continuing in this manner, we find

$$a_j = \frac{1}{j!} \frac{d^j}{dx^j}[(x - \lambda)^r F(x)]\bigg|_{x=\lambda} \tag{B.64b}$$

Observe that $(x - \lambda)^r F(x)$ is obtained from $F(x)$ by omitting the factor $(x - \lambda)^r$ from its denominator. Therefore, the coefficient a_j is obtained by concealing the factor $(x - \lambda)^r$ in $F(x)$, taking the jth derivative of the remaining expression, and then letting $x = \lambda$ (while dividing by $j!$).

■ **Example B.16**

Expand $F(x)$ into partial fractions if

$$F(x) = \frac{4x^3 + 16x^2 + 23x + 13}{(x + 1)^3(x + 2)}$$

The partial fractions are

$$F(x) = \frac{a_0}{(x + 1)^3} + \frac{a_1}{(x + 1)^2} + \frac{a_2}{x + 1} + \frac{k}{x + 2}$$

The coefficient k is obtained by concealing the factor $(x+2)$ in $F(x)$ and then substituting $x = -2$ in the remaining expression:

$$k = \left. \frac{4x^3 + 16x^2 + 23x + 13}{(x + 1)^3 \boxed{(x + 2)}} \right|_{x=-2} = 1$$

To find a_0, we conceal the factor $(x + 1)^3$ in $F(x)$ and let $x = -1$ in the remaining expression:

$$a_0 = \left. \frac{4x^3 + 16x^2 + 23x + 13}{\boxed{(x + 1)^3}(x + 2)} \right|_{x=-1} = 2$$

To find a_1, we conceal the factor $(x + 1)^3$ in $F(x)$, take the derivative of the remaining expression, and then let $x = -1$:

$$a_1 = \left. \frac{d}{dx} \left[\frac{4x^3 + 16x^2 + 23x + 13}{\boxed{(x + 1)^3}(x + 2)} \right] \right|_{x=-1} = 1$$

Similarly,

$$a_2 = \left. \frac{1}{2!} \frac{d^2}{dx^2} \left[\frac{4x^3 + 16x^2 + 23x + 13}{\boxed{(x + 1)^3}(x + 2)} \right] \right|_{x=-1} = 3$$

Therefore

$$F(x) = \frac{2}{(x + 1)^3} + \frac{1}{(x + 1)^2} + \frac{3}{x + 1} + \frac{1}{x + 2} \quad ■$$

B.8-3 A Hybrid Method: Mixture of the Heaviside "Cover-Up" and Clearing Fractions

For multiple roots, especially of higher order, the Heaviside expansion method, which requires repeated differentiation, can become cumbersome. For a function which contains several repeated and unrepeated roots, a hybrid of the two procedures proves the best. The simpler coefficients are determined by the Heaviside method, and the remaining coefficients are found by clearing fractions or short-cuts, thus incorporating the best of the two methods. We demonstrate this procedure by solving Example B.16 once again by this method:

In Example B.16, coefficients k and a_0 are relatively simple to determine by the Heaviside expansion method. These values were found to be $k_1 = 1$ and $a_0 = 2$. Therefore

$$\frac{4x^3 + 16x^2 + 23x + 13}{(x+1)^3(x+2)} = \frac{2}{(x+1)^3} + \frac{a_1}{(x+1)^2} + \frac{a_2}{x+1} + \frac{1}{x+2}$$

We now multiply both sides of the above equation by $(x+1)^3(x+2)$ to clear the fractions. This yields

$4x^3 + 16x^2 + 23x + 13$

$\qquad = 2(x+2) + a_1(x+1)(x+2) + a_2(x+1)^2(x+2) + (x+1)^3$

$\qquad = (1+a_2)x^3 + (a_1 + 4a_2 + 3)x^2 + (5 + 3a_1 + 5a_2)x + (4 + 2a_1 + 2a_2 + 1)$

Equating coefficients of the third and second powers of x on both sides, we obtain

$$\left.\begin{array}{r} 1 + a_2 = 4 \\ a_1 + 4a_2 + 3 = 16 \end{array}\right\} \implies \begin{array}{l} a_1 = 1 \\ a_2 = 3 \end{array}$$

We may stop here if we wish because the two desired coefficients, a_1 and a_2, are now determined. However, equating the coefficients of the two remaining powers of x yields a convenient check on the answer. Equating the coefficients of the x^1 and x^0 terms, we obtain

$$23 = 5 + 3a_1 + 5a_2$$

$$13 = 4 + 2a_1 + 2a_2 + 1$$

These equations are satisfied by the values $a_1 = 1$ and $a_2 = 3$, found earlier, providing an additional check for our answers. Therefore

$$F(x) = \frac{2}{(x+1)^3} + \frac{1}{(x+1)^2} + \frac{3}{x+1} + \frac{1}{x+2}$$

which agrees with the previous result.

A Mixture of the Heaviside "Cover-Up" and Short Cuts

In the above example after determining the coefficients $a_0 = 2$ and $k = 1$ by the Heaviside method as before, we have

$$\frac{4x^3 + 16x^2 + 23x + 13}{(x+1)^3(x+2)} = \frac{2}{(x+1)^3} + \frac{a_1}{(x+1)^2} + \frac{a_2}{x+1} + \frac{1}{x+2}$$

There are only two unknown coefficients, a_1 and a_2. If we multiply both sides of the above equation by x and then let $x \to \infty$, we can eliminate a_1. This yields

$$4 = a_2 + 1 \implies a_2 = 3$$

Therefore

$$\frac{4x^3 + 16x^2 + 23x + 13}{(x+1)^3(x+2)} = \frac{2}{(x+1)^3} + \frac{a_1}{(x+1)^2} + \frac{3}{x+1} + \frac{1}{x+2}$$

There is now only one unknown a_1, which can be readily found by setting x equal to any convenient value, say $x = 0$. This yields

$$\tfrac{13}{2} = 2 + a_1 + 3 + \tfrac{1}{2} \quad \Longrightarrow \quad a_1 = 1$$

which agrees with our earlier answer.

B.8-4 Improper $F(x)$ with $m = n$

A general method of handling an improper function is indicated in the beginning of this section. However, for a special case where the numerator and denominator polynomials of $F(x)$ are of the same degree $(m = n)$, the procedure is the same as that for proper function. We can show that for

$$F(x) = \frac{b_n x^n + b_{n-1} x^{n-1} + \cdots + b_1 x + b_0}{x^n + a_{n-1} x^{n-1} + \cdots + a_1 x + a_0}$$

$$= b_n + \frac{k_1}{x - \lambda_1} + \frac{k_2}{x - \lambda_2} + \cdots + \frac{k_n}{x - \lambda_n}$$

the coefficients $k_1, k_2, \ldots, k_n$ are computed as if $F(x)$ were proper. Thus

$$k_r = (x - \lambda_r) F(x) \big|_{x = \lambda_r}$$

For quadratic or repeated factors, the appropriate procedures discussed in Secs. B.8-2 or B.8-3 should be used as if $F(x)$ were proper. In other words, when $m = n$, the only difference between the proper and improper case is the appearance of an extra constant b_n in the latter. Otherwise the procedure remains the same. The proof is left as an exercise for the reader.

■ Example B.17

Expand $F(x)$ into partial fractions if

$$F(x) = \frac{3x^2 + 9x - 20}{x^2 + x - 6} = \frac{3x^2 + 9x - 20}{(x - 2)(x + 3)}$$

Here $m = n = 2$ with $b_n = b_2 = 3$. Therefore

$$F(x) = \frac{3x^2 + x - 20}{(x - 2)(x + 3)} = 3 + \frac{k_1}{x - 2} + \frac{k_2}{x + 3}$$

in which

$$k_1 = \frac{3x^2 + 9x - 20}{(x - 2)(x + 3)} \bigg|_{x=2} = \frac{12 + 18 - 20}{(2 + 3)} = \frac{10}{5} = 2$$

and

$$k_2 = \frac{3x^2 + 9x - 20}{(x - 2)(x + 3)} \bigg|_{x=-3} = \frac{27 - 27 - 20}{(-3 - 2)} = \frac{-20}{-5} = 4$$

Therefore

$$F(x) = \frac{3x^2 + 9x - 20}{(x - 2)(x + 3)} = 3 + \frac{2}{x - 2} + \frac{4}{x + 3} \qquad ■$$

B.8-5 Modified Partial Fractions

Often we require partial fractions of the form $kx/(x-\lambda_i)$ rather than $k/(x-\lambda_i)$. This can be achieved by expanding $F(x)/x$ into partial fractions. Consider, for example,

$$F(x) = \frac{5x^2 + 20x + 18}{(x+2)(x+3)^2}$$

Dividing both sides by x yields

$$\frac{F(x)}{x} = \frac{5x^2 + 20x + 18}{x(x+2)(x+3)^2}$$

Expansion of the right-hand side into partial fractions as usual yields

$$\frac{F(x)}{x} = \frac{5x^2 + 20x + 18}{x(x+2)(x+3)^2} = \frac{a_1}{x} + \frac{a_2}{x+2} + \frac{a_3}{(x+3)} + \frac{a_4}{(x+3)^2}$$

Using the procedure discussed earlier, we find $a_1 = 1$, $a_2 = 1$, $a_3 = -2$, and $a_4 = 1$. Therefore

$$\frac{F(x)}{x} = \frac{1}{x} + \frac{1}{x+2} - \frac{2}{x+3} + \frac{1}{(x+3)^2}$$

Now multiplying both sides by x yields

$$F(x) = 1 + \frac{x}{x+2} - \frac{2x}{x+3} + \frac{x}{(x+3)^2}$$

This expresses $F(x)$ as the sum of partial fractions having the form $\frac{kx}{x-\lambda_i}$.

B.9 VECTORS AND MATRICES

An entity specified by n numbers in a certain order (ordered n-tuple) is an n-dimensional **vector**. Thus, if an entity is specified by an ordered n-tuple $(x_1, x_2, \ldots, x_n)$, it represents an n-dimensional vector $\mathbf{x}$. Vectors may be represented as a row (**row vector**):

$$\mathbf{x} = \begin{bmatrix} x_1 & x_2 & \cdots & x_n \end{bmatrix}$$

or as a column (**column vector**):

$$\mathbf{x} = \begin{bmatrix} x_1 \\ x_2 \\ \vdots \\ x_n \end{bmatrix}$$

Simultaneous linear equations can be viewed as the transformation of one vector into another. Consider, for example, the n simultaneous linear equations

$$y_1 = a_{11}x_1 + a_{12}x_2 + \cdots + a_{1n}x_n$$

$$y_2 = a_{21}x_1 + a_{22}x_2 + \cdots + a_{2n}x_n$$

$$\cdots\cdots\cdots\cdots\cdots\cdots\cdots\cdots\cdots\cdots\cdots\cdots$$

$$y_m = a_{m1}x_1 + a_{m2}x_2 + \cdots + a_{mn}x_n \tag{B.65}$$

If we define two column vectors $\mathbf{x}$ and $\mathbf{y}$ as

$$\mathbf{x} = \begin{bmatrix} x_1 \\ x_2 \\ \vdots \\ x_n \end{bmatrix}, \qquad \mathbf{y} = \begin{bmatrix} y_1 \\ y_2 \\ \vdots \\ y_m \end{bmatrix} \tag{B.66}$$

then Eqs. (B.65) may be viewed as the relationship or the function that transforms vector $\mathbf{x}$ into vector $\mathbf{y}$. Such a transformation is called the **linear transformation** of vectors. In order to perform a linear transformation, we need to define the array of coefficients a_{ij} appearing in Eqs. (B.65). This array is called a **matrix** and is denoted by $\mathbf{A}$ for convenience:

$$\mathbf{A} = \begin{bmatrix} a_{11} & a_{12} & \cdots & a_{1n} \\ a_{21} & a_{22} & \cdots & a_{2n} \\ \cdot & \cdot & & \cdot \\ a_{m1} & a_{m2} & \cdots & a_{mn} \end{bmatrix} \tag{B.67}$$

A matrix with m rows and n columns is called a matrix of the order (m, n) or an $(m \times n)$ matrix. For the special case where $m = n$, the matrix is called a **square matrix** of order n.

It should be stressed at this point that a matrix is not a number such as a determinant, but an array of numbers arranged in a particular order. It is convenient to abbreviate the representation of matrix $\mathbf{A}$ in Eq. (B.67) with the form $(a_{ij})_{m \times n}$, implying a matrix of order $m \times n$ with a_{ij} as its ijth element. In practice, when the order $m \times n$ is understood or need not be specified, the notation can be abbreviated to (a_{ij}). Note that the first index i of a_{ij} indicates the row and the second index j indicates the column of the element a_{ij} in matrix $\mathbf{A}$.

The simultaneous equations (B.65) may now be expressed in a symbolic form as

$$\mathbf{y} = \mathbf{A}\mathbf{x} \tag{B.68}$$

or

$$
\begin{bmatrix} y_1 \\ y_2 \\ \vdots \\ y_m \end{bmatrix} = \begin{bmatrix} a_{11} & a_{12} & \cdots & a_{1n} \\ a_{21} & a_{22} & \cdots & a_{2n} \\ \vdots & \vdots & & \vdots \\ a_{m1} & a_{m2} & \cdots & a_{mn} \end{bmatrix} \begin{bmatrix} x_1 \\ x_2 \\ \vdots \\ x_n \end{bmatrix} \tag{B.69}
$$

Equation (B.68) is the symbolic representation of Eq. (B.65). As yet, we have not defined the operation of the multiplication of a matrix by a vector. The quantity $\mathbf{Ax}$ is not meaningful until we define such an operation.

B.9-1   Some Definitions and Properties

A square matrix whose elements are zero everywhere except on the main diagonal is a **diagonal matrix**. An example of a diagonal matrix is

$$
\begin{bmatrix} 2 & 0 & 0 \\ 0 & 1 & 0 \\ 0 & 0 & 5 \end{bmatrix}
$$

A diagonal matrix with unity for all its diagonal elements is called an **identity matrix** or a **unit matrix**, denoted by $\mathbf{I}$. Note that this is a square matrix:

$$
\mathbf{I} = \begin{bmatrix} 1 & 0 & 0 & \cdots & 0 \\ 0 & 1 & 0 & \cdots & 0 \\ 0 & 0 & 1 & \cdots & 0 \\ & & \cdots\cdots\cdots & & \\ 0 & 0 & 0 & \cdots & 1 \end{bmatrix} \tag{B.70}
$$

The order of the unit matrix is sometimes indicated by a subscript. Thus $\mathbf{I}_n$ represents the $n \times n$ unit matrix (or identity matrix). However, we shall omit the subscript. The order of the unit matrix will be understood from the context.

A matrix having all its elements zero is a **zero matrix**.

A square matrix $\mathbf{A}$ is a **symmetric matrix** if $a_{ij} = a_{ji}$ (symmetry about the main diagonal).

Two matrices of the same order are said to be **equal** if they are equal element by element. Thus, if

$$
\mathbf{A} = (a_{ij})_{m \times n} \qquad \text{and} \qquad \mathbf{B} = (b_{ij})_{m \times n}
$$

then $\mathbf{A} = \mathbf{B}$ only if $a_{ij} = b_{ij}$ for all i and j.

If the rows and columns of an $m \times n$ matrix $\mathbf{A}$ are interchanged so that the elements in the ith row now become the elements of the ith column (for $i = 1, 2, \ldots, m$), the resulting matrix is called the **transpose** of $\mathbf{A}$ and is denoted by $\mathbf{A}^T$. It is evident that $\mathbf{A}^T$ is an $n \times m$ matrix. For example, if

$$\mathbf{A} = \begin{bmatrix} 2 & 1 \\ 3 & 2 \\ 1 & 3 \end{bmatrix} \quad \text{then} \quad \mathbf{A}^T = \begin{bmatrix} 2 & 3 & 1 \\ 1 & 2 & 3 \end{bmatrix}$$

Thus, if

$$\mathbf{A} = (a_{ij})_{m \times n}$$

then

$$\mathbf{A}^T = (a_{ji})_{n \times m} \tag{B.71}$$

Note that

$$(\mathbf{A}^T)^T = \mathbf{A} \tag{B.72}$$

B.9-2 Matrix Algebra

We shall now define matrix operations, such as addition, subtraction, multiplication, and division of matrices. The definitions should be formulated so that they are useful in the manipulation of matrices.

1. Addition of Matrices

For two matrices $\mathbf{A}$ and $\mathbf{B}$, both of the same order $(m \times n)$,

$$\mathbf{A} = \begin{bmatrix} a_{11} & a_{12} & \cdots & a_{1n} \\ a_{21} & a_{22} & \cdots & a_{2n} \\ \cdots\cdots\cdots\cdots\cdots\cdots \\ a_{m1} & a_{m2} & \cdots & a_{mn} \end{bmatrix} \quad \text{and} \quad \mathbf{B} = \begin{bmatrix} b_{11} & b_{12} & \cdots & b_{1n} \\ b_{21} & b_{22} & \cdots & b_{2n} \\ \cdots\cdots\cdots\cdots\cdots\cdots \\ b_{m1} & b_{m2} & \cdots & b_{mn} \end{bmatrix}$$

we define the sum $\mathbf{A} + \mathbf{B}$ as

$$\mathbf{A} + \mathbf{B} = \begin{bmatrix} (a_{11} + b_{11}) & (a_{12} + b_{12}) & \cdots & (a_{1n} + b_{1n}) \\ (a_{21} + b_{21}) & (a_{22} + b_{22}) & \cdots & (a_{2n} + b_{2n}) \\ \cdots\cdots\cdots\cdots\cdots\cdots\cdots\cdots\cdots\cdots \\ (a_{m1} + b_{m1}) & (a_{m2} + b_{m2}) & \cdots & (a_{mn} + b_{mn}) \end{bmatrix}$$

or

$$\mathbf{A} + \mathbf{B} = (a_{ij} + b_{ij})_{m \times n}$$

Note that two matrices can be added only if they are of the same order.

2. Multiplication of a Matrix by a Scalar

We define the multiplication of a matrix $\mathbf{A}$ by a scalar c as

$$c\mathbf{A} = c \begin{bmatrix} a_{11} & a_{12} & \cdots & a_{1n} \\ a_{21} & a_{22} & \cdots & a_{2n} \\ \cdots\cdots\cdots\cdots\cdots\cdots \\ a_{m1} & a_{m2} & \cdots & a_{mn} \end{bmatrix} = \begin{bmatrix} ca_{11} & ca_{12} & \cdots & ca_{1n} \\ ca_{21} & ca_{22} & \cdots & ca_{2n} \\ \cdots\cdots\cdots\cdots\cdots\cdots \\ ca_{m1} & ca_{m2} & \cdots & ca_{mn} \end{bmatrix}$$

3. Matrix Multiplication

We define the product

$$\mathbf{AB} = \mathbf{C}$$

in which c_{ij}, the element of $\mathbf{C}$ in the ith row and jth column, is found by adding the products of the elements of $\mathbf{A}$ in the ith row with the corresponding elements of $\mathbf{B}$ in the jth column. Thus

$$c_{ij} = a_{i1}b_{1j} + a_{i2}b_{2j} + \cdots + a_{in}b_{nj}$$

$$= \sum_{k=1}^{n} a_{ik}b_{kj} \tag{B.73}$$

This is shown below:

$$\begin{bmatrix} & & & \\ & & & \\ a_{i1} \ a_{i2} & \cdots a_{in} \\ & & & \\ & & & \end{bmatrix} \underbrace{}_{} \begin{bmatrix} b_{1j} \\ b_{2j} \\ \vdots \\ \cdots \ b_{ij} \ \cdots \\ \vdots \\ b_{nj} \end{bmatrix} = \begin{bmatrix} & & & \\ & & & \\ \cdots \ c_{ij} \ \cdots \\ & & & \\ & & & \end{bmatrix}$$

$$\underbrace{}_{\mathbf{A}(m\times n)} \qquad \underbrace{}_{\mathbf{B}(n\times p)} \qquad \underbrace{}_{\mathbf{C}(m\times p)}$$

Note carefully that the number of columns of $\mathbf{A}$ must be equal to the number of rows of $\mathbf{B}$ if this procedure is to work. In other words, $\mathbf{AB}$, the product of matrices $\mathbf{A}$ and $\mathbf{B}$, is defined only if the number of columns of $\mathbf{A}$ is equal to the number of rows of $\mathbf{B}$. If this condition is not satisfied, the product $\mathbf{AB}$ is not defined and is meaningless. When the number of columns of $\mathbf{A}$ is equal to the number of rows of $\mathbf{B}$, matrix $\mathbf{A}$ is said to be **conformable** to matrix $\mathbf{B}$ for the product $\mathbf{AB}$. Observe that if $\mathbf{A}$ is an $m \times n$ matrix and $\mathbf{B}$ is an $n \times p$ matrix, $\mathbf{A}$ and $\mathbf{B}$ are conformable for the product, and $\mathbf{C}$ is an $m \times p$ matrix.

We demonstrate the use of the rule in Eq. (B.73) with the following examples:

$$
\begin{bmatrix} 2 & 3 \\ 1 & 1 \\ 3 & 1 \end{bmatrix}
\begin{bmatrix} 1 & 3 & 1 & 2 \\ 2 & 1 & 1 & 1 \end{bmatrix} =
\begin{bmatrix} 8 & 9 & 5 & 7 \\ 3 & 4 & 2 & 3 \\ 5 & 10 & 4 & 7 \end{bmatrix}
$$

$$
\begin{bmatrix} 2 & 1 & 3 \end{bmatrix}
\begin{bmatrix} 2 \\ 1 \\ 1 \end{bmatrix} = 8
$$

In both cases above, the two matrices are conformable. However, if we interchange the order of the matrices as follows,

$$
\begin{bmatrix} 1 & 3 & 1 & 2 \\ 2 & 1 & 1 & 1 \end{bmatrix}
\begin{bmatrix} 2 & 3 \\ 1 & 1 \\ 3 & 1 \end{bmatrix}
$$

the matrices are no longer conformable for the product. It is evident that in general,

$$
\mathbf{AB} \neq \mathbf{BA}
$$

Indeed, $\mathbf{AB}$ may exist and $\mathbf{BA}$ may not exist, or vice versa, as in the above examples. We shall see later that for some special matrices,

$$
\mathbf{AB} = \mathbf{BA} \tag{B.74}
$$

When Eq. (B.74) is true, matrices $\mathbf{A}$ and $\mathbf{B}$ are said to **commute**. We must stress here again that in general, matrices do not commute. Operation (B.74) is valid only for some special cases.

In the matrix product $\mathbf{AB}$, matrix $\mathbf{A}$ is said to be **postmultiplied** by $\mathbf{B}$ or matrix $\mathbf{B}$ is said to be **premultiplied** by $\mathbf{A}$. We may also verify the following relationships:

$$
(\mathbf{A} + \mathbf{B})\mathbf{C} = \mathbf{AC} + \mathbf{BC} \tag{B.75}
$$

$$
\mathbf{C}(\mathbf{A} + \mathbf{B}) = \mathbf{CA} + \mathbf{CB} \tag{B.76}
$$

We can verify that any matrix $\mathbf{A}$ premultiplied or postmultiplied by the identity matrix $\mathbf{I}$ remains unchanged

$$
\mathbf{AI} = \mathbf{IA} = \mathbf{A} \tag{B.77}
$$

Of course, we must make sure that the order of $\mathbf{I}$ is such that the matrices are conformable for the corresponding product.

4. Multiplication of a Matrix by a Vector

Consider the matrix Eq. (B.69), which represents Eq. (B.65). The right-hand side of Eq. (B.69) is a product of the $m \times n$ matrix $\mathbf{A}$ and a vector $\mathbf{x}$. If, for the time being, we treat the vector $\mathbf{x}$ as if it were an $n \times 1$ matrix, then the product $\mathbf{Ax}$, according to the matrix multiplication rule, yields the right-hand side of Eq. (B.65). Thus, we may multiply a matrix by a vector by treating the vector as if it were an $n \times 1$ matrix. Note that the constraint of conformability still applies. Thus, in this case, $\mathbf{xA}$ is not defined and is meaningless.

5. Matrix Inversion

To define the inverse of a matrix, let us consider the set of equations

$$
\begin{bmatrix} y_1 \\ y_2 \\ \cdots \\ y_n \end{bmatrix} = \begin{bmatrix} a_{11} & a_{12} & \cdots & a_{1n} \\ a_{21} & a_{22} & \cdots & a_{2n} \\ \cdots\cdots\cdots\cdots\cdots \\ a_{n1} & a_{n2} & \cdots & a_{nn} \end{bmatrix} \begin{bmatrix} x_1 \\ x_2 \\ \cdots \\ x_n \end{bmatrix} \tag{B.78}
$$

We can solve this set of equations for $x_1, x_2, \ldots, x_n$ in terms of $y_1, y_2, \ldots, y_n$ by using Cramer's rule [see Eq. (B.48)]. This yields

$$
\begin{bmatrix} x_1 \\ x_2 \\ \vdots \\ x_n \end{bmatrix} = \begin{bmatrix} \frac{|\mathbf{D}_{11}|}{|\mathbf{A}|} & \frac{|\mathbf{D}_{21}|}{|\mathbf{A}|} & \cdots & \frac{|\mathbf{D}_{n1}|}{|\mathbf{A}|} \\ \frac{\mathbf{D}_{12}}{|\mathbf{A}|} & \frac{|\mathbf{D}_{22}|}{|\mathbf{A}|} & \cdots & \frac{|\mathbf{D}_{n2}|}{|\mathbf{A}|} \\ \cdots\cdots\cdots\cdots\cdots\cdots \\ \frac{|\mathbf{D}_{1n}|}{|\mathbf{A}|} & \frac{|\mathbf{D}_{2n}|}{|\mathbf{A}|} & \cdots & \frac{|\mathbf{D}_{nn}|}{|\mathbf{A}|} \end{bmatrix} \begin{bmatrix} y_1 \\ y_2 \\ \vdots \\ y_n \end{bmatrix} \tag{B.79}
$$

in which $|\mathbf{A}|$ is the determinant of the matrix $\mathbf{A}$ and $|\mathbf{D}_{ij}|$ is the cofactor of element a_{ij} in the matrix $\mathbf{A}$. The cofactor of element a_{ij} is given by $(-1)^{i+j}$ times the determinant of the $(n-1) \times (n-1)$ matrix that is obtained when the ith row and the jth column in matrix $\mathbf{A}$ are deleted.

We can express Eq. (B.78) in matrix form as

$$
\mathbf{y} = \mathbf{Ax} \tag{B.80}
$$

We can now define $\mathbf{A}^{-1}$, the inverse of a square matrix $\mathbf{A}$, with the property

$$
\mathbf{A}^{-1}\mathbf{A} = \mathbf{I} \qquad \text{(unit matrix)} \tag{B.81}
$$

Then, premultiplying both sides of Eq. (B.80) by $\mathbf{A}^{-1}$, we obtain

$$
\mathbf{A}^{-1}\mathbf{y} = \mathbf{A}^{-1}\mathbf{Ax} = \mathbf{Ix} = \mathbf{x}
$$

or

$$
\mathbf{x} = \mathbf{A}^{-1}\mathbf{y} \tag{B.82}
$$

A comparison of Eq. (B.82) with Eq. (B.79) shows that

$$\mathbf{A}^{-1} = \frac{1}{|\mathbf{A}|} \begin{bmatrix} |\mathbf{D}_{11}| & |\mathbf{D}_{21}| & \cdots & |\mathbf{D}_{n1}| \\ |\mathbf{D}_{12}| & |\mathbf{D}_{22}| & \cdots & |\mathbf{D}_{n2}| \\ \cdots\cdots\cdots\cdots\cdots\cdots\cdots \\ |\mathbf{D}_{1n}| & |\mathbf{D}_{2n}| & \cdots & |\mathbf{D}_{nn}| \end{bmatrix} \tag{B.83}$$

One of the conditions necessary for a unique solution of Eq. (B.78) is that the number of equations must equal the number of unknowns. This implies that the matrix $\mathbf{A}$ must be a square matrix. In addition, we observe from the solution as given in Eq. (B.79) that if the solution is to exist, $|\mathbf{A}| \neq 0$.† Therefore, the inverse exists only for a square matrix and only under the condition that the determinant of the matrix be nonzero. A matrix whose determinant is nonzero is a **nonsingular** matrix. Thus, an inverse exists only for a nonsingular (square) matrix. By definition, we have

$$\mathbf{A}^{-1}\mathbf{A} = \mathbf{I} \tag{B.84a}$$

Postmultiplying this equation by $\mathbf{A}^{-1}$ and then premultiplying by $\mathbf{A}$, we can show that

$$\mathbf{A}\mathbf{A}^{-1} = \mathbf{I} \tag{B.84b}$$

Note that the matrices $\mathbf{A}$ and $\mathbf{A}^{-1}$ commute.

■ **Example B.18**

Let us find $\mathbf{A}^{-1}$ if

$$\mathbf{A} = \begin{bmatrix} 2 & 1 & 1 \\ 1 & 2 & 3 \\ 3 & 2 & 1 \end{bmatrix}$$

Here

$$|\mathbf{D}_{11}| = -4, \quad |\mathbf{D}_{12}| = 8, \quad |\mathbf{D}_{13}| = -4$$

$$|\mathbf{D}_{21}| = 1, \quad |\mathbf{D}_{22}| = -1, \quad |\mathbf{D}_{23}| = -1$$

$$|\mathbf{D}_{31}| = 1, \quad |\mathbf{D}_{32}| = -5, \quad |\mathbf{D}_{33}| = 3$$

and $|\mathbf{A}| = -4$. Therefore

$$\mathbf{A}^{-1} = -\frac{1}{4} \begin{bmatrix} -4 & 1 & 1 \\ 8 & -1 & -5 \\ -4 & -1 & 3 \end{bmatrix} \qquad ■$$

---

†These two conditions imply that the number of equations is equal to the number of unknowns and that all the equations are independent.

B.9-3 Derivatives and Integrals of a Matrix

Elements of a matrix need not be constants; they may be functions of a variable. For example, if

$$\mathbf{A} = \begin{bmatrix} e^{-2t} & \sin t \\ e^t & e^{-t} + e^{-2t} \end{bmatrix} \tag{B.85}$$

then the matrix elements are functions of t. Here, it is helpful to denote $\mathbf{A}$ by $\mathbf{A}(t)$. Also, it would be helpful to define the derivative and integral of $\mathbf{A}(t)$.

The derivative of a matrix $\mathbf{A}(t)$ (with respect to t) is defined as a matrix whose ijth element is the derivative (with respect to t) of the ijth element of the matrix $\mathbf{A}$. Thus, if

$$\mathbf{A}(t) = [a_{ij}(t)]_{m \times n}$$

then

$$\frac{d}{dt}[\mathbf{A}(t)] = \left[\frac{d}{dt} a_{ij}(t) \right]_{m \times n} \tag{B.86a}$$

or

$$\dot{\mathbf{A}}(t) = [\dot{a}_{ij}(t)]_{m \times n} \tag{B.86b}$$

Thus the derivative of the matrix in Eq. (B.85) is given by

$$\dot{\mathbf{A}}(t) = \begin{bmatrix} -2e^{-2t} & \cos t \\ e^t & -e^{-t} - 2e^{-2t} \end{bmatrix}$$

Similarly, we define the integral of $\mathbf{A}(t)$ (with respect to t) as a matrix whose ijth element is the integral (with respect to t) of the ijth element of the matrix $\mathbf{A}$:

$$\int \mathbf{A}(t)\, dt = \left(\int a_{ij}(t)\, dt \right)_{m \times n} \tag{B.87}$$

Thus, for the matrix $\mathbf{A}$ in Eq. (B.85), we have

$$\int \mathbf{A}(t)\, dt = \begin{bmatrix} \int e^{-2t}\, dt & \int \sin\, dt \\ \int e^t\, dt & \int (e^{-t} + 2e^{-2t})\, dt \end{bmatrix}$$

We can readily prove the following identities:

$$\frac{d}{dt}(\mathbf{A} + \mathbf{B}) = \frac{d\mathbf{A}}{dt} + \frac{d\mathbf{B}}{dt} \tag{B.88a}$$

$$\frac{d}{dt}(c\mathbf{A}) = c\frac{d\mathbf{A}}{dt} \tag{B.88b}$$

$$\frac{d}{dt}(\mathbf{AB}) = \frac{d\mathbf{A}}{dt}\mathbf{B} + \mathbf{A}\frac{d\mathbf{B}}{dt} = \dot{\mathbf{A}}\mathbf{B} + \mathbf{A}\dot{\mathbf{B}} \tag{B.88c}$$

The proofs of identities (B.88a) and (B.88b) are trivial. We can prove Eq. (B.88c) as follows:

Let $\mathbf{A}$ be an $m \times n$ matrix and $\mathbf{B}$ an $n \times p$ matrix; then, if

$$\mathbf{C} = \mathbf{AB}$$

from Eq. (B.73), we have

$$c_{ik} = \sum_{j-1}^{n} a_{ij} b_{jk}$$

and

$$\dot{c}_{ik} = \underbrace{\sum_{j-1}^{n} \dot{a}_{ij} b_{jk}}_{d_{ik}} + \underbrace{\sum_{j-1}^{n} a_{ij} \dot{b}_{jk}}_{e_{ik}} \tag{B.89}$$

or

$$\dot{c}_{ij} = d_{ij} + e_{ik}$$

It can be seen from Eq. (B.89) and the multiplication rule that d_{ik} is the ikth element of matrix $\dot{\mathbf{A}}\mathbf{B}$ and e_{ik} is the ikth element of matrix $\mathbf{A}\dot{\mathbf{B}}$. Equation (B.88c) then follows.

If we let $\mathbf{B} = \mathbf{A}^{-1}$ in Eq. (B.88c), we obtain

$$\frac{d}{dt}(\mathbf{A}\mathbf{A}^{-1}) = \frac{d\mathbf{A}}{dt}\mathbf{A}^{-1} + \mathbf{A}\frac{d}{dt}\mathbf{A}^{-1}$$

But since

$$\frac{d}{dt}(\mathbf{A}\mathbf{A}^{-1}) = \frac{d}{dt}\mathbf{I} = 0$$

we have

$$\frac{d}{dt}(\mathbf{A}^{-1}) = -\mathbf{A}^{-1}\frac{d\mathbf{A}}{dt}\mathbf{A}^{-1} \tag{B.90}$$

B.9-4   The Characteristic Equation of a Matrix: The Cayley-Hamilton Theorem

For an $(n \times n)$ square matrix $\mathbf{A}$, any vector $\mathbf{x}$ ($\mathbf{x} \neq 0$) that satisfies the equation

$$\mathbf{A}\mathbf{x} = \lambda\mathbf{x} \tag{B.91}$$

is an **eigenvector** (or **characteristic vector**), and λ is the corresponding **eigenvalue** (or **characteristic value**) of $\mathbf{A}$. Equation (B.91) can be expressed as

$$(\mathbf{A} - \lambda\mathbf{I})\mathbf{x} = 0 \tag{B.92}$$

The solution for this set of homogeneous equations exists if and only if

$$|\mathbf{A} - \lambda\mathbf{I}| = |\lambda\mathbf{I} - \mathbf{A}| = 0 \tag{B.93a}$$

or

$$\begin{vmatrix} a_{11} - \lambda & a_{12} & \cdots & a_{1n} \\ a_{21} & a_{22} - \lambda & \cdots & a_{2n} \\ \cdots\cdots\cdots\cdots\cdots\cdots\cdots\cdots \\ a_{n1} & a_{n2} & \cdots & a_{nn} - \lambda \end{vmatrix} = 0 \qquad \text{(B.93b)}$$

Equation (B.93a) [or (B.93b)] is known as the **characteristic equation** of the matrix $\mathbf{A}$ and can be expressed as

$$Q(\lambda) = |\lambda \mathbf{I} - \mathbf{A}| = \lambda^n + a_{n-1}\lambda^{n-1} + \cdots + a_1\lambda + a_0\lambda^0 = 0 \qquad \text{(B.94)}$$

$Q(\lambda)$ is called the **characteristic polynomial** of the matrix $\mathbf{A}$. The n zeros of the characteristic polynomial are the eigenvalues of $\mathbf{A}$, and, corresponding to each eigenvalue, there is an eigenvector that satisfies Eq. (B.91).

The **Cayley-Hamilton theorem** states that every $n \times n$ matrix $\mathbf{A}$ satisfies its own characteristic equation. In other words, Eq. (B.94) is valid if λ is replaced by $\mathbf{A}$:

$$\mathbf{Q(A)} = \mathbf{A}^n + a_{n-1}\mathbf{A}^{n-1} + \cdots + a_1\mathbf{A} + a_0\mathbf{A}^0 = 0 \qquad \text{(B.95)}$$

Functions of a Matrix

The Cayley-Hamilton theorem can be used to evaluate functions of a square matrix $\mathbf{A}$ as shown below.

Consider a function $f(\lambda)$ in the form of an infinite power series:

$$f(\lambda) = \alpha_0 + \alpha_1\lambda + \alpha_2\lambda_2^2 + \cdots + \cdots = \sum_{i=0}^{\infty} \alpha_i \lambda^i \qquad \text{(B.96)}$$

Because λ satisfies the characteristic Eq. (B.94), we can express

$$\lambda^n = -a_{n-1}\lambda^{n-1} - a_{n-2}\lambda^{n-2} - \cdots - a_1\lambda - a_0 \qquad \text{(B.97)}$$

If we multiply both sides by λ, the left-hand side is λ^{n+1}, and the right-hand side contains the terms $\lambda^n, \lambda^{n-1}, \ldots, \lambda$. If we substitute λ^n in terms of $\lambda^{n-1}, \lambda^{n-2}, \ldots$, λ using Eq. (B.97), the highest power on the right-hand side is reduced to $n-1$. Continuing in this way, we see that λ^{n+k} can be expressed in terms of $\lambda^{n-1}, \lambda^{n-2}$, $\ldots, \lambda$ for any k. Hence the infinite series on the right-hand side of Eq. (B.96) can always be expressed in terms of $\lambda^{n-1}, \lambda^{n-2}, \ldots, \lambda$:

$$f(\lambda) = \beta_0 + \beta_1\lambda + \beta_2\lambda^2 + \cdots + \beta_{n-1}\lambda^{n-1} \qquad \text{(B.98)}$$

If we assume that there are n distinct eigenvalues $\lambda_1, \lambda_2, \ldots, \lambda_n$, then Eq. (B.98) holds for these n values of λ. The substitution of these values in Eq. (B.98) yields n simultaneous equations:

$$\begin{bmatrix} f(\lambda_1) \\ f(\lambda_2) \\ \cdots \\ f(\lambda_n) \end{bmatrix} = \begin{bmatrix} 1 & \lambda_1 & \lambda_1^2 & \cdots & \lambda_1^{n-1} \\ 1 & \lambda_2 & \lambda_2^2 & \cdots & \lambda_2^{n-1} \\ \cdots\cdots\cdots\cdots\cdots\cdots\cdots\cdots \\ 1 & \lambda_n & \lambda_n^2 & \cdots & \lambda_n^{n-1} \end{bmatrix} \begin{bmatrix} \beta_0 \\ \beta_1 \\ \cdots \\ \beta_{n-1} \end{bmatrix} \tag{B.99a}$$

and

$$\begin{bmatrix} \beta_0 \\ \beta_1 \\ \cdots \\ \beta_{n-1} \end{bmatrix} = \begin{bmatrix} 1 & \lambda_1 & \lambda_1^2 & \cdots & \lambda_1^{n-1} \\ 1 & \lambda_2 & \lambda_2^2 & \cdots & \lambda_2^{n-1} \\ \cdots\cdots\cdots\cdots\cdots\cdots\cdots\cdots \\ 1 & \lambda_n & \lambda_n^2 & \cdots & \lambda_n^{n-1} \end{bmatrix}^{-1} \begin{bmatrix} f(\lambda_1) \\ f(\lambda_2) \\ \cdots \\ f(\lambda_n) \end{bmatrix} \tag{B.99b}$$

Since $\mathbf{A}$ also satisfies Eq. (B.97), we may use a similar argument to show that if $f(\mathbf{A})$ is a function of a square matrix $\mathbf{A}$ expressed as an infinite power series in $\mathbf{A}$, then

$$f(\mathbf{A}) = \alpha_0 \mathbf{I} + \alpha_1 \mathbf{A} + \alpha_2 \mathbf{A}^2 + \cdots + \cdots = \sum_{i=0}^{\infty} \alpha_i \mathbf{A}^i \tag{B.100a}$$

and

$$f(\mathbf{A}) = \beta_0 \mathbf{I} + \beta_1 \mathbf{A} + \beta_2 \mathbf{A}^2 + \cdots + \beta_{n-1} \mathbf{A}^{n-1} \tag{B.100b}$$

in which the coefficients β_is are found in Eq. (B.99b). If some of the eigenvalues are repeated (multiple roots), the results are somewhat modified.

We shall demonstrate the utility of this result with the following two examples.

B.9-5 Computation of an Exponential and a Power of a Matrix

Let us compute $e^{\mathbf{A}t}$ defined by

$$e^{\mathbf{A}t} = \mathbf{I} + \mathbf{A}t + \frac{\mathbf{A}^2 t^2}{2!} + \cdots + \frac{\mathbf{A}^n t^n}{n!} + \cdots$$

$$= \sum_{k=0}^{\infty} \frac{\mathbf{A}^k t^k}{k!}$$

From Eq. (B.100b), we can express

$$e^{\mathbf{A}t} = \sum_{i=1}^{n-1} \beta_i (\mathbf{A}t)^i$$

in which the β_is are given by Eq. (B.99b), with $f(\lambda_i) = e^{\lambda_i t}$.

■ **Example B.19**

Let us consider the case where

$$\mathbf{A} = \begin{bmatrix} 0 & 1 \\ -2 & -3 \end{bmatrix}$$

The eigenvalues are

$$|\lambda \mathbf{I} - \mathbf{A}| = \begin{vmatrix} \lambda & -1 \\ 2 & \lambda + 3 \end{vmatrix} = \lambda^2 + 3\lambda + 2 = (\lambda + 1)(\lambda + 2) = 0$$

Hence $\lambda_1 = -1$, $\lambda_2 = -2$, and

$$e^{\mathbf{A}t} = \beta_0 \mathbf{I} + \beta_1 \mathbf{A}$$

in which

$$\begin{bmatrix} \beta_0 \\ \beta_1 \end{bmatrix} = \begin{bmatrix} 1 & -1 \\ 1 & -2 \end{bmatrix}^{-1} \begin{bmatrix} e^{-t} \\ e^{-2t} \end{bmatrix}$$

$$= \begin{bmatrix} 2 & -1 \\ 1 & -1 \end{bmatrix} \begin{bmatrix} e^{-t} \\ e^{-2t} \end{bmatrix} = \begin{bmatrix} 2e^{-t} - e^{-2t} \\ e^{-t} - e^{-2t} \end{bmatrix}$$

and

$$e^{\mathbf{A}t} = (2e^{-t} - e^{-2t}) \begin{bmatrix} 1 & 0 \\ 0 & 1 \end{bmatrix} + (e^{-t} - e^{-2t}) \begin{bmatrix} 0 & 1 \\ -2 & -3 \end{bmatrix}$$

$$= \begin{bmatrix} 2e^{-t} - e^{-2t} & (e^{-t} - e^{-2t}) \\ -2e^{-t} + 2e^{-2t} & -e^{-t} + 2e^{-2t} \end{bmatrix} \tag{B.101}$$

■

Computation of $\mathbf{A}^k$

As seen in Eq. (B.100b), we can express $\mathbf{A}^k$ as

$$\mathbf{A}^k = \beta_0 \mathbf{I} + \beta_1 \mathbf{A} + \cdots + \beta_{n-1} \mathbf{A}^{n-1}$$

in which the β_is are given by Eq. (B.99b) with $f(\lambda_i) = \lambda_i^k$. For a completed example of the computation of $\mathbf{A}^k$ by this method, see Example 10.12.

B.10 MISCELLANEOUS

B.10-1 L'Hôpital's Rule

If $\lim f(x)/g(x)$ results in the indeterministic form $0/0$ or ∞/∞, then

$$\lim \frac{f(x)}{g(x)} = \lim \frac{\dot{f}(x)}{\dot{g}(x)}$$

B.10-2 The Taylor and Maclaurin Series

$$f(x) = f(a) + \frac{(x-a)}{1!}\dot{f}(a) + \frac{(x-a)^2}{2!}\ddot{f}(a) + \cdots$$

$$f(x) = f(0) + \frac{x}{1!}\dot{f}(0) + \frac{x^2}{2!}\ddot{f}(0) + \cdots$$

B.10-3 Power Series

$$e^x = 1 + x + \frac{x^2}{2!} + \frac{x^3}{3!} + \cdots + \frac{x^n}{n!} + \cdots$$

$$\sin x = x - \frac{x^3}{3!} + \frac{x^5}{5!} - \frac{x^7}{7!} + \cdots$$

$$\cos x = 1 - \frac{x^2}{2!} + \frac{x^4}{4!} - \frac{x^6}{6!} + \frac{x^8}{8!} - \cdots$$

$$\tan x = x + \frac{x^3}{3} + \frac{2x^5}{15} + \frac{17x^7}{315} + \cdots \qquad x^2 < \pi^2/4$$

$$\tanh x = x - \frac{x^3}{3} + \frac{2x^5}{15} - \frac{17x^7}{315} + \cdots \qquad x^2 < \pi^2/4$$

$$(1+x)^n = 1 + nx + \frac{n(n-1)}{2!}x^2 + \frac{n(n-1)(n-2)}{3!}x^3 + \cdots + \binom{n}{k}x^k + \cdots + x^n$$

$$\approx 1 + nx \qquad |x| << 1$$

$$\frac{1}{1-x} = 1 + x + x^2 + x^3 + \cdots \qquad |x| < 1$$

B.10-4 Sums

$$\sum_{m=0}^{k} r^m = \frac{r^{k+1} - 1}{r - 1} \qquad r \neq 1$$

$$\sum_{m=M}^{N} r^m = \frac{r^{N+1} - r^M}{r - 1} \qquad r \neq 1$$

$$\sum_{m=0}^{k} \left(\frac{a}{b}\right)^m = \frac{a^{k+1} - b^{k+1}}{b^k(a-b)} \qquad a \neq b$$

B.10-5 Complex Numbers

$$e^{\pm j\pi/2} = \pm j$$

$$e^{\pm jn\pi} = \begin{cases} 1 & n \text{ even} \\ -1 & n \text{ odd} \end{cases}$$

$$e^{\pm j\theta} = \cos\theta \pm j\sin\theta$$

$$a + jb = re^{j\theta} \qquad r = \sqrt{a^2 + b^2}, \qquad \theta = \tan^{-1}\left(\tfrac{b}{a}\right)$$

$$(re^{j\theta})^k = r^k e^{jk\theta}$$

$$(r_1 e^{j\theta_1})(r_2 e^{j\theta_2}) = r_1 r_2 e^{j(\theta_1 + \theta_2)}$$

B.10-6 Trigonometric Identities

$$e^{\pm jx} = \cos x \pm j\sin x$$

$$\cos x = \tfrac{1}{2}[e^{jx} + e^{-jx}]$$

$$\sin x = \tfrac{1}{2j}[e^{jx} - e^{-jx}]$$

$$\cos\left(x \pm \tfrac{\pi}{2}\right) = \mp\sin x$$

$$\sin\left(x \pm \tfrac{\pi}{2}\right) = \pm\cos x$$

$$2\sin x\cos x = \sin 2x$$

$$\sin^2 x + \cos^2 x = 1$$

$$\cos^2 x - \sin^2 x = \cos 2x$$

$$\cos^2 x = \tfrac{1}{2}(1 + \cos 2x)$$

$$\sin^2 x = \tfrac{1}{2}(1 - \cos 2x)$$

$$\cos^3 x = \tfrac{1}{4}(3\cos x + \cos 3x)$$

$$\sin^3 x = \tfrac{1}{4}(3\sin x - \sin 3x)$$

$$\sin(x \pm y) = \sin x\cos y \pm \cos x\sin y$$

$$\cos(x \pm y) = \cos x\cos y \mp \sin x\sin y$$

$$\tan(x \pm y) = \frac{\tan x \pm \tan y}{1 \mp \tan x\tan y}$$

$$\sin x\sin y = \tfrac{1}{2}[\cos(x - y) - \cos(x + y)]$$

$$\cos x\cos y = \tfrac{1}{2}[\cos(x - y) + \cos(x + y)]$$

$$\sin x\cos y = \tfrac{1}{2}[\sin(x - y) + \sin(x + y)]$$

$$a\cos x + b\sin x = C\cos(x + \theta)$$

$$\text{in which } C = \sqrt{a^2 + b^2} \quad \text{and} \quad \theta = \tan^{-1}\left(\frac{-b}{a}\right)$$

B.10-7   Indefinite Integrals

$$\int u\,dv = uv - \int v\,du$$

$$\int f(x)\dot{g}(x)\,dx = f(x)g(x) - \int \dot{f}(x)g(x)\,dx$$

$$\int \sin ax\,dx = -\frac{1}{a}\cos ax \qquad\qquad \int \cos ax\,dx = \frac{1}{a}\sin ax$$

$$\int \sin^2 ax\,dx = \frac{x}{2} - \frac{\sin 2ax}{4a} \qquad\qquad \int \cos^2 ax\,dx = \frac{x}{2} + \frac{\sin 2ax}{4a}$$

$$\int x\sin ax\,dx = \frac{1}{a^2}(\sin ax - ax\cos ax)$$

$$\int x\cos ax\,dx = \frac{1}{a^2}(\cos ax + ax\sin ax)$$

$$\int x^2\sin ax\,dx = \frac{1}{a^3}(2ax\sin ax + 2\cos ax - a^2x^2\cos ax)$$

$$\int x^2\cos ax\,dx = \frac{1}{a^3}(2ax\cos ax - 2\sin ax + a^2x^2\sin ax)$$

$$\int \sin ax\sin bx\,dx = \frac{\sin(a-b)x}{2(a-b)} - \frac{\sin(a+b)x}{2(a+b)} \qquad a^2 \neq b^2$$

$$\int \sin ax\cos bx\,dx = -\left[\frac{\cos(a-b)x}{2(a-b)} + \frac{\cos(a+b)x}{2(a+b)}\right] \qquad a^2 \neq b^2$$

$$\int \cos ax\cos bx\,dx = \frac{\sin(a-b)x}{2(a-b)} + \frac{\sin(a+b)x}{2(a+b)} \qquad a^2 \neq b^2$$

$$\int e^{ax}\,dx = \frac{1}{a}e^{ax}$$

$$\int xe^{ax}\,dx = \frac{e^{ax}}{a^2}(ax - 1)$$

$$\int x^2 e^{ax}\,dx = \frac{e^{ax}}{a^3}(a^2x^2 - 2ax + 2)$$

$$\int e^{ax}\sin bx\,dx = \frac{e^{ax}}{a^2 + b^2}(a\sin bx - b\cos bx)$$

$$\int e^{ax}\cos bx\,dx = \frac{e^{ax}}{a^2 + b^2}(a\cos bx + b\sin bx)$$

$$\int \frac{1}{x^2 + a^2}\,dx = \frac{1}{a}\tan^{-1}\frac{x}{a}$$

$$\int \frac{x}{x^2 + a^2}\,dx = \frac{1}{2}\ln(x^2 + a^2)$$

B.10-8   Differentiation Table

$$\frac{d}{dx}f(u) = \frac{d}{du}f(u)\frac{du}{dx} \qquad\qquad \frac{d}{dx}a^{bx} = b(\ln a)a^{bx}$$

$$\frac{d}{dx}(uv) = u\frac{dv}{dx} + v\frac{du}{dx} \qquad\qquad \frac{d}{dx}\sin ax = a\cos ax$$

$$\frac{d}{dx}\left(\frac{u}{v}\right) = \frac{v\frac{du}{dx} - u\frac{dv}{dx}}{v^2} \qquad\qquad \frac{d}{dx}\cos ax = -a\sin ax$$

$$\frac{dx^n}{dx} = nx^{n-1} \qquad\qquad \frac{d}{dx}\tan ax = \frac{a}{\cos^2 ax}$$

$$\frac{d}{dx}\ln(ax) = \frac{1}{x} \qquad\qquad \frac{d}{dx}(\sin^{-1}ax) = \frac{a}{\sqrt{1 - a^2x^2}}$$

$$\frac{d}{dx}\log(ax) = \frac{\log e}{x} \qquad\qquad \frac{d}{dx}(\cos^{-1}ax) = \frac{-a}{\sqrt{1 - a^2x^2}}$$

$$\frac{d}{dx}e^{bx} = be^{bx} \qquad\qquad \frac{d}{dx}(\tan^{-1}ax) = \frac{a}{1 + a^2x^2}$$

B.10-9   Some Useful Constants

$$\pi \approx 3.1415926535$$

$$e \approx 2.7182818284$$

$$\frac{1}{e} \approx 0.3678794411$$

$$\log_{10} 2 = 0.30103$$

$$\log_{10} 3 = 0.47712$$

B.10-10   Solution of Quadratic and Cubic Equations

Any **quadratic** equation can be reduced to the form

$$ax^2 + bx + c = 0$$

The solution of this equation is given by

$$x = \frac{-b \pm \sqrt{b^2 - 4ac}}{2a}$$

A general **cubic** equation

$$y^3 + py^2 + qy + r = 0$$

may be reduced to the **depressed cubic** form

$$x^3 + ax + b = 0$$

by substituting

$$y = x - \tfrac{p}{3}$$

This yields

$$a = \tfrac{1}{3}(3q - p^2) \qquad b = \tfrac{1}{27}(2p^3 - 9pq + 27r)$$

Now let

$$A = \sqrt[3]{-\tfrac{b}{2} + \sqrt{\tfrac{b^2}{4} + \tfrac{a^3}{27}}}, \qquad B = \sqrt[3]{-\tfrac{b}{2} - \sqrt{\tfrac{b^2}{4} + \tfrac{a^3}{27}}}$$

The solution of the depressed cubic is

$$x = A + B, \qquad x = -\tfrac{A+B}{2} + \tfrac{A-B}{2}\sqrt{-3}, \qquad x = -\tfrac{A+B}{2} - \tfrac{A-B}{2}\sqrt{-3}$$

and

$$y = x - \tfrac{p}{3}$$

REFERENCES

1. Asimov, Isaac, *Asimov on Numbers,* Bell Publishing Co., N.Y., 1982.

2. Calinger, R., Ed., *Classics of Mathematics,* Moore Publishing Co.,Oak Park, IL., 1982.

3. Hogben, Lancelot, *Mathematics in the Making,* Doubleday & Co. Inc., New York, 1960.

4. Cajori, Florian, *A History of Mathematics*, 4th ed., Chelsea, New York, 1985.

5. Encyclopaedia Britannica, 15th ed., *Micropaedia*, vol. 11, p. 1043, 1982.

6. Singh, Jagjit, *Great Ideas of Modern Mathematics,* Dover, New York, 1959.

7. Dunham, William, *Journey through Genius,* Wiley, New York, 1990.

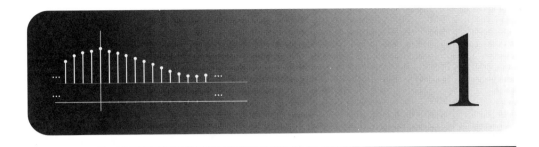

Introduction to Systems

This chapter introduces important concepts and qualitatively explains many of the hows and whys of systems theory, thus building a solid foundation for assimilating of the material in the remainder of the book.

1.1 SIGNALS AND SYSTEMS

Signals

A signal, as the term implies, is a set of information or data. Examples include a telephone or a television signal, monthly sales of a corporation, or the daily closing prices of a stock market (e.g., Dow Jones averages). In all these examples, the signals are functions of the independent variable *time*. This is not always the case, however. When an electrical charge is distributed over a body, for instance, the signal is the charge density, which is a function of *space* rather than time. In this book we deal almost exclusively with signals that are functions of time. The discussion, however, applies equally well to other independent variables.

A signal that is specified for every value of time t (Fig. 1.1a) is a **continuous-time signal**, and a signal that is specified only at discrete values of t (Fig. 1.1b) is a **discrete-time signal**. Telephone and television signals are continuous-time signals, whereas the quarterly GNP, monthly sales of a corporation and stock market daily averages are discrete-time signals.

Systems

Signals may be processed further by **systems**, which may modify them or extract an additional information from them. Thus, a system is an entity that *processes* a set of signals (inputs) to yield another set of signals (outputs). A system may be made up of physical components, as in electrical, mechanical, or hydraulic systems (hardware realization), or may be an algorithm that computes an output from an input signal (software realization).

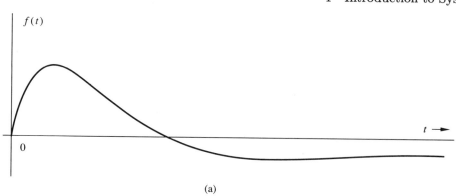

(a)

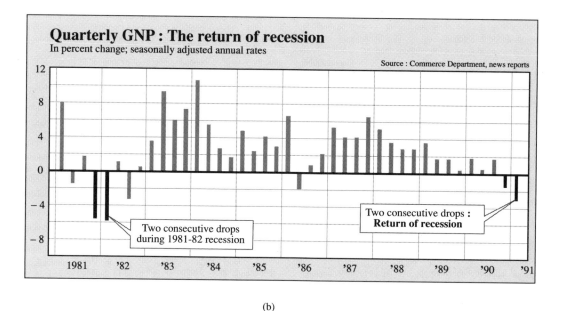

(b)

Fig. 1.1   Continuous-time and discrete-time signals.

A system is characterized by its **inputs**, its **outputs** (or **responses**), and the **rules of operation** (or **laws**) adequate to describe its behavior. For example, in electrical systems, the laws of operation are the familiar voltage-current relationships for the resistors, capacitors, inductors, transformers, transistors, and so on, as well as the laws of interconnection (i.e., Kirchhoff's laws). Using these laws, we derive mathematical equations relating the outputs to the inputs. These equations then represent a **mathematical model** of the system. Thus a system is characterized by its inputs, its outputs, and its mathematical model.

A system can be conveniently illustrated by a "black box" with one set of accessible terminals where the input variables $f_1(t)$, $f_2(t)$, ..., $f_j(t)$ are applied and another set of accessible terminals where the output variables $y_1(t)$, $y_2(t)$, ..., $y_k(t)$ are observed. Note that the direction of the arrows for the variables in Fig. 1.2 is always from cause to effect.

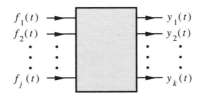

Fig. 1.2 Representation of a system.

In addition to the input and output signals, we may also have intermediate signals, which will appear inside the box. Intermediate signals can be regarded as outputs since they are also the result of input signals; however, because we are not directly interested in these signals, they are conveniently tucked inside the box and are called *suppressed outputs*.[1] Only the outputs of direct interest are shown outside the box at the output terminals.

The study of systems consists of three major areas: mathematical modeling, analysis, and design. Although we shall be dealing with mathematical modeling, our main concern is with analysis and design. The major portion of this book is devoted to the analysis problem—how to determine the system outputs for the given inputs and a given mathematical model of the system (or rules governing the system). To a lesser extent, we will also consider the problem of design or synthesis—how to construct a system which will produce a desired set of outputs for the given inputs.

In general, inputs must be known from $t = -\infty$ (or at least from the instant when the system was created) to the present moment. It is impractical, however, to keep a record of inputs from $t = -\infty$. In practice, we often know the inputs starting only from some $t = t_0$, with total ignorance of their earlier history; in such cases, we must also know auxiliary pieces of information indicating the **state** or situation of the system at $t = t_0$. As an example, consider a simple RC circuit with a current source $f(t)$ as its input (Fig. 1.3). The output voltage $y(t)$ is given by

$$y(t) = Rf(t) + \frac{1}{C} \int_{-\infty}^{t} f(\tau) \, d\tau \tag{1.1a}$$

The limits of the integral on the right-hand side are from $-\infty$ to t because this integral represents the capacitor charge due to the current $f(t)$ flowing in the capacitor, and this charge is the result of the current flowing in the capacitor from $-\infty$. Now, Eq. (1.1a) can be expressed as

$$y(t) = Rf(t) + \frac{1}{C} \int_{-\infty}^{0} f(\tau) \, d\tau + \frac{1}{C} \int_{0}^{t} f(\tau) \, d\tau \tag{1.1b}$$

However, the middle term on the right-hand side is $v_C(0)$, the capacitor voltage at $t = 0$. Therefore

$$y(t) = v_C(0) + Rf(t) + \frac{1}{C} \int_{0}^{t} f(\tau) \, d\tau \tag{1.1c}$$

This equation can be readily generalized as

$$y(t) = v_C(t_0) + Rf(t) + \frac{1}{C} \int_{t_0}^{t} f(\tau) \, d\tau \tag{1.1d}$$

From Eq. (1.1a), the output voltage $y(t)$ at any instant t can be computed if we know the current flowing in the capacitor throughout its entire past $(-\infty$ to $t)$.

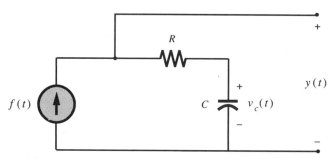

Fig. 1.3   An example of a simple electrical system.

Alternatively, if we know the input current $f(t)$ from some moment t_0 onward, then, using Eq. (1.1d), we can still calculate $y(t)$ for $t \geq t_0$ from a knowledge of the input current, provided we know $v_C(t_0)$, the initial capacitor voltage (voltage at t_0). Thus $v_C(t_0)$ contains all the relevant information about the circuit's entire past ($-\infty$ to t_0) that we need to compute $y(t)$ for $t \geq t_0$. Therefore, the response of a system at $t > t_0$ can be determined from its input(s) during the interval t_0 to t and from certain **initial conditions** at $t = t_0$.

In the preceding example, we needed only one initial condition. In more complex systems, several initial conditions may be necessary. We know, for example, that in passive RLC networks, the initial values of all inductor currents and all capacitor voltages† are needed to determine the outputs for $t \geq 0$ if the inputs are given over the interval $[0, t]$.

1.2   CLASSIFICATION OF SYSTEMS

Systems may be classified broadly in the following categories:‡

1. Linear and nonlinear systems;
2. Constant-parameter and time-varying-parameter systems;
3. Instantaneous (memoryless) and dynamic (with memory) systems;
4. Causal and noncausal systems;
5. Lumped-parameter and distributed-parameter systems;
6. Continuous-time and discrete-time systems;
7. Analog and Digital systems;

1.2-1   Linear and Nonlinear Systems

The Concept of Linearity

A system whose output is proportional to its input is an *example* of a linear system. But linearity implies more than this; it also implies the **superposition** or **additivity property**. This property states that if several causes are acting on a system, then the total effect on the system due to all these causes can be determined by considering each cause separately while assuming all the other causes to be zero.

---

† Strictly speaking, independent inductor currents and capacitor voltages.
‡ Other classifications, such as deterministic and probabilistic systems, are beyond the scope of this text and are not considered.

The total effect is then the sum of all the component effects. This property may be expressed as follows: For a linear system, if a cause c_1 acting alone has an effect e_1, and if another cause c_2 also acting alone has an effect e_2, then, with both causes acting on the system, the total effect will be $e_1 + e_2$. Thus, if

$$c_1 \longrightarrow e_1 \qquad \text{and} \qquad c_2 \longrightarrow e_2 \qquad (1.2)$$

then for all c_1 and c_2

$$c_1 + c_2 \longrightarrow e_1 + e_2 \qquad (1.3)$$

In addition, a linear system must satisfy the **homogeneity property**, which states that if a cause is increased k-fold, the effect also increases k-fold. Thus if

$$c \longrightarrow e$$

then for all k

$$kc \longrightarrow ke \qquad (1.4)$$

Thus, linearity implies two properties: homogeneity and additivity.† Both these properties can be combined into one property (**superposition**), which is expressed as follows: If

$$c_1 \longrightarrow e_1 \qquad \text{and} \qquad c_2 \longrightarrow e_2$$

then for all values of constants k_1 and k_2,

$$k_1 c_1 + k_2 c_2 \longrightarrow k_1 e_1 + k_2 e_2 \qquad (1.5)$$

Response of a Linear System

For the sake of simplicity, we discuss below only **single-input, single-output** (**SISO**) systems. But the discussion can be readily extended to **multiple-input, multiple-output** (**MIMO**) systems.

A system's output for $t \geq 0$ is the result of two independent causes: the initial conditions of the system (or the system state) at $t = 0$ and the input $f(t)$ for $t \geq 0$. If a system is to be linear, the output must be the sum of the two components resulting from these two causes: first, the **zero-input response** component that results only from the initial conditions at $t = 0$ with the input $f(t) = 0$ for $t \geq 0$, and then the **zero-state response** component that results only from the input $f(t)$ for $t \geq 0$ when the initial conditions (at $t = 0$) are assumed to be zero. When all the appropriate initial conditions are zero, the system is said to be in **zero state**. The system output is zero when the input is zero only if the system is in zero state.

In summary, a linear system response can be expressed as the sum of a zero-input and a zero-state component:

$$\textbf{Total response = zero-input response + zero-state response} \qquad (1.6)$$

---

† For all practical purposes, additivity implies homogeneity. It follows from the additivity property (1.3) that if a cause c results in an effect e, then the cause $2c$ results in the effect $2e$, and, in general, the cause kc results in the effect ke for all rational values of k. Homogeneity property ensures that this proportionality holds for irrational values of k as well. A linear system must also satisfy the additional condition of **smoothness**, where small changes in the system's inputs must result in small changes in its outputs.[2] This condition is usually ignored because practical systems satisfying the superposition property generally satisfy this condition as well.

This property of linear systems which permits the separation of an output into components resulting from the initial conditions and from the input is called the **decomposition property**.

For the RC circuit of Fig. 1.3, the response $y(t)$ was found to be [see Eq. (1.1c)]

$$y(t) = \underbrace{v_C(0)}_{z-i \text{ component}} + \underbrace{Rf(t) + \frac{1}{C} \int_0^t f(\tau)\, d\tau}_{z-s \text{ component}} \tag{1.7}$$

From Eq. (1.7), it is clear that if the input $f(t) = 0$ for $t \geq 0$, the output $y(t) = v_C(0)$. Hence $v_C(0)$ is the zero-input component of the response $y(t)$. Similarly, if the system state (the voltage v_C in this case) is zero at $t = 0$, the output is given by the second component on the right-hand side of Eq. (1.7). Clearly this is the zero-state component of the response $y(t)$.

In addition to the decomposition property, linearity implies that both the zero-input and zero-state components must obey the principle of superposition with respect to each of their respective causes. For example, if we increase the initial condition k-fold, the zero-input component must also increase k-fold. Similarly, if we increase the input k-fold, the zero-state component must also increase k-fold. These facts can be readily verified from Eq. (1.7) for the RC circuit in Fig. 1.3. For instance, if we double the initial condition $v_C(0)$, the zero-input component doubles; if we double the input $f(t)$, the zero-state component doubles.

■ **Example 1.1**

Show that the system described by the equation

$$\frac{dy}{dt} + 3y(t) = f(t) \tag{1.8}$$

is linear.

Let the system response to the inputs $f_1(t)$ and $f_2(t)$ be $y_1(t)$ and $y_2(t)$, respectively. Then

$$\frac{dy_1}{dt} + 3y_1(t) = f_1(t)$$

and

$$\frac{dy_2}{dt} + 3y_2(t) = f_2(t)$$

Multiplying the first equation by k_1, the second with k_2, and adding them yields

$$\frac{d}{dt}\left[k_1 y_1(t) + k_2 y_2(t)\right] + 3\left[k_1 y_1(t) + k_2 y_2(t)\right] = k_1 f_1(t) + k_2 f_2(t)$$

But this equation is the system equation [Eq. (1.8)] with

$$f(t) = k_1 f_1(t) + k_2 f_2(t)$$

and

$$y(t) = k_1 y_1(t) + k_2 y_2(t)$$

Therefore, when the input is $k_1 f_1(t) + k_2 f_2(t)$, the system response is $k_1 y_1(t) + k_2 y_2(t)$. Consequently, the system is linear. Using this argument, we can readily generalize the result to show that a system described by a differential equation of the form

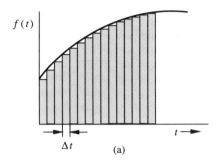

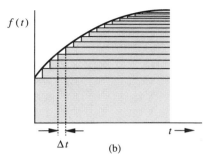

Fig. 1.4 Signal representation in terms of impulse and step components.

$$\frac{d^n y}{dt^n} + a_{n-1}\frac{d^{n-1} y}{dt^{n-1}} + \cdots + a_0 y = b_m\frac{d^m f}{dt^m} + \cdots + b_1\frac{df}{dt} + b_0 f \qquad (1.9)$$

is a linear system. The coefficients a_i and b_i in this equation can be constants or functions of time. ■

△ **Exercise E1.1**
Show that the system described by

$$\frac{dy}{dt} + t^2 y(t) = (2t + 3)f(t)$$

is linear. This is an example of a linear, time-varying parameter system.   ▽

△ **Exercise E1.2**
Show that a system described by the following equation is nonlinear:

$$y(t)\frac{dy}{dt} + 3y(t) = f(t)   ▽$$

More Comments on Linear Systems

Almost all systems observed in practice become nonlinear when large enough signals are applied to them. However, many systems show linear behavior for small signals. The analysis of nonlinear systems is generally difficult; nonlinearities can arise in so many ways that describing them with a common mathematical form is impossible; not only is each system a category in itself, but even for a given system, changes in initial conditions or input amplitudes may change the nature of the problem. On the other hand, the superposition property of linear systems is a powerful unifying principle which allows for a general solution. The superposition property (linearity) greatly simplifies the analysis of linear systems. Because of the decomposition property, we can evaluate separately the two components of the output. The zero-input component can be computed by assuming the input to be zero, and the zero-state component can be computed by assuming zero initial conditions. Moreover, if we express an input $f(t)$ as a sum of simpler functions,

$$f(t) = a_1 f_1(t) + a_2 f_2(t) + \cdots + a_m f_m(t)$$

then, by virtue of linearity, the response $y(t)$ is given by

$$y(t) = a_1 y_1(t) + a_2 y_2(t) + \cdots + a_m y_m(t) \qquad (1.10)$$

where $y_k(t)$ is the zero-state response to an input $f_k(t)$. This apparently trivial observation has profound implications. As we shall see repeatedly in later chapters, it proves extremely useful and opens new avenues for analyzing linear systems.

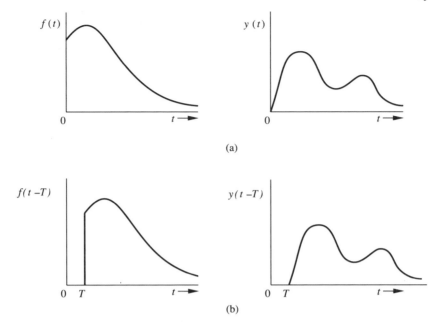

Fig. 1.5 Time-invariance property.

As an example, consider an arbitrary input $f(t)$ such as the one shown in Fig. 1.4a. We can approximate $f(t)$ with a sum of rectangular pulses of width Δt and of varying heights. The approximation improves as $\Delta t \to 0$, when the rectangular pulses become impulses spaced Δt seconds apart (with $\Delta t \to 0$). Thus, an arbitrary input can be replaced by a weighted sum of impulses spaced Δt ($\Delta t \to 0$) seconds apart. Therefore, if we know the system response to a unit impulse, we can immediately determine the system response to an arbitrary input $f(t)$ by adding the system response to each impulse component of $f(t)$. A similar situation is shown in Fig. 1.4b. There $f(t)$ is approximated by a sum of step functions of varying magnitude and spaced Δt seconds apart. The approximation improves as Δt becomes smaller. Therefore, if we know the system response to a unit step input, we can compute the system response to any arbitrary input $f(t)$ with relative ease. Time-domain analysis of linear systems (discussed in Chapters 2 and 3) uses this approach.

In Chapters 4 and 5, we employ the same approach but instead use sinusoids or exponentials as our basic signal components. There we show that any arbitrary input signal can be expressed as a weighted sum of sinusoids (or exponentials) having various frequencies. Thus a knowledge of the system response to sinusoids enables us to determine the system response to an arbitrary input $f(t)$.

1.2-2 Time-Invariant and Time-Varying Parameter Systems

Systems whose parameters do not change with time are **time-invariant** (also **constant-parameter**) systems. For any given initial state in a time-invariant system, the shape of the output will depend only on the shape of the input and not on the instant at which the input is applied. If the input is delayed by T seconds, the output is the same as before but is delayed by T. This property is expressed graphically in Fig. 1.5.

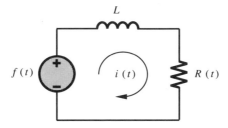

Fig. 1.6 An example of a linear time-varying system.

It is possible to verify that the system in Fig. 1.3 is a time-invariant system. All systems with input-output relationships described by linear differential equations of the form (1.9) are linear time-invariant (LTI) systems when the coefficients of such equations are constants. If the coefficients a_i and b_i in these equations are functions of time, then the system is a linear **time-varying** system. (Networks composed of RLC elements and other commonly used active elements such as transistors are time-invariant systems). A familiar example of a time-varying system is the carbon microphone, in which the resistance R is a function of the mechanical pressure generated by sound waves on the carbon granules of the microphone. An equivalent circuit for the microphone is given in Fig. 1.6. The response is the current $i(t)$, and the equation describing the circuit is

$$L\frac{di(t)}{dt} + R(t)i(t) = f(t)$$

One of the coefficients in this equation, $R(t)$, is time-varying.

1.2-3 Instantaneous and Dynamic Systems

As observed earlier, a system's output at any instant t generally depends upon the entire past input. However, in a special class of systems the output at any instant t depends only on its input at that instant. In resistive networks, for example, any output of the network at some instant t depends only on the input at the instant t. In these systems, past history is irrelevant in determining the response. Such systems are said to be **instantaneous** or **memoryless** systems. More precisely, a system is said to be instantaneous (or memoryless) if its output at any instant t depends, at most, on the strength of its input(s) at the same instant but not on any past or future values of the input(s). Otherwise, the system is said to be **dynamic** (or a system with memory). A system whose response at t is completely determined by the input signals over the past T seconds [interval from $(t - T)$ to t] is a **finite-memory system** with a memory of T seconds. Networks containing inductive and capacitive elements generally have infinite memory because the response of such networks at any instant t is determined by their inputs over the entire past $(-\infty, t)$. This is true for the RC circuit of Fig. 1.3.

In this book we will generally examine dynamic systems. Instantaneous systems are a special case of dynamic systems.

1.2-4 Causal and Noncausal Systems

A **causal** (also known as a **physical** or **non-anticipative**) system is one for which the output at any instant t_0 depends only on the value of the input $f(t)$ for

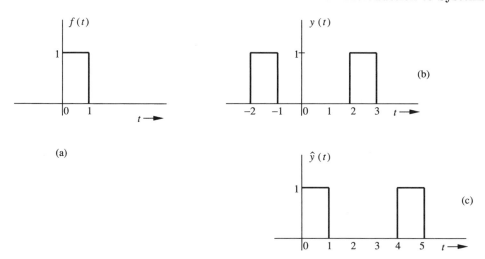

Fig. 1.7 A noncausal system and its realization by a delayed causal system.

$t \le t_0$. In other words, the value of the output at the present instant depends only on the past and present values of the input $f(t)$, not on its future values. A system that violates this condition is called a **noncausal** (or **anticipative**) system. To put it simply, a system is causal if the output does not start before the input is applied; otherwise it is noncausal.

Any practical system that operates in real time† must necessarily be causal. We do not yet know how to build a system that can respond to future inputs (inputs not yet applied). A noncausal system is a prophetic system that knows the future input and acts on it in the present. Thus, if we apply an input starting at $t = 0$ to a noncausal system, the output would begin even before $t = 0$. As an example, consider the system specified by

$$y(t) = f(t - 2) + f(t + 2) \tag{1.11}$$

For the input $f(t)$ shown in Fig. 1.7a, the output $y(t)$ (Fig. 1.7b) starts even before the input is applied. Equation (1.11) shows that $y(t)$, the output at t, is given by the sum of the input values two seconds before and two seconds after t (at $t - 2$ and $t + 2$ respectively). But if we are operating the system in real time at t, we do not know what the value of the input will be two seconds later. Thus it is impossible to implement this system in real time. For this reason, noncausal systems are unrealizable in **real time**.

Why Study Noncausal systems?

From the above discussion it may seem that noncausal systems have no practical purpose. This is not the case; they are valuable in the study of systems for several reasons. First, noncausal systems are realizable when the independent variable is other than "time" (e.g., **space**). Consider, for example, an electric charge of density $q(x)$ placed along the x-axis for $x \ge 0$. This charge density produces an electric field $E(x)$ that is present at every point on the x-axis from $x = -\infty$ to ∞. The

---

†In real-time operations, the response to an input is essentially simultaneous (contemporaneous) with the input itself.

Noncausal systems are realizable with time delay!

input [i.e., the charge density $q(x)$] starts at $x = 0$, but its output [the electric field $E(x)$] begins before $x = 0$. Clearly, this space charge system is noncausal. This discussion shows that only temporal systems (systems with time as independent variable) must be causal in order to be realizable. The terms "before" and "after" have a special connection to causality only when the independent variable is time. This connection is lost for variables other than time. Nontemporal systems, such as those occurring in optics, can be noncausal and still realizable.

Moreover, even for temporal systems, such as those used for signal processing, the study of noncausal systems is important. In such systems we may have all input data prerecorded. (This often happens with speech, geophysical, and meteorological signals, and with space probes.) In such cases, the input's future values are available to us. For example, suppose we had a set of input signal records available for the system described by Eq. (1.11). We can then compute $y(t)$ since, for any t, we need only refer to the records to find the input's value two seconds before and two seconds after t. Thus noncausal systems can be realized, although not in real time. We may therefore be able to realize a noncausal system, provided that we are willing to accept a time delay in the output. Consider a system whose output $\widehat{y}(t)$ is the same as $y(t)$ in Eq. (1.11) delayed by two seconds (Fig 1.7c), so that

$$\widehat{y}(t) = y(t - 2)$$
$$= f(t - 4) + f(t)$$

Here the value of the output $\widehat{y}$ at any instant t is the sum of the values of the input f at t and at the instant four seconds earlier [at $(t - 4)$]. In this case, the output at any instant t does not depend on future values of the input, and the system is causal. The output of this system is identical to that in Eq. (1.11) or Fig. 1.7b except for a delay of two seconds. Thus a noncausal system may be realized or satisfactorily approximated in real time by using a causal system with a delay.

A third reason for studying noncausal systems is that they provide an upper bound on the performance of causal systems. For example, if we wish to design a filter for separating a signal from noise, then the optimum filter is invariably a noncausal system. Although unrealizable, this noncausal system's performance acts as the upper limit on what can be achieved and gives us a standard for evaluating the performance of causal filters.

At first glance, noncausal systems may seem inscrutable. Actually, there is nothing mysterious about these systems and their approximate realization through using physical systems with delay. If we want to know what will happen one year from now, we have two choices: go to a prophet (an unrealizable person), who can give the answers immediately, or go to a wise man and allow him a delay of one year to give us the answer! If the wise man is truly wise, he may even be able to shrewdly guess the future very closely with a delay of less than a year by studying trends. Such is the case with noncausal systems—nothing more and nothing less.

1.2-5 Lumped-Parameter and Distributed-Parameter Systems

In the study of electrical systems, we make use of voltage-current relationships for various components (Ohm's law, for example). In doing so, we implicitly assume that the current in any system component (resistor, inductor, etc.) is the same at every point throughout that component. Thus we assumes that electrical signals are propagated instantaneously throughout the system. In reality, electrical signals are electromagnetic space waves requiring some finite propagation time. An electric current, for example, propagates through a component with a finite velocity and therefore may exhibit different values at different locations in the same component. Thus, an electric current is a function not only of time but also of space. However, if the variation in the current with time is slow compared to the time required for the current to propagate through the component, we may assume that the current is constant throughout the component. This assumption requires the wavelength of the signal to be much greater than the dimensions of the component. This is the assumption made in **lumped-parameter systems**, where each component is regarded as being lumped at one point in space. Such an assumption is justified at lower frequencies (higher wavelength). Therefore, in lumped-parameter models, signals can be assumed to be functions of time alone. For such systems, the system equations require only one independent variable (time) and therefore are ordinary differential equations.

In contrast, for **distributed-parameter systems** such as transmission lines, waveguides, antennas, and microwave tubes, the system dimensions cannot be assumed to be small compared to the wavelengths of the signals; thus the lumped-parameter assumption breaks down. The signals here are functions of space as well as of time, leading to mathematical models consisting of partial differential equations.[3] The discussion in this book will be restricted to lumped-parameter systems only.

1.2-6 Continuous-Time and Discrete-Time Systems

Signals defined or specified over a continuous range of time are *continuous-time signals*, denoted by symbols $f(t)$, $y(t)$, etc. Systems whose inputs and outputs are continuous-time signals are **continuous-time systems**. On the other hand, signals defined only at discrete instants of time t_0, t_1, t_2, ..., t_k, ..., are *discrete-time signals*, denoted by the symbols $f(t_k)$, $y(t_k)$, etc., where k is some integer. Systems whose inputs and outputs are discrete-time signals are **discrete-time systems**. A digital computer is a familiar example of this type of system. We assume that the discrete instants t_0, t_1, t_2, ... are uniformly spaced so that

$$t_{k+1} - t_k = T \qquad \text{for all } k.$$

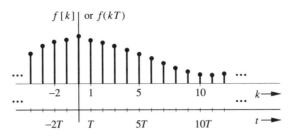

Fig. 1.8 A discrete-time signal.

With uniform spacing, discrete-time signals can be represented by $f(kT)$, $y(kT)$, etc.; for convenience we further simplify this notation to $f[k]$, $y[k]$, ..., where it is understood that $f[k] = f(kT)$ and that k is some integer. A typical discrete-time signal is shown in Fig. 1.8. A discrete-time signal may also be viewed as a sequence of numbered values $f[0]$, $f[1]$, $f[2]$, ... Thus a discrete-time system may be seen as processing a sequence of numbers $f[k]$ and yielding as an output another sequence of numbers $y[k]$.

Discrete-time signals arise naturally in situations which are inherently discrete-time, such as population studies, amortization problems, national income models, or radar tracking. They may also arise as a result of sampling continuous-time signals in sampled data systems, digital filtering, and the like. Digital filtering is a particularly interesting application in which continuous-time signals are processed by using discrete-time systems as shown in Fig. 1.9. A continuous-time signal $f(t)$ is first sampled to convert it into a discrete-time signal $f[k]$, which then is processed by the discrete-time system to yield a discrete-time output $y[k]$. A continuous-time signal $y(t)$ is finally constructed from $y[k]$. In this manner we can process a continuous-time signal with an appropriate discrete-time system such as a digital computer. Because discrete-time systems have several significant advantages over continuous-time systems, there is an accelerating trend toward processing continuous-time signals with discrete-time systems.

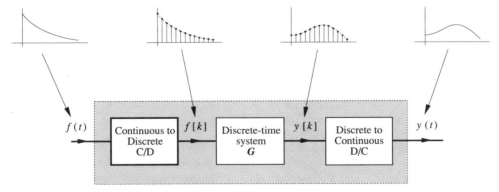

Fig. 1.9 Processing continuous-time signals by discrete-time systems.

1.2-7 Analog and Digital Systems

A signal whose amplitude can take on any value in a continuous range is an **analog signal**. This means that an analog signal amplitude can take on an infinite

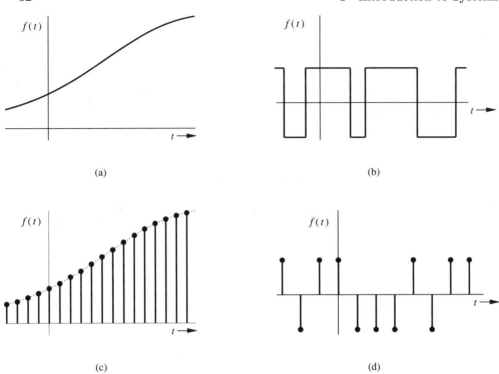

Fig. 1.10   Examples of signals: (a) an analog and continuous-time (b) digital and continuous-time (c) analog and discrete-time (d) digital and discrete-time.

number of values. A **digital signal**, on the other hand, is one whose amplitude can take on only a finite number of values. Signals associated with a digital computer are digital because they take on only two values (binary signals). For a signal to qualify as digital, the number of values need not be restricted to two. It can be any finite number. A digital signal whose amplitudes can take on M values is an **M-ary signal**, of which **binary** ($M = 2$) is a special case. An analog signal can be converted into a digital signal (analog-to-digital or A/D conversion) through quantization (rounding off), as explained in Sec. 8.1-2.

The terms **"continuous-time" and "discrete-time" qualify the nature of a signal along the time axis (horizontal axis). The terms "analog" and "digital," on the other hand, qualify the nature of the signal amplitude (vertical axis).** Figure 1.10 shows examples of various types of signals. It is clear that analog is not necessarily continuous-time and digital need not be discrete-time. Figure 1.10c shows an example of an analog but discrete-time signal.

A system whose input and output signals are analog is an **analog system**; a system whose input and output signals are digital is a **digital system**. A digital computer is an example of a digital (binary) system. Observe that a digital computer is an example of a system that is digital as well as discrete-time.

1.3 SYSTEM MODEL: INPUT-OUTPUT DESCRIPTION

As mentioned earlier, systems theory encompasses a variety of systems, such as electrical, mechanical, hydraulic, acoustic, electromechanical, and chemical, as well as social, political, economic, and biological. The first step in analyzing any system is the construction of a system model, which is a mathematical expression or a rule that satisfactorily approximates the dynamical behavior of the system. In this chapter we shall consider only the continuous-time systems. (Modeling of discrete-time systems is discussed in Chapter 3.) We now describe a procedure for deriving input-output relationships of various types of systems.

1.3-1   Electrical Systems

To construct a system model, we must study the relationships between different variables in the system. In electrical systems, for example, we must determine a satisfactory model for the voltage-current relationship of each element, such as Ohm's law for a resistor. In addition, we must determine the various constraints on voltages and currents when several electrical elements are interconnected. These are laws of interconnection—the well-known Kirchhoff's voltage and current laws (KVL and KCL). From all these equations, we eliminate unwanted variables to obtain equation(s) relating the desired output variable(s) to the input(s). The following examples demonstrate the procedure of deriving input-output relationships for various systems.

■ **Example 1.2**

For the series RLC circuit of Fig. 1.11 find the input-output equation relating the input voltage $f(t)$ to the output current (loop current) $y(t)$.

Application of the Kirchhoff's voltage law around the loop yields

$$v_L(t) + v_R(t) + v_C(t) = f(t) \tag{1.12}$$

By using the voltage-current laws of each element (inductor, resistor, and capacitor), this equation can be expressed as

$$\frac{dy}{dt} + 3y(t) + 2 \int_{-\infty}^{t} y(\tau)\, d\tau = f(t) \tag{1.13}$$

Differentiating both sides of this equation, we obtain

$$\frac{d^2y}{dt^2} + 3\frac{dy}{dt} + 2y(t) = \frac{df}{dt} \tag{1.14}$$

This differential equation is the input-output relationship between the output $y(t)$ and the input $f(t)$. ■

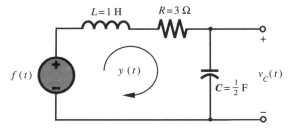

Fig. 1.11   Circuit for Example 1.2.

It proves convenient to use a compact notation D for the differential operator $\frac{d}{dt}$. Thus

$$\frac{dy}{dt} \equiv Dy(t) \tag{1.15}$$

$$\frac{d^2y}{dt^2} \equiv D^2y(t) \tag{1.16}$$

and so on. With this notation, Eq. (1.14) can be expressed as

$$\left(D^2 + 3D + 2\right) y(t) = Df(t) \tag{1.17}$$

Moreover, the differential operator is the inverse of the integral operator, so we can use the operator $1/D$ to represent integration†

$$\int_{-\infty}^{t} y(\tau)\, d\tau \equiv \frac{1}{D}y(t) \tag{1.18}$$

Consequently, the loop equation (1.13) can be expressed as

$$\left(D + 3 + \frac{2}{D}\right) y(t) = f(t) \tag{1.19}$$

Multiplying both sides by D, that is, differentiating Eq. (1.19), we obtain

$$\left(D^2 + 3D + 2\right) y(t) = Df(t) \tag{1.20}$$

which is identical to Eq. (1.17).

Recall that Eq. (1.20) is not an algebraic equation, and $D^2 + 3D + 2$ is not an algebraic term that multiplies $y(t)$; it is an operator that operates on $y(t)$. It means that we must perform the following operations on $y(t)$: take the second derivative of $y(t)$ and add to it 3 times the first derivative of $y(t)$ and 2 times $y(t)$. Clearly, a polynomial in D multiplied by $y(t)$ represents a certain differential operation on $y(t)$.

■ **Example 1.3**

Find the equation relating the input to output for the series RC circuit of Fig. 1.12 if the input is the voltage $f(t)$ and output is **(a)** the loop current $y(t)$ **(b)** the capacitor voltage $v_C(t)$.

The loop equation for the circuit is

$$Ry(t) + \frac{1}{C} \int_{-\infty}^{t} y(\tau)\, d\tau = f(t) \tag{1.21}$$

or

$$y(t) + 5 \int_{-\infty}^{t} y(\tau)\, d\tau = f(t) \tag{1.22}$$

---

† See footnote on p. 91 for a possible pitfall in use of this operator.

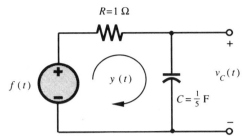

$R=1\ \Omega$

$f(t)$

$y\,(t)$

$v_C(t)$

$C = \frac{1}{5}\,\mathrm{F}$

Fig. 1.12   Circuit for Example 1.3.

With operational notation, this equation can be expressed as

$$y(t) + \frac{5}{D}y(t) = f(t) \tag{1.23}$$

Multiplying both sides of the above equation by D (that is, differentiating the above equation), we obtain

$$(D + 5)y(t) = Df(t) \tag{1.24a}$$

or

$$\frac{dy}{dt} + 5y(t) = \frac{df}{dt} \tag{1.24b}$$

Moreover,

$$y(t) = C\frac{dv_C}{dt}$$

$$= \frac{1}{5}Dv_C(t)$$

Substitution of this result in Eq. (1.24a) yields

$$(D + 5)v_C(t) = 5f(t) \tag{1.25}$$

or

$$\frac{dv_C}{dt} + 5v_C(t) = 5f(t) \tag{1.26}$$

■

△ **Exercise E1.3**
   For the RLC circuit in Fig. 1.11, find the input-output relationship if the output is the inductor voltage $v_L(t)$.

Hint: $v_L(t) = LDy(t) = Dy(t)$.   Answer: $\left(D^2 + 3D + 2\right)v_L(t) = D^2f(t)$ ▽

△ **Exercise E1.4**
   For the RLC circuit in Fig. 1.11, find the input-output relationship if the output is the capacitor voltage $v_C(t)$.

Hint: $v_C(t) = \frac{1}{CD}y(t) = \frac{2}{D}y(t)$.   Answer: $\left(D^2 + 3D + 2\right)v_C(t) = 2f(t)$ ▽

1.3-2   Mechanical Systems

   Planar motion can be resolved into translational (rectilinear) motion and rotational (torsional) motion. Translational motion will be considered first. We shall restrict ourselves to motions in one dimension only.

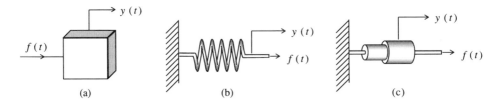

Fig. 1.13   Some elements in translational mechanical systems.

Translational Systems

The basic elements used in modeling translational systems are ideal masses, linear springs, and dashpots providing viscous damping. The laws of various mechanical elements are now discussed.

For a **mass** M (Fig. 1.13a), a force $f(t)$ causes a motion $y(t)$ and acceleration $\ddot{y}(t)$. From Newton's law of motion,

$$f(t) = M\ddot{y}(t) = M\frac{d^2y}{dt^2} = MD^2y(t) \qquad (1.27)$$

The force $f(t)$ required to stretch (or compress) a **linear spring** (Fig. 1.13b) by an amount $y(t)$ is given by

$$f(t) = Ky(t) \qquad (1.28)$$

where K is the **stiffness** of the spring.

For **a linear dashpot** (Fig. 1.13c), which operates by virtue of viscous friction, the force moving the dashpot is proportional to the relative velocity $\dot{y}(t)$ of one surface with respect to the other. Thus

$$f(t) = B\dot{y}(t) = B\frac{dy}{dt} = BDy(t) \qquad (1.29)$$

where B is the **damping coefficient** of the dashpot or the viscous friction.

■**Example 1.4**

Find the input-output relationship for the translational mechanical system shown in Fig. 1.14a or its equivalent in Fig. 1.14b. The input is the force $f(t)$, and the output is the mass position $y(t)$.

In mechanical systems it is helpful to draw a free-body diagram of each junction, which is a point where two or more elements are connected. In Fig. 1.14, the point representing the mass is a junction. The displacement of the mass is denoted by $y(t)$. The spring is also stretched by the amount $y(t)$, and therefore it exerts a force $-Ky(t)$ on the mass. The dashpot exerts a force $-B\dot{y}(t)$ on the mass as shown in the free-body diagram Fig. 1.14c. By Newton's second law, the net force must be $M\ddot{y}(t)$. Therefore

$$M\ddot{y}(t) = -B\dot{y}(t) - Ky(t) + f(t)$$

or

$$\left(MD^2 + BD + K\right)y(t) = f(t) \qquad ■ \qquad (1.30)$$

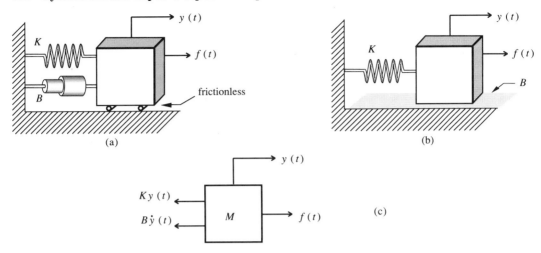

Fig. 1.14   Mechanical system for Example 1.4.

Rotational Systems

In rotational systems, the motion of a body may be defined as its motion about a certain axis. The variables used to describe rotational motion are torque (in place of force), angular position (in place of linear position), angular velocity (in place of linear velocity), and angular acceleration (in place of linear acceleration). The system elements are **rotational mass** or **moment of inertia** (in place of mass) and **torsional springs** and **torsional dashpots** (in place of linear springs and dashpots). The terminal equations for these elements are analogous to the corresponding equations for translational elements. If J is the moment of inertia (or rotational mass) of a rotating body about a certain axis, then the external torque required for this motion is equal to J (rotational mass) times the angular acceleration. If θ is the angular position of the body, $\ddot{\theta}$ is its angular acceleration, and

$$\text{Torque} = J\ddot{\theta} = J\frac{d^2\theta}{dt^2} = JD^2\theta(t) \tag{1.31}$$

Similarly, if K is the stiffness of a torsional spring (per unit angular twist), and θ is the angular displacement of one terminal of the spring with respect to the other, then

$$\text{Torque} = K\theta \tag{1.32}$$

Finally, the torque due to viscous damping of a torsional dashpot with damping coefficient B is

$$\text{Torque} = B\dot{\theta}(t) = BD\theta(t) \tag{1.33}$$

■Example 1.5

The attitude of an aircraft can be controlled by three sets of surfaces (shown shaded in Fig. 1.15): elevators, rudder, and ailerons. By manipulating these surfaces, one can set the aircraft on a desired flight path. The roll angle φ can be controlled by deflecting in the

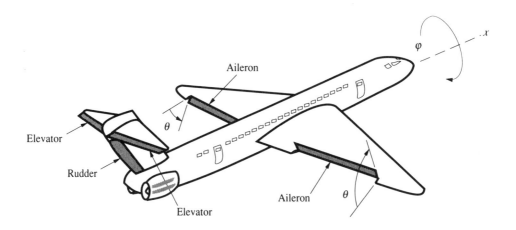

Fig. 1.15 Attitude control of an airplane.

opposite direction the two aileron surfaces as shown in Fig. 1.15. Assuming only rolling motion, find the equation relating the roll angle φ to the input (deflection) θ.

The aileron surfaces generate a torque about the roll-axis proportional to the aileron deflection angle θ. Let this torque be $c\theta$, where c is the constant of proportionality. Air friction dissipates the torque $B\dot{\varphi}(t)$. The torque available for rolling motion is then $c\theta(t) - B\dot{\varphi}(t)$. If J is the moment of inertia of the plane about the x-axis (roll axis), then

$$J\ddot{\varphi}(t) = \text{Net torque}$$
$$= c\theta(t) - B\dot{\varphi}(t) \qquad (1.34)$$

and

$$J\frac{d^2\varphi}{dt^2} + B\frac{d\varphi}{dt} = c\theta(t) \qquad (1.35)$$

or

$$\left(JD^2 + BD\right)\varphi(t) = c\theta(t) \qquad (1.36)$$

This is the desired equation relating the output (roll angle φ) to the input (aileron angle θ).

The roll velocity ω is $\dot{\varphi}(t)$. If the desired output is the roll velocity ω rather than the roll angle φ, then the input-output equation would be

$$J\frac{d\omega}{dt} + B\omega = \theta \qquad (1.37)$$

or

$$(JD + B)\omega(t) = \theta(t) \qquad (1.38)$$

■

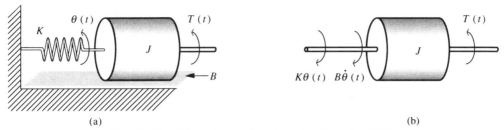

Fig. 1.16 The rotational system for Exercise E1.5.

△ **Exercise E1.5**

Torque $T(t)$ is applied to the rotational mechanical system shown in Fig. 1.16a. The torsional spring stiffness is K; the rotational mass (the cylinder's moment of inertia about the shaft) is J; the viscous damping coefficient between the cylinder and the ground is B. Find the equation relating the output angle θ to the input torque T. Hint: A free-body diagram is shown in Fig. 1.16b.

Answer: $J\frac{d^2\theta}{dt^2} + B\frac{d\theta}{dt} + K\theta(t) = T(t)$ or $\left(JD^2 + BD + K\right)\theta(t) = T(t)\triangledown$

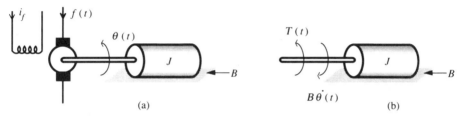

Fig. 1.17 Armature controlled dc motor.

1.3-3 Electromechanical Systems

A wide variety of electromechanical systems convert electrical signals into mechanical motion (mechanical energy) and vice versa. Here we consider a rather simple example of an armature-controlled dc motor driven by a current source $f(t)$, as shown in Fig. 1.17a. The torque $T(t)$ generated in the motor is proportional to the armature current $f(t)$. Therefore

$$T(t) = K_T f(t) \tag{1.39}$$

where K_T is a constant of the motor. This torque drives a mechanical load whose free-body diagram is shown in Fig. 1.17b. The viscous damping (with coefficient B) dissipates a torque $B\dot{\theta}(t)$. If J is the moment of inertia of the load (including the rotor of the motor), then the net torque $T(t) - B\dot{\theta}(t)$ must be equal to $J\ddot{\theta}(t)$;

$$J\ddot{\theta}(t) = T(t) - B\dot{\theta}(t) \tag{1.40}$$

Thus

$$\left(JD^2 + BD\right)\theta(t) = T(t)$$
$$= K_T f(t) \tag{1.41}$$

which in conventional form can be expressed as

$$J\frac{d^2\theta}{dt^2} + B\frac{d\theta}{dt} = K_T f(t) \tag{1.42}$$

1.4 SIMULTANEOUS DIFFERENTIAL EQUATIONS

So far we have intentionally considered rather simple systems. In more complex systems we must deal with simultaneous differential equations in several variables. Such equations can be solved directly, as in the state-variable approach discussed in Chapter 10, or by eliminating variables to obtain differential equations in a single variable only, as shown in the following discussion.

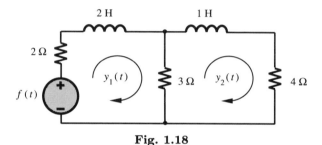

Fig. 1.18

Consider the electrical network in Fig. 1.18. The loop equations are

$$(2D + 5)y_1(t) - 3y_2(t) = f(t) \tag{1.43a}$$

$$-3y_1(t) + (D + 7)y_2(t) = 0 \tag{1.43b}$$

These are two simultaneous differential equations in two unknowns $y_1(t)$ and $y_2(t)$. To find the relationship between the output $y_1(t)$ and the input $f(t)$, we must eliminate $y_2(t)$ from these equations. If these were algebraic equations, this elimination could be achieved by multiplying Eq. (1.43a) by $(D + 7)$ and Eq. (1.43b) by 3 and then adding the resulting equations. However, Eqs. (1.43) are not algebraic equations but differential equations in operator notation. Thus we cannot apply algebraic procedure without testing its validity. Fortunately, this algebraic procedure is also valid for differential equations (1.43) because multiplying an equation by $D + 7$ means that we multiply the equation by D (differentiate the equation) and add 7 times the original equation. This is clearly valid. Therefore, the multiplication of any differential equation by a polynomial in the operator D is a valid operation. Eliminating $y_2(t)$ from Eqs. (1.43) in this manner, we obtain

$$(D + 7)(2D + 5)y_1(t) - 9y_1(t) = (D + 7)f(t)$$

or

$$\left(2D^2 + 19D + 26\right)y_1(t) = (D + 7)f(t) \tag{1.44a}$$

Similarly, to eliminate $y_1(t)$, we multiply Eq. (1.43a) by 3, and Eq. (1.43b) by $(2D + 5)$, and add the resulting equations to obtain

$$-9y_2(t) + (2D + 5)(D + 7)y_2(t) = 3f(t)$$

or

$$\left(2D^2 + 19D + 26\right) y_2(t) = 3f(t) \tag{1.44b}$$

By treating differential equations in operational notation as if they were algebraic equations, we can eliminate variables. Thus we can write these equations in matrix form and use Cramer's rule.† For example, Eqs. (1.43) can be expressed as

$$\begin{bmatrix} 2D + 5 & -3 \\ -3 & D + 7 \end{bmatrix} \begin{bmatrix} y_1(t) \\ y_2(t) \end{bmatrix} = \begin{bmatrix} f(t) \\ 0 \end{bmatrix}$$

Application of Cramer's rule (see Sec. B.7) to this equation yields

$$y_1(t) = \frac{\begin{vmatrix} f(t) & -3 \\ 0 & D + 7 \end{vmatrix}}{\begin{vmatrix} 2D + 5 & -3 \\ -3 & D + 7 \end{vmatrix}}$$

$$= \frac{(D + 7)f(t)}{2D^2 + 19D + 26} \tag{1.45a}$$

or

$$\left(2D^2 + 19D + 26\right) y_1(t) = (D + 7)f(t)$$

Similarly,

$$y_2(t) = \frac{\begin{vmatrix} 2D + 5 & f(t) \\ -3 & 0 \end{vmatrix}}{\begin{vmatrix} 2D + 5 & -3 \\ -3 & D + 7 \end{vmatrix}}$$

$$= \frac{3f(t)}{2D^2 + 19D + 26} \tag{1.45b}$$

---

† Use of operator $1/D$ for integration generates some subtle mathematical difficulties because the operators D and $1/D$ do not commute. For instance, we know that $D(1/D) = 1$ because $\frac{d}{dt}[\int_{-\infty}^{t} y(\tau)\,d\tau] = y(t)$. However, $\frac{1}{D}D$ is not necessarily unity. Use of Cramer's rule in solving simultaneous integro-differential equations will always result in cancellation of operators $1/D$ and D. This may yield erroneous results in those cases where the factor D occurs in the numerator as well as in the denominator. This happens, for instance, in circuits with all-inductor loops or all-capacitor cutsets. To eliminate this problem, the integral operation in system equations should be avoided so that the resulting equations are differential rather than integro-differential. In electrical circuits, this can be done by using charge (instead of current) variables in loops containing capacitors and using current variables for loops without capacitors. In the literature this problem of commutativity of D and $1/D$ is largely ignored. As mentioned earlier, this gives erroneous results only in special systems, such as the circuits with all-inductor loops or all-capacitor cutsets. Fortunately such systems constitute a very small fraction of the systems we deal with. For further discussion of this topic and a correct method of handling problems involving integrals, see Lathi.[4]

or

$$\left(2D^2 + 19D + 26\right) y_2(t) = 3f(t)$$

which confirms our earlier results in Eqs. (1.44).

Observe that in the solution of simultaneous equations, the denominator of every variable is given by the determinant of the system equation matrix. In fact, the denominators of all variables in a given system will be identical. This is clearly seen from Eqs. (1.45a) and (1.45b), and explains why the polynomials in D on the left-hand side of Eqs. (1.44a) and (1.44b) are identical. All differential equations relating the various outputs of a system to the input will have the same polynomial in D on the left-hand side.

1.5 INTERNAL AND EXTERNAL DESCRIPTIONS OF A SYSTEM

With a knowledge of the internal structure of a system, we can write system equations yielding an **internal description** of the system. In contrast, an **external description** is seen from the system's input and output terminals. To understand an external description, suppose that a system is enclosed in a "black box" with only its input(s) and output(s) terminals accessible. In order to describe or characterize such a system, we must perform some measurements at these terminals. For example, we might apply a known input, such as a unit impulse or a unit step, and then measure the system's output. The description provided by such a measurement is an external description of the system. The following example will clarify these concepts.

Let us find the equation relating $v(t)$, the output voltage, to the input voltage $f(t)$ for the RC circuit in Fig. 1.19a. There are three loops with loop currents y_1, y_2, and y_3. The three loop equations are

$$6y_1(t) - 2y_2(t) - 2y_3(t) = f(t) \tag{1.46a}$$

$$-2y_1(t) + 4y_2(t) + \frac{1}{D}\left[y_2(t) - y_3(t)\right] = 0 \tag{1.46b}$$

$$-2y_1(t) + 4y_3(t) + \frac{1}{D}\left[y_3(t) - y_2(t)\right] = 0 \tag{1.46c}$$

or

$$\begin{bmatrix} 6 & -2 & -2 \\ -2 & 4 + \frac{1}{D} & -\frac{1}{D} \\ -2 & -\frac{1}{D} & 4 + \frac{1}{D} \end{bmatrix} \begin{bmatrix} y_1(t) \\ y_2(t) \\ y_3(t) \end{bmatrix} = \begin{bmatrix} f(t) \\ 0 \\ 0 \end{bmatrix} \tag{1.47}$$

The output voltage $v(t)$ is equal to $y_2(t)$. Application of Cramer's rule to Eq. (1.47) yields

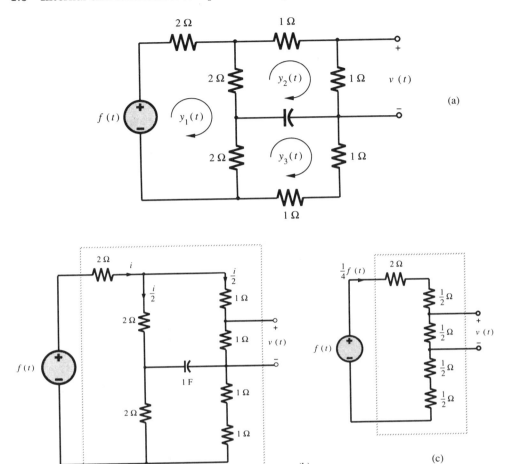

Fig. 1.19   A system that cannot be described by external measurements.

$$v(t) = y_2(t) = \frac{\begin{vmatrix} 6 & f(t) & -2 \\ -2 & 0 & -\frac{1}{D} \\ -2 & 0 & 4+\frac{1}{D} \end{vmatrix}}{\begin{vmatrix} 6 & -2 & -2 \\ -2 & 4+\frac{1}{D} & -\frac{1}{D} \\ -2 & -\frac{1}{D} & 4+\frac{1}{D} \end{vmatrix}}$$

$$= \frac{4(2+\frac{1}{D})f(t)}{32(2+\frac{1}{D})}$$

$$= \frac{1}{8}\frac{(2D+1)f(t)}{2D+1} \tag{1.48}$$

Equation (1.48) can be expressed in conventional form as

$$(2D+1)v(t) = \frac{1}{8}(2D+1)f(t) \tag{1.49a}$$

or

$$\left[2\frac{dv}{dt} + v(t)\right] = \frac{1}{8}\left[2\frac{df}{dt} + f(t)\right] \tag{1.49b}$$

If Eq. (1.49a) were an algebraic equation, we could have canceled the common factor $(2D+1)$ from both sides of the equation. But because this is a differential equation in operational notation and not an algebraic equation, we cannot cancel this common factor. Recall that if

$$Dx(t) = Dy(t) \tag{1.50a}$$

then we cannot assert that

$$x(t) = y(t) \tag{1.50b}$$

Extending this result, if

$$(D+a)x(t) = (D+a)y(t) \tag{1.51a}$$

then we cannot assert that

$$x(t) = y(t) \tag{1.51b}$$

In other words, the cancellation of common factors in differential equations expressed in operational notation is not permitted.

Note that we derived Eq. (1.49) from the knowledge of the system's internal structure. For this reason it is the *internal description* of the system. Now suppose that we are totally ignorant of the system's internal structure, and that the system is enclosed in a "black box" with only the input and output terminals accessible. Under these conditions the only way of describing or specifying the system is with external measurements. We can, for example, apply a known voltage $f(t)$ at the input terminals and measure the resulting output voltage $v(t)$. From this information we can describe or characterize the system. This is the *external description*.

In order to obtain the external description of this system, we redraw the circuit in Fig. 1.19a in a slightly different way, as shown in Fig. 1.19b, to emphasize its balanced nature. We then apply a voltage $f(t)$ at the input, assuming that the initial capacitor voltage is zero. This voltage produces the current i (Fig. 1.19b), which will divide equally between the two branches because of the balanced nature of the circuit. Thus voltage across the capacitor continues to remain zero. Therefore, for the purpose of computing the current i, the capacitor may be removed or replaced by a short. The resulting circuit is equivalent to that shown in Fig. 1.19c. It is clear from Fig. 1.19c that $f(t)$ sees a net resistance of $4\,\Omega$, and

$$i(t) = \frac{1}{4}f(t)$$

Also, because $v(t) = i/2$,

$$v(t) = \frac{1}{8}f(t) \tag{1.52}$$

The equivalent system as seen from the system's external terminals is shown in Fig. 1.19c. Clearly, for the external description, the capacitor does not exist.

The external description of the system is the same as the internal description [Eq. (1.49a)] except for the common factor $2D+1$, which canceled out. For most systems, the external and internal descriptions are identical, but there are a few exceptions, as in the present case, where the external description gives an inadequate picture of the systems. This happens when the system is **uncontrollable** and/or **unobservable**. Figures 1.20a and 1.20b show a structural representation of simple uncontrollable and unobservable systems respectively. In Fig. 1.20a we note that part of the system (subsystem S_2) inside the box cannot be controlled by the input $f(t)$. In Fig. 1.20b some of the system outputs (those in subsystem S_2) cannot be observed from the output terminals. If we try to describe either of these systems by applying an external input $f(t)$ and then measuring the output $y(t)$, the measurement will not characterize the complete system but only the part of the system (here S_1) that is both controllable and observable (linked to both the input and output). Such systems are undesirable in practice and should be avoided in any system design. The system in Fig. 1.19a can be shown to be neither controllable nor observable. It can be represented structurally as a combination of the systems in Figs. 1.20a and 1.20b. The presence of common factors on both sides of a system equation (in operational form) is symptomatic of system uncontrollability and/or unobservability.

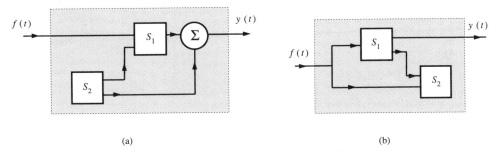

(a)   (b)

Fig. 1.20 Structures of uncontrollable and unobservable systems.

1.5-1 Large Systems

Until now we have discussed only very simple systems. In practice, systems tend to be much larger and more complex than those we have discussed so far. Anyone who has seen a circuit diagram of a television set knows this, and a TV receiver, though quite complicated, is relatively simple compared to many systems that electrical engineers are required to handle.

Clearly the approach discussed here so far(e.g., loop or node analysis), would lead us nowhere if we decided to analyze a large system. For large systems, it is necessary to break the system into a number of subsystems, each with a specific function. Fortunately, decomposing large systems in this manner is usually possible. Each subsystem, being smaller and performing a specific simple task, can be characterized in terms of its own terminal relationship (input-output relationship). Thus each subsystem can be represented by a block diagram characterized by this relationship. The analysis of a very large system then is reduced to the examination

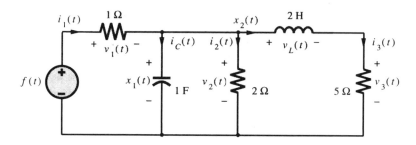

Fig. 1.21 Choosing suitable initial conditions in a network.

of the input-output relationships of a few interconnected subsystems.

1.6 STATE-SPACE DESCRIPTION OF A SYSTEM

The mathematical equations describing a system (its mathematical model) can be expressed in more than one way. The reader is familiar with the nth-order differential equation (Eq. 1.9) used to describe an nth-order dynamic system. It is possible to replace this equation with n simultaneous first-order differential equations, known collectively as the **normal form** of the differential equation.

Consider, for example, the following third-order differential equation:

$$\dddot{y} + a_2\ddot{y} + a_1\dot{y} + a_0 y = f \tag{1.53}$$

This equation can be expressed by three simultaneous first-order differential equations in the **state variables** x_1, x_2 and x_3 defined as

$$x_1 = y$$
$$x_2 = \dot{y}$$
$$x_3 = \ddot{y} \tag{1.54}$$

From Eqs. (1.53) and (1.54), we obtain our first-order state equations:

$$\dot{x}_1 = x_2$$
$$\dot{x}_2 = x_3$$
$$\dot{x}_3 = -a_0 x_1 - a_1 x_2 - a_2 x_3 + f \tag{1.55}$$

These three state equations are equivalent to Eq. (1.53). The procedure illustrated in this example for the third-order case extends readily to the nth-order case (see Chapter 10). Thus, instead of analyzing one nth-order equation, we analyze n simultaneous first-order equations.

We must emphasize that Eqs. (1.54) show only one possible choice of state variables. There is, in fact, an unlimited number of ways of defining state variables, although they cannot be selected arbitrarily. For example, in passive linear RLC circuits, a possible choice of state variables consists of all independent inductor currents and all independent capacitor voltages; however, the set of all capacitor and inductor voltages cannot generally be chosen as state variables.

One property of a valid set of state variables is that every possible output (including suppressed outputs) can be represented as a linear combination of the state variables and the inputs. To illustrate this point, consider the network in Fig. 1.21. Let us choose the capacitor voltage x_1 and the inductor current x_2 as the state variables. If the values of x_1, x_2, and the input $f(t)$ are known at some instant t, we can demonstrate that every possible signal (current or voltage) in the circuit can be determined at t. For example, if $x_1 = 10$, $x_2 = 1$, and $f = 20$ at some instant, the remaining voltages and currents at that instant will be

$$
\begin{aligned}
i_1 &= (20 - 10)/1 = 10\,\text{A} \\
v_1 &= i_1 = 10\,\text{V} \\
v_2 &= x_1 = 10\,\text{V} \\
i_2 &= v_2/2 = 5\,\text{A} \\
i_C &= i_1 - i_2 - x_2 = 4\,\text{A} \\
i_3 &= x_2 = 1\,\text{A} \\
v_3 &= 5i_3 = 5\,\text{V} \\
v_L &= v_2 - v_3 = 10 - 5 = 5\,\text{V}
\end{aligned}
\tag{1.56}
$$

Thus all signals in this circuit are determined. Clearly, state variables consist of the *key variables* in a system; a knowledge of the state variables allows one to determine every possible output of the system. Note that the **state-variable description is an internal description** of a system because it is capable of describing all possible input-output relationships of the system.

■ Example 1.6

This example illustrates how state equations may be natural and easier to determine than other descriptions, such as loop or node equations. Consider again the network in Fig. 1.21 with x_1 and x_2 as the state variables. The state equations, being first-order differential equations in x_1 and x_2, can be written by simple inspection of Fig. 1.21. Since $\dot{x}_1$ is the current through the capacitor,

$$
\begin{aligned}
\dot{x}_1 &= i_1 - i_2 - x_2 \\
&= (f - x_1) - 0.5\,x_1 - x_2 \\
&= -1.5\,x_1 - x_2 + f
\end{aligned}
$$

Also $2\dot{x}_2$, the voltage across the inductor, is given by

$$
\begin{aligned}
2\dot{x}_2 &= x_1 - v_3 \\
&= x_1 - 5x_2
\end{aligned}
$$

or

$$
\dot{x}_2 = 0.5\,x_1 - 2.5\,x_2
$$

Thus the state equations are

$$
\begin{aligned}
\dot{x}_1 &= -1.5\,x_1 - x_2 + f \\
\dot{x}_2 &= 0.5\,x_1 - 2.5\,x_2
\end{aligned}
$$

Once these equations are solved for x_1 and x_2, everything else in the circuit can be determined. ■

Chapter 10 deals with the state-space analysis of continuous-time and discrete-time systems. State-space techniques are useful for several reasons, including the following:

1. Time-varying parameter systems and nonlinear systems can be characterized effectively with state-space descriptions.
2. State equations lend themselves readily to accurate simulation on analog or digital computers.
3. For second-order systems ($n = 2$), a graphical method called **phase-plane analysis** can be used on state equations, whether they are linear or nonlinear.
4. State equations can yield a great deal of information about a system even when they are not solved explicitly.

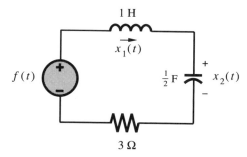

Fig. 1.22   Circuit for Exercise 1.6.

△ **Exercise E1.6**
   Write the state equations for the series RLC circuit shown in Fig. 1.22, using the inductor current $x_1(t)$ and the capacitor voltage $x_2(t)$ as state variables. Express every voltage and current in this circuit as a linear combination of x_1, x_2, and $f(t)$.

Answer: $\dot{x}_1 = -3\,x_1 - x_2 + f$, $\dot{x}_2 = 2\,x_1$ ▽

1.7 SUMMARY

   A system processes input signals to produce output signals (response). The input is the cause and the output is its effect. In general, the output is affected by two causes; the internal conditions of the system (such as the initial conditions) and the external input.
   Systems can be classified in several ways.

1. Linear systems are characterized by the linearity property, which implies superposition; if several causes (such as various inputs and initial conditions) are acting on a linear system, the total effect (response) is the sum of the responses from each cause, assuming that all the remaining causes are absent. A system is nonlinear if it is not linear.
2. Time-invariant systems are characterized by the fact that system parameters do not change with time. The parameters of time-varying parameter systems change with time.
3. For memoryless (or instantaneous) systems, the system response at any instant t depends only on the present value of the input (value at t). For systems with

memory (also known as dynamic systems), the system response at any instant t depends not only on the present value of the input, but also on the past values of the input (values before t).

4. In contrast, if a system response at t also depends on the future values of the input (values of input beyond t), the system is noncausal. In causal systems, the response does not depend on the future values of the input. Because of the dependence of the response on the future values of input, the effect (response) of noncausal systems occurs before cause. When the independent variable is time (temporal systems), the noncausal systems are prophetic systems because they anticipate the cause (effect occurs before the cause). Such systems are unrealizable, although close approximation is possible with some time delay in the response. Noncausal systems with independent variables other than time (e.g., space) are realizable.

5. If dimensions of system elements are small compared to the wavelengths of the signals, we may assume that each element is lumped at a single point in space, and the system may be considered as a lumped-parameter system. The signals under this assumption are functions of time only. If this assumption does not hold, the signals are functions of space and time; such a system is a distributed-parameter system.

6. A signal defined for a continuum of values of the independent variable (such as time) is a continuous-time signal. Similarly, a signal defined only at discrete instants is a discrete-time signal. Systems whose inputs and outputs are continuous-time signals are continuous-time systems; systems whose inputs and outputs are discrete-time signals are discrete-time systems. If a continuous-time signal is sampled, the resulting signal is a discrete-time signal. We can process a continuous-time signal by processing samples of this signal with a discrete-time system.

7. A signal whose amplitudes can take on any value in a continuous range is an analog signal. A digital signal, on the other hand, is one whose amplitude can take on only a finite number of values. The terms *discrete-time* and *continuous-time* qualify the nature of a signal along the time axis (horizontal axis). The terms *analog* and *digital*, on the other hand, qualify the nature of the signal amplitude (vertical axis). Systems whose inputs and outputs are analog signals are analog systems; those whose inputs and outputs are digital signals are digital systems.

The system model derived from a knowledge of the internal structure of the system is its internal description. In contrast, an external description of a system is its description as seen from system's input and output terminals; it can be obtained by applying a known input and measuring the resulting output. In the majority of practical systems, an external description of a system so obtained is equivalent to its internal description. In some cases, however, the external description fails to give adequate information about the system. Such is the case with the so-called uncontrollable or unobservable systems.

A system may also be described in terms of certain set of key variables called state variables. This description consists of a set of simultaneous first-order differential equations in state variables. State equations of a system represent an internal description of that system.

REFERENCES

1. Zadeh, L., and C. Desoer, *Linear System Theory*, McGraw-Hill, New York, 1963.

2. Kailath, T., *Linear Systems*, Prentice-Hall, Englewood Cliffs, New Jersey, 1980.

3. Lathi, B.P., *Signals, Systems, and Communication*, Wiley, New York, 1965.

4. Lathi, B.P., *Signals and Systems*, Berkeley-Cambridge Press, Carmichael, California, 1987.

PROBLEMS

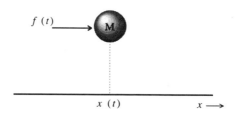

Fig. P1.1-1

1.1-1 A force $f(t)$ acts on a ball of mass M (Fig. P1.1). Show that the velocity $v(t)$ of the ball at any instant $t > 0$ can be determined if we know the force $f(t)$ over the interval $(0, t)$ and the ball's initial velocity $v(0)$.

1.1-2 Write the input-output relationship for an ideal integrator. Determine the zero-input and zero-state components of the response.

1.2-1 For the systems described by the equations below, with the input $f(t)$ and output $y(t)$, determine which of the systems are linear and which are nonlinear.

(a) $\dfrac{dy}{dt} + 2y(t) = f^2(t)$ (e) $\left(\dfrac{dy}{dt}\right)^2 + 2y(t) = f(t)$

(b) $\dfrac{dy}{dt} + 3ty(t) = t^2 f(t)$ (f) $\dfrac{dy}{dt} + (\sin t)y(t) = \dfrac{df}{dt} + 2f(t)$

(h) $3y(t) + 2 = f(t)$ (g) $\dfrac{dy}{dt} + 2y(t) = f(t)\dfrac{df}{dt}$

(d) $\dfrac{dy}{dt} + y^2(t) = f(t)$

Hint: If $y_1(t)$ and $y_2(t)$ are the responses to the inputs $f_1(t)$ and $f_2(t)$ respectively, then for a linear system, $y_1(t) + y_2(t)$ should be the response to input $f_1(t) + f_2(t)$.

1.3-1 For the circuit shown in Fig. P1.3-1, find the differential equations relating outputs $y_1(t)$ and $y_2(t)$ to the input $f(t)$.

1.3-2 Repeat Prob. 1.3-1 for the circuit in Fig. P1.3-2.

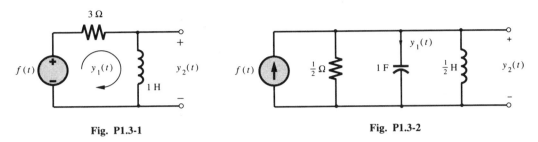

Fig. P1.3-1   Fig. P1.3-2

1.3-3 A simplified (one-dimensional) model of an automobile suspension system is shown in Fig. P1.3-3. In this case, the input is not a force but a displacement $x(t)$ (the road contour). Find the differential equation relating the output $y(t)$ (auto body displacement) to the input $x(t)$ (the road contour).

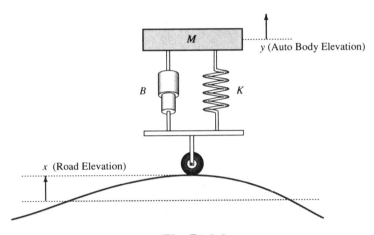

Fig. P1.3-3

1.3-4 Water flows into a tank at a rate of q_i units/s and flows out through the outflow valve at a rate of q_0 units/s (Fig. P1.3-4). Determine the equation relating the outflow q_0 to the input q_i. The outflow rate is proportional to the head h. Thus $q_0 = Rh$ where R is the valve resistance. Determine also the differential equation relating the head h to the input q_i.

(Hint: The net inflow of water in time $\triangle t$ is $(q_i - q_0)\triangle t$. This inflow is also $A\triangle h$ where A is the cross section of the tank.)

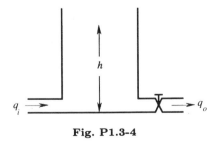

Fig. P1.3-4

1.3-5 A field-controlled dc motor is shown in Fig. P1.3-5. Its armature current i_a is maintained constant. The torque generated by this motor is proportional to the field current i_f (Torque$=K_f i_f$). Find the differential equation relating the output position θ to the input voltage $f(t)$. The motor and load together have a moment of inertia J.

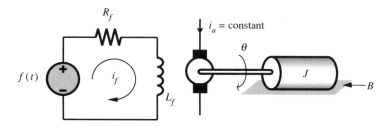

Fig. P1.3-5

1.4-1   For the circuit shown in Fig. P1.4-1, find the differential equations relating the loop currents $y_1(t)$ and $y_2(t)$ to the input $f(t)$.

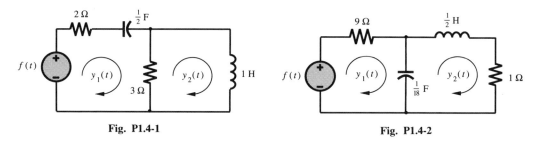

Fig. P1.4-1   **Fig. P1.4-2**

1.4-2   Repeat Prob. 1.4-1 for the circuit in Fig. P1.4-2.

1.5-1   Find the differential equations relating the loop currents $y_1(t)$ and $y_2(t)$ to the input voltage $f(t)$ in Fig. P1.5-1.

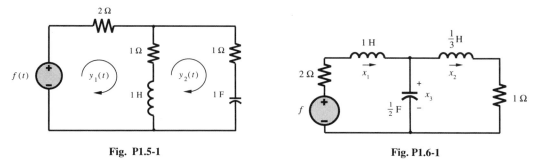

Fig. P1.5-1   **Fig. P1.6-1**

1.6-1   Write state equations for the third-order circuit shown in Fig. P1.6-1, using the inductor currents x_1, x_2 and the capacitor voltage x_3 as state variables. Show that every possible voltage or current in this circuit can be expressed as a linear combination of x_1, x_2, x_3, and $f(t)$. Also, at some instant t it was found that $x_1 = 5$, $x_2 = 1$, $x_3 = 2$, and $f = 10$. Determine the voltage across and the current through every element in this circuit.

II

TIME-DOMAIN ANALYSIS OF LTI SYSTEMS

Time-Domain Analysis: Continuous-Time Systems

In this book we consider two methods of analysis of linear time-invariant (LTI) systems: the time-domain method and frequency-domain method. In this chapter we discuss the **time-domain analysis** of linear, time-invariant, continuous-time (LTIC) systems. Response of a linear system can be expressed as a sum of zero-input and zero-state components. The zero-input response is the result of initial conditions alone with zero input, whereas the zero-state response is the result of the input alone with zero initial conditions. We now develop a procedure for determining both components of the response.

2.1 INTRODUCTION

For the class of LTIC systems discussed in Chapter 1 (differential systems), the input $f(t)$ and the output $y(t)$ are related by linear differential equations of the form

$$\frac{d^n y}{dt^n} + a_{n-1} \frac{d^{n-1} y}{dt^{n-1}} + \cdots + a_1 \frac{dy}{dt} + a_0 y(t) =$$

$$b_m \frac{d^m f}{dt^m} + b_{m-1} \frac{d^{m-1} f}{dt^{m-1}} + \cdots + b_1 \frac{df}{dt} + b_0 f(t) \qquad (2.1a)$$

where $a_0, a_1, a_2, \ldots, a_{n-1}, b_0, b_1, b_2, \ldots, b_{m-1}, b_m$ are constants.

In differential equations, the symbol d/dt represents the operation of differentiation. We replace this notation by D for the sake of compactness. The second derivative is represented by D^2, and the kth derivative is represented by D^k. With this operational notation, Eq. (2.1a) can be expressed as

$$(D^n + a_{n-1} D^{n-1} + \cdots + a_1 D + a_0) y(t)$$

$$= b_m D^m + b_{m-1} D^{m-1} + \cdots + b_1 D + b_0) f(t) \qquad (2.1b)$$

or

$$Q(D)y(t) = P(D)f(t) \qquad\qquad (2.1c)$$

where the polynomials $Q(D)$ and $P(D)$ are

$$Q(D) = D^n + a_{n-1}D^{n-1} + \cdots + a_1 D + a_0 \qquad\qquad (2.2a)$$

$$P(D) = b_m D^m + b_{m-1}D^{m-1} + \cdots + b_1 D + b_0 \qquad\qquad (2.2b)$$

Theoretically the powers m and n in the above equations can take on any value. Practical noise considerations, however, require $m \le n$ Noise can be any undesirable signal, natural or manmade, which interferes with the desired signals in the system. Sources of noise include such things as the electromagnetic radiation from stars, random motion of electrons in system components, interference from nearby radio and television stations, transients produced by automobile ignition systems, fluorescent lighting, and so on. We show in Chapter 4 that a system specified by Eq. (2.1) behaves as an $(m-n)$th-order differentiator of high-frequency signals if $m > n$. Unfortunately, noise is a wideband signal containing components of all frequencies from 0 to ∞. For this reason noise contains a significant amount of rapidly varying components with derivatives that are, consequently, very large. Therefore, any system specified by Eq. (2.1) in which $m > n$ will magnify the high-frequency components of noise. It is entirely possible for noise to be magnified so much that it swamps the desired system output even if the noise signal at the system's input is tolerably small. For this reason in practical systems it is desirable that $m \le n$. For the rest of this text we will assume implicitly that $m \le n$.

We now turn to the **time-domain method** for solving differential equations in the form of Eq. (2.1) with $m = n$ for generality. The time-domain method is one of two basic techniques considered in this book for the analysis of linear time-invariant (LTI) systems. With this method we are able to solve system equations directly, using t as the independent variable. The other technique, known as the **frequency-domain method**, will be discussed in later chapters.

To demonstrate that any system described by Eq. (2.1) is linear, let the input $f_1(t)$ to the system generate the output $y_1(t)$, and another input $f_2(t)$ generate the output $y_2(t)$. From Eq. (2.1c) it follows that

$$Q(D)y_1(t) = P(D)f_1(t)$$

and

$$Q(D)y_2(t) = P(D)f_2(t)$$

The addition of these two equations yields

$$Q(D)\,[y_1(t) + y_2(t)] = P(D)\,[f_1(t) + f_2(t)]$$

This equation shows that the input $f_1(t) + f_2(t)$ generates the response $y_1(t) + y_2(t)$. Therefore, the system has a superposition property, which makes it linear.† In Chapter 1 we showed that the response of a linear system can be expressed as the

---

† We can readily incorporate the homogeneity property by showing that the input $k_1 f_1(t) + k_2 f_2(t)$ generates the response $k_1 y_1(t) + k_2 y_2(t)$.

sum of two components: the zero-input component and the zero-state component (decomposition property).‡ Therefore

Total response = zero-input response + zero-state response   (2.3)

The zero-input component is the system response when the input $f(t) = 0$ so that it is the result of internal system conditions (such as energy storages) alone. It is independent of the external input $f(t)$. In contrast, the zero-state component is the system response to the external input $f(t)$ when the system is relaxed, that is, when the system is in zero state. This statement implies that a system in zero state cannot generate any response in the absence of $f(t)$. Therefore, the term "zero state" implies the absence of all internal energy storages, that is, all initial conditions are zero.

This discussion shows that the two components of the response are independent of each other, and the one can be computed independently of the other.

2.2 SYSTEM RESPONSE TO INTERNAL CONDITIONS: THE ZERO-INPUT RESPONSE

The zero-input response $y_0(t)$ is the solution of Eq. (2.1) when the input $f(t) = 0$ so that

$$Q(D)y_0(t) = 0 \tag{2.4a}$$

or

$$\left(D^n + a_{n-1}D^{n-1} + \cdots + a_1 D + a_0\right) y_0(t) = 0 \tag{2.4b}$$

To gain an intuitive sense of the problem, we first consider the simplest case of the first-order system given by the equation

$$(D + a_0)y_0(t) = 0 \tag{2.5a}$$

or

$$\frac{dy_0}{dt} = -a_0 y_0(t) \tag{2.5b}$$

---

‡ We can verify readily that the system described by Eq. (2.1) has the decomposition property. If $y_0(t)$ is the zero-input response, then, by definition,

$$Q(D)y_0(t) = 0$$

If $y(t)$ is the zero-state response, then $y(t)$ is the solution of

$$Q(D)y(t) = P(D)f(t)$$

subject to zero initial conditions (zero-state). The addition of these two equations yields

$$Q(D)\left[y_0(t) + y(t)\right] = P(D)f(t)$$

Clearly $y_0(t) + y(t)$ is the general solution of Eq. (2.1).

This equation can readily be solved in a systematic way,† but by solving it with heuristic reasoning, we gain more intuitive insight. Equation (2.5b) shows that $y_0(t)$ is a function whose derivative has the same form as itself (within a multiplicative constant). Only the exponential function has this property, so let us assume that the solution is

$$y_0(t) = ce^{\lambda t} \tag{2.6}$$

Differentiating this equation yields

$$\frac{dy_0}{dt} = c\lambda e^{\lambda t} \tag{2.7}$$

If $y_0(t)$ in Eq. (2.6) is the solution of Eq. (2.5), $y_0(t)$ must satisfy Eq. (2.5b). Substituting Eq. (2.6) and (2.7) in Eq. (2.5b), we obtain

$$c\lambda e^{\lambda t} = -a_0 c e^{\lambda t} \tag{2.8}$$

so that

$$\lambda = -a_0 \tag{2.9}$$

This result means that Eq. (2.6) is a solution of Eq. (2.5), provided that $\lambda = -a_0$. Thus the solution of

$$(D + a_0)\, y_0(t) = 0$$

is

$$y_0(t) = ce^{-a_0 t} \tag{2.10}$$

Note that c is an arbitrary constant; the solution in Eq. (2.10) satisfies Eq. (2.5) for any value of c. Consequently, there is an infinite number of possible solutions to Eq. (2.5). The unique solution can be determined only if an additional constraint is placed on the solution. This constraint is usually in the form of an initial condition.

Let us now consider the nth-order homogeneous equation (2.4):

$$Q(D)y_0(t) = 0 \tag{2.11a}$$

or

$$\left(D^n + a_{n-1}D^{n-1} + \cdots + a_1 D + a_0\right) y_0(t) = 0 \tag{2.11b}$$

A solution to this equation can be found in a systematic way.[1] However, we will take a short cut by using heuristic reasoning, as before. Equation (2.11b) shows that the linear combination of $y_0(t)$ and its n successive derivatives is zero. We know that an exponential function $e^{\lambda t}$ has this property, so let us assume that

---

† This can be done by rewriting Eq. (2.5b) as

$$\frac{dy_0}{y_0} = -a_0 dt$$

Integrating both sides of this equation yields

$$\ln y_0(t) = -a_0 t + k$$

or

$$y_0(t) = e^{-a_0 t_0 + k} = ce^{-a_0 t_0}$$

where c is an arbitrary constant.

$$y_0(t) = ce^{\lambda t}$$

is a solution to Eq. (2.11b). Then

$$Dy_0(t) = \frac{dy_0}{dt} = c\lambda e^{\lambda t}$$

$$D^2 y_0(t) = \frac{d^2 y_0}{dt^2} = c\lambda^2 e^{\lambda t}$$

$$\cdots\cdots\cdots\cdots\cdots\cdots\cdots\cdots$$

$$D^n y_0(t) = \frac{d^n y_0}{dt^n} = c\lambda^n e^{\lambda t}$$

Substituting these equations in Eq. (2.11b), we obtain

$$c\left(\lambda^n + a_{n-1}\lambda^{n-1} + \cdots + a_1\lambda + a_0\right)e^{\lambda t} = 0$$

For a nontrivial solution of this equation,

$$\lambda^n + a_{n-1}\lambda^{n-1} + \cdots + a_1\lambda + a_0 = 0 \qquad (2.12a)$$

This result means that $ce^{\lambda t}$ is indeed a solution of Eq. (2.11), provided that λ satisfies Eq. (2.12a). Note that the polynomial in Eq. (2.12a) is identical to the polynomial $Q(D)$ in Eq. (2.11b), with λ replacing D. Therefore, Eq. (2.12a) can be expressed as

$$Q(\lambda) = 0 \qquad (2.12b)$$

When $Q(\lambda)$ is expressed in factorized form, Eq. (2.12b) can be represented as

$$Q(\lambda) = (\lambda - \lambda_1)(\lambda - \lambda_2)\cdots(\lambda - \lambda_n) = 0 \qquad (2.12c)$$

Clearly λ has n solutions: $\lambda_1, \lambda_2, \ldots, \lambda_n$. Consequently, Eq. (2.11) has n possible solutions: $c_1 e^{\lambda_1 t}, c_2 e^{\lambda_2 t}, \ldots, c_n e^{\lambda_n t}$, with $c_1, c_2, \ldots, c_n$ as arbitrary constants. We can readily show that a general solution is given by the sum of these n solutions†, so that

$$y_0(t) = c_1 e^{\lambda_1 t} + c_2 e^{\lambda_2 t} + \cdots + c_n e^{\lambda_n t} \qquad (2.13)$$

where $c_1, c_2, \ldots, c_n$ are arbitrary constants determined by n constraints (the auxiliary conditions) on the solution.

---

† To prove this fact, assume that $y_1(t), y_2(t), \ldots, y_n(t)$ are all solutions of Eq. (2.11). Then

$$Q(D)y_1(t) = 0$$
$$Q(D)y_2(t) = 0$$
$$\cdots\cdots\cdots\cdots\cdots$$
$$Q(D)y_n(t) = 0$$

Multiplying these equations by $c_1, c_2, \ldots, c_n$, respectively, and adding them together yields
$$Q(D)\left[c_1 y_1(t) + c_2 y_2(t) + \cdots + c_n y_n(t)\right] = 0$$
This result shows that $c_1 y_1(t) + c_2 y_2(t) + \cdots + c_n y_n(t)$ is also a solution of the homogeneous equation (2.11).

Observe that the polynomial $Q(\lambda)$, which is characteristic of the system, has nothing to do with the input. For this reason the polynomial $Q(\lambda)$ is called the **characteristic polynomial** of the system. The equation

$$Q(\lambda) = 0 \qquad (2.14)$$

is called the **characteristic equation** of the system. From Eq. (2.12c), it is clear that λ_1, λ_2, ..., λ_n are the roots of the characteristic equation; consequently, they are called the **characteristic roots** of the system. The terms **characteristic values**, **eigenvalues**, and **natural frequencies** are also used for characteristic roots.‡ The exponentials $e^{\lambda_i t}(i = 1, 2, \ldots, n)$ in the zero-input response are the **characteristic modes** (also known as **modes** or **natural modes**) of the system. There is a characteristic mode for each characteristic root of the system, and the **zero-input response is a linear combination of the characteristic modes of the system.**

The single most important attribute of an LTIC system is its characteristic modes. Characteristic modes not only determine the zero-input response but also play an important role in determining the zero-state response. In other words, the entire behavior of a system is dictated primarily by its characteristic modes. In the rest of this chapter we shall see the pervasive presence of characteristic modes in every aspect of system behavior.

Repeated Roots

The solution of Eq. (2.11) as given in Eq. (2.13) assumes that the n characteristic roots λ_1, λ_2, ... , λ_n are distinct. If there are repeated roots (same root occurring more than once), the form of the solution is modified slightly. By direct substitution we can show that the solution of the equation

$$(D - \lambda)^2 y_0(t) = 0$$

is given by

$$y_0(t) = (c_1 + c_2 t)e^{\lambda t}$$

In this case the root λ repeats twice. Observe that the characteristic modes in this case are $e^{\lambda t}$ and $te^{\lambda t}$. Continuing this pattern, we can show that for the differential equation

$$(D - \lambda)^r y_0(t) = 0 \qquad (2.15)$$

the characteristic modes are $e^{\lambda t}$, $te^{\lambda t}$, $t^2 e^{\lambda t}$, ..., $t^{r-1} e^{\lambda t}$, and the solution is

$$y_0(t) = \left(c_1 + c_2 t + \cdots + c_r t^{r-1}\right) e^{\lambda t} \qquad (2.16)$$

Consequently, for a system with the characteristic polynomial

$$Q(\lambda) = (\lambda - \lambda_1)^r (\lambda - \lambda_{r+1}) \cdots (\lambda - \lambda_n)$$

the characteristic modes are $e^{\lambda_1 t}$, $te^{\lambda_1 t}$, ..., $t^{r-1} e^{\lambda t}$, $e^{\lambda_{r+1} t}$, ..., $e^{\lambda_n t}$. and the solution is

---

‡ The term *eigenvalue* is German for characteristic value.

$$y_0(t) = (c_1 + c_2 t + \cdots + c_r t^{r-1})e^{\lambda_1 t} + c_{r+1}e^{\lambda_{r+1} t} + \cdots + c_n e^{\lambda_n t}$$

■ **Example 2.1**

For an LTIC system specified by the differential equation

$$\left(D^2 + 6D + 9\right)y(t) = (3D + 5)f(t) \tag{2.17}$$

determine $y_0(t)$, the zero-input component of the response if the initial conditions are $y_0(0) = 3$ and $\dot{y}_0(0) = -7$.

The characteristic polynomial of the system is $\lambda^2 + 6\lambda + 9$. Its characteristic equation is $\lambda^2 + 6\lambda + 9 = (\lambda + 3)^2 = 0$, and its characteristic roots are $\lambda_1 = -3$, $\lambda_2 = -3$ (repeated roots). Consequently, the characteristic modes of the system are e^{-3t} and te^{-3t}. The zero-input response, being a linear combination of the characteristic modes, is given by

$$y_0(t) = (c_1 + c_2 t)e^{-3t} \tag{2.18}$$

This is the general form of the solution. To find the unique solution, we must determine the arbitrary constants c_1 and c_2. For this purpose we differentiate Eq. (2.18) to obtain

$$\dot{y}_0(t) = [-3(c_1 + c_2 t) + c_2]\, e^{-3t} \tag{2.19}$$

Setting $t = 0$ in Eqs. (2.18) and (2.19), and substituting the given initial conditions $y_0(0) = 3$ and $\dot{y}_0(0) = -7$, we obtain

$$3 = c_1$$

$$-7 = -3c_1 + c_2$$

Solving these two simultaneous equations yields

$$c_1 = 3 \quad \text{and} \quad c_2 = 2$$

Therefore

$$y_0(t) = (3 + 2t)e^{-3t} \qquad t \geq 0 \qquad ■$$

In Example 2.1 the initial conditions $y_0(0)$ and $\dot{y}_0(0)$ were supplied. In practical problems we must derive such conditions from the physical situation, as demonstrated in the next example.

■ **Example 2.2**

For the RLC circuit shown in Fig. 2.1a, the initial inductor current is zero, that is, $y(0) = 0$, and the initial capacitor voltage is 5 volts, that is, $v_C(0) = 5$. Find the loop current $y(t)$ if the input $f(t) = 10e^{-3t}u(t)$.

The differential (loop) equation relating $y(t)$ to $f(t)$ was derived in Eq. (1.20) as

$$\left(D^2 + 3D + 2\right)y(t) = Df(t)$$

The loop current, being a response of an LTI system to an input $f(t)$, consists of two components—the zero-input component and the zero-state component. The zero-input component of the loop current $y(t)$ is $y_0(t)$ that results from the initial conditions when the input $f(t) = 0$, that is, when the input is shorted as shown in Fig. 2.1b. The second (zero-state) component of the loop current $y(t)$ is the loop current resulting from the input $f(t)$, assuming that all initial conditions are zero; that is, $y(0) = v_C(0) = 0$. In

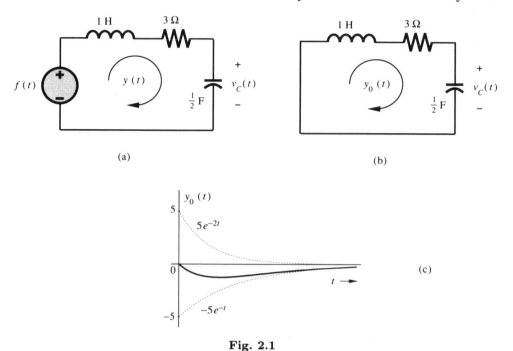

Fig. 2.1

this example we shall find the first (zero-input) component $y_0(t)$. The second (zero-state) component is found in Example 2.8.

The characteristic polynomial of the system is $\lambda^2 + 3\lambda + 2$. The characteristic equation of the system is therefore $\lambda^2 + 3\lambda + 2 = (\lambda + 1)(\lambda + 2) = 0$. The characteristic roots of the system are $\lambda_1 = -1$ and $\lambda_2 = -2$, and the characteristic modes of the system are e^{-t} and e^{-2t}. Consequently, the zero-input component of the loop current is

$$y_0(t) = c_1 e^{-t} + c_2 e^{-2t} \tag{2.20}$$

To determine the arbitrary constants c_1 and c_2, we need two constraints on $y_0(t)$, such as the initial conditions $y_0(0)$ and $\dot{y}_0(0)$. These conditions can be derived from the given initial conditions, $y(0) = 0$ and $v_C(0) = 5$, as follows. Recall that $y_0(t)$ is the loop current when the input terminals are shorted at $t = 0$, so that the input $f(t) = 0$ as shown in Fig. 2.1b. We now compute $y_0(0)$ and $\dot{y}_0(0)$, the values of the loop current and its derivative at $t = 0$, from the initial values of the inductor current and the capacitor voltage. Remember that the inductor current cannot change instantaneously in the absence of an impulsive voltage. Similarly, the capacitor voltage cannot change instantaneously in the absence of an impulsive current. Therefore, when the input terminals are shorted at $t = 0$, the inductor current is still zero and the capacitor voltage is still 5 volts. This means

$$y_0(0) = 0$$

To determine $\dot{y}_0(0)$, we use the loop equation for the circuit in Fig. 2.1b. Because the voltage across the inductor is $L(dy_0/dt)$ or $\dot{y}_0(t)$, this equation can be written as follows:

$$\dot{y}_0(t) + 3y_0(t) + v_C(t) = 0$$

Setting $t = 0$, we obtain

$$\dot{y}_0(0) + 3y_0(0) + v_C(0) = 0$$

But $y_0(0) = 0$ and $v_C(0) = 5$. Consequently,

$$\dot{y}_0(0) = -5$$

Therefore the desired initial conditions are

$$y_0(0) = 0 \qquad \text{and} \qquad \dot{y}_0(0) = -5$$

We can now determine the arbitrary constants c_1 and c_2 using the initial conditions $y_0(0)$ and $\dot{y}_0(0)$. Differentiating Eq. (2.20), we obtain

$$\dot{y}_0(t) = -c_1 e^{-t} - 2c_2 e^{-2t} \tag{2.21}$$

Setting $t = 0$ in Eqs. (2.20) and (2.21), and substituting the initial conditions $y_0(0) = 0$ and $\dot{y}_0(0) = -5$ we obtain

$$0 = c_1 + c_2$$

$$-5 = -c_1 - 2c_2$$

Solving these two simultaneous equations in two unknowns for c_1 and c_2 yields

$$c_1 = -5, \qquad c_2 = 5$$

Therefore

$$y_0(t) = -5e^{-t} + 5e^{-2t} \tag{2.22}$$

Note that the initial conditions are given at $t = 0$. Hence this solution is valid for $t \geq 0$. Figure 2.1c shows the plot of $y_0(t)$. ■

Independence of Zero-Input and Zero-State Response

In this example we computed the zero-input component without using the input $f(t)$. The zero-state component can be computed from the knowledge of the input $f(t)$ alone; the initial conditions are assumed to be zero (system in zero state). The two components of the system response (the zero-input and zero-state components) are independent of each other. The two worlds of zero-input response and zero-state response coexist side by side, neither of them knowing or caring what the other is doing. For each component the other is totally irrelevant.

Complex Roots

The procedure for handling complex roots is the same as that for real roots. For complex roots the usual procedure leads to complex characteristic modes and the complex form of solution. Nevertheless, it is possible to avoid the complex form altogether by selecting a real form of solution, as described below.

For a real system, complex roots must occur in pairs of conjugates if the coefficients of the characteristic polynomial $Q(\lambda)$ are to be real. Therefore, if $\alpha + j\beta$ is a characteristic root, $\alpha - j\beta$ must also be a characteristic root. The zero-input response corresponding to this pair of complex conjugate roots is

$$y_0(t) = c_1 e^{(\alpha+j\beta)t} + c_2 e^{(\alpha-j\beta)t} \tag{2.23a}$$

For a real system, the response $y_0(t)$ must also be real. This is possible only if c_1 and c_2 are conjugates. Let

$$c_1 = \frac{c}{2}e^{j\theta} \qquad \text{and} \qquad c_2 = \frac{c}{2}e^{-j\theta}$$

This yields

$$y_0(t) = \frac{c}{2}e^{j\theta}e^{(\alpha+j\beta)t} + \frac{c}{2}e^{-j\theta}e^{(\alpha-j\beta)t}$$

$$= \frac{c}{2}e^{\alpha t}\left[e^{j(\beta t+\theta)} + e^{-j(\beta t+\theta)}\right]$$

$$= ce^{\alpha t}\cos\left(\beta t + \theta\right) \qquad (2.23b)$$

Therefore the zero-input response corresponding to complex conjugate roots $\alpha \pm j\beta$ can be expressed in a complex form (2.23a) or a real form (2.23b). The latter is more convenient from computational viewpoint because it avoids dealing with complex numbers.

■ **Example 2.3**

In the RLC circuit in Fig. 2.1a, if the circuit parameters are $L = 1$, $R = 4$, and $C = 1/40$, find the zero-input loop current $y_0(t)$ if the initial conditions are $y_0(0) = 2$ and $\dot{y}_0(0) = 16.78$.

In this case the loop equation is

$$\left(D^2 + 4D + 40\right)y_0(t) = 0$$

The characteristic polynomial is

$$\lambda^2 + 4\lambda + 40 = (\lambda + 2 - j6)(\lambda + 2 + j6)$$

The characteristic roots are $-2 \pm j6$.† The solution can be written either in the complex form (2.23a) or in the real form (2.23b). We select the latter form in order to avoid complex algebra. In this case $\alpha = -2$ and $\beta = 6$. Therefore [see Eq. (2.23b)]

$$y_0(t) = ce^{-2t}\cos\left(6t + \theta\right) \qquad (2.24)$$

where c and θ are arbitrary constants to be determined from the initial conditions $y_0(0) = 2$ and $\dot{y}_0(0) = 16.78$. Differentiation of Eq. (2.24) yields

$$\dot{y}_0(t) = -2ce^{-2t}\cos\left(6t + \theta\right) - 6ce^{-2t}\sin\left(6t + \theta\right) \qquad (2.25)$$

Setting $t = 0$ in Eqs. (2.24) and (2.25), and then substituting initial conditions, we obtain

$$2 = c\cos\theta$$

$$16.78 = -2c\cos\theta - 6c\sin\theta$$

Solution of these two simultaneous equations in two unknowns $c\cos\theta$ and $c\sin\theta$ yields

$$c\cos\theta = 2 \qquad (2.26a)$$

$$c\sin\theta = -3.463 \qquad (2.26b)$$

Dividing $c\sin\theta$ by $c\cos\theta$ gives

$$\tan\theta = \frac{-3.463}{2}$$

---

†   The complex conjugate roots of a second-order polynomial can be determined by using the formula in Sec. B.10-10 or by expressing the polynomial as a sum of two squares. This is done readily by completing the square with the first two terms, as shown below:

$$\lambda^2 + 4\lambda + 40 = (\lambda + 2)^2 + (6)^2 = (\lambda + 2 - j6)(\lambda + 2 + j6)$$

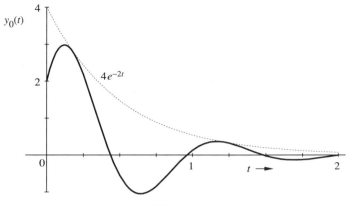

Fig. 2.2

and

$$\theta = \tan^{-1}\left(\tfrac{-3.463}{2}\right) = -\tfrac{\pi}{3}$$

Substituting this value of θ in Eq. (2.26a) yields

$$c\cos\left(-\tfrac{\pi}{3}\right) = \tfrac{c}{2} = 2 \implies c = 4$$

Therefore

$$y_0(t) = 4e^{-2t}\cos\left(6t - \tfrac{\pi}{3}\right) \tag{2.27}$$

The response $y_0(t)$ is shown in Fig. 2.2. ∎

⊙ **Computer Example C2.1**

In the RLC circuit in Fig. 2.1a, if the circuit parameters are **(a)** $L = 1, R = 2, C = 1/50$, **(b)** $L = 1, R = 2, C = 1/4$, and **(c)** $L = 1, R = 5, C = 1/6$, find the zero-input loop current $y_0(t)$ if the initial conditions are $y_0(0) = 2$, and $\dot{y}_0(0) = 16.78$

(a) The loop equation is $(D^2 + 2D + 50)y_0(t) = 0$. The characteristic polynomial for this system must be entered into the computer.

```
p = [1 2 50] % Representation of the characteristic polynomial.
lambda = roots(p)
a = [0 1; -50 -2]; % Enter the characteristic eqn.
b = [0;0];         % Enter input coefficients.
t = 0:.025:2;
[m,k] = size(t);
[n, nb] = size(b);
y0 = [2;16.78];    % Initial condition vector.
tt=0;
    for j=1:k
    z=expm(a*tt);
    A=z(1:nb,:);
    y(1,j)=A*y0;
    tt=tt+.025;
    end
    y = y(1:nb,:);
    plot(t,y),grid,
    title('plot for C2.1a'),pause
```

(b) The loop equation is $(D^2 + 2D + 4)y_0(t) = 0$. The characteristic polynomial for this system must be entered into the computer.

```
p = [1 2 4]          % Representation of the characteristic polynomial.
lambda = roots(p)
a = [0 1; -4 -2];    % Enter the characteristic eqn.
b = [0;0];           % Enter input coefficients.
t = 0:.025:2;
[m,k] = size(t);
[n, nb] = size(b);
y0 = [2;16.78];      % Initial condition vector.
tt=0;
    for j=1:k
    z=expm(a*tt);
    A=z(1:nb,:);
    y(1,j)=A*y0;
    tt=tt+.025;
    end
    y = y(1:nb,:);
    plot(t,y),grid,
    title('plot for C2.1b'),pause
```

(c) The loop equation is $(D^2 + 5D + 6)y_0(t) = 0$. The characteristic polynomial for this system must be entered into the computer.

```
p = [1 5 6] % Representation of the characteristic polynomial.
lambda = roots(p)
a = [0 1; -6 -5];    % Enter the characteristic eqn.
b = [0;0];           % Enter input coefficients.
t = 0:.025:2;
[m,k] = size(t);
[n, nb] = size(b);
y0 = [2;16.78];      % Initial condition vector.
tt=0;
    for j=1:k
    z=expm(a*tt);
    A=z(1:nb,:);
    y(1,j)=A*y0;
    tt=tt+.025;
    end
    y = y(1:nb,:);
    plot(t,y),grid,
    title('plot for C2.1c')     ⊙
```

△ **Exercise E2.1**

The series RC circuit in Fig. 2.3 has the initial capacitor voltage $v_C(0) = 10$ volts. Find $y_0(t)$, the zero-input component of the loop current for $t \geq 0$. Hint: Recognize that $y_0(0)$ is the initial value of the loop current under the condition that $f(t) = 0$. Find $y_0(0)$ by shorting the input.

Answer: $y_0(t) = -5e^{-5t} \qquad t \geq 0 \quad \bigtriangledown$

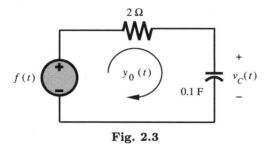

Fig. 2.3

△ **Exercise E2.2**
Solve

$$\left(D^2 + 2D\right) y_0(t) = 0$$

if $y_0(0) = 1$ and $\dot{y}_0(0) = 4$. Hint: The characteristic roots are 0 and -2. The characteristic modes are e^{0t} and e^{-2t}.

Answer: $y_0(t) = 3 - 2e^{-2t}$ $\quad t \geq 0$ $\quad \triangledown$

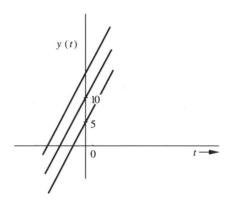

Fig. 2.4 There are infinite solutions of a differential equation.

Why Arbitrary Constants? A Physical Explanation

The discussion in this section shows that a differential equation does not, in general, have a unique solution unless some additional constraints (or conditions) on the solution are known. This fact should not come as a surprise. A function $y(t)$ has a unique derivative $\dot{y}(t)$, but for a given derivative $\dot{y}(t)$ there are infinite possible functions $y(t)$. If we are given $\dot{y}(t)$, it is impossible to determine $y(t)$ uniquely unless an additional piece of information about $y(t)$ is given. For example, the solution of a differential equation

$$\frac{dy}{dt} = 2 \tag{2.28}$$

obtained by integrating both sides of the above equation is

$$y(t) = 2t + c \tag{2.29}$$

for any value of c. Equation (2.28) specifies a function whose slope is 2 for all t. Any straight line with a slope of 2 satisfies this equation, as Fig. 2.4 shows. Clearly the solution is not unique, but if we place an additional constraint on the solution $y(t)$, then we specify a unique solution. For example, suppose we require that $y(0) = 5$; then out of all the possible solutions available, only one function has a slope of 2 and an intercept with the vertical axis at 5. By setting $t = 0$ in Eq. (2.29) and substituting $y(0) = 5$ in the same equation, we obtain $y(0) = 5 = c$ and

$$y(t) = 2t + 5$$

which is the unique solution satisfying both Eq. (2.28) and the constraint (or initial condition).

In conclusion, differentiation is an irreversible operation during which certain information is lost. To reverse this operation, one piece of information about $y(t)$ must be provided to restore the original $y(t)$. Using a similar argument, we can show that, given $\ddot{y}(t)$, we can determine $y(t)$ uniquely only if two additional pieces of information (constraints) about $y(t)$ are given. In general, to determine $y(t)$ uniquely from its nth derivative, we need n additional pieces of information (constraints) about $y(t)$. These constraints are also called *auxiliary conditions*, and they are generally available in the form of *initial* or *boundary conditions*.

2.2-1   Some Insights into the Zero-Input Behavior of a System

By definition the zero-input response is the system response to its internal conditions, assuming that its input is zero. Understanding this phenomenon provides interesting insight into system behavior. If a system is disturbed momentarily from its rest position and if the disturbance is then removed, the system will not come back to rest instantaneously. In general, it will come back to rest over a period of time and only through a special type of motion that is characteristic of the system.† For example, if we press on an automobile fender momentarily and then release it at $t = 0$, there is no external force on the automobile for $t > 0$.‡ The auto-body will eventually come back to its rest (equilibrium) position, but not through any arbitrary motion. It must do so using only a form of response which is sustainable by the system on its own without any external source, because the input is zero. Only characteristic modes satisfy this condition. **The system uses a proper combination of characteristic modes to come back to the rest position while satisfying appropriate boundary (or initial) conditions.**

If the shock absorbers of the automobile are in good condition (high damping coefficient), the characteristic modes will be monotonically decaying exponentials, and the auto-body will come to rest rapidly without oscillation. On the other hand, for poor shock absorbers (low damping coefficients), the characteristic modes will be exponentially decaying sinusoids, and the body will come to rest through oscillatory motion. In a series RC circuit such as the one in Fig. 2.3 with the input $f(t) = 0$, we can place a charge on the capacitor by passing a momentary current pulse through it. When left to itself, the capacitor will start to discharge exponentially through the resistor. This response of the RC circuit is caused entirely by its internal conditions and is sustained by this system without the aid of any external input.

---

† This assumes that the system will eventually come back to its original rest (or equilibrium) position.
‡ We ignore the force of gravity, which merely causes a constant displacement of the auto-body without affecting the other motion.

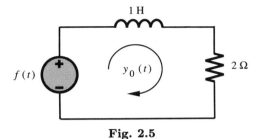

Fig. 2.5

The exponential current waveform is therefore the characteristic mode of this RC circuit.

Mathematically we know that **any combination of characteristic modes can be sustained by the system alone without requiring an external input.** This fact can be readily verified for the series RL circuit shown in Fig. 2.5. The loop equation for this system is

$$(D + 2)y(t) = f(t)$$

It has a single characteristic root $\lambda = -2$, and the characteristic mode is e^{-2t}. We now verify that a loop current $y(t) = ce^{-2t}$ can be sustained through this circuit without any input voltage. The input voltage $f(t)$ required to drive a loop current $y(t) = ce^{-2t}$ is given by

$$f(t) = L\frac{dy}{dt} + Ry(t)$$

$$= \frac{d}{dt}(ce^{-2t}) + 2ce^{-2t}$$

$$= -2ce^{-2t} + 2ce^{-2t}$$

$$= 0$$

Clearly the loop current $y(t) = ce^{-2t}$ is sustained by the RL circuit on its own, without the necessity of an external input.

The Resonance Phenomenon

We have seen that any signal consisting of a system's characteristic modes is sustained by the system on its own; the system offers no obstacle to such signals. Imagine what would happen if we actually drove the system with an external input that is one of its characteristic modes. This would be like pouring gasoline on a fire in a dry forest or hiring an alcoholic to taste liquor. An alcoholic would gladly do the job without pay. Think what will happen if he were paid by the amount of liquor he tasted! The system response to characteristic modes would naturally be very high. We call this behavior the **resonance phenomenon.** An intelligent discussion of this important phenomenon requires an understanding of the zero-state response; for this reason we postpone this topic to Sec. 2.7-7.

Uniqueness of the Characteristic Roots of a System

So far we have examined the case of a system with only a single output. A system may have several outputs, generally with a different input-output equation for each output. Even so, $Q(D)$, which is the determinant of the system equations' matrix, remains the same for every possible output of a given system, as we saw in Sec. 1.4. Consequently the characteristic roots of a system are the same regardless of the output. This observation means that the form of the zero-input response is the same for each and every output. Therefore a system with a set of initial conditions and zero input will give rise to signals that have the same form (characteristic modes) throughout the system.

(a)   (b)

Fig. 2.6 A unit impulse and its approximation.

2.3   UNIT IMPULSE FUNCTION

The unit impulse function $\delta(t)$ is one of the most important functions in the study of signals and systems. This function was first defined by P.A.M Dirac as

$$\delta(t) = 0 \qquad t \neq 0$$

$$\int_{-\infty}^{\infty} \delta(t)\, dt = 1 \tag{2.30}$$

We can visualize an impulse as a tall, narrow rectangular pulse of unit area, as shown in Fig. 2.6b. The width of this rectangular pulse is some very small value ϵ; its height is a very large value $1/\epsilon$ in the limit as $\epsilon \to 0$. The unit impulse therefore can be regarded as a rectangular pulse with a width that has become infinitesimally small, a height that has become infinitely large, and an overall area that has been maintained at unity. Thus $\delta(t) = 0$ everywhere except at $t = 0$, where it is undefined. For this reason a unit impulse is represented by the spear-like symbol in Fig. 2.6a.

We need not use a rectangular pulse in approximating the unit impulse function. Other pulses, such as exponential pulse, triangular pulse, or Gaussian pulse, may also be used. The important feature of the unit impulse function is not its shape but the fact that its effective duration (pulse width) approaches zero while its area remains at unity. For example, the exponential pulse $\alpha e^{-\alpha t} u(t)$ in Fig. 2.7a becomes taller and narrower as α increases. In the limit as α approaches ∞, the pulse height $\to \infty$, and its width or duration $\to 0$. The area under the pulse is unity regardless of the value of α because

$$\int_{0}^{\infty} \alpha e^{-\alpha t}\, dt = 1 \tag{2.31}$$

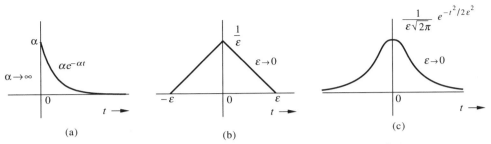

Fig. 2.7 Other possible approximations to a unit impulse.

The pulses in Figs. 2.7b and 2.7c behave in a similar fashion.

Multiplication of a Function by an Impulse

Let us now consider what happens when we multiply the unit impulse $\delta(t)$ by a function $\phi(t)$ that is known to be continuous at $t = 0$. Since the impulse exists only at $t = 0$, and the value of $\phi(t)$ at $t = 0$ is $\phi(0)$, we obtain

$$\phi(t)\delta(t) = \phi(0)\delta(t) \tag{2.32a}$$

Similarly, if $\phi(t)$ is multiplied by an impulse $\delta(t - T)$ (impulse located at $t = T$), then

$$\phi(t)\delta(t - T) = \phi(T)\delta(t - T) \tag{2.32b}$$

provided $\phi(t)$ is continuous at $t = T$.

Sampling Property of the Unit Impulse Function

From Eq. (2.32a) it follows that

$$\int_{-\infty}^{\infty} \phi(t)\delta(t)\, dt = \phi(0) \int_{-\infty}^{\infty} \delta(t)\, dt$$

$$= \phi(0) \tag{2.33a}$$

provided $\phi(t)$ is continuous at $t = 0$. This result means that **the area under the product of a function with an impulse $\delta(t)$ is equal to the value of that function at the instant where the unit impulse is located.** This property is very important and useful, and is known as the **sampling (or sifting) property** of the unit impulse.

From Eq. (2.32b) it follows that

$$\int_{-\infty}^{\infty} \phi(t)\delta(t - T)\, dt = \phi(T) \tag{2.33b}$$

Equation (2.33b) is just another form of sampling or sifting property. In the case of Eq. (2.33b), the impulse $\delta(t - T)$ is located at $t = T$. Therefore the area under $\phi(t)\delta(t - T)$ is $\phi(T)$, the value of $\phi(t)$ at the instant where the impulse is located (at $t = T$). In these derivations we have assumed that the function is continuous at the instant where the impulse is located.

The Unit Impulse as a Generalized Function

The definition of the unit impulse function given in Eq. (2.30) is not mathematically rigorous. Some difficulties are associated with it. First, it does not define a unique function: for example, it can be shown that $\delta(t) + \dot{\delta}(t)$ also satisfies Eq. (2.30).[2] Moreover, $\delta(t)$ is not even a true function in the ordinary sense. An ordinary function is specified by its values for all time t. The impulse function is zero everywhere except at $t = 0$, and at this only interesting part of its range it is undefined. These difficulties are resolved by defining the impulse as a generalized function rather than an ordinary function. A **generalized function** is defined by its effect on other functions instead of by its value at every instant of time.

In this approach the impulse function is defined by the sampling property [Eq. (2.33)]. We say nothing about what the impulse function is or what it looks like. Instead, it is defined in terms of the effect it has on a test function $\phi(t)$. We define a unit impulse as a function for which the area under its product with a function $\phi(t)$ is equal to the value of the function $\phi(t)$ at the instant where the impulse is located. It is assumed that $\phi(t)$ is continuous at the location of the impulse. Therefore either Eq. (2.33a) or (2.33b) can serve as a definition of the impulse function in this approach. Recall that the sampling property [Eq. (2.33)] is the consequence of the classical (Dirac) definition of impulse in Eq. (2.30). In contrast, **the sampling property [Eq. (2.33)] defines the impulse function in the generalized function approach.**

We now give an interesting application of the generalized function definition of an impulse. Because the unit step function $u(t)$ is discontinuous at $t = 0$, its derivative du/dt does not exist at $t = 0$ in the ordinary sense. Now, however, we show that this derivative exists in the generalized sense, and it is given by $\delta(t)$. To show this, let us evaluate the integral of $(du/dt)\phi(t)$, using integration by parts:

$$\int_{-\infty}^{\infty} \frac{du}{dt} \phi(t)\, dt = u(t)\phi(t)\Big|_{-\infty}^{\infty} - \int_{-\infty}^{\infty} u(t)\dot{\phi}(t)\, dt \qquad (2.34)$$

$$= \phi(\infty) - 0 - \int_{0}^{\infty} \dot{\phi}(t)\, dt$$

$$= \phi(\infty) - \phi(t)\big|_{0}^{\infty}$$

$$= \phi(0) \qquad (2.35)$$

This result shows that du/dt satisfies the definition of $\delta(t)$; therefore it is an impulse $\delta(t)$ in the generalized sense:

$$\frac{du}{dt} = \delta(t) \qquad (2.36)$$

Consequently

$$\int_{-\infty}^{t} \delta(\tau)\, d\tau = u(t) \qquad (2.37)$$

These results can also be obtained graphically from Fig. 2.6b. We observe that the area from $-\infty$ to t under the limiting form of $\delta(t)$ in Fig. 2.6b is zero if $t < 0$ and unity if $t \geq 0$. Consequently

$$\int_{-\infty}^{t} \delta(\tau)\, d\tau = \begin{cases} 0 & t < 0 \\ 1 & t \geq 0 \end{cases}$$

$$= u(t) \tag{2.38}$$

The derivatives of impulse function can also be defined as generalized functions (see Prob. 2.3-5).

△ **Exercise E2.3**
   Show that

   (a) $(t^3 + 3)\delta(t) = 3\delta(t)$ **(b)** $\left[\sin\left(t^2 - \frac{\pi}{2}\right)\right]\delta(t) = -\delta(t)$

   (c) $e^{-2t}\delta(t) = \delta(t)$ **(d)** $\dfrac{\omega^2 + 1}{\omega^2 + 9}\delta(\omega - 1) = \dfrac{1}{5}\delta(\omega - 1)$

Hint: Use Eqs. (2.32).   ▽

△ **Exercise E2.4**
   Show that

   (a) $\displaystyle\int_{-\infty}^{\infty} \delta(t)e^{-j\omega t}\, dt = 1$ **(b)** $\displaystyle\int_{-\infty}^{\infty} \delta(t - 2)\cos\left(\frac{\pi t}{4}\right) dt = 0$

   (c) $\displaystyle\int_{-\infty}^{\infty} e^{-2(x-t)}\delta(2 - t)\, dt = e^{-2(x-2)}$

Hint: In part **c** recall that $\delta(x)$ is located at $x = 0$. Therefore $\delta(2 - t)$ is located at $2 - t = 0$; that is at $t = 2$.   ▽

2.3-1   The Unit Impulse Response h(t)

The impulse function $\delta(t)$ is also used in determining the response of a linear system to an arbitrary input $f(t)$. In Chapter 1 we explained how a system response to an input $f(t)$ may be found by breaking this input into narrow rectangular pulses, as shown in Fig. 1.4a and then summing the system response to all the components. The rectangular pulses become impulses in the limit as their widths approach zero. Therefore the system response is the sum of its response to various impulse components. This discussion shows that if we know the system response to an impulse input, we can determine the system response to an arbitrary input $f(t)$. We now discuss a method of determining $h(t)$, the unit impulse response of an LTIC system described by the nth-order differential equation

$$Q(D)y(t) = P(D)f(t) \tag{2.39a}$$

in which $Q(D)$ and $P(D)$ are the polynomials shown in Eq. (2.2). Recall that noise considerations restrict practical systems to $m \leq n$. Under this constraint, the most general case is $m = n$. Therefore Eq. (2.39a) can be expressed as

$$(D^n + a_{n-1}D^{n-1} + \cdots + a_1 D + a_0)y(t) =$$
$$(b_n D^n + b_{n-1}D^{n-1} + \cdots + b_1 D + b_0)\, f(t) \tag{2.39b}$$

Before deriving the general expression for the unit impulse response $h(t)$, it is illuminating to understand qualitatively the nature of $h(t)$. The impulse response $h(t)$ is the system response to an impulse input $\delta(t)$ applied at $t = 0$ with all the initial conditions zero at $t = 0^-$. An impulse input $\delta(t)$ is like lightning, which strikes instantaneously and then vanishes. But in its wake, in that single moment, it rearranges things where it strikes. Similarly, an impulse input $\delta(t)$ appears momentarily at $t = 0$, and then it is gone forever. But in that moment it generates energy storages; that is, it creates nonzero initial conditions instantaneously within the system at $t = 0^+$. Although the impulse input $\delta(t)$ vanishes for $t > 0$ so that the system has no input after the impulse has been applied, the system will still have a response generated by these newly created initial conditions. The impulse response $h(t)$ therefore must consist of the system's characteristic modes for $t \geq 0^+$. As a result

$$h(t) = \text{characteristic mode terms} \qquad t \geq 0^+$$

This response is valid for $t > 0$. But what happens at $t = 0$? At a single moment $t = 0$, there can at most be an impulse,† so the form of the complete response $h(t)$ is given by

$$h(t) = A_0\delta(t) + \text{characteristic mode terms} \qquad t \geq 0 \qquad (2.40)$$

The detailed derivation of $h(t)$ is neither illuminating nor necessary for our future development, so to prevent needless distraction, this derivation is placed in Appendix 2.1 at the end of the chapter. There it is shown that for an LTIC system specified by Eq. (2.39), the unit impulse response $h(t)$ is given by

$$h(t) = b_n\delta(t) + [P(D)y_0(t)]u(t) \qquad (2.41)$$

where b_n is the coefficient of the nth-order term in $P(D)$ [see Eq. (2.39b)], and $y_0(t)$ is the zero-input response of the system subject to the following initial conditions:

$$y_0^{(n-1)}(0) = 1, \quad \text{and} \quad y_0(0) = \dot{y}_0(0) = \ddot{y}_0(0) = \cdots = y_0^{(n-2)}(0) = 0 \qquad (2.42)$$

where $y_0^{(k)}(0)$ is the value of the kth derivative of $y_0(t)$ at $t = 0$.

If the order of $P(D)$ is less than the order of $Q(D)$, $b_n = 0$, and the impulse term $b_n\delta(t)$ in $h(t)$ is zero.

■ **Example 2.4**

Determine the unit impulse response $h(t)$ for a system specified by the equation

$$\left(D^2 + 3D + 2\right)y(t) = Df(t) \qquad (2.43)$$

This is a second-order system ($n = 2$) having the characteristic polynomial

---

† It might be possible for the derivatives of $\delta(t)$ to appear at the origin. However, if $m \leq n$, it is impossible for $h(t)$ to have any derivatives of $\delta(t)$. This conclusion follows from Eq. (2.39b) with $f(t) = \delta(t)$ and $y(t) = h(t)$. The coefficients of the impulse and all of its derivatives must be matched on both sides of this equation. If $h(t)$ contains $\delta^{(1)}(t)$, the first derivative of $\delta(t)$, the left-hand side of Eq. (2.39b) will contain a term $\delta^{(n+1)}(t)$. But the highest-order derivative term on the right-hand side is $\delta^{(n)}(t)$. Therefore, the two sides cannot match. Similar arguments can be made against the presence of the impulse's higher-order derivatives in $h(t)$.

$$\left(\lambda^2 + 3\lambda + 2\right) = (\lambda + 1)(\lambda + 2) \tag{2.44}$$

The characteristic roots of this system are $\lambda = -1$ and $\lambda = -2$. Therefore

$$y_0(t) = c_1 e^{-t} + c_2 e^{-2t} \tag{2.45a}$$

Differentiation of this equation yields

$$\dot{y}_0(t) = -c_1 e^{-t} - 2c_2 e^{-2t} \tag{2.45b}$$

The initial conditions are [see Eq. (2.42)]

$$\dot{y}_0(0) = 1, \qquad \text{and} \qquad y_0(0) = 0$$

Setting $t = 0$ in Eqs. (2.45a) and (2.45b), and substituting the above initial conditions, we obtain

$$0 = c_1 + c_2$$
$$1 = -c_1 - 2c_2 \tag{2.46}$$

Solution of these two simultaneous equations yields

$$c_1 = 1 \qquad \text{and} \quad c_2 = -1$$

Therefore
$$y_0(t) = e^{-t} - e^{-2t} \tag{2.47}$$

Moreover, from Eq. (2.43), $P(D) = D$, so that

$$P(D)y_0(t) = Dy_0(t) = \dot{y}_0(t) = -e^{-t} + 2e^{-2t}$$

Also in this case, $b_n = b_2 = 0$ [the second-order term is absent in $P(D)$]. Therefore

$$h(t) = b_n \delta(t) + [P(D)y_0(t)]u(t) = (-e^{-t} + 2e^{-2t})u(t) \tag{2.48}$$

This response is shown in Fig. 2.8.   ■

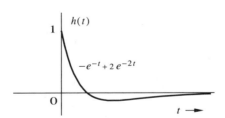

Fig. 2.8

■ **Example 2.5**

Determine the unit impulse response $h(t)$ for a system specified by the equation

$$\left(D^2 + 6D + 9\right)y(t) = (3D^2 + 6D + 3)f(t) \tag{2.49}$$

This is a second-order system ($n = 2$) having the characteristic polynomial

$$\left(\lambda^2 + 6\lambda + 9\right) = (\lambda + 3)^2 \tag{2.50}$$

The system has two repeated characteristic roots at -3. Therefore

$$y_0(t) = (c_1 + c_2 t)e^{-3t} \tag{2.51a}$$

Differentiation of this equation yields

$$\dot{y}_0(t) = (-3c_1 + c_2 - 3c_2 t)e^{-3t} \tag{2.51b}$$

The initial conditions are [see Eq. (2.42)]

$$\dot{y}_0(0) = 1, \quad \text{and} \quad y_0(0) = 0$$

Setting $t = 0$ in Eqs. (2.51a) and (2.51b), and substituting the above initial conditions, we obtain

$$0 = c_1$$

$$1 = -3c_1 + c_2$$

Solution of these two simultaneous equations yields $c_1 = 0$, and $c_2 = 1$. Therefore

$$y_0(t) = te^{-3t} \tag{2.52}$$

Moreover, from Eq. (2.49), $P(D) = 3D^2 + 6D + 3$, so that

$$P(D)y_0(t) = (3D^2 + 6D + 3)y_0(t) = 3\ddot{y}_0(t) + 6\dot{y}_0(t) + 3y_0(t) = 12(t-1)e^{-3t}$$

Also in this case, $b_n = b_2 = 3$. Therefore

$$h(t) = b_n\delta(t) + [P(D)y_0(t)]u(t) = 3\delta(t) + 12(t-1)e^{-3t}u(t) \tag{2.53}$$

∎

Comment

In the above discussion, we have assumed $m \leq n$, as specified by Eq. (2.39). In Appendix 2.1 it is shown that the expression for $h(t)$ applicable to all possible values of m and n is given by

$$h(t) = P(D)[y_0(t)u(t)]$$

where $y_0(t)$ is the zero-input response of the system subject to initial conditions (2.42). This expression reduces to Eq. (2.41) when $m \leq n$.

Determination of the impulse response $h(t)$ using the procedure in this section is relatively simple. However, in Chapter 4 we shall discuss another, even simpler, method using the Laplace transform.

△ **Exercise E2.5**

Determine the unit impulse response of LTIC systems described by the equations:

    (a) $(D+2)y(t) = (3D+5)f(t)$

    (b) $D(D+2)y(t) = (D+4)f(t)$

    (c) $(D^2 + 2D + 1)y(t) = Df(t)$

Answers: **(a)** $3\delta(t) - e^{-2t}u(t)$ **(b)** $(2 - e^{-2t})u(t)$ **(c)** $(1-t)e^{-t}u(t)$ ▽

System Response to Delayed Impulse

If $h(t)$ is the response of an LTIC system to the input $\delta(t)$, then $h(t - \tau)$ is the response of this same system to the input $\delta(t - \tau)$. This conclusion follows from the time-invariance property of LTIC systems. Thus, by knowing the unit impulse response $h(t)$, we can determine the system response to a delayed impulse $\delta(t - \tau)$.

2.4 SYSTEM RESPONSE TO EXTERNAL INPUT: ZERO-STATE RESPONSE

Although the zero-input component (response due to initial conditions) is important in determining the total response and in theoretical understanding of the system behavior, from a practical viewpoint it is much less important than the zero-state response (response due to external input). The reason is that in almost all practical systems used for signal processing, the zero-input component decays with time and soon becomes negligible. Moreover most of the practical systems have zero initial conditions, so that this component is zero. Systems are generally designed to yield a certain response to a given input, and it is the zero-state component (response due to external input) in which we are most interested.

This section is devoted to the determination of the zero-state response of an LTIC system. Recall that the zero-state response is the system response $y(t)$ to an input $f(t)$ when the system is in zero state, that is, all initial conditions are zero. **We shall assume that the systems discussed in this section are in zero state unless mentioned otherwise.** Under these conditions, the zero-state response will be the total response of the system. It alone will be the solution of the equation

$$Q(D)y(t) = P(D)f(t) \tag{2.54}$$

subject to zero initial conditions (zero state).

Superposition is one of the most powerful properties of linear systems. We shall use the superposition principle here to derive a linear system's response to some arbitrary input $f(t)$. We can express an input $f(t)$ as the sum of simpler components, so that

$$f(t) = f_1(t) + f_2(t) + \cdots + f_m(t) = \sum_{i=1}^{m} f_i(t) \tag{2.55a}$$

The response $y(t)$ of a linear system to the input $f(t)$ can be expressed as the sum of its responses to each of these m input components. Therefore

$$y(t) = y_1(t) + y_2(t) + \cdots + y_m(t) = \sum_{i=1}^{m} y_i(t) \tag{2.55b}$$

where $y_i(t)$ is the system response to the input component $f_i(t)$. There are many (in fact, infinite) families of functions $f_i(t)$ which can be used to express $f(t)$, as in Eq. (2.55a). Not all of these possible choices lead to simpler results, nor do all of them provide any special insight into system behavior. Two possibilities stand out for reasons that will become clear as our discussion unfolds. In the first approach, we express $f(t)$ in terms of impulses. We begin by approximating $f(t)$ with narrow rectangular pulses, as shown in Fig. 2.9a. This procedure gives us a staircase approximation of $f(t)$ that improves as pulse width is reduced. In the limit as pulse width approaches zero, this representation becomes exact, and the rectangular pulses become impulses delayed by various amounts. The system response to the input $f(t)$ is then given by the sum of the system's responses to each (delayed) impulse component of $f(t)$. In other words, we can determine $y(t)$, the system response to any arbitrary input $f(t)$, if we know the impulse response of the system. We now discuss this approach. The second approach (frequency-domain analysis) where an input is expressed as a sum of sinusoids or exponentials, is discussed in Chapters 4 and 5.

For the sake of generality, we place no restriction on $f(t)$ as to where it starts and where it ends. It is therefore assumed to exist for all time, starting at $t = -\infty$. The system's total response to this input will then be given by the sum of its responses to each of these impulse components. This process is illustrated in Fig. 2.9.

Figure 2.9a shows $f(t)$ as a sum of rectangular pulses, each of width $\triangle\tau$. In the limit as $\triangle\tau \rightarrow 0$, each pulse approaches an impulse having a strength equal to the area under that pulse. For example, as $\triangle\tau \rightarrow 0$, the shaded rectangular pulse located at $t = n\triangle\tau$ in Fig. 2.9a will approach an impulse at the same location with strength $f(n\triangle\tau)\triangle\tau$ (the shaded area under the rectangular pulse). This impulse can therefore be represented by $[f(n\triangle\tau)\triangle\tau]\delta(t - n\triangle\tau)$, as shown in Fig. 2.9d.

If the system's response to a unit impulse $\delta(t)$ is $h(t)$ (Fig. 2.9b), its response to a delayed impulse $\delta(t - n\triangle\tau)$ will be $h(t - n\triangle\tau)$ (Fig. 2.9c). Consequently the system's response to $[f(n\triangle\tau)\triangle\tau]\delta(t - n\triangle\tau)$ will be $[f(n\triangle\tau)\triangle\tau]h(t - n\triangle\tau)$, as shown in Fig. 2.9d.

We have now found the response $\triangle y(t)$ to one of the impulse components of $f(t)$, where

$$\triangle y(t) = \lim_{\triangle\tau\to0} [f(n\triangle\tau)\triangle\tau]h(t - n\triangle\tau) \qquad (2.56)$$

The desired zero-state response $y(t)$ is then the sum of the system's responses to each of these impulses, as shown in Fig. 2.9e. Consequently

$$y(t) = \lim_{\triangle\tau\to0} \sum_{n=-\infty}^{\infty} f(n\triangle\tau)h(t - n\triangle\tau)\triangle\tau \qquad (2.57)$$

The right-hand side, by definition, is an integral, that is,†

$$y(t) = \int_{-\infty}^{\infty} f(\tau)h(t - \tau)\, d\tau \qquad (2.58)$$

This is the result we seek. We have obtained the system response $y(t)$ to input $f(t)$ in terms of the unit impulse response $h(t)$. Knowing $h(t)$, we can determine

---

† In deriving this result we have assumed a time-invariant system. If the system is time-varying, then the system response to the input $\delta(t - n\triangle\tau)$ cannot be expressed as $h(t - n\triangle\tau)$, but instead has the form $h(t, n\triangle\tau)$. Using this form in Eq. (2.57), we see that

$$y(t) = \int_{-\infty}^{\infty} f(\tau)h(t, \tau)\, d\tau \qquad (2.58\text{n})$$

where $h(t, \tau)$ is the system response at instant t to a unit impulse input located at τ.

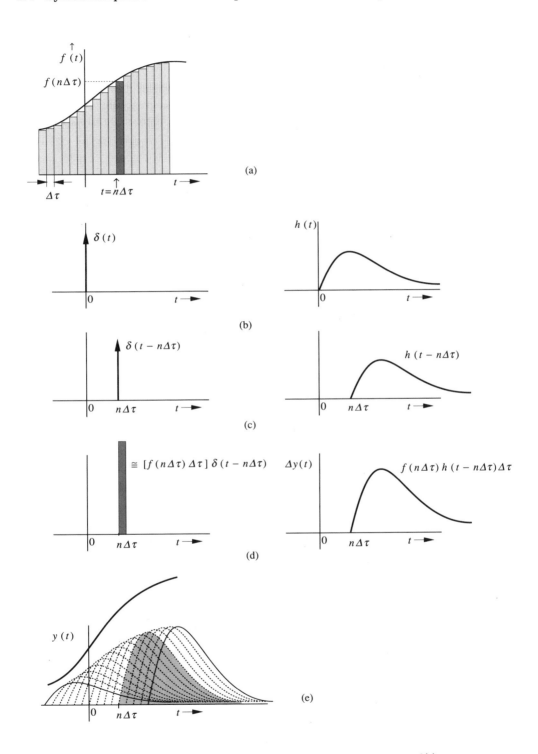

Fig. 2.9 Finding the system response to an arbitrary input $f(t)$.

the response $y(t)$ to any input. **Observe once again the all-pervasive nature of the system's characteristic modes. The system response to any input is determined by the impulse response, which in turn is made up of characteristic modes of the system.**

It is important to keep in mind the assumptions used in deriving Eq. (2.58). We assumed a linear, time-invariant (LTI) system. Linearity allowed us to use the principle of superposition, and time-invariance made it possible to express the system's response to $\delta(t - n\triangle\tau)$ as $h(t - n\triangle\tau)$.

2.4-1 The Convolution Integral

The zero-state response $y(t)$ obtained in Eq. (2.58) is given by an integral that occurs frequently in the physical sciences, engineering, and mathematics. For this reason this integral is given a special name: the **convolution integral**. The convolution integral of two functions $f_1(t)$ and $f_2(t)$ is denoted symbolically by $f_1(t) * f_2(t)$ and is defined as

$$f_1(t) * f_2(t) \equiv \int_{-\infty}^{\infty} f_1(\tau) f_2(t - \tau) \, d\tau \qquad (2.59)$$

Some important properties of the convolution integral are given below.

1. **The Commutative Property**: Convolution operation is commutative; that is, $f_1(t) * f_2(t) = f_2(t) * f_1(t)$. This property can be proved by a change of variable. In Eq. (2.59), if we let $x = t - \tau$ so that $\tau = t - x$ and $d\tau = -dx$, we obtain

$$f_1(t) * f_2(t) = -\int_{\infty}^{-\infty} f_2(x) f_1(t - x) \, dx$$

$$= \int_{-\infty}^{\infty} f_2(x) f_1(t - x) \, dx$$

$$= f_2(t) * f_1(t) \qquad (2.60)$$

2. **The Distributive Property**: This property states that

$$f_1(t) * [f_2(t) + f_3(t)] = f_1(t) * f_2(t) + f_1(t) * f_3(t) \qquad (2.61)$$

3. **The Associative Property**: This property states that

$$f_1(t) * [f_2(t) * f_3(t)] = [f_1(t) * f_2(t)] * f_3(t) \qquad (2.62)$$

The proofs of (2.61) and (2.62) follow directly from the definition of the convolution integral. They are left as an exercise for the reader.

4. **The Shift Property**: If

$$f_1(t) * f_2(t) = c(t)$$

then

$$f_1(t) * f_2(t - T) = c(t - T) \qquad (2.63a)$$

$$f_1(t - T) * f_2(t) = c(t - T) \qquad (2.63b)$$

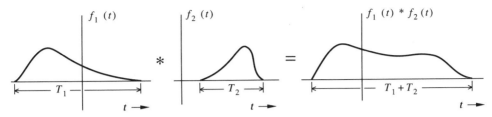

Fig. 2.10 Width property of convolution.

and

$$f_1(t - T_1) * f_2(t - T_2) = c(t - T_1 - T_2) \tag{2.63c}$$

Proof:
We are given

$$f_1(t) * f_2(t) = \int_{-\infty}^{\infty} f_1(\tau) f_2(t - \tau) \, d\tau = c(t)$$

Therefore

$$f_1(t) * f_2(t - T) = \int_{-\infty}^{\infty} f_1(\tau) f_2(t - T - \tau) \, d\tau$$

$$= c(t - T)$$

Equation (2.63b) follows from (2.63a) and the commutative property of convolution; Eq. (2.63c) follows directly from (2.63a) and (2.63b).

5. **Convolution with an Impulse**: Convolution of a function $f(t)$ with a unit impulse results in the function $f(t)$ itself. By definition

$$f(t) * \delta(t) = \int_{-\infty}^{\infty} f(\tau) \delta(t - \tau) \, d\tau$$

Note that $\delta(t - \tau)$ is located at $\tau = t$. According to the sampling property of the impulse [Eq. (2.33)], the integral in the above equation is the value of $f(\tau)$ at $\tau = t$, that is, $f(t)$. Therefore†

$$f(t) * \delta(t) = f(t) \tag{2.64}$$

6. **The Width Property**: If $f_1(t)$ and $f_2(t)$ have the durations (or widths) T_1 and T_2 respectively, then the duration (or width) of $f_1(t) * f_2(t)$ is $T_1 + T_2$ (Fig. 2.10). The proof of this property follows readily from the graphical considerations discussed later in Sec. 2.4-2.

---

† This result also follows from Fig. 2.9a, where we approximate $f(t)$ using a sum of rectangular pulses, each of width $\triangle \tau$. A typical pulse located at $t = n\triangle \tau$ has an area of $f(n\triangle \tau)\triangle \tau$. As $\triangle \tau \to 0$, this pulse approaches an impulse having a strength of $f(n\triangle \tau)\triangle \tau$. Therefore

$$f(t) = \lim_{\triangle \tau \to 0} \sum_{n=-\infty}^{\infty} [f(n\triangle \tau)\triangle \tau]\delta(t - n\triangle \tau) = \int_{-\infty}^{\infty} f(\tau)\delta(t - \tau) \, d\tau = f(t) * \delta(t)$$

Zero-State Response and Causality

From Eqs. (2.58) and (2.59), it follows that the (zero-state) response $y(t)$ of an LTIC system is

$$y(t) = \int_{-\infty}^{\infty} f(\tau)h(t - \tau)\,d\tau \qquad (2.65a)$$

$$= f(t) * h(t) \qquad (2.65b)$$

In deriving Eq. (2.65), we assumed the system to be linear and time-invariant. There were no other restrictions either on the system or on the input signal $f(t)$. In practice, most systems are causal, so that their response cannot begin before the input starts. Furthermore, most inputs start at $t = 0$. Such signals are called **causal signals**. To repeat, a signal $f(t)$ is said to be causal if

$$f(t) = 0 \qquad\qquad t < 0 \qquad (2.66)$$

If a signal $f(t) \neq 0$ for $t < 0$, the signal is **noncausal**.

Causality restrictions on both signals and systems further simplify the limits of integration in Eq. (2.65). By definition the response of a causal system cannot begin before its input begins. Consequently the causal system's response to a unit impulse $\delta(t)$ (which is located at $t = 0$) cannot begin before $t = 0$. Therefore a **causal system's unit impulse response $h(t)$ is a causal signal**.

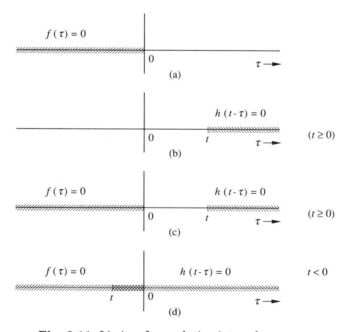

Fig. 2.11 Limits of convolution integral.

It is important to remember that the integration in Eq. (2.65) is performed with respect to τ (not t). If the input $f(t)$ is causal, $f(x) = 0$ for $x < 0$. Therefore, $f(\tau) = 0$ for $\tau < 0$, as shown in Fig. 2.11a. Similarly, if $h(t)$ is causal, $h(t - \tau) = 0$

for $t - \tau < 0$; that is, for $\tau > t$, as shown in Fig. 2.11b. Therefore the product $f(\tau)h(t - \tau) = 0$ everywhere except over the interval $0 \leq \tau \leq t$, as shown in Fig. 2.11c (assuming $t \geq 0$). Observe that if t is negative, $f(\tau)h(t - \tau) = 0$ for all τ, as shown in Fig. 2.11d. Therefore

$$y(t) = f(t) * h(t) = \int_{0^-}^{t} f(\tau)h(t - \tau)\, d\tau \qquad t \geq 0 \qquad (2.67)$$
$$= 0 \qquad\qquad t < 0$$

This result shows that if $f(t)$ and $h(t)$ are both causal, the response $y(t)$ is also causal. The lower limit of integration in Eq. (2.67) is taken as 0^- to avoid the difficulty in integration that can arise if $f(t)$ contains an impulse at the origin. In subsequent discussion, the lower limit will be shown as 0 with the understanding that it means 0^-.

Because of the convolution's commutative property [Eq. (2.60)], we can also express Eq. (2.67) as

$$y(t) = \int_{0}^{t} h(\tau)f(t - \tau)\, d\tau \qquad t \geq 0 \qquad (2.68)$$

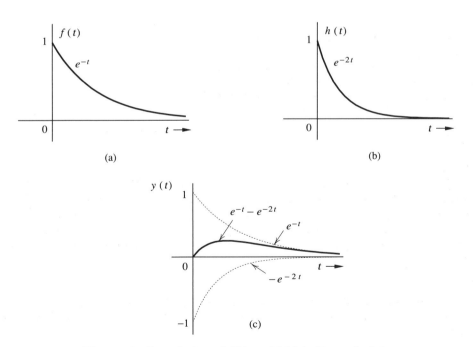

Fig. 2.12 Convolution of $f(t)$ and $h(t)$ in Example 2.6.

■ **Example 2.6**

For an LTIC system with the unit impulse response $h(t) = e^{-2t}u(t)$, determine the response $y(t)$ for the input

$$f(t) = e^{-t}u(t) \qquad (2.69)$$

Here both $f(t)$ and $h(t)$ are causal, and we need only to perform the convolution's integration over the range $(0, t)$ [see Eq. (2.67)]. The system response is therefore given by

$$y(t) = \int_0^t f(\tau)h(t - \tau)\, d\tau \qquad t \geq 0$$

Because $f(t) = e^{-t}u(t)$ and $h(t) = e^{-2t}u(t)$

$$f(\tau) = e^{-\tau}u(\tau) \quad \text{and} \quad h(t - \tau) = e^{-2(t-\tau)}u(t - \tau)$$

Remember that the integration is performed with respect to τ (not t), and the region of integration is $0 \leq \tau \leq t$. In other words, τ lies between 0 and t. Therefore, if $t \geq 0$, then $\tau \geq 0$ and $t - \tau \geq 0$, so that $u(\tau) = 1$ and $u(t - \tau) = 1$; consequently

$$y(t) = \int_0^t e^{-\tau}e^{-2(t-\tau)}\, d\tau \qquad t \geq 0$$

Because this integration is with respect to τ, we can pull e^{-2t} outside the integral, giving us

$$y(t) = e^{-2t}\int_0^t e^{\tau}\, d\tau = e^{-2t}(e^t - 1) = e^{-t} - e^{-2t} \qquad t \geq 0 \tag{2.70}$$

Also, $y(t) = 0$ when $t < 0$ [see Eq. (2.67)]. This result, along with Eq. (2.70), yields

$$y(t) = (e^{-t} - e^{-2t})u(t) \tag{2.71}$$

The response is shown in Fig. 2.12c. ■

△ **Exercise E2.6**

For an LTIC system with the impulse response $h(t) = 6e^{-t}u(t)$, determine the system response to the input: (a) $2u(t)$, and (b) $3e^{-3t}u(t)$.

Answer: (a) $12(1 - e^{-t})u(t)$ (b) $9(e^{-t} - e^{-3t})u(t)$ ▽

△ **Exercise E2.7**

Repeat Prob. E2.6 if the input $f(t) = e^{-t}u(t)$.

Answer: $6te^{-t}u(t)$ ▽

The Convolution Table

The task of convolution is considerably simplified by the use of a ready-made convolution table (Table 2.1). This table, which lists several pairs of signals $[f_1(t)$ and $f_2(t)]$ and their resulting convolution $f_1(t) * f_2(t)$, can be used conveniently to determine $y(t)$, a system response to an input $f(t)$, without having to perform the tedious job of integration. The following examples demonstrate the utility of this table.

■ **Example 2.7**

Rework Example 2.6 using the convolution table (Table 2.1).

Here $f(t) = e^{-t}u(t)$ and $h(t) = e^{-2t}u(t)$. The use of Pair 4 in Table 2.1 with $\lambda_1 = -1$ and $\lambda_2 = -2$ yields

$$y(t) = e^{-t}u(t) * e^{-2t}u(t)$$

$$= \frac{1}{-1 - (-2)}\left(e^{-t} - e^{-2t}\right)u(t)$$

$$= \left(e^{-t} - e^{-2t}\right)u(t) \tag{2.72}$$

which agrees with the result in Example 2.6. ■

TABLE 2.1: Convolution Table

No	$f_1(t)$	$f_2(t)$	$f_1(t) * f_2(t) = f_2(t) * f_1(t)$
1	$f(t)$	$\delta(t-T)$	$f(t-T)$
2	$e^{\lambda t}u(t)$	$u(t)$	$\dfrac{-1}{\lambda}(1-e^{\lambda t})u(t)$
3	$u(t)$	$u(t)$	$tu(t)$
4	$e^{\lambda_1 t}u(t)$	$e^{\lambda_2 t}u(t)$	$\dfrac{1}{\lambda_1-\lambda_2}[e^{\lambda_1 t}-e^{\lambda_2 t}]u(t) \qquad \lambda_1 \neq \lambda_2$
5	$e^{\lambda t}u(t)$	$e^{\lambda t}u(t)$	$te^{\lambda t}u(t)$
6	$te^{\lambda t}u(t)$	$e^{\lambda t}u(t)$	$\dfrac{1}{2}t^2 e^{\lambda t}u(t)$
7	$t^n u(t)$	$e^{\lambda t}u(t)$	$\dfrac{n!}{\lambda^{n+1}}e^{\lambda t}u(t) - \displaystyle\sum_{j=o}^{n}\dfrac{n!}{\lambda^{j+1}(n-j)!}t^{n-j}u(t)$
8	$t^m u(t)$	$t^n u(t)$	$\dfrac{m!n!}{(m+n+1)!}t^{m+n+1}u(t)$
9	$te^{\lambda_1 t}u(t)$	$e^{\lambda_2 t}u(t)$	$\dfrac{1}{(\lambda_1-\lambda_2)^2}[e^{\lambda_2 t}-e^{\lambda_1 t}+(\lambda_1-\lambda_2)te^{\lambda_1 t}]u(t)$
10	$t^m e^{\lambda t}u(t)$	$t^n e^{\lambda t}u(t)$	$\dfrac{m!\,n!}{(n+m+1)!}t^{m+n+1}e^{\lambda t}u(t)$
11	$t^m e^{\lambda_1 t}u(t)$	$t^n e^{\lambda_2 t}u(t)$	$\displaystyle\sum_{j=0}^{m}\dfrac{(-1)^j m!(n+j)!}{j!(m-j)!(\lambda_1-\lambda_2)^{n+j+1}}t^{m-j}e^{\lambda_1 t}u(t)$
	$\lambda_1 \neq \lambda_2$		$+\displaystyle\sum_{k=0}^{n}\dfrac{(-1)^k n!(m+k)!}{k!(n-k)!(\lambda_2-\lambda_1)^{m+k+1}}t^{n-k}e^{\lambda_2 t}u(t)$
12	$e^{-\alpha t}\cos(\beta t+\theta)u(t)$	$e^{\lambda t}u(t)$	$\dfrac{\cos(\theta-\phi)e^{\lambda t}-e^{-\alpha t}\cos(\beta t+\theta-\phi)}{\sqrt{(\alpha+\lambda)^2+\beta^2}}u(t)$
			$\phi=\tan^{-1}[-\beta/(\alpha+\lambda)]$
13	$e^{\lambda_1 t}u(t)$	$e^{\lambda_2 t}u(-t)$	$\dfrac{1}{\lambda_2-\lambda_1}[e^{\lambda_1 t}u(t)+e^{\lambda_2 t}u(-t)] \quad \operatorname{Re}\lambda_2 > \operatorname{Re}\lambda_1$
14	$e^{\lambda_1 t}u(-t)$	$e^{\lambda_2 t}u(-t)$	$\dfrac{1}{\lambda_2-\lambda_1}(e^{\lambda_1 t}-e^{\lambda_2 t})u(-t)$

Example 2.8

Find the loop current $y(t)$ of the RLC circuit in Example 2.2 for the input $f(t) = 10e^{-3t}u(t)$, when all the initial conditions are zero.

The loop equation for this circuit [see also Example 2.2 or Eq. (1.20)] is

$$\left(D^2 + 3D + 2\right) y(t) = Df(t)$$

The impulse response $h(t)$ for this system was found in Example 2.4 to be

$$h(t) = \left(2e^{-2t} - e^{-t}\right) u(t)$$

The input is $f(t) = 10e^{-3t}u(t)$. The response $y(t)$ is therefore given by

$$y(t) = f(t) * h(t)$$
$$= 10e^{-3t}u(t) * \left[2e^{-2t} - e^{-t}\right] u(t)$$

Using the distributive property of the convolution [Eq. (2.61)], we obtain

$$y(t) = 10e^{-3t}u(t) * 2e^{-2t}u(t) - 10e^{-3t}u(t) * e^{-t}u(t)$$
$$= 20 \left[e^{-3t}u(t) * e^{-2t}u(t)\right] - 10 \left[e^{-3t}u(t) * e^{-t}u(t)\right]$$

Now the use of Pair 4 in Table 2.1 yields

$$y(t) = \frac{20}{-3 - (-2)} \left[e^{-3t} - e^{-2t}\right] u(t) - \frac{10}{-3 - (-1)} \left[e^{-3t} - e^{-t}\right] u(t)$$
$$= -20 \left(e^{-3t} - e^{-2t}\right) u(t) + 5 \left(e^{-3t} - e^{-t}\right) u(t)$$
$$= \left(-5e^{-t} + 20e^{-2t} - 15e^{-3t}\right) u(t) \qquad (2.73)$$

This response is shown in Fig. 2.13. ■

Fig. 2.13 The zero-state response for the RLC circuit in Fig. 2.1a (Example 2.8).

Total Response

The total response of a linear system can be expressed as the sum of its zero-input and zero-state components:

$$\text{Total Response} = \underbrace{\sum_{j=1}^{n} c_j e^{\lambda_j t}}_{\text{zero-input component}} + \underbrace{f(t) * h(t)}_{\text{zero-state component}}$$

We have developed procedures for determining these two components. From the system equation we find the system's characteristic roots and hence its characteristic modes. The zero-input response is a linear combination of these characteristic modes; the arbitrary constants $c_1, c_2, \ldots, c_n$ in this response are determined from n auxiliary conditions. From the system equation we can determine $h(t)$, the system's impulse response, using the procedure described in Sec. 2.3-1. Knowing $h(t)$ and the input $f(t)$, we can determine the zero-state response as the convolution of $f(t)$ and $h(t)$.

As an example, consider the series RLC circuit in Fig. 2.1a with the input $f(t) = 10e^{-3t}u(t)$; the initial inductor current $y(0) = 0$, and the initial capacitor voltage $v_C(0) = 5$. We determined the zero-input current of this circuit (due to initial conditions alone) in Example 2.2. We also determined the system's unit impulse response $h(t)$ in Example 2.4. Knowing $h(t)$ and $f(t)$, we found the zero-state current (due to the input alone, with zero initial conditions) in Example 2.8. Using the results in Examples 2.2 and 2.8, we obtain

$$\text{Total current} = \underbrace{\left(-5e^{-t} + 5e^{-2t}\right)}_{\text{zero-input current}} + \underbrace{\left(-5e^{-t} + 20e^{-2t} - 15e^{-3t}\right)}_{\text{zero-state current}} \qquad t \geq 0 \quad (2.74a)$$

Figure 2.14a shows the zero-input, the zero-state, and the total response.

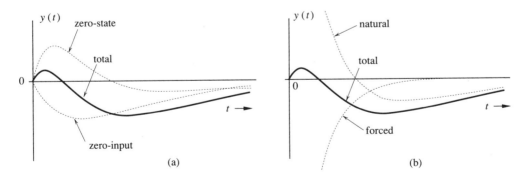

Fig. 2.14 Total response of the RLC circuit in Fig. 2.1a (Example 2.2).

Natural and Forced Response

For the RLC circuit of Figure 2.1a, the characteristic modes were found to be e^{-t} and e^{-2t}. As we expected, the zero-input response is composed exclusively of characteristic modes. Note, however, that even the zero-state response [Eq. (2.73)] contains characteristic mode terms. This observation is generally true of LTI systems. We can now lump together all the characteristic mode terms in the total response, giving us a component known as the **natural response**. The remainder, consisting entirely of noncharacteristic mode terms, is known as the **forced response**. The total response of the RLC circuit in Fig. 2.1a can be expressed in terms of natural and forced components by regrouping the terms in Eq. (2.74a). This gives us

$$\text{Total current} = \underbrace{\left(-10e^{-t} + 25e^{-2t}\right)}_{\text{natural response}} + \underbrace{\left(-15e^{-3t}\right)}_{\text{forced response}} \qquad t \geq 0 \qquad (2.74b)$$

Figure 2.14b shows the natural, forced, and total response.

△ **Exercise E2.8**
   Rework Probs. E2.6 and E2.7, using the convolution table.   ▽

△ **Exercise E2.9**
   Using the convolution table, determine

$$e^{-2t}u(t) * \left(1 - e^{-t}\right)u(t)$$

Answer: $\left(\frac{1}{2} - e^{-t} + \frac{1}{2}e^{-2t}\right)u(t)$ ▽

△ **Exercise E2.10**
   For an LTIC system with the unit impulse response $h(t) = e^{-2t}u(t)$, determine the zero-state response $y(t)$ if the input $f(t) = \sin 3t\, u(t)$. Hint: Use the convolution table (Pair 12).

Answer: $\frac{1}{13}\left[3e^{-2t} + \sqrt{13}\cos\left(3t - 146.32°\right)\right]u(t)$

or $\frac{1}{13}\left[3e^{-2t} - \sqrt{13}\cos\left(3t + 33.68°\right)\right]u(t)$ ▽

Multiple Inputs

   Multiple inputs to LTI systems can be treated by applying the superposition principle. Each input is considered separately, with all other inputs assumed to be zero. The sum of all these individual system responses constitutes the total system output due to all the inputs acting simultaneously.

2.4-2   Graphical Understanding of Convolution

   To have a proper grasp of convolution operation, it is helpful to understand graphical interpretation of convolution. Such a comprehension also helps in evaluating the convolution integral of more complicated signals. In addition, graphical convolution allows us to grasp visually or mentally the convolution integral's result, which can be of great help in sampling, filtering, and many other problems. Finally, many signals have no exact mathematical description, so that they can be described only graphically. If two such signals are to be convolved, we have no choice but to perform their convolution graphically.

   We shall now explain the convolution operation by convolving the signals $f(t)$ and $g(t)$, shown in Figs. 2.15a and 2.15b respectively. We have intentionally selected simple functions in order to focus attention on the process of convolution

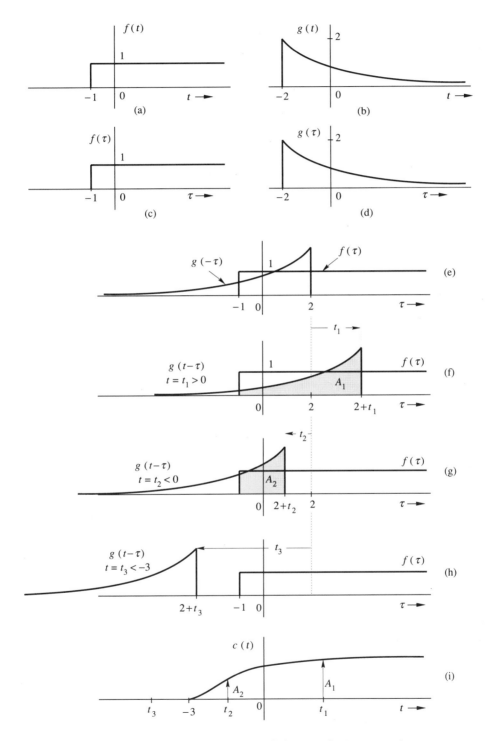

Fig. 2.15 Graphical explanation of the convolution operation.

without being distracted by the complexities of $f(t)$ and $g(t)$. In later examples more complex functions will be considered.

If $c(t)$ is the convolution of $f(t)$ with $g(t)$, then

$$c(t) = \int_{-\infty}^{\infty} f(\tau)g(t - \tau)\, d\tau \qquad (2.75)$$

One of the crucial things to remember here is that this integration is performed with respect to τ, so that t is just a parameter (like a constant). This is especially important when we sketch the graphical representations of the functions $f(\tau)$ and $g(t - \tau)$ appearing in the integrand of Eq. (2.75). Both of these functions should be sketched as functions of τ, not of t.

The function $f(\tau)$ is identical to $f(t)$, with τ replacing t (Fig. 2.15c). Therefore $f(t)$ and $f(\tau)$ will have the same graphical representations. Similar remarks apply to $g(t)$ and $g(\tau)$ (Fig. 2.15d).

The function $g(t - \tau)$ is not as easy to comprehend. To understand what this function looks like, let us start with the function $g(\tau)$ (Fig. 2.15d). Time-inversion (or reflection) of this function about the vertical axis ($\tau = 0$) yields $g(-\tau)$ (Fig. 2.15e). Let us denote this function by $\phi(\tau)$.

$$\phi(\tau) = g(-\tau)$$

Now $\phi(\tau)$ shifted by t seconds is $\phi(\tau - t)$, given by

$$\phi(\tau - t) = g[-(\tau - t)] = g(t - \tau)$$

Therefore we first time-invert $g(\tau)$ to obtain $g(-\tau)$ and then time-shift $g(-\tau)$ by t to obtain $g(t - \tau)$. For positive t, the shift is to the right (Fig. 2.15f); for negative t, the shift is to the left (Fig. 2.15g).

This gives us a graphical interpretation of the functions $f(\tau)$ and $g(t - \tau)$. The convolution $c(t)$ is the area under the product of these two functions. Thus, to compute $c(t)$ at some positive instant $t = t_1$, we first obtain $g(-\tau)$ by inverting $g(\tau)$ about the vertical axis. Next, we right-shift or delay $g(-\tau)$ by t_1 to obtain $g(t_1 - \tau)$ (Fig. 2.15f), and then we multiply this function by $f(\tau)$, giving us the product $f(\tau)g(t_1 - \tau)$ (Fig. 2.15f). The area A_1 under this product is $c(t_1)$, the value of $c(t)$ at $t = t_1$. We can therefore plot $c(t_1) = A_1$ on a curve describing $c(t)$, as shown in Fig. 2.15i. Observe that the area under the product $f(\tau)g(-\tau)$ in Fig. 2.15e is $c(0)$, the value of the convolution for $t = 0$ (at the origin).

A similar procedure is followed in computing the value of $c(t)$ at $t = t_2$, where t_2 is negative (Fig. 2.15g). In this case, the function $g(-\tau)$ is shifted by a negative amount (that is, left-shifted) to obtain $g(t_2 - \tau)$. Multiplication of this function with $f(\tau)$ yields the product $f(\tau)g(t_2 - \tau)$. The area under this product is $c(t_2) = A_2$, giving us another point on the curve $c(t)$ at $t = t_2$ (Figure 2.15i). This procedure can be repeated for all values of t, from $-\infty$ to ∞. The result will be a curve describing $c(t)$ for all time t. Note that when $t \leq -3$, $f(\tau)$ and $g(t - \tau)$ do not overlap (see Fig. 2.15h); therefore $c(t) = 0$ for $t \leq -3$.

Summary of the Graphical Procedure

The procedure for graphical convolution can be summarized as follows:

1. Keep the function $f(\tau)$ fixed.

2. Visualize the function $g(\tau)$ as a rigid wire frame, and rotate (or invert) this frame about the vertical axis ($\tau = 0$) to obtain $g(-\tau)$.

3. Shift the inverted frame along the τ axis by t_0 seconds. The shifted frame now represents $g(t_0 - \tau)$.

4. The area under the product of $f(\tau)$ and $g(t_0 - \tau)$ (the shifted frame) is $c(t_0)$, the value of the convolution at $t = t_0$.

5. Repeat this procedure, shifting the frame by different values (positive and negative) to obtain $c(t)$ for all values of t.

The graphical procedure discussed here appears very complicated and discouraging at first reading. Actually, its bark is worse than its bite! (We assure the reader that there is absolutely no truth to the rumor that convolution has, for decades, driven electrical engineering undergraduates to contemplate theology either for salvation or as an alternative career. For the source of rumor, see Paul Nahin *IEEE Spectrum*, March 1991, p.60). In graphical convolution, we need to determine the area under the product $f(\tau)g(t - \tau)$ for all values of t from $-\infty$ to ∞. However, a mathematical description of $f(\tau)g(t - \tau)$ is generally valid over a range of t. Therefore repeating the procedure for every value of t amounts to repeating it only a few times for different ranges of t.

Convolution: its bark is worse than its bite!

We can also use the commutative property of convolution to our advantage by computing $f(t) * g(t)$ or $g(t) * f(t)$, whichever is simpler. As a rule of thumb, **convolution computations are simplified if we choose to invert the simpler of the two functions**. For example, if the mathematical description of $g(t)$ is simpler than that of $f(t)$, then $f(t) * g(t)$ will be easier to compute than $g(t) * f(t)$. On the other hand, if the mathematical description of $f(t)$ is simpler, the reverse will be true.

We shall demonstrate graphical convolution with the following examples. Let us start by reworking Example 2.6, using this graphical method.

■ **Example 2.9**

Determine graphically $y(t) = f(t) * h(t)$ for $f(t) = e^{-t}u(t)$ and $h(t) = e^{-2t}u(t)$.

Figures 2.16a and 2.16b show $f(t)$ and $h(t)$ respectively, and Fig. 2.16c shows $f(\tau)$ and $h(-\tau)$ as functions of τ. The function $h(t-\tau)$ is now obtained by shifting $h(-\tau)$ by t. If t is positive, the shift is to the right (delay); if t is negative, the shift is to the left (advance). Figure 2.16d shows that for negative t, $h(t-\tau)$ [obtained by left-shifting $h(-\tau)$] does not overlap $f(\tau)$, and the product $f(\tau)h(t-\tau) = 0$, so that

$$y(t) = 0 \qquad t < 0$$

Figure 2.16e shows the situation for $t \geq 0$. Here $f(\tau)$ and $h(t-\tau)$ do overlap, but the product is nonzero only over the interval $0 \leq \tau \leq t$ (shaded interval). Therefore

$$y(t) = \int_0^t f(\tau)h(t-\tau)\, d\tau \qquad t \geq 0$$

All we need to do now is substitute correct expressions for $f(\tau)$ and $h(t-\tau)$ in this integral. From Figs. 2.16a and 2.16b, it is clear that the segments of $f(t)$ and $g(t)$ to be used in this convolution (Fig. 2.16e) are described by

$$f(t) = e^{-t} \qquad \text{and} \qquad h(t) = e^{-2t}$$

Therefore

$$f(\tau) = e^{-\tau} \qquad \text{and} \qquad h(t-\tau) = e^{-2(t-\tau)}$$

Consequently

$$y(t) = \int_0^t e^{-\tau} e^{-2(t-\tau)}\, d\tau$$

$$= e^{-2t} \int_0^t e^{\tau}\, d\tau$$

$$= e^{-t} - e^{-2t} \qquad t \geq 0$$

Moreover, $y(t) = 0$ for $t < 0$, so that

$$y(t) = \left(e^{-t} - e^{-2t} \right) u(t) \quad ■$$

■ **Example 2.10**

Find $c(t) = f(t) * g(t)$ for the signals shown in Figs. 2.17a and 2.17b.

Since $f(t)$ is simpler than $g(t)$, it is easier to evaluate $g(t) * f(t)$ than $f(t) * g(t)$. However, we shall intentionally take the more difficult route and evaluate $f(t) * g(t)$ to clarify some of the finer points of convolution.

Figures 2.17a and 2.17b show $f(t)$ and $g(t)$ respectively. Observe that $g(t)$ is composed of two segments. As a result, it can be described as

$$g(t) = \begin{cases} 2e^{-t} & \text{segment A} \\ -2e^{2t} & \text{segment B} \end{cases}$$

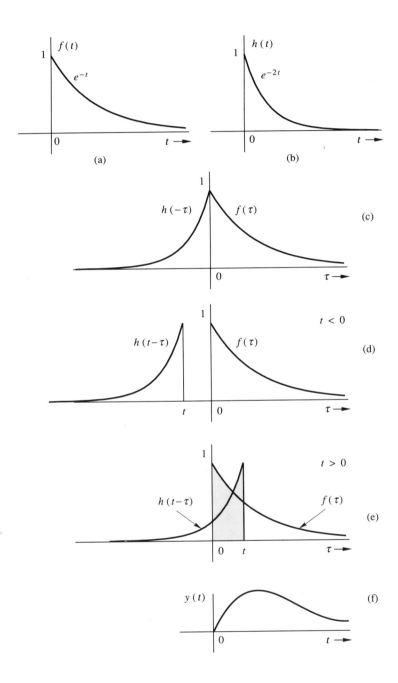

Fig. 2.16 Convolution of $f(t)$ and $h(t)$ in example 2.9.

Therefore

$$g(t - \tau) = \begin{cases} 2e^{-(t-\tau)} & \text{segment A} \\ -2e^{2(t-\tau)} & \text{segment B} \end{cases} \tag{2.76}$$

The segment of $f(t)$ that is used in convolution is $f(t) = 1$, so that $f(\tau) = 1$. Figure 2.17c shows $f(\tau)$ and $g(-\tau)$.

To compute $c(t)$ for $t \geq 0$, we right-shift $g(-\tau)$ to obtain $g(t - \tau)$, as shown in Fig. 2.17d. Clearly, $g(t - \tau)$ overlaps with $f(\tau)$ over the shaded interval, that is, over the range $\tau \geq 0$; Segment A overlaps with $f(\tau)$ over the interval $(0, t)$, while Segment B overlaps with $f(\tau)$ over (t, ∞). Remembering that $f(\tau) = 1$, we have

$$c(t) = \int_0^\infty f(\tau)g(t - \tau)\, d\tau$$

$$= \int_0^t 2e^{-(t-\tau)}\, d\tau + \int_t^\infty -2e^{2(t-\tau)}\, d\tau$$

$$= 2\left(1 - e^{-t}\right) - 1$$

$$= 1 - 2e^{-t} \qquad\qquad t \geq 0$$

Figure 2.17e shows the situation for $t < 0$. Here the overlap is over the shaded interval, that is, over the range $\tau \geq 0$, where only the segment B of $g(t)$ is involved. Therefore

$$c(t) = \int_0^\infty f(\tau)g(t - \tau)\, d\tau$$

$$= \int_0^\infty g(t - \tau)\, d\tau$$

$$= \int_0^\infty -2e^{2(t-\tau)}\, d\tau$$

$$= -e^{2t} \qquad\qquad t < 0$$

Therefore

$$c(t) = \begin{cases} 1 - 2e^{-2t} & t \geq 0 \\ -e^{2t} & t < 0 \end{cases}$$

Figure 2.17f shows a plot of $c(t)$. ■

■ Example 2.11

Find $f(t) * g(t)$ for the functions $f(t)$ and $g(t)$ shown in Figs. 2.18a and 2.18b respectively.

In this case $f(t)$ has a simpler mathematical description than that of $g(t)$, so it is preferable to invert $f(t)$. For this reason we shall determine $g(t) * f(t)$ rather than $f(t) * g(t)$. Therefore

$$c(t) = g(t) * f(t)$$

$$= \int_{-\infty}^\infty g(\tau)f(t - \tau)\, d\tau$$

First, we determine the expressions for the segments of $f(t)$ and $g(t)$ that are used in finding $c(t)$. From Figs. 2.18a and 2.18b, these segments can be expressed as

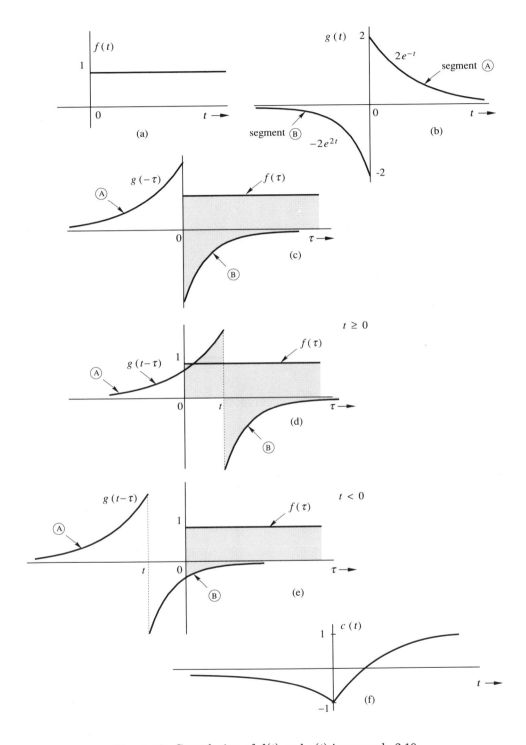

Fig. 2.17 Convolution of $f(t)$ and $g(t)$ in example 2.10.

$$f(t) = 1 \qquad \text{and} \qquad g(t) = \tfrac{1}{3}t$$

Therefore

$$f(t - \tau) = 1 \qquad \text{and} \qquad g(\tau) = \tfrac{1}{3}\tau$$

Fig. 2.18c shows $g(\tau)$ and $f(-\tau)$, whereas Fig. 2.18d shows $g(\tau)$ and $f(t - \tau)$ valid for $-1 \leq t \leq 1$. For $t < -1$ the situation is as shown in Fig. 2.18h; for $t > 1$ the situation is as shown in Fig. 2.18e. Note that the edges of $f(-\tau)$ are at -1 and 1 (Fig. 2.18c). Because $f(t - \tau)$ is $f(-\tau)$ time-shifted by t, the edges of $f(t - \tau)$ are at $-1 + t$ and $1 + t$, as shown in Fig. 2.18d. The two functions overlap over the interval $(0, 1 + t)$ (shaded interval), so that

$$
\begin{aligned}
c(t) &= \int_0^{1+t} g(\tau) f(t - \tau)\, d\tau \\
&= \int_0^{1+t} \tfrac{1}{3}\tau \, d\tau \\
&= \tfrac{1}{6}(t + 1)^2 \qquad\qquad -1 \leq t \leq 1
\end{aligned}
\tag{2.77a}
$$

This situation, shown in Fig. 2.18d, is valid only for $-1 \leq t \leq 1$. For $t > 1$ but < 2, the situation is as shown in Fig. 2.18e. The two functions overlap only over the range $-1 + t$ to $1 + t$ (shaded interval). Note that the expressions for $g(\tau)$ and $f(t - \tau)$ do not change; only the range of integration changes. Therefore

$$
\begin{aligned}
c(t) &= \int_{-1+t}^{1+t} \tfrac{1}{3}\tau \, d\tau \\
&= \tfrac{2}{3}t \qquad\qquad 1 \leq t \leq 2
\end{aligned}
\tag{2.77b}
$$

Also note that the expressions in Eqs. (2.77a) and (2.77b) both apply at $t = 1$, the transition point between their respective ranges. We can readily verify that both expressions yield a value of $2/3$ at $t = 1$, so that $c(1) = 2/3$. The continuity of $c(t)$ at transition points indicates a high probability of a right answer.† For $t \geq 2$ but < 4 the situation is as shown in Fig. 2.18f. The functions $g(\tau)$ and $f(t - \tau)$ overlap over the interval from $-1 + t$ to 3 (shaded interval), so that

$$
\begin{aligned}
c(t) &= \int_{-1+t}^{3} \tfrac{1}{3}\tau \, d\tau \\
&= -\tfrac{1}{6}\left(t^2 - 2t - 8\right)
\end{aligned}
\tag{2.77c}
$$

Again, both Eqs. (2.77b) and (2.77c) apply at the transition point $t = 2$. We can readily verify that $c(2) = 4/3$ when either of these expressions is used.

     For $t \geq 4$, $f(t - \tau)$ has been shifted so far to the right that it no longer overlaps with $g(\tau)$ as shown in Fig. 2.18g. Consequently

$$c(t) = 0 \qquad\qquad t \geq 4 \tag{2.77d}$$

---

† Even if $c(t)$ is continuous at the transition, the answer could be wrong in the unlikely event of two or more errors canceling out their effects. In our discussion we are assuming that there are no impulses in $f(t - \tau)$ and $g(\tau)$ after the transition which were not present before.

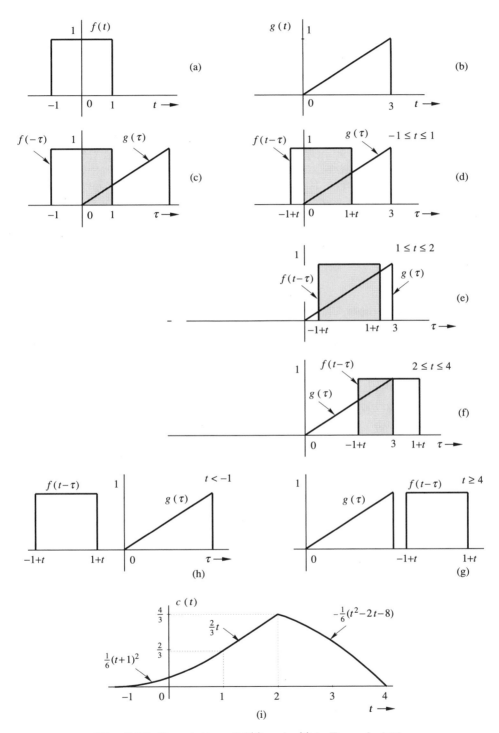

Fig. 2.18 Convolution of $f(t)$ and $g(t)$ in Example 2.11.

We now turn our attention to negative values of t. We have already determined $c(t)$ up to $t = -1$. For $t < -1$ there is no overlap between the two functions, as shown in Fig. 2.18h, so that

$$c(t) = 0 \qquad\qquad t < -1 \qquad\qquad (2.77\text{e})$$

Figure 2.18i shows $c(t)$ plotted according to Eqs. (2.77a) through (2.77e). ■

Width of the Convolved Function

The widths (durations) of $f(t)$, $g(t)$, and $c(t)$ in Example 2.11 (Fig. 2.18) are 2, 3, and 5 respectively. Note that the width of $c(t)$ in this case is the sum of the widths of $f(t)$ and $g(t)$. This is not a coincidence. Using the concept of graphical convolution, we can readily show that if $f(t)$ and $g(t)$ have the finite widths of T_1 and T_2 respectively, then the width of $c(t)$ is equal to $T_1 + T_2$. This conclusion follows from the fact that the time it takes for a signal of width (duration) T_1 to completely pass another signal of width (duration) T_2 so that they become nonoverlapping is $T_1 + T_2$.

⊙ **Computer Example C2.2**

Find $c(t) = f(t) * g(t)$ for the signals shown in Fig. 2.17.

```
t1 = -10:.1:0;        % Create a vector for negative time.
g1 = -2*exp(2*t1);    % Create g for negative time.
plot(t1,g1),grid, title('segment B'),pause,
t2 = 0:.1:10;         % Create a  vector for positive time.
g2 = 2*exp(-t2);      % Create g for positive time.
plot(t2,g2),grid, title('segment A'),pause,
g = [g1';g2'];   t3=[t1';t2'];
plot(t3,g),grid, title('segments A and B'),pause,
f = [zeros(g1)   ones(g2)];
plot(t3,f),grid, title('plot of f(t)'),pause,
c =0.1*conv(f,g);   % Convolve the two vectors f and g.
t = -20:0.1:7.4;
plot(t,c(1:length(t))),grid,title('c(t)')   ⊙
```

△ **Exercise E2.11**

Rework Example 2.10 by evaluating $g(t) * f(t)$ ▽

△ **Exercise E2.12**

Use graphical convolution to show that $f(t) * g(t) = g(t) * f(t)$ in Fig. 2.19. ▽

△ **Exercise E2.13**

Repeat Prob. E2.12 for the functions in Fig. 2.20. ▽

△ **Exercise E2.14**

Repeat Prob. E2.12 for the functions in Fig. 2.21. ▽

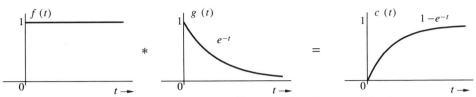

Fig. 2.19 Convolution of $f(t)$ and $g(t)$ in Exercise E2.12.

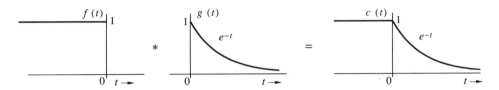

Fig. 2.20 Convolution of $f(t)$ and $g(t)$ in Exercise E2.13.

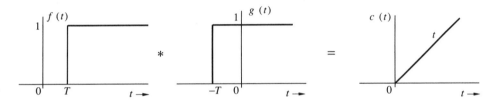

Fig. 2.21 Convolution of $f(t)$ and $g(t)$ in Exercise E2.14.

2.4-3   Why Convolution? An Intuitive Explanation of System Response

On the surface it appears rather strange that the response of linear systems (those gentlest of the gentle systems) should be given by such a tortuous operation of convolution, where one signal is fixed and the other is inverted and shifted. To understand this odd behavior, consider a hypothetical impulse response $h(t)$ that decays linearly with time (Fig. 2.22a). This response is strongest at $t = 0$, the moment the impulse is applied, and it decays linearly at future instants, so that one second later (at $t = 1$ and beyond), it ceases to exist. This means that the closer the impulse input is to an instant t, the stronger is its response at t.

Now consider the input $f(t)$ shown in Fig. 2.22b. To compute the system response, we break the input into rectangular pulses and approximate these pulses with impulses. Generally the response of a causal system at some instant t will be determined by all the impulse components of the input before t. Each of these impulse components will have different weight in determining the response at the instant t, depending on its proximity to t. As seen earlier, the closer the impulse is to t, the stronger is its influence at t. The impulse component at t has the greatest weight (unity) in determining the response at t. The weight decreases linearly for all impulses before t until the instant $t - 1$. The input before $t - 1$ has no influence (zero weight). Thus, to determine the system response at t, we must assign a linearly decreasing weight to impulses occurring before t, as shown in Fig. 2.22b. This weighting function is precisely the function $h(t - \tau)$. The system response at t is then determined not by the input $f(\tau)$ but by the weighted input $f(\tau)h(t - \tau)$, and the summation of all these weighted inputs is the convolution integral.

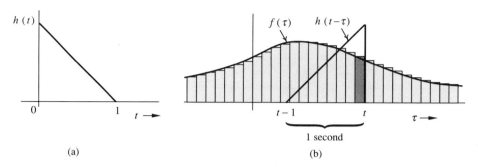

Fig. 2.22 Intuitive explanation of convolution.

2.4-4   Some Reflections on the Use of Impulse Function

In the study of signals and systems we often come across signals such as impulses, which cannot be generated in practice. One wonders why we even consider such signals. The answer should be clear from our discussion so far in this chapter. Even if the impulse function has no physical existence, we can compute the impulse response $h(t)$ of a system according to the procedure in Sec. 2.3-1, and knowing $h(t)$, we can compute the system response to any arbitrary input. The concept of impulse response therefore provides an effective intermediary for computing system response to an arbitrary input. In addition, the impulse response $h(t)$ itself provides a great deal of information and insight about the system behavior. In Sec. 2.6 we show that the knowledge of impulse response provides much valuable information, such as the response time, pulse dispersion, and filtering properties of the system. If the system is used to transmit information through pulses, $h(t)$ indicates the system capability in terms of rate of information transmission. Many other useful insights about the system behavior can be obtained by inspection of $h(t)$.

We have a similar situation in frequency-domain analysis (discussed in later chapters) where we use an *everlasting sinusoid* (a sinusoid starting at $t = -\infty$) to determine system response. The everlasting sinusoid has no physical existence, but it provides another effective intermediary for computing system response to an arbitrary input. Moreover, the system response to everlasting sinusoid provides valuable information and insight regarding the system's behavior.

2.5   NUMERICAL CONVOLUTION

Convolution can be performed conveniently with digital computers, which are routinely used for signal processing and for the analysis and simulation of systems. Since convolution is basically an integration, numerical convolution in a direct form is actually a simple numerical integration based on a staircase approximation of the signals convolved.

For the sake of simplicity we start with causal signals. As an example, consider

$$c(t) = f(t) * g(t) \tag{2.78a}$$

$$= \int_0^t f(\tau)g(t - \tau)\, d\tau \tag{2.78b}$$

But by definition an integral is a sum in the limit, so that Eq. (2.78b) can be expressed [see Eq. (2.57)] as

$$c(t) = \lim_{\triangle\tau\to 0} \sum_{m\triangle\tau=0}^{t} f(m\triangle\tau)g(t-m\triangle\tau)\triangle\tau \qquad (2.78\text{c})$$

Eq. (2.78c) expresses the convolution $c(t)$ in terms of the samples of $f(t)$ and $g(t)$ taken every $\triangle\tau$ seconds. For the sake of compactness we will denote the sampling interval $\triangle\tau$ by T. This leads to

$$c(t) = \lim_{T\to 0} \sum_{mT=0}^{t} f(mT)g(t-mT)T$$

$$= \lim_{T\to 0} T \sum_{mT=0}^{t} f(mT)g(t-mT) \qquad (2.78\text{d})$$

Setting $t = kT$ in this equation yields

$$c(kT) = \lim_{T\to 0} T \sum_{mT=0}^{kT} f(mT)g(kT-mT)$$

It is convenient to use compact notation $c[k]$ to represent the kth sample of $c(t)$; that is, $c[k] = c(kT)$. Similarly, the mth sample of $f(t)$ is $f[m] = f(mT)$, and so on. Using this compact notation, we can express the above equation as

$$c[k] = \lim_{T\to 0} T \sum_{m=0}^{k} f[m]g[k-m] \qquad (2.79)$$

For a general case where both $f(t)$ and $g(t)$ are noncausal, the limits for the sum are from $m = -\infty$ to ∞.

$$c[k] = \lim_{T\to 0} T \sum_{m=-\infty}^{\infty} f[m]g[k-m] \qquad (2.80)$$

Observe that we have expressed the samples of $c(t)$ in terms of the samples of $f(t)$ and $g(t)$ taken every T seconds ($T \to 0$). The relationship in Eq. (2.79) or Eq. (2.80) is exact only when $T \to 0$. In practical computations, we can make T small, but we can never make it 0. Therefore the numerical computation of convolution using these equations will yield approximate results; the approximation improves as the value of T is reduced. (The choice of a suitable value for T can be discussed intelligently only after a study of sampling theorem in Chapter 8).

To demonstrate how convolution is evaluated numerically, let us consider Eq. (2.79), where both the system and the input are causal. We compute $c[k]$ at $k = 0$ by setting $k = 0$ in Eq. (2.79). The summation on the right-hand side has only one term. Similarly, for $c[k]$ at $k = 1$, we set $k = 1$ in Eq. (2.79), giving us two terms on the right-hand side. If we continue in this manner, substituting $k = 0, 1, 2, \ldots$ in Eq. (2.79), we obtain

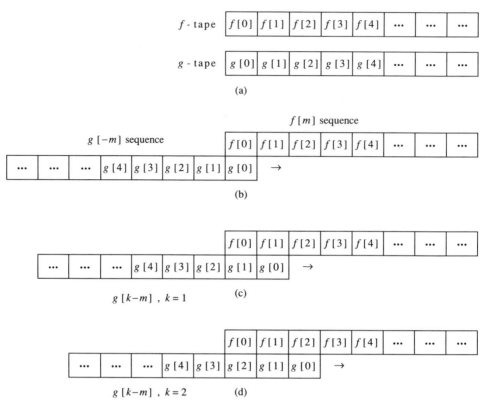

Fig. 2.23 Sliding tape algorithm for numerical convolution.

$$c[0] = Tf[0]g[0]$$

$$c[1] = T\{f[0]g[1] + f[1]g[0]\}$$

$$c[2] = T\{f[0]g[2] + f[1]g[1] + f[2]g[0]\} \tag{2.81}$$

We can also obtain this result graphically by using the "sliding-tape" algorithm† illustrated in Fig. 2.23. The samples of $f(t)$ and $g(t)$ are written sequentially on two tapes. The tapes therefore contain the sequences $f[0]$, $f[1]$, $f[2]$, $f[3]$, ..., and $g[0]$, $g[1]$, $g[2]$, $g[3]$, ... respectively, as shown in Fig. 2.23a. The sequence $g[k-m]$ is obtained in two steps:

1. First, we obtain $g[-m]$ by inverting or folding the $g[m]$ tape about the $g[0]$ slot so that the $g[0]$ slot remains aligned with the $f[0]$ slot.

2. Shifting the $g[-m]$ tape by k units (slots) yields $g[k-m]$ sequence. Figure 2.23b shows $f[m]$ and $g[-m]$. Figures 2.23c and 2.23d show $g[k-m]$ for $k=1$ and 2 respectively.

To compute $c[k]$, we multiply the sample values lying in adjacent slots on the $f[m]$ and $g[k-m]$ tapes, and then add the resulting products and multiply by T. Figure 2.23b shows the case for $k=0$ (no shift). In this case, only one slot overlaps,

---

† This development closely follows Lathi.[1, 3]

yielding the product $f[0]g[0]$. Therefore $c[0] = Tf[0]g[0]$, which is verified in Eq. (2.81). Figure 2.23c shows the case for $k = 1$ ($g[-m]$ right-shifted by one slot). Here two slots overlap, and the sum of their products is

$$c[1] = T\{f[0]g[1] + f[1]g[0]\}$$

This result is verified in Eq. (2.81). Figure 2.23d shows the case for $k = 2$ ($g[-m]$ right-shifted by two slots). Now we have three overlapping slots and

$$c[2] = T\{f[0]g[2] + f[1]g[1] + f[2]g[0]\}$$

which is verified in Eq. (2.81). We can continue this graphical procedure for all values of k (positive and negative). Observe that the sliding-tape technique is basically the same as the graphical procedure in Sec. 2.4-2 applied to staircase-approximated signals $f(t)$ and $g(t)$.

Jump Discontinuities

When jump discontinuities are present in $f(t)$ and/or $g(t)$, the error in numerical convolution is minimized if the sample at a point of discontinuity is assigned a value that is the mean of the values on two sides of the discontinuity. In the following example we consider convolution of signals with jump discontinuities.

■ Example 2.12

Using the numerical procedure of convolution, determine $c[k]$, the samples of $c(t) = f(t) * g(t)$, where $f(t)$ and $g(t)$ are shown in Fig. 2.24a and 2.24b respectively. Use $T = 0.1$ for the sampling interval.

The signal $f(t)$ has jump discontinuities at $t = 0$ and 1. At $t = 0$, the signal values on the two sides of the discontinuity are 0 and 2, and the mean value is 1. Therefore the sample at $t = 0$ is assigned a value 1. With the same argument, the sample at $t = 1$ is assigned a value 1. Each of the remaining samples in the interval $t < 1$ has a value 2. Similarly, the samples of $g(t)$ at $t = 0$ and 0.5 are assigned values 0.5, and each of the remaining samples in the interval $t < 0.5$ has a unit value. All the remaining sample values are zero.

First, we write in the sequences $f[m]$ and $g[m]$ on the tapes shown in Fig. 2.24c. Next, we keep the $f[m]$ tape fixed and invert the $g[m]$ tape about $m = 0$ to obtain $g[-m]$, as shown in Fig. 2.24d. Observe that the $f[m]$ tape has 11 samples ranging from $m = 0$ to 10, and the $g[m]$ tape has six samples ranging from $m = 0$ to $m = 5$. Samples for all other values of m (not shown in the figure) are zero.

We now shift $g[-m]$ by k slots. Figure 2.24d shows the case for $k = 0$. For this case there is only one overlapping slot, and the product of two overlapping values is 0.5. Hence

$$c[0] = T(0.5) = 0.05$$

Figure 2.24e shows the case for $k = 1$, where two slots overlap and

$$c[1] = T(1 + 1) = 0.2$$

Continuing in this way, we find

$$c[2] = 4T = 0.4 \qquad c[3] = 6T = 0.6$$
$$c[4] = 8T = 0.8 \qquad c[5] = 9.5T = 0.95$$
$$c[6] = c[7] = c[8] = c[9] = 10T = 1$$
$$c[10] = 9.5T = 0.95$$

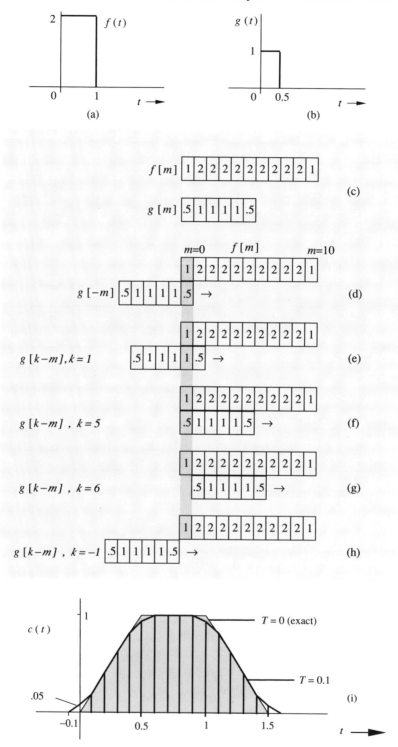

Fig. 2.24 Numerical convolution for Example 2.12 using sliding tape algorithm.

Figures 2.24f and 2.24g show the cases for $k = 5$ and $k = 6$. For $k > 10$, the number of overlapping slots decreases linearly. It can be seen that

$$c[11] = 8T = 0.8 \qquad c[12] = 6T = 0.6$$
$$c[13] = 4T = 0.4 \qquad c[14] = 2T = 0.2$$
$$c[15] = 0.5T = 0.05 \qquad c[16] = 0$$

For $k > 15$ the two strips are nonoverlapping. Similarly, for $k < 0$ the strips are nonoverlapping and

$$c[k] = 0 \qquad k > 15 \quad \text{and} \quad k < 0$$

Figure 2.24i shows both the numerical convolution and the exact convolution. The error in the numerical convolution can be made as small as desired by choosing a sufficiently small value of T. ■

The sliding tape method is conceptually quite valuable in understanding the convolution mechanism. Numerical convolution can also be performed from the arrays by using the sets $f[0]$, $f[1]$, $f[2]$, ..., and $g[0]$, $g[1]$, $g[2]$, ... This method, although convenient for computation, fails to give proper understanding of the convolution mechanism. It is explained in Prob. 3.5-16.

⊙ **Computer Example C2.3**
Solve Example 2.12 using a computer. Determine $c[k]$.
Here, $c(t) = f(t) * g(t)$, where $f(t)$ and $g(t)$ are shown in Fig.2.24a and b, respectively. Use $T = 0.1$ for the sampling interval.

```
f = [1;2*ones(9,1);1];      % Create the vector representing f.
g = [.5;ones(4,1);.5];      % Create the vector representing g.
T=0.1;
c = T*conv(f,g);
t = T*[0:length(c)-1]';
plot(t,c),grid,
xlabel('t'),ylabel('c(t)'),  % Label x and y axes.
title('plot of c(t)')    ⊙
```

△ **Exercise E2.15**
Show that the numerical convolution of $f(t)$ and $g(t)$ in Fig. 2.25 with $T = 0.2$ yields $c(t)$. Caution: Both $f(t)$ and $g(t)$ have jump discontinuities. ▽

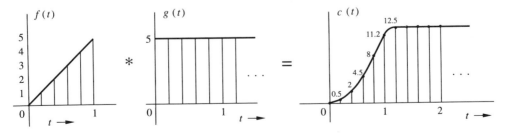

Fig. 2.25 Numerical convolution of $f(t)$ and $g(t)$ using sliding tape algorithm.

2.6 SYSTEM STABILITY

Because of the great variety of possible system behaviors, there are several definitions of stability in the literature. Here we shall consider a definition that is suitable for causal, linear, time-invariant (LTI) systems.

In order to understand system stability intuitively, let us examine the stability concept as applied to a right circular cone. Such a cone can be made to stand forever on its circular base, on its apex, or on its side. For this reason these three states of the cone are said to be **equilibrium states**. Qualitatively, however, the three states show very different behavior. If this cone, standing on its circular base, were to be disturbed slightly, it would eventually return to its original equilibrium state if left to itself. In this case the cone is said to be in **stable equilibrium**. On the other hand, if the cone stands on its apex, then the slightest disturbance will cause the cone to move farther and farther away from its equilibrium state. The cone in this case is said to be in an **unstable equilibrium**. The cone lying on its side, if disturbed, will neither go back to the original state nor continue to move farther away from the original state. The cone in this case is said to be in a **neutral equilibrium**.

Let us apply these observations to systems in general. If, in the absence of an external input, a system remains in a particular state (or condition) indefinitely, then that state is said to be an **equilibrium state of the system**. For an LTI system this equilibrium state is the zero state, in which all initial conditions are zero. Now suppose an LTI system is in equilibrium or zero state and we change this state by creating some nonzero initial conditions. By analogy with the cone, if the system is stable, it should eventually return to zero state. This means that, when left to itself, the system's output due to the nonzero initial conditions should approach 0 as $t \to \infty$. But the system output generated by initial conditions (zero-input response) is made up of its characteristic modes. For this reason we define stability as follows: a system is **(asymptotically) stable** if, and only if, all its characteristic modes $\to 0$ as $t \to \infty$. If any of the modes grows without bound as $t \to \infty$, the system is **unstable**. There is also a borderline situation in which the zero-input response remains bounded (approaches neither zero nor infinity), approaching a constant or oscillating with a constant amplitude as $t \to \infty$. For this borderline situation, the system is said to be **marginally stable** or just stable.

If an LTIC system has n distinct characteristic roots $\lambda_1, \lambda_2, \ldots, \lambda_n$, the zero-input response is given by

$$y_0(t) = \sum_{j=1}^{n} c_j e^{\lambda_j t} \tag{2.82}$$

We have shown elsewhere [see Eq. (B.14)]

$$\lim_{t \to \infty} e^{\lambda t} = \begin{cases} 0 & \text{Re } \lambda < 0 \\ \infty & \text{Re } \lambda > 0 \end{cases} \tag{2.83}$$

It is helpful to study system stability in terms of the location of the system's characteristic roots in the complex plane. Let us first assume that the system has distinct roots only. If a characteristic root λ is located in the left half of the complex plane (LHP), its real part is negative (Re $\lambda < 0$). Similarly, if a root λ is located in the right half of the complex plane (RHP), its real part is positive (Re $\lambda > 0$). Along the imaginary axis, the real part is zero (Re $\lambda = 0$). These regions are delineated in Fig. 2.26. From Eq. (2.83) it is clear that the characteristic modes corresponding to roots in LHP vanish as $t \to \infty$, while the modes corresponding to roots in RHP grow without bound as $t \to \infty$. However, the modes corresponding to simple (unrepeated) roots on the imaginary axis are of the form $e^{j\beta t}$; these are bounded (neither vanish nor grow without limit) as $t \to \infty$.

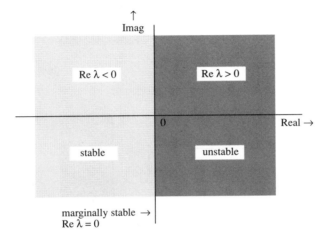

Fig. 2.26 Characteristic roots location and system stability.

From this discussion it follows that a system is asymptotically stable if, and only if, all of its characteristic roots lie in the left half of the complex plane. If any of the roots—even one—lies in RHP, the system is unstable. If none of the roots lie in RHP, but if some unrepeated (simple) roots lie on the imaginary axis, then the system is marginally stable (Fig. 2.26).

So far we have assumed all of the system's n roots to be distinct. The modes corresponding to a root λ repeated r times are $e^{\lambda t}, te^{\lambda t}, t^2 e^{\lambda t}, \cdots, t^{r-1} e^{\lambda t}$. But as $t \to \infty$, $t^k e^{\lambda t} \to 0$, if Re $\lambda < 0$ (λ in LHP). Therefore repeated roots in LHP do not cause instability. But when the repeated roots are on the imaginary axis ($\lambda = j\omega$), the corresponding modes $t^k e^{j\omega t}$ approach infinity as $t \to \infty$. Therefore repeated roots on the imaginary axis cause instability. Figure 2.27 shows characteristic modes corresponding to characteristic roots at various location in the complex plane. Observe the central role played by the characteristic roots or characteristic modes in determining the system's stability.

To summarize:

1. An LTIC system is asymptotically stable if, and only if, all the characteristic roots are in the LHP. The roots may be simple (unrepeated) or repeated.

2. An LTIC system is unstable if, and only if, either one or both of the following conditions exist: (i) at least one root is in the RHP, (ii) there are repeated roots on the imaginary axis.

3. An LTIC system is marginally stable if, and only if, there are no roots in the RHP, and there are some unrepeated roots on the imaginary axis.

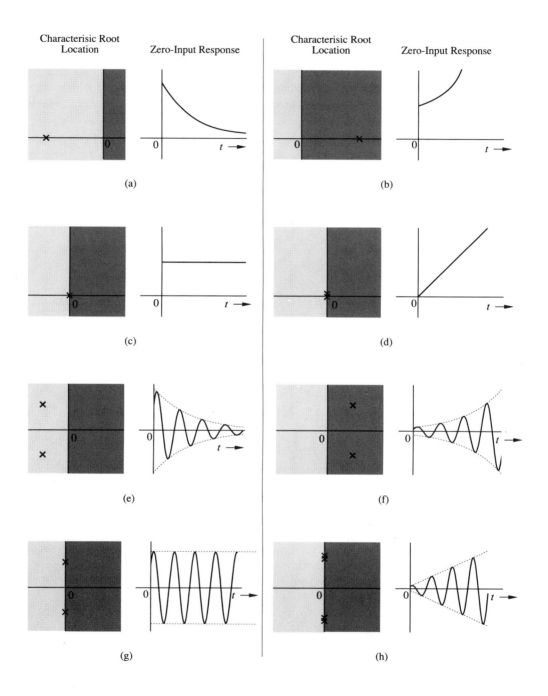

Fig. 2.27 Location of characteristic roots and the corresponding characteristic modes.

■ **Example 2.13**

Investigate the stability of LTIC system described by the following equations:

(a) $(D+1)\left(D^2+4D+8\right)y(t)=(D-3)f(t)$

(b) $(D-1)\left(D^2+4D+8\right)y(t)=(D+2)f(t)$

(c) $(D+2)(D^2+4)y(t)=\left(D^2+D+1\right)f(t)$

(d) $(D+1)(D^2+4)^2y(t)=\left(D^2+2D+8\right)f(t)$

The characteristic polynomials of these systems are

(a) $(\lambda+1)\left(\lambda^2+4\lambda+8\right)=(\lambda+1)(\lambda+2-j2)(\lambda+2+j2)$

(b) $(\lambda-1)\left(\lambda^2+4\lambda+8\right)=(\lambda-1)(\lambda+2-j2)(\lambda+2+j2)$

(c) $(\lambda+2)(\lambda^2+4)=(\lambda+2)(\lambda-j2)(\lambda+j2)$

(d) $(\lambda+1)(\lambda^2+4)^2=(\lambda+2)(\lambda-j2)^2(\lambda+j2)^2$

Consequently the characteristic roots of the systems above are (see Fig. 2.28):

(a) $-1,\ -2\pm j2$ (b) $1,\ -2\pm j2$ (c) $-2,\ \pm j2$ (d) $-1,\ \pm j2,\ \pm j2$.

System (a) is asymptotically stable (all roots in LHP), (b) is unstable (one root in RHP), (c) is marginally stable (unrepeated roots on imaginary axis) and no roots in RHP, and (d) is unstable (repeated roots on the imaginary axis.) ■

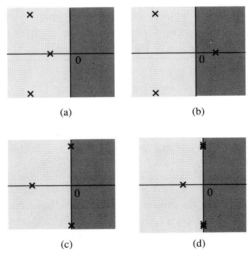

Fig. 2.28

△ **Exercise E2.16**

For each of the systems specified by the equations below, plot its characteristic roots in the complex plane and determine whether it is asymptotically stable, marginally stable, or unstable.

(a) $D(D+2)y(t)=3f(t)$

(b) $D^2(D+3)y(t)=(D+5)f(t)$

(c) $(D+1)(D+2)y(t)=(2D+3)f(t)$

(d) $(D^2+1)(D^2+9)y(t)=\left(D^2+2D+4\right)f(t)$

(e) $(D+1)\left(D^2-4D+9\right)y(t)=(D+7)f(t)$

Answer: (a) marginally stable (b) unstable (c) stable (d) marginally stable (e) unstable.   ▽

2.6-1 System Response to Bounded Inputs

From the example of the right circular cone, it appears that when a system is in stable equilibrium, application of a small force (input) produces a small response.

On the other hand, when the system is in unstable equilibrium, a small force (input) produces an unbounded response. Intuitively we feel that every bounded input should produce a bounded response in a stable system, whereas in an unstable system this would not be the case. We shall now verify this hunch and show that it is indeed true.

Recall that for an LTIC system

$$y(t) = h(t) * f(t)$$
$$= \int_{-\infty}^{\infty} h(\tau)f(t-\tau)\, d\tau \tag{2.84}$$

Therefore

$$|y(t)| \le \int_{-\infty}^{\infty} |h(\tau)||f(t-\tau)|\, d\tau$$

Moreover, if $f(t)$ is bounded, then $|f(t-\tau)| < K_1 < \infty$, and

$$|y(t)| \le K_1 \int_{-\infty}^{\infty} |h(\tau)|\, d\tau$$

Because $h(t)$ contains terms of the form $e^{\lambda_j t}$ or $t^k e^{\lambda_j t}$, $h(t)$ decays exponentially with time if $\operatorname{Re} \lambda_j < 0$. Consequently for an asymptotically stable system†

$$\int_{-\infty}^{\infty} |h(\tau)|\, d\tau < K_2 < \infty \tag{2.85}$$

and

$$|y(t)| \le K_1 K_2 < \infty$$

Thus, for an asymptotically stable system, a bounded input always produces a bounded output. Moreover, we can show that for an unstable or a marginally stable system, the output $y(t)$ is unbounded for some bounded input (see Problem 2.6-4). These results lead to the formulation of an alternative definition of stability known as **bounded-input, bounded-output (BIBO) stability**: a system is BIBO stable if, and only if, a bounded input produces a bounded output. Observe that an asymptotically stable system is always BIBO stable.‡ However, a marginally stable system is BIBO unstable.

---

† This can be shown as follows. If $\lambda_i = \alpha_i + j\beta_i$, then $e^{\lambda_i t} = e^{\alpha_i t} e^{j\beta_i t}$ and $\left| e^{\lambda_i t} \right| = e^{\alpha_i t}$. Therefore

$$\int_{-\infty}^{\infty} \left| e^{\lambda_i \tau} u(\tau) \right| d\tau = \int_{0}^{\infty} e^{\alpha_i \tau}\, d\tau = -\frac{1}{\alpha_i} \qquad \text{if } \operatorname{Re} \lambda_i = \alpha_i < 0$$

and Eq. (2.85) follows. This conclusion is also valid when the integrand is of the form $|t^k e^{\lambda_i t}| u(t)$.

‡ However, a BIBO stable system is not necessarily asymptotically stable because BIBO stability is determined from the system's impulse response, which is an external description of the system, while asymptotic stability is determined from the internal description of the system obtained from system equations. In certain systems (e.g., uncontrollable or unobservable systems), the two descriptions may not be the same. Remember that the external description describes only that part of the system which is coupled to both the input and the output. Hence a system may be internally unstable while appearing stable from the system's external terminals (BIBO stable)[4].

Implications of Stability

All practical signal processing systems must be stable. Unstable systems are useless from the viewpoint of signal processing because any set of intended or unintended initial conditions leads to an unbounded response that either destroys the system or (more likely) leads it to some saturation conditions that change the nature of the system. Even if the discernible initial conditions are zero, stray voltages or thermal noise signals generated within the system will act as initial conditions. Because of exponential growth, a stray signal, no matter how small, will eventually cause an unbounded output in an unstable system.

Marginally stable systems do have one important application in the oscillator. The oscillator is a system that generates a signal on its own without the application of an external input. Consequently the oscillator output is a zero-input response. If such a response is to be a sinusoid of frequency ω_0, the system should be marginally stable with characteristic roots at $\pm j\omega_0$. Thus, to design an oscillator of frequency ω_0, we should pick a system with the characteristic polynomial $(\lambda - j\omega_0)(\lambda + j\omega_0) = \lambda^2 + \omega_0^2$. A system described by the differential equation

$$\left(D^2 + \omega_0^2\right) y(t) = f(t)$$

or

$$\frac{d^2 y}{dt^2} + \omega_0^2 y(t) = f(t)$$

will do the job.

2.7 INTUITIVE INSIGHTS IN SYSTEM BEHAVIOR

This section attempts to provide an understanding of what determines system behavior. Because of its intuitive nature, the following discussion will be more or less qualitative. We shall now show that the most important attributes of a system are its characteristic roots or characteristic modes because they determine not only the zero-input response but also the entire behavior of the system.

2.7-1 Dependence of System Behavior on Characteristic Modes

Recall that the zero-input response of a system consists of the system's characteristic modes. For a stable system, these characteristic modes decay exponentially and eventually vanish. This may give the impression that these modes do not substantially affect system behavior in general and system response in particular. This impression is totally wrong! We shall now see that the system's characteristic modes leave their imprint on every aspect of the system behavior. **We may compare the system's characteristic modes (or roots) to a seed which eventually dissolves in the ground; however, the plant that springs from it is totally determined by the seed. The imprint of the seed exists on every cell of the plant.** In order to understand this interesting phenomenon, recall that the characteristic modes of a system are very special to that system because it can sustain these signals without the application of an external input. In other words, the system offers a free ride and ready access to these signals. Now imagine what would happen

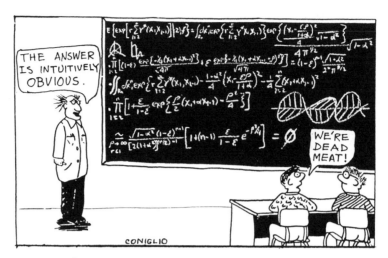

Intuition can cut the math jungle instantly!

if we actually drove the system with an input having the form of a characteristic mode! We would expect the system to respond strongly (this is, in fact, the resonance phenomenon discussed later in this section). If the input is not exactly a characteristic mode but is close to such a mode, we would still expect the system response to be strong. However, if the input is very different from any of the characteristic modes, we would expect the system to respond poorly. We shall now show that these intuitive deductions are indeed true.

When we say that an input is close to a characteristic mode signal, we must have some measure of the closeness or similarity of the two signals. (This measure is developed fully in Chapter 6.) At this point it is sufficient to devise a rough measure of similarity by restricting the system's inputs only to exponentials of the form $e^{\zeta t}$, where ζ is generally a complex number. The similarity of two exponential signals $e^{\zeta t}$ and $e^{\lambda t}$ will then be measured by the closeness of ζ and λ. If the difference $\zeta - \lambda$ is small, the signals are similar; if $\zeta - \lambda$ is large, the signals are dissimilar.

Now consider a first-order system with a single characteristic mode $e^{\lambda t}$ and the input $e^{\zeta t}$. The impulse response of this system is then given by $Ae^{\lambda t}$, where the exact value of A is not important for this qualitative discussion. The system response $y(t)$ is given by

$$y(t) = h(t) * f(t)$$
$$= Ae^{\lambda t}u(t) * e^{\zeta t}u(t)$$

From the convolution table (Table 2.1), we find that

$$y(t) = \frac{A}{\zeta - \lambda}\left[e^{\zeta t} - e^{\lambda t}\right]u(t) \tag{2.86}$$

Clearly, if the input $e^{\zeta t}$ is similar to $e^{\lambda t}$, $\zeta - \lambda$ is small, and the system response is large. **The closer the input $f(t)$ is to the characteristic mode, the stronger is the system response.** On the other hand, if the input is very different from the natural mode, $\zeta - \lambda$ is large, and the system responds poorly. This is precisely what we set out to prove.

We have proved the desired result for a single-mode (first-order) system. It can be generalized to an nth-order system, which has n characteristic modes. The impulse response $h(t)$ of such a system is a linear combination of its n modes. Therefore if $f(t)$ is similar to any one of the modes, the corresponding response will be high; if it is similar to none of the modes, the response will be small. Clearly the characteristic modes are very influential in determining system response to a given input.

It would be tempting to conclude from Eq. (2.86) that if the input is identical to the characteristic mode, so that $\zeta = \lambda$, then the response goes to infinity. Remember, however, that if $\zeta = \lambda$, the numerator on the right-hand side of Eq. (2.86) also goes to zero. We shall study this complex behavior (resonance phenomenon) later in this section.

We shall now show that **by mere inspection of the impulse response $h(t)$ (which is composed of characteristic modes), we can learn a great deal about system behavior.**

2.7-2 Response Time of a System: The System Time Constant

Like human beings, systems have a certain response time. In other words, when an input (stimulus) is applied to a system, a certain amount of time elapses before the system fully responds to that input. This time lag or response time is called the system **time constant**. As we shall see, a system's time constant is equal to the width of its impulse response $h(t)$.

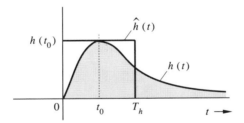

Fig. 2.29 Effective duration of an impulse response.

An input $\delta(t)$ to a system is instantaneous (zero duration), but its response $h(t)$ has a duration T_h. Therefore the system requires a time T_h to respond fully to this input, and we are justified in viewing T_h as the system's response time or time constant. We arrive at the same conclusion using another argument. The output is a convolution of the input with $h(t)$. If an input is a pulse of width T_f, then the output pulse width is $T_f + T_h$ according to the width property of convolution. This shows that the system requires T_h seconds to respond fully to any input. **The system time constant indicates how fast the system is. A system with a smaller time constant is a faster system that responds quickly to an input. A system with a relatively large time constant is a sluggish system that cannot respond well to rapidly varying signals.**

Strictly speaking, the duration of the impulse response $h(t)$ is ∞ because the characteristic modes approach zero asymptotically as $t \to \infty$. However, beyond some value of t, $h(t)$ becomes negligible. It is therefore necessary to use some suitable measure of the impulse response's effective width.

There is no single satisfactory definition of effective signal duration (or width) that is applicable to every situation. For the situation shown in Fig. 2.29, a reasonable definition of the duration $h(t)$ would be T_h, the width of the rectangular pulse $\hat{h}(t)$. This rectangular pulse $\hat{h}(t)$ has an area identical to that of $h(t)$ and a height identical to that of $h(t)$ at some suitable instant $t = t_0$. In Fig. 2.29, t_0 is chosen as the instant at which $h(t)$ is maximum. According to this definition,[†]

$$T_h h(t_0) = \int_{-\infty}^{\infty} h(t)\, dt$$

or

$$T_h = \frac{\int_{-\infty}^{\infty} h(t)\, dt}{h(t_0)} \tag{2.87}$$

Now if a system has a single mode

$$h(t) = A e^{\lambda t} u(t)$$

with λ negative and real, then $h(t)$ is maximum at $t = 0$ with value $h(0) = A$. Therefore from Eq. (2.87)

$$T_h = \frac{1}{A} \int_0^{\infty} A e^{\lambda t}\, dt = -\frac{1}{\lambda} \tag{2.88}$$

Thus the time constant in this case is simply (the negative of the) reciprocal of the system's characteristic root. For the multimode case, $h(t)$ is a weighted sum of the system's characteristic modes, and T_h is a weighted average of the time constants associated with the n modes of the system.

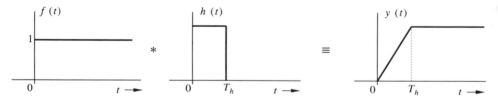

Fig. 2.30 Rise time of a system.

2.7-3 Time Constant and Rise Time of a System

The system time constant also may be viewed from a different perspective. The unit step response $y(t)$ of a system is the convolution of $u(t)$ with $h(t)$. Let the impulse response $h(t)$ be a rectangular pulse of width T_h, as shown in Fig. 2.30. This simplifies the discussion, yet gives satisfactory results for qualitative discussion. The result of this convolution is shown in Fig. 2.30. Note that the output does not rise from zero to a final value instantaneously as the input rises; instead, the output

---

[†] This definition is satisfactory when $h(t)$ is a single, mostly positive (or mostly negative) pulse. Such systems are lowpass systems. This definition should not be applied indiscriminately to all systems.

takes T_h seconds to accomplish this. Hence the rise time T_r of the system is equal to the system time constant:‡

$$T_r = T_h \tag{2.89}$$

This result and Fig. 2.30 show clearly that a system generally does not respond to an input instantaneously. Instead, it takes time T_h for the system to respond fully.

2.7-4 Time Constant and Filtering

A larger time constant implies a sluggish system because the system takes a longer time to respond fully to an input. Such a system cannot respond effectively to rapid variations in the input. On the other hand, a smaller time constant indicates that a system is capable of responding to rapid variations in the input. Thus there is a direct connection between a system's time constant and its filtering properties.

Consider a high-frequency sinusoid that varies rapidly with time. A system with a large time constant will not be able to respond well to this input. Therefore such a system will suppress rapidly varying (high-frequency) sinusoids and other high-frequency signals, thereby acting as a lowpass filter (a filter allowing the transmission of low-frequency signals only). We shall now show that a system with a time constant T_h acts as a lowpass filter having a cutoff frequency of $\mathcal{F}_c = 1/T_h$ Hz, so that sinusoids with frequencies below $\mathcal{F}_c$ Hz are transmitted reasonably well, while those with frequencies above $\mathcal{F}_c$ Hz are suppressed.

To demonstrate this fact, let us determine the system response to a sinusoidal input $f(t)$ by convolving this input with the effective impulse response $h(t)$ in Fig. 2.31a. Figures 2.31b and 2.31c show the process of convolution of $h(t)$ with the sinusoidal inputs of two different frequencies. The sinusoid in Fig. 2.31b has a relatively high frequency, while the frequency of the sinusoid in Fig. 2.31c is low. Recall that the convolution of $f(t)$ and $h(t)$ is equal to the area under the product $f(\tau)h(t - \tau)$. This area is shown shaded in Figs. 2.31b and 2.31c for the two cases. For the high-frequency sinusoid it is clear from Fig. 2.31b that the area under $f(\tau)h(t - \tau)$ is very small because its positive and negative areas nearly cancel each other out. In this case the output $y(t)$ remains periodic but has a rather small amplitude. This happens when the period of the sinusoid is much smaller than the system time-constant T_h. On the other hand, for the low-frequency sinusoid, the period of the sinusoid is larger than T_h, so that the partial cancellation of area under $f(\tau)h(t - \tau)$ is less effective. Consequently the output $y(t)$ is much larger, as shown in Fig. 2.31c.

Between these two possible extremes in system behavior, a transition point occurs when the period of the sinusoid is equal to the system time constant T_h. The frequency at which this transition occurs is known as the **cutoff frequency** $\mathcal{F}_c$ of the system. Because T_h is the period of cutoff frequency $\mathcal{F}_c$,

$$\mathcal{F}_c = \frac{1}{T_h} \tag{2.90}$$

The frequency $\mathcal{F}_c$ is also known as the bandwidth of the system because the system transmits or passes sinusoidal components with frequencies below $\mathcal{F}_c$ while attenuating components with frequencies above $\mathcal{F}_c$. Of course, the transition in system

---

‡ Because of varying definitions of rise time, the reader may find different results in the literature. The qualitative and intuitive nature of this discussion should always be kept in mind.

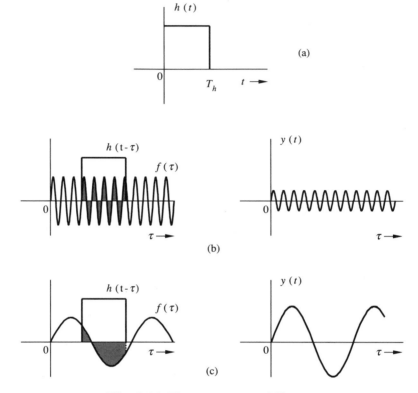

Fig. 2.31 Time constant and filtering.

behavior is gradual. There is no dramatic change in system behavior at $\mathcal{F}_c = 1/T_h$. Moreover, these results are based on an idealized (rectangular pulse) impulse response; in practice these results will vary somewhat, depending on the exact shape of $h(t)$. Remember that the "feel" of general system behavior is more important than exact system response for this qualitative discussion.

Since the system time constant is equal to its rise time, we have

$$T_r = \frac{1}{\mathcal{F}_c} \qquad (2.91a)$$

or

$$\mathcal{F}_c = \frac{1}{T_r} \qquad (2.91b)$$

Thus a system's bandwidth is the reciprocal of its rise time. An experienced engineer often can quickly determine the bandwidth of an unknown system by simply observing the system response to a step input on an oscilloscope.

2.7-5   Time Constant and Pulse Dispersion (Spreading)

In general the transmission of a pulse through a system causes pulse dispersion (or spreading). Therefore the output pulse is generally wider than the input pulse. This system behavior can have serious consequences in communication systems where information is transmitted by pulse amplitudes. Dispersion (or spreading) causes interference or overlap with neighboring pulses, thereby distorting pulse amplitudes and introducing errors in the received information.

Earlier we saw that if an input $f(t)$ is a pulse of width T_f, then T_y, the width of the output $y(t)$, is

$$T_y = T_f + T_h \qquad (2.92)$$

This result shows that an input pulse spreads out (disperses) as it passes through a system. Since T_h is also the system's time constant or rise time, the amount of spread in the pulse is equal to the time constant (or rise time) of the system.

2.7-6 Time Constant and Rate of Information Transmission

In pulse communications systems where information is conveyed through pulse amplitudes, the rate of information transmission is proportional to the rate of pulse transmission. We shall show that to avoid the destruction of information caused by dispersion of pulses during their transmission through the channel (transmission medium), the rate of information transmission should not exceed the bandwidth of the communications channel.

Since an input pulse spreads out by T_h seconds, the consecutive pulses should be spaced T_h seconds apart in order to avoid interference between pulses. Thus the rate of pulse transmission should not exceed $1/T_h$ pulses/second. But $1/T_h = \mathcal{F}_c$, the channel's bandwidth, so that we can transmit pulses through a communications channel at a rate of $\mathcal{F}_c$ pulses/second and still avoid significant interference between the pulses. The rate of information transmission is therefore proportional to the channel's bandwidth (or to the reciprocal of its time constant).†

The previous discussion (Secs. 2.7-2 through 2.7-6) showed that the system time constant determines much of a system's behavior—its filtering characteristics, rise time, pulse dispersion, and so on. In turn, the time constant is determined by the system's characteristic roots. Clearly the characteristic roots and their relative amounts in the impulse response $h(t)$ determine the behavior of a system.

2.7-7 The Resonance Phenomenon

Finally, we come to this fascinating phenomenon of resonance. As we have mentioned already several times, this phenomenon is observed when the input signal is identical or is very similar to a characteristic mode of the system. For the sake of simplicity and clarity, we consider a first-order system which has only a single mode, $e^{\lambda t}$. Let the impulse response of this system be‡

$$h(t) = Ae^{\lambda t} \qquad (2.93)$$

and let the input be

$$f(t) = e^{(\lambda - \epsilon)t}$$

---

† Theoretically, a channel of bandwidth $\mathcal{F}_c$ can transmit up to $2\mathcal{F}_c$ pulse amplitudes per second correctly.[5] Our derivation here, being very simple and qualitative, yields only half the theoretical limit. In practice it is difficult to attain the upper theoretical limit; transmission rates of $\mathcal{F}_c$ pulses per second are more common.

‡ For convenience we omit multiplying $f(t)$ and $h(t)$ by $u(t)$. Throughout this discussion, it is assumed that they are causal.

The system response $y(t)$ is then given by

$$y(t) = Ae^{\lambda t} * e^{(\lambda - \epsilon)t}$$

From the convolution table we obtain

$$y(t) = \frac{A}{\epsilon} \left[e^{\lambda t} - e^{(\lambda - \epsilon)t} \right]$$

$$= Ae^{\lambda t} \left(\frac{1 - e^{-\epsilon t}}{\epsilon} \right) \tag{2.94}$$

Now, as $\epsilon \to 0$, both the numerator and the denominator of the term in the parentheses approach zero. Applying L'Hôpital's rule to this term yields

$$\lim_{\epsilon \to 0} y(t) = At e^{\lambda t} \tag{2.95}$$

Clearly the response does not go to infinity as $\epsilon \to 0$, but it acquires a factor t, which approaches ∞ as $t \to \infty$. If λ has a negative real part (so that it lies in LHP), $e^{\lambda t}$ decays faster than t, and $y(t) \to 0$ as $t \to \infty$. The resonance phenomenon in this case is present, but its manifestation is aborted by the signal's own exponential decay.

This discussion shows that **resonance is a cumulative phenomenon**, not instantaneous. It builds up linearly‡ with t. When the mode decays exponentially, the signal decays at a rate too fast for resonance to counteract the decay; as a result, the signal vanishes before resonance has a chance to build it up. However, if the mode were to decay at a rate less than $1/t$, we should see the resonance phenomenon clearly. This would be possible if Re $\lambda = 0$, so that λ lies on the imaginary axis of the complex plane. In this case

$$\lambda = j\omega$$

and Eq. (2.95) becomes

$$y(t) = At e^{j\omega t} \tag{2.96}$$

In this case, the response does go to infinity linearly with t.

For a real system, if $\lambda = j\omega$ is a root, $\lambda^* = -j\omega$ must also be a root; the impulse response is of the form $Ae^{j\omega t} + Ae^{-j\omega t} = 2A \cos \omega t$. The response of this system to input $A \cos \omega t$ is $2A \cos \omega t * \cos \omega t$. The reader can show that this convolution contains a term of the form $At \cos \omega t$. The resonance phenomenon is clearly visible. The system response to its characteristic mode increases linearly with time, eventually reaching ∞, as shown in Fig. 2.32.

Recall that when $\lambda = j\omega$, the system is marginally stable. As we have seen, the full effect of resonance cannot be seen for an asymptotically stable system; only in a marginally stable system does the resonance phenomenon boost the system's response to infinity when the system's input is a characteristic mode. But even for an asymptotically stable system, we see a manifestation of resonance if its characteristic roots are close to the imaginary axis, so that Re λ is very small. We can

---

‡ If the characteristic root in question repeats r times, resonance effect increases as t^{r-1}. However, $t^{r-1}e^{\lambda t} \to 0$ as $t \to \infty$ for any value of r, provided Re $\lambda < 0$ (λ in LHP).

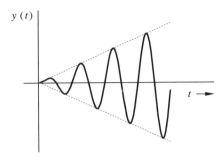

Fig. 2.32 Build up of system response in resonance.

show that when the characteristic roots of a system are $\sigma \pm j\omega_0$, the system response to the input $e^{j\omega_0 t}$ or the sinusoid $\cos \omega_0 t$ can become very large if σ is small enough. The response drops off rapidly as the input signal's frequency is moved away from ω_0. This frequency-selective behavior of asymptotically stable systems can be studied more profitably after an understanding of frequency-domain analysis is acquired. For this reason we postpone full discussion of this subject to Chapter 4.

Importance of the Resonance Phenomenon

The resonance phenomenon is very important because it allows us to design frequency-selective systems by choosing their characteristic roots properly. Low-pass, bandpass, highpass, and bandstop filters are all examples of frequency selective networks. In mechanical systems, the inadvertent presence of resonance can cause signals of such tremendous magnitudes that the system may fall apart. A musical note (periodic vibrations) of proper the frequency can shatter a glass if the the frequency is matched to the characteristic root of the glass, which acts as a mechanical system. Similarly, a company of soldiers marching in step across a bridge amounts to applying a periodic force to the bridge. If the frequency of this input force happens to be close to a characteristic root of the bridge, the bridge may respond (vibrate) violently and collapse, even though it would have been strong enough to carry many soldiers marching out of step. The Tacoma Narrow Bridge failure of 1940 is a case in point. This bridge was opened to traffic in July 1940. Within four months of opening (on 7 November 1940), it collapsed in a mild gale, not because of the wind's brute force but because the frequency of wind-generated vortices matched the natural frequency (characteristic roots) of the bridge, causing resonance.

Because of the great damage which may occur, mechanical resonance is generally something to be avoided, especially in structures or vibrating mechanisms. If an engine with periodic force (such as piston motion) is mounted on a platform, the platform with its mass and springs should be designed so that their characteristic roots are not close to the engine's frequency of vibration. Proper design of this platform can not only avoid resonance, but also attenuate vibrations if the system roots are placed far away from the frequency of vibration.

2.8 CLASSICAL SOLUTION OF DIFFERENTIAL EQUATIONS

The classical method is relatively simple compared to the method discussed so far. In this method we determine the natural and forced components rather than the zero-input and zero-state components of the response.

The Natural and Forced Response

Recall that the zero-input component of the system response consists entirely of characteristic modes, while the zero-state component contains both characteristic and noncharacteristic mode terms; we illustrated this observation on p. 138. When all of the characteristic mode terms of the total system response are lumped together, they form the system's **natural response** (also known as the **homogeneous solution** or **complementary solution**). The remaining portion of the response consists entirely of noncharacteristic mode terms and is called the system's **forced response** (also known as the **particular solution**). Equation (2.74b) shows these two components for the loop current in the RLC circuit of Fig. 2.1a.

If $y_n(t)$ and $y_\phi(t)$ denote the natural and forced response respectively, then the total system response is $y_n(t) + y_\phi(t)$. Since this total system response must satisfy the system equation [Eq. (2.1)],

$$Q(D)\,[y_n(t) + y_\phi(t)] = P(D)f(t) \qquad (2.97)$$

or

$$Q(D)y_n(t) + Q(D)y_\phi(t) = P(D)f(t)$$

But $y_n(t)$ is composed entirely of characteristic modes. Therefore

$$Q(D)y_n(t) = 0$$

so that

$$Q(D)y_\phi(t) = P(D)f(t) \qquad (2.98)$$

Because the natural response is a linear combination of the system's characteristic modes, it has the same form as that of the zero-input response; only its arbitrary constants are different. These constants are determined from auxiliary conditions, as explained later. We shall now discuss a method of determining the forced response.

2.8-1 Forced Response: Method of Undetermined Coefficients

It is a relatively simple task to determine $y_\phi(t)$, the forced response of an LTIC system, when the input $f(t)$ is such that it yields only a finite number of independent derivatives. Inputs having the form $e^{\zeta t}$ or t^r fall into this category. For example, $e^{\zeta t}$ has only one independent derivative; the repeated differentiation of $e^{\zeta t}$ yields the same form as this input, that is, $e^{\zeta t}$. Similarly, the repeated differentiation of t^r yields only r independent derivatives. The forced response to such an input can be expressed as a linear combination of the input and its independent derivatives. Consider, for example, the input $at^2 + bt + c$. The successive derivatives of this input are $2at + b$ and $2a$. In this case, the input has only two independent derivatives. Therefore the forced response can be assumed to be a linear combination of $f(t)$ and its two derivatives. The suitable form for $y_\phi(t)$ in this case is therefore

$$y_\phi(t) = \beta_2 t^2 + \beta_1 t + \beta_0$$

The undetermined coefficients β_0, β_1, and β_2 are determined by substituting this expression for $y_\phi(t)$ in Eq. (2.98):

$$Q(D)y_\phi(t) = P(D)f(t)$$

and then equating coefficients of similar terms on both sides of the resulting expression.

Although this method can be used only for inputs with a finite number of derivatives, this class of inputs includes a wide variety of the most commonly encountered signals in practice. Table 2.2 shows a variety of such inputs and the form of the forced response corresponding to each input. We shall demonstrate this procedure with an example.

TABLE 2.2

	Input $f(t)$		Forced Response
1.	$e^{\zeta t}$	$\zeta \neq \lambda_i \ (i = 1, 2, \cdots, n)$	$\beta e^{\zeta t}$
2.	$e^{\zeta t}$	$\zeta = \lambda_i$	$\beta t e^{\zeta t}$
3.	k		β
4.	$\cos(\omega t + \theta)$		$\beta \cos(\omega t + \phi)$
5.	$\left(t^r + \alpha_{r-1}t^{r-1} + \cdots + \alpha_1 t + \alpha_0\right)e^{\zeta t}$		$(\beta_r t^r + \beta_{r-1}t^{r-1} + \cdots + \beta_1 t + \beta_0)e^{\zeta t}$

Note: By definition, $y_\phi(t)$ cannot have any characteristic mode terms. If any term shown in the right-hand column for the forced response is also a characteristic mode of the system, the correct form of the forced response must be modified to $t^i y_\phi(t)$, where i is the smallest possible integer that can be used and still can prevent $t^i y_\phi(t)$ from having a characteristic mode term. For example, when the input is $e^{\zeta t}$, the forced response (right-hand column) has the form $\beta e^{\zeta t}$. But if $e^{\zeta t}$ happens to be a characteristic mode of the system, the correct form of the forced response is $\beta t e^{\zeta t}$ (see Pair 2). If $t e^{\zeta t}$ also happens to be a characteristic mode of the system, the correct form of the forced response is $\beta t^2 e^{\zeta t}$, and so on.

■ **Example 2.14**

Solve the differential equation

$$\left(D^2 + 3D + 2\right)y(t) = Df(t)$$

if the input

$$f(t) = t^2 + 5t + 3$$

and the initial conditions are $y(0^+) = 2$ and $\dot{y}(0^+) = 3$.

The characteristic polynomial of the system is

$$\lambda^2 + 3\lambda + 2 = (\lambda + 1)(\lambda + 2)$$

Therefore the characteristic modes are e^{-t} and e^{-2t}. The natural response is then a linear combination of these modes, so that

$$y_n(t) = K_1 e^{-t} + K_2 e^{-2t} \qquad t \geq 0$$

Here the arbitrary constants K_1 and K_2 must be determined from the system's initial conditions.

The forced response to the input $t^2 + 5t + 3$ is found from Table 2.2 (Pair 5 with $\zeta = 0$) to be

$$y_\phi(t) = \beta_2 t^2 + \beta_1 t + \beta_0$$

Moreover, $y_\phi(t)$ satisfies the system equation [Eq. (2.98)]; that is,

Now

$$\left(D^2 + 3D + 2\right) y_\phi(t) = Df(t) \qquad\qquad (2.99)$$

$$Dy_\phi(t) = \frac{d}{dt}\left(\beta_2 t^2 + \beta_1 t + \beta_0\right) = 2\beta_2 t + \beta_1$$

$$D^2 y_\phi(t) = \frac{d^2}{dt^2}\left(\beta_2 t^2 + \beta_1 t + \beta_0\right) = 2\beta_2$$

and

$$Df(t) = \frac{d}{dt}\left[t^2 + 5t + 3\right] = 2t + 5$$

Substituting these results in Eq. (2.99) yields

$$2\beta_2 + 3(2\beta_2 t + \beta_1) + 2(\beta_2 t^2 + \beta_1 t + \beta_0) = 2t + 5$$

or

$$2\beta_2 t^2 + (2\beta_1 + 6\beta_2)t + (2\beta_0 + 3\beta_1 + 2\beta_2) = 2t + 5$$

Equating coefficients of similar powers on both sides of this expression yields

$$2\beta_2 = 0$$

$$2\beta_1 + 6\beta_2 = 2$$

$$2\beta_0 + 3\beta_1 + 2\beta_2 = 5$$

Solving these three equations for their unknowns, we obtain $\beta_0 = 1$, $\beta_1 = 1$, and $\beta_2 = 0$. Therefore

$$y_\phi(t) = t + 1 \qquad\qquad t > 0$$

The total system response $y(t)$ is the sum of the natural and forced solutions. Therefore

$$y(t) = y_n(t) + y_\phi(t)$$
$$= K_1 e^{-t} + K_2 e^{-2t} + t + 1 \qquad\qquad t > 0$$

so that

$$\dot{y}(t) = -K_1 e^{-t} - 2K_2 e^{-2t} + 1$$

Setting $t = 0$ and substituting $y(0) = 2$ and $\dot{y}(0) = 3$ in these equations, we have

$$2 = K_1 + K_2 + 1$$

$$3 = -K_1 - 2K_2 + 1$$

The solutions to these two simultaneous equations are $K_1 = 4$ and $K_2 = -3$. Therefore

$$y(t) = 4e^{-t} - 3e^{-2t} + t + 1 \qquad t \geq 0 \quad \blacksquare$$

Comments on Initial Conditions

In the classical method, the initial conditions are required at $t = 0^+$. This is because at $t = 0^-$, only the zero-input component exists, and the initial conditions at $t = 0^-$ can be applied to the zero-input component only. In the classical method, the zero-input and zero-state components cannot be separated. Consequently the initial conditions must be applied to the total response, which begins at $t = 0^+$.

$\triangle$ **Exercise E2.17**

An LTIC system is specified by the equation

$$\left(D^2 + 5D + 6\right) y(t) = (D + 1)f(t)$$

The input is $f(t) = 6t^2$. Find **(a)** the forced response $y_\phi(t)$ **(b)** the total response $y(t)$ if the initial conditions are $y(0^+) = \frac{25}{18}$ and $\dot{y}(0^+) = -\frac{2}{3}$.

Answers: **(a)** $y_\phi(t) = t^2 + \frac{1}{3}t - \frac{11}{18}$ **(b)** $y(t) = 5e^{-2t} - 3e^{-3t} + \left(t^2 + \frac{1}{3}t - \frac{11}{18}\right)$. $\triangledown$

The Exponential Input $e^{\zeta t}$

The exponential signal is the most important signal in the study of LTI systems. Interestingly, the forced response for an exponential input signal turns out to be very simple. From Table 2.2 we see that the forced response for the input $e^{\zeta t}$ has the form $\beta e^{\zeta t}$. We now show that $\beta = Q(\zeta)/P(\zeta)$.† To determine the constant β, we substitute $y_\phi(t) = \beta e^{\zeta t}$ in the system equation [Eq. (2.98)], which gives us

$$Q(D) \left[\beta e^{\zeta t}\right] = P(D)e^{\zeta t} \qquad (2.100)$$

Now observe that

$$De^{\zeta t} = \frac{d}{dt}\left(e^{\zeta t}\right) = \zeta e^{\zeta t}$$

$$D^2 e^{\zeta t} = \frac{d^2}{dt^2}\left(e^{\zeta t}\right) = \zeta^2 e^{\zeta t}$$

$$\dotfill$$

$$D^r e^{\zeta t} = \zeta^r e^{\zeta t}$$

Consequently

$$Q(D)e^{\zeta t} = Q(\zeta)e^{\zeta t} \qquad \text{and} \qquad P(D)e^{\zeta t} = P(\zeta)e^{\zeta t}$$

Therefore Eq. (2.98) becomes

$$\beta Q(\zeta)e^{\zeta t} = P(\zeta)e^{\zeta t}$$

and

$$\beta = \frac{P(\zeta)}{Q(\zeta)}$$

Thus, for the input $f(t) = e^{\zeta t}u(t)$, the forced response is given by

---

† This is true only if ζ is not a natural frequency of the system.

$$y_\phi(t) = H(\zeta)e^{\zeta t} \qquad t > 0 \tag{2.101}$$

where

$$H(\zeta) = \frac{P(\zeta)}{Q(\zeta)} \tag{2.102}$$

This is an interesting and significant result. It states that for an exponential input $e^{\zeta t}$ the forced response $y_\phi(t)$ is the same exponential multiplied by $H(\zeta) = P(\zeta)/Q(\zeta)$. The total system response $y(t)$ to an exponential input $e^{\zeta t}$ is then given by

$$y(t) = \sum_{j=1}^{n} K_j e^{\lambda_j t} + H(\zeta)e^{\zeta t} \tag{2.103}$$

where the arbitrary constants K_1, K_2, ..., K_n are determined from auxiliary conditions.

Recall that the exponential signal includes a large variety of signals, such as a constant ($\zeta = 0$), a sinusoid ($\zeta = \pm j\omega$), and an exponentially growing or decaying sinusoid ($\zeta = \sigma \pm j\omega$). Let us consider the forced response for some of these cases.

The Constant Input $f(t) = C$

Because $C = Ce^{0t}$, the constant input is a special case of the exponential input $Ce^{\zeta t}$ with $\zeta = 0$. The forced response to this input is then given by

$$y_\phi(t) = CH(\zeta)e^{\zeta t} \qquad \text{with} \qquad \zeta = 0$$
$$= CH(0) \tag{2.104}$$

The Exponential Input $e^{j\omega t}$

Here $\zeta = j\omega$, and

$$y_\phi(t) = H(j\omega)e^{j\omega t} \tag{2.105}$$

The Sinusoidal Input $f(t) = \cos \omega_0 t$

We know that the forced response for the input $e^{\pm j\omega t}$ is $H(\pm j\omega)e^{\pm j\omega t}$. Since $\cos \omega t = (e^{j\omega t} + e^{-j\omega t})/2$, the forced response to $\cos \omega t$ is

$$y_\phi(t) = \tfrac{1}{2}\left[H(j\omega)e^{j\omega t} + H(-j\omega)e^{-j\omega t}\right]$$

Because the two terms on the right-hand side are conjugates,

$$y_\phi(t) = \text{Re}\left[H(j\omega)e^{j\omega t}\right]$$

But

$$H(j\omega) = |H(j\omega)|e^{j\angle H(j\omega)}$$

so that

$$y_\phi(t) = \text{Re}\left\{|H(j\omega)|e^{j[\omega t + \angle H(j\omega)]}\right\}$$
$$= |H(j\omega)| \cos\left[\omega t + \angle H(j\omega)\right] \tag{2.106}$$

This result can be generalized for the input $f(t) = \cos(\omega t + \theta)$. The forced response in this case is

$$y_\phi(t) = |H(j\omega)| \cos [\omega t + \theta + \angle H(j\omega)] \qquad (2.107)$$

■ **Example 2.15**

For the RLC circuit of Fig. 2.1a, find the forced response $y_\phi(t)$ for the following inputs:

(a) $10e^{-3t}$ (b) 5 (c) e^{-2t} (d) $10\cos(3t + 30°)$.

The loop equation for this system [see Example 2.2 or Eq. (1.20)] is

$$\left(D^2 + 3D + 2\right) y(t) = Df(t)$$

Therefore

$$H(\zeta) = \frac{P(\zeta)}{Q(\zeta)} = \frac{\zeta}{\zeta^2 + 3\zeta + 2}$$

(a) For input $f(t) = 10e^{-3t}$, $\zeta = -3$, and

$$y_\phi(t) = 10H(-3)e^{-3t}$$

$$= 10 \left[\frac{-3}{(-3)^2 + 3(-3) + 2} \right] e^{-3t}$$

$$= -15e^{-3t} \qquad t > 0$$

(b) For input $f(t) = 5 = 5e^{0t}$, $\zeta = 0$, and

$$y_\phi(t) = 5H(0) = 0 \qquad t > 0$$

(c) Here $\zeta = -2$, which is also a characteristic root of the system. Hence (see Pair 2, Table 2.2, or the comment at the bottom of the table)

$$y_\phi(t) = \beta t e^{-2t}$$

To find β, we substitute $y_\phi(t)$ in the system equation, giving us

$$\left(D^2 + 3D + 2\right) y_\phi(t) = Df(t)$$

or

$$\left(D^2 + 3D + 2\right) \left[\beta t e^{-2t}\right] = De^{-2t}$$

But

$$D\left[\beta t e^{-2t}\right] = \beta(1 - 2t)e^{-2t}$$

$$D^2\left[\beta t e^{-2t}\right] = 4\beta(t - 1)e^{-2t}$$

$$De^{-2t} = -2e^{-2t}$$

Consequently

$$\beta(4t - 4 + 3 - 6t + 2t)e^{-2t} = -2e^{-2t}$$

or

$$-\beta e^{-2t} = -2e^{-2t}$$

This means that $\beta = 2$, so that

$$y_\phi(t) = 2te^{-2t}$$

(d) For the input $f(t) = 10\cos(3t + 30°)$, the forced response [see Eq. (2.107)] is

$$y_\phi(t) = 10|H(j3)|\cos[3t + 30° + \angle H(j3)]$$

where

$$H(j3) = \frac{P(j3)}{Q(j3)} = \frac{j3}{(j3)^2 + 3(j3) + 2} = \frac{j3}{-7 + j9} = \frac{27 - j21}{130} = 0.263e^{-j37.9°}$$

Therefore

$$|H(j3)| = 0.263, \qquad \angle H(j3) = -37.9°$$

and

$$y_\phi(t) = 10(0.263)\cos(3t + 30° - 37.9°)$$
$$= 2.63\cos(3t - 7.9°) \quad\blacksquare$$

■ **Example 2.16**
    For the *RLC* circuit of Fig. 2.1a, find the loop current $y(t)$ if the input voltage $f(t) = 10e^{-3t}$; the initial conditions are $y(0) = 0$ and $v_C(0) = 5$.
    The zero-input and zero-state responses for this problem are found in Examples 2.2 and 2.8. The natural and forced responses are found in Eq. 2.74b. Here we shall solve this problem by the classical method.
    First, we determine the initial conditions $y(0^+)$ and $\dot{y}(0^+)$ from the given auxiliary conditions $y(0) = 0$ and $v_C(0) = 5$. The inductor current cannot change instantaneously in the absence of impulsive voltage, and the capacitor voltage cannot change instantaneously in the absence of impulsive current. Therefore $y(0^+) = 0$ and $v_C(0^+) = 5$. Also, the loop equation of the circuit in Fig. 2.1 with $f(t) = 10e^{-3t}$ is

$$\dot{y}(t) + 3y(t) + v_C(t) = 10e^{-3t}$$

Setting $t = 0^+$ and using $y(0^+) = 0, v_C(0^+) = 5$ in this equation yields

$$\dot{y}(0^+) + 3(0^+) + 5 = 10 \quad\Longrightarrow\quad \dot{y}(0^+) = 5$$

Therefore the initial conditions are

$$y(0^+) = 0 \qquad \text{and} \qquad \dot{y}(0^+) = 5$$

The loop equation for this system [see Example 2.2 or Eq. (1.20)] is

$$\left(D^2 + 3D + 2\right)y(t) = Df(t)$$

The characteristic polynomial is $\lambda^2 + 3\lambda + 2 = (\lambda + 1)(\lambda + 2)$. Therefore the natural response is

$$y_n(t) = K_1 e^{-t} + K_2 e^{-2t}$$

The forced response, already found in Example 2.15 **(a)**, is

$$y_\phi(t) = -15e^{-3t}$$

The total response is

$$y(t) = K_1 e^{-t} + K_2 e^{-2t} - 15 e^{-3t}$$

Differentiation of this equation yields

$$\dot{y}(t) = -K_1 e^{-t} - 2K_2 e^{-2t} + 45 e^{-3t}$$

Setting $t = 0^+$ and substituting $y(0^+) = 0$, $\dot{y}(0^+) = 5$ in these equations yields

$$\left. \begin{array}{r} 0 = K_1 + K_2 - 15 \\ 5 = -K_1 - 2K_2 + 45 \end{array} \right\} \implies \begin{array}{l} K_1 = -10 \\ K_2 = 25 \end{array}$$

Therefore

$$y(t) = -10 e^{-t} + 25 e^{-2t} - 15 e^{-3t}$$

which agrees with the solution found previously in Eq. 2.74b. ∎

Assessment of the Classical Method

The development in this section shows that the classical method is relatively simple compared to the method of finding the response as a sum of the zero-input and zero-state components. Unfortunately, the classical method has a serious draw-back because it yields the total response, which cannot be separated into components arising from the internal conditions and the external input. In the study of systems it is important to be able to express the system response to an input $f(t)$ as an explicit function of $f(t)$. This is not possible in the classical method. Moreover, the classical method is restricted to a certain class of inputs; it cannot be applied to any input. Another minor problem is that because the classical method yields total response, the auxiliary conditions must be on the total response which exists only for $t \geq 0^+$. In practice we are most likely to know the conditions at $t = 0^-$ (before the input is applied). Therefore we need to derive a new set of auxiliary conditions at $t = 0^+$ from the known conditions at $t = 0^-$.

If we must solve a particular linear differential equation or find a response of a particular LTI system, the classical method may be the best. In the theoretical study of linear systems, however, it is practically useless.

2.9 APPENDIX 2.1: DETERMINING THE IMPULSE RESPONSE

We now derive the unit impulse response of an LTIC system S specified by the nth-order differential equation

$$Q(D)y(t) = P(D)f(t) \tag{2.108a}$$

or

$$(D^n + a_{n-1}D^{n-1} + \cdots + a_1 D + a_0)y(t)$$
$$= (b_n D^n + b_{n-1}D^{n-1} + \cdots + b_1 D + b_0)f(t) \tag{2.108b}$$

In Sec. 2.3 we showed that the impulse response $h(t)$ is given by

$$h(t) = A_0 \delta(t) + \text{characteristic modes} \tag{2.109}$$

We now show that in the above equation $A_0 = b_n$ where b_n is the coefficient of the nth-order term on the right-hand side of Eq. (2.108b). When the input $f(t) = \delta(t)$, the response $y(t) = h(t)$. Therefore from Eq. (2.108b), we obtain

$$(D^n + a_{n-1}D^{n-1} + \cdots + a_1 D + a_0)h(t) = (b_n D^n + b_{n-1}D^{n-1} + \cdots + b_1 D + b_0)\delta(t)$$

In this equation we substitute $h(t)$ from Eq. (2.109) and compare the coefficients of similar impulsive terms on both sides. The highest order of the derivative of impulse on both sides is n, with its coefficient value as A_0 on the left-hand side and b_n on the right-hand side. The two values must be matched. Therefore $A_0 = b_n$, and

$$h(t) = b_n \delta(t) + \text{characteristic modes} \tag{2.110}$$

To determine the characteristic mode terms in the above equation, let us consider a system S_0 whose input $f(t)$ and the corresponding output $x(t)$ are related by

$$Q(D)x(t) = f(t) \tag{2.111}$$

Observe that both the systems, S and S_0, have the same characteristic polynomial, namely, $Q(\lambda)$, and consequently the same characteristic modes. Moreover, S_0 is the same as S with $P(D) = 1$, that is, $b_n = 0$. Therefore according to Eq. (2.110), the impulse response of S_0 consists of characteristic mode terms only without an impulse at $t = 0$. Let us denote this impulse response of S_0 by $y_0(t)$. Observe that $y_0(t)$ consists of characteristic modes of S, and therefore may be viewed as a zero-input response of S. Now $y_0(t)$ is the response of S_0 to input $\delta(t)$. Therefore from Eq. (2.111),

$$Q(D)y_0(t) = \delta(t) \tag{2.112a}$$

or

$$(D^n + a_{n-1}D^{n-1} + \cdots + a_1 D + a_0)y_0(t) = \delta(t) \tag{2.112b}$$

or

$$y_0^{(n)}(t) + a_{n-1}y_0^{(n-1)}(t) + \cdots + a_1 y_0^{(1)}(t) + a_0 y_0(t) = \delta(t) \tag{2.112c}$$

where $y_0^{(k)}(t)$ represents the kth derivative of $y_0(t)$. The right-hand side contains a single impulse term $\delta(t)$. This is possible only if $y_0^{(n-1)}(t)$ has a unit jump discontinuity at $t = 0$, so that $y_0^{(n)}(t) = \delta(t)$. Moreover, the lower-order terms can not have any jump discontinuity. This means that $y_0(0) = y_0^{(1)}(0) = \cdots = y_0^{(n-2)}(0) = 0$. Therefore the n initial conditions on $y_0(t)$ are

$$y_0^{(n-1)}(0) = 1$$
$$y_0(0) = y_0^{(1)}(0) = \cdots = y_0^{(n-2)}(0) = 0 \tag{2.113}$$

This discussion means that $y_0(t)$ is the zero-input response of the system S subject to initial conditions (2.113).

We now show that for the same input $f(t)$ to both systems, S and S_0, their respective outputs $y(t)$ and $x(t)$ are related by

$$y(t) = P(D)x(t) \tag{2.114}$$

To prove this result, we operate on both sides of Eq. (2.111) by $P(D)$ to obtain

$$Q(D)P(D)x(t) = P(D)f(t)$$

Comparison of this equation with Eq. (2.108a) leads immediately to Eq. (2.114).

Now if the input $f(t) = \delta(t)$, the output of S_0 is $y_0(t)$, and the output of S, according to Eq. (2.114), is $P(D)y_0(t)$. This output is $h(t)$, the unit impulse response of S. Note, however, that because it is an impulse response of a causal system S_0, the function $y_0(t)$ is causal. To incorporate this fact we must represent this function as $y_0(t)u(t)$. Now it follows that $h(t)$, the unit impulse of the system S, is given by

$$h(t) = P(D)[y_0(t)u(t)] \tag{2.115}$$

where $y_0(t)$ is the zero-input response of the system subject to initial conditions (2.113).

The right-hand side of Eq. (2.115) is a linear combination of the derivatives of $y_0(t)u(t)$. Evaluating these derivatives is clumsy and inconvenient because of the presence of $u(t)$. The derivatives will generate an impulse and its derivatives at the origin. Fortunately when $m \leq n$ [Eq. (2.108)], we can avoid this difficulty by using the observation in Eq. (2.110), which shows that at $t = 0$ (the origin), $h(t) = b_n\delta(t)$. Therefore we need not bother to find $h(t)$ at the origin. This means that instead of deriving $P(D)[y_0(t)u(t)]$, we can derive $P(D)y_0(t)$ and add to it the term $b_n\delta(t)$, so that

$$h(t) = b_n\delta(t) + P(D)y_0(t) \qquad t \geq 0$$
$$= b_n\delta(t) + [P(D)y_0(t)]u(t) \tag{2.116}$$

This expression is valid when $m \leq n$ [the form given in Eq. (2.108b)]. When $m > n$, Eq. (2.115) should be used.

2.10 SUMMARY

This chapter discusses time-domain analysis of LTIC systems. The total response of a linear system is a sum of the zero-input response and zero-state response. The zero-input response is the system response generated only by the internal conditions (initial conditions) of the system, assuming that the external input is zero; hence the term "zero-input." The zero-state response is the system response generated by the external input, assuming that all initial conditions are zero; that is, when the system is in zero state.

Every system can sustain certain forms of response on its own with no external input (zero input). These forms are intrinsic characteristics of the system; that is, they do not depend on any external input. For this reason they are called characteristic modes of the system. Needless to say, the zero-input response is made up of characteristic modes chosen in a suitable combination so as to satisfy the initial conditions of the system. For an nth-order system, there are n distinct modes.

Unit impulse function is an idealized mathematical model of a signal that cannot be generated in practice. Nevertheless, introduction of such a signal as an intermediary is very helpful in analysis of signals and systems. The impulse function has a sampling or sifting property, which states that the area under the product of a function with a unit impulse is given by the value of the function where the impulse is located (assuming the function to be continuous at the impulse location). The unit impulse response of a system is a combination of the characteristic modes of the system† because the impulse $\delta(t) = 0$ for $t > 0$. Therefore the system response for $t > 0$ must necessarily be a zero-input response, which, as seen earlier, is a combination of characteristic modes.

The zero-state response (response due to external input) of a linear system can be found by breaking the input into simpler components and then adding the responses to all the components. There are many ways of separating an input signal into simpler components. In this chapter we consider representing an arbitrary input signal $f(t)$ as a sum of narrow rectangular pulses [staircase approximation of $f(t)$]. In the limit as the pulse width $\rightarrow 0$, the rectangular pulse components approach impulses. Knowing the impulse response of the system, we can find the system response to all the impulse components and add them to yield the system response to the input $f(t)$. The sum of the responses to the impulse components is in the form of an integral, known as the convolution integral. The system response is obtained as the convolution of the input $f(t)$ with the system's impulse response $h(t)$. Therefore the knowledge of the system's impulse response allows us to determine the system response to any arbitrary input. The impulse response of a system can be found from external measurements on a system (applying an impulse at the input and measuring the resulting response). It is therefore an external description of a system.

A linear system is in a zero state if all initial conditions are zero. A system in a zero state is incapable of generating any response in the absence of an external input. When some initial conditions are applied to a system, then, if the system eventually goes to zero state in the absence of any external input, the system is said to be asymptotically stable. On the other hand, if the system's response increases without bound, it is unstable. If neither the system goes to zero state nor the response increases indefinitely, the system is marginally stable. For LTIC systems, the system response in the absence of an external input is a combination of the characteristic modes of the system. The stability criterion in terms of the location of the characteristic roots of a system can be summarized as follows:

1. An LTIC system is asymptotically stable if, and only if, all the characteristic roots are in the LHP. The roots may be repeated or unrepeated.

2. An LTIC system is unstable if, and only if, either one or both of the following conditions exist: (i) at least one root is in the RHP; (ii) there are repeated roots on the imaginary axis.

3. An LTIC system is marginally stable if, and only if, there are no roots in the RHP, and there are some unrepeated roots on the imaginary axis.

According to an alternative definition of stability— bounded-input bounded-output (BIBO) stability—a system is stable if every bounded input produces a bounded output. Otherwise the system is (BIBO-) unstable. Asymptotically stable system is always BIBO-stable. The converse is not necessarily true, however.

Characteristic behavior of a system is extremely important because it deter-

---

† There is the possibility of an impulse in addition to characteristic modes.

mines not only the system response to internal conditions (zero-input behavior), but also the system response to external inputs (zero-state behavior) and the system stability. The system response to external inputs is determined by the impulse response, which itself is made up of characteristic modes. The width of the impulse response is called the time constant of the system, which indicates how fast the system can respond to an input. The time constant plays an important role in determining such diverse system behavior as the response time and filtering properties of the system, dispersion of pulses, and the rate of pulse transmission through the system.

Differential equations of LTIC systems can also be solved by the classical method, where the response is found as a sum of natural and forced response. These are not the same as the zero-input and zero-state components, although they satisfy the same equations respectively. Although simple, this method suffers from the fact that it is applicable to a restricted class of input signals, and the system response cannot be set forth as an explicit function of the input. This makes it useless in the theoretical study of systems.

REFERENCES

1. Lathi, B.P., *Signals and Systems*, Berkeley-Cambridge Press, Carmichael, Ca. 1987.

2. Papoulis A., *The Fourier Integral and Its Applications*, McGraw-Hill, New York, 1962.

3. Lathi, B.P., *Signals, Systems, and Communication*, Wiley, New York, 1965.

4. Kailath, T., *Linear System*, Prentice-Hall, Englewood Cliffs, N.J., 1980.

5. Lathi, B.P., *Modern Digital and Analog Communication Systems*, Second Ed., Holt, Rinehart and Winston, New York, 1989.

PROBLEMS

2.2-1 An LTIC system is specified by the equation

$$\left(D^2 + 5D + 6\right) y(t) = (D + 1)f(t)$$

(a) Find the characteristic polynomial, characteristic equation, characteristic roots, and characteristic modes of this system.
(b) Find $y_0(t)$, the zero-input component of the response $y(t)$ for $t \geq 0$, if the initial conditions are $y_0(0) = 2$ and $\dot{y}_0(0) = -1$.

2.2-2 Repeat Prob. 2.2-1 if
$$\left(D^2 + 4D + 4\right) y(t) = Df(t)$$

and $y_0(0) = 3$, $\dot{y}_0(0) = -4$.

2.2-3 Repeat Prob. 2.2-1 if
$$D(D + 1)y(t) = (D + 2)f(t)$$

and $y_0(0) = \dot{y}_0(0) = 1$.

2.2-4 Repeat Prob. 2.2-1 if
$$\left(D^2 + 9\right) y(t) = (3D + 2)f(t)$$

and $y_0(0) = 0$, $\dot{y}_0(0) = 6$.

2.2-5 Repeat Prob. 2.2-1 if

$$\left(D^2 + 4D + 13\right) y(t) = 4(D + 2)f(t)$$

with $y_0(0) = 5$, $\dot{y}_0(0) = 15.98$.

2.2-6 Repeat Prob. 2.2-1 if
$$D^2(D + 1)y(t) = (D^2 + 2)f(t)$$

with $y_0(0) = 4$, $\dot{y}_0(0) = 3$ and $\ddot{y}_0(0) = -1$.

2.2-7 Repeat Prob. 2.2-1 if

$$(D + 1)\left(D^2 + 5D + 6\right) y(t) = Df(t)$$

with $y_0(0) = 2$, $\dot{y}_0(0) = -1$ and $\ddot{y}_0(0) = 5$.

2.3-1 Simplify the following expressions:

(a) $\left(\dfrac{\sin t}{t^2 + 2}\right) \delta(t)$

(d) $\left(\dfrac{\sin \frac{\pi}{2}(t - 2)}{t^2 + 4}\right) \delta(t - 1)$

(b) $\left(\dfrac{j\omega + 2}{\omega^2 + 9}\right) \delta(\omega)$

(e) $\left(\dfrac{1}{j\omega + 2}\right) \delta(\omega + 3)$

(c) $\left(e^{-t} \cos (3t - 60°)\right) \delta(t)$

(f) $\left(\dfrac{\sin k\omega}{\omega}\right) \delta(\omega)$

Hint: Use Eq. (2.32). For part (f) use L'Hôpital's rule.

2.3-2 Evaluate the following integrals:

(a) $\displaystyle\int_{-\infty}^{\infty} f(\tau)\delta(t - \tau) \, d\tau$

(b) $\displaystyle\int_{-\infty}^{\infty} \delta(\tau)f(t - \tau) \, d\tau$

(c) $\displaystyle\int_{-\infty}^{\infty} \delta(t)e^{-j\omega t} \, dt$

(d) $\displaystyle\int_{-\infty}^{\infty} \delta(t - 2) \sin \pi t \, dt$

(e) $\displaystyle\int_{-\infty}^{\infty} \delta(t + 3)e^{-t} \, dt$

(f) $\displaystyle\int_{-\infty}^{\infty} (t^3 + 4)\delta(1 - t) \, dt$

(g) $\displaystyle\int_{-\infty}^{\infty} f(2 - t)\delta(3 - t) \, dt$

(h) $\displaystyle\int_{-\infty}^{\infty} e^{(x-1)} \cos \frac{\pi}{2}(x - 5)\delta(x - 3) \, dx$

Hint: $\delta(x)$ is located at $x = 0$. For example, $\delta(1 - t)$ is located at $1 - t = 0$; that is, at $t = 1$, and so on.

2.3-3 Using the generalized function definition, show that $\delta(t)$ is an even function of t.

Hint: Start with Eq. (2.33a) as the definition of $\delta(t)$. Now change variable $t = -x$ to show that

$$\int_{-\infty}^{\infty} \phi(t)\delta(-t) \, dt = \phi(0)$$

2.3-4 Prove that

$$\delta(at) = \frac{1}{|a|}\delta(t)$$

Hint: Show that

$$\int_{-\infty}^{\infty} \phi(t)\delta(at)\, dt = \frac{1}{|a|}\phi(0)$$

2.3-5 Show that

$$\int_{-\infty}^{\infty} \dot{\delta}(t)\phi(t)\, dt = -\dot{\phi}(0)$$

where $\phi(t)$ and $\dot{\phi}(t)$ are continuous at $t = 0$, and $\phi(t) \to 0$ as $t \to \pm\infty$. This integral defines $\dot{\delta}(t)$ as a generalized function. Hint: Use integration by parts.

2.3-6 Find the unit impulse response of a system specified by the equation

$$\left(D^2 + 4D + 3\right) y(t) = (D + 5)f(t)$$

2.3-7 Repeat Prob. 2.3-6 if

$$\left(D^2 + 5D + 6\right) y(t) = \left(D^2 + 7D + 11\right) f(t)$$

2.3-8 Repeat Prob. 2.3-6 for the first-order allpass filter specified by the equation

$$(D + 1)y(t) = -(D - 1)f(t)$$

Hint: For a first-order system $(n = 1)$, condition (2.42) implies $y_0^{(n-1)}(0) = y_0(0) = 1$.

2.3-9 **(a)** Find the unit impulse response of a system specified by the equation

$$\left(D^2 + 5D + 6\right) y(t) = (D + 2)f(t)$$

(b) Show that the unit impulse response found in part **(a)** is identical to that of a system specified by the equation

$$(D + 3)y(t) = f(t)$$

This proof will show that common factors in the polynomials $Q(D)$ and $P(D)$ can be canceled for the purpose of computing $h(t)$. This result occurs because $h(t)$ is an external description of the system (see Sec. 1.5).

2.3-10 Find the unit impulse response of an LTIC system specified by the equation

$$\left(D^2 + 4D + 4\right) y(t) = (2D + 9)\, f(t)$$

2.3-11 Find the unit impulse response of an LTIC system specified by the equation

$$(D^2 + 10D + 34)y(t) = (D + 9)f(t)$$

Hint: The characteristic roots are complex conjugate, and $y_0(t)$ is of the form $ce^{\alpha t} \cos(\beta t + \theta)$.

2.3-12 Find the unit impulse response of an LTIC system specified by the equation

$$(D + 1)\left(D^2 + 5D + 6\right) y(t) = (5D + 9)f(t)$$

2.4-1 Using direct integration, find $e^{-at}u(t) * e^{-bt}u(t)$.

2.4-2 Using direct integration, find $u(t) * u(t)$, $e^{-at}u(t) * e^{-at}u(t)$, and $tu(t) * u(t)$.

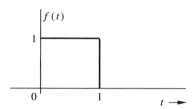

Fig. P2.4-8

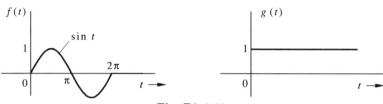

Fig. P2.4-11

2.4-3 Using direct integration, find $\sin t\, u(t) * u(t)$ and $\cos t\, u(t) * u(t)$.

2.4-4 The unit impulse response of an LTIC system is $h(t) = e^{-t}u(t)$. Find this system's (zero-state) response $y(t)$ if the input $f(t)$ is:

(a) $u(t)$ (b) $e^{-t}u(t)$ (c) $e^{-2t}u(t)$ (d) $\sin 3t\, u(t)$.

Use the convolution table to find your answers.

2.4-5 Repeat Prob. 2.4-4 if

$$h(t) = \left[2e^{-3t} - e^{-2t}\right]u(t)$$

and if the input $f(t)$ is: (a) $u(t)$ (b) $e^{-t}u(t)$ (c) $e^{-2t}u(t)$.

2.4-6 Repeat Prob. 2.4-4 if

$$h(t) = (1 - 2t)e^{-2t}u(t)$$

and if the input $f(t) = u(t)$.

2.4-7 Repeat Prob. 2.4-4 if $h(t) = 4e^{-2t}\cos 3t\, u(t)$ and if the input $f(t)$ is: (a) $u(t)$ (b) $e^{-t}u(t)$.

2.4-8 Repeat Prob. 2.4-4 if

$$h(t) = e^{-t}u(t)$$

and if the input $f(t)$ is: (a) $e^{-2t}u(t)$ (b) $e^{-2(t-3)}u(t)$ (c) $e^{-2t}u(t-3)$ (d) the gate pulse shown in Fig. P2.4-8. For (d), sketch $y(t)$.

Hint: $e^{-2(t-3)} = e^6 e^{-2t}$, and $e^{-2t}u(t-3) = e^{-6}e^{-2(t-3)}u(t-3)$. Also, the input in (d) can be expressed as $u(t) - u(t-1)$. For parts (c) and (d), use the shift property (2.63) of convolution. (Alternatively, you may want to invoke the system's time-invariance and superposition properties).

2.4-9 A first-order allpass filter impulse response is given by

$$h(t) = -\delta(t) + 2e^{-t}u(t)$$

(a) Find the zero-state response of this filter for the input $e^t u(-t)$.

(b) Sketch the input and the corresponding zero-state response.

2.4-10 Sketch the functions $f(t) = \frac{1}{t^2+1}$ and $u(t)$. Now find $f(t) * u(t)$ and sketch the result.

2.4-11 Figure P2.4-11 shows $f(t)$ and $g(t)$. Find and sketch $c(t) = f(t) * g(t)$.

Fig. P2.4-12

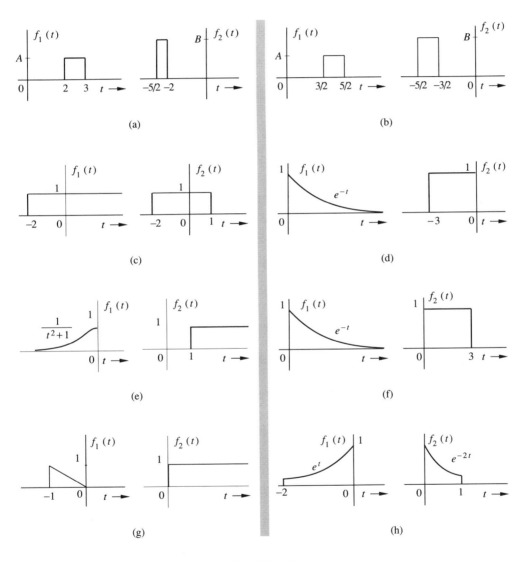

(a)

(b)

(c)

(d)

(e)

(f)

(g)

(h)

Fig. P2.4-13

2.4-12 Find and sketch $c(t) = f(t) * g(t)$ for the functions shown in Fig. P2.4-12.

2.4-13 Find and sketch $c(t) = f_1(t) * f_2(t)$ for the pairs of functions shown in Fig. P2.4-13.

2.4-14 A line charge is located along the x axis with a charge density $f(x)$. Show that the electric field $E(x)$ produced by this line charge at a point x is given by

$$E(x) = f(x) * h(x)$$

where

$$h(x) = \frac{1}{4\pi\epsilon x^2}$$

Hint: the charge over an interval $\triangle\tau$ located at $\tau = n\triangle\tau$ is $f(n\triangle\tau)\triangle\tau$. Also by Coulomb's law, the electric field $E(r)$ at a distance r from a charge q is given by

$$E(r) = \frac{q}{4\pi\epsilon r^2}$$

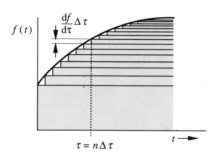

Fig. P2.4-15

2.4-15 As mentioned in Chapter 1 (Fig. 1.4b), it is possible to express an input in terms of its step components, as shown in Fig. P2.4-15. If $g(t)$ is the unit step response of an LTIC system, show that the (zero-state) response $y(t)$ of the system to an input $f(t)$ can be expressed as

$$y(t) = \int_{-\infty}^{\infty} \dot{f}(\tau)g(t - \tau)\,d\tau = \dot{f}(t) * g(t)$$

Hint: From Figure P2.4-15, the shaded step component of the input is given by $(\triangle f)u(t - n\triangle\tau) \simeq [\dot{f}(\tau)\triangle\tau]u(t - n\triangle\tau)$, and the system response corresponding to this input component is $[\dot{f}(\tau)\triangle\tau]g(t - n\triangle\tau)$.

2.4-16 For an LTIC system, if the (zero-state) response to an input $f(t)$ is $y(t)$, show that the (zero-state) response to the input $\dot{f}(t)$ is $\dot{y}(t)$ and that for the input $\int_0^t f(\tau)\,d\tau$ is $\int_0^t y(\tau)\,d\tau$.

Hint: Recognize that $\dot{f}(t) = \lim_{T\to 0}\frac{1}{T}[f(t) - f(t - T)]$. Now use linearity and time invariance to find the response to $\dot{f}(t)$. Also recognize that $\int_0^t f(\tau)\,d\tau = f(t) * u(t)$, so that the response to this input is $[f(t) * u(t)] * h(t) = [f(t) * h(t)] * u(t)$.

2.5-1 Find $f_1(t) * f_2(t)$ numerically, using $T = 0.1$ for the pair shown in Fig. P2.4-13a.

2.5-2 Repeat Prob. 2.5-1 for the pair in Fig. P2.4-13b.

2.5-3 Repeat Prob. 2.5-1 for the pair in Fig. P2.4-13c.

2.6-1 Explain with reasons, whether the LTIC systems described by the following equations are asymptotically stable, marginally stable, or unstable.

(a) $(D^2 + 8D + 12)y(t) = (D - 1)f(t)$

(b) $D(D^2 + 3D + 2)y(t) = (D + 5)f(t)$

(c) $D^2(D^2 + 2)y(t) = f(t)$

(d) $(D + 1)(D^2 - 6D + 5)y(t) = (3D + 1)f(t)$

2.6-2 Repeat Prob. 2.6-1 if

(a) $(D + 1)(D^2 + 2D + 5)^2 y(t) = f(t)$

(b) $(D + 1)(D^2 + 9)y(t) = (2D + 9)f(t)$

(c) $(D + 1)(D^2 + 9)^2 y(t) = (2D + 9)f(t)$

(d) $(D^2 + 1)(D^2 + 4)(D^2 + 9)y(t) = 3Df(t)$

2.6-3 For a certain LTIC system, the impulse response $h(t) = u(t)$.
(a) Determine the characteristic root(s) of this system.
(b) Is this system asymptotically or marginally stable, or is it unstable?
(c) Is this system BIBO stable?
(d) What can this system be used for?

2.6-4 In Sec. 2.6 we demonstrated that for an LTIC system, Condition (2.85) is sufficient for BIBO stability. Show that this is also a necessary condition for BIBO stability in such systems. In other words, show that if Eq. (2.85) is not satisfied, then there exists a bounded input that produces an unbounded output.

Hint: Assume that a system exists for which $h(t)$ violates Eq. (2.85) and yet produces an output that is bounded for every bounded input. Establish contradiction in this statement by considering an input $f(t)$ defined by $f(t_1 - \tau) = 1$ when $h(\tau) \geq 0$ and $f(t_1 - \tau) = -1$ when $h(\tau) < 0$, where t_1 is some fixed instant.

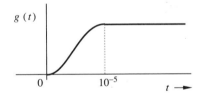

Fig. P2.7-1

2.7-1 Data at a rate of 1 million pulses per second are to be transmitted over a certain communications channel. The unit step response $g(t)$ for this channel is shown in Fig. P2.7-1.
(a) Explain whether this channel is capable of transmitting the data at the required rate.
(b) Can an audio signal consisting of components with frequencies up to 15kHz be transmitted over this channel with reasonable fidelity?

2.7-2 A certain communication channel has a bandwidth of 10 kHz. If a pulse of 0.5 ms. duration is transmitted over this channel, determine the width (duration) of the received pulse. Find the maximum rate at which these pulses can be transmitted over this channel without interference between the successive pulses.

2.8-1 Using the classical method, solve

$$\left(D^2 + 7D + 12\right)y(t) = (D + 2)f(t)$$

if the initial conditions are $y(0^+) = 0$, $\dot{y}(0^+) = 1$ and if the input $f(t)$ is:

(a) $u(t)$ **(b)** $e^{-t}u(t)$ **(c)** $e^{-2t}u(t)$.

2.8-2 Using the classical method, solve

$$\left(D^2 + 6D + 25\right) y(t) = (D+3)f(t)$$

if the initial conditions are $y(0^+) = 0$, $\dot{y}(0^+) = 2$ and if the input $f(t) = u(t)$.

2.8-3 Using the classical method, solve

$$\left(D^2 + 4D + 4\right) y(t) = (D+1)f(t)$$

if the initial conditions are $y(0^+) = \frac{9}{4}$, $\dot{y}(0^+) = 5$ and if the input $f(t)$ is:

(a) $e^{-3t}u(t)$ **(b)** $e^{-t}u(t)$.

2.8-4 Using the classical method, solve

$$(D^2 + 2D)y(t) = (D+1)f(t)$$

if the initial conditions are $y(0^+) = 2$, $\dot{y}(0^+) = 1$ and if the input is $f(t) = u(t)$.

Hint: The input $u(t)$ is itself a characteristic mode; hence the form of the forced response is $\beta t u(t)$.

2.8-5 Repeat Prob. 2.8-1 if the input

$$f(t) = e^{-3t}u(t)$$

Hint: Note that e^{-3t} is a characteristic mode of the system. See Pair 2 in Table 2.2.

COMPUTER PROBLEMS

C2-1 In the RLC circuit in Fig. 2.1a, if the circuit parameters are (a)$L = 2, R = 4, C = 1/40$, (b) $L = 1, R = 4, C = 1/50$ (c) $L = 1, R = 8, C = 1/12$, find the zero-input loop current $y_0(t)$ if the initial conditions are $y_0(0) = 4$, and $\dot{y}(t) = 10$.

C2-2 Find $c(t) = f(t) * g(t)$ if $f(t) = tu(t)$ and

$$g(t) = \begin{cases} te^{-t} & \geq 0 \\ e^t & < 0 \end{cases}$$

C2-3 Use numerical convolution to determine $c[k] = c(kT)$, where $c(t) = f(t) * g(t)$ and $f(t) = tu(t)$ and $g(t) = 2u(t)$. Use $T = 0.1$ for the sampling interval.

Time-Domain Analysis:
Discrete-Time Systems

In this chapter we discuss time-domain analysis of LTID (linear time-invariant discrete-time systems). As we shall see, the procedure here is parallel to that for continuous-time systems, with minor differences.

3.1 DISCRETE-TIME SYSTEMS

Signals defined or specified over a continuous range of t are **continuous-time signals**, denoted by the symbols $f(t), y(t)$, etc. Systems whose inputs and outputs are continuous-time signals are **continuous-time systems**. On the other hand, signals defined only at discrete instants of time $t_0, t_1, t_2, \ldots, t_k, \ldots$, are **discrete-time signals**, which we denote by the symbols $f(t_k), y(t_k), \ldots$ (k, an integer). Systems whose input and output signals are discrete-time signals are called **discrete-time systems**. A digital computer is a familiar example of this type of system. We shall assume that the discrete instants $t_0, t_1, t_2, \ldots$ are uniformly spaced; that is

$$t_{k+1} - t_k = T \qquad \text{for all} \quad k$$

With uniform spacing, discrete-time signals may be represented in the form $f(kT), y(kT)$, etc. We further simplify this notation to $f[k], y[k], \ldots$, etc., where it is understood that $f[k] = f(kT)$ and that k is an integer. A typical discrete-time signal, shown in Fig. 3.1, is therefore a sequence of numbers. This signal may be denoted by $f(kT)$ and viewed as a function of time where signal values are specified at $t = kT$ (k, integer). It may also be denoted by $f[k]$ and viewed as a function of k (k, integer). The latter representation is more convenient and will be followed throughout this book. A discrete-time system may be seen as processing a sequence of numbers $f[k]$ and yielding as output another sequence of numbers $y[k]$.

Discrete-time signals arise naturally in situations which are inherently discrete-time, such as population studies, amortization problems, national income models, and radar tracking.. They may also arise as a result of sampling continuous-time

189

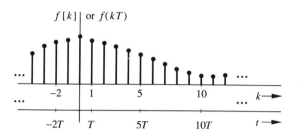

Fig. 3.1 A discrete-time signal.

signals in sampled data systems, digital filtering, and such. Digital filtering is a particularly interesting application in which continuous-time signals are processed by using discrete-time systems, as shown in Fig. 3.2. A continuous-time signal $f(t)$ is first sampled to convert it into a discrete-time signal $f[k]$, which is then processed by a discrete-time system to yield the output $y[k]$. A continuous-time signal $y(t)$ is finally constructed from $y[k]$. We shall use the notations C/D and D/C for continuous-to-discrete-time and discrete-to-continuous-time conversion. Using the interfaces in this manner, we can process a continuous-time signal with an appropriate discrete-time system. As we shall see later in our discussion, discrete-time systems have several advantages over continuous-time systems. For this reason, there is an accelerating trend toward processing continuous-time signals with discrete-time systems. We shall give here three examples of discrete-time systems. In the first two examples, the signals are inherently discrete-time. In the third example, a continuous-time signal is processed by a discrete-time system, as shown in Fig. 3.2, by discretizing the signal through sampling.

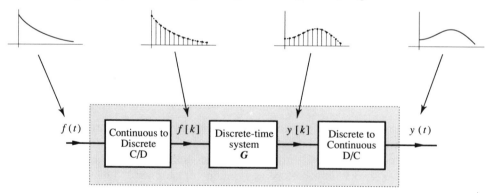

Fig. 3.2 Processing a continuous-time signal by a discrete-time system.

■ **Example 3.1**

A person makes a deposit (the input) in a bank regularly every period T. The bank pays a certain interest on the account balance during the period T and mails out a periodic statement of the account balance (the output) to the depositor. Find the equation relating the output $y[k]$ (the balance) to the input $f[k]$ (the deposit).†

---

†A withdrawal is a negative deposit. Therefore this formulation can handle deposits as well as withdrawals. It also applies to a mortgage payment problem with the initial value $y[0] = -M$, where M is the amount of the mortgage. The mortgage is an initial deposit with a negative value.

Here the signals are inherently discrete-time. Let

$f[k]$ = the deposit made at the kth discrete instant.

$y[k]$ = the account balance at the kth instant computed

immediately after the kth deposit $f[k]$ is received.

r = interest per dollar per period T.

The balance $y[k]$ is the sum of (i) the previous balance $y[k-1]$, (ii) the interest on $y[k-1]$ during the period T, and (iii) the deposit $f[k]$:

$$y[k] = y[k-1] + ry[k-1] + f[k]$$
$$= (1+r)y[k-1] + f[k] \tag{3.1}$$

or

$$y[k] - \gamma y[k-1] = f[k] \qquad \gamma = 1+r \tag{3.2a}$$

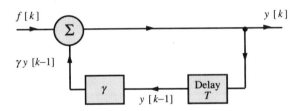

Fig. 3.3 Realization of the savings account system.

In this example the deposit $f[k]$ is the input (cause) and the balance $y[k]$ is the output (effect).

We can express Eq. (3.2a) in an alternate form. The choice of index k in Eq. (3.2a) is completely arbitrary, so we can substitute $k+1$ for k to obtain

$$y[k+1] - \gamma y[k] = f[k+1] \tag{3.2b}$$

We also could have obtained Eq. (3.2b) directly by realizing that $y[k+1]$, the balance at the $(k+1)$st instant, is the sum of $y[k]$ plus $ry[k]$ (the interest on $y[k]$) plus the input, $f[k+1]$ at the $(k+1)$st instant.

For a hardware realization of such a system, we rewrite Eq. (3.2a) as

$$y[k] = \gamma y[k-1] + f[k] \tag{3.2c}$$

Figure 3.3 shows the hardware realization of this equation using a single time delay of T units.† To understand this realization, assume that $y[k]$ is available. Delaying it by T, we generate $y[k-1]$. Next, we generate $y[k]$ from $f[k]$ and $y[k-1]$ according to Eq. (3.2c).

■

■ **Example 3.2**

In the kth semester, $f[k]$ number of students enroll in a course requiring a certain textbook. The publisher sells $y[k]$ new copies of the book in the kth semester. On the average, one quarter of the students with books in salable condition resell their books at the end of the semester, and the book life is three semesters. Write the equation relating

---

†The time delay in Fig. 3.3 need not be T. The use of any other value will result in a time-scaled output.

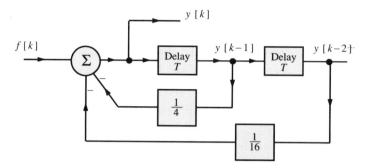

Fig. 3.4 Realization of a second-order discrete-time system in Example 3.2.

$y[k]$, the new books sold by the publisher, to $f[k]$, the number of students enrolled in the kth semester, assuming that every student buys a book.

In the kth semester, the total books $f[k]$ sold to the students must be equal to $y[k]$ (the new books from the publisher) plus the used books from the students enrolled in the two previous semesters (because the book life is only three semesters). There are $y[k-1]$ new books sold in the $(k-1)$st semester, and one quarter of these books, that is, $\frac{1}{4}y[k-1]$, will be resold in the kth semester. Also, $y[k-2]$ new books are sold in the $(k-2)$nd semester, and one quarter of these, that is, $\frac{1}{4}y[k-2]$, will be resold in the $(k-1)$st semester. Again a quarter of these, that is, $\frac{1}{16}y[k-2]$ will be resold in the kth semester. Therefore $f[k]$ must be equal to the sum of $y[k]$, $\frac{1}{4}y[k-1]$, and $\frac{1}{16}y[k-2]$.

$$y[k] + \tfrac{1}{4}y[k-1] + \tfrac{1}{16}y[k-2] = f[k] \tag{3.3a}$$

Equation (3.3a) can also be expressed in an alternative form by realizing that this equation is valid for any value of k. Therefore, replacing k by $k+2$, we obtain

$$y[k+2] + \tfrac{1}{4}y[k+1] + \tfrac{1}{16}y[k] = f[k+2] \tag{3.3b}$$

This is the alternative form of Eq. (3.3a).

For a realization of a system with this input-output equation, we rewrite Eq. (3.3a) as

$$y[k] = -\tfrac{1}{4}y[k-1] - \tfrac{1}{16}y[k-2] + f[k] \tag{3.3c}$$

Figure 3.4 shows a hardware realization of Eq. (3.3c) using two time delays (here the time delay T is a semester). To understand this realization, assume that $y[k]$ is available. Then, by delaying it successively, we generate $y[k-1]$ and $y[k-2]$. Next we generate $y[k]$ from $f[k]$, $y[k-1]$, and $y[k-2]$ according to Eq. (3.3c). ■

Equations (3.2) and (3.3) are examples of difference equations; the former is a first-order and the latter is a second-order difference equation. Difference equations also arise in numerical solution of differential equations.

■ **Example 3.3: Digital Differentiator**
Design a discrete-time system, such as in Fig. 3.2, to differentiate continuous-time signals.

In this case, the output $y(t)$ is required to be the derivative of the input $f(t)$. The discrete-time processor (system) G processes the samples of $f(t)$ to produce the discrete-time output $y[k]$. If $f[k]$ and $y[k]$ represent the samples T seconds apart of the signals $f(t)$ and $y(t)$ respectively—that is,

$$f[k] = f(kT) \qquad \text{and} \qquad y[k] = y(kT) \tag{3.4}$$

The signals $f[k]$ and $y[k]$ are the input and output for the discrete-time system G. Now, we require that

$$y(t) = \frac{df}{dt} \tag{3.5}$$

Therefore at $t = kT$ (see Fig. 3.5a)

$$y(kT) = \frac{df}{dt}\bigg|_{t=kT}$$

$$= \lim_{T \to 0} \frac{1}{T}\left[f(kT) - f[(k-1)T]\right]$$

By using the notation in (3.4), the above equation can be expressed as

$$y[k] = \lim_{T \to 0} \frac{1}{T}\left\{f[k] - f[k-1]\right\}$$

This is the input-output relationship for G required to achieve our objective. In practice the sampling interval T cannot be zero. Assuming T to be sufficiently small, the above equation can be expressed as

$$y[k] \simeq \frac{1}{T}\left\{f[k] - f[k-1]\right\} \tag{3.6}$$

The approximation improves as T approaches 0. A discrete-time processor G to realize Eq. (3.6) is shown inside the box in Fig. 3.5b. The system in Fig. 3.5b acts as a differentiator.

Observe that we have achieved the operation of differentiation of a continuous-time signal entirely by operating on its samples. The discrete-time system G that performs this operation consists of a time delay, a summer and a scalar multiplier. We shall see in Chapter 5 that these are all the elements we need to build a general discrete-time system.

To gain some insight into this method of signal processing, let us consider the differentiator in Fig. 3.5b with a ramp input $f(t) = t$, shown in Fig. 3.5c. If the system were to act as a differentiator, then the output $y(t)$ of the system should be the unit step function $u(t)$. Let us investigate how the system performs this particular operation and how well it achieves the objective.

The samples of the input $f(t) = t$ at the interval of T seconds act as the input to the discrete-time system G. These samples, denoted by a compact notation $f[k]$, are therefore

$$f[k] = f(t)|_{t=kT} = t|_{t=kT} \qquad t \geq 0$$

$$= kT \qquad k \geq 0$$

Figure 3.5d shows the sampled signal $f[k]$. This signal acts as an input to the discrete-time system G. Figure 3.5b shows that the operation of G consists of subtracting a sample from the previous (delayed) sample and then multiplying the difference with $1/T$. From Fig. 3.5d, it is clear that the difference between the successive samples is a constant $kT - (k-1)T = T$ for all samples. The output of G is $1/T$ times the difference T, which is unity for all values of k. Therefore the output $y[k]$ of G consists of samples of unit values for $k \geq 1$, as shown in Fig. 3.5e. The D/C (discrete-time to continuous-time) converter converts these samples into a continuous-time signal $y(t)$, as shown in Fig. 3.5f. Ideally the output should have been $y(t) = u(t)$. This deviation from the ideal is caused by the fact that we have used a nonzero sampling interval T. As T approaches zero, the output $y(t)$ approaches the desired output $u(t)$. ■

These examples show that the basic elements required in the realization of discrete-time systems are time delays, scalar multipliers, and adders (summers).

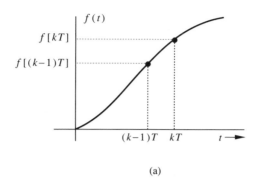

(a)

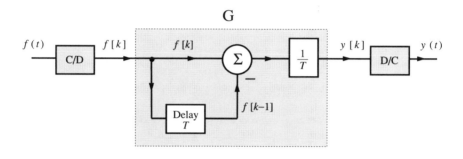

(b)

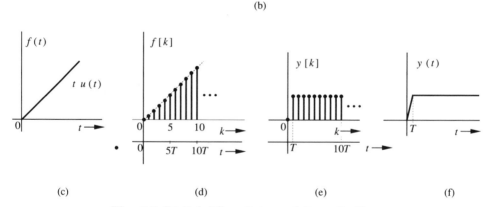

(c)   (d)   (e)   (f)

Fig. 3.5 Digital differentiator and its realization.

We show in Chapter 5 that this is generally true of discrete-time systems. The discrete-time systems can be realized in two ways:

1. By using digital computers which readily perform the operations of adding, multiplying, and delaying. Minicomputers and microprocessors are well suited for this purpose, especially for signals with frequencies below 100 kHz.

2. By using special-purpose time-delay devices that have been developed in the last two decades. These include monolithic MOS charge-transfer devices (CTD) such as charge-coupled devices (CCD) and bucket brigade devices (BBD), which are implemented on silicon substrate as integrated circuit elements. In addition, there are surface acoustic wave (SAW) devices built on piezoelectric

substrates. Systems using these devices are less expensive but are not as reliable or as accurate as the digital systems. Digital systems are preferable for signals below 100 kHz. Systems using CTD are suitable and competitive with those using SAW devices in the frequency range 1 kHz to 20 MHz. At frequencies higher than 20 MHz, SAW devices are preferred and are the only realistic choice for frequencies higher than 50 MHz. Systems using SAW devices with frequencies in the range of 10 MHz to 1 GHz are implemented routinely.[1]

Digital Filters

There is a basic difference between continuous-time systems and analog systems. The same is true of discrete-time and digital systems, explained fully in Secs. 1.2-6 and 1.2-7.† For historical reasons, digital computers (rather than time-delay elements, such as CCD or SAW devices) were used in the realization of early discrete-time systems. Because of this, the terms *digital filters* and *discrete-time systems* are used synonymously in the literature. This distinction is irrelevant in the analysis of discrete-time systems. For this reason, in this book, the term *digital filters* implies *discrete-time systems*, and *analog filters* means *continuous-time systems*. Moreover, the terms C/D (continuous-to-discrete-time) and D/C will be used interchangeably with terms A/D (analog-to-digital) and D/A respectively.

Advantages of Digital Signal Processing

1. Digital filters have a greater degree of precision and stability.
2. Digital filters are more flexible. Their characteristics can be easily altered simply by changing the program.
3. A greater variety of filters can be realized by digital systems.
4. Very low frequency filters, if realized by continuous-time systems, require prohibitively bulky components. Such is not the case with digital filters.
5. Digital signals can be stored easily on magnetic tapes or disks without deterioration of signal quality.
6. More sophisticated signal processing algorithms can be used to process digital signals.
7. Digital filters can be time-shared and therefore can serve a number of inputs simultaneously.
8. Because of these advantages, the overall economic cost of digital systems is comparable (or lower in many cases) to the corresponding analog systems.

3.2 DISCRETE-TIME SYSTEM EQUATIONS

Difference Equations

Equations (3.2), (3.3), and (3.6) are examples of difference equations. Equations (3.2) and (3.6) are first-order difference equations, and Eq. (3.3) is a second-order difference equation. All these equations are linear, with constant (not time-varying) coefficients. Before giving a general form of nth-order difference equation,

---

†The terms *discrete-time* and *continuous-time* qualify the nature of a signal along the time axis (horizontal axis). The terms *analog* and *digital*, on the other hand, qualify the nature of the signal amplitude (vertical axis).

we recall that a difference equation can be written in two forms; the first form uses delay terms such as $y[k-1]$, $y[k-2]$, $f[k-1]$, $f[k-2]$, ..., etc., and the alternate form uses advance terms such as $y[k+1]$, $y[k+2]$, ..., etc. Both forms are useful. We start here with a general nth-order difference equation, using advance operator form:

$$y[k+n] + a_{n-1}y[k+n-1] + \cdots + a_1 y[k+1] + a_0 y[k] =$$
$$b_m f[k+m] + b_{m-1}f[k+m-1] + \cdots + b_1 f[k+1] + b_0 f[k] \qquad (3.7)$$

Observe that the coefficient of $y[k+n]$ can be taken to be equal to unity without loss of generality. If this coefficient is other than unity, we can divide the equation throughout by the value of the coefficient of $y[k+n]$ to normalize the equation to the form in (3.7).

Causality Condition

The left-hand side of Eq. (3.7) consists of the output at instants $k+n$, $k+n-1$, $k+n-2$, and so on. The right-hand side of Eq. (3.7) consists of the input at instants $k+m$, $k+m-1$, $k+m-2$, and so on. For a causal system the output cannot depend on future input values. This shows that when the system equation is in the advance operator form (3.7), the causality requires $m \le n$. For a general causal case, $m = n$, and Eq. (3.7) can be expressed as

$$y[k+n] + a_{n-1}y[k+n-1] + \cdots + a_1 y[k+1] + a_0 y[k] =$$
$$b_n f[k+n] + b_{n-1}f[k+n-1] + \cdots + b_1 f[k+1] + b_0 f[k] \qquad (3.8a)$$

where some of the coefficients on both sides can be zero. However, the coefficient of $y[k+n]$ is normalized to unity. Equation (3.8a) is valid for all values of k. Therefore the equation is still valid if we replace k by $k-n$ throughout the equation [see Eqs. (3.2a) and (3.2b)]. This yields the alternative form (the delay operator form) of Eq. (3.8a):

$$y[k] + a_{n-1}y[k-1] + \cdots + a_1 y[k-n+1] + a_0 y[k-n] =$$
$$b_n f[k] + b_{n-1}f[k-1] + \cdots + b_1 f[k-n+1] + b_0 f[k-n] \qquad (3.8b)$$

We shall designate Form (3.8a) the **advance operator form**, and Form (3.8b) the **delay operator form**.

3.2-1   Initial Conditions and Iterative Solution of Difference Equations

Equation (3.8b) can be expressed as

$$y[k] = -a_{n-1}y[k-1] - a_{n-2}y[k-2] - \cdots - a_0 y[k-n]$$
$$+ b_n f[k] + b_{n-1}f[k-1] + \cdots + b_0 f[k-n] \qquad (3.8c)$$

This equation shows that $y[k]$, the output at the kth instant, is computed from $2n+1$ pieces of information. These are the past n values of the output: $y[k-1]$, $y[k-2]$, ..., $y[k-n]$ and the present and past n values of the input: $f[k]$, $f[k-1]$, $f[k-2]$, ..., $f[k-n]$. If the input $f[k]$ is given for $k = 0, 1, 2, \ldots$, then the output $y[k]$ for $k = 0, 1, 2, \ldots$ can be computed from the $2n$ initial conditions $y[-1], y[-2], \ldots, y[-n]$ and $f[-1], f[-2], \ldots, f[-n]$. If the input is causal, then $f[-1] = f[-2] = \ldots = f[-n] = 0$, and we need only n initial conditions $y[-1], y[-2], \ldots, y[-n]$. This

allows us to compute iteratively or recursively the output $y[0]$, $y[1]$, $y[2]$, $y[3]$, ...,
and so on.† For instance, to find $y[0]$ we set $k = 0$ in Eq. (3.8c). The left-hand
side is $y[0]$, and the right-hand side contains terms $y[-1]$, $y[-2]$, ..., $y[-n]$, and
the inputs $f[0]$, $f[-1]$, $f[-2]$, ..., $f[-n]$. Therefore, to begin with, we must know
the n initial conditions $y[-1]$, $y[-2]$, ..., $y[-n]$. Knowing these conditions and the
input $f[k]$, we can iteratively find the response $y[0]$, $y[1]$, $y[2]$, ..., and so on. The
following examples demonstrate this procedure. This method basically reflects the
manner in which a computer would solve a difference equation, given the input and
initial conditions.

■ **Example 3.4**
Solve iteratively

$$y[k] - 0.5y[k-1] = f[k] \tag{3.9a}$$

with initial condition $y[-1] = 16$ and causal input $f[k] = k^2$ (starting at $k = 0$). This
equation can be expressed as

$$y[k] = 0.5y[k-1] + f[k] \tag{3.9b}$$

If we set $k = 0$ in this equation, we obtain

$$y[0] = 0.5y[-1] + f[0]$$
$$= 0.5(16) + 0 = 8$$

Now, setting $k = 1$ in Eq. (3.9b) and using the value $y[0] = 8$ (computed in the first step)
and $f[1] = (1)^2 = 1$, we obtain

$$y[1] = 0.5(8) + (1)^2 = 5$$

Next, setting $k = 2$ in Eq. (3.9b) and using the value $y[1] = 5$ (computed in the previous
step) and $f[2] = (2)^2$, we obtain

$$y[2] = 0.5(5) + (2)^2 = 6.5$$

Continuing in this way iteratively, we obtain

$$y[3] = 0.5(6.5) + (3)^2 = 12.25$$
$$y[4] = 0.5(12.25) + (4)^2 = 22.125$$

....

The output $y[k]$ is shown in Fig. 3.6.   ■

Note carefully the iterative or recursive nature of the computations. From the
n initial conditions (and input) we found $y[0]$ first. Then, using this value of $y[0]$

---

†For this reason Eq. (3.8) is called a **recursive difference equation**. However in Eq. (3.8)
if $a_0 = a_1 = a_2 = \cdots = a_{n-1} = 0$, then it follows from Eq. (3.8c) that determination of the
present output $y[k]$ does not require the past values $y[k-1], y[k-2], \ldots$, etc. For this reason
when $a_i = 0$, $(i = 0, 1, \ldots, n-1)$, the difference Eq. (3.8) is **nonrecursive**. This classification is
important in designing and realizing digital filters. In this chapter, however, this classification is
not important. The analysis techniques developed here (and in Chap. 5) apply to general recursive
and nonrecursive systems. Observe that a nonrecursive system is a special case of recursive system
with $a_0 = a_1 = \ldots = a_{n-1} = 0$.

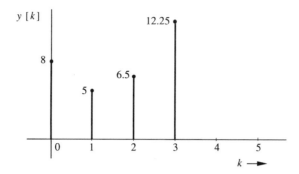

Fig. 3.6 Iterative solution of a difference equation in Example 3.4.

and the previous $n-1$ initial conditions (along with the input), we found $y[1]$. Next, using $y[0]$, $y[1]$ along with the initial conditions and input, we found $y[2]$, and so on. This method is general and can be applied to a difference equation of any order. It is interesting to note that the hardware realization of Eq. (3.9a) shown in Fig. 3.3 (with $\gamma = 0.5$) generates the solution precisely in this fashion.

We now present one more example of iterative solution—this time for a second-order equation. Iterative method can be applied to a difference equation in delay or advance operator form. In Example 3.4 we considered the former. Let us now apply it to the advance operator form.

■ **Example 3.5**
Solve iteratively

$$y[k+2] - y[k+1] + 0.24y[k] = f[k+2] - 2f[k+1] \tag{3.10}$$

with initial conditions $y[-1] = 2$, $y[-2] = 1$ and a causal input $f[k] = k$ (starting at $k = 0$). The system equation can be expressed as

$$y[k+2] = y[k+1] - 0.24y[k] + f[k+2] - 2f[k+1] \tag{3.11}$$

Setting $k = -2$ and then substituting $y[-1] = 2$, $y[-2] = 1$, $f[0] = f[-1] = 0$ (recall that $f[k] = k$ starting at $k = 0$), we obtain

$$y[0] = 2 - 0.24(1) + 0 - 0 = 1.76$$

Setting $k = -1$ in Eq. (3.11) and then substituting $y[0] = 1.76$, $y[-1] = 2$, $f[1] = 1$, $f[0] = 0$, we obtain

$$y[1] = 1.76 - 0.24(2) + 1 - 0 = 2.28$$

Setting $k = 0$ in Eq. (3.11) and then substituting $y[0] = 1.76$, $y[1] = 2.28$, $f[2] = 2$ and $f[1] = 1$ yields

$$y[2] = 2.28 - 0.24(1.76) + 2 - 2(1) = 1.8576$$

and so on. ■

⊙ **Computer Example C3.1**
Do Example 3.5 using a computer.
The system equation can be expressed as

$$y[k+2] = y[k+1] - 0.24y[k] + f[k+2] - 2f[k+1]$$

```
Y = [1 2];      % Initial condition vector for the output.
f0=[0 0 0];     % Initial condition vector for the input.
f=[f0 1:12];    % Input f[k] = k, for k > 0.
    for k=1:12
        y=Y(k+1)-.24*Y(k)+f(k+2)-2*f(k+1);
        Y=[Y y];
    end
k=0:1:11;
plot(k,Y(3:14),'o'),grid,
pause
Y(3:14)'    ⊙
```

We shall see in the future that the solution of a difference equation obtained in this direct (iterative) way is useful in many situations. This procedure is available only for discrete-time systems; it cannot be applied to continuous-time systems. Despite the many uses of this method, a closed-form solution of a difference equation is far more useful in study of the system behavior and its dependence on the input and the various system parameters. For this reason we shall develop a systematic procedure to analyze discrete-time systems along lines similar to those used for continuous-time systems.

△ **Exercise E3.1**
Using the iterative method, find the first three terms of $y[k]$ if
$$y[k+1] - 2y[k] = f[k]$$
The initial condition is $y[-1] = 10$ and the input $f[k] = 2$ starting at $k = 0$.
Answer: $y[0] = 20, y[1] = 42,$ and $y[2] = 86$.   ▽

Operational Notation
    In difference equations it is convenient to use operational notation similar to that used in differential equations for the sake of compactness. In continuous-time systems we used the operator D to denote the operation of differentiation. For discrete-time systems we shall use the operator E to denote the operation for advancing the sequence by one time interval. Thus

$$Ef[k] \equiv f[k+1]$$
$$E^2 f[k] \equiv f[k+2]$$
$$\cdots\cdots\cdots\cdots$$
$$E^n f[k] \equiv f[k+n] \tag{3.12}$$

    The first-order difference equation of the savings account problem was found to be [see Eq. (3.2b)]:

$$y[k+1] - \gamma y[k] = f[k+1] \tag{3.13}$$

Using the operational notation, this equation can be expressed as

$$Ey[k] - \gamma y[k] = Ef[k]$$

or

$$(E - \gamma)y[k] = Ef[k] \qquad (3.14)$$

The second-order difference equation (3.3b)

$$y[k + 2] + \tfrac{1}{4}y[k + 1] + \tfrac{1}{16}y[k] = f[k + 2]$$

can be expressed in operational notation as

$$\left(E^2 + \tfrac{1}{4}E + \tfrac{1}{16}\right) y[k] = E^2 f[k]$$

A general nth-order difference Eq. (3.8a) can be expressed as

$$(E^n + a_{n-1}E^{n-1} + \cdots + a_1 E + a_0)y[k] =$$
$$(b_n E^n + b_{n-1}E^{n-1} + \cdots + b_1 E + b_0)f[k] \qquad (3.15a)$$

or

$$Q[E]y[k] = P[E]f[k] \qquad (3.15b)$$

where $Q[E]$ and $P[E]$ are nth-order polynomial operators

$$Q[E] = E^n + a_{n-1}E^{n-1} + \cdots + a_1 E + a_0 \qquad (3.16)$$
$$P[E] = b_n E^n + b_{n-1}E^{n-1} + \cdots + b_1 E + b_0 \qquad (3.17)$$

Response of Linear Discrete-Time Systems

Following the procedure used for continuous-time systems on p. 106, we can show that Eq. (3.15) is a linear equation (with constant coefficients). A system described by such an equation is a linear time-invariant discrete-time (LTID) system. We can verify, as in the case of LTIC systems (see footnote on p. 107), that the general solution of Eq. (3.15) consists of zero-input and zero-state components.

3.3 SYSTEM RESPONSE TO INTERNAL CONDITIONS: ZERO-INPUT RESPONSE

The zero-input response $y_0[k]$ is the solution of Eq. (3.15) with $f[k] = 0$; that is,

$$Q[E]y_0[k] = 0 \qquad (3.18a)$$

or

$$(E^n + a_{n-1}E^{n-1} + \cdots + a_1 E + a_0)y_0[k] = 0 \qquad (3.18b)$$

Let us start with the first-order case

$$(E - \gamma)y_0[k] = 0 \qquad (3.19a)$$

that is,

$$y_0[k+1] - \gamma y_0[k] = 0 \qquad (3.19b)$$

or

$$y_0[k+1] = \gamma y_0[k] \qquad (3.19c)$$

Therefore

$$\frac{y_0[k+1]}{y_0[k]} = \gamma$$

This means that the ratio of $y_0[k+1]$ to $y_0[k]$ is γ. Clearly, the sequence $y_0[k]$ is a geometrical progression with a common ratio γ. Thus if $y_0[0] = c$, $y_0[1] = c\gamma$, $y_0[2] = c\gamma^2$, $y_0[3] = c\gamma^3$, and in general

$$y_0[k] = c\gamma^k \qquad (3.20)$$

where c is an arbitrary constant to be determined from an additional constraint such as an initial (or auxiliary) condition. Consider now the zero-input solution of the nth-order Eq. (3.15)

$$(E^n + a_{n-1}E^{n-1} + \cdots + a_1 E + a_0)y_0[k] = 0 \qquad (3.21)$$

As in the case of the first-order system, let us try a solution

$$y_0[k] = c\gamma^k \qquad (3.22)$$

Then

$$Ey_0[k] = y_0[k+1] = c\gamma^{k+1}$$

$$E^2 y_0[k] = y_0[k+2] = c\gamma^{k+2}$$

$$\cdots\cdots\cdots\cdots\cdots\cdots\cdots\cdots\cdots\cdots\cdots$$

$$E^n y_0[k] = y_0[k+n] = c\gamma^{k+n}$$

Substitution of this in Eq. (3.21) yields

$$c(\gamma^n + a_{n-1}\gamma^{n-1} + \cdots + a_1\gamma + a_0)\gamma^k = 0$$

For a nontrivial solution of this equation

$$(\gamma^n + a_{n-1}\gamma^{n-1} + \cdots + a_1\gamma + a_0) = 0 \qquad (3.23a)$$

or

$$Q[\gamma] = 0 \qquad (3.23b)$$

Our solution $c\gamma^k$ [Eq. (3.23)] is correct, provided that γ satisfies Eq. (3.23). Equation (3.21) is an nth-order polynomial and can be expressed in the factorized form (assuming all distinct roots):

$$(\gamma - \gamma_1)(\gamma - \gamma_2) \cdots (\gamma - \gamma_n) = 0 \qquad (3.23c)$$

Clearly γ has n solutions $\gamma_1, \gamma_2, \cdots, \gamma_n$ and therefore Eq. (3.18) also has n solutions $c_1\gamma_1^k, c_2\gamma_2^k, \cdots, c_n\gamma_n^k$. In such a case we have shown (see footnote on p. 109) that the general solution is a linear combination of the n solutions. Thus

$$y_0[k] = c_1 \gamma_1^k + c_2 \gamma_2^k + \cdots + c_n \gamma_n^k \qquad (3.24)$$

where $\gamma_1, \gamma_2, \cdots, \gamma_n$ are the roots of Eq. (3.23) and $c_1, c_2, \ldots, c_n$ are arbitrary constants determined from n auxiliary conditions, generally given in the form of initial conditions. The polynomial $Q[\gamma]$ is called the **characteristic polynomial** of the system, and

$$Q[\gamma] = 0$$

is the **characteristic equation** of the system. Moreover, $\gamma_1, \gamma_2, \cdots, \gamma_n$, the roots of the characteristic equation, are called **characteristic roots** or **characteristic values** (also **eigenvalues**) of the system. The exponentials $\gamma_i^k (i = 1, 2, \ldots, n)$ are the **characteristic modes** or **natural modes** of the system. A characteristic mode corresponds to each characteristic root of the system, and the **zero-input response is a linear combination of the characteristic modes of the system.**

Repeated Roots

In the discussion so far we have assumed the system to have n distinct characteristic roots $\gamma_1, \gamma_2, \ldots, \gamma_n$ with corresponding characteristic modes $\gamma_1^k, \gamma_2^k, \ldots, \gamma_n^k$. If two or more roots coincide (repeated roots), the form of characteristic modes is modified. It can be shown by direct substitution that if a root γ repeats r times (root of multiplicity r), the characteristic modes corresponding to this root are $\gamma^k, k\gamma^k, k^2\gamma^k, \ldots, k^{r-1}\gamma^k$. Thus if the characteristic equation of a system is

$$Q[\gamma] = (\gamma - \gamma_1)^r (\gamma - \gamma_{r+1})(\gamma - \gamma_{r+2}) \cdots (\gamma - \gamma_n) \qquad (3.25)$$

the zero-input response of the system is

$$y_0[k] = (c_1 + c_2 k + c_3 k^2 + \cdots + c_r k^{r-1})\gamma_1^k + c_{r+1}\gamma_{r+1}^k + c_{r+2}\gamma_{r+2}^k + \cdots + c_n \gamma_n^k \quad (3.26)$$

The Nature of Discrete-Time Exponentials

Until now the development of discrete-time systems has been parallel to that of the continuous-time systems, with one minor difference. In continuous-time systems the characteristic modes were exponentials with a natural base (e.g., $e^{\lambda_i t}$), whereas for discrete-time systems, the characteristic modes are also exponential but their base is not necessarily natural (e.g., γ_i^k).† Observe that γ^k is also an exponential as is e^k; the only difference is in the base. In fact, γ^k is a more general form and e^k is a special case with $\gamma = e$. Because of unfamiliarity with exponentials with bases other than e, the characteristic modes of the form γ_i^k may seem inconvenient and confusing. The reader is urged to plot some exponentials to acquire a sense of these functions. Figure 3.7 shows $(0.8)^k, (-0.8)^k, (0.5)^k, (1.2)^k, k(0.8)^k$, and $(0.9)^k \cos\left(\frac{\pi}{6}k + \theta\right)$. Observe that γ^k is a decaying exponential if $|\gamma| < 1$, and γ^k is a growing exponential if $|\gamma| > 1$. A signal $(-\gamma)^k$ changes sign successively for real γ. Also, a signal $|\gamma|^k \cos(\beta k + \theta)$ is a sinusoid with exponentially varying amplitude. Its envelope is $|\gamma|^k$. There are on average $2\pi/\beta$ samples per period of the sinusoid.

---

†We could express γ_i^k as $e^{\lambda_i k}$, where $\lambda_i = \ln \gamma_i$. Thus, we could define $\lambda_1, \lambda_2, \cdots, \lambda_n$ as the characteristic roots with the corresponding characteristic modes $e^{\lambda_1 k}, e^{\lambda_2 k}, \cdots, e^{\lambda_n k}$.

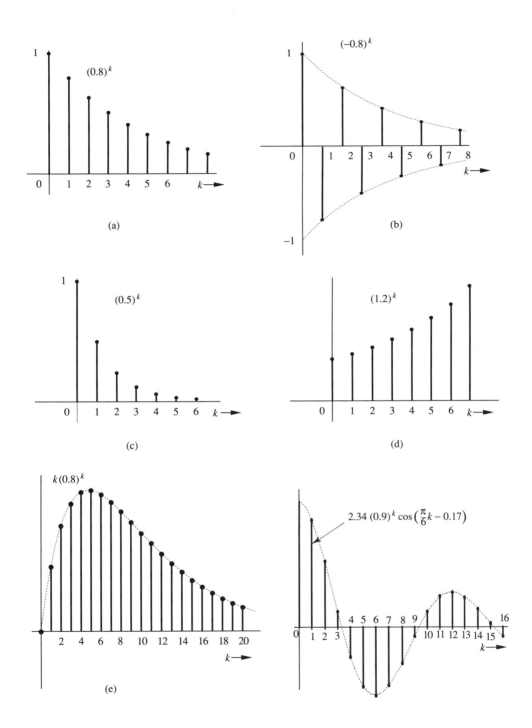

Fig. 3.7 Nature of discrete-time exponentials.

It is convenient to define a unit step function $u[k]$, shown in Fig. 3.8, as

$$u[k] = \begin{cases} 1 & \text{for } k \geq 0 \\ 0 & \text{for } k < 0 \end{cases} \tag{3.27}$$

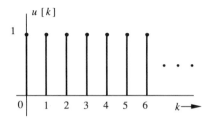

Fig. 3.8 A discrete-time unit step function $u[k]$.

<h3>■ Example 3.6</h3>

For an LTID system described by the difference equation

$$y[k+2] - 0.6y[k+1] - 0.16y[k] = 5f[k+2] \tag{3.28}$$

find the total response if the initial conditions are $y[-1] = 0$ and $y[-2] = \frac{25}{4}$, and if the input $f[k] = 4^{-k}u[k]$. In this example we shall determine the zero-input component $y_0[k]$ only. The zero-state component is determined in Example 3.13.

The system equation in operational notation is

$$(E^2 - 0.6E - 0.16)y[k] = 5E^2 f[k] \tag{3.29}$$

The characteristic polynomial is

$$\gamma^2 - 0.6\gamma - 0.16 = (\gamma + 0.2)(\gamma - 0.8)$$

The characteristic equation is

$$(\gamma + 0.2)(\gamma - 0.8) = 0 \tag{3.30}$$

The characteristic roots are $\gamma_1 = -0.2$ and $\gamma_2 = 0.8$. The zero-input response is

$$y_0[k] = c_1(-0.2)^k + c_2(0.8)^k \tag{3.31}$$

The initial conditions $y[-1]$ and $y[-2]$ are the conditions given on the total response, but the input does not start until $k = 0$. Therefore the zero-state response is zero for $k < 0$. Hence at $k = -1$ and -2 the total response consists of only the zero-input component so that $y[-1] = y_0[-1]$ and $y[-2] = y_0[-2]$. To determine arbitrary constants c_1 and c_2, we set $k = -1$ and -2 in Eq. (3.31), then substitute $y_0[-1] = 0$, and $y_0[-2] = \frac{25}{4}$ to obtain

$$\left. \begin{array}{l} 0 = -5c_1 + \frac{5}{4}c_2 \\ \frac{25}{4} = 25c_1 + \frac{25}{16}c_2 \end{array} \right\} \implies \begin{array}{l} c_1 = \frac{1}{5} \\ c_2 = \frac{4}{5} \end{array}$$

Therefore

$$y_0[k] = \frac{1}{5}(-0.2)^k + \frac{4}{5}(0.8)^k \tag{3.32}$$

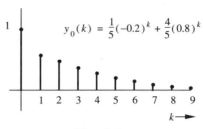

$$y_0(k) = \frac{1}{5}(-0.2)^k + \frac{4}{5}(0.8)^k$$

Fig. 3.9

The reader can verify this solution by computing the first few terms using the iterative method (see Examples 3.4 and 3.5). Figure 3.9 shows the plot of $y_0[k]$. ∎

Example 3.7

For an LTID system specified by difference equation

$$(E^2 + 6E + 9)y[k] = (2E^2 + 6E)f[k]$$

find the zero-input response $y_0[k]$ if the initial conditions are $y_0[-1] = -\frac{1}{3}$ and $y_0[-2] = -\frac{2}{9}$. The characteristic equation of the system is

$$\gamma^2 + 6\gamma + 9 = (\gamma + 3)^2 = 0$$

Here the root $\gamma = -3$ repeats twice. Hence the zero-input response is

$$y_0[k] = (c_1 + c_2 k)(-3)^k$$

To determine arbitrary constants c_1 and c_2, we set $k = -1$ and -2 in this equation, then substitute $y_0[-1] = -\frac{1}{3}$, $y_0[-2] = -\frac{2}{9}$ to obtain

$$\left. \begin{array}{l} -\frac{1}{3} = -\frac{1}{3}(c_1 - c_2) \\ -\frac{2}{9} = \frac{1}{9}(c_1 - 2c_2) \end{array} \right\} \implies \begin{array}{l} c_1 = 4 \\ c_2 = 3 \end{array}$$

Therefore

$$y_0[k] = (4 + 3k)(-3)^k$$

The reader can verify this solution by computing the first few terms using the iterative method (see Examples 3.4 and 3.5). ∎

Complex Roots

As in the case of continuous-time systems, the complex roots of a discrete-time system must occur in pairs of conjugates so that the system response is a real function of time. Complex roots can be treated exactly as we would treat real roots. However, just as in the case of continuous-time systems, we can eliminate dealing with complex numbers by using the real form of solution. This is clarified in the following example.

Example 3.8

For an LTID system specified by the difference equation

$$(E^2 - 1.56E + 0.81)y[k] = (E + 3)f[k] \qquad (3.33)$$

find the zero-input response $y_0[k]$ if the initial conditions are $y[-1] = 2$ and $y[-2] = 1$. The characteristic equation is

$$(\gamma^2 - 1.56\gamma + 0.81) = (\gamma - 0.78 - j0.45)(\gamma - 0.78 + j0.45) = 0$$

There are two complex (conjugate) roots: $0.78 \pm j0.45$; that is,

$$\gamma_1 = 0.78 + j0.45 = 0.9e^{j\frac{\pi}{6}}$$
$$\gamma_2 = 0.78 - j0.45 = 0.9e^{-j\frac{\pi}{6}}$$

The zero-input response is

$$y_0[k] = c_1(0.78 - j0.45)^k + c_2(0.78 + j0.45)^k \qquad (3.34)$$

As before, using the two initial conditions, we can determine the constants c_1 and c_2, which are complex (conjugate) in general. This yields the answer for $y_0[k]$ in the complex form, which can be converted to real form, requiring tedious and error-prone manipulation of complex numbers. All of this can be avoided by starting with the real form of solution, as explained below.

First we express the complex conjugate roots γ and γ^* in polar form. If $|\gamma|$ is the magnitude and β is the angle of γ, then

$$\gamma = |\gamma|e^{j\beta} \qquad \text{and} \qquad \gamma^* = |\gamma|e^{-j\beta}$$

The zero-input response is given by

$$y_0[k] = c_1\gamma^k + c_2(\gamma^*)^k$$
$$= c_1|\gamma|^k e^{j\beta k} + c_2|\gamma|^k e^{-j\beta k}$$

For a real system, c_1 and c_2 must be conjugates so that $y_0[k]$ is a real function of k. Let

$$c_1 = \frac{c}{2}e^{j\theta} \qquad \text{and} \qquad c_2 = \frac{c}{2}e^{-j\theta}$$

Then

$$y_0[k] = \frac{c}{2}|\gamma|^k \left[e^{j(\beta k + \theta)} + e^{-j(\beta k + \theta)} \right]$$
$$= c|\gamma|^k \cos(\beta k + \theta)$$

where c and θ are arbitrary constants determined from the auxiliary conditions. This is the solution in real form, which avoids dealing with complex numbers.

In the present case, the roots are $0.9e^{\pm j\frac{\pi}{6}}$ ($|\gamma| = 0.9$ and $\beta = \frac{\pi}{6}$). Therefore

$$y_0[k] = c(0.9)^k \cos\left(\frac{\pi}{6}k + \theta\right)$$

Setting $k = -1$ and -2 in this equation and substituting the initial conditions $y_0[-1] = 2$, $y_0[-2] = 1$, we obtain

$$2 = \frac{c}{0.9}\cos\left(-\frac{\pi}{6} + \theta\right) = \frac{c}{0.9}\left[\frac{\sqrt{3}}{2}\cos\theta + \frac{1}{2}\sin\theta\right]$$
$$1 = \frac{c}{(0.9)^2}\cos\left(-\frac{\pi}{3} + \theta\right) = \frac{c}{0.81}\left[\frac{1}{2}\cos\theta + \frac{\sqrt{3}}{2}\sin\theta\right]$$

or

$$\frac{\sqrt{3}}{1.8}c\cos\theta + \frac{1}{1.8}c\sin\theta = 2$$

$$\tfrac{1}{1.62} c \cos \theta + \tfrac{\sqrt{3}}{1.62} c \sin \theta = 1$$

These are two simultaneous equations in two unknowns $c \cos \theta$ and $c \sin \theta$. Solution of these equations yields

$$c \cos \theta = 2.308$$

$$c \sin \theta = -0.397$$

Dividing $c \sin \theta$ by $c \cos \theta$ yields

$$\tan \theta = \frac{-0.397}{2.308} = \frac{-0.172}{1}$$

$$\theta = \tan^{-1}(-0.172) = -0.17 \text{ rad.}$$

Substituting $\theta = -0.17$ radian in $c \cos \theta = 2.308$ yields $c = 2.34$ and

$$y_0[k] = 2.34(0.9)^k \cos\left(\tfrac{\pi}{6}k - 0.17\right) \qquad k \geq 0$$

Observe that here we have used radian unit for both β and θ. We also could have used the degree unit, although it is not recommended. The important thing is to be consistent and to use the same units for both β and θ. Figure 3.10 shows the plot of $y_0[k]$. ■

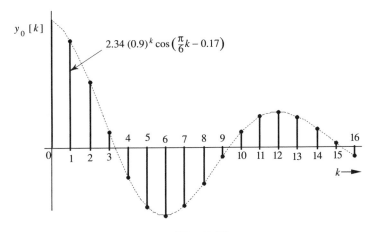

Fig. 3.10

⊙ **Computer Example C3.2**
An LTID system is described by the following difference equation:

$$y[k+2] - 0.6y[k+1] - 0.16y[k] = 5f[k+2]$$

Find the zero-input response if the initial conditions are $y[-1] = 0, y[-2] = \frac{25}{4}$.

```
Y = [25/4 0];           % Y initial conditions.
f = [0 0 0 ];           % Input initial conditions.
char = [1 -0.6 -0.16];  % Enter characteristic eqn.
r = roots(char)         % Find roots of CE.
    for k=1:12
    y = 0.6*Y(k+1)+0.16*Y(k);
```

```
      Y = [Y y];
   end
k=0:1:11;
plot(k,Y(3:14),'o'),grid, % Plot the response to Y initial conditions.
xlabel('k'),ylabel('y[k]'),   % Label x and y axes.
pause
Y(3:14)'   ⊙
```

△ **Exercise E3.2**
Find and sketch the zero-input response for the systems described by the following equations:
(a) $y[k+1] - 0.8y[k] = 3f[k+1]$ (b) $y[k+1] + 0.8y[k] = 3f[k+1]$.
In each case the initial condition is $y[-1] = 10$. Verify the solutions by computing the first three terms using iterative method.
Answer: (a) $8(0.8)^k$ (b) $-8(-0.8)^k$.

See Fig. 3.7 for the plots of $(0.8)^k$ and $(-0.8)^k$.   ▽

△ **Exercise E3.3**
Find the zero-input response of a system described by the equation

$$y[k] + 0.3y[k-1] - 0.1y[k-2] = f[k] + 2f[k-1]$$

The initial conditions are $y_0[-1] = 1$ and $y_0[-2] = 33$. Verify the solution by computing the first three terms iteratively.
Hint: First express the equation in the advanced operator form (3.8a).
Answer: $y_0[k] = (0.2)^k + 2(-0.5)^k$ ▽

△ **Exercise E3.4**
Find the zero-input response of a system described by the equation

$$y[k] + 4y[k-2] = 2f[k]$$

The auxiliary conditions are $y_0[0] = -\frac{1}{\sqrt{2}}$ and $y_0[1] = \sqrt{2}$. Verify the solution by computing the first three terms iteratively.
Answer: $y_0[k] = (2)^k \cos\left(\frac{\pi}{2}k - \frac{3\pi}{4}\right)$ ▽

3.4 UNIT IMPULSE RESPONSE h[k]

The discrete-time counterpart of continuous-time impulse function $\delta(t)$ is $\delta[k]$ defined by

$$\delta[k] = \begin{cases} 1 & k = 0 \\ 0 & k \neq 0 \end{cases} \qquad (3.35)$$

This function, also called unit impulse sequence, is shown in Fig. 3.11a. The time-shifted impulse sequence $\delta[k-m]$ is shown in Fig. 3.11b. Unlike its continuous-time counterpart $\delta(t)$, this is a very simple function without any mystery.

Later we shall express an arbitrary input $f[k]$ in terms of impulse components. The (zero-state) system response to input $f[k]$ is found as the sum of system responses to impulse components of $f[k]$. For this reason we need to find $h[k]$, the system response to input $\delta[k]$ when the system is in zero state (initially relaxed system). Recall that when the system is in zero state, all outputs are zero in the absence of input(s). Iterative solution of Eq. (3.8c) shows that when the input

(a)

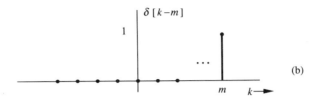

(b)

Fig. 3.11 Discrete-time impulse function.

$f[k] = 0$, the output $y[k]$ is zero if $y[-1] = y[-2] = \ldots = y[-n] = 0$. Therefore zero state implies initial conditions $y[-1] = y[-2] = \ldots = y[-n] = 0$.

3.4-1 Determining Impulse Response h[k].

Consider an nth-order system specified by the equation

$$(E^n + a_{n-1}E^{n-1} + \cdots + a_1 E + a_0)y[k] =$$
$$(b_n E^n + b_{n-1}E^{n-1} + \cdots + b_1 E + b_0)f[k] \qquad (3.36a)$$

or

$$Q[E]y[k] = P[E]f[k] \qquad (3.36b)$$

The impulse response $h[k]$ is the solution of this equation for the input $\delta[k]$ and when all the initial conditions are zero; that is,

$$Q[E]h[k] = P[E]\delta[k] \qquad (3.37)$$

subject to initial conditions

$$h[-1] = h[-2] = \cdots = h[-n] = 0 \qquad (3.38)$$

Equation (3.37) can be solved to determine $h[k]$ iteratively or in a closed form. The following example demonstrates the iterative solution.

■**Example 3.9: Iterative Determination of h[k]**

Find $h[k]$, the unit impulse response of a system described by the equation

$$y[k] - 0.6y[k-1] - 0.16y[k-2] = 5f[k] \qquad (3.39)$$

To determine impulse response, we must let the input $f[k] = \delta[k]$ and the output $y[k] = h[k]$ in the above equation. The resulting equation is

$$h[k] - 0.6h[k-1] - 0.16h[k-2] = 5\delta[k] \qquad (3.40)$$

subject to zero initial state; that is, $h[-1] = h[-2] = 0$.

Setting $k = 0$ in this equation yields

$$h[0] - 0.6(0) - 0.16(0) = 5(1) \implies h[0] = 5$$

Next, setting $k = 1$ in eq. (3.40) and using $h[0] = 5$, we obtain

$$h[1] - 0.6(5) - 0.16(0) = 5(0) \implies h[1] = 3 \quad \blacksquare$$

Continuing this way, we can determine any number of terms of $h[k]$. Unfortunately, such a solution does not yield a closed-form expression for $h[k]$. Nevertheless, determining a few values of $h[k]$ can be useful in determining the closed-form solution, as the following development shows.

The Closed-Form Solution of h[k]

Recall that $h[k]$ is the system response to input $\delta[k]$, which is zero for $k > 0$. Therefore $h[k]$ must have a form of zero-input response $y_0[k]$, consisting of characteristic modes for $k > 0$. At $k = 0$, it may have some nonzero value A_0, so that $h[k]$ can be expressed as‡

$$h[k] = A_0 \delta[k] + y_0[k] \tag{3.41}$$

In Appendix 3.1 we show that

$$A_0 = \frac{b_0}{a_0} \tag{3.42}$$

Moreover, because $h[k]$ is causal, we must multiply $y_0[k]$ by $u[k]$. Therefore

$$h[k] = \frac{b_0}{a_0} \delta[k] + y_0[k] u[k] \tag{3.43}$$

The zero-input response $y_0[k]$ on the right-hand side contains n unknown coefficients, which can be determined from a knowledge of n values of $h[k]$. Fortunately, it is a straightforward task to determine values of $h[k]$ iteratively, as demonstrated in Example 3.9. We compute n values $h[0], h[1], h[2], \cdots, h[n-1]$ iteratively. Now, setting $k = 0, 1, 2, \cdots, n-1$ in Eq. (3.43), we can determine the n unknowns in $y_0[k]$. This will become clear from the following examples.

■**Example 3.10**

Determine the unit impulse response $h[k]$ for a system in Example 3.9 specified by the equation

$$y[k] - 0.6y[k-1] - 0.16y[k-2] = 5f[k]$$

This equation can be expressed in the advance operator form as

$$y[k+2] - 0.6y[k+1] - 0.16y[k] = 5f[k+2] \tag{3.44}$$

---

‡We showed that $h[k]$ consists of characteristic modes only for $k > 0$. Hence the characteristic mode terms in $h[k]$ start at $k = 1$. To reflect this behavior, they should be expressed in the form $\gamma_j^k u[k-1]$. But because $u[k-1] = u[k] - \delta[k]$, $A_j \gamma_j^k u[k-1] = A_j \gamma_j^k u[k] - A_j \delta[k]$, and $h[k]$ can be expressed in terms of exponentials $\gamma_j^k u[k]$ (which start at $k = 0$), plus an impulse at $k = 0$.

or

$$(E^2 - 0.6E - 0.16)y[k] = 5E^2 f[k] \qquad (3.45)$$

The characteristic polynomial is

$$\gamma^2 - 0.6\gamma - 0.16 = (\gamma + 0.2)(\gamma - 0.8)$$

The characteristic modes are $(-0.2)^k$ and $(0.8)^k$. Therefore the zero-input response of the system is

$$y_0[k] = c_1(-0.2)^k + c_2(0.8)^k \qquad (3.46)$$

Also, from Eq. (3.45), we have $a_0 = -0.16$ and $b_0 = 0$. Therefore from Eq. (3.43)

$$h[k] = [c_1(-0.2)^k + c_2(0.8)^k]u[k] \qquad (3.47)$$

To determine c_1 and c_2, we need to find two values of $h[k]$ iteratively. This is already done in Example 3.9, where we found that $h[0] = 5$ and $h[1] = 3$. Now, setting $k = 0$ and 1 in Eq. (3.47) and using the fact that $h[0] = 5$ and $h[1] = 3$, we obtain

$$\left.\begin{array}{l} 5 = c_1 + c_2 \\ 3 = -0.2c_1 + 0.8c_2 \end{array}\right\} \implies \begin{array}{l} c_1 = 1 \\ c_2 = 4 \end{array}$$

Therefore

$$h[k] = \left[(-0.2)^k + 4(0.8)^k\right]u[k] \qquad (3.48)$$

■

■ **Example 3.11**

Find the unit impulse response $h[k]$ of a system described by the equation

$$(E^2 - 6E + 9)y[k] = (E + 18)f[k] \qquad (3.49)$$

The characteristic polynomial is $\gamma^2 - 6\gamma + 9 = (\gamma - 3)^2$. This is a case of repeated characteristic roots (root at 3 repeating twice) so that the zero-input response is $y_0[k] = (c_1 + c_2 k)3^k$. Moreover, $b_0 = 18$ and $a_0 = 9$. Therefore according to Eq. (3.43),

$$h[k] = 2\delta[k] + (c_1 + c_2 k)3^k u[k] \qquad (3.50)$$

To determine c_1 and c_2, we find two values of $h[k]$ from the iterative solution of the system equation (3.49) with $f[k] = \delta[k]$ and $y[k] = h[k]$; that is,

$$h[k + 2] - 6h[k + 1] + 9h[k] = \delta[k + 1] + 18\delta[k] \qquad (3.51)$$

subject to zero initial state; that is, $h[-1] = h[-2] = 0$. Setting $k = -2$ and $k = -1$ in succession yields

$$h[0] - 6(0) + 9(0) = 0 + 18(0) \implies h[0] = 0$$

$$h[1] - 6(0) + 9(0) = 1 + 18(0) \implies h[1] = 1 \qquad (3.52)$$

Now, setting $k = 0$ and 1 in Eq. (3.50) and using the fact that $h[0] = 0$ and $h[1] = 1$, we obtain

$$\left. \begin{array}{l} 0 = 2 + c_1 \\ 1 = (c_1 + c_2)3 \end{array} \right\} \implies \begin{array}{l} c_1 = -2 \\ c_2 = \frac{7}{3} \end{array}$$

Therefore

$$h[k] = 2\delta[k] - (2 - \tfrac{7}{3}k)(3)^k u[k] \qquad (3.53)$$

■

Comment

For a special case where $a_0 = 0$, A_0 cannot be determined from Eq. (3.42). It is shown in Appendix 3.1 that in such a case the impulse response contains an additional impulse at $k = 1$, and $h[k]$ can be expressed as

$$h[k] = A_0\delta[k] + A_1\delta[k-1] + y_0[k]u[k]$$

Now $h[k]$ contains $n + 2$ unknowns (A_0, A_1, and n coefficients of $y_0[k]$), which can be determined from $n + 2$ values of $h[k]$ found iteratively. See Appendix 3.1 for development of this special case with a worked example.

Although determination of the impulse response $h[k]$ using the procedure in this section is relatively simple, in Chapter 5 we shall discuss another much simpler method of z-transform.

⊙ **Computer Example C3.3**
Find the unit impulse response $h[k]$ of a system described by the equation

$$(E^2 - 6E + 9)y[k] = (E + 18)f[k]$$

```
Y = [0 0];        % Initial condition vector for the output.
f0= [0 0];        % Initial condition vector for the input.
f = [f0 1 0 0 0 0 0 0 0 0 0];  % Unit impulse;
char = [1 -6 9];               % Enter characteristic eqn.
r = roots(char)                % Find roots of CE.
    for k=1:12
        y = 6*Y(k+1)-9*Y(k)+f(k+1)+18*f(k);
        Y = [Y y];
    end
k=0:1:11;
plot(k,Y(3:14),'o'),grid,
pause,
Y(3:14)'    ⊙
```

△ **Exercise E3.5**
Find $h[k]$, the unit impulse response of the LTID systems specified by the following equations:
(a) $y[k+2] - 5y[k+1] + 6y[k] = 8f[k+1] - 19f[k]$
(b) $y[k] - 1.6y[k-1] + 0.8y[k-2] = f[k] - f[k-1]$
(c) $y[k+2] - 4y[k+1] + 4y[k] = 2f[k+2] - 2f[k+1]$

Answers:

(a) $h[k] = -\frac{19}{6}\delta[k] + [\frac{3}{2}(2)^k + \frac{5}{3}(3)^k]u[k]$

(b) $h[k] = \frac{\sqrt{5}}{2}(\frac{2}{\sqrt{5}})^k \cos(0.464k + 0.464)u[k]$ ▽

(c) $h[k] = (2 + k)2^k u[k]$

3.5   SYSTEM RESPONSE TO EXTERNAL INPUT: ZERO-STATE RESPONSE

Zero-state response is the system response $y[k]$ to an input $f[k]$ when the system is in zero state. In this section we shall assume that systems are in zero state unless mentioned otherwise, so that the zero-state response will be the total response of the system. Here we follow the procedure parallel to that used in the continuous-time case by expressing an arbitrary input $f[k]$ as a sum of impulse components. Figure 3.12 shows how a signal $f[k]$ in Fig. 3.12a can be expressed as a sum of impulse components shown in Figs. 3.12 b, c, d, e, f, etc. The component of $f[k]$ at $k = m$ is $f[m]\delta[k - m]$, and $f[k]$ is the sum of all these components summed from $m = -\infty$ to ∞. Therefore

$$f[k] = f[0]\delta[k] + f[1]\delta[k - 1] + f[2]\delta[k - 2] + \cdots$$
$$+ f[-1]\delta[k + 1] + f[-2]\delta[k + 2] + \cdots$$
$$= \sum_{m=-\infty}^{\infty} f[m]\delta[k - m] \qquad (3.54)$$

If we knew the system response to impulse $\delta[k]$, the system response to any arbitrary input could be found by summing the system response to various impulse components. Let $h[k]$ be the system response to impulse input $\delta[k]$. We shall use the notation

$$f[k] \rightarrow y[k]$$

to indicate the input and the corresponding response of the system. Thus if

$$\delta[k] \rightarrow h[k]$$

then because of time invariance

$$\delta[k - m] \rightarrow h[k - m]$$

and because of linearity

$$f[m]\delta[k - m] \rightarrow f[m]h[k - m]$$

and again because of linearity

$$\sum_{m=-\infty}^{\infty} f[m]\delta[k - m] \rightarrow \sum_{m=-\infty}^{\infty} f[m]h[k - m]$$

The left hand-side is $f[k]$ [see Eq. (3.54)], and the right-hand side is the system response $y[k]$ to input $f[k]$. Therefore

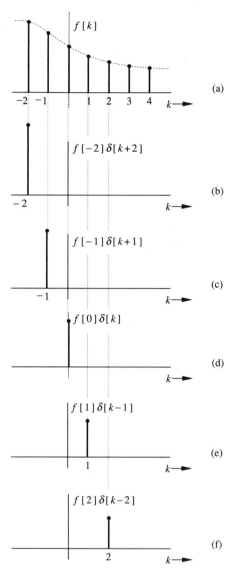

Fig. 3.12 Representation of an arbitrary signal $f[k]$ in terms of impulse components.

$$y[k] = \sum_{m=-\infty}^{\infty} f[m]h[k-m] \tag{3.55}$$

The summation on the right-hand side is known as the **convolution sum** of $f[k]$ and $h[k]$, and is represented symbolically by $f[k] * h[k]$

$$f[k] * h[k] = \sum_{m=-\infty}^{\infty} f[m]h[k-m] \tag{3.56}$$

Properties of the Convolution Sum

The structure of the convolution sum is similar to that of the convolution integral. Moreover, the properties of the convolution sum are similar to those of the convolution integral. We shall enumerate these properties here without proof. The proofs are similar to those for the convolution integral and may be derived by the reader.

1. **The Commutative Property**

$$f_1[k] * f_2[k] = f_2[k] * f_1[k] \tag{3.57}$$

2. **The Distributive Property**

$$f_1[k] * (f_2[k] + f_3[k]) = f_1[k] * f_2[k] + f_1[k] * f_3[k] \tag{3.58}$$

3. **The Associative Property**

$$f_1[k] * (f_2[k] * f_3[k]) = (f_1[k] * f_2[k]) * f_3[k] \tag{3.59}$$

4. **The Shifting Property**
   If

   then
$$f_1[k] * f_2[k] = c[k]$$

$$f_1[k - m] * f_2[k - n] = c[k - m - n] \tag{3.60}$$

5. **The Convolution with an Impulse**

$$f[k] * \delta[k] = f[k] \tag{3.61}$$

6. **The Width Property**

   If $f_1[k]$ and $f_2[k]$ have lengths of m and n elements respectively, then the length of $c[k]$ is $m + n - 1$ elements†.

Causality and Zero-State Response

In deriving Eq. (3.56) we assumed the system to be linear and time-invariant. There were no other restrictions on either the input signal or the system. In practice almost all of the input signals are causal, and a majority of the systems are also causal. These restrictions further simplify the limits of summation in Eq. (3.55). If the input $f[k]$ is causal, $f[m] = 0$ for $m < 0$. Similarly, if the system is causal (that is, if $h[k]$ is causal), then $h[x] = 0$ for negative x, so that $h[k - m] = 0$ when $m > k$. Therefore if $f[k]$ and $h[k]$ are both causal, the product $f[m]h[k - m] = 0$ for $m < 0$ and for $m > k$, and it is nonzero only for the range $0 \le m \le k$. Therefore Eq. (3.55) in this case reduces to

$$y[k] = \sum_{m=0}^{k} f[m]h[k - m] \tag{3.62}$$

We shall evaluate the convolution sum first by an analytical method and later with graphical aid.

---

†The width of a signal is one less than the number of its elements (length). For instance, the signal in Fig. 3.13h has 6 elements, but has a width of only 5. Thus if $f_1[k]$ and $f_2[k]$ have widths of W_1 and W_2, respectively, then the width of $c[k]$ is $W_1 + W_2$.

∎ **Example 3.12**

Determine $c[k] = f[k] * g[k]$ for

$$f[k] = (0.8)^k u[k] \qquad \text{and} \qquad g[k] = (0.3)^k u[k]$$

We have

$$c[k] = \sum_{m=-\infty}^{\infty} f[m]g[k-m] \tag{3.63}$$

Note that

$$f[m] = (0.8)^m u[m] \qquad \text{and} \qquad g[k-m] = (0.3)^{k-m}u[k-m] \tag{3.64}$$

Both $f[k]$ and $g[k]$ are causal. Therefore [see Eq. (3.62)]

$$c[k] = \sum_{m=0}^{k} f[m]g[k-m]$$

$$= \sum_{m=0}^{k} (0.8)^m u[m] \, (0.3)^{k-m} u[k-m] \tag{3.65}$$

In the above summation, m lies between 0 and k ($0 \le m \le k$). Therefore if $k \ge 0$, then $m \ge 0$ and $k - m \ge 0$, so that $u[m] = u[k-m] = 1$. If $k < 0$, m is negative because m lies between 0 and k, and $u[m] = 0$. Therefore Eq. (3.65) becomes

$$c[k] = \sum_{m=0}^{k} (0.8)^m \, (0.3)^{k-m} \qquad k \ge 0$$

$$= 0 \qquad\qquad\qquad k < 0$$

and

$$c[k] = (0.3)^k \sum_{m=0}^{k} \left(\frac{0.8}{0.3} \right)^m u[k]$$

This is a geometric progression with common ratio $(0.8/0.3)$. From Sec. B.10-4 we have

$$c[k] = (0.3)^k \frac{(0.8)^{k+1} - (0.3)^{k+1}}{(0.3)^k(0.8 - 0.3)} u[k]$$

$$= 2 \left[(0.8)^{k+1} - (0.3)^{k+1} \right] u[k] \tag{3.66}$$

∎

△ **Exercise E3.6**

Show that $(0.8)^k u[k] * u[k] = 5[1 - (0.8)^{k+1}]u[k]$ ▽

Convolution Sum from a Table

Just as in the continuous-time case, we have prepared a table of convolution sums (Table 3.1) from which convolution sums may be found directly for a variety of signal pairs. For example the convolution in Example 3.12 can be read directly from this table (Pair 4) as

$$(0.8)^k u[k] * (0.3)^k u[k] = \frac{(0.8)^{k+1} - (0.3)^{k+1}}{0.8 - 0.3} u[k] = 2[(0.8)^{k+1} - (0.3)^{k+1}]u[k]$$

TABLE 3.1: Convolution Sums

No.	$f_1[k]$	$f_2[k]$	$f_1[k] * f_2[k] = f_2[k] * f_1[k]$
1	$\delta[k-j]$	$f[k]$	$f[k-j]$
2	$\gamma^k u[k]$	$u[k]$	$\left[\dfrac{1 - \gamma^{k+1}}{1 - \gamma}\right] u[k]$
3	$u[k]$	$u[k]$	$(k+1)u[k]$
4	$\gamma_1^k u[k]$	$\gamma_2^k u[k]$	$\left[\dfrac{\gamma_1^{k+1} - \gamma_2^{k+1}}{\gamma_1 - \gamma_2}\right] u[k] \quad \gamma_1 \neq \gamma_2$
5	$\gamma_1^k u[k]$	$\gamma_2^k u[-(k+1)]$	$\dfrac{\gamma_1}{\gamma_2 - \gamma_1}\gamma_1^k u[k] + \dfrac{\gamma_2}{\gamma_2 - \gamma_1}\gamma_2^k u[-(k+1)] \qquad |\gamma_2| > |\gamma_1|$
6	$k\gamma_1^k u[k]$	$\gamma_2^k u[k]$	$\dfrac{\gamma_1\gamma_2}{(\gamma_1 - \gamma_2)^2}\left[\gamma_2^k - \gamma_1^k + \dfrac{\gamma_1 - \gamma_2}{\gamma_2}k\gamma_1^k\right] u[k] \qquad \gamma_1 \neq \gamma_2$
7	$ku[k]$	$ku[k]$	$\dfrac{1}{6}k(k-1)(k+1)u[k]$
8	$\gamma^k u[k]$	$\gamma^k u[k]$	$(k+1)\gamma^k u[k]$
9	$\gamma^k u[k]$	$ku[k]$	$\left[\dfrac{\gamma(\gamma^k - 1) + k(1 - \gamma)}{(1 - \gamma)^2}\right] u[k]$
10	$|\gamma_1|^k \cos(\beta k + \theta)u[k]$	$\gamma_2^k u[k]$	$\dfrac{1}{R}\left[|\gamma_1|^{k+1}\cos(\beta(k+1) + \theta - \phi) - \gamma_2^{k+1}\cos(\theta - \phi)\right] u[k] \qquad \gamma_2 \text{ real}$

$$R = \left[|\gamma_1|^2 + \gamma_2^2 - 2|\gamma_1|\gamma_2\cos\beta\right]^{1/2}$$

$$\phi = \tan^{-1}\left[\frac{(|\gamma_1|\sin\beta)}{(|\gamma_1|\cos\beta - \gamma_2)}\right]$$

We shall demonstrate the use of the convolution table in the following examples.

■ **Example 3.13**

Find the (zero-state) response $y[k]$ of an LTID system described by the equation

$$y[k+2] - 0.6y[k+1] - 0.16y[k] = 5f[k+2]$$

if the input $f[k] = 4^{-k}u[k]$.

The unit impulse response of this system is found in Example 3.10.

$$h[k] = [(-0.2)^k + 4(0.8)^k]u[k] \tag{3.67}$$

Therefore

$$
\begin{aligned}
y[k] &= f[k] * h[k] \\
&= (4)^{-k}u[k] * \left[(-0.2)^k u[k] + 4(0.8)^k u[k]\right] \\
&= (4)^{-k}u[k] * (-0.2)^k u[k] + (4)^{-k}u[k] * 4(0.8)^k u[k]
\end{aligned} \tag{3.68}
$$

Note that

$$(4)^{-k}u[k] = \left(\tfrac{1}{4}\right)^k u[k] = (0.25)^k u[k]$$

Therefore

$$y[k] = (0.25)^k u[k] * (-0.2)^k u[k] + 4(0.25)^k u[k] * (0.8)^k u[k]$$

We use Pair 4 (Table 3.1) to find the above convolution sums.

$$
\begin{aligned}
y[k] &= \left[\frac{(0.25)^{k+1} - (-0.2)^{k+1}}{0.25 - (-0.2)} + 4\frac{(0.25)^{k+1} - (0.8)^{k+1}}{0.25 - 0.8}\right]u[k] \\
&= \left(2.22\left[(0.25)^{k+1} - (-0.2)^{k+1}\right] - 7.27\left[(0.25)^{k+1} - (0.8)^{k+1}\right]\right)u[k] \\
&= \left[-5.05(0.25)^{k+1} - 2.22(-0.2)^{k+1} + 7.27(0.8)^{k+1}\right]u[k]
\end{aligned}
$$

Recognizing that

$$\gamma^{k+1} = \gamma(\gamma)^k$$

We can express $y[k]$ as

$$
\begin{aligned}
y[k] &= \left[-1.26(0.25)^k + 0.444(-0.2)^k + 5.81(0.8)^k\right]u[k] \\
&= \left[-1.26(4)^{-k} + 0.444(-0.2)^k + 5.81(0.8)^k\right]u[k]
\end{aligned} \tag{3.69}
$$

■

⊙ **Computer Example C3.4**

An LTID system is described by the following difference equation:

$$y[k+2] + 6y[k+1] + 9y[k] = 2f[k+2] + 6f[k+1]$$

Find the system response $y[k]$ if all the initial conditions are zero (system is in zero state) and the input is $f[k] = 4^{-k}u[k]$.

```
Y = [0 0];              % Initial condition vector for the output.
f = [0 0 4^(-0)];       % Initial condition vector for the input.
char=[1 6 9];           % Enter characteristic eqn.
r = roots(char)         % Find roots of CE.
disp('strike any key to see the plot')
pause
clc                     % clear command window
    for k=1:12
        f=[f 4^(-k)];   % Input f[k] = 4^{-k}u[k].
        y = -6*Y(k+1)-9*Y(k)+2*f(k+2)+6*f(k+1);
        Y = [Y y];
    end
k=0:1:11;
plot(k,Y(3:14),'o'),grid,xlabel('k'),ylabel('y[k]'),
text(0.5,-3*1e5,'Strike any key to see more...')
pause
Y(3:14)'    ⊙
```

Total Response

The total response of an LTID can be expressed as a sum of the zero-input and zero-state components:

$$\text{Total response} = \underbrace{\sum_{j=1}^{n} c_j \gamma_j^k}_{\text{Zero-input component}} + \underbrace{f[k] * h[k]}_{\text{Zero-state component}}$$

We have developed procedures to determine these two components. From the system equation we find the characteristic roots and characteristic modes. The zero-input response is a linear combination of the characteristic modes. From the system equation we also determine $h[k]$, the impulse response, as discussed in Sec. 3.4-1. Knowing $h[k]$ and the input $f[k]$, we find the zero-state response as the convolution of $f[k]$ and $h[k]$. The arbitrary constants $c_1, c_2, \ldots, c_n$ in the zero-input response are determined from the n auxiliary conditions. For the system described by the equation

$$y[k+2] - 0.6y[k+1] - 0.16y[k] = 5f[k+2]$$

with initial conditions $y[-1] = 0, y[-2] = \frac{25}{4}$ and input $f[k] = (4)^{-k}u[k]$, we have found the two components of the response in Examples 3.6 and 3.13 respectively . From the results in these examples, the total response for $k \geq 0$ is

$$\text{Total response} = \underbrace{0.2(-0.2)^k + 0.8(0.8)^k}_{\text{Zero-input component}} \underbrace{-1.26(4)^{-k} + 0.444(-0.2)^k + 5.81(0.8)^k}_{\text{Zero-state component}}$$

$$(3.70a)$$

Natural and Forced Response

The characteristic modes of this system are $(-0.2)^k$ and $(0.8)^k$. The zero-input component is made up of characteristic modes exclusively as expected, but the characteristic modes also appear in the zero-state response. When all the characteristic mode terms in the total response are lumped together, the resulting component is the **natural response**. The remaining part of the total response that is made up of noncharacteristic modes is the **forced response**. For the present case, from Eq. (3.70a), we have

$$\text{Total response} = \underbrace{0.644(-0.2)^k + 6.61(0.8)^k}_{\text{Natural response}} - \underbrace{1.26(4)^{-k}}_{\text{Forced response}} \qquad k \geq 0 \qquad (3.70b)$$

⊙ **Computer Example C3.5**
An LTID system is described by the following difference equation:

$$y[k+2] - 1.56y[k+1] + 0.81y[k] = f[k+1] + 3f[k]$$

Find the (total) response if the initial conditions are $y[-1] = 0, y[-2] = \frac{25}{4}$ and the input is $f[k] = 4^{(-k)}u[k]$.

```
Y = [25/4 0];          % Initial condition vector for the output.
f = [0 0 4^(-0)];      % Initial condition vector for the input.
char = [1 -1.56 0.81]; %  Enter characteristic eqn.
r = roots(char)        % Find roots of CE.
disp('Strike any key to see more...')
pause
   for k=1:12
      f = [f 4^(-k)];      % Input f[k] = 4^(-k)u[k].
      y = 1.56*Y(k+1)-0.81*Y(k)+f(k+1)+3*f(k);
      Y = [Y y];
   end
k=0:1:11;
plot(k,Y(3:14),'o'),grid,xlabel('k'),ylabel('y[k]'),
text(4,-6,'Strike any key')
pause
Y(3:14)'    ⊙
```

△ **Exercise E3.7**
Show that $(0.8)^{k+1}u[k] * u[k] = 4[1 - 0.8(0.8)^k]u[k]$
Use convolution table. Recognize that $(0.8)^{k+1} = 0.8(0.8)^k$ ▽

△ **Exercise E3.8**
Show that $k3^{-k}u[k] * (0.2)^k u[k] = \frac{15}{4}[(0.2)^k - (1 - \frac{2}{3}k)3^{-k}]u[k]$
Hint: Use convolution table. Recognize that $3^{-k} = (\frac{1}{3})^k$. ▽

△ **Exercise E3.9**
Using convolution table, show that $e^{-k}u[k] * 2^{-k}u[k] = \frac{2}{2-e}[e^{-k} - (\frac{e}{2})2^{-k}]u[k]$
Hint: $e^{-k} = (\frac{1}{e})^k$ and $2^{-k} = (0.5)^k$. ▽

3.5-1 Graphical Procedure for the Convolution Sum

The steps in evaluating the convolution sum are parallel to those followed in evaluating the convolution integral. The convolution sum of causal signals $f[k]$ and $g[k]$ is given by

$$c[k] = \sum_{m=0}^{k} f[m]g[k - m]$$

We first plot $f[m]$ and $g[k - m]$ as functions of m (not k), because the summation is over m. Functions $f[m]$ and $g[m]$ are the same as $f[k]$ and $g[k]$, plotted respectively as functions of m (see Fig. 3.13). The convolution operation can be performed as follows:

1. Invert $g[m]$ about the vertical axis $(m = 0)$ to obtain $g[-m]$ (Fig. 3.13d). Figure 3.13e shows both $f[m]$ and $g[-m]$.

2. Time-shift $g[-m]$ by k units to obtain $g[k - m]$. For $k > 0$, the shift is to the right (delay); for $k < 0$, the shift is to the left (advance). Figures 3.13f and 3.13g show $g[k - m]$ for $k > 0$ and for $k < 0$, respectively.

3. Next we multiply $f[m]$ and $g[k - m]$ and add all the products to obtain $c[k]$. The procedure is repeated for each value of k over the range $-\infty$ to ∞.

We shall demonstrate by an example the graphical procedure for finding the convolution sum. Although both the functions in this example are causal, the procedure is applicable to the general case.

■ **Example 3.14**
Find

$$c[k] = f[k] * g[k]$$

where $f[k]$ and $g[k]$ are shown in Figs. 3.13a and 3.13b, respectively.
We are given

$$f[k] = (0.8)^k \text{ and } g[k] = (0.3)^k$$

Therefore

$$f[m] = (0.8)^m \quad \text{and} \quad g[k - m] = (0.3)^{k-m}$$

Figure 3.13f shows the general situation for $k \geq 0$. The two functions $f[m]$ and $g[k - m]$ overlap over the interval $0 \leq m \leq k$. Therefore

$$c[k] = \sum_{m=0}^{k} f[m]g[k - m]$$

$$= \sum_{m=0}^{k} (0.8)^m (0.3)^{k-m}$$

$$= (0.3)^k \sum_{m=0}^{k} \left(\frac{0.8}{0.3}\right)^m$$

$$= 2 \left[(0.8)^{k+1} - (0.3)^{k+1}\right] u[k] \qquad \text{(see Sec. B.10-4)}$$

For $k < 0$, there is no overlap between $f[m]$ and $g[k - m]$, as shown in Fig. 3.13g so that

$$c[k] = 0 \qquad k < 0$$

and

$$c[k] = 2 \left[(0.8)^{k+1} - (0.3)^{k+1}\right] u[k]$$

which agrees with the earlier result in Eq. (3.66). ■

△ **Exercise E3.10**
Find $(0.8)^k u[k] * u[k]$ graphically and sketch the result.
Answer: $5(1 - (0.8)^{k+1})u[k]$ ▽

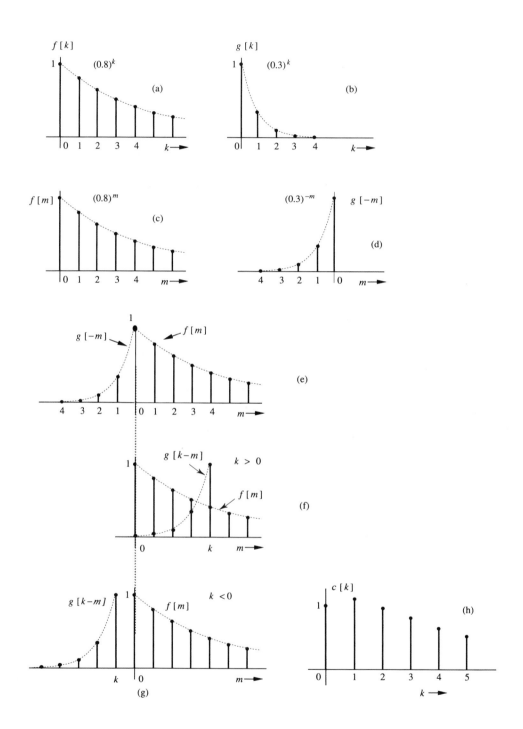

Fig. 3.13 Graphical understanding of convolution of $f[k]$ and $g[k]$ for Example 3.14.

An Alternative Form of Graphical Procedure: The Sliding Tape Method

The expression for the convolution sum is identical to that in Eq. (2.80) used to compute convolution integral numerically. Therefore we can use the sliding-tape algorithm discussed in Sec. 2.5, especially when the sequences $f[k]$ and $g[k]$ are short or when they are available only in graphical form. This algorithm is basically the same as the graphical procedure in Fig. 3.13. The only difference is that instead of presenting the data as graphical plots, we display it as sequence of numbers on tapes. Otherwise the procedure is the same, as will become clear from the following example.

■ **Example 3.15**

Using the sliding tape method, convolve the two sequences $f[k]$ and $g[k]$ shown in Fig. 3.14a and 3.14b, respectively. In this procedure we write the sequences $f[k]$ and $g[k]$ in the slots of two tapes: f tape and g tape (Fig. 3.14c). Now leave the f tape stationary (to correspond to $f[m]$). The $g[-m]$ tape is obtained by time-inverting the $g[m]$ tape about the origin ($k = 0$) so that the slots corresponding to $f[0]$ and $g[0]$ remain aligned (Fig. 3.14d). We now shift the inverted tape by k slots, multiply values on two tapes in adjacent slots, and add all the products to find $c[k]$. Figures 3.14d, e, f, g, h, i and j show the cases for $k = 0, 1, 2, 3, 4, 5$, and 6, respectively. For the case of $k = 0$, for example (Fig. 3.14d)

$$c[0] = 0 \times 1 = 0$$

For $k = 1$ (Fig. 3.14e)

$$c[1] = (0 \times 1) + (1 \times 1) = 1$$

Similarly,

$$c[2] = (0 \times 1) + (1 \times 1) + (2 \times 1) = 3$$
$$c[3] = (0 \times 1) + (1 \times 1) + (2 \times 1) + (3 \times 1) = 6$$
$$c[4] = (0 \times 1) + (1 \times 1) + (2 \times 1) + (3 \times 1) + (4 \times 1) = 10$$
$$c[5] = (0 \times 1) + (1 \times 1) + (2 \times 1) + (3 \times 1) + (4 \times 1) + (5 \times 1) = 15$$
$$c[6] = (0 \times 1) + (1 \times 1) + (2 \times 1) + (3 \times 1) + (4 \times 1) + (5 \times 1) = 15$$

Figure 3.14j shows that $c[k] = 15$ for $k \geq 5$. Moreover, the two tapes are nonoverlapping for $k < 0$, so that $c[k] = 0$ for $k < 0$. Figure 3.14k shows the plot of $c[k]$ versus k.

■

An Array Form of Graphical Procedure

The convolution sum can also be found from the array formed by sequences $f[k]$ and $g[k]$. This procedure, although convenient from a computational viewpoint, fails to give proper understanding of the convolution mechanism. It is explained in Prob. 3.5-16.

△ **Exercise E3.11**

Using the graphical procedure of Example 3.15 (sliding-tape technique), show that $f[k] * g[k] = c[k]$ in Fig. 3.15. Verify the width property of convolution.   ▽

3.6 SYSTEM STABILITY

Just as in a continuous-time system, we define a discrete-time system to be **asymptotically stable** if, and only if, the zero-input response approaches zero as

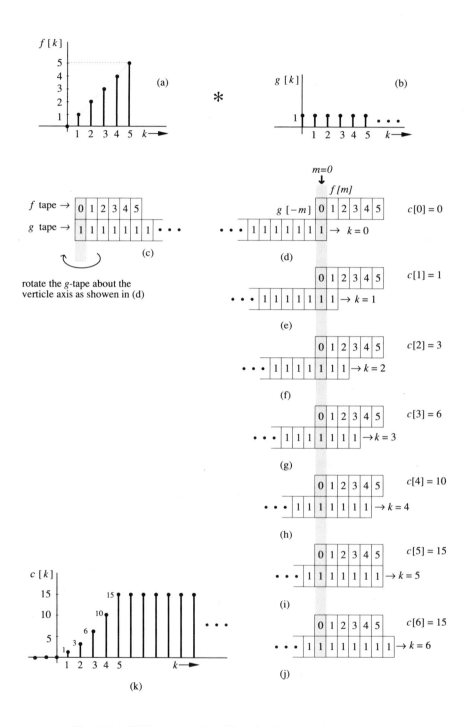

Fig. 3.14 Sliding tape algorithm for discrete-time convolution.

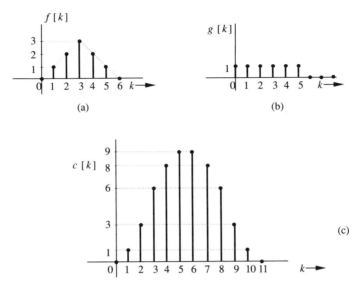

Fig. 3.15 Signals for Exercise E3.11.

$k \to \infty$. If the zero-input response grows without bound as $k \to \infty$, the system is **unstable**. If the zero-input response neither approaches zero nor grows without bound, but remains within a finite limit as $k \to \infty$, the system is **marginally stable**. In the last case, the zero-input response approaches a constant or oscillates with a constant amplitude. Recall that the zero-input response consists of characteristic modes of the system. The mode corresponding to a characteristic root γ is γ^k. To be more general, let γ be complex so that

$$\gamma = |\gamma|e^{j\beta} \qquad \text{and} \qquad \gamma^k = |\gamma|^k e^{j\beta k}$$

Since the magnitude of $e^{j\beta k}$ is always unity regardless of the value of k, the magnitude of γ^k is $|\gamma|^k$. Therefore

$$\text{if } |\gamma| < 1, \quad \gamma^k \to 0 \qquad \text{as } k \to \infty$$

$$\text{if } |\gamma| > 1, \quad \gamma^k \to \infty \qquad \text{as } k \to \infty$$

$$\text{and if } |\gamma| = 1, \quad |\gamma|^k = 1 \qquad \text{for all } k$$

It is clear that a system is asymptotically stable if and only if

$$|\gamma_i| < 1 \qquad i = 1, 2, \cdots, n \tag{3.71}$$

These results can be grasped more effectively in terms of the location of characteristic roots in the complex plane. Figure 3.16 shows a circle of unit radius, centered at the origin in a complex plane . From our discussion it is clear that if all characteristic roots of the system lie inside this circle (**unit circle**), $|\gamma_i| < 1$ for all i and the system is asymptotically stable. On the other hand, even if one characteristic root lies outside the unit circle, the system is unstable. If none of the characteristic roots lie outside the unit circle, but some simple (unrepeated) roots

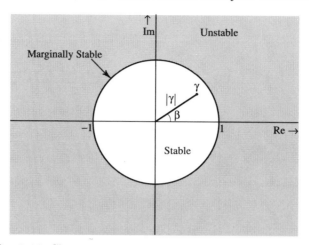

Fig. 3.16 Characteristic roots location and system stability.

lie on the circle itself, the system is marginally stable. If two or more characteristic roots coincide on the unit circle (repeated roots), the system is unstable. This is because for repeated roots, the zero-input response is of the form $k^{r-1}\gamma^k$, and if $|\gamma| = 1$, then $|k^{r-1}\gamma^k| = k^{r-1} \to \infty$ as $k \to \infty$.† Note, however, that repeated roots inside the unit circle do not cause instability. Figure 3.17 shows the characteristic modes corresponding to characteristic roots at various location in the complex plane.

To summarize:

1. An LTID system is asymptotically stable if and only if all the characteristic roots are inside the unit circle. The roots may be simple or repeated.

2. An LTID system is unstable if and only if either one or both of the following conditions exist: (i) at least one root is outside the unit circle; (ii) there are repeated roots on the unit circle.

3. An LTID system is marginally stable if and only if there are no roots outside the unit circle and there are some unrepeated roots on the unit circle.

■ **Example 3.16**
Determine whether the systems specified by the following equations are asymptotically stable, marginally stable, or unstable. In each case plot the characteristic roots in the complex plane.

    (a) $y[k + 2] + 2.5y[k + 1] + y[k] = f[k + 1] - 2f[k]$
    (b) $y[k] - y[k - 1] + 0.21y[k - 2] = 2f[k - 1] + 3f[k - 2]$
    (c) $y[k + 3] + 2y[k + 2] + \frac{3}{2}y[k + 1] + \frac{1}{2}y[k] = f[k + 1]$
    (d) $(E^2 - E + 1)^2 y[k] = (3E + 1)f[k]$

---

†If the development of discrete-time systems is parallel to that of continuous-time systems, we wonder why the parallel breaks down here. Why, for instance, aren't LHP and RHP the regions demarcating stability and instability? The reason lies in the form of the characteristic modes. In continuous-time systems we chose the form of characteristic mode as $e^{\lambda_i t}$. In discrete-time systems we choose the form (for computational convenience) to be γ_i^k. Had we chosen this form to be $e^{\lambda_i k}$ where $\gamma_i = e^{\lambda_i}$, then LHP and RHP (for the location of λ_i) again would demarcate stability and instability. This is because if $\gamma = e^\lambda$, $|\gamma| = 1$ implies $|e^\lambda| = 1$, and therefore $\lambda = j\omega$. This shows that the unit circle in γ plane maps into the imaginary axis in the λ plane.

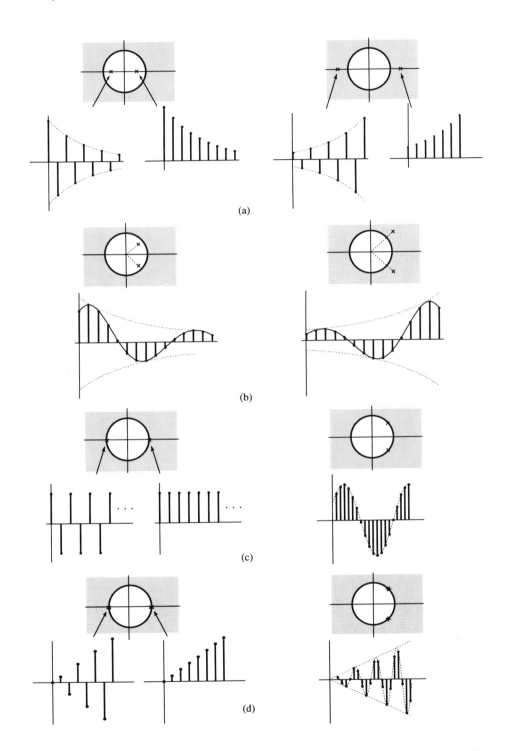

Fig. 3.17 Characteristic roots location and the corresponding characteristic modes.

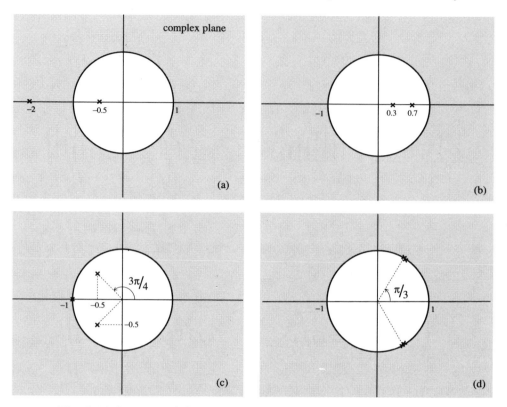

Fig. 3.18 Location of characteristic roots for systems in Example 3.16.

(a) The characteristic polynomial is

$$\gamma^2 + 2.5\gamma + 1 = (\gamma + 0.5)(\gamma + 2)$$

The characteristic roots are -0.5 and -2. Because $|-2| > 1$ (-2 lies outside the unit circle), the system is unstable (Fig. 3.18a).

(b) The characteristic polynomial is

$$\gamma^2 - \gamma + 0.21 = (\gamma - 0.3)(\gamma - 0.7)$$

The characteristic roots are 0.3 and 0.7, both of which lie inside the unit circle. The system is asymptotically stable (Fig. 3.18b).

(c) The characteristic polynomial is

$$\gamma^3 + 2\gamma^2 + \tfrac{3}{2}\gamma + \tfrac{1}{2} = (\gamma + 1)(\gamma^2 + \gamma + \tfrac{1}{2}) = (\gamma + 1)(\gamma + 0.5 - j0.5)(\gamma + 0.5 + j0.5)$$

The characteristic roots are -1, $-0.5 \pm j0.5$ (Fig. 3.18c). One of the characteristic roots is on the unit circle and the remaining two roots are inside the unit circle (Fig. 3.18c). The system is marginally stable.

(d) The characteristic polynomial is

$$(\gamma^2 - \gamma + 1)^2 = \left(\gamma - \tfrac{1}{2} - j\tfrac{\sqrt{3}}{2}\right)^2 \left(\gamma - \tfrac{1}{2} + j\tfrac{\sqrt{3}}{2}\right)^2$$

The characteristic roots are $\tfrac{1}{2} \pm j\tfrac{\sqrt{3}}{2} = 1e^{\pm j\frac{\pi}{3}}$ repeated twice, and they lie on the unit circle (Fig. 3.18d). The system is unstable. ∎

△ **Exercise E3.12**
    Find and sketch the location in the complex plane of the characteristic roots of the system specified by the following equation:

$$(E + 1)(E^2 + 6E + 25)y[k] = 3Ef[k]$$

Is this a stable, unstable, or marginally stable system? Answer: unstable.   ▽

△ **Exercise E3.13**
    Repeat Prob. E3.12 for

$$(E - 1)^2(E + 0.5)y[k] = (E^2 + 2E + 3)f[k]$$

Answer: unstable.   ▽

3.6-1 System Response to Bounded Inputs

As in the case of continuous-time systems, asymptotically stable discrete-time systems have the property that every bounded input produces a bounded output. On the other hand, for a marginally stable and unstable system it is possible to find a bounded input that produces unbounded output. To prove this result we recall that

$$y[k] = h[k] * f[k]$$
$$= \sum_{m=-\infty}^{\infty} h[m]f[k - m]$$

and

$$|y[k]| = \left| \sum_{m=-\infty}^{\infty} h[m]f[k - m] \right|$$
$$\leq \sum_{m=-\infty}^{\infty} |h[m]| \, |f[k - m]|$$

If $f[k]$ is bounded, then $|f[k - m]| < K_1 < \infty$, and

$$|y[k]| \leq K_1 \sum_{m=-\infty}^{\infty} |h[m]|$$

Clearly the output is bounded if the summation on the right-hand side is bounded; that is, if

$$\sum_{k=-\infty}^{\infty} |h[k]| < K_2 < \infty \qquad (3.72)$$

For an asymptotically stable system

$$h[k] = (c_1\gamma_1^k + c_2\gamma_2^k + \cdots + c_n\gamma_n^k)u[k]$$

with the characteristic roots $\gamma_1, \gamma_2, \ldots \gamma_n$ lying within the unit circle. Under these conditions $h[k]$ satisfies Eq. (3.72).† Therefore the response of an asymptotically stable system to a bounded input is also bounded. Moreover, we can show that

---

†This follows from the fact that (see Sec. B.10-4)

for an unstable or marginally stable system, the response is unbounded for some bounded input. As seen in Sec. 2.6, these results lead to the formulation of an alternate definition of system stability. A system is said to be stable in the sense of having bounded output for every bounded input (BIBO stability) if and only if its impulse response $h[k]$ satisfies Eq. (3.72). Condition (3.72) is sufficient for a system to produce bounded output for any bounded input. It is relatively easy to show that it is also a necessary condition; that is, for a system that violates (3.72), there exists a bounded input that produces unbounded output (see Prob. 3.6-2). Note that an asymptotically stable system is always BIBO-stable.‡

3.7 INTUITIVE INSIGHTS INTO SYSTEM BEHAVIOR

The intuitive insights into the behavior of continuous-time systems and their qualitative proofs, discussed in Sec. 2.7, also apply to discrete-time systems. For this reason we shall merely mention here some of these insights without discussion or proof. The interested reader should read Sec. 2.7 for more explanation.

The system's behavior is strongly influenced by the characteristic roots (or modes) of the system. The system responds strongly to input signals similar to its characteristic modes, and responds poorly to inputs very different from its characteristic modes. In fact, when the input is a characteristic mode of the system, the response goes to infinity provided that the mode is a nondecaying signal. This is the resonance phenomenon. The width of an impulse response $h[k]$ indicates the response time (time required to respond fully to an input) of the system. It is the time constant of the system.† Discrete-time pulses are generally dispersed when passed through a discrete-time system. The amount of dispersion (or spreading out) is equal to the system time constant (or width of $h[k]$).

3.8 CLASSICAL SOLUTION OF LINEAR DIFFERENCE EQUATIONS

As in the case of LTIC systems, we can analyze LTID systems by using the classical method, where the response is found as a sum of natural and forced components of the response.

$$\sum_{k=-\infty}^{\infty} |\gamma_i^k| u[k] = \sum_{k=0}^{\infty} |\gamma_i|^k = \frac{1}{1 - |\gamma_i|} \qquad |\gamma_i| < 1$$

Therefore if $|\gamma_i| < 1$ for $i = 1, 2, \ldots, n$,

$$\sum_{k=-\infty}^{\infty} |h[k]| \leq \sum_{i=1}^{n} \frac{1}{1 - |\gamma_i|} < \infty$$

Although we have assumed all the roots to be distinct in this derivation, it is also valid for repeated roots provided that they lie inside the unit circle.

‡The converse is not true. See footnote on p. 160.

†This part of the discussion applies to systems with impulse response $h[k]$ that is a mostly positive (or mostly negative) pulse.

Natural and Forced Response

The zero-input response consists exclusively of characteristic mode terms; the zero-state response consists of characteristic and noncharacteristic mode terms. We lump together the characteristic mode terms of the two components and call it the *natural response* of the system. The remaining noncharacteristic mode terms form the *forced response*. If $y_n[k]$ and $y_\phi[k]$ denote the natural and the forced response respectively, then the total response is given by

$$\text{Total response} = \underbrace{y_n[k]}_{\text{modes}} + \underbrace{y_\phi[k]}_{\text{nonmodes}}$$

The total response $y_n[k] + y_\phi[k]$ is a solution of the system equation (3.15); that is,

$$Q[E]\left(y_n[k] + y_\phi[k]\right) = P[E]f[k] \tag{3.73}$$

But since $y_n[k]$ is made up of characteristic modes,

$$Q[E]y_n[k] = 0 \tag{3.74}$$

Substitution of this equation in Eq. (3.73) yields

$$Q[E]y_\phi[k] = P[E]f[k] \tag{3.75}$$

The natural response is a linear combination of characteristic modes. The arbitrary constants (multipliers) are determined from suitable auxiliary conditions usually given as $y[0], y[1], \ldots, y[n-1]$. We now turn our attention to the forced response.

Forced Response

We have shown that the forced response $y_\phi[k]$ satisfies the system Equation (3.75)

$$Q[E]y_\phi[k] = P[E]f[k] \tag{3.76}$$

By definition, the forced response contains only nonmode terms. To determine the forced response, we shall use a method of undermined coefficients, the same method used for continuous-time system. However, rather than retracing all the steps of continuous-time system, we shall present a table (Table 3.2) listing the inputs and the corresponding forms of forced function with undetermined coefficients. These coefficients can be determined by substituting $y_\phi[k]$ in Eq. (3.75) and equating the coefficients of similar terms.

■ **Example 3.17**

Solve

$$(E^2 - 5E + 6)y[k] = (E - 5)f[k] \tag{3.77}$$

If the input $f[k] = (3k + 5)u[k]$ and the auxiliary conditions are $y[0] = 4, y[1] = 13$.

The characteristic equation is

$$\gamma^2 - 5\gamma + 6 = (\gamma - 2)(\gamma - 3) = 0$$

Therefore the natural response is

$$y_n[k] = B_1(2)^k + B_2(3)^k$$

Table 3.2

Input $f[k]$		Forced Response $y_\phi[k]$
1.	$r^k \quad r \neq \gamma_i \ (i = 1, 2, \cdots, n)$	cr^k
2.	$r^k \quad r = \gamma_i$	ckr^k
3.	$\cos(\beta k + \theta)$	$c\cos(\beta k + \phi)$
4.	$\left(\displaystyle\sum_{i=0}^{m} \alpha_i k^i \right) r^k$	$\left(\displaystyle\sum_{i=0}^{m} c_i k^i \right) r^k$

Note: By definition, $y_\phi[k]$ cannot have any characteristic mode terms. If any terms shown in the right-hand column for the forced response should also be a characteristic mode of the system, the correct form of the forced response must be modified to $k^i y_\phi[k]$, where i is the smallest integer that will prevent $k^i y_\phi[k]$ from having a characteristic mode term. For example, when the input is r^k, the forced response in the right-hand column is of the form cr^k. But if r^k happens to be a natural mode of the system, the correct form of the forced response is ckr^k (see Pair 2).

To find the form of forced response $y_\phi[k]$ we use Table 3.2, Pair 4 with $r = 1$, $m = 1$. This yields

$$y_\phi[k] = c_1 k + c_0$$

Therefore

$$y_\phi[k+1] = c_1(k+1) + c_0 = c_1 k + c_1 + c_0$$

$$y_\phi[k+2] = c_1(k+2) + c_0 = c_1 k + 2c_1 + c_0$$

Also

$$f[k] = 3k + 5$$

and

$$f[k+1] = 3(k+1) + 5 = 3k + 8$$

Substitution of the above results in Eq. (3.75) yields

$$c_1 k + 2c_1 + c_0 - 5(c_1 k + c_1 + c_0) + 6(c_1 k + c_0) = 3k + 8 - 5(3k + 5)$$

or

$$2c_1 k - 3c_1 + 2c_0 = -12k - 17$$

Comparison of similar terms on two sides yields

$$\left. \begin{array}{r} 2c_1 = -12 \\ -3c_1 + 2c_0 = -17 \end{array} \right\} \implies \begin{array}{l} c_1 = -6 \\ c_2 = -\frac{35}{2} \end{array}$$

This means

$$y_\phi[k] = -6k - \tfrac{35}{2}$$

The total response is

$$\begin{aligned} y[k] &= y_n[k] + y_\phi[k] \\ &= B_1(2)^k + B_2(3)^k - 6k - \tfrac{35}{2} \qquad k \geq 0 \end{aligned} \tag{3.78}$$

To determine arbitrary constants B_1 and B_2 we set $k = 0$ and 1 and substitute the initial conditions $y[0] = 4, y[1] = 13$ to obtain

$$\left. \begin{array}{l} 4 = B_1 + B_2 - \frac{35}{2} \\[2mm] 13 = 2B_1 + 3B_2 - \frac{47}{2} \end{array} \right\} \implies \begin{array}{l} B_1 = 28 \\[2mm] B_2 = \frac{-13}{2} \end{array}$$

Therefore

$$y_n[k] = 28(2)^k - \tfrac{13}{2}(3)^k \tag{3.79}$$

and

$$y[k] = \underbrace{28(2)^k - \tfrac{13}{2}(3)^k}_{y_n[k]} \underbrace{- 6k - \tfrac{35}{2}}_{y_\phi[k]} \tag{3.80}$$

∎

A Comment on Initial Conditions

This method requires auxiliary conditions $y[0], y[1], \ldots, y[n-1]$ for the reasons explained in Sec. 2.8. If we are given the initial conditions $y[-1], y[-2], \ldots, y[-n]$, we can derive the conditions $y[0], y[1], \ldots, y[n-1]$ using iterative procedure.

Exponential Input

As in the case of continuous-time systems, we can show that for a system specified by the equation

$$Q[E]y[k] = P[E]f[k] \tag{3.81}$$

the forced response for the exponential input $f[k] = r^k$ is given by

$$y_\phi[k] = H[r]r^k \qquad r \neq \gamma_i \tag{3.82}$$

where

$$H[r] = \frac{P[r]}{Q[r]} \tag{3.83}$$

The proof follows from the fact that if the input $f[k] = r^k$, then from Table 3.2 (Pair 4), $y_\phi[k] = cr^k$. Therefore

$$E^i f[k] = f[k+i] = r^{k+i} = r^i r^k \qquad \text{and} \qquad P[E]f[k] = P[r]r^k$$

$$E^j y_\phi[k] = cr^{k+j} = cr^j r^k \qquad \text{and} \qquad Q[E]y[k] = cQ[r]r^k$$

so that Eq. (3.81) reduces to

$$cQ[r]r^k = P[r]r^k$$

which yields $c = P[r]/Q[r] = H[r]$.

This result is valid only if r is not a characteristic root of the system. If r is a characteristic root, the forced response is ckr^k where c is determined by substituting $y_\phi[k]$ in the system equation and equating coefficients of similar terms on the two sides. Observe that the exponential r^k includes a wide variety of signals such as a constant C, a sinusoid $\cos{(\beta k + \theta)}$, and an exponentially growing or decaying sinusoid $|\gamma|^k \cos{(\beta k + \theta)}$.

1. A Constant Input $f[k] = C$

This is a special case of exponential Cr^k with $r = 1$. Therefore from Eq. (3.82) we have

$$y_\phi[k] = C\frac{P[1]}{Q[1]} = CH[1] \tag{3.84}$$

2. A Sinusoidal Input

The input $e^{j\beta k}$ is an exponential r^k with $r = e^{j\beta}$. Hence

$$y_\phi[k] = H[e^{j\beta}]e^{j\beta k} = \frac{P[e^{j\beta}]}{Q[e^{j\beta}]}e^{j\beta k}$$

Similarly for the input $e^{-j\beta k}$

$$y_\phi[k] = H[e^{-j\beta}]e^{-j\beta k}$$

Consequently, if the input $f[k] = \cos{\beta k} = \frac{1}{2}(e^{j\beta k} + e^{-j\beta k})$,

$$y_\phi[k] = \frac{1}{2}\left\{ H[e^{j\beta}]e^{j\beta k} + H[e^{-j\beta}]e^{-j\beta k} \right\}$$

Since the two terms on the right-hand side are conjugates

$$y_\phi[k] = \mathrm{Re}\left\{ H[e^{j\beta}]e^{j\beta k} \right\}$$

If

$$H[e^{j\beta}] = |H[e^{j\beta}]|e^{j\angle H[e^{j\beta}]}$$

$$y_\phi[k] = \mathrm{Re}\left\{ |H[e^{j\beta}]|\, e^{j(\beta k + \angle H[e^{j\beta}])} \right\}$$
$$= |H[e^{j\beta}]|\cos{\left(\beta k + \angle H[e^{j\beta}] \right)} \tag{3.85}$$

Using a similar argument, we can show that for the input

$$f[k] = \cos{(\beta k + \theta)}$$

$$y_\phi[k] = |H[e^{j\beta}]|\cos{\left(\beta k + \theta + \angle H[e^{j\beta}] \right)} \tag{3.86}$$

■ **Example 3.18**

For a system specified by the equation

$$(E^2 - 3E + 2)y[k] = (E + 2)f[k]$$

Find the forced response for the input $f[k] = (3)^k u[k]$.

In this case

$$H[r] = \frac{P[r]}{Q[r]} = \frac{r + 2}{r^2 - 3r + 2}$$

and the forced response to input $(3)^k u[k]$ is $H[3](3)^k$; that is,

$$y_\phi[k] = \frac{3+2}{(3)^2 - 3(3) + 2}(3)^k = \frac{5}{2}(3)^k \qquad k \geq 0 \quad \blacksquare$$

■ **Example 3.19**

For an LTID system described by the equation

$$(E^2 - E + 0.16)y[k] = (E + 0.32)f[k]$$

determine the forced response $y_\phi[k]$ if the input is

$$f[k] = \cos\left(2k + \tfrac{\pi}{3}\right)u[k]$$

Here

$$H[r] = \frac{P[r]}{Q[r]} = \frac{r + 0.32}{r^2 - r + 0.16}$$

for the input $\cos\left(2k + \tfrac{\pi}{3}\right)u[k]$, the forced response is

$$y_\phi[k] = \left|H[e^{j2}]\right| \cos\left(2k + \tfrac{\pi}{3} + \angle H[e^{j2}]\right)u[k]$$

where

$$H[e^{j2}] = \frac{e^{j2} + 0.32}{(e^{j2})^2 - e^{j2} + 0.16} = \frac{(-0.416 + j0.909) + 0.32}{(-0.654 - j0.757) - (-0.416 + j0.909) + 0.16}$$

$$= \frac{-0.0.96 + j0.909}{-0.078 - j1.66} = \frac{0.914e^{j1.676}}{1.668e^{-j1.618}} = 0.548e^{j3.294}$$

Therefore

$$|H[e^{j2}]| = 0.548 \qquad \text{and} \qquad \angle H[e^{j2}] = 3.294$$

so that

$$y_\phi[k] = 0.548 \cos\left(2k + \tfrac{\pi}{3} + 3.294\right)u[k]$$

$$= 0.548 \cos\left(2k + 4.34\right)u[k] \quad \blacksquare$$

Assessment of the Classical Method

The remarks in Chapter 2 concerning the classical method for solving differential equations (p. 177) also apply to difference equations.

3.9   APPENDIX 3.1: DETERMINING THE IMPULSE RESPONSE

For a discrete-time system specified by Eq. (3.36), we have shown that

$$h[k] = A_0\delta[k] + y_0[k]u[k] \tag{3.87}$$

We now show that

$$A_0 = \frac{b_0}{a_0} \tag{3.88}$$

To prove this, we substitute Eq. (3.41) in Eq. (3.37) to obtain

$$Q[E]\left(A_0\delta[k] + y_0[k]u[k]\right) = P[E]\delta[k] \qquad (3.89)$$

Observe that $y_0[k]u[k]$ is a sum of characteristic modes of the system. Therefore†
[see Eq. (3.18a)]

$$Q[E]\left(y_0[k]u[k]\right) = 0 \quad k \geq 0 \qquad (3.90)$$

Equation (3.89) now reduces to

$$A_0 Q[E]\delta[k] = P[E]\delta[k] \qquad k \geq 0$$

or

$$A_0(\delta[k+n] + a_{n-1}\delta[k+n-1] + \cdots + a_1\delta[k+1] + a_0\delta[k]) = b_n\delta[k+n]$$
$$+ b_{n-1}\delta[k+n-1] + \cdots + b_1\delta[k+1] + b_0\delta[k] \quad k \geq 0$$

If we set $k = 0$ in this equation and recognize that $\delta[0] = 1$, and, $\delta[m] = 0 \quad m \neq 0$,
all but the last terms vanish on both sides, yielding

$$A_0 a_0 = b_0$$

and

$$A_0 = \frac{b_0}{a_0}$$

A Special Case: $a_0 = 0$

In this case $A_0 = \frac{b_0}{a_0}$ becomes indeterminate. The procedure has to be modified
slightly. When $a_0 = 0$, $Q[E]$ can be expressed as $E\hat{Q}[E]$, and Eq. (3.37) can be
expressed as

$$E\hat{Q}[E]h[k] = P[E]\delta[k]$$

If we recognize that $\frac{1}{E}$ is a delay operator, this equation can be rearranged as

$$\hat{Q}[E]h[k] = P[E]\left(\frac{1}{E}\delta[k]\right) = P[E]\delta[k-1]$$

In this case the input vanishes not at $k = 1$, but at $k = 2$. Therefore the response
consists not only of the zero-input term and an impulse $A_0\delta[k]$ (at $k = 0$), but also
of an impulse $A_1\delta[k-1]$ (at $k = 1$). Therefore

$$h[k] = A_0\delta[k] + A_1\delta[k-1] + y_0[k]u[k]$$

---

†Note that

$$Q[E]\left(y_0[k]\right) = 0$$

for all k. But if we restrict the mode terms to be causal, the equation is valid only for $k \geq 0$; that
is,

$$Q[E]\left(y_0[k]u[k]\right) = 0 \qquad k \geq 0$$

We can determine the unknowns A_0, A_1, and the n coefficients in $y_0[k]$ from the $n + 2$ number of initial values $h[0]$, $h[1]$, $\cdots$, $h[n + 1]$, determined as usual from the iterative solution of the equation $Q[E]h[k] = P[E]\delta[k]$. The following example demonstrates the procedure.

■ **Example 3.20**

Determine the unit impulse response for a system specified by the equation

$$y[k] - \tfrac{1}{2}y[k - 1] + \tfrac{1}{18}y[k - 2] = -f[k] + 2f[k - 3]$$

or

$$y[k + 3] - \tfrac{1}{2}y[k + 2] + \tfrac{1}{18}y[k + 1] = -f[k + 3] + 2f[k]$$

In operational notation

$$(E^3 - \tfrac{1}{2}E^2 + \tfrac{1}{18}E)y[k] = (-E^3 + 2)f[k]$$

or

$$E(E^2 - \tfrac{1}{2}E + \tfrac{1}{18})y[k] = (-E^3 + 2)f[k]$$

Note that $\hat{Q}[E] = (E^2 - \tfrac{1}{2}E + \tfrac{1}{18})$. The characteristic polynomial is $\gamma^2 - \tfrac{1}{2}\gamma + \tfrac{1}{18} = (\gamma - \tfrac{1}{3})(\gamma - \tfrac{1}{6})$. Therefore

$$h[k] = A_0\delta[k] + A_1\delta[k - 1] + c_1(\tfrac{1}{3})^k + c_2(\tfrac{1}{6})^k \tag{3.91}$$

There are four unknowns in this equation. Therefore we need to determine four values $h[0]$, $h[1]$, $h[2]$, and $h[3]$, iteratively from the Eq. (3.37) which in this case is

$$h[k + 3] - \tfrac{1}{2}h[k + 2] + \tfrac{1}{18}h[k + 1] = -\delta[k + 3] + 2\delta[k]$$

subject to zero initial state; that is $h[-1] = h[-2] = h[-3] = 0$. Setting $k = -3, -2, -1$, and 0, iteratively, we obtain $h[0] = -1$, $h[1] = -0.5$, $h[2] = -\tfrac{7}{36}$, and $h[3] = \tfrac{139}{72}$. Finally, setting $k = 0, 1, 2$, and 3 in Eq. (3.91), and solving the four simultaneous equations, we obtain $A_0 = 324$, $A_1 = 36$, $c_1 = 106$, and $c_2 = -431$. Therefore

$$h[k] = 324\delta[k] + 36\delta[k - 1] + 106(\tfrac{1}{3})^k - 431(\tfrac{1}{6})^k$$

This discussion can be readily extended to the case in which $Q[E]$ is expressed as $E^2\hat{Q}[E]$. In this case

$$h[k] = A_0\delta[k] + A_1\delta[k - 1] + A_2\delta[k - 2] + y_0[k]u[k]$$

and so on. ■

3.10 SUMMARY

This chapter discusses time-domain analysis of LTID (linear, time-invariant, discrete-time) systems. The analysis is parallel to that of LTIC systems, with some minor differences. Discrete-time systems can be realized by using scalar multipliers, summers, and time delays. These operations can be readily performed by digital computers. Time delays also can be obtained from charge coupled devices (CCD), bucket brigade devices (BBD), and surface acoustic wave devices (SAW).

Several advantages of discrete-time systems over continuous-time systems are discussed in Sec. 3.1. Because of these advantages, discrete-time systems are replacing continuous-time systems in several applications.

Discrete-time systems are described by difference equations. For an nth-order system, n auxiliary conditions must be specified for a unique solution. Characteristic modes are discrete-time exponentials of the form γ^k corresponding to a unrepeated root γ, and the modes are of the form $k^i \gamma^k$ corresponding to a repeated root γ.

The unit impulse function $\delta[k]$ is a sequence of a single number of unit value at $k = 0$. The unit impulse response $h[k]$ of a discrete-time system is a linear combination of its characteristic modes.†

The zero-state response (response due to external input) of a linear system is found by breaking the input into impulse components and then adding the system response to all the impulse components. The sum of the system response to the impulse components is in the form of a sum, known as the convolution sum, whose structure and properties are similar to the convolution integral. The system response is obtained as the convolution sum of the input $f[k]$ with the system's impulse response $h[k]$. Therefore the knowledge of the system's impulse response allows us to determine the system response to any arbitrary input.

The stability criterion in terms of the location of characteristic roots of the system can be summarized as follows:

1. An LTID system is asymptotically stable if and only if all the characteristic roots are inside the unit circle. The roots may be repeated or unrepeated.

2. An LTID system is unstable if and only if either one or both of the following conditions exist: (i) at least one root is outside the unit circle, (ii) there are repeated roots on the unit circle.

3. An LTID system is marginally stable if and only if there are no roots outside the unit circle, and there are some unrepeated roots on the unit circle.

According to an alternate definition of stability—the bounded-input bounded-output (BIBO) stability—a system is stable if and only if every bounded input produces a bounded output. Otherwise the system is unstable. An asymptotically stable system is always BIBO-stable. The converse is not necessarily true.

Difference equations of LTID systems also can be solved by the classical method, whereby the response is found as a sum of natural and forced responses. These are not the same as the zero-input and the zero-state components, although they satisfy the same equations respectively. Although simple, this method suffers from the fact that it is applicable to a restricted class of input signals, and the system response cannot be set forth as an explicit function of the input. This makes it useless in the theoretical study of systems.

REFERENCES

1. Milstein, L. B., and P.K. Das, "Surface Acoustic wave Devices", IEEE Communication Society Magazine, vol. 17, No. 5, pp 25-33, September 1979.

---

†There is a possibility of an impulse $\delta[k]$ in addition to characteristic modes.

PROBLEMS

3.1-1 A cash register output $y[k]$ represents the total cost of k items rung up by an operator. The input $f[k]$ is the cost of the kth item.

    (a) Write the difference equation relating $y[k]$ to $f[k]$.

    (b) Realize this system using a time-delay element.

3.1-2 Let $p[k]$ be the population of a certain country at the beginning of the kth year. The birth and death rates of the population during any year are 3.3% and 1.3%, respectively. If $i[k]$ is the total number of immigrants entering the country during the kth year, write the difference equation relating $p[k+1]$, $p[k]$, and $i[k]$.

Hint: Assume that the immigrants enter the country throughout the year uniformly, so that their average birth and death rates are only half the normal rates.

3.1-3 A discrete-time differentiator is realized in Fig. 3.5b. Using a similar approach, realize a discrete-time integrator. The equation for an integrator is

$$\lim_{T \to 0} y(kT) - y\left[(k-1)T\right] = Tf\left[(k-1)T\right]$$

or

$$y[k] = y[k-1] + Tf[k-1]$$

If an input $u(t)$ is applied to such an integrator show that the output is a ramp.

3.1-4 In Example 3.1, if the bank pays 1% interest per month ($r = 0.01$) on the money that is in the account for a month, and 1.5% ($r = 0.015$) per month on the money that is in the account for more than a month, write the equation relating the bank balance $y[k]$ to the deposit $f[k]$. Assume that there are no withdrawals and that the interest is compounded monthly.

3.1-5 A moving average is used to detect a trend of a rapidly fluctuating variable such as the stock market average. A variable may fluctuate (up and down) daily, masking its long-term trend. We can discern the long-term trend by smoothing or averaging the past N values of the variable. For the stock market average, we may consider a five-day moving average $y[k]$ to be the mean of the past five days' market closing values $f[k], f[k-1], \ldots, f[k-4]$.

    (a) Write the difference equation relating $y[k]$ to the input $f[k]$.

    (b) Using time-delay elements, realize the five-day moving average filter.

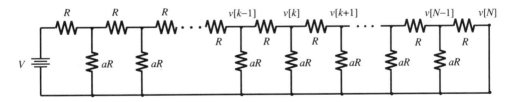

Fig. P3.1-6

3.1-6 The voltage at the kth node of a resistive ladder in Fig. P3.1-6 is $v[k]$ ($k = 0, 1, 2, \ldots, N$). Show that the $v[k]$ satisfies the second-order difference equation

$$v[k+2] - Av[k+1] + v[k] = 0 \qquad A = 2 + \tfrac{1}{a}$$

with auxiliary conditions $v[0] = V$ and $v[N] = 0$.

Hint: Consider the node equation at the kth node with voltage $v[k]$.

3.2-1 Solve iteratively (first three terms only):

(a) $y[k+1] - 0.5y[k] = 0$, with $y[-1] = 10$

(b) $y[k+1] + 2y[k] = f[k+1]$, with $f[k] = e^{-k}u[k]$ and $y[-1] = 0$

3.2-2 Solve the following equation iteratively (first three terms only):

$$y[k] - 0.6y[k-1] - 0.16y[k-2] = 0 \text{ with } y[-1] = -25,\ y[-2] = 0$$

3.2-3 Solve the second-order difference Eq. (3.3b) iteratively (first three terms only), assuming $y[-1] = y[-2] = 0$ and $f[k] = 100u[k]$.

3.2-4 Solve the following equation iteratively (first three terms only):

$$y[k+2] + 3y[k+1] + 2y[k] = f[k+2] + 3f[k+1] + 3f[k]$$

with $f[k] = (3)^k u[k]$, $y[-1] = 3$, and $y[-2] = 2$

3.2-5 Repeat Prob. 3.2-4 if

$$y[k] + 2y[k-1] + y[k-2] = 2f[k] - f[k-1]$$

with $f[k] = (3)^{-k}u[k]$, $y[-1] = 2$, and $y[-2] = 3$.

3.3-1 Solve

$$y[k+2] + 3y[k+1] + 2y[k] = 0 \text{ if } y[-1] = 0,\ y[-2] = 1$$

3.3-2 Solve

$$y[k+2] + 2y[k+1] + y[k] = 0 \text{ if } y[-1] = 1,\ y[-2] = 1$$

3.3-3 Solve

$$y[k+2] - 2y[k+1] + 2y[k] = 0 \text{ if } y[-1] = 1,\ y[-2] = 0$$

3.3-4 Solve

$$(E^2 + 2E + 2)y[k] = 0 \text{ if } y[0] = 0,\ y[1] = 2$$

3.3-5 Solve

$$y[k] + 4y[k-2] = 0 \text{ if } y[0] = 1,\ y[1] = 2$$

3.3-6 Find $v[k]$, the voltage at the kth node of the resistive ladder shown in Fig. P3.1-6, if $V = 100$ volts and $a = 2$.

Hint: The difference equation for $v[k]$ is given in Prob. 3.1-6. The auxiliary conditions are $v[0] = 100$, and $v[N] = 0$.

3.4-1 Find the unit impulse response $h[k]$ of a system specified by the equation

$$y[k+1] + 2y[k] = f[k]$$

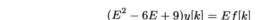

3.4-2 Repeat Prob. 3.4-1 if

$$y[k+1] + 2y[k] = f[k+1]$$

3.4-3 Repeat Prob. 3.4-1 if

$$(E^2 - 6E + 9)y[k] = Ef[k]$$

3.4-4 Repeat Prob. 3.4-1 if

$$y[k] - 6y[k-1] + 25y[k-2] = 2f[k] - 4f[k-1]$$

Hint: The characteristic roots are $3 \pm j4 = 5e^{\pm j0.927}$ and
$y_0[k] = c(5)^k \cos(0.927k + \theta)$.

3.4-5   For the general nth-order difference Eq. (3.15) if

$$a_0 = a_1 = a_2 = \cdots = a_{n-1} = 0$$

the resulting equation is called a **nonrecursive** difference equation.
(a) Show that the impulse response of a system described by this equation is

$$h[k] = b_n\delta[k] + b_{n-1}\delta[k-1] + \cdots + b_1\delta[k-n+1] + b_0\delta[k-n]$$

Hint: Express the system equation in delay-operator form and let $f[k] = \delta[k]$, $y[k] = h[k]$.
(b) Find the impulse response of an LTID system described by the equation

$$y[k] = 3f[k] - 5f[k-1] - 2f[k-3]$$

Observe that the impulse response has only a finite (n) number of nonzero elements.
For this reason such systems are said to be **finite impulse response (FIR)** systems.
For a general recursive case [Eq. (3.15)], the impulse response has an infinite number
of nonzero elements, and such systems are said to be **infinite impulse response
(IIR)** systems.

3.5-1   Find the (zero-state) response $y[k]$ of an LTID system whose unit impulse response
is

$$h[k] = (-2)^k u[k]$$

and the input is $f[k] = e^{-k}u[k]$. Hint: Here $f[k] = \gamma^k u[k]$ with $\gamma = 1/e$.

3.5-2   Repeat Prob. 3.5-1 if

$$h[k] = \tfrac{1}{2}[\delta[k] - (-2)^k]u[k]$$

3.5-3   Repeat Prob. 3.5-1 if the input $f[k] = (3)^{k+2}u[k]$, and

$$h[k] = [(2)^k + 3(-5)^k]u[k]$$

Hint: $(3)^{k+2} = 9(3)^k$.

3.5-4   Repeat Prob. 3.5-1 if the input $f[k] = (3)^{-k}u[k]$, and

$$h[k] = 3k(2)^k u[k]$$

3.5-5   Repeat Prob. 3.5-1 if the input $f[k] = (2)^k u[k]$, and

$$h[k] = (3)^k \cos\left(\tfrac{\pi}{3}k - 0.5\right)u[k]$$

3.5-6   Find the total response of a system specified by the equation

$$y[k+1] + 2y[k] = f[k+1]$$

if $y[-1] = 10$, and the input $f[k] = e^{-k}u[k]$.

3.5-7   Repeat Prob. 3.5-1 if the impulse response $h[k] = (0.5)^k u[k]$, and the input $f[k]$ is
(a) $2^k u[k]$ **(b)** $2^{(k-3)}u[k]$ **(c)** $2^k u[k-2]$.
Hint: $2^{(k-3)} = \tfrac{1}{8}(2)^k$. For part **(c)**, $2^k u[k-2] = 4(2)^{(k-2)}u[k-2]$. Use shift property
(3.60) of the convolution.

3.5-8   In the savings account problem worked out in Example 3.1, a person deposits \$500
at the beginning of every month, starting at $k = 0$ with the exception at $k = 4$, when

instead of depositing \$500, she withdraws \$1000. Find $y[k]$ if the interest rate is 1% per month ($r = 0.01$).

Hint: Because the deposit starts at $k = 0$, the initial condition $y[-1] = 0$. Withdrawal is a deposit of negative amount. Recognize that the input is $500u[k] - 1500\delta[k - 4]$.

3.5-9 To pay off a loan of M dollars in N months using a fixed monthly payment of P dollars, show that

$$P = \frac{r\gamma^N}{\gamma^N - 1} M \qquad \gamma = 1 + r$$

where r is the interest rate per dollar per month.

Hint: This problem can be modeled by Eq. (3.2) with the initial balance $y[0] = -M$. The first payment begins at $k = 1$. Therefore $f[k] = Pu[k - 1]$. Solve Eq. (3.2) to determine $y[k]$. Because there is a nonzero initial condition and an input, both the zero-input and the zero-state components should be determined. Because the loan is paid off in N payments, set $y[N] = 0$. Remember also that from shifting property of convolution, $f[k] * u[k - 1] = c[k - 1]$ where $c[k] = f[k] * u[k]$.

3.5-10 A person receives an automobile loan of \$10,000 from a bank at the interest rate of 1.5% per month. His monthly payment is \$500, with the first payment due one month after he receives the loan. Compute the number of payments required to pay off the loan. Note that the last payment may not be exactly \$500.

Hint: Follow the procedure in Prob. 3.5-9 to determine the balance $y[k]$. To determine N, the number of payments, set $y[N] = 0$. In general N will not be an integer. The number of payments K is the largest integer $\leq N$. The last payment is $y[K]$.

3.5-11 Using the sliding-tape algorithm, show that

    (a) $u[k] * u[k] = (k + 1)u[k]$

    (b) $(u[k] - u[k - m]) * u[k] = (k + 1)u[k] - (k - m + 1)u[k - m]$

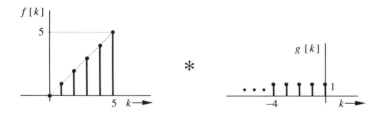

Fig. P3.5-12

3.5-12 Using the sliding-tape algorithm find $f[k] * g[k]$ for the signals shown in Fig. P3.5-12.

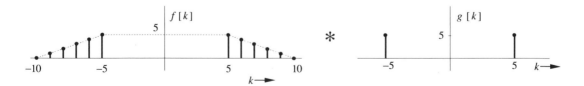

Fig. P3.5-13

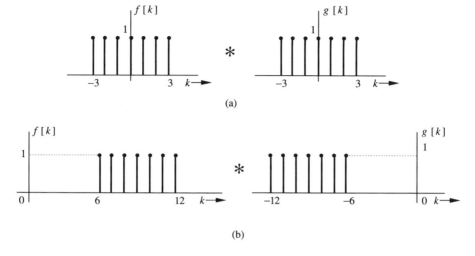

Fig. P3.5-14

3.5-13   Repeat Prob. 3.5-12 for signals shown in Fig. P3.5-13.

3.5-14   Repeat Prob. 3.5-12 for signals shown in Fig. P3.5-14.

3.5-15   The convolution sum in Eq. (3.62) can be expressed in a matrix form as

$$\underbrace{\begin{bmatrix} y[0] \\ y[1] \\ \cdots \\ y[k] \end{bmatrix}}_{\mathbf{y}} \underbrace{\begin{bmatrix} h[0] & 0 & 0 & \cdots & 0 \\ h[1] & h[0] & 0 & \cdots & 0 \\ \cdots\cdots\cdots\cdots\cdots\cdots\cdots\cdots\cdots \\ h[k] & h[k-1] & \cdots & \cdots & h[0] \end{bmatrix}}_{\mathbf{H}} \underbrace{\begin{bmatrix} f[0] \\ f[1] \\ \cdots \\ f[k] \end{bmatrix}}_{\mathbf{f}}$$

or

$$\mathbf{y} = \mathbf{Hf}$$

and

$$\mathbf{f} = \mathbf{H}^{-1}\mathbf{y}$$

Knowing $h[k]$ and the output $y[k]$, we can determine the input $f[k]$. This operation is the reverse of the convolution and is known as the **deconvolution**. Moreover, knowing $f[k]$ and $y[k]$, we can determine $h[k]$. This can be done by expressing the above matrix equation as $k + 1$ simultaneous equations in terms of $k + 1$ unknowns $h[0]$, $h[1]$, ... , $h[k]$. These equations can readily be solved iteratively. Thus we can synthesize a system that yields a certain output $y[k]$ for a given input $f[k]$.
(a) Design a system (that is, determine $h[k]$) that will yield the output sequence (8, 12, 14, 15, 15.5, 15.75, $\cdots$) for the input sequence (1, 1, 1, 1, 1, 1, $\cdots$).
(b) For a system with the impulse response sequence (1, 2, 4, $\cdots$), the output sequence was (1, 7/3, 43/9, $\cdots$). Determine the input sequence.

3.5-16   The sliding-tape method is conceptually quite valuable in understanding the convolution mechanism. Numerical convolution can also be performed from the arrays using the sets $f[0]$, $f[1]$, $f[2]$, ..., and $g[0]$, $g[1]$, $g[2]$, ..., as shown in Fig. 3.5-16. The ijth element (element in the ith row and jth column) is given by $g[i]f[j]$. We add the

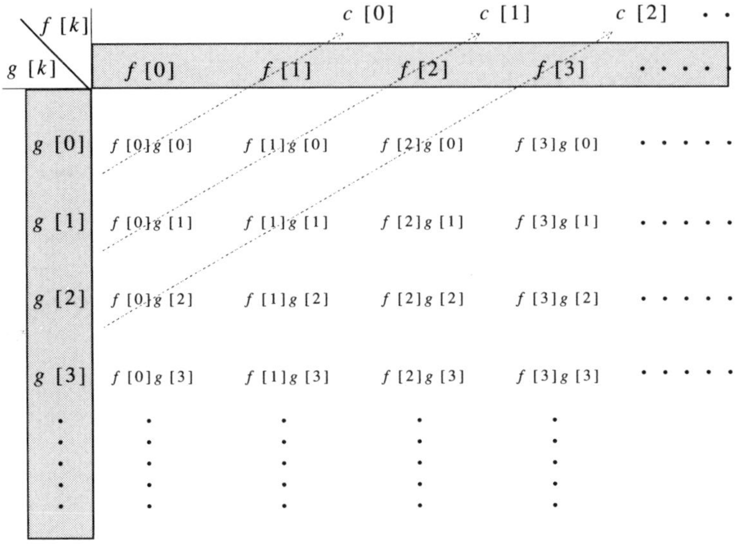

Fig. P3.5-16

elements of the array along its diagonals to produce $c[k] = f[k] * g[k]$. For example, if we sum the elements corresponding to the first diagonal of the array, we obtain $c[0]$. Similarly, if we sum along the second diagonal, we obtain $c[1]$, and so on. Draw the array for the signals $f[k]$ and $g[k]$ in Example 3.15, and find $f[k] * g[k]$.

3.6-1 The following equations specify several LTID systems. Determine whether each of these systems is asymptotically stable, unstable, or marginally stable.

(a) $y[k + 2] + 0.6y[k + 1] - 0.16y[k] = f[k + 1] - 2f[k]$

(b) $(E^2 + 1)(E^2 + E + 1)y[k] = Ef[k]$

(c) $(E - 1)^2(E + \frac{1}{2})y[k] = (E + 2)f[k]$

(d) $y[k] + 2y[k - 1] + 0.96y[k - 2] = 2f[k - 1] + 3f[k - 3]$

(e) $(E^2 - 1)(E^2 + 1)y[k] = f[k]$

3.6-2 In Sec. 3.6 we showed that for BIBO stability in an LTID system, it is sufficient for its impulse response $h[k]$ to satisfy Eq. (3.72). Show that this is also a necessary condition for the system to be BIBO-stable. In other words, show that if Eq. (3.72) is not satisfied, there exists a bounded input that produces unbounded output.
Hint: Assume that a system exists for which $h[k]$ violates Eq. (3.72); yet its output is bounded for every bounded input. Establish contradiction in this statement by considering an input $f[k]$ defined by $f[k_1 - m] = 1$ when $h[m] > 0$ and $f[k_1 - m] = -1$ when $h[m] < 0$, where k_1 is some fixed integer.

3.6-3 Show that a marginally stable system is BIBO unstable. Verify your result by considering a system with characteristic roots on the unit circle and show that for the input of the form of the natural mode (which is bounded), the response is unbounded.

3.8-1 Using the classical method, solve

$$y[k + 1] + 2y[k] = f[k + 1]$$

with the input $f[k] = e^{-k}u[k]$, and the auxiliary condition $y[0] = 1$.

3.8-2 Using the classical method, solve

$$y[k] + 2y[k-1] = f[k-1]$$

with the input $f[k] = e^{-k}u[k]$ and the auxiliary condition $y[-1] = 0$. Hint: You will have to determine the auxiliary condition $y[0]$ using the iterative method.

3.8-3 (a) Using the classical method, solve

$$y[k+2] + 3y[k+1] + 2y[k] = f[k+2] + 3f[k+1] + 3f[k]$$

with the input $f[k] = (3)^k$ and the auxiliary conditions $y[0] = 1, y[1] = 3$.
(b) Repeat (a) if the auxiliary conditions are $y[-1] = y[-2] = 1$. Hint: Using the iterative method, determine $y[0]$ and $y[1]$.

3.8-4 Using the classical method, solve

$$y[k] + 2y[k-1] + y[k-2] = 2f[k] - f[k-1]$$

with the input $f[k] = (3)^{-k}u[k]$ and the auxiliary conditions $y[0] = 2$ and $y[1] = -\frac{13}{3}$. Hint: The input is $r^k u[k]$ with $r = \frac{1}{3}$.

3.8-5 Using the classical method, solve

$$(E^2 - E + 0.16)y[k] = Ef[k]$$

with the input $f[k] = (0.2)^k u[k]$ and the auxiliary conditions $y[0] = 1, y[1] = 2$. Hint: The input is a natural mode of the system. Hence the forced response is $ck(0.2)^k u[k]$.

3.8-6 Using the classical method, solve

$$(E^2 - E + 0.16)y[k] = Ef[k]$$

with the input $f[k] = \cos\left(\frac{\pi k}{2} + \frac{\pi}{3}\right)u[k]$ and the initial conditions $y[-1] = y[-2] = 0$. Hint: Find $y[0]$ and $y[1]$ iteratively.

COMPUTER PROBLEMS

C3-1 Solve iteratively

$$y[k+2] - y[k+1] + 0.35y[k] = 2f[k+2] - f[k+1]$$

with initial conditions $y[-1] = 2, y[-2] = 1, f[0] = f[-1] = 0$, and $f[k] = k$ for $k > 0$.

C3-2 For the LTID systems described by the following difference equation, find the zero-input response, using a computer, if the initial conditions are $y[-1] = 0.5$ and $y[-2] = 0.75$.

$$y[k+2] - y[k+1] - 0.09y[k] = 5f[k+2]$$

C3-3 For the system described by the difference equation below find the (zero-state) response to the input $f[k] = u[k]$.

$$(E^2 - 0.6E + 0.09)y[k] = (E + 18)f[k]$$

C3-4 For the LTID systems described by the following difference equation, find the total response, using a computer, if the initial conditions are $y[-1] = 0.5$ and $y[-2] = 0.75$, and the input $f[k] = 4e^k u[k]$.

$$y[k+2] - 0.5y[k+1] - 0.25y[k] = 5f[k+2]$$

III

FREQUENCY-DOMAIN (TRANSFORM) ANALYSIS OF LTI SYSTEMS

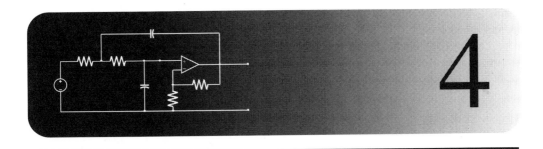

Continuous-Time Systems: Laplace Transform Analysis

Because of the linearity (superposition) property of linear time-invariant systems, we can find the response of these systems by breaking the input $f(t)$ into several components and then summing the system response to all the components of $f(t)$. We have already used this procedure in time-domain analysis, in which the input $f(t)$ is broken into impulsive components. In the **frequency-domain analysis** developed in this chapter, we break up the input $f(t)$ into sinusoids or exponentials of the form e^{st}, where the parameter s (which is complex in general) is the complex frequency of the signal e^{st} (see Sec. B.2-3 for discussion of complex frequency). This method offers an insight into the system behavior which is complementary to that seen in the time-domain analysis. In fact, the time-domain and the frequency-domain methods are duals of one another.

The tool that makes it possible to represent arbitrary input $f(t)$ in terms of exponential components is the **Laplace transform**, which is discussed in the following section.

4.1 THE LAPLACE TRANSFORM

For a signal $f(t)$, the Laplace transform $F(s)$ is defined by

$$F(s) \equiv \int_{-\infty}^{\infty} f(t)e^{-st}\, dt \qquad (4.1)$$

The signal $f(t)$ is said to be the **inverse Laplace transform** of $F(s)$. It will be shown later in Sec. 7.2-1 that

$$f(t) = \frac{1}{2\pi j} \int_{c-j\infty}^{c+j\infty} F(s)e^{st}\, ds \qquad (4.2)$$

where c is a constant chosen to ensure the convergence of the integral in Eq. (4.1), as explained later.

This pair of equations is known as the **bilateral Laplace transform pair**, where $F(s)$ is the direct Laplace transform of $f(t)$ and $f(t)$ is the inverse Laplace transform of $F(s)$. Symbolically,

$$F(s) = \mathcal{L}\left[f(t)\right] \qquad \text{and} \qquad f(t) = \mathcal{L}^{-1}\left[F(s)\right] \tag{4.3}$$

Note that

$$\mathcal{L}^{-1}\left\{\mathcal{L}[f(t)]\right\} = f(t) \qquad \text{and} \qquad \mathcal{L}\left\{\mathcal{L}^{-1}[F(s)]\right\} = F(s)$$

It is also common practice to use a bidirectional arrow to indicate a Laplace transform pair, as

$$f(t) \Longleftrightarrow F(s)$$

This means that $F(s)$ is the direct transform of $f(t)$ and $f(t)$ is the inverse transform of $F(s)$. This Laplace transform can handle signals existing over the entire time interval from $-\infty$ to ∞ (causal and noncausal signals). For this reason it is called the **bilateral** (or **two-sided**) Laplace transform. Later we shall consider a special case—the **unilateral** or **one-sided** Laplace transform—which can handle only causal signals.

Linearity of the Laplace Transform

We now prove that the Laplace transform is a linear operator by showing that the principle of superposition holds. We now show that if

$$f_1(t) \Longleftrightarrow F_1(s) \quad \text{and} \quad f_2(t) \Longleftrightarrow F_2(s)$$

then

$$a_1 f_1(t) + a_2 f_2(t) \Longleftrightarrow a_1 F_1(s) + a_2 F_2(s)$$

The proof is simple. By definition

$$\mathcal{L}\left[a_1 f_1(t) + a_2 f_2(t)\right] = \int_{-\infty}^{\infty} \left[a_1 f_1(t) + a_2 f_2(t)\right] e^{-st}\, dt$$

$$= a_1 \int_{-\infty}^{\infty} f_1(t) e^{-st}\, dt + a_2 \int_{-\infty}^{\infty} f_2(t) e^{-st}\, dt$$

$$= a_1 F_1(s) + a_2 F_2(s) \tag{4.4}$$

This result can be extended to any finite number of terms.

The Region of Convergence

The integral in Eq. (4.1) defining the direct Laplace transform $F(s)$ may not converge for all values of s. The values of s (the region) for which the integral in Eq. (4.1) converges is called the **region of convergence** (or **region of existence**) for $F(s)$. This concept will become clear from the following example.

■**Example 4.1**

For the signal $f(t) = e^{-at}u(t)$, find the Laplace transform $F(s)$ and its region of convergence.

By definition

$$F(s) = \int_{-\infty}^{\infty} e^{-at}u(t)e^{-st}\,dt$$

Because $u(t) = 0$ for $t < 0$ and $u(t) = 1$ for $t \geq 0$,

$$F(s) = \int_{0}^{\infty} e^{-at}e^{-st}\,dt = \int_{0}^{\infty} e^{-(s+a)t}\,dt = -\left.\frac{1}{s+a}e^{-(s+a)t}\right|_{0}^{\infty} \qquad (4.5)$$

Note that s is complex and as $t \to \infty$, the term $e^{-(s+a)t}$ does not necessarily vanish. Here we recall that for a complex number $z = \alpha + j\beta$,

$$e^{-zt} = e^{-(\alpha+j\beta)t} = e^{-\alpha t}e^{-j\beta t}$$

Now $|e^{-j\beta t}| = 1$ regardless of the value of βt (see p. 10). Therefore as $t \to \infty$, $e^{-zt} \to 0$ if $\alpha > 0$, and $e^{-zt} \to \infty$ if $\alpha < 0$. Thus

$$\lim_{t\to\infty} e^{-zt} = \begin{cases} 0 & \text{Re } z > 0 \\ \infty & \text{Re } z < 0 \end{cases} \qquad (4.6)$$

Clearly

$$\lim_{t\to\infty} e^{-(s+a)t} = \begin{cases} 0 & \text{Re}\,(s+a) > 0 \\ \infty & \text{Re}\,(s+a) < 0 \end{cases}$$

Use of this result in Eq. (4.5) yields

$$F(s) = \frac{1}{s+a} \qquad \text{Re}\,(s+a) > 0 \qquad (4.7a)$$

or

$$e^{-at}u(t) \Longleftrightarrow \frac{1}{s+a} \qquad \text{Re } s > -a \qquad (4.7b)$$

The region of convergence of $F(s)$ is Re $s > -a$, as shown shaded in Fig. 4.1a. This means that the integral defining $F(s)$ in Eq. (4.5) exists only for the values of s in the shaded region in Fig. 4.1a. For other values of s, the integral in Eq. (4.5) does not converge. For this reason the shaded region is called the *region of convergence* (or the *region of existence*) for $F(s)$.   ■

The region of convergence is required for evaluating the inverse Laplace transform $f(t)$ from $F(s)$ as defined by Eq. (4.2). The operation of finding the inverse transform requires an integration in the complex plane, which needs some explanation. The path of integration is along $c + j\omega$, with ω varying from $-\infty$ to ∞. Moreover, the path of integration must lie in the region of convergence (or existence) for $F(s)$. For the signal e^{-at}, this is possible if $c > -a$. One possible path of integration is shown (dotted) in Fig. 4.1a. Thus, to obtain $f(t)$ from $F(s)$, the integration in (4.2) is performed along this path. When we integrate $1/(s+a)$ along this path, the result is $e^{-at}u(t)$. Such integration in complex plane requires a background in the theory of functions of complex variables. We can avoid this integration by compiling a table of Laplace transform (Table 4.1), where Laplace transform pairs are tabulated for a variety of signals. To find the inverse Laplace

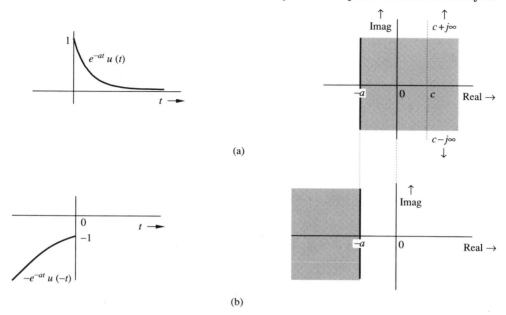

(a)

(b)

Fig. 4.1 Signals $e^{-at}u(t)$ and $-e^{at}u(-t)$ have the same Laplace transform but different regions of convergence.

transform of say, $1/(s+a)$, instead of using the complex integral (4.2), we look up the table and find the inverse Laplace transform to be $e^{-at}u(t)$. Although the table given here is rather short, it comprises the functions of most practical interest. A more comprehensive table can be found in Doetsch.[1]

Unilateral Laplace Transform

In order to understand the need for defining unilateral transform, let us find the Laplace transform of signal $f(t)$ shown in Fig. 4.1b:

$$f(t) = -e^{-at}u(-t)$$

The Laplace transform of this signal is

$$F(s) = \int_{-\infty}^{\infty} -e^{-at}u(-t)e^{-st}\,dt$$

Because $u(-t) = 1$ for $t < 0$ and $u(-t) = 0$ for $t > 0$,

$$F(s) = \int_{-\infty}^{0} -e^{-at}e^{-st}\,dt = -\int_{-\infty}^{0} e^{-(s+a)t}\,dt = \frac{1}{s+a}e^{-(s+a)t}\bigg|_{-\infty}^{0}$$

From Eq. (4.6), it follows that

$$\lim_{t \to -\infty} e^{-(s+a)t} = 0 \qquad \text{Re}\,(s+a) < 0$$

Hence

$$F(s) = \frac{1}{s+a} \qquad \text{Re}\,s < -a$$

The signal $-e^{-at}u(-t)$ and its region of convergence (Re $s < -a$) are shown in Fig. 4.1b. Note that the Laplace transforms for the signals $e^{-at}u(t)$ and $-e^{-at}u(-t)$ are identical except for their regions of convergence. Therefore, for a given $F(s)$ there may be more than one inverse transform depending on the region of convergence. In other words, there is no one-to-one correspondence between $F(s)$ and $f(t)$ unless the region of convergence is specified. This fact causes additional complexity in using the Laplace transform. The complexity is the result of trying to handle causal as well as noncausal signals. If we restrict all our signals to the causal type such an ambiguity does not arise. There is only one inverse transform of $F(s) = 1/(s + a)$, namely $e^{-at}u(t)$. To find $f(t)$ from $F(s)$, we need not even specify the region of convergence. In summary, if all signals are restricted to be causal, then, for a given $F(s)$, there is only one inverse transform $f(t)$.†

The unilateral Laplace transform is a special case of the bilateral Laplace transform, where all signals are restricted to be causal; consequently the limits of integration for integral in Eq. (4.1) can be taken from 0 to ∞. Therefore the unilateral Laplace transform $F(s)$ of a signal $f(t)$ is defined as

$$F(s) \equiv \int_{0^-}^{\infty} f(t)e^{-st}\, dt \tag{4.8}$$

The lower limit of integration is chosen to be 0^- (rather than 0) to avoid any ambiguity that may arise if $f(t)$ contains an impulse at the origin. The lower limit of 0^- ensures the inclusion of $\delta(t)$ in the interval of integration.

The unilateral Laplace transform simplifies the system analysis problem considerably, but the price for this simplification is that we cannot analyze noncausal systems or use noncausal inputs. However, in most practical problems this is of little consequence. For this reason we shall first consider the unilateral Laplace transform and its application to system analysis. (The bilateral Laplace transform is discussed later in Sec. 4.8.)

Observe that basically there is no difference between the unilateral and the bilateral Laplace transform. Unilateral transform is the bilateral transform that deals with a subclass of signals starting at $t = 0$ (causal signals). Therefore the expression (4.2) for the inverse Laplace transform remains unchanged. In practice the term *Laplace transform* means *the unilateral Laplace transform*.

Existence of the Laplace Transform

The variable s in the Laplace transform is complex in general, and it can be expressed as $s = \sigma + j\omega$. By definition

$$F(s) = \int_{0^-}^{\infty} f(t)e^{-st}\, dt$$

$$= \int_{0^-}^{\infty} \left[f(t)e^{-\sigma t}\right] e^{-j\omega t}\, dt$$

---

†Actually, $F(s)$ specifies $f(t)$ within a null function $n(t)$ which has the property that the area under $|n(t)|^2$ is zero over any finite interval 0 to t ($t > 0$) (Lerch's theorem). For example, if two functions are identical everywhere except at points of discontinuity, they differ by a null function. For all practical purposes , two functions differing by a null function are the same.

Because $|e^{j\omega t}| = 1$ (see p. 10), the integral on the right-hand side of this equation converges if

$$\int_{0-}^{\infty} \left| f(t)e^{-\sigma t} \right| \, dt \tag{4.9}$$

is finite. Hence the existence of the Laplace transform is guaranteed if the integral (4.9) is finite for some value of σ. Any signal that grows no faster than an exponential signal $Me^{\sigma_0 t}$ for some M and σ_0 satisfies condition (4.9). Thus, if for some M and σ_0,

$$|f(t)| \leq Me^{\sigma_0 t} \tag{4.10}$$

we can choose $\sigma > \sigma_0$ to make integral (4.9) finite.† The signal e^{t^2}, on the other hand, grows at a rate faster than $e^{\sigma_0 t}$, and consequently e^{t^2} is not Laplace transformable. Fortunately such signals (which are not Laplace transformable) are of little consequence from a practical and theoretical viewpoint. If σ_0 is the smallest value of σ for which the integral in (4.9) is finite, σ_0 is called the **abscissa of convergence** and the region of convergence of $F(s)$ is Re $s > \sigma_0$. The abscissa of convergence for $e^{-at}u(t)$ is $-a$ (the region of convergence is Re $s > -a$).

■ **Example 4.2**

Determine the Laplace transform of **(a)** $\delta(t)$ **(b)** $u(t)$ **(c)** $\cos \omega_0 t \, u(t)$.

  (a)

$$\mathcal{L}\left[\delta(t)\right] = \int_{0-}^{\infty} \delta(t)e^{-st} \, dt$$

Using the sampling property [Eq. (2.33a)], we obtain

$$\mathcal{L}\left[\delta(t)\right] = 1 \qquad \text{for all } s$$

that is

$$\delta(t) \Longleftrightarrow 1 \qquad \text{for all } s \tag{4.11}$$

  (b) To find the Laplace transform of $u(t)$, recall that $u(t) = 1$ for $t \geq 0$. Therefore

$$\mathcal{L}\left[u(t)\right] = \int_{0-}^{\infty} u(t)e^{-st} \, dt = \int_{0-}^{\infty} e^{-st} \, dt = -\left. \frac{1}{s}e^{-st} \right|_{0-}^{\infty}$$

$$= \frac{1}{s} \qquad \text{Re } s > 0 \tag{4.12}$$

We also could have obtained this result from (4.7b) by letting $a = 0$.

  (c) Because

$$\cos \omega_0 t \, u(t) = \tfrac{1}{2}\left[e^{j\omega_0 t} + e^{-j\omega_0 t}\right]u(t) \tag{4.13}$$

$$\mathcal{L}\left[\cos \omega_0 t \, u(t)\right] = \tfrac{1}{2}\mathcal{L}\left[e^{j\omega_0 t}u(t) + e^{-j\omega_0 t}u(t)\right]$$

From Eq. (4.7a), it follows that

---

†Condition (4.10) is sufficient but not necessary for the existence of the Laplace transform. For example $f(t) = 1/\sqrt{t}$ is infinite at $t = 0$ and (4.10) cannot be satisfied, but the transform of $1/\sqrt{t}$ exists and is given by $\sqrt{\pi/s}$.

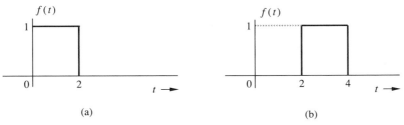

Fig. 4.2 Signals for Exercise E4.1.

$$\mathcal{L}\left[\cos \omega_0 t\, u(t)\right] = \frac{1}{2}\left[\frac{1}{s-j\omega_0} + \frac{1}{s+j\omega_0}\right] \qquad \mathrm{Re}\,(s \pm j\omega) = \mathrm{Re}\,s > 0$$

$$= \frac{s}{s^2 + \omega_0{}^2} \qquad \mathrm{Re}\,s > 0 \tag{4.14}$$

∎

For the unilateral Laplace transform, there is a unique inverse transform of $F(s)$; consequently there is no need to specify the region of convergence explicitly. For this reason we shall generally ignore any mention of the region of convergence for unilateral transforms.

In unilateral Laplace transform (discussed so far) it is understood that every signal $f(t)$ is zero for $t < 0$; the practice of multiplying $f(t)$ by $u(t)$ (as done here) is redundant and unnecessary. However, such a precaution may save us from some potential pitfalls, as pointed out later.

△ **Exercise E4.1**

By direct integration, find the Laplace transform $F(s)$ and the region of convergence of $F(s)$ for signals shown in Fig. 4.2.

Answer: **(a)** $\frac{1}{s}(1 - e^{-2s})$ for all s.   **(b)** $\frac{1}{s}(e^{-2s} - e^{-4s})$ for all s.   ▽

4.1-1   Finding the Inverse Transform

Finding the inverse Laplace transform by using the definition (4.2) requires integration in the complex plane. This is beyond the scope of this book.[2] For our purpose, we can find the inverse transforms from the transform table. All we need to do is to express $F(s)$ as a sum of simpler functions of the form listed in the table. Most of the transforms $F(s)$ of practical interest are **rational functions**, that is, ratios of polynomials in s. Such functions can be expressed as a sum of simpler functions by using partial fraction expansion (see Sec. B.8). Values of s for which $F(s) = 0$ are called the zeros of $F(s)$; the values of s for which $F(s) \to \infty$ and are called the **poles** of $F(s)$. If $F(s)$ is a rational function of the form $P(s)/Q(s)$, the roots of $P(s)$ are the zeros and the roots of $Q(s)$ are the poles of $F(s)$.

■ **Example 4.3**

Find the inverse Laplace transforms of:

(a) $\dfrac{7s - 6}{s^2 - s - 6}$ **(b)** $\dfrac{2s^2 + 5}{s^2 + 3s + 2}$ **(c)** $\dfrac{6(s + 34)}{s(s^2 + 10s + 34)}$ **(d)** $\dfrac{8s + 10}{(s + 1)(s + 2)^3}$

The inverse transform of none of the above functions is directly available in Table 4.1. We need to expand these functions into partial fractions. Readers should familiarize themselves thoroughly with the techniques of partial fraction expansion discussed in Sec. B.8.

Table 4.1

A Short Table of Laplace Transforms

	$f(t)$	$F(s)$
1	$\delta(t)$	1
2	$u(t)$	$\dfrac{1}{s}$
3	$tu(t)$	$\dfrac{1}{s^2}$
4	$t^n u(t)$	$\dfrac{n!}{s^{n+1}}$
5	$e^{\lambda t} u(t)$	$\dfrac{1}{s - \lambda}$
6	$te^{\lambda t} u(t)$	$\dfrac{1}{(s - \lambda)^2}$
7	$t^n e^{\lambda t} u(t)$	$\dfrac{n!}{(s - \lambda)^{n+1}}$
8a	$\cos bt\, u(t)$	$\dfrac{s}{s^2 + b^2}$
8b	$\sin bt\, u(t)$	$\dfrac{b}{s^2 + b^2}$
9a	$e^{-at} \cos bt\, u(t)$	$\dfrac{s + a}{(s + a)^2 + b^2}$
9b	$e^{-at} \sin bt\, u(t)$	$\dfrac{b}{(s + a)^2 + b^2}$
10a	$re^{-at} \cos(bt + \theta)\, u(t)$	$\dfrac{(r\cos\theta)s + (ar\cos\theta - br\sin\theta)}{s^2 + 2as + (a^2 + b^2)}$
10b	$re^{-at} \cos(bt + \theta)\, u(t)$	$\dfrac{0.5re^{j\theta}}{s + a - jb} + \dfrac{0.5re^{-j\theta}}{s + a + jb}$
10c	$re^{-at} \cos(bt + \theta)\, u(t)$	$\dfrac{As + B}{s^2 + 2as + c}$
	$r = \sqrt{\dfrac{A^2 c + B^2 - 2ABa}{c - a^2}},\ \theta = \tan^{-1}\dfrac{Aa - B}{A\sqrt{c - a^2}}$	
	$b = \sqrt{c - a^2}$	
10d	$e^{-at}\left[A\cos bt + \dfrac{B - Aa}{b}\sin bt\right] u(t)$	$\dfrac{As + B}{s^2 + 2as + c}$
	$b = \sqrt{c - a^2}$	

(a)

$$F(s) = \frac{7s - 6}{(s + 2)(s - 3)}$$

$$= \frac{k_1}{s + 2} + \frac{k_2}{s - 3}$$

To determine k_1, corresponding to the term $(s + 2)$, we cover the term $(s + 2)$ in $F(s)$ and substitute $s = -2$ (the value of s that makes $s + 2 = 0$) in the remaining expression (see Sec. B.8-2):

$$k_1 = \frac{7s - 6}{(s + 2)(s - 3)}\bigg|_{s=-2} = \frac{-14 - 6}{-2 - 3} = 4$$

Similarly, to determine k_2 corresponding to the term $(s - 3)$, we cover up the term $(s - 3)$ in $F(s)$ and substitute $s = 3$ in the remaining expression

$$k_2 = \frac{7s - 6}{(s + 2)(s - 3)}\bigg|_{s=3} = \frac{21 - 6}{3 + 2} = 3$$

Therefore

$$F(s) = \frac{7s - 6}{(s + 2)(s - 3)} = \frac{4}{s + 2} + \frac{3}{s - 3} \tag{4.15a}$$

Checking the answer

It is easy to make a mistake in partial fraction computations. Fortunately it is simple to check the answer by recognizing that $F(s)$ and its partial fractions must be equal for every value of s if the partial fractions are correct. Let us verify this in Eq. (4.15a) for some convenient value, say $s = 1$. Substitution of $s = 1$ in Eq. (4.15a) yields†

$$-\frac{1}{6} = \frac{4}{3} - \frac{3}{2} = -\frac{1}{6}$$

We can now be sure of our answer with a high margin of confidence. Using Pair 5 (Table 4.1) in Eq. (4.15a), we obtain

$$f(t) = \mathcal{L}^{-1}\left(\frac{4}{s + 2} + \frac{3}{s - 3}\right) = \left(4e^{-2t} + 3e^{3t}\right)u(t) \tag{4.15b}$$

(b)

$$F(s) = \frac{2s^2 + 5}{s^2 + 3s + 2} = \frac{2s^2 + 5}{(s + 1)(s + 2)}$$

Observe that $F(s)$ is an improper function with $m = n$. In such a case we can express $F(s)$ as a sum of the coefficient b_n (the coefficient of the highest power in the numerator) plus partial fractions corresponding to the poles of $F(s)$ (see Sec.B.8-4). In the present case $b_n = 2$. Therefore

$$F(s) = 2 + \frac{k_1}{s + 1} + \frac{k_2}{s + 2}$$

where

$$k_1 = \frac{2s^2 + 5}{(s + 1)(s + 2)}\bigg|_{s=-1} = \frac{2 + 5}{-1 + 2} = 7$$

---

†Because $F(s) = \infty$ at its poles, we should avoid the pole values (-2 and 3 in the present case) for checking. It is possible that the answers may check even if partial fractions are wrong. This can happen when two or more errors cancel their effects. But the chances of this happening for a randomly selected s are extremely small.

and

$$k_2 = \left. \frac{2s^2 + 5}{(s+1)(s+2)} \right|_{s=-2} = \frac{8+5}{-2+1} = -13$$

Therefore

$$F(s) = 2 + \frac{7}{s+1} - \frac{13}{s+2}$$

From Table 4.1, Pairs 1 and 5, we obtain

$$f(t) = 2\delta(t) + \left(7e^{-t} - 13e^{-2t}\right) u(t) \qquad (4.16)$$

(c)

$$F(s) = \frac{6(s+34)}{s(s^2 + 10s + 34)}$$

$$= \frac{6(s+34)}{s(s+5-j3)(s+5+j3)}$$

$$= \frac{k_1}{s} + \frac{k_2}{s+5-j3} + \frac{k_2{}^*}{s+5+j3}$$

Note that the coefficients (k_2 and $k_2{}^*$) of the conjugate terms must also be conjugate (see Sec. B.8). Now

$$k_1 = \left. \frac{6(s+34)}{s(s^2 + 10s + 34)} \right|_{s=0} = \frac{6 \times 34}{34} = 6$$

$$k_2 = \left. \frac{6(s+34)}{s(s+5-j3)(s+5+j3)} \right|_{s=-5+j3} = \frac{29 + j3}{-3 - j5} = -3 + j4$$

Therefore

$$k_2{}^* = -3 - j4$$

In order to be able to use Pair 10b (Table 4.1), we need to express k_2 and $k_2{}^*$ in polar form

$$-3 + j4 = \left(\sqrt{3^2 + 4^2}\right) e^{j \tan^{-1}(4/-3)} = 5e^{j \tan^{-1}(4/-3)}$$

Observe that $\tan^{-1}(\frac{4}{-3}) \neq \tan^{-1}(-\frac{4}{3})$. This is evident from Fig. 4.3. Remember that electronic calculators can give answers only for the angles in the first and the fourth quadrant. For this reason it is important to plot the point (e.g. $-3 + j4$) in the complex plane, as shown in Fig. 4.3, and determine the angle. For further discussion of this topic, see Example B.1.

From Fig. 4.3, we observe that

$$k_2 = -3 + j4 = 5e^{j126.9°}$$

so that

$$k_2{}^* = 5e^{-j126.9°}$$

Therefore

$$F(s) = \frac{6}{s} + \frac{5e^{j126.9°}}{s+5-j3} + \frac{5e^{-j126.9°}}{s+5+j3}$$

From Table 4.1 (Pairs 2 and 10b), we obtain

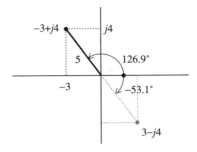

Fig. 4.3 $\tan^{-1}\left(\frac{-4}{3}\right) \neq \tan^{-1}\left(\frac{4}{-3}\right)$.

$$f(t) = \left[6 + 10e^{-5t}\cos\left(3t + 126.9°\right)\right]u(t) \qquad (4.17)$$

Alternative Method Using Quadratic Factors

The above procedure involves considerable manipulation of complex numbers. As indicated by Pair 10c (Table 4.1), the inverse transform of quadratic terms (with complex conjugate poles) can be found directly without having to find first-order partial fraction. We now show the alternative method (also discussed in Sec. B.8-2), which exploits this observation. For this purpose we shall express $F(s)$ as

$$F(s) = \frac{6(s+34)}{s(s^2+10s+34)} = \frac{k_1}{s} + \frac{As+B}{s^2+10s+34}$$

We have already found $k_1 = 6$ by the (Heaviside) "cover-up" method. Therefore

$$\frac{6(s+34)}{s(s^2+10s+34)} = \frac{6}{s} + \frac{As+B}{s^2+10s+34}$$

Clearing the fractions by multiplying both sides with $s(s^2+10s+34)$ yields

$$6(s+34) = 6\left(s^2+10s+34\right) + s(As+B)$$
$$= (6+A)s^2 + (60+B)s + 204$$

Now equating the coefficients of s^2 and s on both sides yields

$$0 = (6+A) \implies A = -6$$
$$6 = 60+B \implies B = -54$$

and

$$F(s) = \frac{6}{s} + \frac{-6s-54}{s^2+10s+34}$$

We now use Pairs 2 and 10c to find the inverse Laplace transform. The parameters for Pair 10c are $A = -6$, $B = -54$, $a = 5$, $c = 34$, and $b = \sqrt{c-a^2} = 3$, and

$$r = \sqrt{\frac{A^2c+B^2-2ABa}{c-a^2}} = 10, \quad \theta = \tan^{-1}\frac{Aa-B}{A\sqrt{c-a^2}} = 126.9°$$

Therefore

$$f(t) = [6 + 10e^{-5t}\cos\left(3t + 126.9°\right)]u(t)$$

which agrees with previous result.

Short-Cuts

The partial fractions with quadratic terms also can be found by using short cuts. We have

$$F(s) = \frac{6(s+34)}{s(s^2 + 10s + 34)} = \frac{6}{s} + \frac{As + B}{s^2 + 10s + 34}$$

We can determine A by eliminating B on the right-hand side. This can be done by multiplying both sides of the above equation by s and then letting $s \to \infty$. This yields

$$0 = 6 + A \implies A = -6$$

Therefore

$$\frac{6(s+34)}{s(s^2 + 10s + 34)} = \frac{6}{s} + \frac{-6s + B}{s^2 + 10s + 34}$$

To find B, we let s take on any convenient value, say $s = 1$, in this equation to yield

$$\frac{210}{45} = 6 + \frac{B - 6}{45}$$

Multiplying both sides of this equation by 45 yields

$$210 = 270 + B - 6 \implies B = -54$$

which agrees with the results we found earlier.

(d)

$$F(s) = \frac{8s + 10}{(s+1)(s+2)^3}$$

$$= \frac{k_1}{s+1} + \frac{a_0}{(s+2)^3} + \frac{a_1}{(s+2)^2} + \frac{a_2}{s+2}$$

where (see Sec. B.8-3)

$$k_1 = \frac{8s + 10}{(s+1)(s+2)^3}\bigg|_{s=-1} = 2$$

$$a_0 = \frac{8s + 10}{(s+1)(s+2)^3}\bigg|_{s=-2} = 6$$

$$a_1 = \left\{\frac{d}{ds}\left[\frac{8s + 10}{(s+1)(s+2)^3}\right]\right\}_{s=-2} = -2$$

$$a_2 = \frac{1}{2}\left\{\frac{d^2}{ds^2}\left[\frac{8s + 10}{(s+1)(s+2)^3}\right]\right\}_{s=-2} = -2$$

Therefore

$$F(s) = \frac{2}{s+1} + \frac{6}{(s+2)^3} - \frac{2}{(s+2)^2} - \frac{2}{s+2}$$

and

$$f(t) = \left[2e^{-t} + (3t^2 - 2t - 2)e^{-2t}\right]u(t) \tag{4.18}$$

Alternative Method: A Hybrid of Heaviside and Clearing Fractions

In this method the simpler coefficients k_1 and a_0 are determined by the Heaviside "cover-up" procedure, as discussed earlier. To determine the remaining coefficients, we use the clearing-fraction method (see Sec. B.8-3). Using the values $k_1 = 2$ and $a_0 = 6$ found earlier by the Heaviside "cover-up" method, we have

$$\frac{8s + 10}{(s + 1)(s + 2)^3} = \frac{2}{s + 1} + \frac{6}{(s + 2)^3} + \frac{a_1}{(s + 2)^2} + \frac{a_2}{s + 2}$$

We now clear fractions by multiplying both sides of the equation by $(s + 1)(s + 2)^3$. This yields†

$$8s + 10 = 2(s + 2)^3 + 6(s + 1) + a_1(s + 1)(s + 2) + a_2(s + 1)(s + 2)^2$$
$$= (2 + a_2)s^3 + (12 + a_1 + 5a_2)s^2 + (30 + 3a_1 + 8a_2)s + (22 + 2a_1 + 4a_2)$$

Equating coefficients of s^3 and s^2 on both sides, we obtain

$$0 = (2 + a_2) \implies a_2 = -2$$
$$0 = 12 + a_1 + 5a_2 = 2 + a_1 \implies a_1 = -2$$

We can stop here if we wish, since the two desired coefficients a_1 and a_2 are already found. However, equating the coefficients of s^1 and s^0 serves as a check on our answers. This yields

$$8 = 30 + 3a_1 + 8a_2$$
$$10 = 22 + 2a_1 + 4a_2$$

Substitution of $a_1 = a_2 = -2$, found earlier, satisfies these equations. This assures the correctness of our answers.

Alternative Method: A Hybrid of Heaviside and Short-Cuts

In this method the simpler coefficients k_1 and a_0 are determined by the Heaviside "cover-up" procedure, as discussed earlier. The usual short-cuts then are used to determine the remaining coefficients. Using the values $k_1 = 2$ and $a_0 = 6$, found earlier by the Heaviside method, we have

$$\frac{8s + 10}{(s + 1)(s + 2)^3} = \frac{2}{s + 1} + \frac{6}{(s + 2)^3} + \frac{a_1}{(s + 2)^2} + \frac{a_2}{s + 2}$$

There are two unknowns, a_1 and a_2. If we multiply both sides by s and then let $s \to \infty$, we eliminate a_1. This yields

$$0 = 2 + a_2 \implies a_2 = -2$$

Therefore

$$\frac{8s + 10}{(s + 1)(s + 2)^3} = \frac{2}{s + 1} + \frac{6}{(s + 2)^3} + \frac{a_1}{(s + 2)^2} - \frac{2}{s + 2}$$

---

†We could have cleared fractions without finding k_1 and a_0. This , however, proves more laborious because it increases the number of unknowns to 4. By predetermining k_1 and a_0, we reduce the unknowns to 2. Moreover, this method provides a convenient check on the solution. This hybrid procedure achieves the best of both methods.

There is now only one unknown, a_1. This can be found readily by setting s equal to any convenient value, say $s = 0$. This yields

$$\frac{10}{8} = 2 + \frac{3}{4} + \frac{a_1}{4} - 1 \implies a_1 = -2 \quad \blacksquare$$

⊙ **Computer Example C4.1**

Find the coefficients of the inverse Laplace transform for the following quadratic function using pair 10c in table 4.1:

$$F(s) = \frac{7s + 6}{s^2 + s + 7}$$

```
num = [7 6];    % Enter the numerator polynomial.
den = [1 1 7];  % Enter the denominator polynomial.
b =sqrt(den(3) - (den(2)/2)^2) % Solve for b
% Solve for theta.
theta = atan((num(1)*den(2)/2-num(2))/(num(1)*b))
r = sqrt((num(1)^2*den(3)+num(2)^2-num(1)*num(2)*den(2))/b^2)
% Calculates the coefficient r.
a = den(2)/2;   % Calculate the coefficient a.
t = 0:.1:10;    % Create a time vector.
f=r*exp(-a*t).*cos(b*t+theta);
disp('Strike any key to see the plot of f(t)')
pause
plot(t,f),grid        % You should see a dampened sinusoid.
xlabel('t'),ylabel('f(t)'),title('plot of f(t)') ⊙
```

△ **Exercise E4.2**

(i) Show that the Laplace transform of $10e^{-3t} \cos(4t + 53.13°)$ is $\dfrac{6s - 14}{s^2 + 6s + 25}$. Use Pair 10a from Table 4.1.

(ii) Find the inverse Laplace transform of: **(a)** $\dfrac{s + 17}{s^2 + 4s - 5}$

(b) $\dfrac{3s - 5}{(s + 1)(s^2 + 2s + 5)}$ **(c)** $\dfrac{16s + 43}{(s - 2)(s + 3)^2}$

Answers: **(a)** $\left(3e^t - 2e^{-5t}\right) u(t)$ **(b)** $\left[-2e^{-t} + \frac{5}{2}e^{-t} \cos(2t - 36.87°)\right] u(t)$

(c) $\left[3e^{2t} + (t - 3)e^{-3t}\right] u(t)$ ▽

A Historical Note: Marquis Pierre-Simon De Laplace (1749-1827)

The Laplace transform is named after the great French mathematician and astronomer Laplace, who first presented the transform and its applications to differential equations in a paper published in 1779.

Laplace developed the foundations of potential theory and made important contributions to special functions, probability theory, astronomy, and celestial mechanics. In his *Exposition du systeme du Monde* (1796), he formulated a nebular hypothesis of cosmic origin and tried to explain the universe as a pure mechanism.

P.S. de Laplace (left) and Oliver Heaviside (right).

In his *traite de Mechanique Celeste (celestial mechanics)*, which completed the work of Newton, he used mathematics and physics to subject the solar system and all heavenly bodies to the laws of motion and the principle of gravitation. Newton was unable to explain the irregularities of some heavenly bodies; in desperation, he concluded that God himself must intervene now and then to prevent some catastrophes, such as Jupiter eventually falling into the sun (and the moon in the earth) as predicted by Newton's calculations. Laplace proposed to show that these irregularities would correct themselves periodically, and that a little patience—in Jupiter's case, 929 years—would see everything returning automatically to order. He concluded that there was no reason why the solar and the stellar systems could not continue to operate by the laws of Newton and Laplace to the end of time.[3]

Laplace presented a copy of *Mechanique Celeste* to Napoleon, who, after reading the book, took Laplace to task for not including God in his scheme: "You have written this huge book on the system of the world without once mentioning the author of the universe." "Sire," Laplace retorted, "I had no need of that hypothesis." Napoleon was not amused, and when he reported this reply to Lagrange, the latter remarked, "Ah, but that is a fine hypothesis. It explains so many things."[4]

Napoleon, following his policy of honoring and promoting scientists, made Laplace the minister of the interior. To Napoleon's dismay, he found the great mathematician-astronomer bringing "the spirit of infinitesimals" into administration, and so had Laplace transferred hastily to the senate.

Oliver Heaviside (1850-1925)

Although Laplace published his transform method to solve differential equations in 1779, it did not catch on until a century later. It was rediscovered independently in a rather awkward form by an eccentric British engineer, Oliver Heaviside (1850-1925), one of the tragic figures in the history of electrical engineering. Despite his prolific contributions to electrical engineering, he was severely criticized during his lifetime, and was neglected later to the point that hardly a textbook today mentions his name or contributions. With the passage of time, Heaviside becomes more distant, although his studies had a major impact on many aspects of modern electrical engineering. It was Heaviside who made transatlantic communication possible by inventing cable loading, but no one ever mentions him as a pioneer or an innovator in telephony. It was Heaviside who suggested the use of inductive cable loading, but the credit is given to M. Pupin, who actually built the first loading coil. In addition Heaviside was[5]

- The first to find a solution to the distortionless transmission line;
- The innovator of lowpass filters;
- The first to write Maxwell's equations in modern form;
- The co-discoverer of rate energy transfer by an electromagnetic field;
- An early champion of the now-common phasor analysis;
- An important contributor to the development of vector analysis. In fact, he essentially created the subject independently of Gibbs.[6]
- An originator of the use of operational mathematics used to solve linear integro-differential equations, which eventually led to rediscovery of the Laplace transform;
- The first to theorize (along with Kennelly of Harvard) that a conducting layer (the Kennelly-Heaviside layer) of atmosphere exists, which allows radio waves to follow earth's curvature instead of traveling off into space in a straight line;
- The first to posit that an electrical charge would increase in mass as its velocity increases; an anticipation of an aspect of Einstein's special theory of relativity.[7] He also forecast the possibility of superconductivity.

Heaviside was a self-made, self-educated man with only an elementary school education, who eventually became a pragmatically successful mathematical physicist. He began his career as a telegrapher, but increasing deafness forced him to retire at the age of 24. He then devoted himself to the study of electricity. His creative work was disdained by many professional mathematicians because of his lack of formal education and his unorthodox methods.

Heaviside's misfortune was that he was criticized by both the mathematicians, who faulted him for lack of rigor, and by men of practice, who faulted him for using too much mathematics and thereby confusing students. Many mathematicians, trying to find solutions to the distortionless transmission line, failed because no rigorous tools were available at the time. Heaviside succeeded because he used mathematics not with rigor, but with insight and intuition. Using his much-maligned operational method, Heaviside successfully attacked problems that the rigid mathematicians could not, problems such as the flow of heat in a body of spatially varying conductivity. Heaviside brilliantly used this method in 1895 to demonstrate a fatal flaw in Lord Kelvin's determination of the geological age of the earth by secular cooling; he used the same flow of heat theory as for his cable analysis. Yet the Royal Society mathematicians remained unmoved and had no sympathy merely because Heaviside found the answer to problems no one else could solve. Many mathematicians who examined his work dismissed it with contempt, asserting that his methods were either complete nonsense or a rehash of already-known ideas.[5]

Sir William Preece, the chief engineer of the British Post Office, a savage critic of Heaviside, ridiculed Heaviside's work as too theoretical and therefore leading to faulty conclusions. Heaviside's work on transmission lines and loading was dismissed by the British Post Office; this work might have remained hidden, had not Lord Kelvin himself publicly expressed admiration for it.[5]

Heaviside's operational calculus may be formally inaccurate, but in fact it anticipated the current operational methods developed in more recent years.[8] Although his method was not fully understood, it provided correct results. When Heaviside was attacked for the vague meaning of his operational calculus, his pragmatic reply was "Shall I refuse my dinner because I do not fully understand the process of digestion?"

Heaviside lived as a bachelor hermit, often in near-squalid conditions, and died largely unnoticed, in poverty. His life demonstrates the arrogance and snobbishness of the intellectual establishment, which does not respects creativity unless it is presented in the strict language of the establishment.

4.2 SOME PROPERTIES OF THE LAPLACE TRANSFORM

Properties of the Laplace transform are useful not only in the derivation of Laplace transform of functions but also in the solutions of linear integro-differential equations. A glance at Eqs. (4.1) and (4.2) shows that there is a certain measure of symmetry in going from $f(t)$ to $F(s)$, and vice versa. This symmetry or duality is also carried over to the properties of the Laplace transform. This will be evident from the following development.

Time Shifting

This property states that if

$$f(t) \Longleftrightarrow F(s)$$

then for $t_0 \geq 0$

$$f(t - t_0) \Longleftrightarrow F(s)e^{-st_0} \qquad (4.19a)$$

Observe that $f(t)$ starts at $t = 0$, and therefore $f(t - t_0)$ starts at $t = t_0$. This fact is implicit, but is not explicitly shown in Eq. (4.19a). This often leads to inadvertent errors. To avoid this pitfall, this property should be restated as follows:

If

$$f(t)u(t) \Longleftrightarrow F(s)$$

then

$$f(t - t_0)u(t - t_0) \Longleftrightarrow F(s)e^{-st_0} \qquad (4.19b)$$

Proof:

$$\mathcal{L}\left[f(t - t_0)u(t - t_0)\right] = \int_0^\infty f(t - t_0)u(t - t_0)e^{-st}\,dt$$

Setting $t - t_0 = x$, we obtain

$$\mathcal{L}\left[f(t - t_0)u(t - t_0)\right] = \int_{-t_0}^\infty f(x)u(x)e^{-s(x+t_0)}\,dx$$

Because $u(x) = 0$ for $x < 0$ and $u(x) = 1$ for $x \geq 0$, the limits of integration may be changed to from 0 to ∞. Thus

$$\mathcal{L}\left[f(t - t_0)u(t - t_0)\right] = \int_0^\infty f(x)e^{-s(x+t_0)}\, dx$$

$$= e^{-st_0} \int_0^\infty f(x)e^{-sx}\, dx$$

$$= F(s)e^{-st_0}$$

Note that $f(t - t_0)u(t - t_0)$ is the signal $f(t)u(t)$ delayed by t_0 seconds. The time-shifting property states that **delaying a signal by t_0 seconds amounts to multiplying its transform e^{-st_0}**.

This property of unilateral Laplace transform holds only for positive t_0 because if t_0 were negative, the signal $f(t - t_0)u(t - t_0)$ would be signal $f(t)u(t)$ shifted to left and therefore not causal.

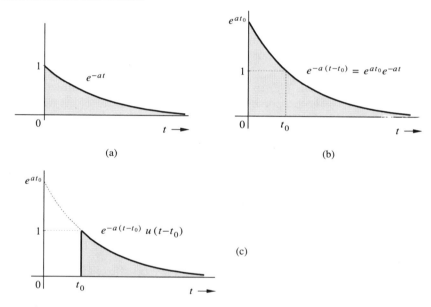

Fig. 4.4 A possible pitfall when a signal is not expressed as $f(t)u(t)$ in the (unilateral) Laplace transform.

Comment

This property illustrates the wisdom of expressing a signal $f(t)$ as $f(t)u(t)$ in (unilateral) Laplace transform analysis. For instance, if we did not use $u(t)$ to indicate the causal nature of a signal, the Laplace transform of a causal exponential e^{-at} would read

$$e^{-at} \Longleftrightarrow \frac{1}{s+a}$$

The time-shifting property yields

$$e^{-a(t-t_0)} \Longleftrightarrow \frac{1}{s+a}e^{-st_0}$$

Now

$$e^{-a(t-t_0)} = e^{at_0}e^{-at}$$

Superficially it appears that $e^{-a(t-t_0)}$ is an exponential e^{-at} multiplied by a constant e^{at_0}, as shown in Fig. 4.4b. This is clearly wrong! We meant $e^{-a(t-t_0)}$ to be $e^{-at}u(t)$ delayed by t_0, as shown in Fig. 4.4c. The correct expression for this signal is $e^{-a(t-t_0)}u(t-t_0)$. Had we expressed the exponential as $e^{-at}u(t)$, the time-delay property would have given the correct expression $e^{-a(t-t_0)}u(t-t_0)$.

The time-shifting property proves very convenient in finding the Laplace transform of piecewise continuous functions, as demonstrated in the following example.

■ Example 4.4

Find the Laplace transform of $f(t)$ shown in Fig. 4.5a.

Describing a piecewise continuous function mathematically is discussed fully in Sec. B.5. The function $f(t)$ in Fig. 4.5a can be described as a sum of two components shown in Fig. 4.5b. The equation for the first component is $t-1$ over $1 \le t \le 2$, so that this part of $f(t)$ can be described by $(t-1)[u(t-1) - u(t-2)]$. The second component can be described by $u(t-2) - u(t-4)$. Therefore

$$f(t) = (t-1)\left[u(t-1) - u(t-2)\right] + \left[u(t-2) - u(t-4)\right]$$
$$= (t-1)u(t-1) - (t-1)u(t-2) + u(t-2) - u(t-4) \qquad (4.20a)$$

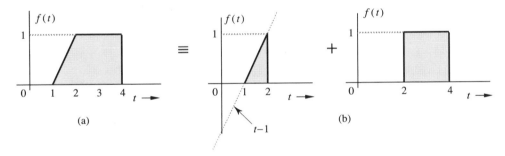

Fig. 4.5 Finding a mathematical description of a piecewise-continuous $f(t)$.

The first term on the right-hand side is the signal $tu(t)$ delayed by 1 second. Also, the third and fourth terms are the signal $u(t)$ delayed by 2 and 4 seconds respectively. The second term, however, cannot be interpreted as a delayed version of any entry in Table 4.1. For this reason, we rearrange it as

$$(t-1)u(t-2) = (t-2+1)u(t-2) = (t-2)u(t-2) + u(t-2)$$

We have now expressed the second term in the desired form as $tu(t)$ delayed by 2 seconds plus $u(t)$ delayed by 2 second. With this result, Eq. (4.20a) can be expressed as

$$f(t) = (t-1)u(t-1) - (t-2)u(t-2) - u(t-4) \qquad (4.20b)$$

Application of the time-shifting property to $tu(t) \Longleftrightarrow 1/s^2$ yields

$$(t-1)u(t-1) \Longleftrightarrow \frac{1}{s^2}e^{-s} \quad \text{and} \quad (t-2)u(t-2) \Longleftrightarrow \frac{1}{s^2}e^{-2s}$$

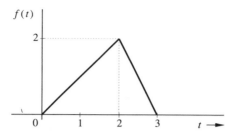

Fig. 4.6 The signal for Exercise E4.3.

Also

$$u(t) \Longleftrightarrow \frac{1}{s} \quad \text{and} \quad u(t-4) \Longleftrightarrow \frac{1}{s}e^{-4s} \tag{4.21}$$

Therefore

$$F(s) = \frac{1}{s^2}e^{-s} - \frac{1}{s^2}e^{-2s} - \frac{1}{s}e^{-4s} \tag{4.22}$$

∎

■**Example 4.5**

Find the inverse Laplace transform of

$$F(s) = \frac{s+3+5e^{-2s}}{(s+1)(s+2)}$$

Observe the exponential term e^{-2s} in the numerator of $F(s)$, indicating time-delay. In such a case we should separate $F(s)$ into terms with and without delay factor, as:

$$F(s) = \underbrace{\frac{s+3}{(s+1)(s+2)}}_{F_1(s)} + \underbrace{\frac{5e^{-2s}}{(s+1)(s+2)}}_{F_2(s)e^{-2s}}$$

where

$$F_1(s) = \frac{s+3}{(s+1)(s+2)} = \frac{2}{s+1} - \frac{1}{s+2}$$

$$F_2(s) = \frac{5}{(s+1)(s+2)} = \frac{5}{s+1} - \frac{5}{s+2}$$

Therefore

$$f_1(t) = \left(2e^{-t} - e^{-2t}\right) u(t)$$

$$f_2(t) = 5\left(e^{-t} - e^{-2t}\right) u(t)$$

Also, because

$$F(s) = F_1(s) + F_2(s)e^{-2s}$$

$$f(t) = f_1(t) + f_2(t-2)$$

$$= \left(2e^{-t} - e^{-2t}\right) u(t) + 5\left[e^{-(t-2)} - e^{-2(t-2)}\right] u(t-2) \quad ■$$

△ **Exercise E4.3**

Find the Laplace transform of the signal shown in Fig. 4.6.

Answer: $\dfrac{1}{s^2}(1 - 3e^{-2s} + 2e^{-3s})$ ▽

△ **Exercise E4.4**

Find the inverse Laplace transform of
$$F(s) = \frac{3e^{-2s}}{(s-1)(s+2)}$$

Answer: $\left[e^{t-2} - e^{-2(t-2)}\right] u(t-2)$ ▽

Frequency Shifting

This property states that if
$$f(t) \Longleftrightarrow F(s)$$
then
$$f(t)e^{s_0 t} \Longleftrightarrow F(s - s_0) \qquad (4.23)$$

Observe the symmetry (or duality) between this property and the time-shifting property (4.19a).

Proof:
$$\mathcal{L}\left[f(t)e^{s_0 t}\right] = \int_{0^-}^{\infty} f(t)e^{s_0 t} e^{-st}\, dt = \int_{0^-}^{\infty} f(t)e^{-(s-s_0)t}\, dt = F(s - s_0)$$

■**Example 4.6**

Derive Pair 9a in Table 4.1 from Pair 8a and the frequency-shifting property.
The Pair 8a is
$$\cos bt\, u(t) \Longleftrightarrow \frac{s}{s^2 + b^2}$$

From the frequency-shifting property [Eq. (4.21)] with $s_0 = -a$ we obtain
$$e^{-at}\cos bt\, u(t) \Longleftrightarrow \frac{s+a}{(s+a)^2 + b^2} \quad ■$$

△ **Exercise E4.5**

Derive Pair 6 in Table 4.1 from Pair 3 and the frequency-shifting property.   ▽

We now come to the two most important properties of the Laplace transform: the time-differentiation and time-integration properties.

The Time-Differentiation Property†

This property states that if
$$f(t) \Longleftrightarrow F(s)$$
then
$$\frac{df}{dt} \Longleftrightarrow sF(s) - f(0^-) \qquad (4.24a)$$

---

†The dual of the time-differentiation property is the frequency-differentiation property, which states that
$$tf(t) \Longleftrightarrow -\frac{d}{ds}F(s)$$

Repeated application of this property yields

$$\frac{d^2 f}{dt^2} \Longleftrightarrow s^2 F(s) - s f(0^-) - \dot{f}(0^-) \tag{4.24b}$$

$$\frac{d^n f}{dt^n} \Longleftrightarrow s^n F(s) - s^{n-1} f(0^-) - s^{n-2} \dot{f}(0^-) - \cdots - f^{(n-1)}(0^-) \tag{4.24c}$$

where $f^{(r)}(0^-)$ is $d^r f/dt^r$ at $t = 0^-$.

Proof:

$$\mathcal{L}\left[\frac{df}{dt}\right] = \int_{0^-}^{\infty} \frac{df}{dt} e^{-st}\, dt$$

Integrating by parts, we obtain

$$\mathcal{L}\left[\frac{df}{dt}\right] = f(t)e^{-st}\Big|_{0^-}^{\infty} + s \int_{0^-}^{\infty} f(t)e^{-st}\, dt$$

Observe that $f(t)e^{-st}$ is absolutely integrable for the values of s in the region of convergence for $F(s)$ [see Eq. (4.9)]. Hence $f(t)e^{-st} \to 0$ as $t \to \infty$, and

$$\mathcal{L}\left[\frac{df}{dt}\right] = -f(0^-) + s F(s)$$

■**Example 4.7**

Find the Laplace transform of the signal $f(t)$ in Fig. 4.7a using Table 4.1 and the time-differentiation and time-shifting properties of the Laplace transform.

Figures 4.7b and 4.7c show the first two derivatives of $f(t)$. Recall that the derivative at a point of jump discontinuity is an impulse of strength equal to the amount of jump [see Eq. (2.36)]. Therefore

$$\frac{d^2 f}{dt^2} = \delta(t) - 3\delta(t-2) + 2\delta(t-3)$$

The Laplace transform of this equation yields

$$\mathcal{L}\left(\frac{d^2 f}{dt^2}\right) = \mathcal{L}\left[\delta(t) - 3\delta(t-2) + 2\delta(t-3)\right]$$

Using the time-differentiation property (4.24b), the time-shifting property (4.19a) and the facts that $f(0^-) = \dot{f}(0^-) = 0$, and $\delta(t) \Longleftrightarrow 1$, we obtain

$$s^2 F(s) - 0 - 0 = 1 - 3e^{-2s} + 2e^{-3s}$$

Therefore

$$F(s) = \frac{1}{s^2}\left(1 - 3e^{-2s} + 2e^{-3s}\right)$$

which confirms the earlier result in Exercise E4.3. ■

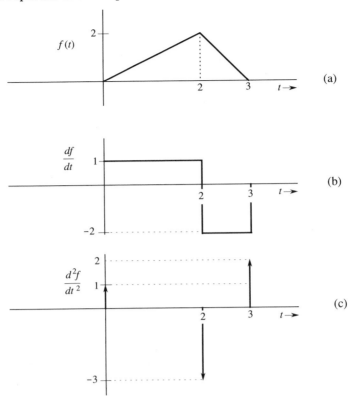

Fig. 4.7 Finding the Laplace transform of a piecewise-linear function using the time-differentiation property.

The Time-Integration Property†

This property states that if

$$f(t) \Longleftrightarrow F(s)$$

then

$$\int_{0^-}^t f(\tau)\,d\tau \Longleftrightarrow \frac{F(s)}{s} \tag{4.25}$$

and

$$\int_{-\infty}^t f(\tau)\,d\tau \Longleftrightarrow \frac{F(s)}{s} + \frac{\int_{-\infty}^{0^-} f(\tau)\,d\tau}{s} \tag{4.26}$$

Proof: We define

$$g(t) \equiv \int_{0^-}^t f(\tau)\,d\tau$$

---

†The dual of the time-integration property is the frequency-integration property, which states that

$$\frac{f(t)}{t} \Longleftrightarrow \int_s^\infty F(x)\,dx$$

so that

$$\frac{d}{dt}g(t) = f(t) \qquad \text{and} \qquad g(0^-) = 0$$

Now if

$$g(t) \Longleftrightarrow G(s)$$

then

$$F(s) = \mathcal{L}\left[\frac{d}{dt}g(t)\right] = sG(s) - g(0^-) = sG(s)$$

Therefore

$$G(s) = \frac{F(s)}{s}$$

or

$$\int_{0^-}^{t} f(\tau)\,d\tau \Longleftrightarrow \frac{F(s)}{s}$$

To prove Eq. (4.26), observe that

$$\int_{-\infty}^{t} f(\tau)\,d\tau = \int_{-\infty}^{0^-} f(\tau)\,d\tau + \int_{0^-}^{t} f(\tau)\,d\tau$$

Note that the first term on the right-hand side is a constant. Taking the Laplace transform of the above equation and using Eq. (4.25), we obtain

$$\int_{-\infty}^{t} f(\tau)\,d\tau \Longleftrightarrow \frac{\int_{-\infty}^{0^-} f(\tau)\,d\tau}{s} + \frac{F(s)}{s}$$

Scaling

This property states that if

$$f(t) \Longleftrightarrow F(s)$$

then for $a > 0$

$$f(at) \Longleftrightarrow \frac{1}{a}F\left(\frac{s}{a}\right) \qquad\qquad (4.27)$$

The proof is given in Chapter 7. Note that a is restricted to only positive values because if $f(t)$ is causal, then $f(at)$ is anticausal (exists only for $t < 0$) for negative a. This is not permitted in the (unilateral) Laplace transform.

Recall that $f(at)$ is the signal $f(t)$ time-compressed by the factor a, and $F(\frac{s}{a})$ is $F(s)$ expanded along the s-scale by the same factor a (see Sec. B.4-2). The scaling property states that *time-compression of a signal by a factor a causes expansion of its Laplace transform in s-scale by the same factor. Similarly, time-expansion $f(t)$ causes compression of $F(s)$ in s-scale by the same factor.*

Time-Convolution and Frequency Convolution

This property states that if

$$f_1(t) \Longleftrightarrow F_1(s) \quad \text{and} \quad f_2(t) \Longleftrightarrow F_2(s)$$

Table 4.2

Laplace Transform Operations

Operation	$f(t)$	$F(s)$
Addition	$f_1(t) + f_2(t)$	$F_1(s) + F_2(s)$
Scalar multiplication	$kf(t)$	$kF(s)$
Time differentiation	$\dfrac{df}{dt}$	$sF(s) - f(0^-)$
	$\dfrac{d^2 f}{dt^2}$	$s^2 F(s) - sf(0^-) - \dot{f}(0^-)$
	$\dfrac{d^3 f}{dt^3}$	$s^3 F(s) - s^2 f(0^-) - s\dot{f}(0^-) - \ddot{f}(0^-)$
Time integration	$\displaystyle\int_{0^-}^{t} f(t)\, dt$	$\dfrac{1}{s} F(s)$
	$\displaystyle\int_{-\infty}^{t} f(t)\, dt$	$\dfrac{1}{s} F(s) + \dfrac{1}{s} \displaystyle\int_{-\infty}^{0^-} f(t)\, dt$
Time shift	$f(t - t_0)u(t - t_0)$	$F(s)e^{-st_0} \quad t_0 \geq 0$
Frequency shift	$f(t)e^{s_0 t}$	$F(s - s_0)$
Frequency differentiation	$-tf(t)$	$\dfrac{dF(s)}{ds}$
Frequency integration	$\dfrac{f(t)}{t}$	$\displaystyle\int_{s}^{\infty} F(s)\, ds$
Scaling	$f(at),\ a \geq 0$	$\dfrac{1}{a} F\left(\dfrac{s}{a}\right)$
Time convolution	$f_1(t) * f_2(t)$	$F_1(s)F_2(s)$
Frequency convolution	$f_1(t)f_2(t)$	$\dfrac{1}{2\pi j} F_1(s) * F_2(s)$
Initial value	$f(0^+)$	$\displaystyle\lim_{s \to \infty} sF(s)$
Final value	$f(\infty)$	$\displaystyle\lim_{s \to 0} sF(s)$ (poles of $sF(s)$ in LHP)

then (**time-convolution property**)

$$f_1(t) * f_2(t) \Longleftrightarrow F_1(s)F_2(s) \tag{4.28}$$

and (**frequency-convolution property**)

$$f_1(t)f_2(t) \Longleftrightarrow \frac{1}{2\pi j}\left[F_1(s) * F_2(s)\right] \tag{4.29}$$

Observe the symmetry (or duality) between the two properties. Proof of these properties is postponed to Chapter 7.

■ **Example 4.8**

Using the time-convolution property of the Laplace transform, determine $e^{at}u(t) * e^{bt}u(t)$.

From Eq. (4.28), it follows that

$$\mathcal{L}\left\{e^{at}u(t) * e^{bt}u(t)\right\} = \frac{1}{(s-a)(s-b)} = \frac{1}{a-b}\left[\frac{1}{s-a} - \frac{1}{s-b}\right]$$

The inverse transform of the above equation yields

$$e^{at}u(t) * e^{bt}u(t) = \frac{1}{a-b}(e^{at} - e^{bt})u(t) \quad ■$$

4.3　LAPLACE TRANSFORM SOLUTION OF DIFFERENTIAL AND INTEGRO-DIFFERENTIAL EQUATIONS

The time-differentiation property of the Laplace transform has set the stage for solving linear differential (or integro-differential) equations with constant coefficients. Because $d^k y/dt^k \leftrightarrow s^k Y(s)$, the Laplace transform of a differential equation is an algebraic equation which can be readily solved for $Y(s)$. Next we take the inverse Laplace transform of $Y(s)$ to find the desired solution $y(t)$. The following examples demonstrate the Laplace transform procedure for solving linear differential equations with constant coefficients.

■ **Example 4.9**

Solve the second-order linear differential equation

$$\left(D^2 + 5D + 6\right) y(t) = (D+1)f(t) \tag{4.30a}$$

if the initial conditions are $y(0^-) = 2$, $\dot{y}(0^-) = 1$, and the input $f(t) = e^{-4t}u(t)$.

The equation is

$$\frac{d^2y}{dt^2} + 5\frac{dy}{dt} + 6y(t) = \frac{df}{dt} + f(t) \tag{4.30b}$$

Let

$$y(t) \Longleftrightarrow Y(s)$$

Then from Eq. (4.24)

$$\frac{dy}{dt} \Longleftrightarrow sY(s) - y(0^-) = sY(s) - 2$$

and

$$\frac{d^2y}{dt^2} \iff s^2Y(s) - sy(0^-) - \dot{y}(0^-) = s^2Y(s) - 2s - 1$$

Moreover, for $f(t) = e^{-4t}u(t)$,

$$F(s) = \frac{1}{s+4}, \quad \text{and} \quad \frac{df}{dt} \iff sF(s) - f(0^-) = \frac{s}{s+4} - 0 = \frac{s}{s+4}$$

Taking the Laplace transform of Eq. (4.30b), we obtain

$$\left[s^2Y(s) - 2s - 1\right] + 5\left[sY(s) - 2\right] + 6Y(s) = \frac{s}{s+4} + \frac{1}{s+4} \qquad (4.31a)$$

Collecting all the terms of $Y(s)$ and the remaining terms separately on the left-hand side, we obtain

$$\left(s^2 + 5s + 6\right)Y(s) - (2s + 11) = \frac{s+1}{s+4} \qquad (4.31b)$$

Therefore

$$\left(s^2 + 5s + 6\right)Y(s) = (2s + 11) + \frac{s+1}{s+4}$$

$$= \frac{2s^2 + 20s + 45}{s+4}$$

and

$$Y(s) = \frac{2s^2 + 20s + 45}{(s^2 + 5s + 6)(s+4)}$$

$$= \frac{2s^2 + 20s + 45}{(s+2)(s+3)(s+4)}$$

Expanding the right-hand side into partial fractions yields

$$Y(s) = \frac{13/2}{s+2} - \frac{3}{s+3} - \frac{3/2}{s+4}$$

The inverse Laplace transform of the above equation yields

$$y(t) = \left(\tfrac{13}{2}e^{-2t} - 3e^{-3t} - \tfrac{3}{2}e^{-4t}\right)u(t) \quad \blacksquare$$

 This example demonstrates the ease with which the Laplace transform can solve linear differential equations with constant coefficients. The method is general and can solve a linear differential equation with constant coefficients of any order.

Zero-Input and Zero-State Components of Response

 The Laplace transform method gives the total response, which includes zero-input and zero-state components. It is possible to separate the two components if we wish to do so. The initial conditions terms in the response give rise to zero-input response. For instance, in Example 4.9, the terms arising due to initial conditions $y(0^-) = 2$ and $\dot{y}(0^-) = 1$ in Eq. (4.31a) generate the zero-input response. These initial condition terms are $-(2s + 11)$, as seen in Eq. (4.31b). The terms on the right-hand side are due exclusively to the input. Equation (4.31b) is reproduced below with the proper labeling of the terms.

$$\left(s^2 + 5s + 6\right) Y(s) - (2s + 11) = \frac{s+1}{s+4} \tag{4.32a}$$

so that

$$\left(s^2 + 5s + 6\right) Y(s) = \underbrace{(2s + 11)}_{\text{initial condition terms}} + \underbrace{\frac{s+1}{s+4}}_{\text{input terms}} \tag{4.32b}$$

Therefore

$$Y(s) = \underbrace{\frac{2s + 11}{s^2 + 5s + 6}}_{\text{zero-input component}} + \underbrace{\frac{s+1}{(s+4)(s^2 + 5s + 6)}}_{\text{zero-state component}}$$

$$= \left[\frac{7}{s+2} - \frac{5}{s+3}\right] + \left[\frac{-1/2}{s+2} + \frac{2}{s+3} - \frac{3/2}{s+4}\right]$$

Taking the inverse transform of this equation yields

$$y(t) = \underbrace{\left(7e^{-2t} - 5e^{-3t}\right)}_{\text{zero-input response}} + \underbrace{\left(-\tfrac{1}{2}e^{-2t} + 2e^{-3t} - \tfrac{3}{2}e^{-4t}\right)}_{\text{zero-state response}}$$

△ **Exercise E4.6**
  Solve

$$\frac{d^2 y}{dt^2} + 4\frac{dy}{dt} + 3y(t) = 2\frac{df}{dt} + f(t)$$

for the input $f(t) = u(t)$. The initial conditions are $y(0^-) = 1$ and $\dot{y}(0^-) = 2$.
Answer: $y(t) = \frac{1}{3}(1 + 9e^{-t} - 7e^{-3t})u(t)$ ▽

■**Example 4.10**
    In the circuit of Fig. 4.8a, the switch is in the closed position for a long time before $t = 0$, when it is opened instantaneously. Find the inductor current $y(t)$ for $t \geq 0$.
    When the switch is in the closed position (for a long time), the inductor current is 2 amps and the capacitor voltage is 10 volts. When the switch is opened, the circuit is equivalent to that shown in Fig. 4.8b, with the initial inductor current $y(0^-) = 2$ and the initial capacitor voltage $v_C(0^-) = 10$. The input voltage is 10 volts, starting at $t = 0$, and therefore can be represented by $10u(t)$.
    The loop equation of the circuit in Fig. 4.8b is

$$\frac{dy}{dt} + 2y(t) + 5\int_{-\infty}^{t} y(\tau)\, d\tau = 10u(t) \tag{4.33}$$

If

$$y(t) \Longleftrightarrow Y(s) \tag{4.34a}$$

then

$$\frac{dy}{dt} \Longleftrightarrow sY(s) - y(0^-) = sY(s) - 2 \tag{4.34b}$$

and [see Eq. (4.26)]

$$\int_{-\infty}^{t} y(\tau)\, d\tau \Longleftrightarrow \frac{Y(s)}{s} + \frac{\int_{-\infty}^{0^-} y(\tau)\, d\tau}{s} \tag{4.34c}$$

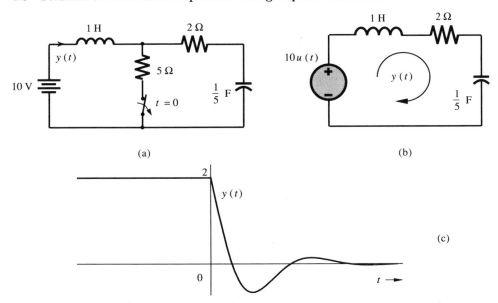

Fig. 4.8 Analysis of a network with a switching action (Example 4.10).

To determine the integral $\int_{-\infty}^{0^-} y(\tau)\,d\tau$, we observe that the capacitor voltage $v_C(t)$ is given by

$$v_C(t) = \frac{1}{C} \int_{-\infty}^{t} y(\tau)\,d\tau$$

Setting $t = 0^-$ in this equation, we find $v_C(0^-)$, the initial capacitor voltage

$$v_C(0^-) = \frac{1}{C} \int_{-\infty}^{0^-} y(\tau)\,d\tau$$

Therefore

$$\int_{-\infty}^{0^-} y(\tau)\,d\tau = C v_C(0^-) = \frac{1}{5}(10) = 2 \tag{4.35}$$

From Eq. (4.34c) it follows that

$$\int_{-\infty}^{t} y(\tau)\,d\tau \iff \frac{Y(s)}{s} + \frac{2}{s} \tag{4.36}$$

Taking the Laplace transform of Eq. (4.33) and using Eqs. (4.34a), (4.34b), and (4.36), we obtain

$$sY(s) - 2 + 2Y(s) + \frac{5Y(s)}{s} + \frac{10}{s} = \frac{10}{s}$$

or

$$\left[s + 2 + \frac{5}{s} \right] Y(s) = 2$$

and

$$Y(s) = \frac{2s}{s^2 + 2s + 5}$$

To find the inverse Laplace transform of $Y(s)$, we use Pair 10c (Table 4.1) with values $A = 2, B = 0, a = 1$, and $c = 5$. This yields

$$r = \sqrt{\tfrac{20}{4}} = \sqrt{5}, \quad b = \sqrt{c - a^2} = 2 \quad \text{and} \quad \theta = \tan^{-1}(\tfrac{2}{4}) = 26.6°$$

Therefore

$$y(t) = \sqrt{5}e^{-t} \cos{(2t + 26.6°)}u(t)$$

This response is shown in Fig. 4.8c. ■

4.3-1   Zero-State Response: The Transfer Function of an LTIC System

Consider an nth-order LTIC system specified by the equation

$$Q(D)y(t) = P(D)f(t)$$

or

$$(D^n + a_{n-1}D^{n-1} + \cdots + a_1 D + a_0)y(t) = $$
$$(b_n D^n + b_{n-1}D^{n-1} + \cdots + b_1 D + b_0)f(t) \qquad (4.37)$$

We shall now find the general expression for the zero-state response of an LTIC system. Zero-state response $y(t)$, by definition, is the system response to an input when the system is initially relaxed (in zero state). Therefore $y(t)$ satisfies the system equation (4.37) with zero initial conditions:

$$y(0^-) = \dot{y}(0^-) = \ddot{y}(0^-) = \cdots = y^{(n-1)}(0^-) = 0$$

Moreover, the input $f(t)$ is causal, so that

$$f(0^-) = \dot{f}(0^-) = \ddot{f}(0^-) = \cdots = f^{(n-1)}(0^-) = 0$$

Let

$$y(t) \Longleftrightarrow Y(s) \quad \text{and} \quad f(t) \Longleftrightarrow F(s)$$

Because of zero initial conditions

$$D^r y(t) = \frac{d^r}{dt^r}y(t) \Longleftrightarrow s^r Y(s)$$

$$D^k f(t) = \frac{d^k}{dt^k}f(t) \Longleftrightarrow s^k F(s)$$

Therefore the Laplace transform of Eq. (4.37) yields

$$\left(s^n + a_{n-1}s^{n-1} + \cdots + a_1 s + a_0\right) Y(s) = \left(b_n s^n + b_{n-1}s^{n-1} + \cdots + b_1 s + b_0\right) F(s)$$

or

$$Y(s) = \frac{b_n s^n + b_{n-1}s^{n-1} + \cdots + b_1 s + b_0}{s^n + a_{n-1}s^{n-1} + \cdots + a_1 s + a_0} F(s) \qquad (4.38a)$$

$$= \frac{P(s)}{Q(s)} F(s) \qquad (4.38b)$$

It is convenient to define $H(s)$, **the transfer function** of an LTIC system, as the ratio of $Y(s)$ to $F(s)$ when all the initial conditions are zero (when the system is in zero state). Thus

$$H(s) \equiv \frac{Y(s)}{F(s)} = \frac{\mathcal{L}[\text{zero-state response}]}{\mathcal{L}[\text{input}]} \tag{4.39}$$

For an LTIC system specified by Eq. (4.37), the transfer function $H(s)$, as seen from Eq. (4.38b), is

$$H(s) = \frac{P(s)}{Q(s)} \tag{4.40}$$

From the definition of the transfer function $H(s)$ in Eq. (4.39), it follows that

$$Y(s) = H(s)F(s) \tag{4.41}$$

This result states that $Y(s)$, the Laplace transform of the zero-state response $y(t)$, is the product of $F(s)$ and $H(s)$, where $F(s)$ is the Laplace transform of the input $f(t)$ and $H(s)$ is the system transfer function (relating the particular output $y(t)$ to the input $f(t)$). Observe that the denominator of $H(s)$ is $Q(s)$, the characteristic polynomial of the system. Therefore **the poles of $H(s)$ are the characteristic roots of the system**. Consequently the system stability criterion can be stated in terms of the poles of the transfer function of a system, as follows:

1. An LTIC system is asymptotically stable if and only if all the poles of its transfer function $H(s)$ are in the LHP. The poles may be repeated or unrepeated.

2. An LTIC system is unstable if and only if either one or both of the following conditions exist: (i) at least one pole of $H(s)$ is in the RHP; (ii) there are repeated poles of $H(s)$ on the imaginary axis.

3. An LTIC system is marginally stable if and only if there are no poles of $H(s)$ in the RHP, and there are some unrepeated poles on the imaginary axis.

We can now represent the transformed version of the system, as shown in Fig. 4.9. The input $F(s)$ is the Laplace transform of $f(t)$, and the output $Y(s)$ is the Laplace transform of $y(t)$. The system is described by the transfer function $H(s)$. The output $Y(s)$ is the product $F(s)H(s)$.

The result $Y(s) = H(s)F(s)$ greatly facilitates derivation of the system response to a given input. We shall demonstrate this by an example.

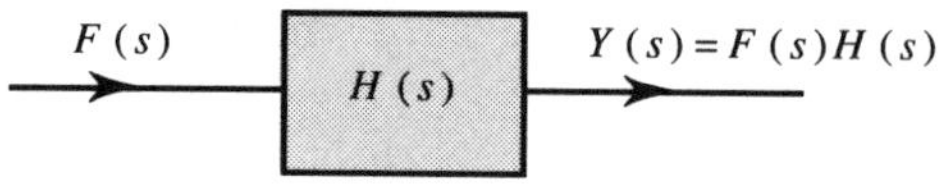

$$F(s) \quad \boxed{H(s)} \quad Y(s) = F(s)H(s)$$

Fig. 4.9 The transformed representation of an LTIC system.

■ **Example 4.11**

Find the response $y(t)$ of an LTIC system described by the equation

$$\frac{d^2y}{dt^2} + 5\frac{dy}{dt} + 6y(t) = \frac{df}{dt} + f(t)$$

if the input $f(t) = 3e^{-5t}u(t)$ and all the initial conditions are zero; that is, the system is in zero-state.

The system equation is

$$\underbrace{\left(D^2 + 5D + 6\right)}_{Q(D)} y(t) = \underbrace{(D+1)}_{P(D)} f(t)$$

Therefore

$$H(s) = \frac{P(s)}{Q(s)} = \frac{s+1}{s^2 + 5s + 6}$$

Also

$$F(s) = \mathcal{L}\left[3e^{-5t}u(t)\right] = \frac{3}{s+5}$$

and

$$Y(s) = F(s)H(s) = \frac{3(s+1)}{(s+5)(s^2 + 5s + 6)}$$

$$= \frac{3(s+1)}{(s+5)(s+2)(s+3)}$$

$$= \frac{-2}{s+5} - \frac{1}{s+2} + \frac{3}{s+3}$$

The inverse Laplace transform of this equation is

$$y(t) = \left(-2e^{-5t} - e^{-2t} + 3e^{-3t}\right)u(t) \quad ■$$

■ **Example 4.12**

Show that the transfer function of

(a) an ideal delay of T seconds is e^{-sT};

(b) an ideal differentiator is s;

(c) an ideal integrator is $1/s$.

(a) Ideal Delay

For an ideal delay of T seconds, the input $f(t)$ and output $y(t)$ are related by

$$y(t) = f(t - T)$$

or

$$Y(s) = F(s)e^{-sT} \qquad \text{[see Eq. (4.19a)]}$$

Therefore

$$H(s) = \frac{Y(s)}{F(s)} = e^{-sT} \tag{4.42}$$

(b) Ideal Differentiator

For an ideal differentiator, the input $f(t)$ and the output $y(t)$ are related by

$$y(t) = \frac{df}{dt}$$

The Laplace transform of this equation yields

$$Y(s) = sF(s) \qquad [f(0^-) = 0 \text{ for causal signal}]$$

and

$$H(s) = \frac{Y(s)}{F(s)} = s \tag{4.43a}$$

(c) Ideal Integrator

For an ideal integrator with zero initial state, that is $y(0^-) = 0$,

$$y(t) = \int_0^t f(\tau)\, d\tau$$

and

$$Y(s) = \frac{1}{s}F(s)$$

Therefore

$$H(s) = \frac{1}{s} \tag{4.43b}$$

■

△ **Exercise E4.7**

For an LTIC system with transfer function
$$H(s) = \frac{s + 5}{s^2 + 4s + 3}$$

(a) Describe the differential equation relating the input $f(t)$ and output $y(t)$.

(b) Find the system response $y(t)$ to the input $f(t) = e^{-2t}u(t)$ if the system is initially in zero state.

Answers: (a) $\dfrac{d^2 y}{dt^2} + 4\dfrac{dy}{dt} + 3y(t) = \dfrac{df}{dt} + 5f(t)$

(b) $y(t) = \left(2e^{-t} - 3e^{-2t} + e^{-3t}\right)u(t)$ ▽

4.3-2 The Relationship between h(t) and H(s)

Consider an LTIC system with transfer function $H(s)$. From Eq. (4.41)

$$\mathcal{L}\left[y(t)\right] = Y(s) = F(s)H(s)$$

For the input $f(t) = \delta(t)$, $F(s) = 1$ and $y(t) = h(t)$. Therefore

$$\mathcal{L}[h(t)] = H(s)$$

or

$$h(t) \Longleftrightarrow H(s) \tag{4.44}$$

This surprising result relates the time-domain and the frequency-domain specifications $h(t)$ and $H(s)$ of an LTIC. We have shown that **the unit impulse response $h(t)$ and the transfer function $H(s)$ are a Laplace transform pair.** This means

$$H(s) = \int_0^\infty h(t)e^{-st}\, dt \tag{4.45}$$

If an LTIC system is described by the differential equation

$$Q(D)y(t) = P(D)f(t)$$

its impulse response can be readily determined by taking the inverse Laplace transform of $H(s)$, which is $P(s)/Q(s)$

$$h(t) = \mathcal{L}^{-1}[H(s)]$$

$$= \mathcal{L}^{-1}\left[\frac{P(s)}{Q(s)}\right]$$

$\triangle$ **Exercise E4.8**

   Find the impulse response $h(t)$ of the systems described by following equations.

   (a) $(D^2 + 3D + 2)y(t) = Df(t)$
   (b) $(D^2 + 4D + 20)y(t) = f(t)$
   (c) $(D^2 + 6D + 9)y(t) = (2D + 9)f(t)$

Answers: **(a)** $h(t) = \left(-e^{-t} + 2e^{-2t}\right)u(t)$ **(b)** $h(t) = \frac{1}{4}e^{-2t}\sin 4t\, u(t)$
(c) $h(t) = (2 + 3t)e^{-3t}u(t)$ $\triangledown$

4.3-3   Frequency-Domain Interpretation of the Transfer Function

   We now show that for a system with transfer function $H(s)$, the system response to an exponential input e^{st} is an exponential $H(s)e^{st}$. Observe that here we are considering an *everlasting* exponential starting at $t = -\infty$, in contrast to a *causal* exponential $e^{st}u(t)$ starting at $t = 0$. The system response to an (everlasting) exponential e^{st} is

$$y(t) = h(t) * e^{st}$$

$$= \int_{-\infty}^\infty h(\tau)e^{s(t-\tau)}\, d\tau$$

$$= e^{st}\int_{-\infty}^\infty h(\tau)e^{-s\tau}\, d\tau$$

For causal $h(t)$, the limits on the integral on the right-hand side would range from 0 to ∞. Clearly, this integral is $H(s)$, the Laplace transform of $h(t)$ [see Eq. (4.45)]. Therefore†

$$y(t) = H(s)e^{st} \tag{4.46}$$

This is an interesting and significant result. It shows that for an LTIC system, the exponential input $f(t) = e^{st}$ yields an exponential response $y(t) = H(s)e^{st}$. We now have an alternate definition of the transfer function of an LTIC system: It is the ratio of the output signal to the input signal when the input signal is an exponential e^{st}:

$$H(s) = \left.\frac{\text{output signal}}{\text{input signal}}\right|_{\text{Input=everlasting exponential } e^{st}}$$

The signal $e^{j\omega t}$ is a special case of e^{st} where $s = \sigma + j\omega$ and $e^{st} = e^{(\sigma + j\omega)t} = e^{\sigma t}e^{j\omega t}$. Therefore e^{st} is a generalized sinusoid with exponentially growing ($\sigma > 0$), exponentially decaying ($\sigma < 0$), or constant ($\sigma = 0$) amplitude. For this reason s is called the *complex frequency* of signal e^{st}. A basic discussion of complex frequency is given in Sec. B.2-3.

The knowledge of $H(s)$, the system response to an (everlasting) exponential e^{st}, enables us to determine the system response to an arbitrary input $f(t)$ [see Eq. (4.41)]. In time-domain analysis (Chapter 2), we saw the crucial role played by an impulse signal, which does not have physical existence. Nevertheless we could use the impulse response as an intermediary to determine the system response to arbitrary inputs. The frequency-domain counterpart of the impulse signal is the everlasting exponential signal. This signal, like the impulse, has no physical existence, but the system response to this signal can be used as an intermediary to determine the system response to arbitrary inputs. In fact, the everlasting exponential is the dual of the impulse (time-frequency duality).

4.3-4 Physical Interpretation of the Laplace Transform

So far we have viewed the Laplace transform as a mathematical tool for solving integro-differential equations. It can also be viewed as a means of signal analysis where an input signal is represented as a sum of (everlasting) exponential (or sinusoidal) components. The Laplace transform method determines the system response by summing the system's response to all the exponential (or sinusoidal) components of the input. To understand this frequency-domain view of Laplace transform, consider an input $f(t)$ with its Laplace transform $F(s)$. From Eq. (4.2),

$$f(t) = \frac{1}{2\pi j}\int_{c-j\infty}^{c+j\infty} F(s)e^{st}\,ds \tag{4.47}$$

where the integration is along a path in the complex plane from $c - j\infty$ to $c + j\infty$ (along the line zz'), as shown in Fig. 4.10. Equation (4.47) can be expressed as

$$f(t) = \lim_{\triangle s \to 0}\sum_{n=-\infty}^{\infty}\left[\frac{F(n\triangle s)\triangle s}{2\pi j}\right]e^{(n\triangle s)t}$$

---

†This result is valid only for the values of s lying in the region of convergence (or existence) of $H(s)$, where the integral $\int_{-\infty}^{\infty} h(\tau)e^{-s\tau}\,d\tau$ exists.

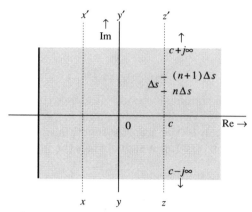

Fig. 4.10 Laplace transform as a means of expressing a signal as a sum of (everlasting) exponentials (or sinusoids).

This shows that $f(t)$ can be synthesized by exponentials $e^{(n\triangle s)t}$ where the complex frequencies $n\triangle s$ are along the path zz' shown in Fig. 4.10. Observe that these exponentials are not multiplied by $u(t)$ and therefore are everlasting. The amount of exponential $e^{(n\triangle s)t}$ is $[F(n\triangle s)\triangle s]/2\pi j$ which is infinitesimal because $\triangle s \to 0$. When all these exponentials are added (from $n = -\infty$ to ∞), the resulting sum is $f(t)$. Therefore $F(s)$ is the spectrum for the signal $f(t)$ of exponential components e^{st}.

The response $y(t)$ is the sum of system response to all these exponential components. We showed in Eq. (4.46) that the system response to exponential e^{st} is $H(s)e^{st}$. Therefore the system response to an exponential $e^{(n\triangle s)t}$ is $H(n\triangle s)e^{(n\triangle s)t}$. We express this fact by an arrow notation:

$$e^{(n\triangle s)t} \longrightarrow H(n\triangle s)e^{(n\triangle s)t}$$

The amount of the exponential component $e^{(n\triangle s)t}$ is $[F(n\triangle s)\triangle s]/2\pi j$. Therefore the response to such an exponential can be expressed as

$$\frac{F(n\triangle s)\triangle s}{2\pi j}e^{(n\triangle s)t} \longrightarrow \frac{F(n\triangle s)H(n\triangle s)\triangle s}{2\pi j}e^{(n\triangle s)t}$$

The response $y(t)$, being the sum of the system responses to all the infinite exponential components, can be expressed as

$$y(t) = \lim_{\triangle s \to 0} \sum_{n=-\infty}^{\infty} \frac{F(n\triangle s)H(n\triangle s)}{2\pi j}e^{(n\triangle s)t}$$

$$= \frac{1}{2\pi j} \int_{c-j\infty}^{c+j\infty} F(s)H(s)e^{st}\, ds \qquad\qquad (4.48)$$

$$= \mathcal{L}^{-1}\left[F(s)H(s)\right]$$

This means

$$Y(s) = F(s)H(s)$$

which is precisely the result obtained earlier [Eq. (4.41)]

We now have an alternative, more physical, view of the Laplace transform. In this view, the Laplace transform is a tool by which a signal is expressed as a sum of its spectral components of the form e^{st}. The relative amount of component e^{st} is $F(s)$; therefore $F(s)$, the Laplace transform of $f(t)$, represents the spectrum of $f(t)$. Also, $H(s)e^{st}$ is the system response to input e^{st}. Therefore $H(s)$ is the system response (or gain) to spectral component e^{st}. The spectrum of the output signal is the input spectrum $F(s)$ times the spectral response (gain) $H(s)$. Consequently, $Y(s)$, the output spectrum [which is the Laplace transform of the output $y(t)$], is $F(s)H(s)$, as shown in Fig. 4.9.

The path of integration for the integral in Eq. 4.47 is any path from $c - j\infty$ to $c + j\infty$, as long as it lies in the region of convergence for $F(s)$ (shown shaded in Fig. 4.10). Three possible paths are shown in Fig. 4.10. The path xx' represents exponentials with complex frequencies in LHP. Therefore choice of this path amounts to spectral representation of $f(t)$ in terms of exponentially decaying sinusoids. The path zz' lies in RHP; the choice of this path amounts to spectral representation of $f(t)$ in terms of exponentially growing sinusoids. The path yy' is the imaginary axis, the choice of this path amounts to spectral representation of $f(t)$ in terms of sinusoids of constant amplitudes. It is clear that there are infinite possible ways to represent a signal spectrally.

An interesting note: when viewed in this physical light, the Laplace transform is not different from the time-domain method in Chapter 2, where the system response is obtained by summing the system's response to impulse components of the input. The only difference is that in the case of Laplace transform, we express the input signal in terms of (everlasting) sinusoidal or exponential components instead of impulse components. The response in both methods is in the form of an integral, which is a sum in the limit. In the time-domain method the response is given by the convolution integral, and in the case of the Laplace transform the response is the Laplace integral (4.48). In both cases we may prepare tables (convolution table and the Laplace transform table) to avoid the integration. Physical view of the Laplace transform therefore unifies the time-domain and frequency-domain approaches.

4.4 ANALYSIS OF ELECTRICAL NETWORKS: THE TRANS-FORMED NETWORK METHOD

Example 4.10 shows how electrical networks may be analyzed by writing the integro-differential equation(s) of the system and then solving these equations by Laplace transform. We now show that it is also possible to analyze electrical networks directly without having to write the integro-differential equations. This procedure is considerably simpler because it permits us to treat an electrical network as if it were a resistive network. This requires representing a network in the "frequency domain" where all the voltages and currents are represented by their Laplace transforms.

For the sake of simplicity let us first discuss the case with zero initial conditions. If $v(t)$ and $i(t)$ are the voltage across and the current through an inductor of L henries, then

$$v(t) = L\frac{di}{dt}$$

The Laplace transform of this equation (assuming zero initial current) is

$$V(s) = LsI(s)$$

Similarly, for a capacitor of C farads, the voltage-current relationship is $i(t) = C\frac{dv}{dt}$ and its Laplace transform, assuming zero initial capacitor voltage, yields $I(s) = CsV(s)$, that is,

$$V(s) = \frac{1}{Cs}I(s)$$

For a resistor of R ohms, the voltage-current relationship is $v(t) = Ri(t)$, and its Laplace transform is

$$V(s) = RI(s)$$

Thus, in the "frequency domain," the voltage-current relationships of an inductor and a capacitor are algebraic; these elements behave like resistors of "resistance" Ls and $1/Cs$, respectively. The generalized "resistance" of an element is called its **impedance** and is given by the ratio $V(s)/I(s)$ for the element (under zero initial conditions). The impedances of a resistor of R ohms, an inductor of L henries, and a capacitance of C farads are R, Ls, and $1/Cs$, respectively.

Also, the interconnection constraints (Kirchhoff's laws) remain valid for voltages and currents in the frequency domain. To show this, let $v_j(t)$ ($j=1, 2, \ldots, k$) be the voltages across k elements in a loop and let $i_j(t)$ ($j = 1, 2, \ldots, m$) be the j currents entering a node. Then

$$\sum_{j=1}^{k} v_j(t) = 0 \quad \text{and} \quad \sum_{j=1}^{m} i_j(t) = 0$$

Now if

$$v_j(t) \Longleftrightarrow V_j(s) \quad \text{and} \quad i_j(t) \Longleftrightarrow I_j(s)$$

then

$$\sum_{j=1}^{k} V_j(s) = 0 \quad \text{and} \quad \sum_{j=1}^{m} I_j(s) = 0$$

This means that if we represent all the voltages and currents in an electrical network by their Laplace transforms, we can treat the network as if it consisted of the "resistances" R, Ls and $1/Cs$ corresponding to a resistor R, an inductor L, and a capacitor C, respectively. The system equations (loop or node) are now algebraic. Moreover, the simplification techniques that have been developed for resistive circuits—equivalent series and parallel impedances, voltage and current divider rules, Thévenin and Norton theorems—can be applied to general electrical networks. The following examples demonstrate these concepts.

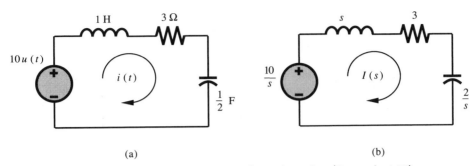

Fig. 4.11 A circuit and its transformed version (Example 4.13).

■Example 4.13

Find the loop current $i(t)$ in the circuit shown in Fig. 4.11a if all the initial conditions are zero.

In the first step, we represent the circuit in the frequency domain, as shown in Fig. 4.11b. All the voltages and currents are represented by their Laplace transforms. The voltage $10u(t)$ is represented by $10/s$ and the (unknown) current $i(t)$ is represented by its Laplace transform $I(s)$. All the circuit elements are represented by their respective impedances. The inductor of 1 henry is represented by s, the capacitor of 1/2 farad is represented by $2/s$, and the resistor of 3 ohms is represented by 3. We now consider the frequency-domain representation of voltages and currents. The voltage across any element is $I(s)$ times its impedance. Therefore the total voltage drop in the loop is $I(s)$ times the total loop impedance, and it must equal to $V(s)$, the (transform of) input voltage. The total impedance in the loop is

$$Z(s) = s + 3 + \frac{2}{s} = \frac{s^2 + 3s + 2}{s}$$

The input "voltage" is $V(s) = 10/s$. Therefore the "loop current" $I(s)$ is

$$I(s) = \frac{V(s)}{Z(s)} = \frac{\frac{10}{s}}{\frac{s^2+3s+2}{s}} = \frac{10}{s^2 + 3s + 2} = \frac{10}{(s+1)(s+2)} = \frac{10}{s+1} - \frac{10}{s+2}$$

The inverse transform of this equation yields the desired result:

$$i(t) = 10(e^{-t} - e^{-2t})u(t) \qquad ■$$

Initial Condition Generators

The above discussion, where we assumed zero initial conditions, can be readily extended to the case of nonzero initial conditions because the initial condition in a capacitor or an inductor can be represented by an equivalent source. We now show that a capacitor C with an initial voltage $v(0)$ (Fig. 4.12a) can be represented in the frequency domain by an uncharged capacitor of impedance $1/Cs$ in series with a voltage source of value $v(0)/s$ (Fig. 4.12b) or as the same uncharged capacitor in parallel with a current source of value $Cv(0)$ (Fig. 4.12c). Similarly, an inductor L with an initial current $i(0)$ (Fig. 4.12d) can be represented in the frequency domain

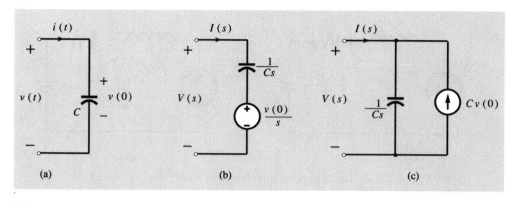

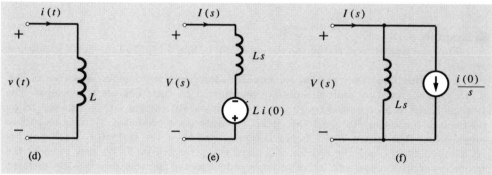

Fig. 4.12 Initial condition generators for a capacitor and an inductor.

by an inductor of impedance Ls in series with a voltage source of value $Li(0)$ (Fig. 4.12e) or by the same inductor in parallel with a current source of value $i(0)/s$ (Fig. 4.12f). To prove this, consider the terminal relationship of the capacitor in Fig. 4.12a:

$$i(t) = C\frac{dv}{dt}$$

The Laplace transform of this equation yields

$$I(s) = C[sV(s) - v(0)]$$

This equation can be rearranged as

$$V(s) = \frac{1}{Cs}I(s) + \frac{v(0)}{s} \tag{4.49a}$$

Observe that $V(s)$ is the voltage (in the frequency domain) across the charged capacitor and $I(s)/Cs$ is the voltage across the same capacitor without any charge. Therefore the above equation shows that the charged capacitor can be represented by the same capacitor (uncharged) in series with a voltage source of value $v(0)/s$, as shown in Fig. 4.12b. Equation (4.49a) can also be rearranged as

$$V(s) = \frac{1}{Cs}[I(s) + Cv(0)] \tag{4.49b}$$

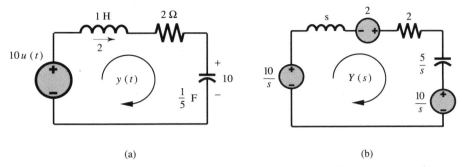

(a)   (b)

Fig. 4.13 A circuit and its transformed version with initial condition generators (Example 4.13).

This equation shows that the charged capacitor voltage $V(s)$ is equal to the uncharged capacitor voltage caused by a current $I(s) + Cv(0)$. This is reflected precisely in Fig. 4.12c, where the current through the uncharged capacitor is $I(s) + Cv(0)$.†

For the inductor in Fig. 4.12d, the terminal equation is

$$v(t) = L\frac{di}{dt}$$

and

$$V(s) = L[sI(s) - i(0)]$$

$$= LsI(s) - Li(0) \tag{4.50a}$$

$$= Ls\left[I(s) - \frac{i(0)}{s}\right] \tag{4.50b}$$

We can verify that Fig. 4.12e satisfies Eq. (4.50a) and that Fig. 4.12f satisfies Eq. (4.50b).

Let us rework Example 4.10 using these ideas. Figure 4.13a shows the circuit in Fig. 4.8b with the initial conditions $y(0) = 2$ and $v_C(0) = 10$. Figure 4.13b shows the frequency-domain representation (transformed circuit) of the circuit in Fig. 4.13a. The resistor is represented by its impedance 2; the inductor with initial current of 2 A is represented according to the arrangement in Fig. 4.12e with a series voltage source $Ly(0) = 2$. The capacitor with initial voltage of 10 V is represented according to arrangement in Fig. 4.12b with a series voltage source $v(0)/s = 10/s$. Note that the impedance of the inductor is s and that of the capacitor is $5/s$. The input of $10u(t)$ is represented by its Laplace transform $10/s$.

The total voltage in the loop is $\frac{10}{s} + 2 - \frac{10}{s} = 2$, and the loop impedance is $s + 2 + \frac{5}{s}$). Therefore

---

†In time domain, a charged capacitor C with initial voltage $v(0)$ can be represented as the same capacitor uncharged in series with a voltage source $v(0)u(t)$, or in parallel with a current source $Cv(0)\delta(t)$. Similarly, an inductor L with initial current $i(0)$ can be represented by the same inductor with zero initial current in series with a voltage source $Li(0)\delta(t)$ or with a parallel current source $i(0)$.

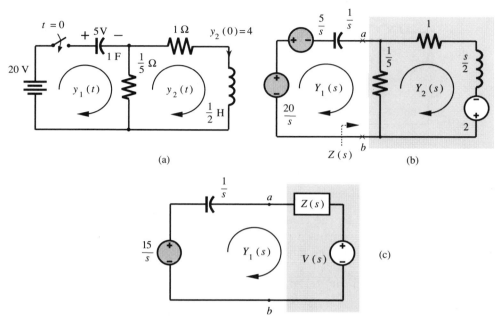

Fig. 4.14 Solving Example 4.13 using initial condition generators and Thèvenin equivalent representation.

$$Y(s) = \frac{2}{s + 2 + \frac{5}{s}}$$

$$= \frac{2s}{s^2 + 2s + 5}$$

which confirms our earlier result in Example 4.10.

■ **Example 4.14**

Find the currents $y_1(t)$ and $y_2(t)$ in the network shown in Fig. 4.14a if the initial capacitor voltage $v_C(0^-) = 5$ volts and the initial inductor current $y_2(0^-) = 4$ amps.

Figure 4.14b shows the transformed version of the circuit in Fig. 4.14a. We have used equivalent sources to account for the initial conditions. The initial capacitor voltage of 5 V is represented by a series voltage of $5/s$ and the initial inductor current of 4 A is represented by a source of value $Ly_2(0) = 2$.

From Fig. 4.14b, the loop equations can be written directly in the frequency domain as

$$\frac{Y_1(s)}{s} + \frac{1}{5}[Y_1(s) - Y_2(s)] = \frac{15}{s}$$

$$-\frac{1}{5}Y_1(s) + \frac{6}{5}Y_2(s) + \frac{s}{2}Y_2(s) = 2$$

$$\begin{bmatrix} \frac{1}{s} + \frac{1}{5} & -\frac{1}{5} \\ -\frac{1}{5} & \frac{6}{5} + \frac{s}{2} \end{bmatrix} \begin{bmatrix} Y_1(s) \\ Y_2(s) \end{bmatrix} = \begin{bmatrix} \frac{15}{s} \\ 2 \end{bmatrix}$$

Application of Cramer's rule to this equation yields

$$Y_1(s) = \frac{79s + 180}{s^2 + 7s + 12}$$

$$= \frac{79s + 180}{(s + 3)(s + 4)} = \frac{-57}{s + 3} + \frac{136}{s + 4}$$

and

$$y_1(t) = \left(-57e^{-3t} + 136e^{-4t}\right) u(t)$$

Similarly, we obtain

$$Y_2(s) = \frac{2(2s + 25)}{s^2 + 7s + 12}$$

$$= \frac{38}{s + 3} - \frac{34}{s + 4}$$

and

$$y_2(t) = \left(38e^{-3t} - 34e^{-4t}\right) u(t)$$

We also could have computed $Y_1(s)$ and $Y_2(s)$ using Thévenin's theorem by replacing the circuit to the right of the capacitor (right of terminals ab) by its Thévenin equivalent, as shown in Fig. 4.14c. From Fig. 4.14b we find the Thévenin impedance $Z(s)$ and the Thévenin source $V(s)$ to be

$$Z(s) = \frac{\frac{1}{5}\left(\frac{s}{2} + 1\right)}{\frac{1}{5} + \frac{s}{2} + 1} = \frac{s + 2}{5s + 12}$$

$$V(s) = \frac{-\frac{1}{5}}{\frac{1}{5} + \frac{s}{2} + 1} 2 = \frac{-4}{5s + 12}$$

From Fig. 4.14c, the current $Y_1(s)$ is given by

$$Y_1(s) = \frac{\frac{15}{s} - V(s)}{\frac{1}{s} + Z(s)}$$

$$= \frac{79s + 180}{s^2 + 7s + 12}$$

which confirms the earlier result. We may determine $Y_2(s)$ in a similar manner. ∎

■**Example 4.15**

The switch in the circuit in Fig. 4.15a is at position a for a long time before $t = 0$, when it is moved instantaneously to position b. Determine the current $y_1(t)$ and the output voltage $v_0(t)$ for $t \geq 0$.

Just before switching, the values of the loop currents are 2 and 1, respectively, that is, $y_1(0^-) = 2$, $y_2(0^-) = 1$.

The equivalent circuits for two types of inductive couplings are shown in Figs. 4.15b and 4.15c. For our situation the circuit in Fig. 4.15c applies. Fig. 4.15d shows the transformed version of circuit in Fig. 4.15a after switching. Note that the inductors $L_1 + M$, $L_2 + M$, and $-M$ are $3, 4$, and -1 henry with impedances $3s, 4s$, and $-s$ respectively. The initial condition voltages in the three branches are $(L_1 + M)y_1(0) = 6$, $(L_2 + M)y_2(0) = 4$,

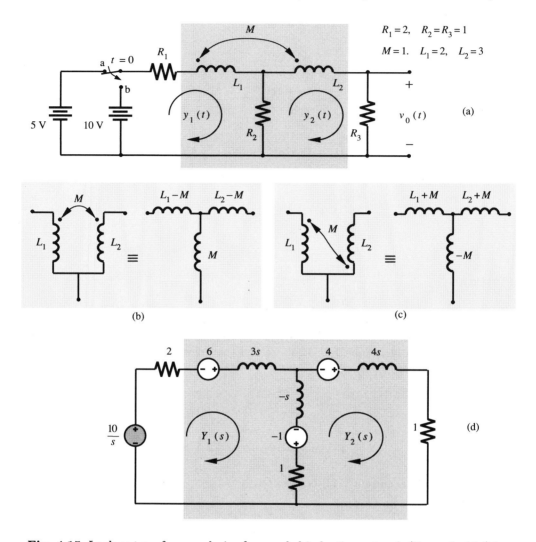

Fig. 4.15 Laplace transform analysis of a coupled inductive network (Example 4.14) by transformed circuit method.

and $-M[y_1(0) - y_2(0)] = -1$, respectively. The two loop equations of the circuit are†

$$(2s + 3)Y_1(s) + (s - 1)Y_2(s) = \frac{10}{s} + 5$$

$$(s - 1)Y_1(s) + (3s + 2)Y_2(s) = 5 \qquad (4.51)$$

---

†The time domain equations (loop equations) are

$$L_1 \frac{dy_1}{dt} + (R_1 + R_2)y_1(t) - R_2 y_2(t) + M \frac{dy_2}{dt} = 10u(t)$$

$$M \frac{dy_1}{dt} - R_2 y_1(t) + L_2 \frac{dy_2}{dt} + (R_2 + R_3)y_2(t) = 0$$

Laplace transform of these equations yields Eq. (4.51).

or

$$\begin{bmatrix} 2s+3 & s-1 \\ s-1 & 3s+2 \end{bmatrix} \begin{bmatrix} Y_1(s) \\ Y_2(s) \end{bmatrix} \begin{bmatrix} \frac{5s+10}{s} \\ 5 \end{bmatrix}$$

and

$$Y_1(s) = \frac{2s^2 + 9s + 4}{s(s^2 + 3s + 1)}$$

$$= \frac{4}{s} - \frac{1}{s+0.382} - \frac{1}{s+2.618}$$

Therefore

$$y_1(t) = \left(4 - e^{-0.382t} - e^{-2.618t}\right) u(t)$$

Similarly

$$Y_2(s) = \frac{s^2 + 2s + 2}{s(s^2 + 3s + 1)}$$

$$= \frac{2}{s} - \frac{1.618}{s+0.382} + \frac{0.618}{s+2.618}$$

and

$$y_2(t) = \left(2 - 1.618e^{-0.382t} + 0.618e^{-2.618t}\right) u(t)$$

The output voltage

$$v_0(t) = y_2(t) = \left(2 - 1.618e^{-0.382t} + 0.618e^{-2.618t}\right) u(t) \quad \blacksquare$$

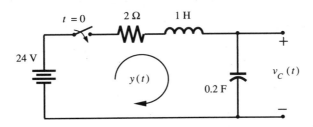

Fig. 4.16 Circuit for Exercise E4.9.

△ **Exercise E4.9**

For the RLC circuit in Fig. 4.16, the input is switched on at $t = 0$. The initial conditions are $y(0^-) = 2$ amps and $v_C(0) = 50$ volts. Find the loop current $y(t)$ and the capacitor voltage $v_C(t)$ for $t \geq 0$.

Answer:

$$y(t) = 10\sqrt{2}e^{-t} \cos(2t + 81.8°)u(t), \quad v_C(t) = (24 + 31.62e^{-t} \cos(2t - 34.7°)u(t) \ \triangledown$$

4.4-1   Analysis of Active Circuits

Although we have considered only examples of passive networks so far, the circuit analysis procedure using Laplace transform is also applicable to active circuits. All that is needed is to replace the active elements with their mathematical models (or equivalent circuits) and proceed as before.

The operational amplifier (shown by the triangular symbol in Fig. 4.17a) is a well-known element in modern electronic circuits. The terminals with the positive and the negative signs correspond to noninverting and inverting terminals, respectively. This means that the polarity of the output voltage v_2 is the same as that of the input voltage at the terminal marked by the positive sign (noninverting). The opposite is true for the inverting terminal, marked by the negative sign.

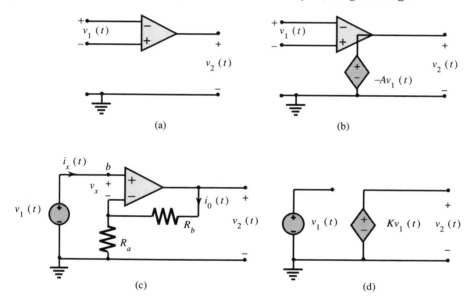

Fig. 4.17 Operational amplifier and its equivalent circuit.

Figure 4.17b shows the model (equivalent circuit) of an operational amplifier (op amp) in Fig. 4.17a. A typical op amp has a very large gain. The output voltage $v_2 = -Av_1$, where A is typically 10^5 to 10^6. The input impedance is very high (typically $10^6\ \Omega$ for BJT to $10^{12}\ \Omega$ for Bi-FET), and the output impedance is very low (50 to $100\,\Omega$). For most applications we are justified in assuming the gain A and the input impedance to be infinite and the output impedance to be zero. For this reason we see an ideal voltage source at the output.

Consider now the operational amplifier with resistors R_a and R_b connected, as shown in Fig. 4.17c. This configuration is known as the **noninverting amplifier**. Observe that the input polarities in this configuration are inverted when compared to those in Fig. 4.17a. We now show that the output voltage v_2 and the input voltage v_1 in this case are related by

$$v_2 = Kv_1 \qquad K = 1 + \tfrac{R_b}{R_a} \qquad (4.52)$$

First, we recognize that because the input impedance and the gain of the operational amplifier approach infinity, the input current i_x and the input voltage v_x in Fig.

4.17c must approach zero. The dependent source in this case is Av_x instead of $-Av_x$ because of the input polarity inversion. The dependent source Av_x (see Fig. 4.17b) at the output will generate current i_0, as shown in Fig. 4.17c. Now

$$v_2 = (R_b + R_a)i_0$$

also

$$v_1 = v_x + R_a i_0$$
$$= R_a i_0$$

Therefore

$$\frac{v_2}{v_1} = \frac{R_b + R_a}{R_a} = 1 + \frac{R_b}{R_a} = K$$

or

$$v_2(t) = Kv_1(t)$$

The equivalent circuit of the noninverting amplifier is shown in Fig. 4.17d.

■ Example 4.16

The circuit in Fig. 4.18a is called the **Sallen-Key** circuit, which is frequently used in filter design. Find the transfer function $H(s)$ relating the output voltage $v_0(t)$ to the input voltage $v_i(t)$.

We are required to find

$$H(s) = \frac{V_0(s)}{V_i(s)}$$

assuming all initial conditions to be zero.

Figure 4.18b shows the transformed version of the circuit in Fig. 4.18a. The noninverting amplifier is replaced by its equivalent circuit. All the voltages are replaced by their Laplace transforms and all the circuit elements are shown by their impedances. All the initial conditions are assumed to be zero, as required for determining $H(s)$.

We shall use node analysis to derive the result. There are two unknown node voltages, $V_a(s)$ and $V_b(s)$, requiring two node equations.

At node aa, $I_{R_1}(s)$, the current in R_1 (leaving the node a), is $[V_a(s) - V_i(s)]/R_1$. Similarly, $I_{R_2}(s)$, the current in R_2 (leaving the node a), is $[V_a(s)-V_b(s)]/R_i$ and $I_{c_1}(s)$, the current in capacitor C_1 (leaving the node a), is $[V_a(s) - V_0(s)]C_1 s = [V_a(s) - KV_b(s)]C_1 s$.

The sum of all the three currents is zero. Therefore

$$\frac{V_a(s) - V_i(s)}{R_1} + \frac{V_a(s) - V_b(s)}{R_2} + [V_a(s) - KV_b(s)]C_1 s = 0$$

or

$$\left(\frac{1}{R_1} + \frac{1}{R_2} + C_1 s\right)V_a(s) - \left(\frac{1}{R_2} + KC_1 s\right)V_b(s) = \frac{1}{R_1}V_i(s) \tag{4.53a}$$

Similarly, the node equation at node b yields

$$\frac{V_b(s) - V_a(s)}{R_2} + C_2 s V_b(s) = 0$$

or

$$-\frac{1}{R_2}V_a(s) + \left(\frac{1}{R_2} + C_2 s\right)V_b(s) = 0 \tag{4.53b}$$

The two node equations (4.53a) and (4.53b) in two unknown node voltages $V_a(s)$ and $V_b(s)$ can be expressed in matrix form as

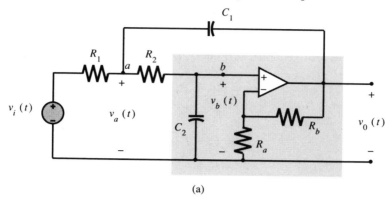

(a)

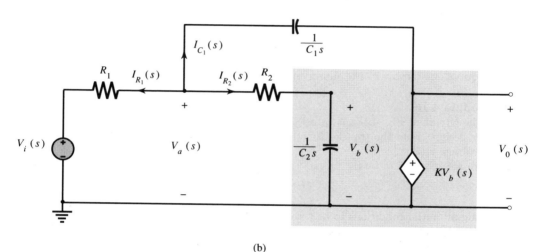

(b)

Fig. 4.18 Sallen-Key circuit and its equivalent.

$$\begin{bmatrix} G_1 + G_2 + C_1 s & -(G_2 + K C_1 s) \\ -G_2 & (G_2 + C_2 s) \end{bmatrix} \begin{bmatrix} V_a(s) \\ V_b(s) \end{bmatrix} = \begin{bmatrix} G_1 V_i(s) \\ 0 \end{bmatrix} \qquad (4.54)$$

where

$$G_1 = \frac{1}{R_1} \qquad \text{and} \qquad G_2 = \frac{1}{R_2}$$

Application of Cramer's rule to Eq. (4.54) yields

$$\frac{V_b(s)}{V_i(s)} = \frac{G_1 G_2}{C_1 C_2 s^2 + [G_1 G_2 + G_2 C_2 + G_2 C_1 (1 - K)] s + G_1 G_2}$$

$$= \frac{\omega_0{}^2}{s^2 + 2\alpha s + \omega_0{}^2}$$

where

$$K = 1 + \frac{R_b}{R_a} \qquad \text{and} \qquad \omega_0{}^2 = \frac{G_1 G_2}{C_1 C_2} = \frac{1}{R_1 R_2 C_1 C_2} \qquad (4.55a)$$

$$2\alpha = \frac{G_1C_2 + G_2C_2 + G_2C_1(1-K)}{C_1C_2} = \frac{1}{R_1C_1} + \frac{1}{R_1C_2} + \frac{1}{R_2C_2}(1-K) \qquad (4.55b)$$

Now

$$V_0(s) = KV_b(s)$$

Therefore

$$H(s) = \frac{V_0(s)}{V_i(s)} = K\frac{V_b(s)}{V_i(s)} = \frac{K\omega_0{}^2}{s^2 + 2\alpha s + \omega_0{}^2} \qquad (4.56)$$

■

4.4-2   Initial and Final Values

In certain applications it is desirable to know the values of $f(t)$ as $t \to 0$ and $t \to \infty$ [initial and final values of $f(t)$] from the knowledge of its Laplace transform $F(s)$. Initial and final value theorems provide such information.

Initial value theorem states that if $f(t)$ and its derivative df/dt are both Laplace transformable, then

$$f(0^+) = \lim_{s \to \infty} sF(s) \qquad (4.57)$$

provided that the limit on the right-hand side of (4.57) exists.

The final value theorem states that if both $f(t)$ and df/dt are Laplace transformable, then

$$\lim_{t \to \infty} f(t) = \lim_{s \to 0} sF(s) \qquad (4.58)$$

provided that $sF(s)$ has no poles in the RHP or on the imaginary axis. To prove these theorems, we use Eq. (4.24a)

$$sF(s) - f(0^-) = \int_{0-}^{\infty} \frac{df}{dt} e^{-st}\, dt$$

$$= \int_{0-}^{0+} \frac{df}{dt} e^{-st}\, dt + \int_{0+}^{\infty} \frac{df}{dt} e^{-st}\, dt$$

$$= f(t)\Big|_{0-}^{0+} + \int_{0+}^{\infty} \frac{df}{dt} e^{-st}\, dt$$

$$= f(0^+) - f(0^-) + \int_{0+}^{\infty} \frac{df}{dt} e^{-st}\, dt$$

Therefore

$$sF(s) = f(0^+) + \int_{0+}^{\infty} \frac{df}{dt} e^{-st}\, dt$$

and

$$\lim_{s \to \infty} sF(s) = f(0^+) + \lim_{s \to \infty} \int_{0+}^{\infty} \frac{df}{dt} e^{-st}\, dt$$

$$= f(0^+) + \int_{0+}^{\infty} \frac{df}{dt} \left(\lim_{s \to \infty} e^{-st} \right) dt$$

$$= f(0^+)$$

To prove the final value theorem, we let $s \to 0$ in Eq. (4.24a) to obtain

$$\lim_{s \to 0} \left[sF(s) - f(0^-) \right] = \lim_{s \to 0} \int_{0-}^{\infty} \frac{df}{dt} e^{-st}\, dt = \int_{0-}^{\infty} \frac{df}{dt}\, dt$$

$$= f(t) \Big|_{0-}^{\infty} = \lim_{t \to \infty} f(t) - f(0^-)$$

which leads to the desired result (4.58)

◼ **Example 4.17**

   Determine the initial and final values of $y(t)$ if its Laplace transform $Y(s)$ is given by

$$Y(s) = \frac{10(2s + 3)}{s(s^2 + 2s + 5)}$$

From Eqs. (4.57) and (4.58), we obtain

$$y(0^+) = \lim_{s \to \infty} sY(s) = \lim_{s \to \infty} \frac{10(2s + 3)}{(s^2 + 2s + 5)} = 0$$

$$y(\infty) = \lim_{s \to 0} sY(s) = \lim_{s \to 0} \frac{10(2s + 3)}{(s^2 + 2s + 5)} = 6 \quad ◼$$

Comment: The initial value theorem should be applied only if $F(s)$ is strictly proper ($m < n$), because for $m \geq n$, $\lim_{s \to \infty} sF(s)$ does not exist, and the theorem does not apply.

4.5 BLOCK DIAGRAMS

   Large systems may consist of a very large number of components or elements. Analyzing such systems all at once could be next to impossible. Anyone who has seen the circuit diagram of a radio or a TV receiver can appreciate this fact. In such cases it is convenient to represent a system by suitably interconnected subsystems, each of which can be analyzed easily. Each subsystem can be characterized in terms of its input-output relationships. A linear (sub)system can be characterized by its transfer function $H(s)$. Figure 4.19a shows a block diagram of a system with a transfer function $H(s)$ and its input and output represented by their frequency domain descriptions $F(s)$ and $Y(s)$ respectively.

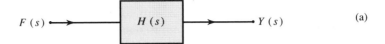

(a)

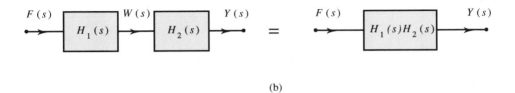

(b)

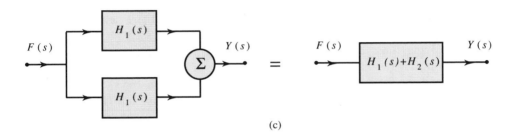

(c)

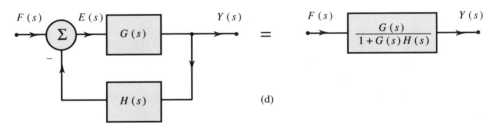

(d)

Fig. 4.19 Elementary connections of blocks and their equivalents.

Subsystems may be interconnected by using three elementary types of inter-connections (Figs. 4.19b, 4.19c, 4.19d): cascade, parallel, and feedback. When two transfer functions appear in cascade, as shown in Fig. 4.19b, the transfer function of the overall system is the product of the two transfer functions. This can be seen from the fact that in Fig. 4.19b

$$\frac{Y(s)}{F(s)} = \frac{W(s)}{F(s)} \frac{Y(s)}{W(s)} = H_1(s)H_2(s)$$

This result can be extended to any number of transfer functions in cascade.

Similarly, when two transfer functions, $H_1(s)$ and $H_2(s)$, appear in parallel, as shown in Fig. 4.19c, the overall transfer function is given by $H_1(s) + H_2(s)$, the sum of the two transfer functions. The proof is trivial. This result can be extended to any number of systems in parallel.

When the output is fed back to the input, as shown in Fig. 4.19d, the overall transfer function $Y(s)/F(s)$ can be computed as follows. The inputs to the summer are $F(s)$ and $-H(s)Y(s)$. Therefore $E(s)$, the output of the summer, is

$$E(s) = F(s) - H(s)Y(s)$$

But

$$Y(s) = G(s)E(s)$$
$$= G(s)[F(s) - H(s)Y(s)]$$

Therefore

$$Y(s)\left[1 + G(s)H(s)\right] = G(s)F(s)$$

so that

$$\frac{Y(s)}{F(s)} = \frac{G(s)}{1 + G(s)H(s)} \tag{4.59}$$

Therefore the feedback loop can be replaced by a single block with the transfer function shown in Eq. (4.59) (see Fig. 4.19d).

In deriving these equations, there is an implicit assumption that when the output of one subsystem is connected to the input of another subsystem, the latter does not load the former. For example, the transfer function $H_1(s)$ in Fig. 4.19b is computed by assuming that the second subsystem $H_2(s)$ was not connected. This is the same as assuming that $H_2(s)$ does not load $H_1(s)$. In other words, the input-output relationship of $H_1(s)$ will remain unchanged regardless of whether $H_2(s)$ is connected or not. Many modern circuits use op amps with high input impedances, and this assumption is justified. When such an assumption is not valid, $H_1(s)$ must be computed under operating conditions (that is, when $H_2(s)$ is connected).

An Example of an Automatic Position Control System

We shall give one example of the efficacy of block diagram representation in simplifying system analysis. Figure 4.20a represents an automatic position control system. This system can be used to control the angular position of any object (e.g., a tracking antenna, an anti-aircraft gun mount, or the position of a ship). The input θ_i is the desired angular position of the object, which can be set at any given value. The actual angular position θ_0 of the object (the output) is measured by a potentiometer whose wiper is mounted on the output shaft. The difference between the output θ_0 and the input θ_i is amplified; the amplified output, which is proportional to $\theta_0 - \theta_i$, is applied to the motor input. If $\theta_0 - \theta_i = 0$ (the output being equal to the desired angle), there is no input to the motor, and the motor stops. But if $\theta_0 \neq \theta_i$, there will be a nonzero input to the motor, which will turn the shaft until $\theta_0 = \theta_i$. It is evident that by setting the input potentiometer at a desired position in this system, we can control the angular position of a heavy remote object.

The block diagram of this system is shown in Fig. 4.20b. The amplifier transfer function is $G_1(s)$. The transfer function of the motor and the load together, which relates the output angle θ_0 to the motor input voltage, is $G_2(s)$. Therefore $T(s)$, the system transfer function relating the output θ_0 to the input θ_i, is

$$T(s) = \frac{G_1(s)G_2(s)}{1 + G_1(s)G_2(s)}$$

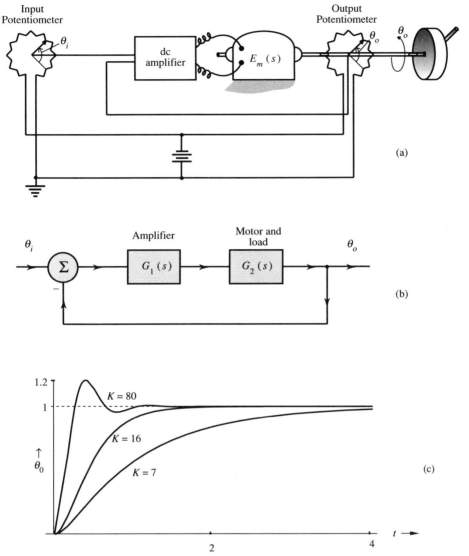

Fig. 4.20 (a) An automatic position control system (b) its block diagram (c) the unit step response.

From this equation, we can investigate the behavior of the automatic position control system in Fig. 4.20a. If, for example, we desire to change the angular position of the object, we may apply a step signal at the input $[\theta_i = u(t)]$. We may then want to know how long the system takes to position itself at the desired angle, whether it reaches the desired angle, and whether it reaches the desired position smoothly (monotonically) or oscillates about the final position. If it oscillates, we may want to know how long it takes for the oscillations to settle down. All this can be readily determined by finding the output $\theta_0(t)$ when the input $\theta_i(t) = u(t)$. A step input implies the instantaneous change in the angle. This input would be one of the most difficult to follow; if the system can perform well for this input, it is likely to give a good account of itself under most other expected situations. This is the reason why we test control systems for a step input.

For the step input $\theta_i(t) = u(t)$, $\Theta_i(s) = 1/s$ and

$$\Theta_0(s) = \frac{1}{s}T(s) = \frac{G_1(s)G_2(s)}{s\left[1 + G_1(s)G_2(s)\right]}$$

The amplifier transfer function can be taken as K (K adjustable), i.e., $G_1(s) = K$. Let the motor (with load) transfer function relating the load angle $\theta_0(t)$ to the motor input voltage be

$$G_2(s) = \frac{1}{s(s + 8)}$$

This yields

$$\Theta_0(s) = \frac{\frac{K}{s(s+8)}}{s\left[1 + \frac{K}{s(s+8)}\right]} = \frac{K}{s(s^2 + 8s + K)}$$

Let us investigate the system behavior for three different values of gain K.

1. $K = 7$

$$\Theta_0(s) = \frac{7}{s(s^2 + 8s + 7)} = \frac{7}{s(s + 1)(s + 7)}$$

$$= \frac{1}{s} - \frac{\frac{7}{6}}{s + 1} + \frac{\frac{1}{6}}{s + 7}$$

and

$$\theta_0(t) = \left(1 - \tfrac{7}{6}e^{-t} + \tfrac{1}{6}e^{-7t}\right)u(t)$$

This response, shown in Fig. 4.20c, appears rather sluggish. To speed up the response let us increase the gain to, say, 80.

2. $K = 80$

$$\Theta_0(s) = \frac{80}{s(s^2 + 8s + 80)} = \frac{80}{s(s + 4 - j8)(s + 4 + j8)}$$

$$= \frac{1}{s} + \frac{\frac{\sqrt{5}}{4}e^{j153°}}{s + 4 - j8} + \frac{\frac{\sqrt{5}}{4}e^{-j153°}}{s + 4 + j8}$$

and

$$\theta_0(t) = \left[1 + \tfrac{\sqrt{5}}{2}e^{-4t}\cos(8t + 153°)\right]u(t)$$

This response, shown in Fig. 4.20c, is certainly faster than the earlier case ($K = 7$), but unfortunately the improvement is achieved at the cost of ringing (oscillations) with high (21%) overshoot. Such behavior may be unacceptable in many applications. Let us try to determine K (the gain) which yields fastest response without oscillations. Complex characteristic roots lead to oscillations; to avoid oscillations, the characteristic roots should be real. In the present case the characteristic polynomial is $s^2 + 8s + K$. For $K > 16$, the characteristic roots are complex; for $K < 16$, the roots are real. The fastest response without oscillations is obtained by choosing $K = 16$. We now consider this case.

3. $K = 16$

$$\Theta_0(s) = \frac{16}{s(s^2 + 8s + 16)} = \frac{16}{s(s + 4)^2}$$

$$= \frac{1}{s} - \frac{1}{s + 4} - \frac{4}{(s + 4)^2}$$

and

$$\theta_0(t) = \left[1 - (4t + 1)e^{-4t}\right] u(t)$$

This response is also shown in Fig. 4.20c. The system with $K > 16$ is said to be **underdamped** (oscillatory response), whereas the system with $K < 16$ is said to be **overdamped**. For $K = 16$, the system is said to be **critically damped**

There is a trade-off between undesirable overshoot and rise time. Reducing overshoots leads to higher rise time (sluggish system). Thus in addition to adjusting gain K, we may need to augment the system with some type of compensator to meet the requirements of smaller overshoot and faster response.

4.6 SYSTEM REALIZATION

We now develop a systematic method for realization (or simulation) of an arbitrary nth-order transfer function. Since realization is basically a synthesis problem, there is no unique way of realizing a system. A given transfer function can be realized in many different ways. We present here three different ways of realization: canonical, cascade and parallel realization. The second form of canonical realization is discussed in Appendix 4.1 at the end of this chapter. A transfer function $H(s)$ can be realized by using integrators or differentiators along with summers and multipliers. For practical reasons we avoid use of differentiators. A differentiator accentuates high-frequency signals, which, by their nature, have large rates of change (large derivative). Signals, in practice, are always corrupted by noise, which happens to be a broad-band signal. That is, it contains components of frequencies ranging from low to very high. In processing desired signals by a differentiator, the high-frequency component of noise is amplified disproportionately, and it may swamp the desired signal. The integrator, on the other hand, tends to suppress a high frequency signal by smoothing it out. In addition, practical differentiators built with op amp circuits tend to be unstable. For these reasons we avoid differentiators in practical realizations.

Consider an LTIC system with a transfer function

$$H(s) = \frac{b_m s^m + b_{m-1} s^{m-1} + \cdots + b_1 s + b_0}{s^n + a_{n-1} s^{n-1} + \cdots + a_1 s + a_0}$$

For large s $(s \to \infty)$

$$H(s) \simeq b_m s^{m-n}$$

Therefore for $m > n$ the system acts as an $(m - n)$th-order differentiator [see Eq. (4.43a)]. For this reason we restrict $m \leq n$ for practical systems. With this restriction, the most general case is $m = n$ with the transfer function

$$H(s) = \frac{b_n s^n + b_{n-1} s^{n-1} + \cdots + b_1 s + b_0}{s^n + a_{n-1} s^{n-1} + \cdots + a_1 s + a_0} \tag{4.60}$$

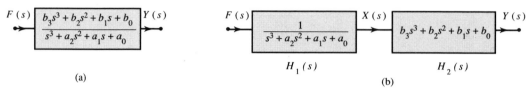

(a)

(b)

Fig. 4.21 Realization of a transfer function in two steps.

4.6-1   Canonical or Direct-Form Realization

Rather than realizing the general nth-order system described by Eq. (4.60), we begin with a specific case of the following third-order system and then extend the results to the nth-order case:

$$H(s) = \frac{b_3 s^3 + b_2 s^2 + b_1 s + b_0}{s^3 + a_2 s^2 + a_1 s + a_0} \tag{4.61}$$

For convenience of realization, we shall express this transfer function as a cascade of two transfer functions $H_1(s)$ and $H_2(s)$, as shown in Fig. 4.21.

$$H(s) = \underbrace{\left(\frac{1}{s^3 + a_2 s^2 + a_1 s + a_0} \right)}_{H_1(s)} \underbrace{\left(b_3 s^3 + b_2 s^2 + b_1 s + b_0 \right)}_{H_2(s)} \tag{4.62}$$

The output of $H_1(s)$ is denoted by $X(s)$, as shown in Fig. 4.21b. Therefore

$$Y(s) = \left(b_3 s^3 + b_2 s^2 + b_1 s + b_0 \right) X(s) \tag{4.63}$$

and

$$X(s) = \frac{1}{s^3 + a_2 s^2 + a_1 s + a_0} F(s) \tag{4.64}$$

We shall first realize $H_1(s)$. From Eq. (4.64), we can write the differential equation relating $x(t)$ to $f(t)$ as

$$(D^3 + a_2 D^2 + a_1 D + a_0) x(t) = f(t) \tag{4.65a}$$

or

$$\dddot{x}(t) + a_2 \ddot{x}(t) + a_1 \dot{x}(t) + a_0 x(t) = f(t) \tag{4.65b}$$

which yields

$$\dddot{x}(t) = -a_2 \ddot{x}(t) - a_1 \dot{x}(t) - a_0 x(t) + f(t) \tag{4.65c}$$

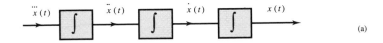

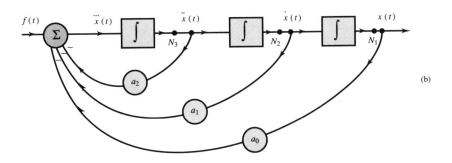

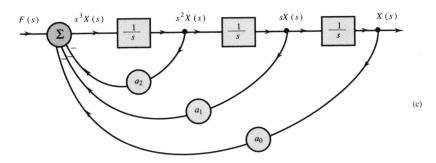

Fig. 4.22 Realization of $H_1(s) = \frac{1}{s^3 + a_2 s^2 + a_1 s + a_0}$

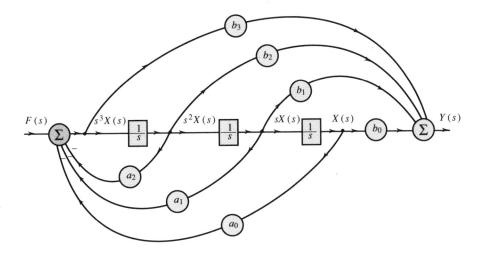

Fig. 4.23 Realization of $H(s) = \frac{b_3 s^3 + b_2 s^2 + b_1 s + b_0}{s^3 + a_2 s^2 + a_1 s + a_0}$

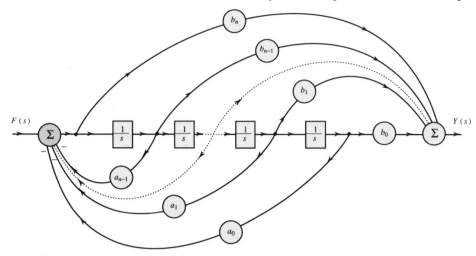

Fig. 4.24 Realization of the nth-order transfer function $H(s)$ in Eq. (4.60).

Our task is to realize a system whose input $f(t)$ and output $x(t)$ satisfy Eq. (4.65c). Let us assume that $\dddot{x}$ is available. Successive integration of $\dddot{x}$ yields $\ddot{x}$, $\dot{x}$, and x (Fig. 4.22a). We can now generate $\dddot{x}$ from $f(t)$, x, $\dot{x}$, and $\ddot{x}$ according to Eq. (4.65c), using the feedback connections, as shown in Fig. 4.22b.† The reader can verify that the system in Fig. 4.22c indeed satisfies Eq. (4.65c). Clearly the transfer function of this system (Fig. 4.22b) is $H_1(s)$ in Eq. (4.62). The signals $x(t)$, $\dot{x}(t)$, and $\ddot{x}(t)$ are available at points N_1, N_2, and N_3, respectively. If the initial conditions $x(0)$, $\dot{x}(t)$, and $\ddot{x}(0)$ are nonzero, they should be added at points N_1, N_2, and N_3, respectively.‡ Figure 4.22c shows the system of Fig. 4.22b in frequency domain. Each integrator is represented by transfer function $1/s$ [see Eq. (4.43b)]. Signals $f(t)$, $x(t)$, $\dot{x}(t)$, $\ddot{x}(t)$, and $\dddot{x}(t)$ are represented by their Laplace transforms $F(s)$, $X(s)$, $sX(s)$, $s^2X(s)$, and $s^3X(s)$, respectively.

Figure 4.22c is a realization of transfer function $H_1(s)$ in Eq. (4.62). To realize the transfer function $H(s)$, we need to augment $H_1(s)$ so that the final output $Y(s)$ is generated from $X(s)$ according to Eq. (4.63):

$$Y(s) = (b_3s^3 + b_2s^2 + b_1s + b_0)X(s)$$
$$= b_3s^3X(s) + b_2s^2X(s) + b_1sX(s) + b_0X(s)$$

Signals $X(s)$, $sX(s)$, $s^2X(s)$, and $s^3X(s)$ are available at various points in Fig. 4.22c so that $Y(s)$ can be generated by using feedforward connections to the output summer, as shown in Fig. 4.23. Therefore the realization in Fig. 4.23 has the desired transfer function $H(s)$ in Eq. (4.61).

---

†It may seem odd that we first assumed the existence of $\dddot{x}$ and generated $\ddot{x}, \dot{x}, x$ by its successive integration, and then in turn generated $\dddot{x}$ from $\ddot{x}, \dot{x}$, and x. This poses a dilemma similar to "which came first, the chicken or egg?" The problem here is satisfactorily resolved by writing the expression for $\dddot{x}$ at the output of the first summer in Fig. 4.22b and verifying that it is indeed the same as Eq. (4.65c).

‡For example, the initial condition $x(0)$ can be incorporated into our realization by adding a constant signal of value $x(0)$ at point N_1. This follows from the fact that $\int_0^t \dot{x}(\tau)\, d\tau = x(t) - x(0)$, and $x(t) = x(0) + \int_0^t \dot{x}(\tau)\, d\tau$. Similarly, the initial condition $\dot{x}(0)$ should be added at point N_2, and so on.

Generalization of this realization for the nth-order transfer function in Eq. (4.60) is shown in Fig. 4.24. This is one of the two **canonical** realizations (also known as the **controller canonical** or **direct-form** realization). The second canonical realization (**observer** canonical realization) is discussed in Appendix 4.1 at the end of this chapter. Observe that n integrators are required for a realization of an nth-order transfer function. The canonical (direct-form) realization procedure of an nth-order transfer function is systematic and straightforward, as shown in Fig. 4.24. From this figure, we can summarize the procedure as follows:

1. Draw an input summer followed by n integrators in cascade.

2. Draw the n feedback connections from the output of each of the n integrators to the input summer. The n feedback coefficients are $a_0, a_1, a_2, \cdots, a_{n-1}$, respectively, and the feedback connections have negative signs (for subtraction) at the input summer.

3. Draw the $n + 1$ feedforward connections to the output summer from the outputs of all the n integrators and the input summer. The $n + 1$ feedforward coefficients are $b_0, b_1, b_2, \cdots, b_n$, respectively, and the feedforward connections have positive signs (for addition) at the output summer.

Note that a_n is assumed to be unity and does not appear explicitly anywhere in the realization. If $a_n \neq 1$, then $H(s)$ should be normalized by dividing both its numerator and its denominator by a_n.

■**Example 4.18**

Find the canonical realization of the following transfer functions.

(a) $\dfrac{5}{s + 2}$ (b) $\dfrac{s + 5}{s + 7}$ (c) $\dfrac{s}{s + 7}$ (d) $\dfrac{4s + 28}{s^2 + 6s + 5}$

All four of these transfer functions are special cases of $H(s)$ in Eq. (4.60).

(a) In this case the transfer function is of the first order ($n = 1$), and therefore we need only one integrator for its realization. The feedback and feedforward coefficients are

$$a_0 = 2 \quad \text{and} \quad b_0 = 5 \quad b_1 = 0$$

The realization is shown in Fig. 4.25. Because $n = 1$, there is a single feedback connection from the output of the integrator to the input summer with coefficient $a_0 = 2$. For $n = 1$, there are $n + 1 = 2$ feedforward connections in general. However, in this case $b_1 = 0$, and there is only one feedforward connection with coefficient $b_0 = 5$ from the output of the integrator to the output summer. Observe that because there is only one signal to be summed at the output summer, a summer is not needed. For this reason the output summer is omitted in Fig. 4.25.

(b)
$$H(s) = \frac{s + 5}{s + 7}$$

The realization is shown in Fig. 4.26. This is also a first order transfer function with $a_0 = 7$ and $b_0 = 5$, $b_1 = 1$. There is a single feedback connection (with coefficient 7) from the integrator output to the input summer. There are two feedforward connections from the outputs of the integrator and the input summer to the output summer.†

---

†When $m = n$ (as in this case), $H(s)$ can also be realized in another way by recognizing that

$$H(s) = 1 - \frac{2}{s + 7}$$

We now realize $H(s)$ as a parallel combination of two transfer functions, as indicated by the above equation.

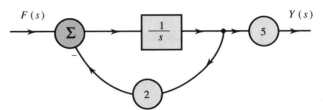

Fig. 4.25 Realization of $\frac{5}{s+2}$

(c)

$$H(s) = \frac{s}{s+7}$$

This first-order transfer function is similar to that in **(b)**, except that $b_0 = 0$. Therefore the realization is similar to that in Fig. 4.26 with the feedforward connection from the output of the integrator missing, as shown in Fig. 4.27a. Also, because there is only one signal to be summed at the output summer, the output summer is omitted. The realization in Fig. 4.27a is redrawn in a more convenient form, as shown in Fig. 4.27b.

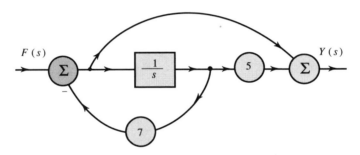

Fig. 4.26 Realization of $\frac{s+5}{s+7}$

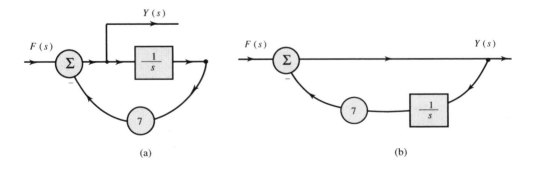

(a)   (b)

Fig. 4.27 Realization of $\frac{s}{s+7}$

(d)

$$H(s) = \frac{4s + 28}{s^2 + 6s + 5}$$

This is a second-order system with $b_0 = 28$, $b_1 = 4$, $b_2 = 0$, $a_0 = 5$, $a_1 = 6$.

Figure 4.28 shows a realization with two feedback connections and two feedforward connections. ∎

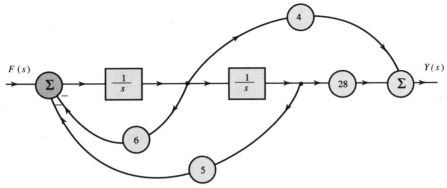

Fig. 4.28 Realization of $\frac{4s+28}{s^2+6s+5}$

△ **Exercise E4.10**
   Realize the transfer function

$$H(s) = \frac{2s}{s^2 + 6s + 25} \qquad \triangledown$$

4.6-2 Cascade and Parallel Realizations

An nth-order transfer function $H(s)$ can be expressed as a product or a sum of n first-order transfer functions. Accordingly, we can also realize $H(s)$ as a cascade (series) or parallel form of these n first-order transfer functions. Consider, for instance, the transfer function in part **(d)** of the last example:

$$H(s) = \frac{4s + 28}{s^2 + 6s + 5}$$

We can express $H(s)$ as

$$H(s) = \frac{4s + 28}{(s+1)(s+5)} = \underbrace{\left(\frac{4s + 28}{s + 1}\right)}_{H_1(s)} \underbrace{\left(\frac{1}{s + 5}\right)}_{H_2(s)} \tag{4.66a}$$

We can also express $H(s)$ as a sum of partial fractions as

$$H(s) = \frac{4s + 28}{(s+1)(s+5)} = \underbrace{\frac{6}{s + 1}}_{H_3(s)} - \underbrace{\frac{2}{s + 5}}_{H_4(s)} \tag{4.66b}$$

From Eqs. (4.66), we now have an option of realizing $H(s)$ as a cascade of $H_1(s)$ and $H_2(s)$, as shown in Fig. 4.29a, or a parallel of $H_3(s)$ and $H_4(s)$, as shown in Fig. 4.29b. Each of the first-order transfer functions in Figs. 4.29a or 4.29b can be realized by using a single integrator, as discussed in Example 4.18.

We have presented here three forms of realization (canonical, cascade, and parallel). The second canonical realization is developed in Appendix 4.1. This discussion by no means exhausts all the possibilities. Moreover, in the cascade form there are various different ways of grouping the factors in the numerator and the denominator of $H(s)$. Accordingly, we will have several possible cascade forms.

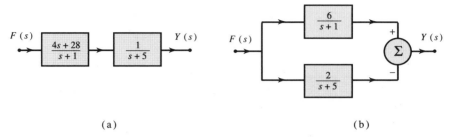

(a)   (b)

Fig. 4.29 Realization of $\frac{4s+28}{(s+1)(s+5)}$: (a) cascade form (b) parallel form.

From a practical viewpoint, parallel and cascade forms are preferable because parallel and certain cascade forms are numerically less sensitive than canonical forms to small parameter variations in the system. Qualitatively this can be explained by the fact that in a canonical realization, all the coefficients interact with each other, and a change in any coefficient will be magnified through its repeated influence from feedback and feedforward connections. In a parallel realization, on the other hand, the change in a coefficient will affect only a localized segment; the case with a cascade realization is similar.

In the above examples of cascade and parallel realization, we have separated $H(s)$ into first-order factors. For $H(s)$ of higher orders, we could group $H(s)$ into factors, not all of which are necessarily of the first order. For example, if $H(s)$ is a third-order transfer function, we could realize it as a cascade (or a parallel) combination of a first-order and a second-order factor.

Realization of Complex-conjugate Poles

The complex poles in $H(s)$ should be realized as a second-order (quadratic) factor because we cannot implement multiplication by complex numbers. Consider, for example,

$$H(s) = \frac{10s + 50}{(s + 3)(s^2 + 4s + 13)}$$

$$= \frac{10s + 50}{(s + 3)(s + 2 - j3)(s + 2 + j3)}$$

$$= \frac{2}{s + 3} - \frac{1 + j2}{s + 2 - j3} - \frac{1 - j2}{s + 2 + j3}$$

We cannot realize first-order transfer functions individually with the poles $-2 \pm j3$ because they require multiplication by complex numbers $-2 \pm j3$ in the feedback path of their realization. Therefore we need to combine the conjugate poles and realize them as a second-order transfer function.† In the present case, we can express $H(s)$ as

---

†It is possible to realize complex conjugate poles indirectly by using a cascade of two first-order transfer functions. A transfer function with poles $-a \pm jb$ can be realized by using a cascade of two identical first-order transfer functions, each having a pole at $-a$. (See Prob. 4.6-7.)

$$H(s) = \left(\frac{10}{s+3}\right)\left(\frac{s+5}{s^2+4s+13}\right) \tag{4.67a}$$

$$= \frac{2}{s+3} - \frac{2s-8}{s^2+4s+13} \tag{4.67b}$$

Now we can realize $H(s)$ in cascade form using Eq. (4.67a) or in parallel form using Eq. (4.67b).

Realization of Repeated Poles

When repeated poles occur, the procedure for canonical and cascade realization is exactly the same as above. In parallel realization, however, the procedure requires a special precaution. This is explained in Example 4.19 below.

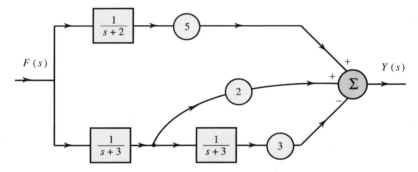

Fig. 4.30 Parallel realization of $\frac{7s^2+37s+51}{(s+2)(s+3)^2}$

■Example 4.19

Determine the parallel realization of

$$H(s) = \frac{7s^2 + 37s + 51}{(s+2)(s+3)^2}$$

$$= \frac{5}{s+2} + \frac{2}{s+3} - \frac{3}{(s+3)^2}$$

This third-order transfer function should require no more than three integrators. But if we try to realize each of the three partial fractions separately, we require four integrators because one of the terms is second-order. This difficulty can be avoided by observing that the terms $1/(s+3)$ and $1/(s+3)^2$ can be realized with a cascade of two subsystems, each having a transfer function $1/(s+3)$, as shown in Fig. 4.30. Each of the three first-order transfer functions in Fig. 4.30 may now be realized as in Fig. 4.25. ■

△ **Exercise E4.11**

Find a canonical, a cascade and a parallel realization of

$$H(s) = \frac{s+3}{s^2+7s+10} = \left(\frac{s+3}{s+2}\right)\left(\frac{1}{s+5}\right) \qquad \bigtriangledown$$

4.6-3   System Realization Using Operational Amplifiers

In this section, we discuss practical implementation of the realizations described in the previous subsection. Earlier we saw that the basic elements required for the synthesis of an LTIC system (or a given transfer function) are (scalar) multipliers, integrators, and summers (or adders). All these elements can be realized by operational amplifier (op amp) circuits, as shown below.

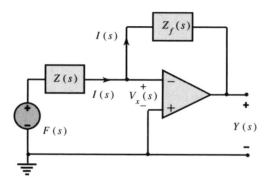

Fig. 4.31 A basic inverting configuration op amp circuit.

Operational Amplifier Circuits

Figure 4.31 shows an op amp circuit in the frequency domain (the transformed circuit). Because the input impedance of the op amp is infinite (very high), all of the current $I(s)$ flows in the feedback path, as shown in Fig. 4.31. Moreover $V_x(s)$, the voltage at the input of the op amp, is zero (very small) because of the infinite (very large) gain of the op amp. Therefore, for all practical purposes,

$$Y(s) = -I(s)Z_f(s)$$

Moreover, because $v_x \approx 0$,

$$I(s) = \frac{F(s)}{Z(s)}$$

Substitution of the second equation in the first yields

$$Y(s) = -\frac{Z_f(s)}{Z(s)}F(s)$$

Therefore the op amp circuit in Fig. 4.31 has the transfer function

$$H(s) = -\frac{Z_f(s)}{Z(s)} \tag{4.68}$$

By properly choosing $Z(s)$ and $Z_f(s)$ we can obtain a variety of transfer functions, as the following development shows.

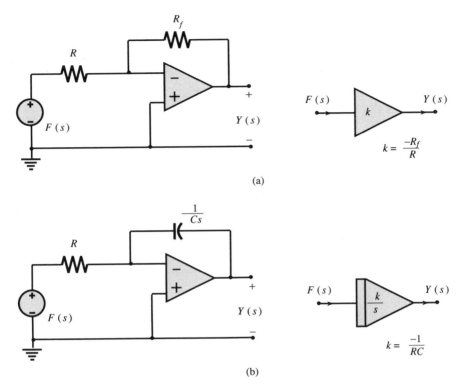

Fig. 4.32 (a) Op amp inverting amplifier (b) integrator.

The Scalar Multiplier

If we use a resistor R_f in the feedback and a resistor R, at the input (Fig. 4.32a), then $Z_f(s) = R_f$, $Z(s) = R$ and

$$H(s) = -\frac{R_f}{R} \tag{4.69a}$$

The system acts as a scalar multiplier (or an amplifier) with a negative gain $\frac{R_f}{R}$. A positive gain can be obtained by using two such multipliers in cascade or by using a single noninverting amplifier, as shown in Fig. 4.17c. Figure 4.32a also shows the compact symbol used in circuit diagrams for a scalar multiplier.

The Integrator

If we use a capacitor C in the feedback and a resistor R at the input (Fig. 4.32b), then $Z_f(s) = 1/Cs$, $Z(s) = R$, and

$$H(s) = \left(-\frac{1}{RC}\right)\frac{1}{s} \tag{4.69b}$$

The system acts as an ideal integrator with a gain $-1/RC$. Figure 4.32b also shows the compact symbol used in circuit diagrams for an integrator.

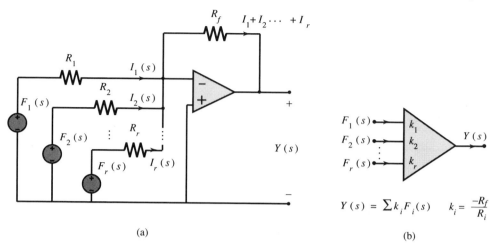

Fig. 4.33 Op amp summing and amplifying circuit.

The Summer

Consider now the circuit in Fig. 4.33a with r inputs $F_1(s)$, $F_2(s)$, ..., $F_r(s)$. As usual, the input voltage $V_x(s) \simeq 0$ because the gain of op amp $\to \infty$. Moreover, the current going into the op amp is very small ($\simeq 0$) because the input impedance $\to \infty$. Therefore the total current in the feedback resistor R_f is $I_1(s) + I_2(s) + \cdots + I_r(s)$. Moreover, because $V_x(s) = 0$,

$$I_j(s) = \frac{F_j(s)}{R_j} \qquad j = 1, 2, \ldots, r$$

Also

$$
\begin{aligned}
Y(s) &= -R_f\left[I_1(s) + I_2(s) + \cdots + I_r(s)\right] \\
&= -\left[\tfrac{R_f}{R_1}F_1(s) + \tfrac{R_f}{R_2}F_2(s) + \cdots + \tfrac{R_f}{R_r}F_r(s)\right] \\
&= k_1 F_1(s) + k_2 F_2(s) + \cdots + k_r F_r(s)
\end{aligned}
\qquad (4.70)
$$

where

$$k_i = \frac{-R_f}{R_i}$$

Clearly, the circuit in Fig. 4.33 serves a summer and an amplifier with any desired gain for each of the input signals. Figure 4.33b shows the compact symbol used in circuit diagrams for a summer with r inputs.

■ Example 4.20

Using op amp circuits, realize the canonical form of the transfer function

$$H(s) = \frac{2s + 5}{s^2 + 4s + 10}$$

The basic canonical realization is shown in Fig. 4.34a. Signals at various points are also indicated in the realization. Op amp elements (multipliers, integrators, and summers)

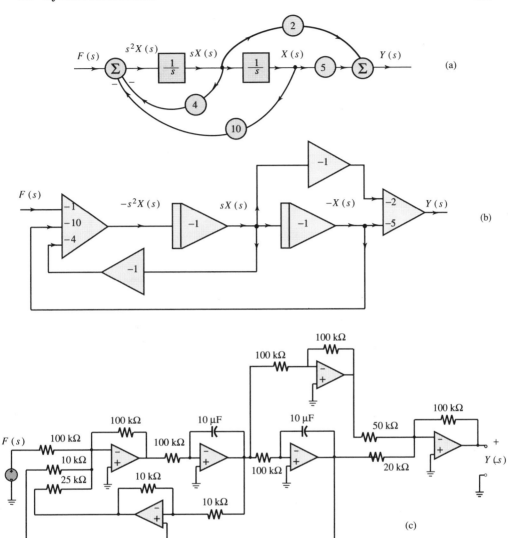

Fig. 4.34 Op amp realization of a second-order transfer function $\frac{2s+5}{s^2+4s+10}$

change the polarity of the output signals. To incorporate this fact, we modify the canonical realization in Fig. 4.34a to that shown in Fig. 4.34b. In Fig. 4.34a, the successive outputs of the summer and the integrators are $s^2X(s)$, $sX(s)$, and $X(s)$ respectively. Because of polarity reversals in op amp circuits, these outputs are $-s^2X(s)$, $sX(s)$, and $-X(s)$ respectively in Fig. 4.34b. This polarity reversal requires corresponding modifications in the signs of feedback and feedforward gains. From Fig. 4.34a, we have

$$s^2X(s) = F(s) - 4sX(s) - 10X(s)$$

Therefore

$$-s^2X(s) = -F(s) + 4sX(s) + 10X(s)$$

Because the summer gains are always negative (see Fig. 4.33b), we rewrite the above equation as

$$-s^2X(s) = -1[F(s)] - 4[-sX(s)] - 10[-X(s)]$$

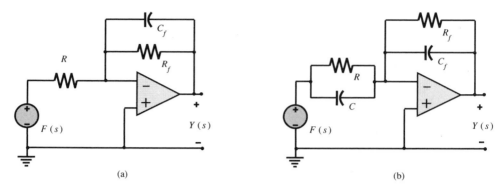

Fig. 4.35 Op amp circuits for exercise E4.12.

Figure 4.34b shows the implementation of this equation. The hardware realization is shown in Fig. 4.34c. Both integrators have a unity gain which requires $RC = 1$. We have used $R=100$ kΩ and $C = 10\mu$F. The gain of 10 in the outer feedback path is obtained in the summer by choosing the feedback resistor of the summer to be 100 kΩ, and an input resistor of 10 kΩ. Similarly, the gain of 4 in the inner feedback path is obtained by using the corresponding input resistor 25 kΩ. The gains of 2 and 5, required in the feedforward connections, are obtained by using a feedback resistor of 100 kΩ and input resistors of 50 kΩ and 20 kΩ respectively.

The op amp realization in Fig. 4.34 is not necessarily the one that uses the fewest op amps. It is possible to avoid the two inverting op amps (with gain -1) in Fig. 4.34 by adding signal $sX(s)$ to the input and output summers directly, using the noninverting amplifier configuration in Fig. 4.17. There are also circuits (such as Sallen-Key) which realize a first- or second-order transfer function using only one op amp. ∎

△ **Exercise E4.12**
   Show that the transfer functions of the op amp circuits in Figs. 4.35a and 4.35b are $H_1(s)$ and $H_2(s)$, respectively, where

$$H_1(s) = \frac{-R_f}{R}\left(\frac{a}{s+a}\right) \qquad a = \frac{1}{R_f C_f}$$

$$H_2(s) = -\frac{C}{C_f}\left(\frac{s+b}{s+a}\right) \qquad a = \frac{1}{R_f C_f} \quad b = \frac{1}{RC}$$

Hint: Use Eq. (4.68). ▽

4.7 SYSTEM RESPONSE TO SINUSOIDAL INPUTS: FREQUENCY RESPONSE OF AN LTIC SYSTEM

In this section we find the system response to sinusoidal inputs. In Sec. 4.3-3 we showed that an LTIC system response to an everlasting (starting at $t = -\infty$) exponential input $f(t) = e^{st}$ is also an everlasting exponential $H(s)e^{st}$. It is helpful to represent this relationship by a directed arrow notation as

$$e^{st} \rightarrow H(s)e^{st} \tag{4.71}$$

Setting $s = \pm j\omega$ this relationship yields

$$e^{j\omega t} \rightarrow H(j\omega)e^{j\omega t} \tag{4.72a}$$

$$e^{-j\omega t} \rightarrow H(-j\omega)e^{-j\omega t} \tag{4.72b}$$

Addition of these two equations yields

$$2\cos \omega t \rightarrow H(j\omega)e^{j\omega t} + H(-j\omega)e^{-j\omega t} = 2\text{Re}\left[H(j\omega)e^{j\omega t}\right] \qquad (4.73)$$

Expressing $H(j\omega)$ in the polar form

$$H(j\omega) = |H(j\omega)|\, e^{j\angle H(j\omega)} \qquad (4.74)$$

Eq. (4.73) can be expressed as

$$\cos \omega t \rightarrow |H(j\omega)|\cos\left[\omega t + \angle H(j\omega)\right]$$

In other words, the system response $y(t)$ to a sinusoidal input $\cos \omega t$ is given by†

$$y(t) = |H(j\omega)|\cos\left[\omega t + \angle H(j\omega)\right] \qquad (4.75a)$$

Using a similar argument, we can show that the system response to a sinusoid $\cos(\omega t + \theta)$ is

$$y(t) = |H(j\omega)|\cos\left[\omega t + \theta + \angle H(j\omega)\right] \qquad (4.75b)$$

This result, which is valid only for asymptotically stable systems,‡ shows that for a sinusoidal input of radian frequency ω, the system response is also a sinusoid of the same frequency ω. **The amplitude of the output sinusoid is $|H(j\omega)|$ times the input amplitude, and the phase of the output sinusoid is shifted by $\angle H(j\omega)$ with respect to the input phase** (see Fig. 4.36b). For instance, if a certain system has $|H(j10)| = 3$ and $\angle H(j10) = -30°$, then the system amplifies a sinusoid of frequency $\omega = 10$ by a factor of 3, and delays its phase by 30°. The system response to an input $5\cos(10t + 50°)$ is $3 \times 5\cos(10t + 50° - 30°) = 15\cos(10t + 20°)$.

Clearly $|H(j\omega)|$ is the system **gain**, and a plot of $|H(j\omega)|$ versus ω shows the system gain as a function of frequency ω. This function is more commonly known as the **amplitude response**. Similarly, $\angle H(j\omega)$ is the **phase response** and a plot of $\angle H(j\omega)$ versus ω shows how the system modifies or changes the phase of the input sinusoid. These two plots together, as functions of ω, are called the **frequency response of the system**. Observe that $H(j\omega)$ has the information of $|H(j\omega)|$ and $\angle H(j\omega)$. For this reason $H(j\omega)$ is also called the **frequency response of the system**. The frequency response plots show at a glance how a system responds to

---

†For a causal sinusoidal input $\cos \omega t\, u(t)$, the system response will also have an additional (natural) component consisting of the characteristic modes. For a stable system, all the modes decay exponentially, and only the sinusoidal component $|H(j\omega)|\cos[\omega t + \angle H(j\omega)]$ persists. For this reason this component is called the **steady-state response** of the system. Thus the steady-state response of the system to an input $\cos \omega t\, u(t)$ is $|H(j\omega)|\cos[\omega t + \angle H(j\omega)]$ (see Prob. 4.7-4).

‡Equation (4.71) is valid only for values of s lying in the region of convergence of $H(s)$. For $s = j\omega$, s lies on the imaginary axis. The region of convergence of $H(s)$ for unstable and marginally stable systems does not include the imaginary axis. Therefore Eqs. (4.74) and (4.75) are valid only for stable systems.

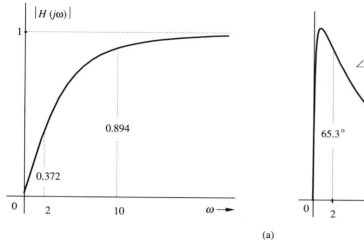
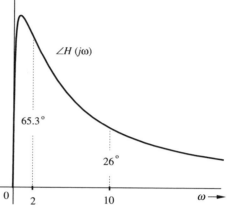

(a)

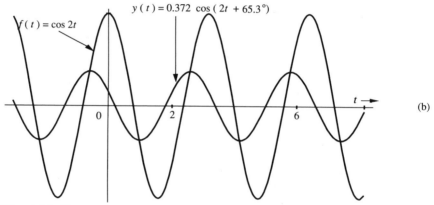

(b)

Fig. 4.36 Frequency response of an LTIC system in Example 4.21.

sinusoids of various frequencies. Thus the frequency response of a system represents its filtering characteristic.

■ Example 4.21

Find the frequency response (amplitude and phase response) of a system whose transfer function is

$$H(s) = \frac{s + 0.1}{s + 5}$$

Also find the system response $y(t)$ if the input $f(t)$ is **(a)** $\cos 2t$ **(b)** $\cos(10t - 50°)$.

In this case

$$H(j\omega) = \frac{j\omega + 0.1}{j\omega + 5}$$

Therefore

$$|H(j\omega)| = \frac{\sqrt{\omega^2 + 0.01}}{\sqrt{\omega^2 + 25}} \quad \text{and} \quad \angle H(j\omega) = \tan^{-1}\left(\frac{\omega}{0.1}\right) - \tan^{-1}\left(\frac{\omega}{5}\right)$$

Both the amplitude and the phase response are shown in Fig. 4.36a as functions of ω. These plot furnish the complete information about the frequency response of the system to sinusoidal inputs.

(a) For the input $f(t) = \cos 2t$, $\omega = 2$, and

$$|H(j2)| = \frac{\sqrt{(2)^2 + 0.01}}{\sqrt{(2)^2 + 25}} = 0.372$$

$$\angle H(j2) = \tan^{-1}\left(\frac{2}{0.1}\right) - \tan^{-1}\left(\frac{2}{5}\right) = 87.1° - 21.8° = 65.3°$$

We also could have read these values directly from the frequency response plots in Fig. 4.36a corresponding to $\omega = 2$. This result means that for a sinusoidal input with frequency $\omega = 2$, the amplitude gain of the system is 0.372, and the phase shift is 65.3°. In other words, the output amplitude is 0.372 times the input amplitude, and the phase of the output is shifted with respect to that of the input by 65.3°. Therefore the system response to an input $\cos 2t$ is

$$y(t) = 0.372 \cos\left(2t + 65.3°\right)$$

The input $\cos 2t$ and the corresponding system response $0.372 \cos\left(2t + 65.34°\right)$ are shown in Fig. 4.36b.

(b) For the input $\cos\left(10t - 50°\right)$, instead of computing the values $|H(j\omega)|$ and $\angle H(j\omega)$ as in part (a), we shall read them directly from the frequency response plots in Fig. 4.36a corresponding to $\omega = 10$. These are

$$|H(j10)| = 0.894 \qquad \text{and} \qquad \angle H(j10) = 26°$$

Therefore, for a sinusoidal input of frequency $\omega = 10$, the output sinusoid amplitude is 0.894 times the input amplitude and the output sinusoid is shifted with respect to the input sinusoid by 26°. Therefore $y(t)$, the system response to an input $\cos\left(10t - 50°\right)$, is

$$y(t) = 0.894 \cos\left(10t - 50° + 26°\right) = 0.894 \cos\left(10t - 24°\right)$$

If the input were $\sin\left(10t - 50°\right)$, the response would be $0.894 \sin\left(10t - 50° + 26°\right) = 0.894 \sin\left(10t - 24°\right)$.

The frequency response plots in Fig. 4.36a show that the system has highpass filtering characteristics; it responds well to sinusoids of higher frequencies (ω well above 5), and suppresses sinusoids of lower frequencies (ω well below 5). ■

■ **Example 4.22**
Find and sketch the frequency response (amplitude and phase response) for
(a) an ideal delay of T seconds;
(b) an ideal differentiator;
(c) an ideal integrator.

(a) **Ideal delay of T seconds**: The transfer function of an ideal delay, as found in Eq. (4.42), is

$$H(s) = e^{-sT}$$

Therefore

$$H(j\omega) = e^{-j\omega T}$$

Consequently

$$|H(j\omega)| = 1 \qquad \text{and} \qquad \angle H(j\omega) = -\omega T \qquad\qquad (4.76)$$

This amplitude and phase response is shown in Fig. 4.37a. The amplitude response is constant (unity) for all frequencies. The phase shift increases linearly with frequency with a slope of $-T$. This result can be explained physically by recognizing that if a sinusoid $\cos \omega t$ is passed through an ideal delay of T seconds, the output is $\cos \omega(t - T)$. The

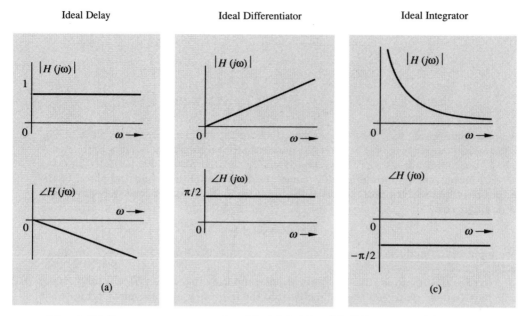

Fig. 4.37 Frequency response of an ideal (a) delay (b) differentiator (c) integrator.

output sinusoid amplitude is the same as that of the input for all values of ω. Therefore the amplitude response (gain) is unity for all frequencies. Moreover, the output $\cos \omega(t-T) = \cos(\omega t - \omega T)$ has a phase shift $-\omega T$ with respect to the input $\cos \omega t$. Therefore the phase response is linearly proportional to the frequency ω with a slope $-T$.

(b) An ideal differentiator: The transfer function of an ideal differentiator, as found in Eq. (4.43a), is
$$H(s) = s$$
Therefore
$$H(j\omega) = j\omega = \omega e^{j\pi/2}$$
Consequently
$$|H(j\omega)| = \omega \qquad \text{and} \qquad \angle H(j\omega) = \tfrac{\pi}{2} \tag{4.77a}$$
This amplitude and phase response is shown in Fig. 4.37b. The amplitude response increases linearly with frequency, and phase response is constant $(\pi/2)$ for all frequencies. This result can be explained physically by recognizing that if a sinusoid $\cos \omega t$ is passed through an ideal differentiator, the output is $-\omega \sin \omega t = \omega \cos(\omega t + \tfrac{\pi}{2})$. Therefore the output sinusoid amplitude is ω times the input amplitude; that is, the amplitude response (gain) increases linearly with frequency ω. Moreover the output sinusoid undergoes a phase shift $\tfrac{\pi}{2}$ with respect to the input $\cos \omega t$. Therefore the phase response is constant $(\pi/2)$ with frequency.

In an ideal differentiator, the amplitude response (gain) is proportional to frequency $[|H(j\omega)| = \omega]$, so that the higher-frequency components are enhanced (see Fig. 4.37b). All practical signals are contaminated with noise, which, by its nature, is a broad-band signal containing components of very high frequencies. A differentiator can increase the noise disproportionately to the point of drowning out the desired signal. This is the reason why ideal differentiators are avoided in practice.

(c) An ideal integrator: The transfer function of an ideal integrator, as found in Eq. (4.43b), is
$$H(s) = \frac{1}{s}$$

Therefore

$$H(j\omega) = \frac{1}{j\omega} = \frac{-j}{\omega} = \frac{1}{\omega}e^{-j\pi/2}$$

Consequently

$$|H(j\omega)| = \frac{1}{\omega} \quad \text{and} \quad \angle H(j\omega) = -\frac{\pi}{2} \tag{4.77b}$$

This amplitude and phase response is shown in Fig. 4.37c. The amplitude response is inversely proportional to frequency, and the phase shift is constant $(-\pi/2)$ with frequency. This result can be explained physically by recognizing that if a sinusoid $\cos \omega t$ is passed through an ideal integrator, the output is $\frac{1}{\omega} \sin \omega t = \frac{1}{\omega} \cos (\omega t - \frac{\pi}{2})$. Therefore the amplitude response is inversely proportional to ω, and the phase response is constant $(-\pi/2)$ with frequency.

Because the gain of an ideal integrator is $1/\omega$, it suppresses higher-frequency components, but enhances lower-frequency components with $\omega < 1$. Consequently noise signals (if they do not contain appreciable amount of very low frequency components) are suppressed (smoothed out) by an integrator. ■

⊙ **Computer Example C4.2**

Compute the frequency response (amplitude and phase) of a system whose transfer function is

$$H(s) = \frac{s + 0.1}{s + 5}$$

There are several ways to compute the solution. The equations for the magnitude and phase can be derived and then programmed, but this is rather problem specific. Here we shall take a general approach.

```
num = [1 0.1];    % Enter the numerator polynomial.
den = [1 5];      % Enter the denominator polynomial.
w=0:.1:100;       % Create a frequency vector.
jw = j*w;         % Create a complex vector w.
H = polyval(num,jw)./polyval(den,jw); % Evaluate the transfer function
% at the given frequencies.
mag = abs(H);              % Calculates |H(jw)|.
phase = angle(H)*180/pi;   % Calculates < H(jw).
subplot(211),plot(w,mag);grid,    % Plot the magnitude response.
subplot(212),plot(w,phase);grid   % Plot the phase response.  ⊙
```

△ **Exercise E4.13**

Find the response of an LTIC system specified by

$$\frac{d^2y}{dt^2} + 3\frac{dy}{dt} + 2y(t) = \frac{df}{dt} + 5f(t)$$

if the input is $20\sin (3t + 35°)$

Answer: $10.23 \sin (3t - 61.91°)$ ▽

Bode Plots

The plotting of amplitude and phase functions is facilitated remarkably by using logarithmic scales. Instead of plotting $|H(j\omega)|$, we plot $20\log_{10}|H(j\omega)|$ as a function of ω. The logarithmic amplitude thus defined is measured in terms of the decibel (dB). The amplitude and phase plots as a function of ω on a logarithmic scale are known as **Bode plots**. By using the asymptotic behavior of the amplitude and phase functions, these plots can be sketched with remarkable ease, even for very complex transfer functions.

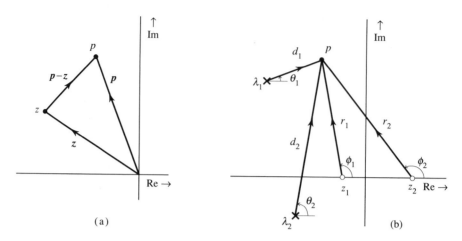

Fig. 4.38 (a)vector representation of complex numbers (b) vector representation of factors of $H(s)$.

4.7-1 Dependence of Frequency Response on Poles and Zeros of H(s)

Frequency response of a system is basically the information about the filtering capability of the system. We now examine the close connection that exists between the pole-zero locations of the system transfer function and its frequency response (or filtering characteristics). A system transfer function can be expressed as

$$H(s) = \frac{P(s)}{Q(s)} = b_n \frac{(s - z_1)(s - z_2) \cdots (s - z_n)}{(s - \lambda_1)(s - \lambda_2) \cdots (s - \lambda_n)} \tag{4.78a}$$

where z_1, z_2, ..., z_n are the zeros of $H(s)$ and the characteristic roots λ_1, λ_2, ..., λ_n are the poles of $H(s)$. Now the value of the transfer function $H(s)$ at $s = p$ is

$$H(s)|_{s=p} = b_n \frac{(p - z_1)(p - z_2) \cdots (p - z_n)}{(p - \lambda_1)(p - \lambda_2) \cdots (p - \lambda_n)} \tag{4.78b}$$

This equation consists of factors of the form $p - z_i$ and $p - \lambda_i$. The factor $p - z$ is a complex number represented by a vector drawn from point z to point p in the complex plane, as shown in Fig. 4.38a. The length of this line segment is $|p - z|$, the magnitude of $p - z$. The angle of this directed line segment (with horizontal axis) is $\angle(p - z)$. To compute $H(s)$ at $s = p$, we draw line segments from all poles and zeros of $H(s)$ to point p, as shown in Fig. 4.38b. The vector connecting a zero z_i to the point p is $p - z_i$. Let the length of this vector be r_i, and let its angle with the horizontal axis be ϕ_i. Then $p - z_i = r_i e^{j\phi_i}$. Similarly the vector connecting a pole λ_i to the point p is $p - \lambda_i = d_i e^{j\theta_i}$, where d_i and θ_i are the length and the angle (with the horizontal axis) respectively of the vector $p - \lambda_i$. Now from Eq. (4.78b) it follows that

$$\begin{aligned}
H(s)|_{s=p} &= b_n \frac{(r_1 e^{j\phi_1})(r_2 e^{j\phi_2}) \cdots (r_n e^{j\phi_n})}{(d_1 e^{j\theta_1})(d_2 e^{j\theta_2}) \cdots (d_n e^{j\theta_1})} \\
&= b_n \frac{r_1 r_2 \cdots r_n}{d_1 d_2 \cdots d_n} e^{j[(\phi_1 + \phi_2 + \cdots + \phi_n) - (\theta_1 + \theta_2 + \cdots + \theta_n)]}
\end{aligned}$$

Therefore

$$|H(s = p)| = b_n \frac{r_1 r_2 \cdots r_n}{d_1 d_2 \cdots d_n}$$

$$= b_n \frac{\text{product of the distances of zeros to } p}{\text{product of distances of poles to } p} \qquad (4.79a)$$

and

$$\angle H(s = p) = (\phi_1 + \phi_2 + \cdots \phi_n) - (\theta_1 + \theta_2 + \cdots + \theta_n)$$

$$= \text{sum of zero angles to } p - \text{sum of pole angles to } p$$

$$(4.79b)$$

Using this procedure, we can determine $H(s)$ for any value of s. To compute the frequency response $H(j\omega)$, we use $s = j\omega$ (a point on the imaginary axis), connect all poles and zeros to the point $j\omega$, and determine $|H(j\omega)|$ and $\angle H(j\omega)$ from Eqs. (4.79). We repeat this procedure for all values of ω from 0 to ∞ to obtain the frequency response.

Gain Enhancement by a Pole

To understand the effect of poles and zeros on the frequency response, consider a hypothetical case of a single pole $-\alpha + j\omega_0$, as shown in Fig. 4.39a. To find the amplitude response $|H(j\omega)|$ for a certain value of ω, we connect the pole to the point $j\omega$ (Fig. 4.39a). If the length of this line is d, then $|H(j\omega)|$ is proportional to $1/d$.

$$|H(j\omega)| = \frac{K}{d} \qquad (4.80)$$

where the exact value of constant K is not important at this point. As ω increases from zero up, d decreases progressively until ω reaches the value ω_0. As ω increases beyond ω_0, d increases progressively. Therefore, according to Eq. (4.80), the amplitude response $|H(j\omega)|$ increases from $\omega = 0$ until $\omega = \omega_0$, and it decreases continuously as ω increases beyond ω_0, as shown in Fig. 4.39b. Therefore a pole at $-\alpha + j\omega_0$ results in a frequency-selective behavior that enhances the gain at frequency ω_0 (resonance). Moreover, as the pole moves closer to the imaginary axis (as α is reduced), this enhancement (resonance) becomes more pronounced. This is because α, the distance between the pole and $j\omega_0$ (d corresponding to $j\omega_0$), becomes smaller, which increases the gain K/d. In the extreme case, when $\alpha = 0$ (pole on the imaginary axis), the gain at ω_0 goes to infinity. Repeated poles further enhance the frequency-selective effect. To summarize, we can enhance a gain at a frequency ω_0 by placing a pole opposite the point $j\omega_0$. The closer the pole is to $j\omega_0$, the higher is the gain at ω_0, and the gain variation is more rapid (more frequency-selective) in the vicinity of frequency ω_0. Note that a pole must be placed in the LHP for stability.

Here we considered the effect of a single complex pole on the system gain. For a real system, a complex pole $-\alpha + j\omega_0$ must accompany its conjugate $-\alpha - j\omega_0$. We can readily show that the presence of the conjugate pole does not appreciably change the frequency-selective behavior in the vicinity of ω_0. This is because the gain in this case is K/dd', where d' is the distance of a point $j\omega$ from the conjugate

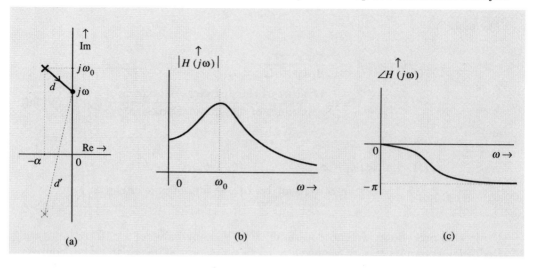

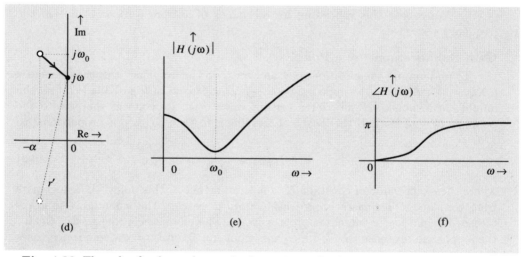

Fig. 4.39 The role of poles and zeros in determining the frequency response of an LTIC system.

pole $-\alpha - j\omega_0$. Because the conjugate pole is far from $j\omega_0$, there is no dramatic change in the length d' as ω varies in the vicinity of ω_0. There is a gradual increase in the value of d' as ω increases, which leaves the frequency-selective behavior as it was originally, with only minor changes.

Gain Suppression by a Zero

Using the same argument, we observe that zeros at $-\alpha \pm j\omega_0$ (Fig. 4.39d) will have exactly the opposite effect of suppressing the gain in the vicinity of ω_0, as shown in Fig. 4.39e. A zero on the imaginary axis at $j\omega_0$ will totally suppress the gain (zero gain) at frequency ω_0. Repeated zeros will further enhance the effect. It is clear that by proper placement of poles and zeros, we can obtain a variety of frequency-selective behaviors. Using these observations, we can design lowpass, highpass, bandpass, and bandstop (or notch) filters.

Phase response can also be computed graphically. In Fig. 4.39a, angles formed

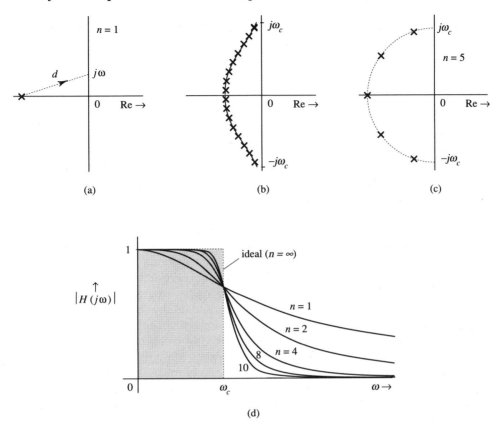

Fig. 4.40 Pole-zero configuration and the amplitude response of a a lowpass (Butterworth) filter.

by the complex conjugate poles $-\alpha \pm j\omega_0$ at $\omega = 0$ (the origin) are equal and opposite. As ω increases from 0 up, the angle θ_1 due to the pole $-\alpha + j\omega_0$, which has a negative value at $\omega = 0$, is reduced in magnitude; the angle θ_2 due to the pole $-\alpha - j\omega_0$, which has a positive value at $\omega = 0$ increases in magnitude. As a result $\theta_1 + \theta_2$, the sum of the two angles, increases continuously, approaching a value π as $\omega \to \infty$. The resulting phase response $\angle H(j\omega) = -(\theta_1 + \theta_2)$ is shown in Fig. 4.39c. Similar arguments apply to zeros at $-\alpha \pm j\omega_0$. The resulting phase response $\angle H(j\omega) = (\phi_1 + \phi_2)$ is shown in Fig. 4.39f.

Observe that a pole and a zero nearby tend to cancel out each other's influence on the frequency response. We now focus on simple filters, using the intuitive insights gained in this discussion. The discussion is essentially qualitative.

Lowpass Filters

A typical lowpass filter has a maximum gain at $\omega = 0$. Therefore we need to place a pole (or poles) on the real axis opposite the origin $(j\omega = 0)$, as shown in Fig. 4.40a. The transfer function of this system is

$$H(s) = \frac{\omega_c}{s + \omega_c}$$

We have chosen the numerator of $H(s)$ to be ω_c in order to normalize the dc gain $H(0)$ to unity. From Fig. 4.40a, we have

$$|H(j\omega)| = \frac{\omega_c}{d}$$

with $H(0) = 1$. As ω increases, r increases and $|H(j\omega)|$ decreases monotonically with ω, as shown in Fig. 4.40d by label $n = 1$. This is clearly a lowpass filter. An ideal lowpass filter characteristic, shown dotted in Fig. 4.40d, has a constant gain of unity up to frequency ω_c. Then the gain drops suddenly to 0 for $\omega > \omega_c$. To achieve the ideal characteristic, we need enhanced gain over the entire frequency band from 0 to ω_c. We know that to enhance a gain at any frequency ω, we need to place a pole opposite ω. To achieve an enhanced gain for all frequencies over the band (0 to ω_c), we need to place a pole opposite every frequency in this band. In other words, we need a **continuous wall of poles** facing the imaginary axis opposite the frequency band 0 to ω_c (and from 0 to $-\omega_c$ for conjugate poles), as shown in Fig. 4.40b. The optimum shape of this wall is not obvious at this point because our arguments are qualitative and intuitive. Yet it is certain that to have enhanced gain (constant gain) at every frequency over this range, we need an infinite number of poles on this wall. It can be shown that for maximally flat† response over the frequency range (0 to ω_c), the wall is a semicircle with infinite number of poles uniformly distributed along the wall[9]. In practice, we compromise by using a finite number (n) of poles with less-than-ideal characteristics. Figure 4.40c shows the pole configuration for a fifth-order ($n = 5$) filter. The amplitude response for various value of n are shown in Fig. 4.40d. As $n \to \infty$, the filter response approaches the ideal. This family of filters is known as the **Butterworth** filters. There are also other families. In **Chebyshev** filters, the wall shape is a semiellipse rather than a semicircle. Chebyshev filter characteristic is inferior to that of Butterworth over the passband $(0, \omega_c)$, where it shows a rippling effect instead of the maximally flat response of Butterworth. But in the stopband ($\omega > \omega_c$), Chebyshev filter gain drops faster than that of the Butterworth filters.

Bandpass Filters

The dotted characteristic in Fig. 4.41b shows the ideal bandpass filter gain. In the bandpass filter, the gain is enhanced over the entire passband. Our earlier discussion indicates that this can be realized by a wall of poles opposite the imaginary axis in front of the passband centered at ω_0. (There is also a wall of conjugate poles opposite $-\omega_0$.) Ideally, an infinite number of poles is required. In practice, we compromise by using a finite number of poles and accepting less-than-ideal characteristics (Fig. 4.41).

Notch (Bandstop) Filter

An ideal notch filter amplitude response is a complement of the amplitude response of an ideal bandpass filter. Its gain is zero over a small band centered at some frequency ω_0 and is unity over the remaining frequencies, as shown by the dotted characteristic in Fig. 4.42b. Realization of such a characteristic requires an infinite number of poles and zeros. Let us consider a practical second-order notch

---

†This means that the first $2n - 1$ derivatives of $|H(j\omega)|$ are zero at $\omega = 0$.

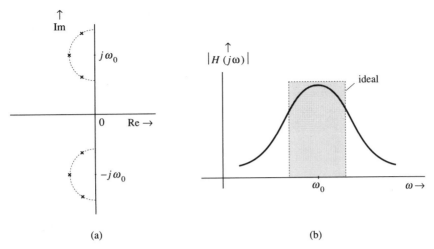

Fig. 4.41 Pole-zero configuration and the amplitude response of a bandpass filter.

filter to obtain zero gain at a frequency $\omega = \omega_0$. For this purpose we must have zeros at $\pm j\omega_0$. The requirement of unity gain at $\omega = \infty$ requires the number of poles to be equal to number of zeros $(m = n)$. This ensures that for very large values of ω, the product of the distances of poles from ω will be equal to the product of distances of zeros from ω. Moreover, unity gain at $\omega = 0$ requires a pole and the corresponding zero to be equidistant from the origin. For example, if we use two (complex conjugate) zeros, we must have two poles; the distance from the origin of the poles and of the zeros should be the same . This can be achieved by placing the two conjugate poles on the semicircle of radius ω_0, as shown in Fig. 4.42a. The poles can be anywhere on the semicircle to satisfy the equidistance condition. Let the two conjugate poles be at angles $\pm\theta$ with respect to the negative real axis. Recall that a pole and a zero in the vicinity tend to cancel out each other's influences. Therefore placing poles closer to zeros (selecting θ closer to $\pi/2$) results in a rapid recovery of the gain from value 0 to 1 as we move away from ω_0 in either direction. Figure 4.42b shows the gain $|H(j\omega)|$ for three different values of θ.

■**Example 4.23**

Design a second-order notch filter to suppress 60 Hz hum in a radio receiver.

We use the poles and zeros in Fig. 4.42a with $\omega_0 = 120\pi$. The zeros are at $s = \pm j\omega_0$. The two poles are at $-\omega_0 \cos \theta \pm j\omega_0 \sin \theta$. The filter transfer function is (with $\omega_0 = 120\pi$)

$$H(s) = \frac{(s - j\omega_0)(s + j\omega_0)}{(s + \omega_0 \cos \theta + j\omega_0 \sin \theta)(s + \omega_0 \cos \theta - j\omega_0 \sin \theta)}$$

$$= \frac{s^2 + \omega_0^2}{s^2 + (2\omega_0 \cos \theta)s + \omega_0^2} = \frac{s^2 + 142122.3}{s^2 + (753.98 \cos \theta)s + 142122.3}$$

and

$$|H(j\omega)| = \frac{-\omega^2 + 142122.3}{\sqrt{(-\omega^2 + 142122.3)^2 + (753.98\omega \cos \theta)^2}}$$

The closer the poles are to the zeros (closer the θ to $\frac{\pi}{2}$), the faster the gain recovery from 0 to 1 on either side of $\omega_0 = 120\pi$. Figure 4.42b shows the amplitude response for three

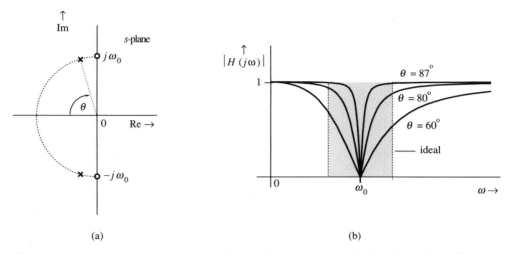

(a)   (b)

Fig. 4.42 Pole-zero configuration and the amplitude response of a bandstop (notch) filter.

different values of θ. This example is a case of very simple design. To achieve zero gain over a band, we need an infinite number of poles as well as of zeros. ∎

Figures 4.41b and 4.42b show that a notch (stopband) filter frequency response is a complement of the bandpass filter frequency response. If $H_{BP}(s)$ and $H_{BS}(s)$ are the transfer functions of a bandpass and a bandstop filter (both centered at the same frequency), then

$$H_{BS}(s) = 1 - H_{BP}(s)$$

Therefore a bandstop filter transfer function may also be obtained from the corresponding bandpass filter transfer function. The case of lowpass and highpass filters is similar. If $H_{LP}(s)$ and $H_{HP}(s)$ are the transfer functions of a lowpass and a highpass filter respectively (both with the same cutoff frequency), then

$$H_{HP}(s) = 1 - H_{LP}(s)$$

Therefore a highpass filter transfer function may also be obtained from the corresponding lowpass filter transfer function.

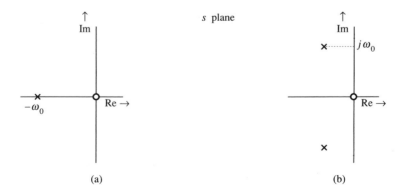

(a)   (b)

Fig. 4.43 Pole-zero configuration of a highpass filter in Exercise E4.14.

⊙ **Computer Example C4.3**
Plot the frequency response of the notch filter in Example 4.23 for $\theta = 80°$.

```
theta=80*pi/180;K=1;
w0=120*pi;
w=0:1:1000;
num=[1 0 w0^2];
den=[1 2*w0*cos(theta) w0^2];
H=polyval(num,j*w)./polyval(den,j*w);
mag=abs(H);
phase=180*angle(H)/pi;
plot(w,mag),grid
xlabel('w'),ylabel('Amplitude')     ⊙
```

⊙ **Computer Example C4.4**
Frequency response using logarithmic scales are very simple and useful. In these plots, known as the Bode plots, $20 \log_{10} |H(j\omega)|$ and $\angle H(j\omega)$ are plotted as functions of ω (also on logarithmic scale). MATLAB has a simple command "bode" for such plots. The transfer function data can be supplied either in terms of its numerator and denominator polynomial coefficients or in terms of its pole-zero locations. Parts a and b of the following example demonstrate this program by plotting the frequency response of

$$H(s) = \frac{20(s+100)}{s^2 + 12s + 20} = \frac{20(s+100)}{(s+2)(s+10)}$$

(a) Bode plots from the numerator and denominator polynomial coefficients:

```
num=[0 20 2000];    den=[1 12 20];
bode(num,den),pause
disp('Strike any key to see part b')
```

(b) Bode plots from the transfer function pole-zero locations:

```
K=20;
zeros=[-100];    poles=[-2 -10];
[num,den]=zp2tf(zeros',poles',K)
bode(num,den),pause   ⊙
```

△ **Exercise E4.14**
Using the qualitative method of sketching the frequency response, show that the system with pole-zero configuration in Fig. 4.43a is a highpass filter, and that with configuration in Fig. 4.43b is a bandpass filter.   ▽

4.8   THE BILATERAL LAPLACE TRANSFORM

Situations involving noncausal signals and/or systems cannot be handled by the (unilateral) Laplace transform discussed so far. These cases can be analyzed by the **bilateral** (or **two-sided**) Laplace transform defined by

$$F(s) = \int_{-\infty}^{\infty} f(t)e^{-st} \, dt \tag{4.81a}$$

We show in Chapter 7 that $f(t)$ can be obtained from $F(s)$ by the inverse transformation

$$f(t) = \frac{1}{2\pi j} \int_{c-j\infty}^{c+j\infty} F(s)e^{st} \, ds \tag{4.81b}$$

Observe that the unilateral Laplace transform discussed so far is a special case of the bilateral Laplace transform, where the signals are restricted to be causal. Basically the two transforms are the same. For this reason we use the same notation for the bilateral Laplace transform.

Earlier we showed that the Laplace transforms of $e^{-at}u(t)$ and of $-e^{at}u(-t)$ are identical. The only difference is between their regions of convergence. The region of convergence for the former is Re $s > -a$; that for the latter is Re $s < -a$, as shown in Fig. 4.1. Clearly the inverse Laplace transform of $F(s)$ is not unique unless the region of convergence is specified. If we restrict all our signals to be causal, however, this ambiguity does not arise. The inverse transform of $1/(s+a)$ is $e^{-at}u(t)$. Thus in unilateral Laplace transform we can ignore the region of convergence in determining the inverse transform of $F(s)$.

We now show that any bilateral transform can be expressed in terms of two unilateral transforms. It is therefore possible to evaluate bilateral transforms from a table of unilateral transforms.

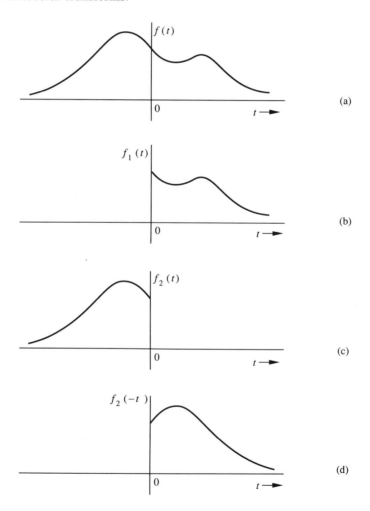

Fig. 4.44 Expressing a signal as a sum of causal and anticausal components.

Consider the function $f(t)$ shown in Fig. 4.44a. We separate $f(t)$ into two components, $f_1(t)$ and $f_2(t)$, representing the positive time (**causal**) component and the negative time (**anticausal**) component of $f(t)$ respectively (Figs. 4.44b and 4.44c):

$$f_1(t) = f(t)u(t) \tag{4.82a}$$

$$f_2(t) = f(t)u(-t) \tag{4.82b}$$

The bilateral Laplace transform of $f(t)$ is given by

$$
\begin{aligned}
F(s) &= \int_{-\infty}^{\infty} f(t)e^{-st}\,dt \\
&= \int_{-\infty}^{0} f_2(t)e^{-st}\,dt + \int_{0}^{\infty} f_1(t)e^{-st}\,dt \\
&= F_2(s) + F_1(s)
\end{aligned}
\tag{4.83}
$$

where $F_1(s)$ is the Laplace transform of the causal component $f_1(t)$ and $F_2(s)$ is the Laplace transform of the anticausal component $f_2(t)$. But $F_2(s)$ is given by

$$
\begin{aligned}
F_2(s) &= \int_{-\infty}^{0} f_2(t)e^{-st}\,dt \\
&= \int_{0}^{\infty} f_2(-t)e^{st}\,dt
\end{aligned}
$$

Therefore

$$F_2(-s) = \int_{0}^{\infty} f_2(-t)e^{-st}\,dt \tag{4.84}$$

It is clear that $F_2(-s)$ is the Laplace transform of $f_2(-t)$, which is causal (Fig. 4.44d), so $F_2(-s)$ can be found from the unilateral transform table. Changing the sign of s in $F_2(-s)$ yields $F_2(s)$.

To summarize, the bilateral transform $F(s)$ in Eq. (4.83) can be computed from the unilateral transforms in two steps:

1) Split $f(t)$ into its causal and anticausal components, $f_1(t)$ and $f_2(t)$, respectively.

2) Note that the signals $f_1(t)$ and $f_2(-t)$ are both causal. Take the (unilateral) Laplace transform of $f_1(t)$ and add to it the (unilateral) Laplace transform of $f_2(-t)$, with s replaced by $-s$. This gives the (bilateral) Laplace transform of $f(t)$.

Since $f_1(t)$ and $f_2(-t)$ are both causal, $F_1(s)$ and $F_2(-s)$ are both unilateral Laplace transforms. Now let σ_{c1} and σ_{c2} be the abscissas of convergence of $F_1(s)$ and $F_2(-s)$, respectively. This means that $F_1(s)$ exists for all s with Re $s > \sigma_{c1}$, $F_2(-s)$ exists for all s with Re $s > \sigma_{c2}$, and $F_2(s)$ exists for all s with Re $s < -\sigma_{c2}$. Because $F(s) = F_1(s) + F_2(s)$, $F(s)$ exists for all s such that

$$\sigma_{c1} < \text{Re } s < -\sigma_{c2} \tag{4.85}$$

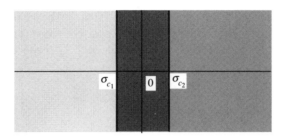

Region of convergence for causal component of $f(t)$.

Region of convergence for anticausal component of $f(t)$.

Region (strip) of convergence for the entire $f(t)$.

Fig. 4.45.

The regions of convergence (or existence) of $F_1(s)$, $F_2(s)$, and $F(s)$ are shown in Fig. 4.45. Because $F(s)$ is finite for all values of s lying in the strip of convergence ($\sigma_{c1} < \text{Re } s < -\sigma_{c2}$), poles of $F(s)$ must lie outside this strip. The poles of $F(s)$ due to the causal component $f_1(t)$ lie to the left of the **strip** (region) **of convergence**, and those due to its anticausal component $f_2(t)$ lie to its right (see Fig. 4.45). This fact is of crucial importance in finding the inverse bilateral transform.

As an example, consider

$$f(t) = e^{bt}u(-t) + e^{at}u(t) \tag{4.86}$$

We already know the Laplace transform of the causal component:

$$e^{at}u(t) \Longleftrightarrow \frac{1}{s-a} \qquad \text{Re } s > a \tag{4.87}$$

For the anticausal component, $f_2(t) = e^{bt}u(-t)$, we have

$$f_2(-t) = e^{-bt}u(t) \Longleftrightarrow \frac{1}{s+b} \qquad \text{Re } s > -b$$

so that

$$F_2(s) = \frac{1}{-s+b} = \frac{-1}{s-b} \qquad \text{Re } s < b$$

Therefore

$$e^{bt}u(-t) \Longleftrightarrow \frac{-1}{s-b} \qquad \text{Re } s < b \tag{4.88}$$

and the Laplace transform of $f(t)$ in Eq. (4.86) is

$$F(s) = -\frac{1}{s-b} + \frac{1}{s-a} \qquad \text{Re } s > a \quad \text{and} \quad \text{Re } s < b$$

$$= \frac{a-b}{(s-b)(s-a)} \qquad a < \text{Re } s < b \tag{4.89}$$

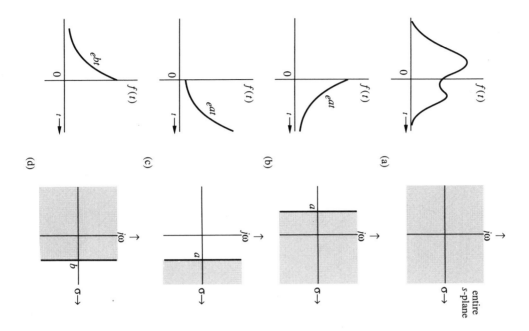

Fig. 4.46 Some signals and their regions of convergence.

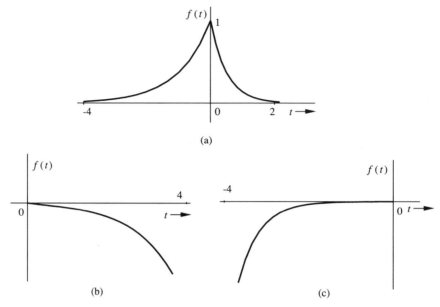

(a)

(b)   (c)

Fig. 4.47 Three possible inverse transforms of $\frac{-3}{(s+2)(s-1)}$.

Figure 4.46 shows $f(t)$ and the region of convergence of $F(s)$ for various values of a and b. From Eq. (4.89), it follows that the region of convergence of $F(s)$ does not exist if $a > b$. This is precisely the case in Fig. 4.46g. Observe that the poles of $F(s)$ are outside (on the edges) of the region of convergence. The poles of $F(s)$ due to the anticausal component of $f(t)$ lie to the right of the region of convergence, and those due to the causal component of $f(t)$ lie to its left.

■**Example 4.24**
Find the inverse Laplace transform of

$$F(s) = \frac{-3}{(s+2)(s-1)}$$

if the region of convergence is (a) $-2 < \text{Re } s < 1$ (b) $\text{Re } s > 1$ (c) $\text{Re } s < -2$

(a) $$F(s) = \frac{1}{s+2} - \frac{1}{s-1}$$

Now, $F(s)$ has poles at -2 and 1. The strip of convergence is $-2 < \text{Re } s < 1$. The pole at -2, being to the left of the strip of convergence, corresponds to the causal signal. The pole at 1, being to the right of the strip of convergence, corresponds to the anticausal signal. From Eqs. (4.87) and (4.88), we have

$$f(t) = e^{-2t}u(t) + e^{t}u(-t)$$

(b) Both poles lie to the left of the region of convergence, so both poles correspond to causal signals. Therefore

$$f(t) = (e^{-2t} - e^{t})u(t)$$

(c) Both poles lie to the right of the region of convergence, so both poles correspond to anticausal signals, and

$$f(t) = (-e^{-2t} + e^t)u(-t)$$

Figure 4.47 shows the three inverse transforms corresponding to the same $F(s)$ but with different regions of convergence. ■

4.8-1   Linear System Analysis Using Bilateral Transform

Since the bilateral Laplace transform can handle noncausal signals, we can analyze noncausal linear systems using the bilateral Laplace transform. We have shown that the (zero-state) output $y(t)$ is given by

$$y(t) = \mathcal{L}^{-1}[F(s)H(s)] \qquad (4.90)$$

This is true only if $F(s)H(s)$ exists. The region of convergence of $F(s)H(s)$ is the region where both $F(s)$ and $H(s)$ exist. In other words, the region of convergence of $F(s)H(s)$ is the region common to the regions of convergence of both $F(s)$ and $H(s)$. These ideas are clarified in the following examples.

■**Example 4.25**

Find the current $y(t)$ for the RC circuit shown in Fig. 4.48a if the voltage $f(t)$ is

$$f(t) = e^t u(t) + e^{2t} u(-t)$$

The transfer function $H(s)$ of the circuit is given by

$$H(s) = \frac{s}{s+1}$$

Because $h(t)$ is a causal function, the region of convergence of $H(s)$ is Re $s > -1$. Next, the bilateral Laplace transform of $f(t)$ is given by

$$F(s) = \frac{1}{s-1} - \frac{1}{s-2} = \frac{-1}{(s-1)(s-2)} \qquad 1 < \text{Re } s < 2$$

The response $y(t)$ is the inverse transform of $F(s)H(s)$

$$y(t) = \mathcal{L}^{-1}\left[\frac{-s}{(s+1)(s-1)(s-2)}\right]$$

$$= \mathcal{L}^{-1}\left[\frac{1}{6}\frac{1}{s+1} + \frac{1}{2}\frac{1}{s-1} - \frac{2}{3}\frac{1}{s-2}\right]$$

The region of convergence of $F(s)H(s)$ is that region of convergence common to both $F(s)$ and $H(s)$. This is $1 < \text{Re } s < 2$. The poles $s = \pm 1$ lie to the left of the region of convergence and therefore correspond to causal signals; the pole $s = 2$ lies to the right of the region of convergence and thus represents an anticausal signal. Hence

$$y(t) = \tfrac{1}{6}e^{-t}u(t) + \tfrac{1}{2}e^t u(t) + \tfrac{2}{3}e^{2t}u(-t)$$

Figure 4.48c shows $y(t)$. Note that in this example, if

$$f(t) = e^{-4t}u(t) + e^{-2t}u(-t)$$

then the region of convergence of $F(s)$ is $-4 < \text{Re } s < -2$. Here no region of convergence exists for $F(s)H(s)$, and the response $y(t)$ goes to infinity. ■

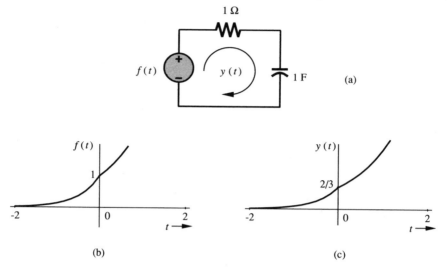

Fig. 4.48 Response of a circuit to a noncausal input (Example 4.25).

■**Example 4.26**

Find the response $y(t)$ of a noncausal system with the transfer function

$$H(s) = \frac{-1}{s-1} \qquad \text{Re } s < 1$$

to the input $f(t) = e^{-2t}u(t)$.

We have

$$F(s) = \frac{1}{s+2} \qquad \text{Re } s > -2$$

and

$$Y(s) = F(s)H(s) = \frac{-1}{(s-1)(s+2)}$$

The region of convergence of $F(s)H(s)$ is therefore the region $-2 < \text{Re } s < 1$. By partial fraction expansion

$$Y(s) = \frac{-1/3}{s-1} + \frac{1/3}{s+2} \qquad -2 < \text{Re } s < 1$$

and

$$y(t) = \tfrac{1}{3}\left[e^t u(-t) + e^{-2t}u(t)\right]$$

Note that the pole of $H(s)$ lies in the RHP at 1. Yet the system is not unstable. The pole(s) in the RHP may indicate instability or noncausality, depending on its location with respect to the region of convergence of $H(s)$. For example, if $H(s) = -1/(s-1)$ with Re $s > 1$, the system is causal and unstable, with $h(t) = -e^t u(t)$. On the other hand, if $H(s) = -1/(s-1)$ with Re $s < 1$, the system is noncausal and stable, with $h(t) = e^t u(-t)$.

■

■**Example 4.27**

Find the response $y(t)$ of a system with the transfer function

$$H(s) = \frac{1}{s+5} \qquad \text{Re } s > -5$$

and the input

$$f(t) = e^{-t}u(t) + e^{-2t}u(-t)$$

The input $f(t)$ is of the type shown in Fig. 4.46g, and the region of convergence for $F(s)$ does not exist. In this case, we must determine separately the system response to each of the two input components, $f_1(t) = e^{-t}u(t)$ and $f_2(t) = e^{-2t}u(-t)$.

$$F_1(s) = \frac{1}{s+1} \qquad \text{Re } s > -1$$

$$F_2(s) = \frac{-1}{s+2} \qquad \text{Re } s < -2$$

If $y_1(t)$ and $y_2(t)$ are the system responses to $f_1(t)$ and $f_2(t)$, respectively, then

$$Y_1(s) = \frac{1}{(s+1)(s+5)} \qquad \text{Re } s > -1$$

$$= \frac{1/4}{s+1} - \frac{1/4}{s+5}$$

so that

$$y_1(t) = \tfrac{1}{4}\left(e^{-t} - e^{-5t}\right)u(t)$$

and

$$Y_2(s) = \frac{-1}{(s+2)(s+5)} \qquad -5 < \text{Re } s < -2$$

$$= \frac{-1/3}{s+2} + \frac{1/3}{s+5}$$

so that

$$y_2(t) = \tfrac{1}{3}\left[e^{-2t}u(-t) + e^{-5t}u(t)\right]$$

Therefore

$$y(t) = y_1(t) + y_2(t)$$

$$= \tfrac{1}{3}e^{-2t}u(-t) + \left(\tfrac{1}{4}e^{-t} + \tfrac{1}{12}e^{-5t}\right)u(t) \qquad \blacksquare$$

4.9   APPENDIX 4.1: SECOND CANONICAL REALIZATION

An nth-order transfer function can also be realized by a second canonical (**observer canonical**) form. As in the case of the first canonical, we begin with a realization of a third-order transfer function in Eq. (4.61):

$$H(s) = \frac{Y(s)}{F(s)} = \frac{b_3 s^3 + b_2 s^2 + b_1 s + b_0}{s^3 + a_2 s^2 + a_1 s + a_0} \qquad (4.91)$$

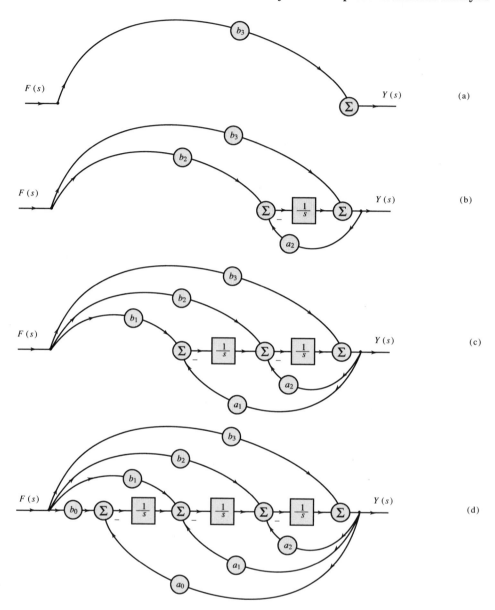

Fig. 4.49 Second (observer) canonical realization of an nth-order transfer function.

Therefore

$$(s^3 + a_2 s^2 + a_1 s + a_0)Y(s) = (b_3 s^3 + b_2 s^2 + b_1 s + b_0)F(s)$$

Transporting all but the first term on the right-hand side yields

$$s^3 Y(s) = [-a_2 Y(s) + b_2 F(s)]s^2 + [-a_1 Y(s) + b_1 F(s)]s$$
$$+ [-a_0 Y(s) + b_0 F(s)] + b_3 s^3 F(s)$$

Dividing throughout by s^3 yields

$$Y(s) = b_3 F(s) + \frac{1}{s}[-a_2 Y(s) + b_2 F(s)] + \frac{1}{s^2}[-a_1 Y(s) + b_1 F(s)]$$

$$+ \frac{1}{s^3}[-a_0 Y(s) + b_0 F(s)] \qquad (4.92)$$

Therefore $Y(s)$ can be generated by adding four signals appearing on the right-hand side of Eq. (4.92). We shall build $Y(s)$ step by step, adding one component at a time. Figure 4.49a shows only the first component; that is, $b_3 F(s)$. Figure 4.49b shows $Y(s)$ formed by the first two components, $b_3 F(s)$ and $\frac{1}{s}[-a_2 Y(s) + b_2 F(s)]$. Observe that the term $a_2 Y(s)$ is obtained from $Y(s)$ itself. We add $-a_2 Y(s)$ to $b_2 F(s)$ and pass it through an integrator to obtain $\frac{1}{s}[-a_2 Y(s) + b_2 F(s)]$. Figure 4.49c shows $Y(s)$ built up from the first three components. Finally Fig. 4.49d shows $Y(s)$ built up from all the components. This is the final form, which represents an alternative realization of $H(s)$ in Eq. (4.91).

This realization can be readily generalized for an nth-order transfer function in Eq. (4.60) using n integrators.

4.10 SUMMARY

This chapter discusses analysis of LTIC (linear, time-invariant, continuous-time) systems by Laplace transform, which transforms integro-differential equations of such systems into algebraic equations. Therefore solving these integro-differential equations reduces to solving algebraic equations. The Laplace transform method cannot be used for time-varying parameter systems or for nonlinear systems in general.

The transfer function of a system is defined as the ratio of the Laplace transform of the output to Laplace transform of the input when all initial conditions are zero (system in zero state). If $F(s)$ is the Laplace transform of the input $f(t)$ and $Y(s)$ is the Laplace transform of the corresponding output $y(t)$ (when all initial conditions are zero), then $Y(s) = F(s)H(s)$, where $H(s)$ is the system transfer function. The system transfer function $H(s)$ is the Laplace transform of the system impulse response $h(t)$. For an LTIC system described by an nth-order differential equation $Q(D)y(t) = P(D)f(t)$, the transfer function is $H(s) = P(s)/Q(s)$. Like the impulse response $h(t)$, the transfer function $H(s)$ is also an external description of the system.

The system response to an everlasting exponential e^{st} is also an everlasting exponential $H(s)e^{st}$. Moreover, the Laplace transform can be viewed as a tool by which a signal is expressed as a sum of everlasting exponentials e^{st}. The relative amount of a component e^{st} is $F(s)$. Therefore $F(s)$, the Laplace transform of $f(t)$, represents the spectrum of exponential components of $f(t)$. Moreover, $H(s)$ is the system response (or gain) to spectral component e^{st}, and the output signal spectrum is the input spectrum $F(s)$ times the spectral response (gain) $H(s)$ $[Y(s) = F(s)H(s)]$.

Electrical circuit analysis can also be carried out by using a transformed circuit method, in which all signals (voltages and currents) are represented by their Laplace transforms, all elements by their impedances (or admittances), and initial conditions by their equivalent sources (initial condition generators). In this method, a network can be analyzed as if it were a resistive circuit.

Large systems can be depicted by suitably interconnected subsystems represented by blocks. Each subsystem, being a smaller system, can be readily analyzed and represented by its input-output relationship, such as its transfer function. Analysis of large systems can be carried out with the knowledge of input-output relationships of its subsystems and the nature of interconnection of various subsystems.

LTIC systems can be realized by scalar multipliers, summers, and integrators. A given transfer function can be synthesized in many different ways. Canonical, cascade, and parallel forms of realization are discussed. In practice, all the building blocks (scalar multipliers, summers, and integrators) can be obtained from operational amplifiers.

Response of an LTIC system to sinusoidal inputs is discussed in Sec. 4.7. The transfer function $H(s)$ is directly related to sinusoidal system response. For a sinusoidal input of unit amplitude and frequency ω, the system response is also a sinusoid of the same frequency (ω) with amplitude $|H(j\omega)|$, and its phase is shifted by $\angle H(j\omega)$ with respect to the input sinusoid. For this reason $|H(j\omega)|$ is called the amplitude response (gain) and $\angle H(j\omega)$ is called the phase response of the system. Amplitude and phase response of a system indicate the filtering characteristics of the system. The general nature of the filtering characteristics of a system can be quickly determined from a knowledge of the location of poles and zeros of the system transfer function.

Most of the input signals and practical systems are causal. Consequently we are required most of the time to deal with causal signals. When all signals are restricted to be causal, the Laplace transform analysis is greatly simplified; the region of convergence of a signal becomes irrelevant to the analysis process. This special case of Laplace transform (which is restricted to causal signals) is called the unilateral Laplace transform. Much of the chapter deals with this variety of Laplace transform. Section 4.9 discusses the general Laplace transform (bilateral Laplace transform), which can handle causal and noncausal signals and systems. In the bilateral transform, the inverse transform of $F(s)$ is not unique but depends on the region of convergence of $F(s)$. Thus the region of convergence plays a very crucial role in the bilateral Laplace transform.

REFERENCES

1. Doetsch, G., *Introduction to the Theory and Applications of the Laplace Transformation with a Table of Laplace Transformations*, Springer Verlag, New York, 1974.

2. LePage, W.R., *Complex Variables and the Laplace Transforms for Engineers*, McGraw-Hill, New York, 1961.

3. Durant, Will, and Ariel, *The Age of Napoleon, The Story of Civilization Series*, Part XI, Simon and Schuster, New York, 1975.

4. Bell, E.T., *Men of Mathematics*, Simon and Schuster, New York, 1937.

5. Nahin, P.J., "Oliver Heaviside: Genius and Carmudgeon," IEEE Spectrum,

vol. 20, pp 63-69, July 1983.

6. Berkey, D., *Calculus*, 2nd ed., Saunder's College Publishing, Philadelphia, Pa. 1988.

7. Encyclopaedia Britannica, *Micropaedia IV*, 15th ed., Chicago, IL. 1982.

8. Churchill, R.V., *Operational Mathematics*, 2nd ed, McGraw-Hill, New York, 1958.

9. Van Valkenburg, M.E., *Analog Filter Design*, Holt, Rinehart, and Winston, New York, 1982.

PROBLEMS

4.1-1 By direct integration [Eq. (4.1)] Find the Laplace transforms and the region of convergence of the following functions:

    (a) $u(t) - u(t-1)$ **(e)** $\cos \omega_1 t \cos \omega_2 t\, u(t)$

    (b) $te^{-t}u(t)$ **(f)** $\cosh(at)\,u(t)$

    (c) $t\cos \omega_0 t\, u(t)$ **(g)** $\sinh(at)\,u(t)$

    (d) $(e^{2t} - 2e^{-t})u(t)$ **(h)** $e^{-2t}\cos(5t+\theta)\,u(t)$

4.1-2 By direct integration find the Laplace transforms of the signals shown in Fig. P4.1-2.

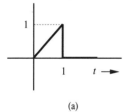
                 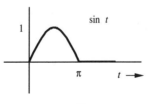   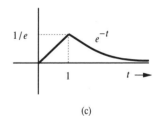

     (a)   (b)   (c)

Fig. P4.1-2

4.1-3 Find the inverse (unilateral) Laplace transforms of the following functions:

    (a) $\dfrac{2s+5}{s^2+5s+6}$ **(f)** $\dfrac{s+2}{s(s+1)^2}$

    (b) $\dfrac{3s+5}{s^2+4s+13}$ **(g)** $\dfrac{1}{(s+1)(s+2)^4}$

    (c) $\dfrac{(s+1)^2}{s^2-s-6}$ **(h)** $\dfrac{s+1}{s(s+2)^2(s^2+4s+5)}$

    (d) $\dfrac{5}{s^2(s+2)}$ **(i)** $\dfrac{s^3}{(s+1)^2(s^2+2s+5)}$

    (e) $\dfrac{2s+1}{(s+1)(s^2+2s+2)}$

4.2-1 Find the Laplace transforms of the following functions using only Table 4.1 and the time-shifting property (if needed) of the unilateral Laplace transform:

(a) $u(t) - u(t-1)$ (e) $te^{-t}u(t-\tau)$

(b) $e^{-(t-\tau)}u(t-\tau)$ (f) $\sin[\omega_0(t-\tau)]\,u(t-\tau)$

(c) $e^{-(t-\tau)}u(t)$ (g) $\sin[\omega_0(t-\tau)]\,u(t)$

(d) $e^{-t}u(t-\tau)$ (h) $\sin\omega_0 t\,u(t-\tau)$

Hint: $e^{-t} = e^{-\tau}e^{-(t-\tau)}$, and $e^{-(t-\tau)} = e^{\tau}e^{-t}$. For part (g) use trigonometric identity in B.10-6 to expand $\sin[\omega_0(t-\tau)] = \sin(\omega_0 t - \omega_0\tau)$, and recognize that $\sin\omega_0\tau$ and $\cos\omega_0\tau$ are constants. For part (h), recognize that $\sin\omega_0 t = \sin[\omega_0(t-\tau) + \omega_0\tau]$. Expand this, using the identity for $\sin(x+y)$.

4.2-2 Using only Table 4.1 and the time-shifting property, determine the Laplace transform of the signals shown in Fig. P4.1-2.

Hint: See Sec. B.5 for discussion of expressing such signals analytically. Recognize that $t = (t-1)+1$ and $e^{-t} = e^{-\tau}e^{-(t-\tau)}$. See also Example 4.4. For (b), show that the signal is $\sin t\,u(t)$ plus the same sine delayed by π.

4.2-3 Find the inverse Laplace transforms of the following functions:

(a) $\dfrac{(2s+5)e^{-2s}}{s^2+5s+6}$ (c) $\dfrac{e^{-(s-1)}+3}{s^2-2s+5}$

(b) $\dfrac{se^{-3s}+2}{s^2+2s+2}$ (d) $\dfrac{e^{-s}+e^{-2s}+1}{s^2+3s+2}$

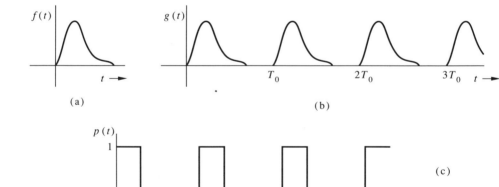

Fig. P4.2-4

4.2-4 The Laplace transform of a causal periodic signal can be found from the knowledge of the Laplace transform of its first cycle (period).

(a) If the Laplace transform of $f(t)$ in Fig. P4.2-4a is $F(s)$, then show that $G(s)$, the Laplace transform of $g(t)$ [Fig. P4.2-4b], is

$$G(s) = \frac{F(s)}{1 - e^{-sT_0}} \qquad \mathrm{Re}\,s > 0$$

(b) Using this result, find the Laplace transform of the signal $p(t)$ shown in Fig. P4.2-4c.

Hint: recognize that $g(t) = f(t) + f(t-T_0) + f(t-2T_0) + \cdots$, and $1 + x + x^2 + x^3 + \cdots = \frac{1}{1-x}$ for $|x| < 1$.

4.2-5 Starting only with the fact that $\delta(t) \Longleftrightarrow 1$, build Pairs 2 through 10b in Table 4.1, using various properties of the Laplace transform. Hint: $u(t)$ is integral of $\delta(t)$, and $tu(t)$ is integral of $u(t)$ (or second integral of $\delta(t)$), and so on.

4.2-6 **(a)** Find the Laplace transform of the pulses in Fig. 4.2 in the text by using only the time-differentiation property, time-shifting property and the fact that $\delta(t) \Longleftrightarrow 1$.

(b) In Example 4.7, the Laplace transform of $f(t)$ is found by finding the Laplace transform of $d^2 f/dt^2$. Find the Laplace transform of $f(t)$ in that example by finding the Laplace transform of df/dt.

Hint for part **(b)**: df/dt can be expressed as a sum of step functions (delayed by various amounts) whose transforms can be determined readily.

4.3-1 Using Laplace transform, solve the following differential equations.

(a) $\left(D^2 + 3D + 2\right) y(t) = Df(t)$ if $y(0^-) = \dot{y}(0^-) = 0$ and $f(t) = u(t)$

(b) $\left(D^2 + 4D + 4\right) y(t) = (D+1)f(t)$ if $y(0^-) = 2$, $\dot{y}(0^-) = 1$ and $f(t) = e^{-t}u(t)$

(c) $\left(D^2 + 6D + 25\right) y(t) = (D+2)f(t)$ if $y(0^-) = \dot{y}(0^-) = 1$ and $f(t) = 25u(t)$

4.3-2 Solve the differential equations in Prob. 4.3-1, using Laplace transform. In each case determine the zero-input and zero-state components of the solution.

4.3-3 Solve the following simultaneous differential equations using Laplace transform, assuming all initial conditions to be zero and the input $f(t) = u(t)$.

$$\textbf{(a)} \ (D+3)y_1(t) - 2y_2(t) = f(t)$$
$$- 2y_1(t) + (2D+4)y_2(t) = 0$$
$$\textbf{(b)} \ (D+2)y_1(t) - (D+1)y_2(t) = 0$$
$$- (D+1)y_1(t) + (2D+1)y_2(t) = f(t)$$

Determine the transfer functions relating outputs $y_1(t)$ and $y_2(t)$ to the input $f(t)$.

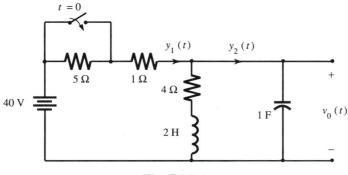

Fig. P4.3-4

4.3-4 For the circuit in Fig. P4.3-4, the switch is in open position for a long time before $t = 0$, when it is closed instantaneously.

(a) Write loop equations (in time domain) for $t \geq 0$.

(b) Solve for $y_1(t)$, and $y_2(t)$ by taking the Laplace transform of loop equations found in part (a).

4.3-5 For each of the systems described by following differential equations, find the system transfer function.

(a) $\dfrac{d^2y}{dt^2} + 11\dfrac{dy}{dt} + 24y(t) = 5\dfrac{df}{dt} + 3f(t)$

(b) $\dfrac{d^3y}{dt^3} + 6\dfrac{d^2y}{dt^2} - 11\dfrac{dy}{dt} + 6y(t) = 3\dfrac{d^2f}{dt^2} + 7\dfrac{df}{dt} + 5f(t)$

(c) $\dfrac{d^4y}{dt^4} + 4\dfrac{dy}{dt} = 3\dfrac{df}{dt} + 2f(t)$

4.3-6 For each of the systems specified by the following transfer functions, find the differential equation relating the output $y(t)$ to the input $f(t)$.

(a) $H(s) = \dfrac{s+5}{s^2 + 3s + 8}$ (b) $H(s) = \dfrac{s^2 + 3s + 5}{s^3 + 8s^2 + 5s + 7}$

(c) $H(s) = \dfrac{5s^2 + 7s + 2}{s^2 - 2s + 5}$

4.3-7 For a system with transfer function

$$H(s) = \frac{s+5}{s^2 + 5s + 6}$$

(a) Find the (zero-state) response if the input $f(t)$ is:

(i) $e^{-3t}u(t)$ (ii) $e^{-4t}u(t)$ (iii) $e^{-4(t-5)}u(t-5)$ (iv) $e^{-4(t-5)}u(t)$ (v) $e^{-4t}u(t-5)$.

Hint: $e^{-4(t-5)} = e^{20}e^{-4t}$, and $e^{-4t}u(t-5) = e^{-20}e^{-4(t-5)}u(t-5)$.

(b) For this system write the differential equation relating the output $y(t)$ to the input $f(t)$.

4.3-8 Repeat Prob. 4.3-7 if

$$H(s) = \frac{2s+3}{s^2 + 2s + 5}$$

and the input $f(t)$ is: (a) $10u(t)$ (b) $u(t-5)$.

4.3-9 Repeat Prob. 4.3-7 if

$$H(s) = \frac{s}{s^2 + 9}$$

and the input $f(t) = (1 - e^{-t})u(t)$

4.3-10 For an LTIC system with zero initial conditions (system initially in zero state), if an input $f(t)$ produces an output $y(t)$, then show that:

(a) the input df/dt produces an output dy/dt, and

(b) the input $\int_0^t f(\tau)\,d\tau$ produces an output $\int_0^t y(\tau)\,d\tau$. Hence show that the unit step response of a system is an integral of the impulse response; that is, $\int_0^t h(\tau)\,d\tau$.

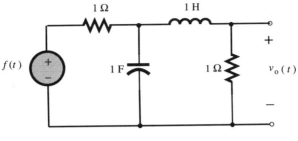

Fig. P4.4-1

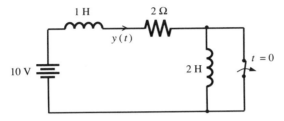

Fig. P4.4-2

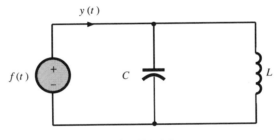

Fig. P4.4-3

4.4-1 Find the zero-state response $v_0(t)$ of the network shown in Fig. P4.4-1 if the input voltage $f(t) = te^{-t}u(t)$. Find the transfer function relating the output $v_0(t)$ to the input $f(t)$. From the transfer function, write the differential equation relating $v_0(t)$ to $f(t)$.

4.4-2 The switch in the circuit of Fig. P4.4-2 is closed for a long time and then opened instantaneously at $t = 0$. Find and sketch the current $y(t)$.

4.4-3 Find the current $y(t)$ for the parallel resonant circuit shown in Fig. P4.4-3 if the input is

(a) $f(t) = A \cos \omega_0 t\, u(t)$

(b) $f(t) = A \sin \omega_0 t\, u(t)$

$$\omega_0^2 = \frac{1}{LC}$$

Assume all initial conditions to be zero.

4.4-4 Find the loop currents $y_1(t)$ and $y_2(t)$ for $t \geq 0$ in the circuit of Fig. P4.4-4a for the input $f(t)$ shown in Fig. P4.4-4b.

4.4-5 For the network in Fig. P4.4-5 the switch is in a closed position for a long time before $t = 0$, when it is opened instantaneously. Find $y_1(t)$ and $v_s(t)$ for $t \geq 0$.

4.4-6 Find the output voltage $v_0(t)$ for $t \geq 0$ for the circuit in Fig. P4.4-6 for the input $f(t) = 100u(t)$. The system is in zero state initially.

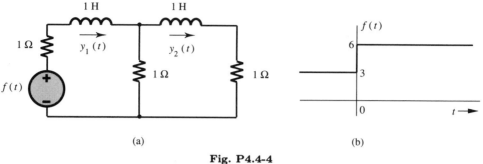

(a)   (b)

Fig. P4.4-4

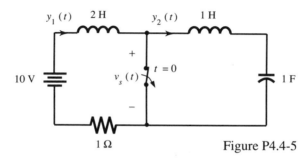

Figure P4.4-5

Fig. P4.4-5

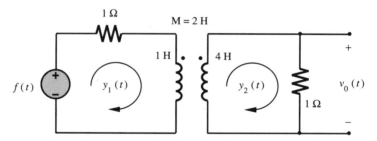

Fig. P4.4-6

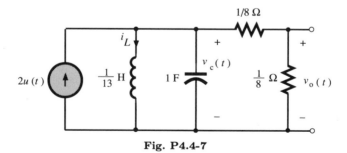

Fig. P4.4-7

4.4-7 Find the output voltage $v_0(t)$ for the network shown in Fig. P4.4-7 if the initial conditions are $i_L(0) = 1$ A and $v_C(0) = 3$ V. (Hint: Use the parallel form of initial condition generators.)

4.4-8 For the network shown in Fig. P4.4-8, the switch is in position a for a long time and then is moved to position b instantaneously at $t = 0$. Determine the current $y(t)$ for $t > 0$. Hint: Use the Thévenin equivalent.

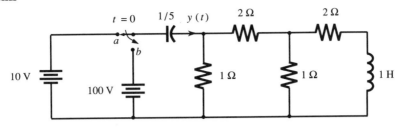

Fig. P4.4-8

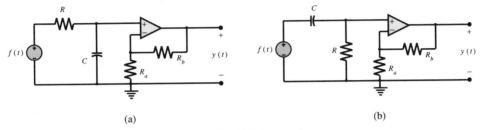

(a)   (b)

Fig. P4.4-9

4.4-9   Show that the transfer function which relates the output voltage $y(t)$ to the input
voltage $f(t)$ for the op amp circuit in Fig. P4.4-9a is given by

$$H(s) = \frac{Ka}{s+a} \quad \text{where} \quad K = 1 + \frac{R_b}{R_a} \quad \text{and} \quad a = \frac{1}{RC}$$

and that the transfer function for the circuit in Fig. P4.4-9b is given by

$$H(s) = \frac{Ks}{s+a}$$

Hint: Use the equivalent circuit for the noninverting op amp equivalent circuit shown
in Fig. 4.17d in the text (p. 294)

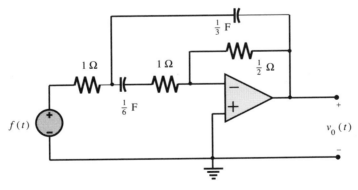

Fig. P4.4-10

4.4-10   For a second-order op amp circuit in Fig. P4.4-10, show that the transfer function
$H(s)$ relating the output voltage $v_0(t)$ to the input voltage $v_i(t)$ is given by

$$H(s) = \frac{-s}{s^2 + 8s + 12}$$

4.4-11 Using the initial and final value theorems, find the initial and final value of the zero-state response of a system with the transfer function

$$H(s) = \frac{6s^2 + 3s + 10}{2s^2 + 6s + 5}$$

and the input **(a)** $u(t)$ **(b)** $e^{-t}u(t)$.

4.5-1 Figure P4.5-1a shows two resistive ladder segments. The transfer function of each segment (ratio of output to input voltage) is $\frac{1}{2}$. Figure P4.5-1b shows these two segments connected in cascade.

(a) Is the transfer function (ratio of output to input voltage) of this cascaded network $\left(\frac{1}{2}\right)\left(\frac{1}{2}\right) = \frac{1}{4}$?

(b) If your answer is affirmative, verify the answer by direct computation of the transfer function. Does this confirm the earlier value $\frac{1}{4}$? If not, why?

(c) Repeat the problem with $R_3 = R_4 = 200$. Does this suggest the answer to the problem in part **(b)**?

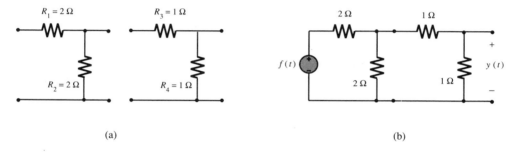

(a)   (b)

Fig. P4.5-1

4.6-1 Realize

$$H(s) = \frac{s(s+2)}{(s+1)(s+3)(s+4)}$$

by canonical, series, and parallel forms.

4.6-2 Repeat Problem 4.6-1 if

(a) $H(s) = \dfrac{3s(s+2)}{(s+1)(s^2+2s+2)}$

(b) $H(s) = \dfrac{2s-4}{(s+2)(s^2+4)}$

4.6-3 Repeat Problem 4.6-1 if

$$H(s) = \frac{2s+3}{5s(s+2)^2(s+3)}$$

Hint: Normalize the highest-power coefficient in the denominator to unity.

4.6-4 Repeat Problem 4.6-1 if

$$H(s) = \frac{s(s+1)(s+2)}{(s+5)(s+6)(s+8)}$$

Hint: Here $m = n = 3$. Take proper precautions in partial fraction expansion.

4.6-5 Repeat Problem 4.6-1 if

$$H(s) = \frac{s^3}{(s+1)^2(s+2)(s+3)}$$

4.6-6 Repeat Problem 4.6-1 if

$$H(s) = \frac{s^3}{(s+1)(s^2 + 4s + 13)}$$

4.6-7 In this problem we show how a pair of complex conjugate poles may be realized using a cascade of two first-order transfer functions. Show that the transfer functions of the block diagrams in Figs. P4.6-7a and b are

(a) $H(s) = \dfrac{1}{(s+a)^2 + b^2} = \dfrac{1}{s^2 + 2as + (a^2 + b^2)}$

(b) $H(s) = \dfrac{s+a}{(s+a)^2 + b^2} = \dfrac{s+a}{s^2 + 2as + (a^2 + b^2)}$

Hence show that the transfer function of the block diagram in Fig. P4.6-7c is

(c) $H(s) = \dfrac{As + B}{(s+a)^2 + b^2} = \dfrac{As + B}{s^2 + 2as + (a^2 + b^2)}$

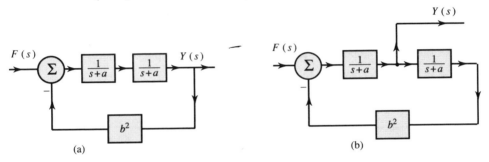

(a)   (b)

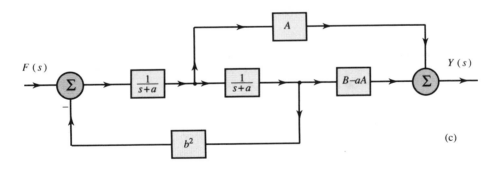

(c)

Fig. P4.6-7

4.6-8 Show op amp realization of the following transfer functions:

(i) $\dfrac{-10}{s+5}$ **(ii)** $\dfrac{10}{s+5}$ **(iii)** $\dfrac{s+2}{s+5}$

4.6-9 Show two different op amp circuit realizations of the transfer function

$$H(s) = \frac{s+2}{s+5} = 1 - \frac{3}{s+5}$$

4.6-10 Show op amp canonical realization of the transfer function

$$H(s) = \frac{3s+7}{s^2 + 4s + 10}$$

4.6-11 Show op amp canonical realization of the transfer function

$$H(s) = \frac{s^2 + 5s + 2}{s^2 + 4s + 13}$$

4.7-1 Find the response of an LTIC system described by the transfer function

$$H(s) = \frac{s + 2}{s^2 + 5s + 4}$$

to a sinusoidal input: **(a)** $5\cos(2t + 30°)$ **(b)** $10\sin(2t + 45°)$ **(c)** $10\cos(3t + 40°)$. Observe that these are everlasting sinusoids.

4.7-2 For an LTIC system described by the transfer function

$$H(s) = \frac{s + 3}{(s + 2)^2}$$

find the system response to the following everlasting inputs:
(a) $\cos(2t + 60°)$ **(b)** $\sin(3t - 45°)$ **(c)** e^{j3t}.

4.7-3 For an allpass filter specified by the transfer function

$$H(s) = \frac{-(s - 10)}{s + 10}$$

find the system response to following everlasting inputs: **(a)** $e^{j\omega t}$ **(b)** $\cos(\omega t + \theta)$
(c) $\cos t$ **(d)** $\sin 2t$ **(e)** $\cos 10t$ **(f)** $\cos 100t$.
Comment on the filter response.

4.7-4 For an asymptotically stable LTIC system, show that the steady-state response to input $e^{j\omega t}u(t)$ is $H(j\omega)e^{j\omega t}$. The steady-state response is that part of the response which does not decay with time and persists forever.
Hint: The total response is $Y(s) = H(s)F(s) = P(s)/Q(s)(s - j\omega)$ where $Q(s) = (s - \lambda_1)(s - \lambda_2)\cdots(s - \lambda_n)$. Show that the partial fraction expansion of $Y(s)$ yields

$$Y(s) = \sum_{i=1}^{n} \frac{k_i}{s - \lambda_i} + \frac{H(j\omega)}{s - j\omega}$$

and

$$y(t) = \sum_{i=1}^{n} k_i e^{\lambda_i t} + H(j\omega)e^{j\omega t}$$

For an asymptotically stable system, $e^{\lambda_i t}$ decays with time.

4.7-5 Using the graphical method of Sec 4.7-1, draw a rough sketch of the amplitude and phase response of an LTIC system described by the transfer function

$$H(s) = \frac{s^2 - 2s + 50}{s^2 + 2s + 50} = \frac{(s - 1 - j7)(s - 1 + j7)}{(s + 1 - j7)(s + 1 + j7)}$$

What kind of filter is this?

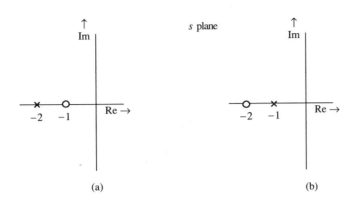

Fig. P4.7-6

4.7-6 Using the graphical method of Sec. 4.7-1, draw a rough sketch of the amplitude and phase response of LTIC systems whose pole-zero plots are shown in Fig. P4.7-6.

4.7-7 Design a second-order bandpass filter with center frequency $\omega = 10$. The gain should be zero at $\omega = 0$ and at $\omega = \infty$. Select poles at $-a \pm j10$. Leave your answer in terms of a. Explain the influence of a on the frequency response.

4.8-1 Find the region of convergence, if it exists, of the (bilateral) Laplace transform of the following signals:

(a) $e^{tu(t)}$

(b) $e^{-tu(t)}$

(c) $\dfrac{1}{1+t^2}$

(d) $\dfrac{1}{1+e^t}$

(e) e^{-kt^2}

4.8-2 Find the (bilateral) Laplace transform and the corresponding region of convergence for the following signals:

(a) $e^{-|t|}$

(b) $e^{-|t|} \cos t$

(c) $e^t u(t) + e^{2t} u(-t)$

(d) $\cos \omega_0 t\, u(t) + e^t\, u(-t)$

(e) $e^{-tu(t)}$

(f) $e^{tu(-t)}$

4.8-3 Find the inverse (bilateral) Laplace transforms of the following functions:

(a) $\dfrac{2s+5}{(s+2)(s+3)}$ $-3 < \sigma < -2$

(b) $\dfrac{2s-5}{(s-2)(s-3)}$ $2 < \sigma < 3$

(c) $\dfrac{2s+3}{(s+1)(s+2)}$ $\sigma > -1$

(d) $\dfrac{2s+3}{(s+1)(s+2)}$ $\sigma < -2$

(e) $\dfrac{3s^2 - 2s - 17}{(s+1)(s+3)(s-5)}$ $-1 < \sigma < 5$

4.8-4 Find

$$\mathcal{L}^{-1}\left[\frac{2s^2 - 2s - 6}{(s+1)(s-1)(s+2)}\right]$$

if the region of convergence is **(a)** Re $s > 1$ **(b)** Re $s < -2$
(c) $-1 <$ Re $s < 1$ **(d)** $-2 <$ Re $s < -1$

4.8-5 For a causal LTIC system having a transfer function

$$H(s) = \frac{1}{s+1}$$

find the output $y(t)$ if the input $f(t)$ is given by

(a) $e^{-|t|/2}$

(b) $e^t u(t) + e^{2t} u(-t)$

(c) $e^{-t/2} u(t) + e^{-t/4} u(-t)$

(d) $e^{2t} u(t) + e^t u(-t)$

(e) $e^{-t/4} u(t) + e^{-t/2} u(-t)$

(f) $e^{-3t} u(t) + e^{-2t} u(-t)$

COMPUTER PROBLEMS

C4-1 Use a computer to solve for the coefficients of the inverse Laplace transform, and to plot f(t) from 0 to 15 seconds with a step size of 0.2 seconds, for the following function:

$$H(s) = \frac{6s + 8.25}{s^2 + 2s + 7.5}$$

C4.2 Compute the frequency response (amplitude and phase) of a system whose transfer function $H(s)$ is

(a) $\dfrac{s + 0.5}{s + 1}$ **(b)** $\dfrac{s + 1}{s + 3}$ **(c)** $\dfrac{s + 2}{s^2 + 4s + 3}$ **(d)** $\dfrac{s^2 + 2s + 1}{s^3 + 4s^2 + 3s + 7}$

5

Discrete-Time Systems: Z-Transform Analysis

The counterpart of the Laplace transform for discrete-time systems is the z-transform, which simplifies the linear time-invariant discrete-time (LTID) system analysis and offers insights into system behavior in terms of frequency-domain concepts such as frequency response and filtering. Laplace transform converts integro-differential equations into algebraic equations. In the same way, the z-transforms changes difference equations into algebraic equations, thereby simplifying the analysis of discrete-time systems. The z-transform method of analyzing of discrete-time systems is parallel to that of the Laplace transform method of analysis of continuous-time systems, with some minor differences. In fact, we shall see that **the z-transform is the Laplace transform in disguise**.

5.1 THE $\mathcal{Z}$-TRANSFORM

We first introduce a simpler version of the z-transform that is capable of analyzing only causal systems with causal inputs (signals starting at $k = 0$). This is the **unilateral z-transform**, or simply the z-transform. The more general version, the **bilateral z-transform**, is capable of handling causal and noncausal signals (inputs) and systems and is discussed in Sec. 5.7.

We define $F[z]$, the **direct z-transform** of a discrete-time signal $f[k]$, as

$$F[z] \equiv \sum_{k=0}^{\infty} f[k] z^{-k} \tag{5.1}$$

where z is complex in general. The signal $f[k]$ is the **inverse z-transform** of $F[z]$. In Sec. 9.6 we show that

$$f[k] = \frac{1}{2\pi j} \oint F[z] z^{k-1} \, dz \tag{5.2}$$

353

where the symbol $\oint$ indicates an integration around a closed path in the complex plane, as will explained later. As in the case of Laplace transform, we need not worry about this integral at this point because inverse z-transforms of many important signals in engineering practice can be found in a z-transform Table. The direct and inverse z-transforms can be expressed symbolically as

$$F[z] = \mathcal{Z}\{f[k]\} \qquad \text{and} \qquad f[k] = \mathcal{Z}^{-1}\{F[z]\} \tag{5.3}$$

or simply as

$$f[k] \Longleftrightarrow F[z] \tag{5.4}$$

Note that

$$\mathcal{Z}^{-1}[\mathcal{Z}\{f[k]\}] = f[k] \qquad \text{and} \qquad \mathcal{Z}\left[\mathcal{Z}^{-1}\{F[z]\}\right] = F[z]$$

Linearity of the $\mathcal{Z}$-Transform

Like the Laplace transform, z-transform is a linear operator. If

$$f_1[k] \Longleftrightarrow F_1[z] \quad \text{and} \quad f_2[k] \Longleftrightarrow F_2[z]$$

then

$$a_1 f_1[k] + a_2 f_2[k] \Longleftrightarrow a_1 F_1[z] + a_2 F_2[z] \tag{5.5}$$

The proof is trivial and follows from the definition of the z-transform. This result can be extended to finite sums.

The Region of Convergence of F[z]

The sum in Eq. (5.1) defining the direct z-transform $F[z]$ may not converge (exist) for all values of z. The values of z (the region in the complex plane) for which the sum in Eq. (5.1) converges (or exists) is called the **region of convergence** (or **region of existence**) of $F[z]$. This concept will become clear from the following example.

■ Example 5.1

Find the z-transform and the corresponding region of convergence for the signal $\gamma^k u[k]$.

By definition

$$F[z] = \sum_{k=0}^{\infty} \gamma^k u[k] z^{-k}$$

Since $u[k] = 1$ for all $k \geq 0$,

$$F[z] = \sum_{k=0}^{\infty} \left(\frac{\gamma}{z}\right)^k$$

$$= 1 + \left(\frac{\gamma}{z}\right) + \left(\frac{\gamma}{z}\right)^2 + \left(\frac{\gamma}{z}\right)^3 + \cdots + \cdots \tag{5.6}$$

It is helpful to remember the following well-known geometric progression and its sum:

$$1 + x + x^2 + x^3 + \cdots = \frac{1}{1-x} \qquad \text{if} \quad |x| < 1 \tag{5.7}$$

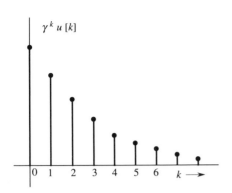

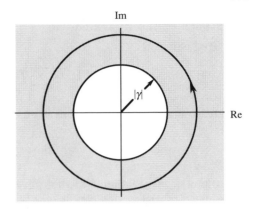

Fig. 5.1 $\gamma^k u[k]$ and the region of convergence of its z-transform.

Use of Eq. (5.7) in Eq. (5.6) yields

$$F[z] = \frac{1}{1 - \dfrac{\gamma}{z}} \qquad \left|\frac{\gamma}{z}\right| < 1$$

$$= \frac{z}{z - \gamma} \qquad |z| > |\gamma| \tag{5.8}$$

Observe that $F[z]$ exists only for $|z| > |\gamma|$. For $|z| < |\gamma|$, the sum in Eq. (5.6) does not converge; it goes to infinity. Therefore the region of convergence (or existence) of $F[z]$ is a region outside the circle of radius $|\gamma|$ centered at the origin in the z-plane, as shown in Fig. 5.1. ∎

   The region of convergence is required for evaluating $f[k]$ from $F[z]$, according to Eq. (5.2). The integral in Eq. (5.2) is a contour integral implying integration in a counterclockwise direction along a closed path centered at the origin and satisfying the condition $|z| > |\gamma|$. Thus any circular path centered at the origin and with a radius greater than $|\gamma|$ (Fig. 5.1) will suffice. It can be shown that the integral in (5.2) along any such path (with a radius greater than $|\gamma|$) yields the same result, namely $f[k]$. Such integration in complex plane requires a background in the theory of functions of complex variables. We can avoid this integration by compiling a table of z-transforms (Table 5.1), where z-transform pairs are tabulated for a variety of signals. To find the inverse z-transform of say, $z/(z - \gamma)$, instead of using the complex integration in (5.2), we consult the table and find the inverse z-transform of $z/(z - \gamma)$ as $\gamma^k u[k]$. Although the table given here is rather short, it comprises the functions of most practical interest. A more comprehensive table can be found in Jury.[1]

   The bilateral z-transform is defined by Eq. (5.1) with the limits of the right-hand sum from $-\infty$ to ∞ instead from 0 to ∞. The situation of the z-transform regarding the uniqueness of the inverse transform is parallel to that of the Laplace transform. For the bilateral case the inverse z-transform is not unique unless the region of convergence is specified. For the unilateral case the inverse transform is unique; the region of convergence need not be specified to determine the inverse z-transform. For this reason we shall ignore the region of convergence in the unilateral z-transform Table 5.1.

Existence of the $\mathcal{Z}$-Transform

By definition

$$F[z] = \sum_{k=0}^{\infty} f[k] z^{-k} = \sum_{k=0}^{\infty} \frac{f[k]}{z^k} \tag{5.9}$$

The existence of the z-transform is guaranteed if

$$|F[z]| \leq \sum_{k=0}^{\infty} \frac{|f[k]|}{|z|^k} < \infty \tag{5.10}$$

for some $|z|$. Any signal $f[k]$ that grows no faster than an exponential signal $r_0{}^k$, for some r_0, satisfies this condition. Thus if

$$|f[k]| \leq r_0{}^k \qquad \text{for some } r_0 \tag{5.11}$$

then

$$|F[z]| \leq \sum_{k=0}^{\infty} \left(\frac{r_0}{|z|} \right)^k = \frac{1}{1 - \frac{r_0}{|z|}} \qquad |z| > r_0$$

Therefore $F[z]$ exists for $|z| > r_0$. All practical signals satisfy (5.11) and are therefore z-transformable. Some signal models (e.g. γ^{k^2}) which grow faster than exponential signal $r_0{}^k$ (for any r_0) do not satisfy (5.11) and therefore are not z-transformable. Fortunately such signals are of little practical or theoretical interest.

■ **Example 5.2**

Find the z-transforms of **(a)** $\delta[k]$ **(b)** $u[k]$ **(c)** $\cos \beta k \, u[k]$ **(d)** signal shown in Fig. 5.2.

Recall that by definition

$$F[z] = \sum_{k=0}^{\infty} f[k] z^{-k}$$

$$= f[0] + \frac{f[1]}{z} + \frac{f[2]}{z^2} + \frac{f[3]}{z^3} + \cdots \tag{5.12}$$

(a) For $f[k] = \delta[k]$, $f[0] = 1$ and $f[2] = f[3] = f[4] = \cdots = 0$. Therefore

$$\delta[k] \Longleftrightarrow 1 \qquad \text{for all } z \tag{5.13}$$

(b) For $f[k] = u[k]$, $f[0] = f[1] = f[3] = \cdots = 1$. Therefore

$$F[z] = 1 + \frac{1}{z} + \frac{1}{z^2} + \frac{1}{z^3} + \cdots$$

From Eq. (5.7) it follows that

$$F[z] = \frac{1}{1 - \frac{1}{z}} \qquad \left| \frac{1}{z} \right| < 1$$

$$= \frac{z}{z - 1} \qquad |z| > 1$$

Therefore

$$u[k] \Longleftrightarrow \frac{z}{z - 1} \qquad |z| > 1 \tag{5.14}$$

(c) Recall that $\cos \beta k = \left(e^{j\beta k} + e^{-j\beta k}\right)/2$. Moreover, From Eq. (5.8)

$$e^{\pm j\beta k} u[k] \Longleftrightarrow \frac{z}{z - e^{\pm j\beta}} \qquad |z| > |e^{\pm j\beta}| = 1$$

Therefore

$$F[z] = \frac{1}{2}\left[\frac{z}{z - e^{j\beta}} + \frac{z}{z - e^{-j\beta}}\right]$$

$$= \frac{z(z - \cos \beta)}{z^2 - 2z\cos \beta + 1} \qquad |z| > 1$$

(d) Here $f[0] = f[1] = f[2] = f[3] = f[4] = 1$ and $f[5] = f[6] = \cdots = 0$. Therefore from Eq. (5.12)

$$F[z] = 1 + \frac{1}{z} + \frac{1}{z^2} + \frac{1}{z^3} + \frac{1}{z^4} \qquad (5.15)$$

$$= \frac{z^4 + z^3 + z^2 + z + 1}{z^4}$$

We can also express this result in a closed form by summing the geometric progression on the right-hand side of Eq. 5.15, using the formula in Sec. B.10-4. Here the common ratio $r = \frac{1}{z}$, $M = 0$, and $N = 4$, so that

$$F[z] = \frac{\left(\frac{1}{z}\right)^5 - \left(\frac{1}{z}\right)^0}{\frac{1}{z} - 1} = \frac{z}{z - 1}(1 - z^{-5}) \quad \blacksquare$$

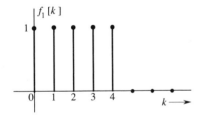
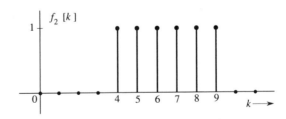

Fig. 5.2   Fig. 5.3

△ **Exercise E5.1**
    (a) Find the z-transform of a signal shown in Fig. 5.3. **(b)** Using Pair 12a (Table 5.1), find the z-transform of $f[k] = 20.65(\sqrt{2})^k \cos\left(\frac{\pi}{4}k - 1.415\right) u[k]$.

Answers: **(a)** $F[z] = \dfrac{z^5 + z^4 + +z^3 + z^2 + z + 1}{z^9}$ or $\dfrac{z}{z - 1}(z^{-4} - z^{-10})$

(b) $\dfrac{z(3.2z + 17.2)}{z^2 - 2z + 2}$ ▽

5.1-1   Finding the Inverse Transform

    As in Laplace transform, we shall avoid the integration in complex plane required to find the inverse z-transform [Eq. (5.2)] by using the transform Table. Many of the transforms $F[z]$ of practical interest are rational functions (ratio of polynomials in z). Such functions can be expressed as a sum of simpler functions using partial fraction expansion. This method works because for every transformable $f[k]$ defined for $k \geq 0$, there is a corresponding unique $F[z]$ defined for $|z| > r_0$ (where r_0 is some constant), and vice versa.

■ **Example 5.3**

Find the inverse z-transform of

(a) $\dfrac{8z - 19}{(z - 2)(z - 3)}$ (b) $\dfrac{z(2z^2 - 11z + 12)}{(z - 1)(z - 2)^3}$ (c) $\dfrac{2z(3z + 17)}{(z - 1)(z^2 - 6z + 25)}$

(a) Expanding $F[z]$ in to partial fractions yields

$$F[z] = \frac{8z - 19}{(z - 2)(z - 3)} = \frac{3}{z - 2} + \frac{5}{z - 3}$$

From Table 5.1, Pair 6, we obtain

$$f[k] = \left[3(2)^{k-1} + 5(3)^{k-1}\right] u[k - 1] \qquad (5.16a)$$

If we expand rational $F[z]$ into partial fractions directly, we shall always obtain an answer that is multiplied by $u[k - 1]$ because of the nature of Pair 6 in Table 5.1. This form is rather awkward as well as inconvenient. We prefer the form that is multiplied by $u[k]$ rather than $u[k - 1]$. A glance at Table 5.1 shows that the z-transform of every signal that is multiplied by $u[k]$ has a factor z in the numerator. This suggests that we expand $F[z]$ into **modified partial fractions**, where each term has a factor z in the numerator. This can be done by expanding $F[z]/z$ into partial fractions and then multiplying both sides by z. We shall demonstrate this by reworking part **(a)** in Example 5.3. For this case

$$\frac{F[z]}{z} = \frac{8z - 19}{z(z - 2)(z - 3)}$$

$$= \frac{(-19/6)}{z} + \frac{(3/2)}{z - 2} + \frac{(5/3)}{z - 3}$$

Multiplying both sides by z yields

$$F[z] = -\frac{19}{6} + \frac{3}{2}\left(\frac{z}{z - 2}\right) + \frac{5}{3}\left(\frac{z}{z - 3}\right)$$

From Pairs 1 and 7 in Table 5.1, it follows that

$$f[k] = -\tfrac{19}{6}\delta[k] + \left[\tfrac{3}{2}(2)^k + \tfrac{5}{3}(3)^k\right] u[k] \qquad (5.16b)$$

The reader can verify that this answer is equivalent to that in Eq. (5.16a) by computing $f[k]$ in both cases for $k = 0, 1, 2, 3, \cdots$, and then comparing the results. The form in Eq. (5.16b) is more convenient than that in Eq. (5.16a). For this reason we shall always expand $F[z]/z$ rather than $F[z]$ into partial fractions and then multiply both sides by z to obtain modified partial fractions of $F[z]$, which have a factor z in the numerator.

(b) $$F[z] = \frac{z(2z^2 - 11z + 12)}{(z - 1)(z - 2)^3}$$

and

$$\frac{F[z]}{z} = \frac{2z^2 - 11z + 12}{(z - 1)(z - 2)^3}$$

$$= \frac{k}{z - 1} + \frac{a_0}{(z - 2)^3} + \frac{a_1}{(z - 2)^2} + \frac{a_2}{(z - 2)}$$

Table 5.1: z-Transform Pairs

	$f[k]$	$F[z]$
1	$\delta[k-j]$	z^{-j}
2	$u[k]$	$\dfrac{z}{z-1}$
3	$ku[k]$	$\dfrac{z}{(z-1)^2}$
4	$k^2u[k]$	$\dfrac{z(z+1)}{(z-1)^3}$
5	$k^3u[k]$	$\dfrac{z(z^2+4z+1)}{(z-1)^4}$
6	$\gamma^{k-1}u[k-1]$	$\dfrac{1}{z-\gamma}$
7	$\gamma^k u[k]$	$\dfrac{z}{z-\gamma}$
8	$k\gamma^k u[k]$	$\dfrac{\gamma z}{(z-\gamma)^2}$
9	$k^2\gamma^k u[k]$	$\dfrac{\gamma z(z+\gamma)}{(z-\gamma)^3}$
10	$\dfrac{k(k-1)(k-2)\cdots(k-m+1)}{\gamma^m m!}\gamma^k u[k]$	$\dfrac{z}{(z-\gamma)^{m+1}}$
11a	$|\gamma|^k \cos\beta k\, u[k]$	$\dfrac{z(z-|\gamma|\cos\beta)}{z^2-(2|\gamma|\cos\beta)z+|\gamma|^2}$
11b	$|\gamma|^k \sin\beta k\, u[k]$	$\dfrac{z|\gamma|\sin\beta}{z^2-(2|\gamma|\cos\beta)z+|\gamma|^2}$
12a	$r|\gamma|^k \cos(\beta k+\theta)u[k]$	$\dfrac{rz[z\cos\theta-|\gamma|\cos(\beta-\theta)]}{z^2-(2|\gamma|\cos\beta)z+|\gamma|^2}$
12b	$r|\gamma|^k \cos(\beta k+\theta)u[k]$	$\dfrac{(0.5re^{j\theta})z}{z-\gamma}+\dfrac{(0.5re^{-j\theta})z}{z-\gamma^*}\qquad \gamma=|\gamma|e^{j\beta}$
12c	$r|\gamma|^k \cos(\beta k+\theta)u[k]$	$\dfrac{z(Az+B)}{z^2+2az+|\gamma|^2}$

$r=\sqrt{\dfrac{A^2|\gamma|^2+B^2-2AaB}{|\gamma|^2-a^2}}$

$\beta=\cos^{-1}\dfrac{-a}{|\gamma|},\ \ \theta=\tan^{-1}\dfrac{Aa-B}{A\sqrt{|\gamma|^2-a^2}}$

where

$$k = \frac{2z^2 - 11z + 12}{(z-1)(z-2)^3}\bigg|_{z=1} = -3$$

$$a_0 = \frac{2z^2 - 11z + 12}{(z-1)(z-2)^3}\bigg|_{z=2} = -2$$

Therefore

$$\frac{F[z]}{z} = \frac{2z^2 - 11z + 12}{(z-1)(z-2)^3} = \frac{-3}{z-1} - \frac{2}{(z-2)^3} + \frac{a_1}{(z-2)^2} + \frac{a_2}{(z-2)} \qquad (5.17)$$

We can determine a_1 and a_2 by clearing fractions or by using the short cuts discussed in Sec. B.8-3. For example, to determine a_2, we multiply both sides of Eq. (5.17) by z and let $z \to \infty$. This yields

$$0 = -3 - 0 + 0 + a_2 \implies a_2 = 3$$

This leaves only one unknown, a_1, which is readily determined by letting z take any convenient value, say $z = 0$, on both sides of Eq. (5.17). This yields

$$\frac{12}{8} = 3 + \frac{1}{4} + \frac{a_1}{4} - \frac{3}{2}$$

Multiplying both sides by 8 yields

$$12 = 24 + 2 + 2a_1 - 12 \implies a_1 = -1$$

Therefore

$$\frac{F[z]}{z} = \frac{-3}{z-1} - \frac{2}{(z-2)^3} - \frac{1}{(z-2)^2} + \frac{3}{z-2}$$

and

$$F[z] = -3\frac{z}{z-1} - 2\frac{z}{(z-2)^3} - \frac{z}{(z-2)^2} + 3\frac{z}{z-2}$$

Now the use of Table 5.1, Pairs 7 and 10 yields

$$f[k] = \left[-3 - 2\frac{k(k-1)}{8}(2)^k - \frac{k}{2}(2)^k + 3(2)^k \right] u[k]$$

$$= -[3 + \tfrac{1}{4}(k^2 + k - 12)2^k]u[k]$$

(c) Complex Poles

$$F[z] = \frac{2z(3z+17)}{(z-1)(z^2 - 6z + 25)} = \frac{2z(3z+17)}{(z-1)(z-3-j4)(z-3+j4)}$$

Poles of $F[z]$ are 1, $3 + j4$, and $3 - j4$. Whenever there are complex conjugate poles, the problem can be worked out in two ways. In the first method we expand $F[z]$ into (modified) first-order partial fractions. In the second method, rather than obtaining one factor corresponding to each complex conjugate pole, we obtain quadratic factors corresponding to each pair of complex conjugate poles. This is explained below.

Method of First-Order Factors

$$\frac{F[z]}{z} = \frac{2(3z+17)}{(z-1)(z^2 - 6z + 25)} = \frac{2(3z+17)}{(z-1)(z-3-j4)(z-3+j4)}$$

We find the partial fraction of $F[z]/z$ using the Heaviside "cover-up" method:

$$\frac{F[z]}{z} = \frac{2}{z-1} + \frac{1.6e^{-j2.246}}{z-3-j4} + \frac{1.6e^{j2.246}}{z-3+j4}$$

and

$$F[z] = 2\frac{z}{z-1} + (1.6e^{-j2.246})\frac{z}{z-3-j4} + (1.6e^{j2.246})\frac{z}{z-3+j4}$$

The inverse transform of the first term on the right-hand side is $2u[k]$. The inverse transform of the remaining two terms (complex conjugate poles) can be found from Pair 12b (Table 5.1) by identifying $\frac{r}{2} = 1.6$, $\theta = -2.246$ rad., $\gamma = 3+j4 = 5e^{j0.927}$, so that $|\gamma| = 5$, $\beta = 0.927$. Therefore

$$f[k] = \left[2 + 3.2(5)^k \cos(0.927k - 2.246)\right] u[k]$$

Method of Quadratic Factors

$$\frac{F[z]}{z} = \frac{2(3z+17)}{(z-1)(z^2-6z+25)} = \frac{2}{z-1} + \frac{Az+B}{z^2-6z+25}$$

Multiplying both sides by z and letting $z \to \infty$, we find

$$0 = 2 + A \Longrightarrow A = -2$$

and

$$\frac{2(3z+17)}{(z-1)(z^2-6z+25)} = \frac{2}{z-1} + \frac{-2z+B}{z^2-6z+25}$$

To find B, we let z take any convenient value, say $z = 0$. This yields

$$\frac{-34}{25} = -2 + \frac{B}{25}$$

Multiplying both sides by 25 yields

$$-34 = -50 + B \Longrightarrow B = 16$$

Therefore

$$\frac{F[z]}{z} = \frac{2}{z-1} + \frac{-2z+16}{z^2-6z+25}$$

and

$$F[z] = \frac{2z}{z-1} + \frac{z(-2z+16)}{z^2-6z+25}$$

We now use Pair 12c where we identify $A = -2$, $B = 16$, $|\gamma| = 5$, $a = -3$. Therefore

$$r = \sqrt{\frac{100+256-192}{25-9}} = 3.2, \quad \beta = \cos^{-1}\left(\frac{3}{5}\right) = 0.927 \text{rad., and}$$

$$\theta = \tan^{-1}\left(\frac{-10}{-8}\right) = -2.246 \text{ rad., so that}$$

$$f[k] = \left[2 + 3.2(5)^k \cos(0.927k - 2.246)\right] u[k] \quad \blacksquare$$

△ **Exercise E5.2**

Find the inverse z-transform of the following functions:

(a) $\dfrac{z(2z-1)}{(z-1)(z+0.5)}$ (b) $\dfrac{1}{(z-1)(z+0.5)}$

(c) $\dfrac{9}{(z+2)(z-0.5)^2}$ (d) $\dfrac{5z(z-1)}{z^2-1.6z+0.8}$

Answer: (a) $\left[\frac{2}{3}+\frac{4}{3}(-0.5)^k\right]u[k]$ (b) $-2\delta[k]+\left[\frac{2}{3}+\frac{4}{3}(-0.5)^k\right]u[k]$

(c) $18\delta[k]-[0.72(-2)^k+17.28(0.5)^k-14.4k(0.5)^k]u[k]$

(d) $\frac{5\sqrt{5}}{2}\left(\frac{2}{\sqrt{5}}\right)^k\cos(0.464k+0.464)u[k]$. Hint: $\sqrt{0.8}=\frac{2}{\sqrt{5}}$.   ▽

Inverse Transform by Expansion of F[z] in Power Series of z^{-1}

By definition

$$F[z]=\sum_{k=0}^{\infty}f[k]z^{-k}$$

$$=f[0]+\frac{f[1]}{z}+\frac{f[2]}{z^2}+\frac{f[3]}{z^3}+\cdots \tag{5.18}$$

$$=f[0]z^0+f[1]z^{-1}+f[2]z^{-2}+f[3]z^{-3}+\cdots$$

This is a power series in z^{-1}. Therefore, if we can expand $F[z]$ into power series in z^{-1}, the coefficients of this power series can be identified as $f[0]$, $f[1]$, $f[2]$, $f[3]$, $\cdots$, and so on. A rational $F[z]$ can be expanded into power series of z^{-1} by dividing its numerator by the denominator. Consider, for example,

$$F[z]=\frac{z^2(7z-2)}{(z-0.2)(z-0.5)(z-1)}$$

$$=\frac{7z^3-2z^2}{z^3-1.7z^2+0.8z-0.1}$$

To obtain a series expansion in powers of z^{-1}, we divide the numerator by the denominator as follows:

$$
\begin{array}{r}
7+9.9\,z^{-1}+11.23z^{-2}+11.87z^{-3}+\cdots \\
z^3-1.7z^2+0.8z-0.1\,\overline{)\,7z^3-2z^2} \\
7z^3-11.9z^2+5.60z-\ 0.7 \\
\hline
9.9z^2-\ 5.60z+\ 0.7 \\
9.9z^2-16.83z+7.92-0.99z^{-1} \\
\hline
11.23z-7.22+0.99z^{-1} \\
11.23z-19.09+8.98z^{-1} \\
\hline
11.87-7.99z^{-1}
\end{array}
$$

Thus

$$F[z]=\frac{z^2(7z-2)}{(z-0.2)(z-0.5)(z-1)}=7+9.9z^{-1}+11.23z^{-2}+11.87z^{-3}+\cdots$$

Therefore

$$f[0] = 7, \ f[1] = 9.9, \ f[2] = 11.23, \ f[3] = 11.87, \ \cdots, \text{ and so on.}$$

Although this procedure yields $f[k]$ directly, it does not provide a closed-form solution. For this reason it is not very useful unless we want to know only the first few terms of the sequence $f[k]$.

△ **Exercise E5.3**

Using long division to find the power series in z^{-1}, show that the inverse z-transform of $z/(z - 0.5)$ is $(0.5)^k u[k]$ or $(2)^{-k} u[k]$.   ▽

5.2   SOME PROPERTIES OF THE Z-TRANSFORM

The z-transform properties are useful in the derivation of z-transforms of many functions and also in the solution of linear difference equations with constant coefficients. Here we consider a few important properties of the z-transform.

Right Shift (Delay)

If

$$f[k]u[k] \Longleftrightarrow F[z]$$

then

$$f[k - 1]u[k - 1] = \frac{1}{z}F[z] \tag{5.19a}$$

and

$$f[k - m]u[k - m] = \frac{1}{z^m}F[z] \tag{5.19b}$$

Moreover

$$f[k - 1]u[k] \Longleftrightarrow \frac{1}{z}F[z] + f[-1] \tag{5.20a}$$

Repeated application of this property yields

$$f[k - 2]u[k] \Longleftrightarrow \frac{1}{z}\left[\frac{1}{z}F[z] + f[-1]\right] + f[-2]$$

$$= \frac{1}{z^2}F[z] + \frac{1}{z}f[-1] + f[-2] \tag{5.20b}$$

and

$$f[k - m]u[k] \Longleftrightarrow z^{-m}F[z] + z^{-m}\sum_{k=1}^{m} f[-k]z^k \tag{5.20c}$$

Proof:

$$\mathcal{Z}\left\{f[k - m]u[k - m]\right\} = \sum_{k=0}^{\infty} f[k - m]u[k - m]z^{-k}$$

Recall that $f[k - m]u[k - m] = 0$ for $k < m$, so that the limits on the summation on the right-hand side can be taken from $k = m$ to ∞. Therefore

$$\mathcal{Z}\left\{f[k-m]u[k-m]\right\} = \sum_{k=m}^{\infty} f[k-m]z^{-k}$$

$$= \sum_{r=0}^{\infty} f[r]z^{-(r+m)}$$

$$= \frac{1}{z^m} \sum_{r=0}^{\infty} f[r]z^{-r}$$

$$= \frac{1}{z^m} F[z]$$

To prove Eq. (5.20c), we have

$$\mathcal{Z}\left\{f[k-m]u[k]\right\} = \sum_{k=0}^{\infty} f[k-m]z^{-k} = \sum_{r=-m}^{\infty} f[r]z^{-(r+m)}$$

$$= z^{-m}\left[\sum_{r=-m}^{-1} f[r]z^{-r} + \sum_{r=0}^{\infty} f[r]z^{-r}\right]$$

$$= z^{-m} \sum_{k=1}^{m} f[-k]z^{k} + z^{-m}F[z] \qquad (5.21)$$

Left Shift (Advance)

If

$$f[k]u[k] \Longleftrightarrow F[z]$$

then

$$f[k+1]u[k] \Longleftrightarrow zF[z] - zf[0] \qquad (5.22a)$$

Repeated application of this property yields

$$f[k+2]u[k] \Longleftrightarrow z\left\{z\left[F[z] - f[0]\right] - f[1]\right\}$$
$$= z^2 F[z] - z^2 f[0] - zf[1] \qquad (5.22b)$$

and

$$f[k+m]u[k] \Longleftrightarrow z^m F[z] - z^m \sum_{k=0}^{m-1} f[k]z^{-k} \qquad (5.22c)$$

Proof: By definition

$$\mathcal{Z}\left\{f[k+m]u[k]\right\} = \sum_{k=0}^{\infty} f[k+m]z^{-k}$$

$$= \sum_{r=m}^{\infty} f[r]z^{-(r-m)}$$

$$= z^{m} \sum_{r=m}^{\infty} f[r]z^{-r}$$

$$= z^{m} \left[\sum_{r=0}^{\infty} f[r]z^{-r} - \sum_{r=0}^{m-1} f[r]z^{-r}\right]$$

$$= z^{m} F[z] - z^{m} \sum_{r=0}^{m-1} f[r]z^{-r}$$

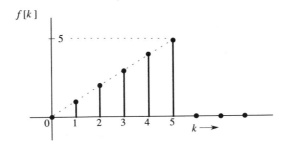

Fig. 5.4 Signal for Example 5.4.

■ **Example 5.4**

Find the z-transform of the signal $f[k]$ shown in Fig. 5.4.

The signal $f[k]$ can be expressed as a product of k and a gate pulse $u[k] - u[k-6]$. Therefore

$$f[k] = k\left\{u[k] - u[k-6]\right\}$$
$$= ku[k] - ku[k-6]$$

We cannot find the z-transform of $ku[k-6]$ directly by using the right-shift property [Eq. (5.19b)], so we rearrange it in terms of $(k-6)u[k-6]$ as follows:

$$f[k] = ku[k] - [(k-6)u[k-6] + 6u[k-6]]$$

We can now find the z-transform of the bracketed term by using the right-shift property [Eq. (5.19b)]. Because $u[k] \Longleftrightarrow \frac{z}{z-1}$

$$u[k-6] \Longleftrightarrow \frac{1}{z^{6}} \frac{z}{z-1} = \frac{1}{z^{5}(z-1)}$$

Also, because $ku[k] \Longleftrightarrow \frac{z}{(z-1)^{2}}$

$$(k-6)u[k-6] \Longleftrightarrow \frac{1}{z^{6}} \frac{z}{(z-1)^{2}} = \frac{1}{z^{5}(z-1)^{2}}$$

Therefore

$$F[z] = \frac{z}{(z-1)^2} - \frac{1}{z^5(z-1)^2} - \frac{6}{z^5(z-1)}$$

$$= \frac{z^6 - 6z + 5}{z^5(z-1)^2} \quad \blacksquare$$

△ **Exercise E5.4**

Using only the fact that $u[k] \iff \frac{z}{z-1}$ and the right-shift property [Eq. (5.19)], find the z-transforms of signals in Figs. 5.2 and 5.3. The answers are given in Example 5.2d and Exercise E5.1a.   ▽

Convolution

If

$$f_1[k] \iff F_1[z] \quad \text{and} \quad f_2[k] \iff F_2[z],$$

then

$$f_1[k] * f_2[k] \iff F_1[z]F_2[z] \tag{5.23}$$

Proof: This property applies to causal as well as noncausal sequences. For this reason, we shall prove it for the more general case of noncausal sequences, where the convolution sum ranges from $-\infty$ to ∞:

$$\mathcal{Z}\left[f_1[k] * f_2[k]\right] = \mathcal{Z}\left[\sum_{m=-\infty}^{\infty} f_1[m]f_2[k-m]\right]$$

$$= \sum_{k=-\infty}^{\infty} z^{-k} \sum_{m=-\infty}^{\infty} f_1[m]f_2[k-m]$$

Interchanging the order of summation,

$$\mathcal{Z}\left[f_1[k] * f_2[k]\right] = \sum_{m=-\infty}^{\infty} f_1[m] \sum_{k=-\infty}^{\infty} f_2[k-m]z^{-k}$$

$$= \sum_{m=-\infty}^{\infty} f_1[m] \sum_{r=-\infty}^{\infty} f_2[r]z^{-(r+m)}$$

$$= \sum_{m=-\infty}^{\infty} f_1[m]z^{-m} \sum_{r=-\infty}^{\infty} f_2[r]z^{-r}$$

$$= F_1[z]F_2[z]$$

Multiplication by γ^k

If

$$f[k]u[k] \iff F[z]$$

then

$$\gamma^k f[k]u[k] \iff F\left[\frac{z}{\gamma}\right] \tag{5.24}$$

Table 5.2

$\mathcal{Z}$- Transform Operations

Operation	$f[k]$	$F[z]$
Addition	$f_1[k] + f_2[k]$	$F_1[z] + F_2[z]$
Scalar multiplication	$af[k]$	$aF[z]$
Right-shift	$f[k-m]u[k-m]$	$\dfrac{1}{z^m}F[z]$
	$f[k-m]u[k]$	$\dfrac{1}{z^m}F[z] + \dfrac{1}{z^m}\displaystyle\sum_{k=1}^{m} f[-k]z^k$
	$f[k-1]u[k]$	$\dfrac{1}{z}F[z] + f[-1]$
	$f[k-2]u[k]$	$\dfrac{1}{z^2}F[z] + \dfrac{1}{z}f[-1] + f[-2]$
	$f[k-3]u[k]$	$\dfrac{1}{z^3}F[z] + \dfrac{1}{z^2}f[-1] + \dfrac{1}{z}f[-2] + f[-3]$
Left-shift	$f[k+m]u[k]$	$z^m F[z] - z^m \displaystyle\sum_{k=0}^{m-1} f[k]z^{-k}$
	$f[k+1]u[k]$	$zF[z] - zf[0]$
	$f[k+2]u[k]$	$z^2 F[z] - z^2 f[0] - zf[1]$
	$f[k+3]u[k]$	$z^3 F[z] - z^3 f[0] - z^2 f[1] - zf[2]$
Multiplication by γ^k	$\gamma^k f[k]u[k]$	$F\left[\dfrac{z}{\gamma}\right]$
Multiplication by k	$kf[k]u[k]$	$-z\dfrac{d}{dz}F[z]$
Convolution	$f_1[k] * f_2[k]$	$F_1[z]F_2[z]$
Initial value	$f[0]$	$\lim_{z\to\infty} F[z]$
Final value	$\lim_{N\to\infty} f[N]$	$\lim_{z\to 1}(z-1)F[z]$ poles of $(z-1)F[z]$ inside the unit circle.

Proof:

$$\mathcal{Z}[\gamma^k f[k]u[k]] = \sum_{k=0}^{\infty} \gamma^k f[k]z^{-k} = \sum_{k=0}^{\infty} f[k]\left(\frac{z}{\gamma}\right)^{-k} = F\left[\frac{z}{\gamma}\right]$$

△ **Exercise E5.5**

Using Eq. (5.24), derive Pairs 7 and 8 in Table 5.1 from Pairs 2 and 3, respectively.   ▽

Multiplication by k

If

$$f[k]u[k] \Longleftrightarrow F[z]$$

then

$$kf[k]u[k] \Longleftrightarrow -z\frac{d}{dz}F[z] \tag{5.25}$$

Proof:

$$-z\frac{d}{dz}F[z] = -z\frac{d}{dz}\sum_{k=0}^{\infty} f[k]z^{-k} = -z\sum_{k=0}^{\infty} -kf[k]z^{-k-1}$$

$$= \sum_{k=0}^{\infty} kf[k]z^{-k} = \mathcal{Z}\{kf[k]u[k]\}$$

△ **Exercise E5.6**

Using Eq. (5.25), derive Pairs 3 and 4 in Table 5.1 from Pair 2. Similarly, derive Pairs 8 and 9 from Pair 7.   ▽

Initial and Final Value

For a causal $f[k]$,

$$f[0] = \lim_{z \to \infty} F[z] \tag{5.26a}$$

This result follows immediately from Eq. (5.18)

We can also show that if $(z-1)F(z)$ has no poles outside the unit circle, then

$$\lim_{N \to \infty} f(N) = \lim_{z \to 1}(z-1)F(z) \tag{5.26b}$$

5.3 $\mathcal{Z}$-TRANSFORM SOLUTION OF LINEAR DIFFERENCE EQUATIONS

The time-shifting (left- or right-shift) property has set the stage for solving linear difference equations with constant coefficients. As in the case of Laplace transform with differential equations, the z-transform converts difference equations into algebraic equations which are readily solved to find the solution in z-domain. Taking the inverse z-transform of the z-domain solution yields the desired time-domain solution. The following examples demonstrate the procedure.

■ **Example 5.5**

Solve

$$y[k+2] - 5y[k+1] + 6y[k] = 3f[k+1] + 5f[k] \tag{5.27}$$

if the initial conditions are $y[-1] = \frac{11}{6}$, $y[-2] = \frac{37}{36}$, and the input $f[k] = (2)^{-k}u[k]$.

As we shall see, difference equations can be solved by using the right-shift or left-shift property. Because the difference equation (5.27) is in advance-operator form, the use of the left-shift property in Eqs. (5.22a) and (5.22b) may seem appropriate for its solution. Unfortunately, as seen from Eqs. (5.22a) and (5.22b), these properties require a knowledge of auxiliary conditions $y[0], y[1], \cdots, y[n-1]$ rather than of the initial conditions $y[-1], y[-2], \cdots, y[-n]$, which are given. This difficulty can be overcome by expressing the difference equation (5.27) in delay operator form (obtained by replacing k with $(k-2)$ and then using the right-shift property.† Equation (5.27) in delay operator form is

$$y[k] - 5y[k-1] + 6y[k-2] = 3f[k-1] + 5f[k-2] \qquad (5.28)$$

We now use the right-shift property to take the z-transform of this equation. But before proceeding, we must be clear about the meaning of a term like $y[k-1]$. Does it mean $y[k-1]u[k-1]$ or $y[k-1]u[k]$? The answer becomes clear when we recognize that the use of unilateral transform implies that we are considering the situation for $k \geq 0$, and that every signal in Eq. (5.28) must be counted from $k = 0$. Therefore the term $y[k-j]$ means $y[k-j]u[k]$. Remember also that although we are considering the situation for $k \geq 0$, $y[k]$ is present even before $k = 0$ (in the form of initial conditions). Now

$$y[k]u[k] \Longleftrightarrow Y[z]$$

$$y[k-1]u[k] \Longleftrightarrow \frac{1}{z}Y[z] + y[-1] = \frac{1}{z}Y[z] + \frac{11}{6}$$

$$y[k-2]u[k] \Longleftrightarrow \frac{1}{z^2}Y[z] + \frac{1}{z}y[-1] + y[-2] = \frac{1}{z^2}Y[z] + \frac{11}{6z} + \frac{37}{36}$$

Also

$$f[k] = (2)^{-k}u[k] = (2^{-1})^k u[k] = (0.5)^k u[k] \Longleftrightarrow \frac{z}{z - 0.5}$$

$$f[k-1]u[k] \Longleftrightarrow \frac{1}{z}F[z] + f[-1] = \frac{1}{z}\frac{z}{z - 0.5} + 0 = \frac{1}{z - 0.5}$$

$$f[k-2]u[k] \Longleftrightarrow \frac{1}{z^2}F[z] + \frac{1}{z}f[-1] + f[-2] = \frac{1}{z^2}F[z] + 0 + 0 = \frac{1}{z(z - 0.5)}$$

Note that for causal input $f[k]$,

$$f[-1] = f[-2] = \cdots = f[-n] = 0$$

Hence

$$f[k-r]u[k] \Longleftrightarrow \frac{1}{z^r}F[z]$$

Taking the z-transform of Eq. (5.28) and substituting the above results, we obtain

$$Y[z] - 5\left[\frac{1}{z}Y[z] + \frac{11}{6}\right] + 6\left[\frac{1}{z^2}Y[z] + \frac{11}{6z} + \frac{37}{36}\right] = \frac{3}{z - 0.5} + \frac{5}{z(z - 0.5)} \qquad (5.29a)$$

or

$$\left(1 - \frac{5}{z} + \frac{6}{z^2}\right)Y[z] - \left(3 - \frac{11}{z}\right) = \frac{3}{z - 0.5} + \frac{5}{z(z - 0.5)} \qquad (5.29b)$$

---

†Another approach is to find $y[0], y[1], y[2], \cdots, y[n-1]$ from $y[-1], y[-2], \cdots, y[-n]$ iteratively, as in Sec. 3.2-1, and then apply the left-shift property to Eq. (5.27)

and

$$\left(1 - \frac{5}{z} + \frac{6}{z^2}\right) Y[z] = \left(3 - \frac{11}{z}\right) + \frac{3z + 5}{z(z - 0.5)}$$

$$= \frac{3z^2 - 9.5z + 10.5}{z(z - 0.5)}$$

Multiplication of both sides by z^2 yields

$$\left(z^2 - 5z + 6\right) Y[z] = \frac{z\left(3z^2 - 9.5z + 10.5\right)}{(z - 0.5)}$$

so that

$$Y[z] = \frac{z(3z^2 - 9.5z + 10.5)}{(z - 0.5)(z^2 - 5z + 6)} \tag{5.30}$$

and

$$\frac{Y[z]}{z} = \frac{3z^2 - 9.5z + 10.5}{(z - 0.5)(z - 2)(z - 3)}$$

$$= \frac{(26/15)}{z - 0.5} - \frac{(7/3)}{z - 2} + \frac{(18/5)}{z - 3}$$

Therefore

$$Y[z] = \frac{26}{15}\left(\frac{z}{z - 0.5}\right) - \frac{7}{3}\left(\frac{z}{z - 2}\right) + \frac{18}{5}\left(\frac{z}{z - 3}\right)$$

and

$$y[k] = \left[\frac{26}{15}(0.5)^k - \frac{7}{3}(2)^k + \frac{18}{5}(3)^k\right] u[k] \tag{5.31}$$

■

This example demonstrates the ease with which linear difference equations with constant coefficients can be solved by z-transform. This method is general; it can be used to solve a single difference equation or a set of simultaneous difference equations of any order as long as the equations are linear with constant coefficients.

Comment

Sometimes auxiliary conditions $y[0], y[1], \cdots, y[n - 1]$ (instead of initial conditions $y[-1], y[-2], \cdots, y[-n]$) are given to solve a difference equation. In this case the equation can be solved by expressing it in the advance operator form and then using the left-shift property (see Exercise E5.8 below).

△   **Exercise E5.7**
Solve the equation below if the initial conditions are $y[-1] = 2$, $y[-2] = 0$, and the input $f[k] = u[k]$:

$$y[k + 2] - \tfrac{5}{6}y[k + 1] + \tfrac{1}{6}y[k] = 5f[k + 1] - f[k]$$

Answer: $y[k] = \left[12 - 15(\tfrac{1}{2})^k + \tfrac{14}{3}(\tfrac{1}{3})^k\right] u[k]$ ▽

△   **Exercise E5.8**
Solve the following equation if the auxiliary conditions are $y[0] = 1$, $y[1] = 2$, and the input $f[k] = u[k]$:

$$y[k + 2] + 3y[k + 1] + 2y[k] = f[k + 1] + 3f[k]$$

Hint: Use left-shift property.   Answer: $y[k] = \left[\tfrac{2}{3} + 2(-1)^k - \tfrac{5}{3}(-2)^k\right] u[k]$ ▽

Zero-Input and Zero-State Components

In Example 5.5 we found the total solution of the difference equation. It is relatively easy to separate the solution into zero-input and zero-state components. All we have to do is to separate the response into terms due to input and terms due to initial conditions. We can separate the response in Eq. (5.29b) as follows:

$$\left(1 - \frac{5}{z} + \frac{6}{z^2}\right) Y[z] - \underbrace{\left(3 - \frac{11}{z}\right)}_{\text{initial condition terms}} = \underbrace{\frac{3}{z - 0.5} + \frac{5}{z(z - 0.5)}}_{\text{terms due to input}} \qquad (5.32)$$

Therefore

$$\left(1 - \frac{5}{z} + \frac{6}{z^2}\right) Y[z] = \underbrace{\left(3 - \frac{11}{z}\right)}_{\text{initial condition terms}} + \underbrace{\frac{(3z + 5)}{z(z - 0.5)}}_{\text{input terms}}$$

Multiplying both sides by z^2 yields

$$\left(z^2 - 5z + 6\right) Y[z] = \underbrace{z(3z - 11)}_{\text{initial condition terms}} + \underbrace{\frac{z(3z + 5)}{z - 0.5}}_{\text{input terms}}$$

and

$$Y[z] = \underbrace{\frac{z(3z - 11)}{z^2 - 5z + 6}}_{\text{zero-input response}} + \underbrace{\frac{z(3z + 5)}{(z - 0.5)(z^2 - 5z + 6)}}_{\text{zero-state response}} \qquad (5.33)$$

We expand both terms on the right-hand side into modified partial fractions to yield

$$Y[z] = \underbrace{\left[5\left(\frac{z}{z - 2}\right) - 2\left(\frac{z}{z - 3}\right)\right]}_{\text{zero-input}} + \underbrace{\left[\frac{26}{15}\left(\frac{z}{z - 0.5}\right) - \frac{22}{3}\left(\frac{z}{z - 2}\right) + \frac{28}{5}\left(\frac{z}{z - 3}\right)\right]}_{\text{zero-state}}$$

and

$$y[k] = \left[\underbrace{5(2)^k - 2(3)^k}_{\text{zero-input}} \underbrace{- \frac{22}{3}(2)^k + \frac{28}{5}(3)^k + \frac{26}{15}(0.5)^k}_{\text{zero-state}}\right] u[k]$$

$$= \left[-\frac{7}{3}(2)^k + \frac{18}{5}(3)^k + \frac{26}{15}(0.5)^k\right] u[k]$$

which agrees with the result in Eq. (5.31).

△ **Exercise E5.9**

Solve

$$y[k + 2] - \tfrac{5}{6}y[k + 1] + \tfrac{1}{6}y[k] = 5f[k + 1] - f[k]$$

if the initial conditions are $y[-1] = 2$, $y[-2] = 0$, and the input $f[k] = u[k]$. Separate the response into zero-input and zero-state components.

Answer:

$$y[k] = \left\{ \underbrace{[3(\tfrac{1}{2})^k - \tfrac{4}{3}(\tfrac{1}{3})^k]}_{\text{zero-input}} + \underbrace{[12 - 18(\tfrac{1}{2})^k + 6(\tfrac{1}{3})^k]}_{\text{zero-state}} \right\} u[k]$$

$$= \left[12 - 15(\tfrac{1}{2})^k + \tfrac{14}{3}(\tfrac{1}{3})^k \right] u[k]$$

$\triangledown$

5.3-1   Zero-State Response of LTID Systems: The Transfer Function

Consider an nth-order LTID system specified by the difference equation

$$Q[E]y[k] = P[E]f[k] \tag{5.34a}$$

or

$$(E^n + a_{n-1}E^{n-1} + \cdots + a_1 E + a_0)y[k] =$$
$$(b_n E^n + b_{n-1}E^{n-1} + \cdots + b_1 E + b_0)f[k] \tag{5.34b}$$

or

$$y[k+n] + a_{n-1}y[k+n-1] + \cdots + a_1 y[k+1] + a_0 y[k]$$
$$= b_n f[k+n] + \cdots + b_1 f[k+1] + b_0 f[k] \tag{5.34c}$$

We now derive the general expression for the zero-state response, that is, the system response to input $f[k]$ when all the initial conditions $y[-1] = y[-2] = \cdots = y[-n] = 0$ (zero state). The input $f[k]$ is assumed to be causal so that $f[-1] = f[-2] = \cdots = f[-n] = 0$.

Equation (5.34c) can be expressed in the delay operator form as

$$y[k] + a_{n-1}y[k-1] + \cdots + a_0 y[k-n]$$
$$= b_n f[k] + b_{n-1}f[k-1] + \cdots + b_0 f[k-n] \tag{5.34d}$$

Because $y[-r] = f[-r] = 0$ for $r = 1, 2, \ldots, n$

$$y[k-m]u[k] \iff \frac{1}{z^m}Y[z]$$
$$f[k-m]u[k] \iff \frac{1}{z^m}F[z] \qquad m = 1, 2, \ldots, n$$

Now the z-transform of Eq. (5.34d) is given by

$$\left(1 + \frac{a_{n-1}}{z} + \frac{a_{n-2}}{z^2} + \cdots + \frac{a_0}{z^n}\right) Y[z] = \left(b_n + \frac{b_{n-1}}{z} + \frac{b_{n-2}}{z^2} + \cdots + \frac{b_0}{z^n}\right) F[z]$$

Multiplication of both sides by z^n yields

$$(z^n + a_{n-1}z^{n-1} + \cdots + a_1 z + a_0)Y[z]$$
$$= (b_n z^n + b_{n-1}z^{n-1} + \cdots + b_1 z + b_0)F[z]$$

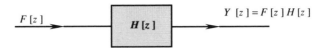

Fig. 5.5 The transformed representation of an LTID system.

Therefore

$$Y[z] = \left(\frac{b_n z^n + b_{n-1} z^{n-1} + \cdots + b_1 z + b_0}{z^n + a_{n-1} z^{n-1} + \cdots + a_1 z + a_0} \right) F[z] \qquad (5.35a)$$

$$= \frac{P[z]}{Q[z]} F[z] \qquad (5.35b)$$

As in the case of continuous-time systems, we define $H[z]$, **the transfer function of an LTID system**, as the ratio of $Y[z]$ to $F[z]$ (assuming all initial conditions zero). Thus

$$H[z] \equiv \frac{Y[z]}{F[z]} = \frac{\mathcal{Z}[\text{zero-state response}]}{\mathcal{Z}[\text{input}]} \qquad (5.36)$$

For an LTID system as specified by Eq. (5.34), the transfer function $H[z]$, as seen in Eq. (5.35b), is

$$H[z] = \frac{P[z]}{Q[z]} = \frac{b_n z^n + b_{n-1} z^{n-1} + \cdots + b_1 z + b_0}{z^n + a_{n-1} z^{n-1} + \cdots + a_1 z + a_0} \qquad (5.37)$$

From the definition of the transfer function $H[z]$ in Eq. (5.37), it follows that

$$Y[z] = H[z]F[z] \qquad (5.38)$$

Therefore $Y[z]$, **the z-transform of the zero-state response $y[k]$, is the product of $F[z]$ and $H[z]$**, as shown in Fig. 5.5, where $F[z]$ is the z-transform of the input $f[k]$ and $H[z]$ is given by Eq. (5.37). We call $H[z]$ the **transfer function** of the (discrete-time) system relating a particular output $y[k]$ to the input $f[k]$. Observe that the denominator of $H[z]$ is $Q[z]$, the characteristic polynomial of the system. **Therefore the poles of $H[z]$ are the characteristic roots of the system.** Consequently the system stability criterion can be stated in terms of the poles of the transfer function of an LTID system as follows:

1. An LTID system is asymptotically stable if and only if all the poles of its transfer function $H[z]$ lie inside a unit circle (centered at the origin) in the complex plane. The poles may be repeated or unrepeated.

2. An LTID system is unstable if and only if either one or both of the following conditions exist: (i) at least one pole of $H[z]$ is outside the unit circle; (ii) there are repeated poles of $H[z]$ on the unit circle.

3. An LTID system is marginally stable if and only if there are no poles of $H[z]$ outside the unit circle, and there are some unrepeated poles on the unit circle.

Just as in continuous-time systems, we can represent discrete-time systems in the transformed manner by representing all signals by their z-transforms and all system components (or elements) by their transfer functions.

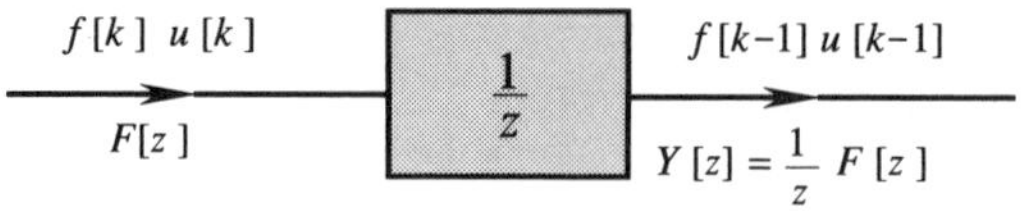

Fig. 5.6 Ideal unit delay and its transfer function.

■ **Example 5.6: The Transfer Function of a Unit Delay**
Show that the transfer function of a unit delay is $1/z$.
If the input to unit delay is $f[k]u[k]$, then its output (Fig. 5.6) is given by

$$y[k] = f[k-1]u[k-1]$$

The z-transform of this equation yields [see Eq. (5.19a)]

$$Y[z] = \frac{1}{z}F[z]$$

$$= H[z]F[z]$$

It follows that the transfer function of unit delay is

$$H[z] = \frac{1}{z} \tag{5.39}$$

■

■ **Example 5.7**
Find the response $y[k]$ of an LTID system described by the difference equation

$$y[k+2] + y[k+1] + 0.16y[k] = f[k+1] + 0.32f[k]$$

or

$$(E^2 + E + 0.16)y[k] = (E + 0.32)f[k]$$

for the input $f[k] = (-2)^{-k}u[k]$ and with all the initial conditions zero (system in zero
state initially).
From the difference equation we find

$$H[z] = \frac{P[z]}{Q[z]} = \frac{z + 0.32}{z^2 + z + 0.16}$$

For the input $f[k] = (-2)^{-k}u[k] = [(-2)^{-1}]^k u(k) = (-0.5)^k u[k]$

$$F[z] = \frac{z}{z + 0.5}$$

and

$$Y[z] = F[z]H[z] = \frac{z(z + 0.32)}{(z^2 + z + 0.16)(z + 0.5)}$$

Therefore

$$\frac{Y[z]}{z} = \frac{(z + 0.32)}{(z^2 + z + 0.16)(z + 0.5)} = \frac{(z + 0.32)}{(z + 0.2)(z + 0.8)(z + 0.5)}$$

$$= \frac{2/3}{z + 0.2} - \frac{8/3}{z + 0.8} + \frac{2}{z + 0.5}$$

so that

$$Y[z] = \frac{2}{3}\left(\frac{z}{z + 0.2}\right) - \frac{8}{3}\left(\frac{z}{z + 0.8}\right) + 2\left(\frac{z}{z + 0.5}\right)$$

and

$$y[k] = \left[\tfrac{2}{3}(-0.2)^k - \tfrac{8}{3}(-0.8)^k + 2(-0.5)^k\right]u[k] \quad ■$$

△ **Exercise E5.10**

A discrete-time system is described by the following transfer function:

$$H[z] = \frac{z - 0.5}{(z + 0.5)(z - 1)}$$

(a) Find the system response to input $f[k] = 3^{-(k+1)}u[k]$ if all initial conditions are zero. **(b)** Write the difference equation relating the output $y[k]$ to input $f[k]$ for this system.

Hint: $3^{-(k+1)} = (3^{-1})(3^{-k}) = \frac{1}{3}\left(\frac{1}{3}\right)^k$

Answers: **(a)** $y[k] = \frac{1}{3}\left[\frac{1}{2} - 0.8(-0.5)^k + 0.3\left(\frac{1}{3}\right)^k\right]u[k]$

(b) $y[k + 2] - 0.5y[k + 1] - 0.5y[k] = f[k + 1] - 0.5f[k]$ ▽

5.3-2 Relationship Between h[k] and H[z]

For an LTID system with transfer function $H[z]$, we have [Eq. (5.38)]

$$Y[z] = H[z]F[z]$$

or

$$\mathcal{Z}\{y[k]\} = H[z]\mathcal{Z}\{f[k]\}$$

When the input $f[k] = \delta[k]$, the response $y[k] = h[k]$. Therefore

$$\mathcal{Z}\{h[k]\} = H[z]\mathcal{Z}\{\delta[k]\} = H[z]$$

or

$$h[k] \Longleftrightarrow H[z] \tag{5.40}$$

This important result relates the impulse response $h[k]$, which is a time-domain specification of a system, to $H[z]$, which is a frequency-domain specification of the system. We have shown that the unit impulse response $h[k]$ and the transfer function $H[z]$ are a z-transform pair. This means

$$H[z] = \sum_{k=0}^{\infty} h[k]z^{-k} \tag{5.41}$$

If an LTID system is described by a difference equation

$$Q[E]y[k] = P[E]f[k]$$

its impulse response is readily determined by taking the inverse z-transform of $H[z]$, which is $P[z]/Q[z]$

$$h[k] = \mathcal{Z}^{-1}\{H[z]\}$$

$$= \mathcal{Z}^{-1}\left\{\frac{P[z]}{Q[z]}\right\} \tag{5.42}$$

△ **Exercise E5.11**

Redo Exercise E3.5 using Eq. (5.42).   ▽

5.3-3   Frequency-Domain Interpretation of the Transfer Function

In Chapter 4 we showed that an LTIC system response to an everlasting exponential input e^{st} is also an everlasting exponential $H(s)e^{st}$. There is a parallel development for LTI discrete-time systems. We now show that **the system response to an everlasting discrete-time exponential z^k is also an everlasting exponential $H[z]z^k$.**

If $h[k]$ is the impulse response of the system, then $y[k]$, the system response to input z^k, is

$$y[k] = h[k] * z^k$$

$$= \sum_{m=-\infty}^{\infty} h[m]z^{k-m}$$

$$= z^k \sum_{m=-\infty}^{\infty} h[m]z^{-m}$$

The sum of the right-hand side defines the (bilateral) z-transform of $h[k]$, that is, $H[z]$. If the system is causal, the limits for the sum are to from 0 to ∞. Therefore

$$y[k] = H[z]z^k \tag{5.43}$$

We now have an alternative definition of the transfer function $H[z]$ as the ratio of the output signal to the input signal when the input is an everlasting exponential z^k:

$$H[z] = \frac{\text{Output signal}}{\text{Input signal}}\bigg|_{\text{Input} = \text{everlasting exponential } z^k} \tag{5.44}$$

5.3-4   Physical Interpretation of the $\mathcal{Z}$-Transform

We now show that the z-transform in reality is a tool for frequency-domain analysis in which an input is represented as a sum of everlasting exponentials of the form z^k. The response is then found by summing the system response to all these exponential components.

If $f[k] \Longleftrightarrow F[z]$, then, according to Eq. (5.2),

$$f[k] = \frac{1}{2\pi j} \oint F[z]z^{k-1}\,dz \tag{5.45}$$

where the contour integration is in the region of convergence of $F[z]$. If $|z| > r_0$ is the region of convergence, then any circle with radius $> r_0$ can be used as a contour of integration (Fig. 5.7). Equation (5.45) can be expressed as

$$f[k] = \lim_{\triangle z_n \to 0} \sum_n \frac{F[z_n]\triangle z_n}{2\pi j}z_n^{k-1} = \lim_{\triangle z_n \to 0} \sum_n \frac{F[z_n]\triangle z_n}{2\pi j z_n}z_n^k$$

This shows that $f[k]$ can be synthesized by exponentials z_n^k where z_n lies on the contour of integration, as shown in Fig. 5.7. The amplitude of exponential z_n^k is $F[z_n]\triangle z_n/2\pi j z_n$ which is infinitesimal because $\triangle z_n \to 0$. The sum of all these exponentials along the entire contour results in $f[k]$. The system response to $f[k]$ is the sum of the system response to all these everlasting exponential components of $f[k]$.

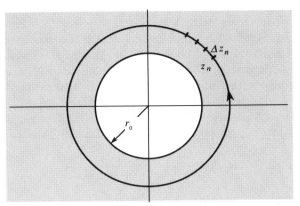

Fig. 5.7   The z-transform as a means of expressing a signal as a sum of (everlasting) exponentials (or sinusoids).

We showed in Eq. (5.43) that the system response to an exponential z^k is also an exponential $H[z]z^k$. Now $f[k]$ is composed of components z_n^k with amplitude $F[z_n]\triangle z_n / 2\pi j z_n$. This component and the corresponding system response can be displayed by using the convenient arrow notation as

$$\frac{F[z_n]\triangle z_n}{2\pi j z_n}z_n^k \longrightarrow \frac{F[z_n]H[z_n]\triangle z_n}{2\pi j z_n}z_n^k$$

The sum of all these response components on the right-hand side is the system output $y[k]$. Therefore

$$y[k] = \lim_{\triangle z \to 0}\sum_n \frac{F[z_n]H[z_n]}{2\pi j z_n}z_n^k \triangle z_n$$

$$= \frac{1}{2\pi j}\oint F[z]H[z]z^{k-1}\,dz$$

$$= \mathcal{Z}^{-1}\left\{F[z]H[z]\right\}$$

This means

$$Y[z] = F[z]H[z]$$

which is precisely the result obtained earlier.

This discussion provides an alternative, more physical view of z-transform. In this view, the z-transform is a tool by which a signal $f[k]$ is expressed as a sum of its spectral components of the form z^k. Also, $H[z]z^k$ is the system response to input z^k. We add the system response to all the exponential components of $f[k]$ to obtain the system response $y[k]$.

5.4   SYSTEM REALIZATION

We now discuss ways to realize an nth-order discrete-time system described by a transfer function

$$H[z] = \frac{b_n z^n + b_{n-1}z^{n-1} + \cdots + b_1 z + b_0}{z^n + a_{n-1}z^{n-1} + \cdots + a_1 z + a_0} \tag{5.46}$$

This transfer function is identical to the general nth-order continuous-time transfer function $H(s)$ in Eq. (4.60) with s replaced by z. It is reasonable to believe that the realization of $H[z]$ in (5.46) would be identical to that of $H(s)$ with s replaced by z. Fortunately this happens to be the case. In realizations of $H(s)$ the basic element used was an integrator with transfer function $1/s$. In realizations of $H[z]$ the basic element is unit delay with transfer function $1/z$. Therefore all the realizations of $H(s)$ studied in Sec. 4.6 are also the realizations of $H[z]$ if we replace integrators by unit delays. To show this, consider a realization of a third-order transfer function.

$$H[z] = \frac{b_3 z^3 + b_2 z^2 + b_1 z + b_0}{z^3 + a_2 z^2 + a_1 z + a_0} \tag{5.47}$$

The output $Y[z]$ and the input $F[z]$ are related by

$$Y[z] = \frac{b_3 z^3 + b_2 z^2 + b_1 z + b_0}{z^3 + a_2 z^2 + a_1 z + a_0} F[z]$$

$$= \left(b_3 z^3 + b_2 z^2 + b_1 z + b_0 \right) \left[\frac{1}{z^3 + a_2 z^2 + a_1 z + a_0} F[z] \right]$$

$$= \left(b_3 z^3 + b_2 z^2 + b_1 z + b_0 \right) X[z] \tag{5.48a}$$

where

$$X[z] = \frac{1}{z^3 + a_2 z^2 + a_1 z + a_0} F[z] \tag{5.48b}$$

We can realize the desired system in two steps. First, we generate $X[z]$ according to Eq. (5.48b). Then $Y[z]$ is generated from $X[z]$ according to Eq. (5.48a).

From Eq. (5.48b) we obtain the difference equation relating the signal $x[k]$ to input $f[k]$:

$$x[k + 3] + a_2 x[k + 2] + a_1 x[k + 1] + a_0 x[k] = f[k] \tag{5.49a}$$

or

$$x[k + 3] = -a_2 x[k + 2] - a_1 x[k + 1] - a_0 x[k] + f[k] \tag{5.49b}$$

Assuming that $x[k + 3]$ is available, we can generate $x[k + 2]$, $x[k + 1]$, and $x[k]$ by passing $x[k + 3]$ successively through unit delays, as shown in Fig. 5.8a. Now we can generate $x[k + 3]$ from $f[k]$, $x[k]$, $x[k + 1]$, and $x[k + 2]$ according to Eq. (5.49b) and can close the loop, as shown in Fig. 5.8b. This realization can be expressed in the transformed version, with each delay represented by a transfer function $1/z$ and the signals $f[k]$, $x[k]$, $x[k + 1]$, $x[k + 2]$, and $x[k + 3]$ represented by their transforms $F[z]$, $X[z]$, $zX[z]$, $z^2 X[z]$, and $z^3 X[z]$, respectively, as shown in Fig. 5.8c. We now generate $Y[z]$ from these signals according to Eq. (5.48a), as shown in Fig. 5.8d. Therefore the system in Fig. 5.8d is a realization of $H[z]$ in Eq. (5.47). Note that this is identical to the first canonical realization of the third-order transfer function $H(s)$ in Fig. 4.23, with $1/s$ replaced by $1/z$. These results are valid for any value of n. Similarly, the cascade and parallel realizations of the continuous-time case are directly applicable to discrete-time systems, with integrators replaced by unit delays. The second canonical realization is developed in Appendix 4.1.

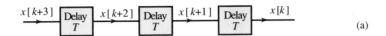

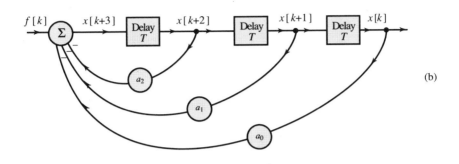

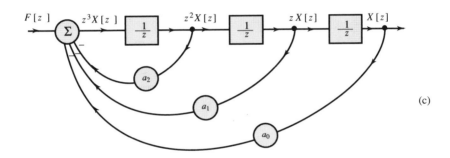

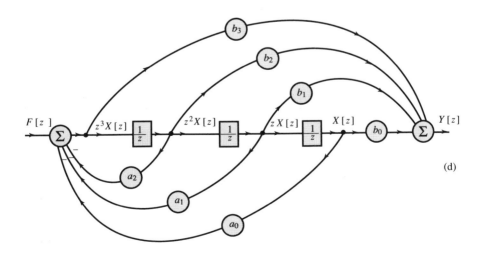

Fig. 5.8 A canonical realization of $H[z]$.

■ **Example 5.8**

Realize the following transfer functions, using only the cascade form for part **a** and using only the parallel form for part **b**.

(a) $H[z] = \dfrac{4z + 28}{z^2 + 6z + 5}$ (b) $H[z] = \dfrac{7z^2 + 37z + 51}{(z + 2)(z + 3)^2}$

Identical transfer functions for continuous-time systems are realized in Figs. 4.29a and 4.30. Replace $1/s$ (integrators) in these realizations by $1/z$ (unit delays). ■

△ **Exercise E5.12**

Give the canonical realization of the following transfer functions. (a) $\dfrac{2}{z + 5}$ (b) $\dfrac{z + 8}{z + 5}$

(c) $\dfrac{z}{z + 5}$ (d) $\dfrac{2z + 3}{z^2 + 6z + 25}$

Answer: See Example 4.18. Replace $1/s$ by $1/z$ and make appropriate changes in coefficients. ▽

5.5 FREQUENCY RESPONSE OF DISCRETE-TIME SYSTEMS

For continuous-time systems we showed that the system response to an input $e^{j\omega t}$ is $H(j\omega)e^{j\omega t}$, and that the response to an input $\cos \omega t$ is $|H(j\omega)| \cos [\omega t + \angle H(j\omega)]$. Similar results hold for discrete-time systems. We now show that for an (asymptotically stable) LTID system, the system response to an input $e^{j\Omega k}$ is $H[e^{j\Omega}]e^{j\Omega k}$ and the response to an input $\cos \Omega k$ is $|H[e^{j\Omega}]| \cos [\Omega k + \angle H[e^{j\Omega}]]$.

The proof is similar to the one used in continuous-time systems. In Sec. 5.3-3 we showed that an LTID system response to an (everlasting) exponential z^k is also an (everlasting) exponential $H[z]z^k$. It is helpful to represent this relationship by a directed arrow notation as

$$z^k \to H[z]z^k \tag{5.50}$$

Setting $z = e^{\pm j\Omega}$ in this relationship yields

$$e^{j\Omega k} \to H[e^{j\Omega}]e^{j\Omega k} \tag{5.51a}$$

$$e^{-j\Omega k} \to H[e^{-j\Omega}]e^{-j\Omega k} \tag{5.51b}$$

Addition of these two equation yields

$$2\cos \Omega k \to H[e^{j\Omega}]e^{j\Omega k} + H[e^{-j\Omega}]e^{-j\Omega k} = 2\mathrm{Re}\left\{H[e^{j\Omega}]e^{j\Omega k}\right\} \tag{5.52}$$

Expressing $H[e^{j\Omega}]$ in the polar form

$$H[e^{j\Omega}] = |H[e^{j\Omega}]|e^{j\angle H[e^{j\Omega}]} \tag{5.53}$$

Eq. (5.52) can be expressed as

$$\cos \Omega k \to |H[e^{j\Omega}]| \cos [\Omega k + \angle H[e^{j\Omega}]] \tag{5.54}$$

In other words, the system response $y[k]$ to a sinusoidal input $\cos \Omega k$ is given by†

$$y[k] = |H[e^{j\Omega}]| \cos [\Omega k + \angle H[e^{j\Omega}]] \tag{5.55a}$$

---

†For a causal sinusoidal input $\cos \Omega k$, the system response will also have an additional natural component consisting of the characteristic modes. For a stable system, all the modes decay exponentially, and only the sinusoidal component $|H[e^{j\Omega}]| \cos [\Omega k + \angle H[e^{j\Omega}]]$ persists. For this reason this component is called the *steady-state* response of the system. Thus, the steady-state response of a system to a causal sinusoidal input $\cos \Omega k$ is $|H[e^{j\Omega}]| \cos [\Omega k + \angle H[e^{j\Omega}]]$.

Using similar argument we can show that the system response to a sinusoid $\cos(\Omega k + \theta)$ is

$$y[k] = |H[e^{j\Omega}]| \cos\left[\Omega k + \theta + \angle H[e^{j\Omega}]\right] \tag{5.55b}$$

Therefore the response of an asymptotically stable‡ LTID system to a discrete-time sinusoidal input of frequency Ω is also a discrete-time sinusoid of the same frequency. **The amplitude of the output sinusoid is $|H[e^{j\Omega}]|$ times the input amplitude, and the phase of the output sinusoid is shifted by $\angle H[e^{j\Omega}]$ with respect to the input phase. Clearly $|H[e^{j\Omega}]|$ is the amplitude gain, and a plot of $|H[e^{j\Omega}]|$ versus Ω is the amplitude response of the discrete-time system. Similarly, $\angle H[e^{j\Omega}]$ is the phase response of the system, and a plot of $\angle H[e^{j\Omega}]$ vs Ω shows how the system modifies or shifts the phase of the input sinusoid.** Note that $H[e^{j\Omega}]$ incorporates the information of both amplitude and phase response and therefore is called the **frequency response** of the system.

These results, although parallel to those for continuous-time systems, differ from them in one significant aspect. In the continuous-time case, the frequency response is $H(j\omega)$. A parallel result for the discrete-time case would lead to frequency response $H[j\Omega]$. Instead we found the frequency response to be $H[e^{j\Omega}]$. This deviation causes some interesting differences between the behavior of continuous-time and discrete-time systems.

System Response to Sampled Continuous-time Sinusoids

So far we have considered the system response of a discrete-time system to a discrete-time sinusoid $\cos \Omega k$ (or exponential $e^{j\Omega k}$). In practice the input may be a sampled continuous-time sinusoid $\cos \omega t$ (or an exponential $e^{j\omega t}$). When a sinusoid $\cos \omega t$ is sampled with sampling interval T, the resulting signal is a discrete-time sinusoid $\cos \omega kT$. Therefore all the results developed in this section apply if we substitute ωT for Ω:

$$\Omega = \omega T$$

■ **Example 5.9**

For a system specified by the equation

$$y[k+1] - 0.8y[k] = f[k+1]$$

Find the system response to the input **(a)** $1^k = 1$ **(b)** $\cos\left[\frac{\pi}{6}k - 0.2\right]$
(c) a sampled sinusoid $\cos 1500t$ with sampling interval $T = 0.001$.

The system equation can be expressed as

$$(E - 0.8)y[k] = Ef[k]$$

Therefore the transfer function of the system is

$$H[z] = \frac{z}{z - 0.8} = \frac{1}{1 - 0.8z^{-1}}$$

---

‡Equation (5.50) is valid only for values of z lying in the region of convergence of $H[z]$. For $z = e^{j\Omega}$, z lies on the unit circle ($|z| = 1$). The region of convergence for unstable and marginally stable systems does not include the unit circle. Therefore Eq. (5.50) is valid only for asymptotically stable systems.

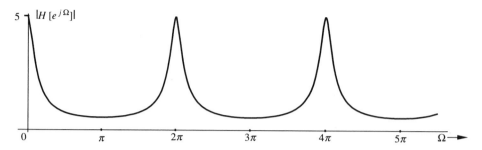

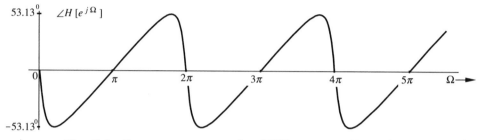

Fig. 5.9   Frequency response of an LTID system in Example 5.9.

The frequency response is

$$H[e^{j\Omega}] = \frac{1}{1 - 0.8e^{-j\Omega}} \tag{5.56}$$

$$= \frac{1}{1 - 0.8(\cos \Omega - j \sin \Omega)}$$

Therefore

$$= \frac{1}{(1 - 0.8\cos \Omega) + j0.8\sin \Omega}$$

$$\left|H[e^{j\Omega}]\right| = \frac{1}{\sqrt{(1 - 0.8\cos \Omega)^2 + (0.8\sin \Omega)^2}}$$

$$= \frac{1}{\sqrt{1.64 - 1.6\cos \Omega}} \tag{5.57a}$$

and

$$\angle H[e^{j\Omega}] = -\tan^{-1}\left[\frac{0.8\sin \Omega}{1 - 0.8\cos \Omega}\right] \tag{5.57b}$$

The amplitude response $|H[e^{j\Omega}]|$ can also be obtained by observing that $|H|^2 = HH^*$. Therefore

$$\left|H[e^{j\Omega}]\right|^2 = H[e^{j\Omega}]H^*[e^{j\Omega}]$$

$$= H[e^{j\Omega}]H[e^{-j\Omega}] \tag{5.58}$$

From Eq. (5.56) it follows that

$$\left|H[e^{j\Omega}]\right|^2 = \left(\frac{1}{1 - 0.8e^{-j\Omega}}\right)\left(\frac{1}{1 - 0.8e^{j\Omega}}\right)$$

$$= \frac{1}{1.64 - 1.6\cos \Omega}$$

which yields the result found earlier in Eq. (5.57a).

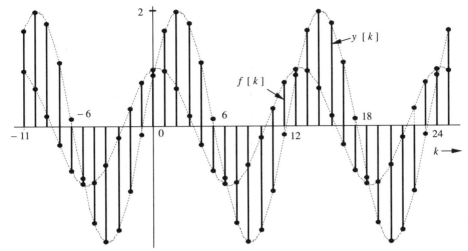

Fig. 5.10 Sinusoidal input and the corresponding output of an LTID system in Example 5.9.

Figure 5.9 shows plots of amplitude and phase response as functions of Ω. Note that **both the amplitude and the phase response are periodic functions of Ω with period 2π.** The reason for this interesting behavior is explained later on p. 385. We now compute the amplitude and the phase response for the various inputs.

(a) $f[k] = 1^k = 1$

Since $1^k = (e^{j\Omega})^k$ with $\Omega = 0$, the amplitude response is $H[e^{j0}]$. From Eq. (5.57a) we obtain

$$H[e^{j0}] = \frac{1}{\sqrt{1.64 - 1.6\cos(0)}} = \frac{1}{\sqrt{0.04}} = 5 = 5\angle 0$$

Therefore

$$|H[e^{j0}]| = 5 \qquad \text{and} \qquad \angle H[e^{j0}] = 0$$

These values also can be read directly from Figs. 5.9a and 5.9b, respectively, corresponding to $\Omega = 0$. Therefore the system response to input 1 is

$$y[k] = 5(1^k) = 5$$

(b) $f[k] = \cos\left[\frac{\pi}{6}k - 0.2\right]$

Here $\Omega = \frac{\pi}{6}$. From Eqs. (5.57)

$$\left|H[e^{j\pi/6}]\right| = \frac{1}{\sqrt{1.64 - 1.6\cos\frac{\pi}{6}}} = 1.983$$

$$\angle H[e^{j\pi/6}] = -\tan^{-1}\left[\frac{0.8\sin\frac{\pi}{6}}{1 - 0.8\cos\frac{\pi}{6}}\right] = -0.916 \text{ rad.}$$

These values also can be read directly from Figs. 5.9a and 5.9b, respectively, corresponding to $\Omega = \frac{\pi}{6}$. Therefore

$$y[k] = 1.983\cos\left(\frac{\pi}{6}k - 0.2 - 0.916\right)$$

$$= 1.983\cos\left(\frac{\pi}{6}k - 1.116\right)$$

Figure 5.10 shows the input $f[k]$ and the corresponding system response.

(c) A sinusoid $\cos 1500t$ sampled every T seconds $(t = kT)$ results in a discrete-time sinusoid

$$f[k] = \cos 1500kT$$

For $T = 0.001$, the input is

$$f[k] = \cos (1.5k)$$

In this case $\Omega = 1.5$. From Eqs. (5.57a) and (5.57b)

$$\left|H[e^{j1.5}]\right| = \frac{1}{\sqrt{1.64 - 1.6\cos(1.5)}} = 0.809$$

$$\angle H[e^{j1.5}] = -\tan^{-1}\left[\frac{0.8\sin(1.5)}{1 - 0.8\cos(1.5)}\right] = -0.702 \text{ rad}$$

These values also could be read directly from Fig. 5.9 corresponding to $\Omega = 1.5$. Therefore

$$y[k] = 0.809\cos(1.5k - 0.702) \quad \blacksquare$$

Comment

Figures 5.9a and 5.9b as well as Eqs. (5.57) indicate that the frequency response of a discrete-time system is a continuous function of frequency Ω. There is no contradiction in this behavior. It is merely an indication of the fact that the frequency variable Ω is continuous (takes on all possible values) and therefore the system response exists at every value of Ω.

⊙ **Computer Example C5.1**

Using a computer, redo Example 5.9. We are required to find the system response to three different inputs. This is done in three parts; a,b, and c.

(a): Input $f[k] = 1^k = 1$ for all k. The input is $e^{j\Omega k}$, with $\Omega = 0$.

```
num=[1 0];
den=[1 -0.8];
z=exp(j*0);
H=polyval(num,z)./polyval(den,z);
mag=abs(H);
phase=angle(H);
k=0:1:10;
y=(mag*k)./k;
plot(k,y,'o'),grid,
xlabel('k'),ylabel('y[k]'),
title('Response y[k] for the input 1^k'),pause
```

(b): Input $f[k] = \cos[\frac{\pi}{6}k - 0.2]$. Here $\Omega = \frac{\pi}{6}$.

```
num=[1 0];
den=[1 -0.8];
z=exp(j*pi/6);
H=polyval(num,z)./polyval(den,z);
mag=abs(H);
phase=angle(H);
k=0:1:20;
y=mag*cos((pi/6)*k-0.2+phase);
```

```
plot(k,y,'o'),grid,
xlabel('k'),ylabel('y[k]'),
title('Response for the input cos((pi/6)k-0.2)'),pause
```

(c): Input $f[k]$ is sampled continuous-time sinusoid $\cos 1500t$ with the sampling interval $T = 0.001$. This results in $\Omega = \omega T = (1500)(0.001) = 1.5$.

```
num=[1 0];
den=[1 -0.8];
z=exp(j*1.5);
H=polyval(num,z)./polyval(den,z);
mag=abs(H);
phase=angle(H);
k=0:1:10;
y=mag*cos(1.5*k+phase);
plot(k,y,'o'),grid,
xlabel('k'),ylabel('y[k]'),
title('Response for the sampled sinusoid cos(1500t)')   ⊙
```

△ **Exercise E5.13**

For a system specified by the equation

$$y[k + 1] - 0.5y[k] = f[k]$$

find the amplitude and phase response. Find the system response to sinusoidal input $\cos\left(1000t - \frac{\pi}{3}\right)$ sampled every $T = 0.5$ ms.
Answer:

$$\left|H[e^{j\Omega}]\right| = \frac{1}{\sqrt{1.25 - \cos\Omega}} \qquad \angle H[e^{j\Omega}] = -\tan^{-1}\left[\frac{\sin\Omega}{\cos\Omega - 0.5}\right]$$

$$y[k] = 1.639\cos\left(0.5k - \frac{\pi}{3} - 0.904\right) = 1.639\cos\left(0.5k - 1.951\right) \quad \triangledown$$

△ **Exercise E5.14**

Show that for an ideal delay ($H[z] = 1/z$), the amplitude response $|H[e^{j\Omega}]| = 1$, and the phase response $\angle H[e^{j\Omega}] = -\Omega$. Thus a pure time-delay does not affect the amplitude gain of sinusoidal input, but it causes a phase shift (delay) of Ω radians in a discrete sinusoid of frequency Ω. This means that the phase shift is proportional to the frequency of the input sinusoid (linear phase shift). $\triangledown$

5.5-1 The Periodic Nature of the Frequency Response

In Example 5.9 and Fig. 5.9, we saw that the frequency response $H[e^{j\Omega}]$ is a periodic function of Ω. This is not a coincidence. Unlike continuous-time systems, the frequency response of all LTID systems is periodic. This is seen clearly from the nature of the expression of the frequency response of an LTID system. Because $e^{j2\pi m} = 1$ for all integral values of m [see Eq. (B.12)],

$$H[e^{j\Omega}] = H\left[e^{j(\Omega + 2\pi m)}\right] \qquad m \text{ integer} \tag{5.59}$$

Therefore the frequency response $H[e^{j\Omega}]$ is a periodic function of Ω with a period 2π. This is the mathematical explanation of the periodic behavior. The physical explanation given below provides a much better insight into the periodic behavior.

Nonuniqueness of Discrete-Time Sinusoids and the Baseband

Until now the development of discrete-time systems has followed almost a parallel course to that of the continuous-time systems. The periodicity of the frequency response of discrete-time systems is a major departure, the reason for which lies in the peculiar behavior of a discrete-time sinusoids (or exponentials). For a continuous-time sinusoid $\cos \omega t$, there is a unique waveform for every value of ω in the range 0 to ∞. Increasing ω results in a sinusoid of ever increasing frequency. Such is not the case for discrete-time sinusoid $\cos \Omega k$ because

$$\cos\left[(\Omega + 2\pi m)k\right] = \cos\left(\Omega k + 2\pi mk\right) = \cos \Omega k \qquad m \text{ integer} \qquad (5.60)$$

This shows that the discrete-time sinusoids $\cos \Omega k$ separated by values of Ω in integral multiples of 2π are identical. The physical reason for the periodic behavior of an LTID system is now clear. Since the sinusoids (or exponentials) separated by values of Ω of 2π are identical, the system response to such sinusoids (or exponentials) is also identical. This discussion shows that the discrete-time sinusoid $\cos \Omega k$ has a unique waveform only for the values of Ω in the range 0 to 2π. This range is further restricted by the fact that

$$\cos\left(\pi \pm \Omega_x\right)k = \cos \Omega_x k \cos \pi k \mp \sin \Omega_x k \sin \pi k$$

$$= \cos \Omega_x k \cos \pi k \qquad (5.61a)$$

Therefore

$$\cos\left(\pi + \Omega_x\right)k = \cos\left(\pi - \Omega_x\right)k \qquad (5.61b)$$

This shows that a sinusoid of frequency $\pi + \Omega_x$ has the same waveform as one with frequency $\pi - \Omega_x$. Therefore a sinusoid with frequency in the range π to 2π is identical to some sinusoid with frequency in the range 0 to π. Consequently a sinusoid with any value of Ω outside the range 0 to π is identical to a sinusoids with Ω in the range 0 to π. Every sinusoid with frequency in the range $(0, \pi)$ has a distinct waveform. This frequency range in which every sinusoid has a distinct waveform is called the *baseband* .

5.5-2 Aliasing and Sampling Rate

On the surface, the periodic repetition of the waveforms of discrete-time sinusoids (or exponentials) may appear innocuous, but in reality it creates a serious problem for processing continuous-time signals by digital filters. A continuous-time sinusoid $\cos \omega t$ sampled every T seconds ($t = kT$) results in a discrete-time sinusoid $\cos \omega kT$, which is $\cos \Omega k$ with $\Omega = \omega T$. Recall that the discrete-time sinusoids $\cos \Omega k$ have unique waveforms only for the values of frequencies in the range $\Omega < \pi$ or $\omega T < \pi$ (baseband). Therefore samples of continuous-time sinusoids of two (or more) different frequencies can generate the same discrete-time signal, as shown in Fig. 5.11. *This phenomenon is known as* **aliasing** *because through sampling, two entirely different analog sinusoids take on the same "discrete-time" identity.*

Aliasing causes ambiguity in digital signal processing, which makes it impossible to determine the true frequency of the sampled signal. Therefore aliasing is highly undesirable and should be avoided. To avoid aliasing, the frequencies of the continuous-time sinusoids to be processed should be kept within the baseband $\omega T \leq \pi$ or $\omega \leq \pi/T$. Under this condition the question of ambiguity or aliasing does not arise because any continuous-time sinusoid of frequency in this range has

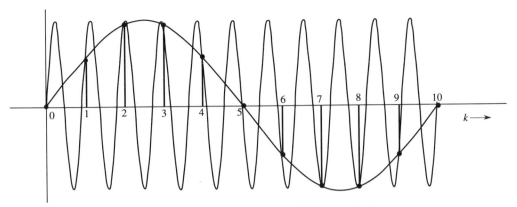

Fig. 5.11   Demonstration of aliasing effect.

a unique waveform when it is sampled. Therefore if ω_h is the highest frequency to be processed, then, to avoid aliasing,

$$\omega_h < \frac{\pi}{T} \qquad (5.62)$$

If $\mathcal{F}_h$ is the highest frequency in Hertz, $\mathcal{F}_h = \omega_h/2\pi$, and, according to Eq. (5.62),

$$\mathcal{F}_h < \frac{1}{2T} \qquad (5.63a)$$

or

$$T < \frac{1}{2\mathcal{F}_h} \qquad (5.63b)$$

This shows that discrete-time signal processing places the limit on the highest frequency $\mathcal{F}_h$ that can be processed for a given value of the sampling interval T according to Eq. (5.63a). But we can process a signal of any frequency (without aliasing) by choosing a suitable value of T according to Eq. (5.63b). The sampling rate or sampling frequency $\mathcal{F}_s$ is the reciprocal of the sampling interval T, and, according to Eq. (5.63b),

$$\mathcal{F}_s = \frac{1}{T} > 2\mathcal{F}_h \qquad (5.64)$$

This result, which is a special case of the well-known **sampling theorem** (to be proved rigorously in Chapter 8), states that to process a continuous-time sinusoid by a discrete-time system, the sampling rate must not be less than twice the frequency (in Hz) of the sinusoid. In short, **a sampled sinusoid must have a minimum of two samples per cycle.** For sampling rate below this minimum value, the output signal will be aliased, which means it will be mistaken for a sinusoid of lower frequency.

■ **Example 5.10**

Determine the maximum sampling interval T that can be used in a discrete-time oscillator which generates a sinusoid of 50kHz.

Here the highest significant frequency $\mathcal{F}_h = 50$ kHz. Therefore from Eq. (5.63b)

$$T < \frac{1}{2\mathcal{F}_h} = 10\,\mu\text{s}$$

The sampling interval must be less than $10\,\mu\text{s}$. The minimum sampling frequency is $\mathcal{F}_s = \frac{1}{T} = 100\,\text{kHz}$. ■

■ **Example 5.11**

A discrete-time amplifier uses a sampling interval $T = 25\,\mu\text{s}$. What is the highest frequency of a signal that can be processed with this amplifier without aliasing?

From Eq. (5.63a)

$$\mathcal{F}_h < \frac{1}{2T} = 20\ \text{kHz}\quad ■$$

5.5-3 Dependence of Frequency Response on Poles and Zeros of H[z]

The frequency response (amplitude and phase response) of a system are determined by pole-zero locations of the transfer function $H[z]$. Just as in continuous-time systems, it is possible to determine quickly the amplitude and the phase response and to obtain physical insight into the filter characteristics of a discrete-time system by using graphical technique. Consider the transfer function

$$H[z] = b_n \frac{(z - z_1)(z - z_2)\cdots(z - z_n)}{(z - \gamma_1)(z - \gamma_2)\cdots(z - \gamma_n)} \tag{5.65}$$

We can compute $H[z]$ graphically using the concepts discussed in Sec. 4.7-1. The directed line segment from z_i to z in the complex plane (Fig. 5.12a) represents the complex number $z - z_i$. The length of this segment is $|z - z_i|$ and its angle with the horizontal axis is $\angle(z - z_i)$.

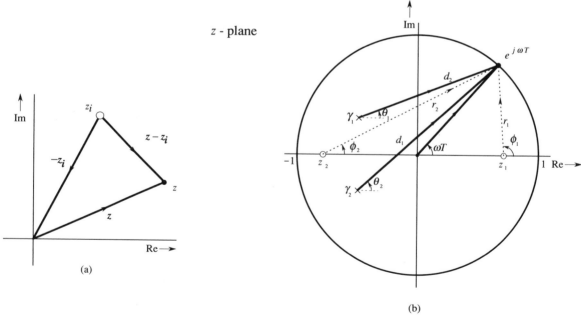

Fig. 5.12 (a) vector representation of complex numbers (b) vector representation of factors of $H[z]$.

In filtering applications, the inputs are often the sampled continuous-time sinusoids. On p. 381 we showed that a sampled continuous-time sinusoid $\cos \omega t$ appears as a discrete-time sinusoid $\cos \Omega k$ $(\Omega = \omega T)$. The appropriate function for computing the frequency response in such a situation, therefore, is $H[e^{j\omega T}]$ $(\Omega = \omega T)$. To compute the frequency response $H[e^{j\omega T}]$ we evaluate $H[z]$ at $z = e^{j\omega T}$. Because for a point $z = e^{j\omega T}$, $|z| = 1$ and $\angle z = \omega T$ so that $z = e^{j\omega T}$ represents a point on the unit circle at an angle ωT with the horizontal. We now connect all zeros (z_1, z_2, ..., z_n) and all poles (γ_1, γ_2, ..., γ_n) to the point $e^{j\omega T}$, as shown in Fig. 5.12b. Let r_1, r_2, ..., r_n be the lengths and ϕ_1, ϕ_2, ..., ϕ_n be the angles, respectively, of the straight lines connecting z_1, z_2, ..., z_n to the point $e^{j\omega T}$. Similarly, let d_1, d_2, ..., d_n be the lengths and θ_1, θ_2, ..., θ_n be the angles, respectively, of the lines connecting γ_1, γ_2, ..., γ_n to $e^{j\omega T}$. Then

$$H[e^{j\omega T}] = H[z]|_{z=e^{j\omega T}} = b_n \frac{(r_1 e^{j\phi_1})(r_2 e^{j\phi_2}) \cdots (r_n e^{j\phi_n})}{(d_1 e^{j\theta_1})(d_2 e^{j\theta_2}) \cdots (d_n e^{j\theta_n})} \tag{5.66}$$

$$= b_n \frac{r_1 r_2 \cdots r_n}{d_1 d_2 \cdots d_n} e^{j[(\phi_1 + \phi_2 + \cdots + \phi_n) - (\theta_1 + \theta_2 + \cdots + \theta_n)]} \tag{5.67}$$

Therefore

$$|H[e^{j\omega T}]| = b_n \frac{r_1 r_2 \cdots r_n}{d_1 d_2 \cdots d_n}$$

$$= b_n \frac{\text{product of the distances of zeros to } e^{j\omega T}}{\text{product of distances of poles to } e^{j\omega T}} \tag{5.68a}$$

and

$$\angle H[e^{j\omega T}] = (\phi_1 + \phi_2 + \cdots \phi_n) - (\theta_1 + \theta_2 + \cdots + \theta_n)$$

$$= \text{sum of zero angles to } e^{j\omega T} - \text{sum of pole angles to } e^{j\omega T} \tag{5.68b}$$

In this manner we can compute the frequency response $H[e^{j\omega T}]$ for any value of ω by selecting the point on the unit circle at an angle ωT corresponding to that value of ω. This point is $e^{j\omega T}$. To compute the frequency response $H[e^{j\omega T}]$ we connect all poles and zeros to this point, and determine $|H[e^{j\omega T}]|$ and $\angle H[e^{j\omega T}]$ using the above equations. We repeat this procedure for all values of ωT from 0 to π to obtain the frequency response.

Controlling Gain by placement of Poles and Zeros

The nature of the influence of pole and zero locations on the frequency response is similar to that observed in continuous-time systems with a minor difference. In place of the imaginary axis of the continuous-time systems, we have a unit circle in the discrete-time case. The nearer the pole (or zero) is to a point $e^{j\omega T}$ (on the unit circle) representing some frequency ω, the more influence that pole (or zero) wields on the amplitude response at that frequency because the length of the vector joining that pole (or zero) to the point $e^{j\omega T}$ is small. This is also true of the phase response. From Eq. (5.68a) it is clear that to enhance the amplitude response at a frequency ω, we should place a pole as close as possible to the point $e^{j\omega T}$ (on the unit circle) representing that frequency ω. Similarly, to suppress the amplitude response at a frequency ω, we should place a zero as close as possible to the point $e^{j\omega T}$ on unit circle. Placing repeated poles or zeros will further enhance their influence.

Total suppression of signal transmission at any frequency can be achieved by placing a zero on the unit circle corresponding to that frequency. This is the principle of the notch (bandstop) filter.

Placing a pole or a zero at the origin does not influence the amplitude response because the length of the vector connecting the origin to any point on the unit circle is unity. However, a pole (a zero) at the origin generates angle $-\omega T$ (ωT) in $\angle H[e^{j\omega T}]$. The phase spectrum $-\omega T$ is a linear function of frequency and therefore represents a pure time-delay of T seconds (see Example 4.22 or Exercise E5.14). Therefore a pole (a zero) at the origin causes a time delay (time advance) of T seconds in the response. There is no change in the amplitude response.

For a stable system, all the poles must be located inside the unit circle. The zeros may lie anywhere. Also for a physically realizable system, $H[z]$ must be a proper fraction, that is, $n \geq m$. If, to achieve a certain amplitude response, we require $m > n$, we can still make the system realizable by placing a sufficient number of poles at the origin. This will not change the amplitude response but it will increase the time delay of the response.

In general a pole at a point has the opposite effect of a zero at that point. Placing a zero closer to a pole tends to cancel the effect of that pole on the frequency response.

Lowpass Filters

A lowpass filter has a maximum gain at $\omega = 0$, which corresponds to point $e^{j0t} = 1$ on the unit circle. Clearly, placing a pole inside the unit circle near point 1 (Fig. 5.13a) would result in a lowpass response. The corresponding amplitude and phase response is also shown in Fig. 5.13a. At smaller values of ω, the point $e^{j\omega T}$ (a point on the unit circle at an angle ωT) is closer to the pole, and consequently the gain is higher. As ω increases, the distance of the point $e^{j\omega T}$ from the pole increases. Consequently the gain decreases, resulting in a lowpass characteristic. Placing a zero at the origin does not change the amplitude response but it does modify the phase response, as shown in Fig. 5.13b. Placing a zero at $z = -1$, however, changes both the amplitude and phase response (Fig. 5.13c). The point $z = -1$ corresponds to frequency $\omega = \pi/T$ ($z = e^{j\omega T} = e^{j\pi} = -1$). Consequently, the amplitude response now becomes more attenuated at higher frequencies, with a zero gain at $\omega T = \pi$. We can approach ideal lowpass characteristics by using more poles staggered near $z = 1$ (but within the unit circle). Figure 5.13d shows a third-order lowpass filter with three poles near $z = 1$ and a third-order zero at $z = -1$, with corresponding amplitude and phase response. For an ideal lowpass filter we need an enhanced gain at every frequency in the band $(0, \omega_c)$. This can be achieved by placing a continuous wall of poles (requiring an infinite number of poles) opposite this band.

Highpass Filters

A highpass filter has a small gain at lower frequencies and a high gain at higher frequencies. Such a characteristic can be realized by placing a pole or poles near $z = -1$ because we want the gain at $\omega T = \pi$ to be the highest. Placing a zero at $z = 1$ further enhances suppression of gain at lower frequencies. Figure 5.13e shows a possible pole-zero configuration of the third-order highpass filter with corresponding amplitude and phase response.

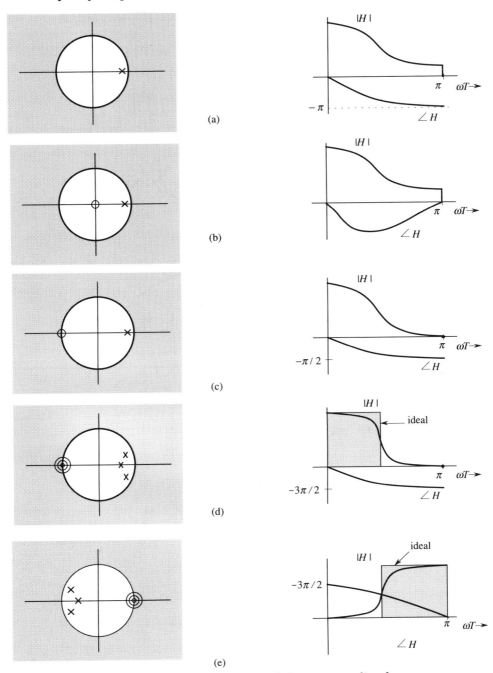

Fig. 5.13 Various pole-zero configurations and the corresponding frequency response.

■ **Example 5.12:   Bandpass Filter**

Using trial-and-error, design a tuned (bandpass) filter with zero transmission at $\omega = 0$ and $\omega = 1000\pi$. The resonant frequency is required to be 250π. The highest frequency to be processed is $\mathcal{F}_h = 400$ Hz.

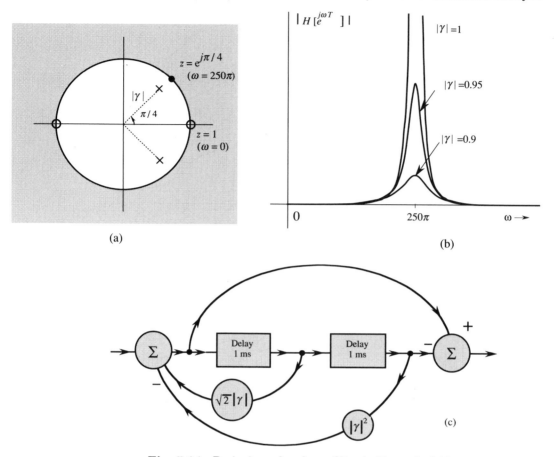

Fig. 5.14   Designing a bandpass filter in Example 5.12.

Because $\mathcal{F}_h = 400$, we require $T < \frac{1}{800}$ [see Eq. (5.63)]. Let us select $T = 10^{-3}$. Since the amplitude response is zero at $\omega = 0$ and $\omega = 1000\pi$, we need to place zeros at $e^{j\omega T}$ corresponding to $\omega = 0$ and $\omega = 1000\pi$. For $\omega = 0, z = e^{j\omega T} = 1$; for $\omega = 1000\pi$ (with $T = 10^{-3}$), $e^{j\omega T} = -1$. Hence there must be zeros at $z = \pm 1$. Moreover we need enhanced frequency response at $\omega = 250\pi$. This frequency (with $\omega T = \pi/4$) corresponds to $z = e^{j\omega T} = e^{j\pi/4}$. Therefore, to enhance the frequency response at this frequency, we place a pole in its vicinity (near $e^{j\pi/4}$). Because this is a complex pole we also need its conjugate near $e^{-j\pi/4}$, as shown in Fig. 5.14a. Let us choose these poles λ_1 and λ_2 as

$$\lambda_1 = |\gamma|e^{j\pi/4} \quad \text{and} \quad \lambda_2 = |\gamma|e^{j\pi/4}$$

where $|\gamma| < 1$ for stability. The closer the value of $|\gamma|$ is to the unit circle, the more sharply peaked is the response around $\omega = 250\pi$. We also have a zeros at ± 1. Hence

$$H[z] = K\frac{(z-1)(z+1)}{(z - |\gamma|e^{j\pi/4})(z - |\gamma|e^{-j\pi/4})} = K\frac{z^2 - 1}{z^2 - \sqrt{2}|\gamma|z + |\gamma|^2} \tag{5.69}$$

For convenience we shall choose $K = 1$. The amplitude response is given by

$$\left|H[e^{j\omega T}]\right| = \frac{|e^{j2\omega T} - 1|}{|e^{j\omega T} - |\gamma|e^{j\pi/4}| \, |e^{j\omega T} - |\gamma|e^{-j\pi/4}|}$$

Now, using Eq.(5.58), we obtain

$$\left|H[e^{j\omega T}]\right|^2 = \frac{2(1 - \cos 2\omega T)}{\left[1 + |\gamma|^2 - 2|\gamma| \cos\left(\omega T - \frac{\pi}{4}\right)\right]\left[1 + |\gamma|^2 - 2|\gamma| \cos\left(\omega T + \frac{\pi}{4}\right)\right]} \tag{5.70}$$

Figure 5.14b shows the amplitude response for values of $|\gamma| = 0.9$, 0.95, and 1. As expected, the gain is zero at $\omega = 0$ and at $\omega = 1000\pi$. The gain peaks at about $\omega = 250\pi$. The resonance (peaking) becomes pronounced as $|\gamma|$ approaches 1. Fig. 5.14c shows a canonical realization of this filter [see Eq. (5.69)]. ■

■ **Example 5.13:   Notch (Bandstop) Filter**
    Design a second-order notch filter to have zero transmission at 250 Hz and a sharp recovery of gain to unity on both sides of 250 Hz. The highest significant frequency to be processed is $\mathcal{F}_h = 500$ Hz.
    In this case $T \leq 1/2\mathcal{F}_h = 10^{-3}$. Let us choose $T = 10^{-3}$. For the frequency 250 Hz, $\omega T = 2\pi(250)T = \pi/2$. This corresponds to the point $e^{j\omega T} = e^{j\pi/2} = j$ on the unit circle, as shown in Fig. 5.15a. Since we need zero transmission at this frequency, we need to place a zero at $z = e^{j\pi/2} = j$ and its conjugate at $z = e^{-j\pi/2} = -j$. We also require a sharp recovery of gain on both sides of frequency 250 Hz. This is achieved by placing two poles close to the two zeros in order to cancel out the effect of the zeros as we move away from the point j (corresponding to frequency 250 Hz). For this reason let us use poles at $\pm ja$ with $a < 1$ for stability. The closer the poles are to the zeros (the closer the a to 1), the faster is the gain recovery on either sides of 250 Hz. The resulting transfer function is

$$H[z] = K\frac{(z - j)(z + j)}{(z - ja)(z + ja)} = K\frac{z^2 + 1}{z^2 + a^2}$$

The dc gain (gain at $\omega = 0$, or $z = 1$) of this filter is

$$H[1] = K\frac{2}{1 + a^2}$$

Because we require a dc gain of unity, we must select $K = \frac{1+a^2}{2}$. The transfer function is therefore

$$H[z] = \frac{(1 + a^2)(z^2 + 1)}{2(z^2 + a^2)} \tag{5.71}$$

and from Eq. (5.58)

$$\left|H[e^{j\omega T}]\right|^2 = \frac{(1 + a^2)^2}{4}\frac{(e^{j2\omega T} + 1)(e^{-j2\omega T} + 1)}{(e^{j2\omega T} + a^2)(e^{-j2\omega T} + a^2)} = \frac{(1 + a^2)^2(1 + \cos 2\omega T)}{2(1 + a^4 + 2a^2 \cos 2\omega T)}$$

Figure 5.15b shows $|H[e^{j\omega T}]|$ for values of $a = 0.3$, 0.6, and 0.95. Figure 5.15c shows a realization of this filter. ■

⊙ **Computer Example C5.2**
    Using MATLAB, we can find the frequency response of a given transfer function either from its numerator and denominator polynomial coefficients or from its pole-zero locations. We demonstrate a procedure by plotting the frequency response of a second-order digital lowpass Butterworth filter, whose transfer function is given by

$$H[z] = \frac{0.333(z + 1)^2}{z^2 + 0.1567z + 0.1759} = \frac{0.333(z + 1)^2}{(z + 0.0783z + j0.412)(z + 0.0783z - j0.412)}$$

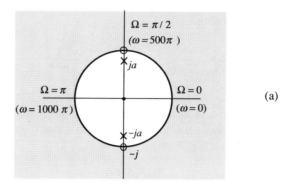

(a)

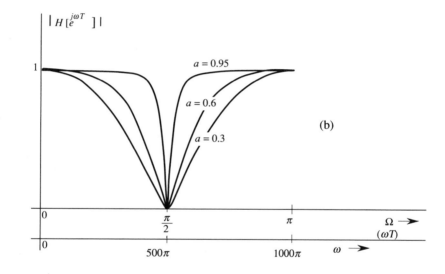

(b)

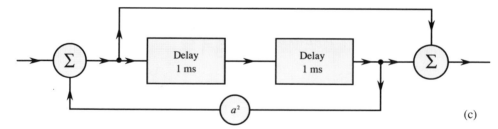

(c)

Fig. 5.15   Designing a notch (bandstop) filter in Example 5.13.

In part (a), we use the numerator and the denominator polynomial coefficients to compute the frequency response. In part (b), the same frequency response is computed by using the pole-zero locations of the transfer functions. In this part we use the fact that $(z+a)(z+b) = z^2 + (a+b)z + ab$, and the coefficients 1, $(a+b)$, ab can be obtained by convolution of the sequences $(1, a)$ and $(1, b)$. Thus from the pole-zero locations, we can determine the coefficients of the numerator and denominator polynomials by a simple command of convolution for a second-order system. Once this is done, we can proceed exactly as in part (a). We shall plot the amplitude response $|H[e^{j\Omega}]|$ and the phase response $\angle H[e^{j\Omega}]$ as functions of Ω over the range $0 \le \Omega \le \pi$.

(a) Frequency response from the numerator and denominator polynomial coefficients:

```
W=0:pi/100:pi;
w=exp(j*W);
num=0.333*[1 2 1];
den=[1 0.1567 0.1759];
H=polyval(num,w)./polyval(den,w);
mag=abs(H);
phase=angle(H)*180/pi;
plot(W,mag),grid,xlabel('W'),ylabel('|H|'),pause
plot(W,phase),grid,xlabel('W'),ylabel('< H'),pause
disp('strike any key for part b'),pause
```

(b) Frequency response from the pole-zero locations:

```
W=0:pi/100:pi;
w=exp(j*W);
num=0.333*conv([1 1],[1 1]);
den=conv([1 .0783+.412*i],[1 .0783-.412*i]);
H=polyval(num,w)./polyval(den,w);
mag=abs(H);
phase=angle(H)*180/pi;
plot(W,mag),grid,xlabel('W'),ylabel('|H|'),pause
plot(W,phase),grid,xlabel('W'),ylabel('< H')   ⊙
```

⊙ **Computer Example C5.3**

Plot the amplitude and phase response for the bandpass filter in Example 5.12 for $|\gamma| = 0.95$ in two ways: (a) using the numerator and the denominator polynomial coefficients of the transfer function and (b) using the pole-zero locations of the transfer function.

(a) Frequency response from the numerator and the denominator polynomial coefficients:

```
gamma=0.95;
w=0:10*pi:1000*pi;
z=exp(j*0.001*w);
num=[1 0 -1];
den=[1 -sqrt(2)*gamma gamma^2];
H=polyval(num,z)./polyval(den,z);
mag=abs(H);
phase=angle(H)*180/pi;
plot(w,mag),grid,xlabel('w'),ylabel('|H|'),pause
plot(w,phase),grid,xlabel('w'),ylabel('< H'),pause
disp('strike any key for part b'),pause
```

(b) Frequency response from the pole-zero locations:

```
gamma=0.95;
w=0:10*pi:1000*pi;
z=exp(j*0.001*w);
num=conv([1 -1],[1 1]);
den=conv([1 -gamma*exp(j*pi/4)],[1 -gamma*exp(-j*pi/4)]);
rz=roots(num),rp=roots(den),
H=polyval(num,z)./polyval(den,z);
mag=abs(H);
phase=angle(H)*180/pi;
plot(w,mag),grid,xlabel('w'),ylabel('|H|'),pause
plot(w,phase),grid,xlabel('w'),ylabel('< H')   ⊙
```

⊙ **Computer Example C5.4**
Find the frequency response of the notch filter in Example 5.13 for the value of
$a = 0.6$.

```
a=0.6;
w=0:10*pi:1000*pi;
z=exp(j*0.001*w);
num=[.5*(1+a^2) 0 0.5*(1+a^2)];
den=[1 0 a^2];
H=polyval(num,z)./polyval(den,z);
mag=abs(H);
phase=angle(H);
plot(w,mag),grid,xlabel('w'),ylabel('|H|'),
title('Amplitude Response for a=0.6'),pause,
plot(w,phase),grid,xlabel('w'),ylabel('< H'),
title('Phase Response for a=0.6')  ⊙
```

⊙ **Computer Example C5.5**
In this example we compute the frequency response of the following transfer function
with a variable parameter G.

$$H[z] = \frac{z^2 + 0.8z + 0.2}{z^2 + 0.09z + 0.6G}$$

where G is variable taking on values 1,3, and 5.

```
W=0:pi/100:pi;
w=exp(j*W);
for G=1:2:5
num=[1 .8 .2];   den=[1 .09 .6*G];
zeros=roots(num);  poles=roots(den);
H=polyval(num,w)./polyval(den,w);
mag=abs(H);       phase=angle(H);
plot(W,mag),grid,xlabel('W'),ylabel('|H|'),
title('Amplitude Response'),pause
plot(W,phase),grid,xlabel('W'),ylabel('< H'),
title('Phase Response'),pause
end  ⊙
```

△ **Exercise E5.15**
Using a graphical argument, show that a filter with transfer function

$$H[z] = \frac{z - 0.9}{z}$$

acts as a highpass filter. Make a rough sketch of the amplitude and phase response. ▽

5.6 THE CONNECTION BETWEEN THE $\mathcal{Z}$-TRANSFORM AND THE LAPLACE TRANSFORM

We now show that discrete-time systems also can be analyzed using Laplace
transform. In fact, we shall see that **the z-transform is the Laplace transform in disguise** and that discrete-time systems can be analyzed as if they were
continuous-time systems.

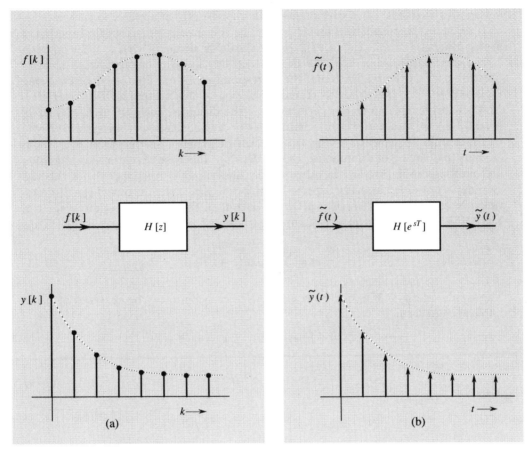

Fig. 5.16 Connection between the Laplace and z-transform.

So far we have considered the discrete-time signal as a sequence of numbers and not as an electrical signal (voltage or current). Similarly, we considered a discrete-time system as a mechanism that processes a sequence of numbers (input) to yield another sequence of numbers (output). The system was built by using delays (along with adders and multipliers) that delay sequences of numbers, not electrical signals (voltages or currents). A digital computer is a perfect example: every signal is a sequence of numbers, and the processing involves delaying sequences of numbers (along with addition and multiplication).

Now suppose we have a discrete-time system with transfer function $H[z]$ and input $f[k]$. We can imagine (or generate, for that matter) a corresponding continuous-time signal $\tilde{f}(t)$ consisting of impulses spaced T seconds apart with the kth impulse of strength $f[k]$. Thus

$$\tilde{f}(t) = \sum_{k=0}^{\infty} f[k]\delta(t - kT) \tag{5.72}$$

Figure 5.16 shows $f[k]$ and corresponding $\tilde{f}(t)$. We also imagine a system identical in structure to the discrete-time system with transfer function $H[z]$ except that the delays in $H[z]$ are replaced by elements that delay continuous-time signals (such as voltages or currents). If a continuous-time impulse $\delta(t)$ is applied to such a delay of T seconds, the output will be $\delta(t-T)$. The continuous-time transfer function of such a delay is e^{-sT} [see Eq. (4.42)]. Hence the delay elements with transfer function $1/z$ in the realization of $H[z]$ will be replaced by the delay elements with transfer function e^{-sT} in the realization of the corresponding $\tilde{H}(s)$. This is same as z being replaced by e^{sT}. Therefore the transfer function of this system is $\tilde{H}(s) = H[e^{sT}]$. Now whatever operations are performed by the discrete-time system $H[z]$ on $f[k]$ (Fig. 5.16a) are also performed by the corresponding continuous-time system $H[e^{sT}]$ on the impulse sequence $\tilde{f}(t)$ (Fig. 5.16b). The delaying of a sequence in $H[z]$ would amount to delaying of an impulse train in $H[e^{sT}]$. The case of operations of adding and multiplying is similar. In other words, one-to-one correspondence of the two systems is preserved in every aspect. Therefore if $y[k]$ is the output of the discrete-time system in Fig. 5.16a, then $\tilde{y}(t)$, the output of the continuous-time system in Fig. 5.16b, would be a sequence of impulse whose kth impulse strength is $y[k]$. Thus

$$\tilde{y}(t) = \sum_{k=0}^{\infty} y[k]\delta(t - kT) \tag{5.73}$$

The system in Fig. 5.16b, being a continuous-time system, can be analyzed by using Laplace transform. If

$$\tilde{f}(t) \Longleftrightarrow \tilde{F}(s) \quad \text{and} \quad \tilde{y}(t) \Longleftrightarrow \tilde{Y}(s)$$

then

$$\tilde{Y}(s) = H[e^{sT}]\tilde{F}(s) \tag{5.74}$$

Also

$$\tilde{F}(s) = \mathcal{L}\left[\sum_{k=0}^{\infty} f[k]\delta(t - kT)\right]$$

Now because the Laplace transform of $\delta(t - kT)$ is e^{-skT}

$$\tilde{F}(s) = \sum_{k=0}^{\infty} f[k]e^{-skT} \tag{5.75}$$

and

$$\tilde{Y}(s) = \sum_{k=0}^{\infty} y[k]e^{-skT} \tag{5.76}$$

Substitution of Eqs. (5.75) and (5.76) in Eq. (5.74) yields

$$\sum_{k=0}^{\infty} y[k]e^{-skT} = H[e^{sT}]\left[\sum_{k=0}^{\infty} f[k]e^{-skT}\right]$$

By introducing a new variable $z = e^{sT}$, this equation can be expressed as

$$\sum_{k=0}^{\infty} y[k]z^{-k} = H[z]\sum_{k=0}^{\infty} f[k]z^{-k}$$

or

$$Y[z] = H[z]F[z]$$

where

$$F[z] = \sum_{k=0}^{\infty} f[k]z^{-k} \quad \text{and} \quad Y[z] = \sum_{k=0}^{\infty} y[k]z^{-k}$$

It is clear from this discussion that the z-transform can be considered as a Laplace transform with a change of variable $z = e^{sT}$ or $s = (1/T)\ln z$. On the other hand, we may consider the z-transform as an independent transform in its own right. Note that the transformation $z = e^{sT}$ transforms the imaginary axis in the s-plane ($s = j\omega$) into a unit circle in the z-plane ($z = e^{sT} = e^{j\omega T}$, or $|z| = 1$). The LHP and RHP in the s-plane map into the inside and the outside, respectively, of the unit circle in the z-plane.

5.7 BILATERAL $\mathcal{Z}$-TRANSFORM

Situations involving noncausal signals or systems cannot be handled by the (unilateral) z-transform discussed so far. These cases, however, can be analyzed by the **bilateral** (or two-sided) z-transform defined by

$$F[z] = \sum_{k=-\infty}^{\infty} f[k]z^{-k} \tag{5.77}$$

The inverse z-transform is given by

$$f[k] = \frac{1}{2\pi j} \oint F[z]z^{k-1}\, dz \tag{5.78}$$

These equations define the bilateral z-transform. The unilateral z-transform discussed so far is a special case, where the input signals are restricted to be causal. Restricting signals in this way results in considerable simplification in the region of convergence. Earlier we showed that

$$\gamma^k u[k] \Longleftrightarrow \frac{z}{z-\gamma} \qquad |z| > |\gamma| \tag{5.79}$$

Now the z-transform of the signal $-\gamma^k u[-(k+1)]$ shown in Fig. 5.17a is

$$\mathcal{Z}\left\{-\gamma^k u[-(k+1)]\right\} = \sum_{-\infty}^{-1} -\gamma^k z^{-k} = \sum_{-\infty}^{-1} -\left(\frac{\gamma}{z}\right)^k$$

$$= -\left[\frac{z}{\gamma} + \left(\frac{z}{\gamma}\right)^2 + \left(\frac{z}{\gamma}\right)^3 + \cdots\right]$$

$$= 1 - \left[1 + \frac{z}{\gamma} + \left(\frac{z}{\gamma}\right)^2 + \left(\frac{z}{\gamma}\right)^3 + \cdots\right]$$

$$= 1 - \frac{1}{1 - \frac{z}{\gamma}} \qquad \left|\frac{z}{\gamma}\right| < 1$$

$$= \frac{z}{z-\gamma} \qquad |z| < |\gamma|$$

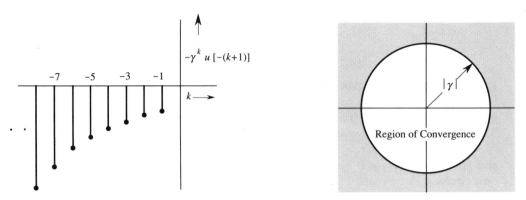

Fig. 5.17 $-\gamma^k u[-(k+1)]$ and the region of convergence of its z-transform.

Therefore

$$\mathcal{Z}\left\{-\gamma^k u[-(k+1)]\right\} = \frac{z}{z-\gamma} \qquad |z| < |\gamma| \tag{5.80}$$

A comparison of Eqs. (5.79) with (5.80) shows that the z-transform of $\gamma^k u[k]$ is identical to that of $-\gamma^k u[-(k+1)]$. The regions of convergence, however, are different. In the former case, $F[z]$ converges for $|z| > |\gamma|$; in the latter, $F[z]$ converges for $|z| < |\gamma|$ (see Fig. 5.17b). Clearly the inverse transform of $F[z]$ is not unique unless the region of convergence is specified. If we restrict all our signals to be causal, however, this ambiguity does not arise. The inverse transform of $z/(z-\gamma)$ is $\gamma^k u[k]$ even without specifying the region of convergence. Thus in unilateral transform we can ignore the region of convergence in determining the inverse z-transform of $F[z]$.

■ **Example 5.14**
Determine the z-transform of

$$f[k] = (0.9)^k u[k] + (1.2)^k u[-(k+1)]$$
$$= f_1[k] + f_2[k]$$

From the results in Eqs. (5.79) and (5.80), we have

$$F_1[z] = \frac{z}{z-0.9} \qquad |z| > 0.9$$

$$F_2[z] = \frac{-z}{z-1.2} \qquad |z| < 1.2$$

The common region where both $F_1[z]$ and $F_2[z]$ converge is $0.9 < |z| < 1.2$ (Fig. 5.18). Hence

$$F[z] = F_1[z] + F_2[z]$$
$$= \frac{z}{z-0.9} - \frac{z}{z-1.2}$$
$$= \frac{-0.3z}{(z-0.9)(z-1.2)} \qquad 0.9 < |z| < 1.2 \tag{5.81}$$

The sequence $f[k]$ and the region of convergence of $F[z]$ are shown in Fig. 5.18.   ■

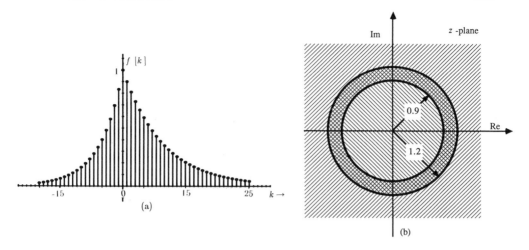

Fig. 5.18   Signal $f[k]$ for Example 5.14.

■ **Example 5.15**

Find the inverse z-transform of

$$F[z] = \frac{-z(z+0.4)}{(z-0.8)(z-2)}$$

if the region of convergence is **(a)** $|z| > 2$ **(b)** $|z| < 0.8$ **(c)** $0.8 < |z| < 2$.

(a)
$$\frac{F[z]}{z} = \frac{-(z+0.4)}{(z-0.8)(z-2)}$$

$$= \frac{1}{z-0.8} - \frac{2}{z-2}$$

and

$$F[z] = \frac{z}{z-0.8} - 2\frac{z}{z-2}$$

Since the region of convergence is $|z| > 2$, both terms correspond to causal sequences and

$$f[k] = \left[(0.8)^k - 2(2)^k\right] u[k]$$

This sequence is shown in Fig. 5.19a.

(b) In this case, $|z| < 0.8$, which is less than the magnitudes of both poles. Hence both terms correspond to anticausal sequences:

$$f[k] = \left[-(0.8)^k + 2(2)^k\right] u\left[-(k+1)\right]$$

This sequence is shown in Fig. 5.19b.

(c) In this case, $0.8 < |z| < 2$; the part of $F[z]$ corresponding to the pole at 0.8 is a causal sequence, and the part corresponding to the pole at 2 is an anticausal sequence:

$$f[k] = (0.8)^k u[k] + 2(2)^k u[-(k+1)]$$

This sequence is shown in Fig. 5.19c.   ■

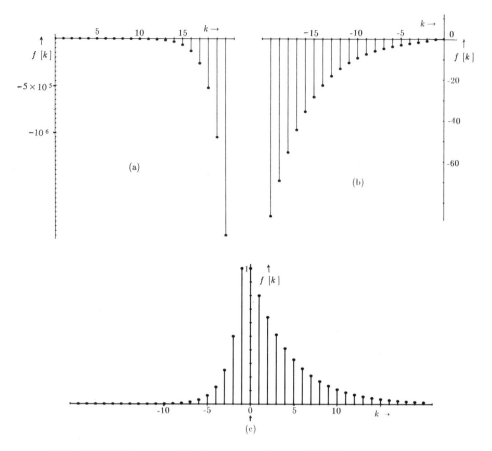

Fig. 5.19 Three possible inverse transforms of $F[z]$ in Example 5.15.

△ **Exercise E5.16**

Find the inverse z-transform of

$$F[z] = \frac{z}{z^2 + \frac{5}{6}z + \frac{1}{6}} \qquad \frac{1}{2} > |z| > \frac{1}{3}$$

Answer: $6(-\frac{1}{3})^k u[k] + 6(-\frac{1}{2})^k u[-(k+1)]$ ▽

5.7-1 Analysis of LTID Systems Using Bilateral $\mathcal{Z}$-Transform

Because bilateral z-transform can handle noncausal signals, we can analyze noncausal linear systems using this transform. The zero-state response $y[k]$ is given by

$$y[k] = \mathcal{Z}^{-1}\left\{F[z]H[z]\right\}$$

provided that $F[z]H[z]$ exists. The region of convergence of $F[z]H[z]$ is the region where both $F[z]$ and $H[z]$ exist, which means that the region is common to the convergence of both $F[z]$ and $H[z]$.

■ **Example 5.16**

For a causal system specified by the transfer function

$$H[z] = \frac{z}{z - 0.5}$$

find the zero-state response to input

$$f[k] = (0.8)^k u[k] + 2(2)^k u[-(k+1)]$$

The z-transform of this signal is found from Example 5.15 (part **c**) as

$$F[z] = \frac{-z(z+0.4)}{(z-0.8)(z-2)} \qquad 0.8 < |z| < 2$$

Therefore

$$Y[z] = F[z]H[z] = \frac{-z^2(z+0.4)}{(z-0.5)(z-0.8)(z-2)}$$

Since the system is causal, the region of convergence of $H[z]$ is $|z| > 0.5$. The region of convergence of $F[z]$ is $0.8 < |z| < 2$. The common region of convergence for $F[z]$ and $H[z]$ is $0.8 < |z| < 2$. Therefore

$$Y[z] = \frac{-z^2(z+0.4)}{(z-0.5)(z-0.8)(z-2)} \qquad 0.8 < |z| < 2$$

Expanding $Y[z]$ into modified partial fractions yields

$$Y[z] = -\frac{z}{z-0.5} + \frac{8}{3}\left(\frac{z}{z-0.8}\right) - \frac{8}{3}\left(\frac{z}{z-2}\right) \qquad 0.8 < |z| < 2$$

The poles at 0.5 and 0.8 are enclosed within the ring of convergence and therefore correspond to the causal part, and the pole at 2 is outside the ring of convergence and corresponds to the anticausal part of $Y[z]$. Therefore

$$y[k] = \left[-(0.5)^k + \tfrac{8}{3}(0.8)^k\right]u[k] + \tfrac{8}{3}(2)^k u[-(k+1)] \qquad ■$$

■ **Example 5.17**

For the system in Example 5.16 find the zero-state response to input

$$f[k] = \underbrace{(0.8)^k u[k]}_{f_1[k]} + \underbrace{(0.6)^k u[-(k+1)]}_{f_2[k]}$$

The z-transforms of the causal and anticausal components $f_1[k]$ and $f_2[k]$ of the output are

$$F_1[z] = \frac{z}{z-0.8} \qquad |z| > 0.8$$

$$F_2[z] = \frac{-z}{z-0.6} \qquad |z| < 0.6$$

Observe that common region of convergence for $F_1[z]$ and $F_2[z]$ does not exist. Therefore $F[z]$ does not exist. In such a case we take advantage of the superposition principle and find $y_1[k]$ and $y_2[k]$, the system response to $f_1[k]$ and $f_2[k]$, separately. The desired response $y[k]$ is the sum of $y_1[k]$ and $y_2[k]$. Now

$$H[z] = \frac{z}{z-0.5} \qquad |z| > 0.5$$

$$Y_1[z] = F_1[z]H[z] = \frac{z^2}{(z-0.5)(z-0.8)} \qquad |z| > 0.8$$

$$Y_2[z] = F_2[z]H[z] = \frac{-z^2}{(z-0.5)(z-0.6)} \qquad 0.5 < |z| < 0.6$$

Expanding $Y_1[z]$ and $Y_2[z]$ into modified partial fractions yields

$$Y_1[z] = -\frac{5}{3}\left(\frac{z}{z-0.5}\right) + \frac{8}{3}\left(\frac{z}{z-0.8}\right) \qquad |z| > 0.8$$

$$Y_2[z] = 5\left(\frac{z}{z-0.5}\right) - 6\left(\frac{z}{z-0.6}\right) \qquad 0.5 < |z| < 0.6$$

Therefore

$$y_1[k] = \left[-\tfrac{5}{3}(0.5)^k + \tfrac{8}{3}(0.8)^k\right]u[k]$$

$$y_2[k] = 5(0.5)^k u[k] + 6(0.6)^k u[-(k+1)]$$

and

$$y[k] = y_1[k] + y_2[k]$$
$$= \left[\tfrac{10}{3}(0.5)^k + \tfrac{8}{3}(0.8)^k\right]u[k] + 6(0.6)^k u[-(k+1)] \qquad \blacksquare$$

△ **Exercise E5.17**

For a causal system in Example 5.16, find the zero-state response to input

$$f[k] = \left(\tfrac{1}{4}\right)^k u[k] + 5(3)^k u[-(k+1)]$$

Answer: $\left[-(\tfrac{1}{4})^k + 3(\tfrac{1}{2})^k\right]u[k] + 6(3)^k u[-(k+1)]$ ▽

5.8 SUMMARY

In this chapter we discuss the analysis of linear, time-invariant, discrete-time (LTID) systems by z-transform, which transforms the difference equations of such systems into algebraic equations. Therefore solving these difference equations reduces to solving algebraic equations.

The transfer function $H[z]$ of an LTID system is defined as the ratio of the z-transform of the output to z-transform of the input when all initial conditions are zero. Therefore if $F[z]$ is the z-transform of the input $f[k]$ and $Y[z]$ is the z-transform of the corresponding output $y[k]$ (when all initial conditions are zero), then $Y[z] = H[z]F[z]$. For a system specified by the difference equation $Q[E]y[k] = P[E]f[k]$, the transfer function $H[z] = P[z]/Q[z]$. Moreover, $H[z]$ is the z-transform of the system impulse response $h[k]$. We also showed that the system response to an everlasting exponential z^k is $H[z]z^k$.

In an alternative and more physical view, the z-transform is a tool by which a signal is expressed as the sum of its spectral components of the form z^k. Also, $H[z]z^k$ is the system response to input z^k. Therefore $H[z]$ is the system response (or gain) to spectral component z^k. The output signal spectrum $Y[z]$ is the input spectrum $F[z]$ times the spectral response (gain) $H[z]$.

LTID systems can be realized by scalar multipliers, summers, and time delays. A given transfer function can be synthesized in many different ways. Canonical, cascade and parallel forms of realization are discussed. The realization procedure is identical to that for continuous-time systems.

Response of an LTID system to sinusoidal inputs is discussed in Sec. 5.5. The transfer function $H[z]$ is directly related to the system response for a sinusoidal input. For an input $\cos \Omega k$ (a sinusoid of unit amplitude and frequency Ω), the system response is also a sinusoid of the same frequency (Ω), with amplitude $|H[e^{j\Omega}]|$, and phase shifted by $\angle H[e^{j\Omega}]$ with respect to the input sinusoid. For this reason $|H[e^{j\Omega}]|$ is called the amplitude response (gain) and $\angle H[e^{j\Omega}]$ is called the phase response of the system. Amplitude and phase response indicate the filtering characteristics of a system. The frequency response of an LTIC system to input $\cos \omega t$ is $H(j\omega)$, whereas the frequency response of an LTID system to input $\cos \Omega k$ is $H[e^{j\Omega}]$ instead of $H[j\Omega]$. This is a major point of departure in the behavior of the two types of systems. This fact also causes the frequency response of an LTID system to be a periodic function of Ω with a period 2π; in general, the frequency response of an LTIC system is not a periodic function of ω. Moreover, continuous-time sinusoid $\cos \omega t$ has a unique waveform for every value of ω. In contrast, discrete-time sinusoid $\cos \Omega k$ has a unique waveform only for values of Ω in the range $(0, \pi)$.† Because of this nonuniqueness of discrete-time sinusoids, two continuous-time sinusoids of entirely different frequencies, when sampled, may yield the same discrete-time sinusoid. This phenomenon is known as *aliasing* because through sampling, two entirely different continuous-time sinusoids take on the same "discrete-time" identity. Aliasing causes ambiguity in digital signal processing, which makes it impossible to determine the true frequency of the sampled signal. To avoid aliasing, a continuous-time sinusoid should be sampled at a rate no less than twice the frequency (in Hz) of the sinusoid; that is, a sampled sinusoid must have a minimum of two samples per cycle. For sampling rates below this minimum value, the sinusoid will be aliased, which means that it will be mistaken for a sinusoid of lower frequency.

In Sec. 5.6, we showed that discrete-time systems can be analyzed by the Laplace transform as if they were continuous-time systems. In fact, we showed that z-transform is Laplace transform with a change in variable.

The majority of the input signals and practical systems are causal. Consequently we are required to deal with causal signals most of the time. When all signals are restricted to be causal, the z-transform analysis is greatly simplified; the region of convergence of a signal becomes irrelevant to the analysis process. This special case of z-transform (which is restricted to causal signals) is called the unilateral z-transform. Much of the chapter deals with this transform. Section 5.7 discusses the general variety of the z-transform (bilateral z-transform), which can handle causal and noncausal signals and systems. In the bilateral transform, the inverse transform of $F[z]$ is not unique, but depends on the region of convergence of $F[z]$. Thus the region of convergence plays a crucial role in the bilateral z-transform.

---

†In fact, in the range $(\Omega_0, \pi + \Omega_0)$ for any Ω_0.

REFERENCES

1.   Juri, E.I., *Theory and Application of the Z-Transform Method*, Wiley, New York, 1964.

PROBLEMS

5.1-1 Using the definition of z-transform, show that

$$\textbf{(a)}\; \gamma^{k-1}u[k-1] \Longleftrightarrow \frac{1}{z-\gamma} \qquad\qquad \textbf{(c)}\; \frac{\gamma^k}{k!}u[k] \Longleftrightarrow e^{\gamma/z}$$

$$\textbf{(b)}\; u[k-m] \Longleftrightarrow \frac{z}{z^m(z-1)} \qquad\qquad \textbf{(d)}\; \frac{(\ln\alpha)^k}{k!}u[k] \Longleftrightarrow \alpha^{1/z}$$

Hint: For parts **(c)** and **(d)** recall that $e^x = \sum_{n=0}^{n=\infty} \frac{x^n}{n!}$. Also $\alpha = e^{\ln\alpha}$

5.1-2 Using only z-transform Table 5.1, show that

(a) $2^{k+1}u[k-1] + e^{k-1}u[k] \Longleftrightarrow \frac{4}{z-2} + \frac{z}{e(z-e)}$
Hint: $2^{k+1} = (4)2^{k-1}$ and $e^{k-1} = (\frac{1}{e})e^k$

(b) $k\gamma^k u[k-1] \Longleftrightarrow \frac{\gamma z}{(z-\gamma)^2}$
Hint: Recognize that $u[k-1] = u[k] - \delta[k]$, and $f[k]\delta[k] = f[0]\delta[k]$. Alternatively recognize that $k\gamma^k u[k-1] = (k-1+1)\gamma^{k-1+1}u[k-1]$.

(c) $\left[2^{-k}\cos\left(\frac{\pi}{3}k\right)\right]u[k-1] \Longleftrightarrow \frac{0.25(z-1)}{z^2-0.5z+0.25}$
Hint: See the hint for part **b**.

(d) $k(k-1)(k-2)2^{k-3}u[k-m] \Longleftrightarrow \frac{6z}{(z-2)^4}$ for m=0, 1, 2, or 3.
Hint: Show that $k(k-1)(k-2)u[k-m] = k(k-1)(k-2)u[k]$ for $m = 0, 1, 2,$ or 3.
This can be done by checking the values of both sides for $k = 0, 1, 2, 3, 4, \cdots$.

5.1-3 Find the inverse z-transform of

$$\textbf{(a)}\; \frac{z(z-4)}{z^2-5z+6} \qquad\qquad \textbf{(g)}\; \frac{z(z-2)}{z^2-z+1}$$

$$\textbf{(b)}\; \frac{z-4}{z^2-5z+6} \qquad\qquad \textbf{(h)}\; \frac{2z^2-0.3z+0.25}{z^2+0.6z+0.25}$$

$$\textbf{(c)}\; \frac{(e^{-2}-2)z}{(z-e^{-2})(z-2)} \qquad\qquad \textbf{(i)}\; \frac{2z(3z-23)}{(z-1)(z^2-6z+25)}$$

$$\textbf{(d)}\; \frac{z(2z+3)}{(z-1)(z^2-5z+6)} \qquad\qquad \textbf{(j)}\; \frac{z(3.83z+11.34)}{(z-2)(z^2-5z+25)}$$

$$\textbf{(e)}\; \frac{z(-5z+22)}{(z+1)(z-2)^2} \qquad\qquad \textbf{(k)}\; \frac{z^2(-2z^2+8z-7)}{(z-1)(z-2)^3}$$

$$\textbf{(f)}\; \frac{z(1.4z+0.08)}{(z-0.2)(z-0.8)^2}$$

5.1-4 Find the first three terms of $f[k]$ if

$$F[z] = \frac{2z^3+13z^2+z}{z^3+7z^2+2z+1}$$

Find your answer by expanding $F[z]$ as a power series in z^{-1}.

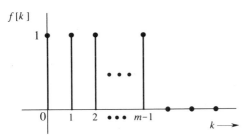

Fig. P5.2-1

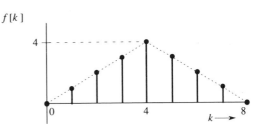

Fig. P5.2-2

5.1-5 By expanding

$$F[z] = \frac{\gamma z}{(z - \gamma)^2}$$

as a power series in z^{-1}, show that $f[k] = k\gamma^k u[k]$.

5.2-1 For a discrete-time signal shown in Fig. P5.2-1 show that

$$F[z] = \frac{1 - z^{-m}}{1 - z^{-1}}$$

Hint: $f[k] = u[k] - u[k - m]$

5.2-2 Find the z-transform of the signal shown in Fig. P5.2-2. Solve this problem in two ways, as in Examples 5.2d and 5.4. The two answers may appear different, but they are equivalent.

5.2-3 Using only the fact that $\gamma^k u[k] \Longleftrightarrow \frac{z}{z-\gamma}$ and properties of the z-transform, find the z-transform of

        (a) $k^2 \gamma^k u[k]$ **(c)** $a^k [u[k] - u[k - m]]$

        (b) $k^3 u[k]$ **(d)** $ke^{-2k} u[k - m]$

Hint for parts **(a)** and **(b)** apply Eq. (5.25) (repeatedly if necessarily) to $\gamma^k u[k] \Longleftrightarrow \frac{z}{z-\gamma}$. Also $a^k = a^{k-m+m} = a^m a^{k-m}$ and $ke^{-2k} = (k - m + m)(e^{-2})^{k-m+m}$.

5.2-4 Using only Pair 1 in Table 5.1 and appropriate properties of the z-transform, derive iteratively pairs 2 through 9. In other words, first derive Pair 2. Then use Pair 2 (and Pair 1, if needed) to derive Pair 3, and so on.

5.3-1 Solve Prob. 3.5-9 by the z-transform method.
Hint: A loan of M dollars is an initial deposit of $-M$ dollars. Solve Eq. (3.2c) subject to the initial condition $y[0] = -M$ and the input $f[k] = Pu[k - 1]$. If the loan is paid off in N payments, $y[N] = 0$.

5.3-2 Solve

$$y[k + 1] + 2y[k] = f[k + 1]$$

with $y[0] = 1$ and $f[k] = e^{-(k-1)} u[k]$

5.3-3 Find the output $y[k]$ of an LTID system specified by the equation

$$2y[k + 2] - 3y[k + 1] + y[k] = 4f[k + 2] - 3f[k + 1]$$

if the initial conditions are $y[-1] = 0$, $y[-2] = 1$, and the input $f[k] = (4)^{-k} u[k]$.

5.3-4 Solve Prob. 5.3-3 if instead of initial conditions $y[-1], y[-2]$ you are given the auxiliary conditions $y[0] = \frac{3}{2}$ and $y[1] = \frac{35}{4}$.

5.3-5 Solve

$$4y[k+2] + 4y[k+1] + y[k] = f[k+1]$$

with $y[-1] = 0$, $y[-2] = 1$, and $f[k] = u[k]$.

5.3-6 Solve

$$y[k+2] - 3y[k+1] + 2y[k] = f[k+1]$$

if $y[-1] = 2$, $y[-2] = 3$, and $f[k] = (3)^k u[k]$.

5.3-7 Solve

$$y[k+2] - 2y[k+1] + 2y[k] = f[k]$$

with $y[-1] = 1$, $y[-2] = 0$, and $f[k] = u[k]$.

5.3-8 Solve

$$y[k] + 2y[k-1] + 2y[k-2] = f[k-1] + 2f[k-2]$$

with $y[0] = 0, y[1] = 1$, and $f[k] = e^k$.

5.3-9 **(a)** Find the zero-state response of an LTID system with transfer function

$$H[z] = \frac{z}{(z+0.2)(z-0.8)}$$

and the input $f[k] = e^{(k+1)} u[k]$.

(b) Write the difference equation relating the output $y[k]$ to input $f[k]$.

5.3-10 Repeat Prob. 5.3-9 if $f[k] = u[k]$, and

$$H[z] = \frac{2z+3}{(z-2)(z-3)}$$

5.3-11 Repeat Prob. 5.3-9 if

$$H[z] = \frac{6(5z-1)}{6z^2 - 5z + 1}$$

and the input $f[k]$ is **(a)** $(4)^{-k} u[k]$ **(b)** $(4)^{-(k-2)} u[k-2]$
(c) $(4)^{-(k-2)} u[k]$ **(d)** $(4)^{-k} u[k-2]$.
Hint: $(4)^{-(k-2)} = 16(4)^{-k}$ and $(4)^{-k} u[k-2] = \frac{1}{16}(4)^{-(k-2)} u[k-2]$.

5.3-12 Repeat Prob. 5.3-9 if $f[k] = u[k]$, and

$$H[z] = \frac{2z-1}{z^2 - 1.6z + 0.8}$$

5.3-13 Find the transfer functions corresponding to each of the systems specified by difference equations in Probs. 5.3-2, 5.3-3, and 5.3-5.

5.3-14 Find the transfer functions for each of the systems specified by difference equations in Probs. 5.3-6 and 5.3-7.

5.3-15 Find $h[k]$, the unit impulse response of the systems described by the following equations:

(a) $y[k] + 3y[k-1] + 2y[k-2] = f[k] + 3f[k-1] + 3f[k-2]$
(b) $y[k+2] + 2y[k+1] + y[k] = 2f[k+2] - f[k+1]$
(c) $y[k] - y[k-1] + 0.5y[k-2] = f[k] + 2f[k-1]$

5.3-16 Find $h[k]$, the unit impulse response of the systems in Probs. 5.3-9, 5.3-10, and 5.3-12.

5.4-1 Show a canonical, a cascade and a parallel realization of the following transfer functions:

(a) $H[z] = \dfrac{z(3z - 1.8)}{z^2 - z + 0.16}$

(b) $H[z] = \dfrac{5z + 2.2}{z^2 + z + 0.16}$

(c) $H[z] = \dfrac{3.8z - 1.1}{(z - 0.2)(z^2 - 0.6z + 0.25)}$

5.4-2 Give cascade and parallel realizations of the following transfer functions:

(a) $\dfrac{z(1.6z - 1.8)}{(z - 0.2)(z^2 + z + 0.5)}$ (b) $\dfrac{z(2z^2 + 1.3z + 0.96)}{(z + 0.5)(z - 0.4)^2}$

5.4-3 Realize a system whose transfer function is

$$H[z] = \frac{2z^4 + z^3 + 0.8z^2 + 2z + 8}{z^4}$$

5.4-4 Realize a system whose transfer function is given by

$$H[z] = \sum_{k=0}^{6} k z^{-k}$$

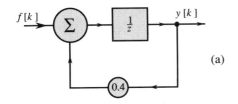

(a)

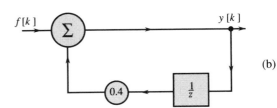

(b)

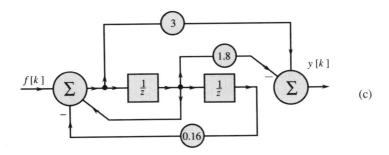

(c)

Fig. P5.5-1.

5.5-1 Find the amplitude and phase response of the digital filters shown in Fig. P5.5-1.

5.5-2 Find the amplitude and phase response of the filters shown in Fig. P5.5-2.
Hint: Express $|H[e^{j\Omega}]|$ as $e^{-j2.5\Omega}\hat{H}[e^{j\Omega}]$.

5.5-3 For an LTID system specified by the equation

$$y[k + 1] - 0.5y[k] = f[k + 1] + 0.8f[k]$$

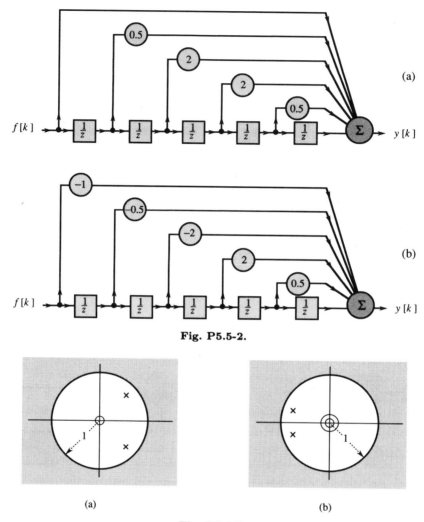

Fig. P5.5-2.

Fig. P5.5-5.

(a) Find the amplitude and phase response.

(b) Find the system response $y[k]$ for the input $f[k] = \cos\left(0.5k - \frac{\pi}{3}\right)$.

5.5-4 **(a)** Realize a digital filter whose transfer function is given by

$$H[z] = K\frac{z+1}{z-a}$$

(b) Choose a value of K such that $H[1] = 1$. The 3-dB (or half-power) bandwidth is the bandwidth over which the amplitude response varies no more than 3-dB (ratio of 0.707). In the present case the amplitude response has a maximum value at $\Omega = 0$, and it decreases monotonically with frequency until $\Omega = \pi$. The 3-dB bandwidth is the frequency where the amplitude response drops to 0.707. Determine the 3-dB bandwidth of this filter when $a = 0.2$.

5.5-5 Pole-zero configurations of certain filters are shown in Fig. P5.5-5. Sketch roughly the amplitude and the phase response of these filters.

5.5-6 Design a digital notch filter to reject frequency 5000 Hz completely, and to have a sharp recovery on either sides of 5000 Hz to a gain of unity. The highest frequency to be processed is 20 kHz ($\mathcal{F}_h = 20,000$). Hint: See Example 5.13. The zeros should be at $e^{\pm j\omega T}$ for ω corresponding to 5000 Hz, and the poles are at $ae^{\pm j\omega T}$ with $a < 1$. Leave your answer in terms of a. Realize this filter using the canonical form. Find the amplitude response of the filter.

5.5-7 Show that a first-order LTID system with a pole at $z = r$ and a zero at $z = \frac{1}{r}$ ($r \leq 1$) is an allpass filter. In other words, show that the amplitude response $|H[e^{j\Omega}]|$ of a system with the transfer function

$$H[z] = \frac{z - \frac{1}{r}}{z - r} \qquad r \leq 1$$

is constant with frequency. This is a first-order allpass filter.

Hint: show that the ratio of the distances of any point on the unit circle from the zero (at $z = \frac{1}{r}$) and the pole (at $z = r$) is a constant $\frac{1}{r}$.

Generalize this result to show that an LTID system with two poles at $z = re^{\pm j\theta}$ and two zeros at $z = \frac{1}{r}e^{\pm j\theta}$ ($r \leq 1$) is an allpass filter. In other words, show that the amplitude response of a system with the transfer function

$$H[z] = \frac{(z - \frac{1}{r}e^{j\theta})(z - \frac{1}{r}e^{-j\theta})}{(z - re^{j\theta})(z - re^{-j\theta})} \qquad r \leq 1$$

is constant with frequency.

5.5-8 For an asymptotically stable LTID system, show that the steady-state response to input $e^{j\Omega k}u[k]$ is $H[e^{j\Omega}]e^{j\Omega k}u[k]$. The steady-state response is that part of the response which does not decay with time and persists forever.

Hint: The total response is $Y[z] = H[z]F[z] = zP[z]/Q[z](z - e^{j\Omega})$ where $Q[z] = (z - \gamma_1)(z - \gamma_2) \cdots (z - \gamma_n)$. Show that the modified partial fraction expansion of $Y[z]$ yields

$$Y[z] = \sum_{i=1}^{n} k_i \frac{z}{z - \gamma_i} + H[e^{j\Omega}]\frac{z}{z - e^{j\Omega}}$$

and

$$y[k] = \sum_{i=1}^{n} k_i \gamma_i^k + H[e^{j\Omega}]e^{j\Omega k}u[k]$$

For an asymptotically stable system, γ_i^k decays with time.

5.7-1 Find the z-transform (if it exists) and the corresponding region of convergence for each of the following signals:

(a) $(0.8)^k u[k] + 2^k u[-(k+1)]$

(b) $2^k u[k] - 3^k u[-(k+1)]$

(c) $(0.8)^k u[k] + (0.9)^k u[-(k+1)]$

(d) $\left[(0.8)^k + 3(0.4)^k\right] u[-(k+1)]$

(e) $\left[(0.8)^k + 3(0.4)^k\right] u[k]$

(f) $(0.8)^k u[k] + 3(0.4)^k u[-(k+1)]$

5.7-2 Find the inverse z-transform of

$$F[z] = \frac{(e^{-2} - 2)z}{(z - e^{-2})(z - 2)}$$

when the region of convergence is

(a) $|z| > 2$ (b) $e^{-2} < |z| < 2$ (c) $|z| < e^{-2}$.

5.7-3 Determine the zero-state response of a system having a transfer function

$$H[z] = \frac{z}{(z+0.2)(z-0.8)} \qquad |z| > 0.8$$

and an input $f[k]$ given by

(a) $f[k] = e^k u[k]$

(b) $f[k] = 2^k u[-(k+1)]$

(c) $f[k] = e^k u[k] + 2^k u[-(k+1)]$

5.7-4 For the system in Problem 5.7-3, determine the zero-state response if the input

$$f[k] = 2^k u[k] + u[-(k+1)]$$

5.7-5 For the system in Problem 5.7-3, determine the zero-state response if the input

$$f[k] = e^{-2k}[-(k+1)]$$

COMPUTER PROBLEMS

C5-1 For a system specified by the equation

$$y[k+1] - 0.9y[k] = 2f[k+1]$$

Find the system response to the inputs: (a) 1 (b) $\cos(\frac{\pi}{4}k - 0.5)$ (c) a sampled sinusoid $\sin 900t$ with a sampling interval $T = 0.002$.

C5-2 Compute the frequency response (amplitude and phase) of a system whose transfer function $H[z]$ is

(a) $\dfrac{z + 0.5}{z + 0.9}$ (b) $\dfrac{z + 0.1}{z + 0.3}$ (c) $\dfrac{z + 0.2}{z^2 + 0.4z + 0.3}$ (d) $\dfrac{z^2 + 0.2z + 0.1}{z^3 + 4z^2 + 3z + 7}$

C5-3 Repeat computer Example C5.3 for $a^2 = 0.3, 0.6, 1.0, 0.9$.

C5-4 Design a digital notch filter to have zero transmission at 60 Hz, and a sharp gain recovery to unity on both sides of 60 Hz. The highest significant frequency to be processed is 1kHz. Plot the amplitude and phase response in each case. Use Eq. 5.71.

IV

SIGNAL ANALYSIS

6

Continuous-Time Signal Analysis: The Fourier Series

Electrical engineers instinctively think of signals in terms of their frequency spectra and think of systems in terms of their frequency response. Even teenagers know about audio signals having a bandwidth of 20 kHz and good quality speakers responding up to 20 kHz. This is basically thinking in frequency domain. In Chapters 4 and 5 we discussed extensively the frequency-domain representation of systems and their spectral response (system response to signals of various frequencies). In the next three Chapters we discuss spectral representation of signals, where signals are expressed as a sum of sinusoids or exponentials. Actually, we touched upon this topic in Chapters 4 and 5. Recall that the Laplace transform of a continuous-time signal is its spectral representation in terms of exponentials (or sinusoids) of complex frequencies. Similarly the z-transform of a discrete-time signal is its spectral representation in terms of discrete-time exponentials. However, in those chapters we were concerned mainly with system representation, and the spectral representation of signals was incidental to the system analysis. Spectral analysis of signals is an important topic in its own right, and the next three chapters are devoted to this subject.

In this chapter we show that a periodic signal can be represented as a sum of sinusoids (or exponentials) of various frequencies. These results are extended to nonperiodic signals in Chapter 7 and to discrete-time signals in Chapter 9.

6.1 REPRESENTATION OF PERIODIC SIGNALS BY TRIGONOMETRIC FOURIER SERIES

A signal $f(t)$ is said to be **periodic** if for some positive constant T_0

$$f(t) = f(t + T_0) \qquad \text{for all } t \tag{6.1}$$

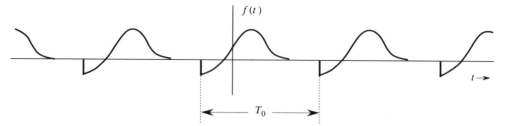

Fig. 6.1 A periodic signal of period T_0.

The **smallest** value of T_0 which satisfies the periodicity condition (6.1) is the **period** of $f(t)$.

By definition, a periodic signal $f(t)$ remains unchanged when time-shifted by one period. This means that a periodic signal must start at $t = -\infty$ because if it starts at some finite instant, say, $t = 0$, the time-shifted signal $f(t+T_0)$ will start at $t = -T_0$ and $f(t+T_0)$ would not be the same as $f(t)$. Therefore a **periodic signal, by definition, must be an everlasting signal** (starting at $t = -\infty$), as shown in Fig. 6.1. Observe that a periodic signal shifted by an integral multiple of T_0 remains unchanged. Therefore $f(t)$ may be considered a periodic signal with period mT_0 where m is any integer. However, by definition, the period is the smallest interval that satisfies periodicity condition (6.1). Therefore T_0 is the period.

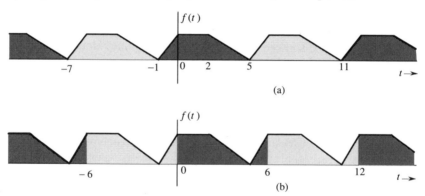

Fig. 6.2 Generation of a periodic signal by periodic extension of its segment of one period duration.

The second important property of a periodic signal $f(t)$ is that $f(t)$ can be generated by periodic extension of any segment of $f(t)$ of duration T_0 (the period). This means that we can generate $f(t)$ from any segment of $f(t)$ with a duration of one period by placing this segment and the reproduction thereof end to end ad infinitum on either side. Figure 6.2 shows a periodic signal $f(t)$ of period $T_0 = 6$. The shaded portion of Fig. 6.2a shows a segment of $f(t)$ starting at $t = -1$ and having a duration of one period (6 secs). This segment, when repeated forever in either direction, results in the periodic signal $f(t)$. Figure 6.2b shows another shaded segment of $f(t)$ of duration T_0 starting at $t = 0$. Again we see that this segment, when repeated forever on either side, results in $f(t)$. The reader can verify that this is possible with any segment of $f(t)$ starting at any instant as long as the segment duration is one period.

The third important property of a periodic signal $f(t)$ **of period** T_0
is that the area under $f(t)$ **over any interval of duration** T_0 **is the same;**
that is, for any real numbers a and b

$$\int_a^{a+T_0} f(t)\, dt = \int_b^{b+T_0} f(t)\, dt \tag{6.2}$$

This result is obvious from Fig. 6.2. **The area under** $f(t)$ **over any contiguous**
interval of duration T_0 **(shaded area in Fig. 6.2) is the same because of the**
periodic nature of $f(t)$. For convenience, the area under $f(t)$ over any interval
of duration T_0 will be denoted by

$$\int_{T_0} f(t)\, dt$$

A sinusoid of frequency $\mathcal{F}_0$ Hz with amplitude C and phase θ can be expressed
as $C\cos(2\pi\mathcal{F}_0 t + \theta) = C\cos(\omega_0 t + \theta)$, where $\omega_0 = 2\pi\mathcal{F}_0$ is the radian frequency
(see Sec. B.2). The period of this sinusoid is $T_0 = 1/\mathcal{F}_0$ sec. A sinusoid of frequency
$n\mathcal{F}_0$ is said to be the nth **harmonic** of the sinusoid of frequency $\mathcal{F}_0$.

Let us consider a signal $f(t)$ made up of a sinusoid of frequency ω_0 and any
number of its harmonics (including the zero-th harmonic; that is, dc) with arbitrary
amplitudes and phases:

$$f(t) = C_0 + C_1\cos(\omega_0 t + \theta_1) + C_2\cos(2\omega_0 t + \theta_2) + C_3\cos(3\omega_0 t + \theta_3) + \cdots + \cdots$$

We can include any number of harmonics. For the sake of generality, all harmonics
up to infinity will be included. This yields

$$f(t) = C_0 + \sum_{n=1}^{\infty} C_n\cos(n\omega_0 t + \theta_n) \tag{6.3}$$

The frequency ω_0 ($\omega_0 = 2\pi\mathcal{F}_0$) is the **fundamental frequency**, and $k\omega_0$ is its kth
harmonic.

We now prove an extremely important property: $f(t)$ in Eq. (6.3) is a periodic
signal with the same period as that of the fundamental, regardless of the values of
the amplitudes C_ks and phases θ_ks. Note that the period T_0 of the fundamental is

$$T_0 = \frac{1}{\mathcal{F}_0} = \frac{2\pi}{\omega_0} \tag{6.4}$$

and

$$\omega_0 T_0 = 2\pi \tag{6.5}$$

To prove the periodicity of $f(t)$, all we need is to show that $f(t) = f(t + T_0)$. From
Eq. (6.3)

$$f(t + T_0) = C_0 + \sum_{n=1}^{\infty} C_n\cos[n\omega_0(t + T_0) + \theta_n]$$

$$= C_0 + \sum_{n=1}^{\infty} C_n\cos[(n\omega_0 t + n\omega_0 T_0) + \theta_n]$$

From Eq. (6.5), $n\omega_0 T_0 = 2\pi n$. Therefore

$$f(t + T_0) = C_0 + \sum_{n=1}^{\infty} C_n \cos\left[(n\omega_0 t + 2\pi n) + \theta_n\right]$$

$$= C_0 + \sum_{n=1}^{\infty} C_n \cos(n\omega_0 t + \theta_n)$$

$$= f(t)$$

Therefore any combination of sinusoids of frequencies 0, $\mathcal{F}_0$, $2\mathcal{F}_0$, ..., $k\mathcal{F}_0$ is a periodic signal of period $T_0 = \frac{1}{\mathcal{F}_0}$ regardless of the amplitudes C_ks and phases θ_ks of these sinusoids. By changing the values of C_ks and θ_ks in Eq. (6.3) we can construct a variety of periodic signals, all of the same period T_0 ($T_0 = \frac{1}{\mathcal{F}_0}$ or $\frac{2\pi}{\omega_0}$).

The converse of this result is also true. **A periodic signal $f(t)$ with a period T_0 can be expressed as a sum of a sinusoid of frequency $\mathcal{F}_0$ ($\mathcal{F}_0 = \frac{1}{T_0}$) and all its harmonics, as shown in Eq. (6.3)**. This is proved in Sec. 6.3.

We can express a sinusoid $C_n \cos(n\omega_0 t + \theta_n)$ in terms of sines and cosines by using an identity

$$C_n \cos(n\omega_0 t + \theta_n) = C_n \cos\theta_n \cos n\omega_0 t - C_n \sin\theta_n \sin n\omega_0 t$$

$$= a_n \cos n\omega_0 t + b_n \sin n\omega_0 t$$

where

$$a_n = C_n \cos\theta_n \qquad \text{and} \qquad b_n = -C_n \sin\theta_n \tag{6.6}$$

Therefore Eq. (6.3) can be expressed as

$$f(t) = a_0 + \sum_{n=1}^{\infty} a_n \cos n\omega_0 t + b_n \sin n\omega_0 t \tag{6.7a}$$

$$= C_0 + \sum_{n=1}^{\infty} C_n \cos(n\omega_0 t + \theta_n) \tag{6.7b}$$

where

$$C_0 = a_0 \tag{6.8a}$$

and from Eq. (6.6)

$$C_n = \sqrt{a_n^2 + b_n^2} \tag{6.8b}$$

$$\theta_n = \tan^{-1}\left(\frac{-b_n}{a_n}\right) \tag{6.8c}$$

The infinite series on the right-hand side of Eq. (6.7) is known as the **Fourier series** of a periodic signal $f(t)$. The series in Eq. (6.7a) is the **trigonometric Fourier series** and that in Eq. (6.7b) is the **compact trigonometric Fourier series** for $f(t)$.

Computing the Coefficients of a Fourier Series

In order to determine the coefficients of a Fourier series, consider an integral I defined by

$$I = \int_{T_0} \cos n\omega_0 t \, \cos m\omega_0 t \, dt \qquad (6.9a)$$

where $\int_{T_0}$ stands for integration over any contiguous interval of T_0 seconds. By using a trigonometric identity (see Sec. B.10-6), Eq. (6.9a) can be expressed as

$$I = \frac{1}{2} \left[\int_{T_0} \cos (n+m)\omega_0 t \, dt + \int_{T_0} \cos (n-m)\omega_0 t \, dt \right] \qquad (6.9b)$$

Since $\cos \omega_0 t$ executes one complete cycle during any interval of duration T_0, $\cos (n+m)\omega_0 t$ executes $(n+m)$ complete cycles during any interval of duration T_0. Therefore the first integral in Eq. (6.9b), which represents the area under $(n+m)$ complete cycles of a sinusoid, equals zero. The same argument shows that the second integral in Eq. (6.9b) is also zero, except when $n = m$. Hence I in Eq. (6.9) is zero for all $n \neq m$. When $n = m$, the first integral in Eq. (6.9b) is still zero, but the second integral yields

$$I = \frac{1}{2} \int_{T_0} dt = \frac{T_0}{2}$$

Thus

$$\int_{T_0} \cos n\omega_0 t \, \cos m\omega_0 t \, dt = \begin{cases} 0 & n \neq m \\ \frac{T_0}{2} & m = n \neq 0 \end{cases} \qquad (6.10a)$$

Using similar arguments, we can show that

$$\int_{T_0} \sin n\omega_0 t \, \sin m\omega_0 t \, dt = \begin{cases} 0 & n \neq m \\ \frac{T_0}{2} & n = m \neq 0 \end{cases} \qquad (6.10b)$$

and

$$\int_{T_0} \sin n\omega_0 t \, \cos m\omega_0 t \, dt = 0 \quad \text{for all } n \text{ and } m \qquad (6.10c)$$

To determine a_0 in Eq. (6.7a) we integrate both sides of Eq. (6.7a) over one period T_0 to yield

$$\int_{T_0} f(t) \, dt = a_0 \int_{T_0} dt + \sum_{n=1}^{\infty} \left[a_n \int_{T_0} \cos n\omega_0 t \, dt + b_n \int_{T_0} \sin n\omega_0 t \, dt \right]$$

Recall that T_0 is the period of a sinusoid of frequency ω_0. Therefore functions $\cos n\omega_0 t$ and $\sin n\omega_0 t$ execute n complete cycles over any interval of T_0 seconds, so that the area under these functions over an interval T_0 is zero, and the last two integrals on the right-hand side of the above equation are zero. This yields

$$\int_{T_0} f(t) \, dt = a_0 \int_{T_0} dt = a_0 T_0$$

and

$$a_0 = \frac{1}{T_0} \int_{T_0} f(t)\,dt \tag{6.11a}$$

Next we multiply both sides of Eq. (6.7a) by $\cos m\omega_0 t$ and integrate the resulting equation over an interval T_0:

$$\int_{T_0} f(t)\cos m\omega_0 t\,dt = a_0 \int_{T_0} \cos m\omega_0 t\,dt + \sum_{n=1}^{\infty} \left[a_n \int_{T_0} \cos n\omega_0 t \cos m\omega_0 t\,dt \right.$$
$$\left. + b_n \int_{T_0} \sin n\omega_0 t \cos m\omega_0 t\,dt \right]$$

The first integral on the right-hand side is zero because it is an area under m integral number of cycles of a sinusoid. Also, the last integral on the right-hand side vanishes because of Eq. (6.10c). This leaves only the middle integral, which is also zero for all $n \neq m$ because of Eq. (6.10a). But n takes on all values from 1 to ∞, including m. When $n = m$, this integral is $\frac{T_0}{2}$, according to Eq. (6.10a). Therefore, from the infinite number of terms on the right-hand side, only one term survives to yield $\frac{a_n T_0}{2} = \frac{a_m T_0}{2}$ (recall that $n = m$). Therefore

$$\int_{T_0} f(t)\cos m\omega_0 t\,dt = \frac{a_m T_0}{2}$$

and

$$a_m = \frac{2}{T_0} \int_{T_0} f(t)\cos m\omega_0 t\,dt \tag{6.11b}$$

Similarly, by multiplying both sides of Eq. (6.7a) by $\sin n\omega_0 t$ and then integrating over an interval T_0, we obtain

$$b_m = \frac{2}{T_0} \int_{T_0} f(t)\sin m\omega_0 t\,dt \tag{6.11c}$$

To sum up our discussion, we have shown that a periodic signal $f(t)$ with period T_0 can be expressed as a sum of a sinusoid of period T_0 and its harmonics:

$$f(t) = a_0 + \sum_{n=1}^{\infty} a_n \cos n\omega_0 t + b_n \sin n\omega_0 t \tag{6.12a}$$

$$= C_0 + \sum_{n=1}^{\infty} C_n \cos(n\omega_0 t + \theta_n) \tag{6.12b}$$

where

$$\omega_0 = 2\pi \mathcal{F}_0 = \frac{2\pi}{T_0} \tag{6.13}$$

$$a_0 = \frac{1}{T_0} \int_{T_0} f(t)\,dt \tag{6.14a}$$

$$a_n = \frac{2}{T_0} \int_{T_0} f(t)\cos n\omega_0 t\,dt \tag{6.14b}$$

$$b_n = \frac{2}{T_0} \int_{T_0} f(t)\sin n\omega_0 t\,dt \tag{6.14c}$$

Moreover,

$$C_0 = a_0 \tag{6.15a}$$

$$C_n = \sqrt{a_n{}^2 + b_n{}^2} \tag{6.15b}$$

$$\theta_n = \tan^{-1}\left(\frac{-b_n}{a_n}\right) \tag{6.15c}$$

These results are summarized in Table 6.2 (p. 440).

Equation 6.14a shows that a_0 (or C_0) is the average value of $f(t)$ (averaged over one period). This value can often be determined by inspection of $f(t)$.

Existence of the Fourier Series: Dirichlet Conditions

For the existence of the Fourier series, coefficients a_0, a_n, and b_n in Eq. (6.14) must be finite. From Eqs. (6.14a), (6.14b), and (6.14c) it follows that the existence of these coefficients is guaranteed if $f(t)$ is absolutely integrable over one period; that is,

$$\int_{T_0} |f(t)|\, dt < \infty \tag{6.16}$$

This is known as the **weak Dirichlet condition. If a function $f(t)$ satisfies the weak Dirichlet condition, the existence of a Fourier series is guaranteed, but the series may not converge at every point.** For example, if a function $f(t)$ is infinite at some point, then obviously the series representing the function will be nonconvergent at that point. Similarly, if a function has an infinite number of maxima and minima in one period, then the function contains an appreciable amount of infinite frequency component and the higher coefficients in the series do not decay rapidly, so that the series will not converge rapidly or uniformly. **Thus, for a convergent Fourier series, in addition to condition (6.16), the function $f(t)$ must remain finite and must have only a finite number of maxima and minima in one period. It may have a finite number of finite discontinuities in one period.** These requirements [along with Eq. (6.16)] are known as the **strong Dirichlet conditions.** We note here that any periodic waveform that can be generated in a laboratory satisfies strong Dirichlet conditions, and hence possesses a convergent Fourier series. Thus a physical possibility of a periodic waveform is a valid and sufficient condition for the existence of a convergent series.

6.1-1   The Fourier Spectrum

The compact trigonometric Fourier series in Eq. (6.12b) indicates that a periodic signal $f(t)$ can be expressed as a sum of sinusoids of frequencies 0 (dc), ω_0, $2\omega_0$, $\cdots$, $n\omega_0$, $\cdots$, whose amplitudes are C_0, C_1, C_2, $\ldots$, C_n, $\cdots$, and whose phases are 0, θ_1, θ_2, $\cdots$, θ_n, $\cdots$, respectively. We can readily plot amplitude C_n vs. ω (**amplitude spectrum**) and θ_n vs. ω (**phase spectrum**). These two plots together are the **frequency spectra** of $f(t)$. These spectra show at a glance the frequency contents of the signal $f(t)$ with their amplitudes and phases. Knowing these spectra, we can reconstruct or synthesize the signal $f(t)$ according to Eq.

(6.12b). Therefore frequency spectra, which are an alternative way of describing a periodic signal $f(t)$, are in every way equivalent to the plot of $f(t)$ as a function of t. The frequency spectra of a signal constitute the **frequency-domain description** of $f(t)$, in contrast to the **time-domain description**, where $f(t)$ is specified as a function of time.

In computing θ_n, the phase of the nth harmonic from Eq. (6.15c), the quadrant in which θ_n lies should be determined from the signs of a_n and b_n. For example, if $a_n = -1$ and $b_n = 1$, θ_n lies in the third quadrant, and

$$\theta_n = \tan^{-1}\left(\frac{-1}{-1}\right) = -135°$$

Observe that

$$\tan^{-1}\left(\frac{-1}{-1}\right) \neq \tan^{-1}(1) = 45°$$

Although C_n, the amplitude of the nth harmonic as defined in Eq. (6.15b), is positive, we shall find it convenient to allow C_n to take on negative values when $b_n = 0$. This will become clear in later examples.

■ **Example 6.1**

Find the compact trigonometric Fourier series for the periodic signal $f(t)$ shown in Fig. 6.3a. Sketch the amplitude and phase spectra for $f(t)$.

In this case the period $T_0 = \pi$ and the fundamental frequency $\mathcal{F}_0 = 1/T_0 = 1/\pi$ Hz, and

$$\omega_0 = \frac{2\pi}{T_0} = 2\,\text{rad/s}$$

Therefore

$$f(t) = a_0 + \sum_{n=1}^{\infty} a_n \cos 2nt + b_n \sin 2nt$$

where

$$a_0 = \frac{1}{\pi} \int_{T_0} f(t)\, dt$$

In this example the obvious choice for the interval of integration is from 0 to π. Hence

$$a_0 = \frac{1}{\pi} \int_0^\pi e^{-t/2}\, dt = 0.504$$

$$a_n = \frac{2}{\pi} \int_0^\pi e^{-t/2} \cos 2nt\, dt = 0.504 \left(\frac{2}{1 + 16n^2}\right)$$

and

$$b_n = \frac{2}{\pi} \int_0^\pi e^{-t/2} \sin 2nt\, dt = 0.504 \left(\frac{8n}{1 + 16n^2}\right)$$

Therefore

$$f(t) = 0.504 \left[1 + \sum_{n=1}^{\infty} \frac{2}{1 + 16n^2} (\cos 2nt + 4n \sin 2nt)\right] \tag{6.17}$$

Also from Eq. (6.15)

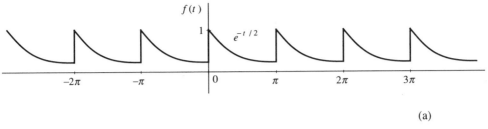

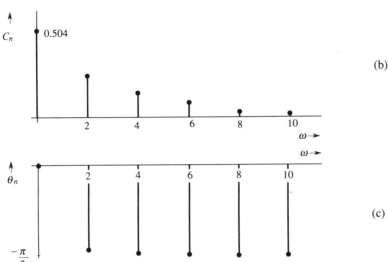

Fig. 6.3 A periodic signal and its Fourier spectra.

$$C_0 = a_0 = 0.504$$

$$C_n = \sqrt{a_n^2 + b_n^2} = 0.504\sqrt{\frac{4}{(1+16n^2)^2} + \frac{64n^2}{(1+16n^2)^2}} = 0.504\left(\frac{2}{\sqrt{1+16n^2}}\right)$$

$$\theta_n = \tan^{-1}\left(\frac{-b_n}{a_n}\right) = \tan^{-1}(-4n) = -\tan^{-1}4n \tag{6.18}$$

Amplitude and phases of the dc and the first seven harmonics are computed from the above equation and displayed in Table 6.1. Using these numerical values, we can express $f(t)$ as

$$f(t) = 0.504 + 0.504\sum_{n=1}^{\infty}\frac{2}{\sqrt{1+16n^2}}\cos\left(2nt - \tan^{-1}4n\right) \tag{6.19a}$$

$$= 0.504 + 0.244\cos\left(2t - 75.96°\right) + 0.125\cos\left(4t - 82.87°\right)$$

$$+ 0.084\cos\left(6t - 85.24°\right) + 0.063\cos\left(8t - 86.42°\right) + \cdots \tag{6.19b}$$

Table 6.1

n	0	1	2	3	4	5	6	7
C_n	0.504	0.244	0.125	0.084	0.063	0.0504	0.042	0.036
θ_n	0	−75.96	−82.87	−85.24	−86.42	−87.14	−87.61	−87.95

⊙ **Computer Example C6.1**

Using a computer, compute and plot the Fourier coefficients for the periodic signal in Fig. 6.3a (Example 6.1).

In this example, $T_0 = \pi$ and $\omega_0 = 2$. We use equations derived in Example 6.1 to compute coefficients a_0, a_n, b_n, C_n, and θ_n.

```
n=1:20;
a0=0.504;        b0=0.504*(8*0/(1+16*0^2));        Cn=a0;
theta0=atan(-b0/a0);    thetan=theta0;
den=(1+16*n.^2);
N=length(den);
for i=1:N
an(i)=0.504*2/den(i);
bn(i)=0.504*8*n(i)/den(i);
cn=sqrt(an(i)^2+bn(i)^2);
Cn=[Cn cn];
theta=atan(-bn(i)/an(i));
thetan=[thetan theta];
end
n=0:1:20;
plot(n,Cn,'o'),grid,xlabel('n'),ylabel('Cn'),pause
plot(n,thetan,'o'),grid,xlabel('n'),ylabel('theta n')    ⊙
```

Figures 6.3b and 6.3c show the amplitude and phase spectra for $f(t)$. These spectra tell us at a glance the frequency composition of $f(t)$; that is, the amplitudes and phases of various sinusoidal components of $f(t)$. Knowing the frequency spectra, we can reconstruct $f(t)$, as shown on the right-hand side of Eq. (6.19). Therefore the frequency spectra in Figs. 6.3b and 6.3c provide an alternative description— the frequency-domain description of $f(t)$. The time-domain description of $f(t)$ is shown in Fig. 6.3a. **A signal, therefore, has a dual identity: the time-domain identity $f(t)$ and the frequency-domain identity (Fourier spectra). The two identities complement each other; taken together, they provide a better understanding of a signal.**

An interesting aspect of Fourier series is that whenever there is a jump discontinuity in $f(t)$, the series at the point of discontinuity converges to an average of the left-hand and right-hand limits of $f(t)$ at the instant of discontinuity.[†] In the present example, for instance, $f(t)$ is discontinuous at $t = 0$ with $f(0^+) = 1$ and $f(0^-) = e^{-\pi/2} = 0.208$. The corresponding Fourier series converges to a value $(1 + 0.208)/2 = 0.604$ at $t = 0$. This is easily verified from Eq. (6.19b) by setting $t = 0$.

■ **Example 6.2**

Find the compact trigonometric Fourier series for the triangular periodic signal $f(t)$ shown in Fig. 6.4a, and sketch the amplitude and phase spectra for $f(t)$.

In this case the period $T_0 = 2$. Hence

and

$$\omega_0 = \frac{2\pi}{2} = \pi$$

$$f(t) = a_0 + \sum_{n=1}^{\infty} a_n \cos n\pi t + b_n \sin n\pi t$$

[†]This behavior of the Fourier series is dictated by its mean-squared property, discussed in Sec. 6.3.

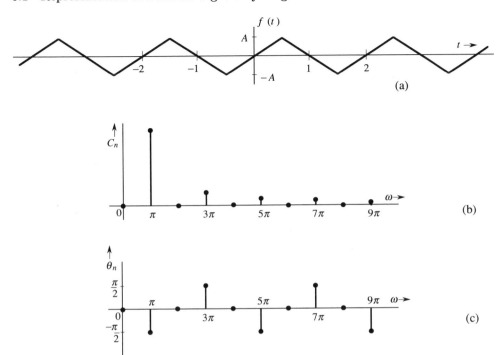

Fig. 6.4 A triangular periodic signal and its Fourier spectra.

where

$$f(t) = \begin{cases} 2At & |t| \le \frac{1}{2} \\ 2A(1-t) & \frac{1}{2} < t \le \frac{3}{2} \end{cases}$$

Here it will be advantageous to choose the interval of integration from $-\frac{1}{2}$ to $\frac{3}{2}$ rather than 0 to 2.

A glance at Fig. 6.4a shows that the average value (dc) of $f(t)$ is zero, so that $a_0 = 0$. Also

$$a_n = \frac{2}{2} \int_{-1/2}^{3/2} f(t) \cos n\pi t \, dt$$

$$= \int_{-1/2}^{1/2} 2At \cos n\pi t \, dt + \int_{1/2}^{3/2} 2A(1-t) \cos n\pi t \, dt$$

The detailed evaluation of the above integrals shows that both have a value of zero. Therefore

$$a_n = 0 \tag{6.20a}$$

$$b_n = \int_{-1/2}^{1/2} 2At \sin n\pi t \, dt + \int_{1/2}^{3/2} 2A(1-t) \sin n\pi t \, dt$$

The detailed evaluation of these integrals yields

$$b_n = \frac{8A}{n^2\pi^2} \sin\left(\frac{n\pi}{2}\right)$$

$$= \begin{cases} 0 & n \text{ even} \\ \frac{8A}{n^2\pi^2} & n = 1, \, 5, \, 9, \, 13, \, \cdots \\ -\frac{8A}{n^2\pi^2} & n = 3, \, 7, \, 11, \, 15, \, \cdots \end{cases} \tag{6.20b}$$

Therefore

$$f(t) = \frac{8A}{\pi^2} \sum_{n=1,3,5,\cdots}^{\infty} \frac{\sin\left(\frac{n\pi}{2}\right)}{n^2} \sin n\pi t \tag{6.21a}$$

$$= \frac{8A}{\pi^2}\left[\sin \pi t - \frac{1}{9}\sin 3\pi t + \frac{1}{25}\sin 5\pi t - \frac{1}{49}\sin 7\pi t + \cdots\right] \tag{6.21b}$$

In order to plot Fourier spectra, the series must be converted into compact trigonometric form as in Eq. (6.12b). In this case this is readily done by converting sine terms into cosine terms with a suitable phase shift. For example,

$$\sin kt = \cos\left(kt - 90°\right)$$

$$-\sin kt = \cos\left(kt + 90°\right)$$

By using these identities, Eq. (6.21b) can be expressed as

$$f(t) = \frac{8A}{\pi^2}\left[\cos\left(\pi t - 90°\right) + \frac{1}{9}\cos\left(3\pi t + 90°\right) + \frac{1}{25}\cos\left(5\pi t - 90°\right)\right.$$

$$\left. + \frac{1}{49}\cos\left(7\pi t + 90°\right) + \cdots\right] \tag{6.22}$$

In this series all the even harmonics are missing. The phases of odd harmonics alternate from $-90°$ to $90°$. Figure 6.4 shows amplitude and phase spectra for $f(t)$.   ■

■ **Example 6.3**

A periodic signal $f(t)$ is represented by a trigonometric Fourier series

$$f(t) = 2 + 3\cos 2t + 4\sin 2t + 2\sin\left(3t + 30°\right) - \cos\left(7t + 150°\right)$$

Express this series as a compact trigonometric Fourier series and sketch amplitude and phase spectra for $f(t)$.

In compact trigonometric Fourier series, the sine and cosine terms of the same frequency are combined into a single term and all terms are expressed as cosine terms with positive amplitudes. From Eqs. (6.8b) and (6.8c), we have

$$3\cos 2t + 4\sin 2t = 5\cos\left(2t - 53.13°\right)$$

Also

$$\sin\left(3t + 30°\right) = \cos\left(3t + 30° - 90°\right) = \cos\left(3t - 60°\right)$$

and

$$-\cos\left(7t + 150°\right) = \cos\left(7t + 150° - 180°\right) = \cos\left(7t - 30°\right)$$

Therefore

$$f(t) = 2 + 5\cos\left(2t - 53.13°\right) + 2\cos\left(3t - 60°\right) + \cos\left(7t - 30°\right)$$

In this case only four components (including dc) are present. The amplitude of dc is 2. The remaining three components are of frequencies $\omega = 2$, 3, and 7 with amplitudes 5, 2, and 1 and phases $-53.13°$, $-60°$, and $-30°$, respectively. The amplitude and phase spectra for this signal are shown in Figs.6.5a and 6.5b, respectively.   ■

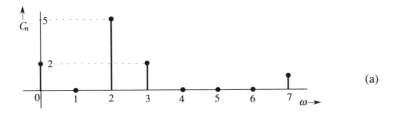

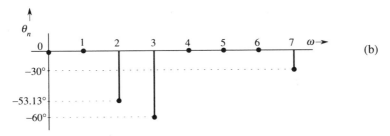

Fig. 6.5 Fourier spectra of the signal in Example 6.3.

■ **Example 6.4**

Find the compact trigonometric Fourier series for the square-pulse periodic signal shown in Fig. 6.6a and sketch its amplitude and phase spectra.

Here the period is $T_0 = 2\pi$ and $\omega_0 = 2\pi/T_0 = 1$. Therefore

$$f(t) = a_0 + \sum_{n=1}^{\infty} a_n \cos nt + b_n \sin nt$$

where

$$a_0 = \frac{1}{T_0} \int_{T_0} f(t)\, dt$$

From Fig. 6.6a, it is clear that a proper choice of region of integration is from $-\pi$ to π. But since $f(t) = 1$ only over $(-\frac{\pi}{2}, \frac{\pi}{2})$ and $f(t) = 0$ over the remaining segment,

$$a_0 = \frac{1}{2\pi} \int_{-\pi/2}^{\pi/2} dt = \frac{1}{2} \tag{6.23a}$$

We could have found a_0, the average value of $f(t)$, to be $\frac{1}{2}$ merely by inspection of $f(t)$ in Fig. 6.6a. Also,

$$a_n = \frac{1}{\pi} \int_{-\pi/2}^{\pi/2} \cos nt\, dt = \frac{2}{n\pi} \sin\left(\frac{n\pi}{2}\right)$$

$$= \begin{cases} 0 & n \text{ even} \\ \frac{2}{\pi n} & n = 1, 5, 9, 13, \cdots \\ -\frac{2}{\pi n} & n = 3, 7, 11, 15, \cdots \end{cases} \tag{6.23b}$$

$$b_n = \frac{1}{\pi} \int_{-\pi/2}^{\pi/2} \sin nt\, dt = 0 \tag{6.23c}$$

Therefore

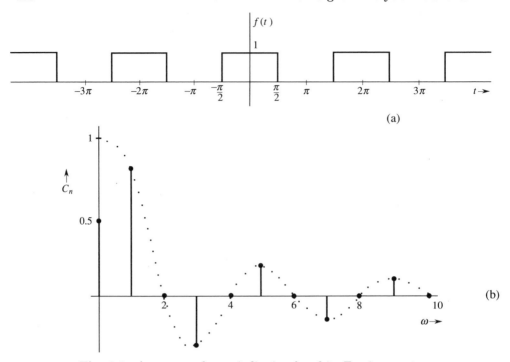

Fig. 6.6 A square pulse periodic signal and its Fourier spectra.

$$f(t) = \frac{1}{2} + \frac{2}{\pi}\left(\cos t - \frac{1}{3}\cos 3t + \frac{1}{5}\cos 5t - \frac{1}{7}\cos 7t + \cdots\right) \qquad (6.24)$$

Observe that $b_n = 0$ and all the sine terms are zero. Only the cosine terms appear in the trigonometric series. The series is therefore already in the compact form except that the amplitudes of alternating harmonics are negative. Now by definition, amplitudes C_n are positive [see Eq. (6.15b)]. The negative sign can be accommodated by a proper phase. This can be seen from the trigonometric identity†

$$-\cos x = \cos(x - \pi)$$

Using this fact, we can express the series in (6.24) as

$$f(t) = \tfrac{1}{2} + \tfrac{2}{\pi}\left[\cos \omega_0 t + \tfrac{1}{3}\cos(3\omega_0 t - \pi) + \tfrac{1}{5}\cos 5\omega_0 t\right.$$
$$\left. + \tfrac{1}{7}\cos(7\omega_0 t - \pi) + \tfrac{1}{9}\cos 9\omega_0 t + \cdots\right]$$

This is the desired form of the compact trigonometric Fourier series. The amplitudes are

$$C_0 = \tfrac{1}{2}$$

$$C_n = \begin{cases} 0 & n \text{ even} \\ \frac{2}{\pi n} & n \text{ odd} \end{cases}$$

---

†Because $\cos(x \pm \pi) = -\cos x$, we could have chosen the phase π or $-\pi$. In fact, $\cos(x \pm N\pi) = -\cos x$ for any odd integral value of N. Therefore the phase can be chosen as $\pm N\pi$ where N is any convenient odd integer.

$$\theta_n = \begin{cases} 0 & \text{for all } n \neq 3, 7, 11, 15, \cdots \\ -\pi & n = 3, 7, 11, 15, \cdots \end{cases}$$

We could plot amplitude and phase spectra using the above values. We can, however, simplify our task in this special case if we allow amplitude C_n to take on negative values. If this is allowed, we do not need a phase of $-\pi$ to account for the sign. This means phases of all components are zero, and we can discard the phase spectrum and manage with only the amplitude spectrum, as shown in Fig. 6.6b. Observe that there is no loss of information in doing so and that the amplitude spectrum in Fig. 6.6b has the complete information about the Fourier series in (6.24). **Therefore, whenever all sine terms vanish ($b_n = 0$), it is convenient to allow C_n to take on negative values.** This permits the spectral information to be conveyed by a single spectrum—the amplitude spectrum. Because C_n can be positive as well as negative, the spectrum is called the *amplitude spectrum* rather than the *magnitude spectrum*. ■

6.1-2 The Effect of Symmetry

The Fourier series for the signal $f(t)$ in Fig. 6.3a (Example 6.1) consists of sine and cosine terms, but the series for the signal $f(t)$ in Fig. 6.4a (Example 6.2) consists of sine terms only and the series for the signal $f(t)$ in Fig. 6.6a (Example 6.4) consists of cosine terms only. This is no accident. We can show that the Fourier series of any even periodic function $f(t)$ consists of cosine terms only and the series for any odd periodic function $f(t)$ consists of sine terms only. Moreover, because of symmetry (even or odd), the information of one period of $f(t)$ is implicit in only half the period, as seen in Figs. 6.4a and 6.6a. In these cases, knowing the signal over a half period and knowing the kind of symmetry (even or odd), we can determine the signal waveform over a complete period. For this reason, the Fourier coefficients in these cases can be computed by integrating over only half the period rather than a complete period. To prove this result, recall that

$$a_0 = \frac{1}{T_0} \int_{-T_0/2}^{T_0/2} f(t) \, dt \tag{6.25a}$$

$$a_n = \frac{2}{T_0} \int_{-T_0/2}^{T_0/2} f(t) \cos n\omega_0 t \, dt \tag{6.25b}$$

$$b_n = \frac{2}{T_0} \int_{-T_0/2}^{T_0/2} f(t) \sin n\omega_0 t \, dt \tag{6.25c}$$

Recall also that $\cos n\omega_0 t$ is an even function and $\sin n\omega_0 t$ is an odd function of t. If $f(t)$ is an even function of t, then $f(t) \cos n\omega_0 t$ is also an even function and $f(t) \sin n\omega_0 t$ is an odd function of t (see Sec. B.6). Therefore, according to Eqs. (B.43a) and (B.43b),

$$a_0 = \frac{2}{T_0} \int_0^{T_0/2} f(t) \, dt \tag{6.26a}$$

$$a_n = \frac{4}{T_0} \int_0^{T_0/2} f(t) \cos n\omega_0 t \, dt \tag{6.26b}$$

$$b_n = 0 \tag{6.26c}$$

Similarly, if $f(t)$ is an odd function of t, then $f(t)\cos n\omega_0 t$ is an odd function of t and $f(t)\sin n\omega_0 t$ is an even function of t. Therefore

$$a_n = 0 \qquad n = 0, 1, 2, 3, \cdots \tag{6.27a}$$

$$b_n = \frac{4}{T_0} \int_0^{T_0/2} f(t)\sin n\omega_0 t\, dt \tag{6.27b}$$

Observe that because of symmetry, the integration required to compute the coefficients need be performed over only half the period.

If a periodic signal $f(t)$ shifted by half the period remains unchanged except for a sign—that is, if

$$f\left(t - \tfrac{T_0}{2}\right) = -f(t)$$

the signal is said to have a **half-wave** symmetry. It can be shown that for a signal with a half-wave symmetry, all the even-numbered harmonics vanish (see Prob. 6.1-2). The signal in Fig. 6.4a is an example of such a symmetry. The signal in Fig. 6.6a also has this symmetry, although it is not obvious. If we subtract the dc component of 0.5 from this signal, the remaining signal has a half-wave symmetry. For this reason this signal has a dc component 0.5 and only the odd harmonics.

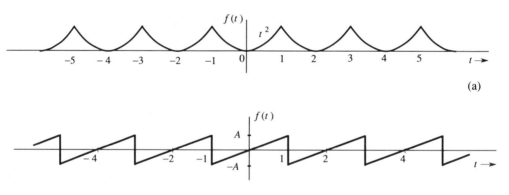

Fig. 6.7 Periodic signals for Exercise E6.1.

△ **Exercise E6.1**

Find the compact trigonometric Fourier series for periodic signals shown in Figs. 6.7a and 6.7b. Sketch their amplitude and phase spectra. Allow C_n to take on negative values if $b_n = 0$ so that the phase spectrum can be eliminated. Hint: Use symmetry conditions (6.26) and (6.27). Answers:

(a) $f(t) = \frac{1}{3} - \frac{4}{\pi^2}\left(\cos \pi t - \frac{1}{4}\cos 2\pi t + \frac{1}{9}\cos 3\pi t - \frac{1}{16}\cos 4\pi t + \cdots\right)$

$\qquad = \frac{1}{3} + \frac{4}{\pi^2}\sum_{n=1}^{\infty} \frac{(-1)^n}{n^2}\cos n\pi t$

(b) $f(t) = \frac{2A}{\pi}\left[\sin \pi t - \frac{1}{2}\sin 2\pi t + \frac{1}{3}\sin 3\pi t - \frac{1}{4}\sin 4\pi t + \cdots\right]$

$\qquad = \frac{2A}{\pi}\left[\cos(\pi t - 90°) + \frac{1}{2}\cos(2\pi t + 90°) + \frac{1}{3}\cos(3\pi t - 90°)\right.$

$\qquad \left. + \frac{1}{4}\cos(4\pi t + 90°) + \cdots\right]$ ▽

6.1-3 Determining the Fundamental Frequency and Period

We have seen that every periodic signal can be expressed as a sum of sinusoids of a fundamental frequency ω_0 and its harmonics. One may ask whether a sum of sinusoids of **any** frequencies represents a periodic signal. If so, how does one determine the period? Consider the following three functions:

$$f_1(t) = 2 + 7\cos\left(\tfrac{1}{2}t + \theta_1\right) + 3\cos\left(\tfrac{2}{3}t + \theta_2\right) + 5\cos\left(\tfrac{7}{6}t + \theta_3\right)$$

$$f_2(t) = 2\cos\left(2t + \theta_1\right) + 5\sin\left(\pi t + \theta_2\right)$$

$$f_3(t) = 3\sin\left(3\sqrt{2}t + \theta\right) + 7\cos\left(6\sqrt{2}t + \phi\right)$$

Recall that every frequency in a periodic signal is an integral multiple of the fundamental frequency ω_0. Therefore the ratio of any two frequencies is of the form m/n where m and n are integers. This means that the ratio of any two frequencies is a rational number. When the ratio of two frequencies is a rational number, they are said to be **harmonically** related. Therefore, for a signal to be periodic, all frequencies in its spectrum must be harmonically related; that is, **the ratio of any two frequencies is a rational number**.

The largest positive number of which all the frequencies are integral multiples is the fundamental frequency. The frequencies in the spectrum of $f_1(t)$ are $\frac{1}{2}$, $\frac{2}{3}$, and $\frac{7}{6}$ (we do not consider dc). The ratios of the successive frequencies are $\frac{3}{4}$ and $\frac{4}{7}$, respectively. Because both these numbers are rational, all the three frequencies in the spectrum are harmonically related and the signal $f_1(t)$ is periodic. The largest number of which $\frac{1}{2}$, $\frac{2}{3}$, and $\frac{7}{6}$ are integral multiples is $\frac{1}{6}$.† Moreover, $3(\frac{1}{6}) = \frac{1}{2}$, $4(\frac{1}{6}) = \frac{2}{3}$, and $7(\frac{1}{6}) = \frac{7}{6}$. Therefore the fundamental frequency is $\frac{1}{6}$. The three frequencies in the spectrum are the third, fourth, and seventh harmonics. Observe that the fundamental frequency component is absent in this Fourier series.

The signal $f_2(t)$ is not periodic because the ratio of two frequencies in the spectrum is $2/\pi$, which is not a rational number. The signal $f_3(t)$ is periodic because the ratio of frequencies $3\sqrt{2}$ and $6\sqrt{2}$ is $1/2$, a rational number. The greatest common divisor of $3\sqrt{2}$ and $6\sqrt{2}$ is $3\sqrt{2}$. Therefore the fundamental frequency $\omega_0 = 3\sqrt{2}$, and the period

$$T_0 = \frac{2\pi}{(3\sqrt{2})} = \frac{\sqrt{2}}{3}\pi$$

△ **Exercise E6.2**
   Determine whether the signal

$$f(t) = \cos\left(\tfrac{2}{3}t + 30°\right) + \sin\left(\tfrac{4}{5}t + 45°\right)$$

---

†The largest number of which $\frac{a_1}{b_1}$, $\frac{a_2}{b_2}$, $\cdots$, $\frac{a_m}{b_m}$ are integral multiples is the ratio of the GCF (greatest common factor) of the numerators set $(a_1, a_2, \cdots, a_m)$ to the LCM (least common multiple) of the denominator set $(b_1, b_2, \cdots, b_m)$. For instance, for the set $(\frac{2}{3}, \frac{6}{7}, 2)$, the GCF of the numerator set $(2, 6, 2)$ is 2; the LCM of the denominator set $(3, 7, 1)$ is 21. Therefore $\frac{2}{21}$ is the largest number of which $\frac{2}{3}, \frac{6}{7}$, and 2 are integral multiples.

is periodic. If it is periodic, find the fundamental frequency and the period. What harmonics are present in $f(t)$?

Answer: Periodic with $\omega_0 = \frac{2}{15}$ and period $T_0 = 15\pi$. The fifth and sixth harmonics.   ▽

6.1-4   The Role of Amplitude and Phase Spectra in Wave Shaping

The trigonometric Fourier series of a signal $f(t)$ shows explicitly the sinusoidal components of $f(t)$. We can synthesize $f(t)$ by adding the sinusoids in the spectrum of $f(t)$. Let us synthesize the square-pulse periodic signal $f(t)$ of Fig. 6.6a by adding successive harmonics in its spectrum step by step and observing the similarity of the resulting signal to $f(t)$. The Fourier series for this function as found in Example 6.4 is

$$f(t) = \frac{1}{2} + \frac{2}{\pi}\left(\cos t - \frac{1}{3}\cos 3t + \frac{1}{5}\cos 5t - \frac{1}{7}\cos 7t + \cdots\right)$$

We start the synthesis with only the first term in the series($n = 0$), a constant $\frac{1}{2}$ (dc); this is a gross approximation of the square wave, as shown in Fig. 6.8a. In the next step we add the dc ($n = 0$) and the first harmonic (fundamental), which results in a signal shown in Fig. 6.8b. Observe that the synthesized signal somewhat resembles $f(t)$. It is a smoothed out version of $f(t)$. The sharp corners in $f(t)$ are not reproduced in this signal because sharp corners mean rapid changes and their reproduction requires rapidly varying (that is, higher frequency) components, which are excluded. Figure 6.8c shows the sum of dc, first, and third harmonics (even harmonics are absent). As we increase the number of harmonics progressively, as shown in Figs. 6.8d (sum up to the fifth harmonic), and 6.8e (sum up to the nineteenth harmonic), the edges of the pulses become sharper and the signal resembles $f(t)$ more closely.

Asymptotic Rate of Amplitude Spectrum Decay

The amplitude spectrum indicates the amounts (amplitudes) of various frequency components of $f(t)$. If $f(t)$ is a smooth function, its variations are less rapid. Synthesis of such a function requires predominantly lower-frequency sinusoids and relatively small amounts of rapidly varying (higher frequency) sinusoids. The amplitude spectrum of such a function would decay swiftly with frequency. To synthesize such a function we require fewer terms in the Fourier series for a good approximation. On the other hand, a signal with sharp changes, such as jump discontinuities, contains rapid variations and its synthesis requires relatively large amount of high-frequency components. The amplitude spectrum of such a signal would decay slowly with frequency, and to synthesize such a function, we require many terms in its Fourier series for a good approximation The square wave $f(t)$ is a discontinuous function with jump discontinuities, and therefore its amplitude spectrum decays rather slowly, as $1/n$ [see Eq. (6.24)]. On the other hand, the triangular pulse periodic signal in Fig. 6.4a is smoother because it is a continuous function (no jump discontinuities). Its spectrum decays rapidly with frequency as $1/n^2$ [see Eq. (6.21b)].

It can be shown[1] that if the first $k - 1$ derivatives of a periodic signal $f(t)$ are continuous and the kth derivative is discontinuous, then its amplitude spectrum C_n decays with frequency at least as rapidly as $1/n^{k+1}$. This result provides a simple

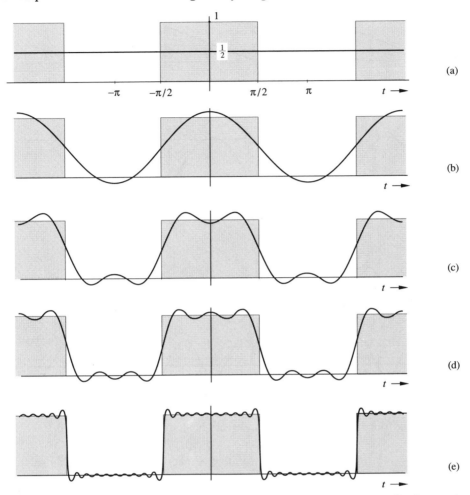

Fig. 6.8 Synthesis of a square pulse periodic signal by successive addition of its harmonics.

and useful means for predicting the asymptotic rate of convergence of the Fourier series. In the case of the square-wave signal (Fig. 6.6a), the zero-th derivative of the signal (the signal itself) is discontinuous, so that $k = 0$. For the triangular periodic signal in Fig. 6.4a, the first derivative is discontinuous; that is, $k = 1$. For this reason the spectra of these signals decay as $1/n$ and $1/n^2$, respectively.

The Role of the Phase Spectrum

The relationship between the amplitude spectrum and the waveform $f(t)$ is reasonably clear. The relationship between the phase spectrum and the periodic signal waveform is not so direct, however. Yet the phase spectrum plays an equally important role in waveshaping. We can explain this role by considering a signal $f(t)$ that has rapid changes such as jump discontinuities. To synthesize an instantaneous change at a jump discontinuity, the phases of the various sinusoidal components in its spectrum must be such that all (or most) of the harmonic components will have one sign before the discontinuity and the opposite sign after the discontinuity. This will result in a sharp change in $f(t)$ at the point of discontinuity. We can verify this

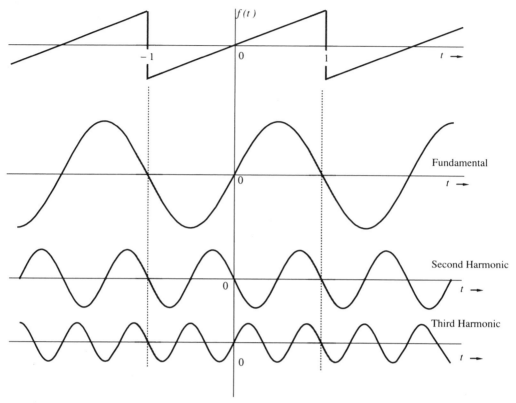

Fig. 6.9 Role of the phase spectrum in shaping a periodic signal.

fact in any waveform with jump discontinuity. Consider, for example, the sawtooth waveform in Fig. 6.7b. This waveform has a discontinuity at $t = 1$. The Fourier series for this waveform as given in Exercise E6.1b is

$$f(t) = \frac{2A}{\pi}\left[\cos(\pi t - 90°) + \tfrac{1}{2}\cos(2\pi t + 90°) + \tfrac{1}{3}\cos(3\pi t - 90°)\right.$$
$$\left. + \tfrac{1}{4}\cos(4\pi t + 90°) + \cdots\right]$$

Figure 6.9 shows the first three components of this series. The phases of all the (infinite) components are such that all the components are positive just before $t = 1$ and turn negative just after $t = 1$, the point of discontinuity. The same behavior is also observed at $t = -1$, where similar discontinuity occurs. This sign change in all the harmonics adds up to produce the jump discontinuity. The role of the phase spectrum is crucial in achieving a sharp change in the waveform. If we try to reconstruct this signal while ignoring the phase spectrum, the result will be a smeared and spread-out waveform. In general the phase spectrum is just as crucial in determining the waveform as is the amplitude spectrum. **The synthesis of any signal $f(t)$ is achieved by using a proper combination of amplitudes and phases of various sinusoids. This unique combination is the Fourier spectrum of $f(t)$.**

⊙ **Computer Example C6.2**

This example illustrates the synthesis of a square wave by successively adding its Fourier components as shown in Fig. 6.8.

```
t11=-3*pi/2:3*pi/200:-pi/2;t11=t11';y11=0*t11;
t22=-pi/2:3*pi/200:pi/2;t22=t22';y22=0*t22+1;
t33=pi/2:3*pi/200:3*pi/2;t33=t33';y33=0*t33;
tt=[t11;t22;t33];yy=[y11;y22;y33];plot(tt,yy),grid,pause
n1=1:4:17;   n1=n1';       n2=3:4:19;   n2=n2';
for k=1:length(n1)
t=-3*(pi/2):3*pi/200:3*pi/2;    t=t';
f1=(2/(pi*n1(k)))*cos(n1(k).*t);
f2=-(2/(pi*n2(k)))*cos(n2(k).*t);
t1=[t1 f1];  t2=[t2 f2];
end,
for c=1:length(t),  fc=0.5;  tc=[tc fc];  end,   tc=tc';
plot(tt,yy,t,tc),grid,xlabel('t'),ylabel('amplitude'),pause,  t33=0.5;
for m=1:length(t1(1,:))
t11=t1(:,m)+t33;
plot(tt,yy,t,t11),grid,xlabel('t'),ylabel('amplitude'),pause,
t22=t11+t2(:,m);
plot(tt,yy,t,t22),grid,xlabel('t'),ylabel('amplitude'),pause
t33=t22;
end   ⊙
```

Abrupt Truncation of the Fourier Series: The Gibbs Phenomenon

Fig. 6.8 shows the square function $f(t)$ and its approximation by a truncated trigonometric Fourier series that includes only the first n harmonics for $n=1$, 3, 5, and 19. The plot of the truncated series approximates closely the function $f(t)$ as n increases, and we expect that the series will converge exactly to $f(t)$ as $n \to \infty$. Yet the curious fact, as seen from Fig. 6.8, is that even for large n, the truncated series exhibits an oscillatory behavior and an overshoot approaching a value of about 9% in the vicinity of the discontinuity at the first peak of oscillation. As n increases, the frequency of oscillation increases, and the overshoot occurs closer to the point of discontinuity. However, no matter how large an n we choose, the plot exhibits oscillations, and the overshoot remains at about 9% of the discontinuity. This property of the Fourier series, known as the **Gibbs phenomenon**, arises because of the abrupt termination of the series after n terms where the first n terms of the series are given a weight of 1 and all the remaining terms (greater than n) are given a weight of 0. An explanation of this seemingly irrational phenomenon is given in Sec. 6.3-5. More discussion on the Gibbs phenomenon and its ramifications is found in Sec. 7.10.

△ **Exercise E6.3**

By inspection of signals in Figs. 6.3a, 6.7a, and 6.7b, determine the asymptotic rate of decay of their amplitude spectra.

Answer: $1/n, 1/n^2$, and $1/n$, respectively. ▽

A Historical Note: Baron Jean-Baptiste-Joseph Fourier (1768-1830)

The Fourier series and integral is a most beautiful and fruitful development, which serves as an indispensable instrument in the treatment of many problems in mathematics, science, and engineering. Maxwell was so taken by the beauty of the Fourier series, that he called it a great mathematical poem. In electrical engineering,

Joseph Fourier (left) and **Napoleon** (right).

it is central to the areas of communication, signal processing and several other fields, including antennas, but it was not received enthusiastically by the scientific world when it was presented. In fact, Fourier could not get his results published as a paper.

Fourier, a tailor's son, was orphaned at age 8 and educated at a local military college (run by Benedictine monks), where he excelled in mathematics. The Benedictines prevailed upon the young genius to choose the priesthood as his profession, but the revolution broke out before he could take his vows. Fourier joined the people's party. But in its early days, the French Revolution, like most revolutions of its kind, liquidated a large segment of the intelligentsia, including prominent scientists such as Lavosier. This caused many intellectuals to leave France to save themselves from a rapidly rising tide of barbarism. Fourier, who was an early enthusiast of the evolution, narrowly escaped the guillotine twice. It was to the everlasting credit of Napoleon that he stopped the persecution of the intelligentsia and founded new schools to replenish their ranks. The 26-year old Fourier was appointed chair of mathematics at the newly created school École Normale in 1794.[2]

Napoleon was the first modern ruler with a scientific education, and he was one of the rare persons who was equally comfortable on the battlefield and in discussing science. The age of Napoleon was one of the most fruitful in the history of science. Napoleon liked to sign himself as a "member of *institut de France*" (a fraternity of scientists), and he once expressed to Laplace his regret that "force of circumstances has led me so far from the career of a scientist."[3] Many great figures in science and mathematics, including Fourier and Laplace, were honored and promoted by Napoleon. In 1798, he took a group of scientists, artists, and scholars—Fourier among them—on his Egyptian expedition, with the promise of an exciting and historic union of adventure and research. Fourier proved to be a capable administrator of the newly formed Institut d'Egypte, which, incidentally, was responsible for the discovery of the Rosetta Stone. The inscription on this stone in two languages and three scripts (hieroglyphic, demotic, and Greek) enabled Thomas Young and Jean-Francois Champollion, a protege of Fourier, to establish a method of translating hieroglyphic Egyptian writings. This is said to be the chief—and only significant—result of Napoleon's Egyptian expedition.

Back in France in 1801, Fourier briefly served in his former position as professor of mathematics at the École Polytechnique in Paris. In 1802 Napoleon appointed him the prefect of Isère (with its headquarters in Grenoble), a position in which Fourier served with distinction. Fourier was created Baron of the Empire by Napoleon in 1809. Later, when Napoleon was exiled to Elba, his route was to take him through Grenoble. Fourier had the route changed to avoid meeting Napoleon, which would have displeased Fourier's new master, the Bourbon King Louis XVIII. Within a year, Napoleon escaped from Elba. On his way home, at Grenoble, Fourier was brought before him in chains. Napoleon scolded Fourier for his ungrateful behavior but reappointed him the prefect of Rhone at Lyons. Within four months Napoleon was defeated at Waterloo, and was exiled to St. Helena, never to return. Fourier once again was in disgrace as a Bonapartist, and had to pawn his effects to keep himself alive. But through the intercession of a former student, who was now a prefect of Paris, he was appointed director of the statistical bureau of the Seine, a position that allowed him ample time for scholarly pursuits. Later, in 1827, he was elected to the powerful position of perpetual secretary of the Paris Academy of Science, a section of the institute.[4]

While serving as the prefect of Grenoble, Fourier carried on his elaborate investigation of propagation of heat in solid bodies, which led him to the Fourier series and the Fourier integral. On 21 December 1807, he announced these results in a prize paper on the theory of heat. Fourier claimed that an arbitrary function (continuous or with discontinuities) defined in a finite interval by an arbitrarily capricious graph can always be expressed as a sum of sinusoids (Fourier series). The judges, who included the three great French mathematicians Laplace, Lagrange, and Legendre admitted the novelty and importance of Fourier's work, but criticized it for lack of mathematical rigor and generality. Lagrange thought it incredible that a sum of sines and cosines could add up to anything but an infinitely differentiable function, which has the property that its knowledge over an arbitrarily small interval determines the behavior of such functions over the entire range (the Taylor-Maclaurin series). Such a function is far from an arbitrary or a capriciously drawn graph.[5] Fourier thought the criticism unjustified but was unable to prove his claim because the tools required for operations with infinite series were not available at the time. Posterity proved Fourier to be closer to the truth than his critics. This is the classic conflict between pure mathematicians and physicists or engineers, as we have seen in the life of Oliver Heaviside (p. 264). In 1829 Dirichlet proved Fourier's claim for capriciously drawn functions with few restrictions (Dirichlet conditions).

Although three of the four judges were in favor of publication, this paper was rejected because of vehement opposition by Lagrange. Fifteen years later, after several attempts and disappointments, Fourier published the results in expanded form as a text, *Theorie analytique de la chaleur*, which is now a classic.

6.2 EXPONENTIAL FOURIER SERIES

Since sinusoids can be expressed in terms of exponentials, the trigonometric Fourier series can be expressed in terms of exponentials. This follows from the Euler's formula

$$C_n \cos\left(n\omega_0 t + \theta_n\right) = \frac{C_n}{2}\left[e^{j(n\omega_0 t + \theta_n)} + e^{-j(n\omega_0 t + \theta_n)}\right]$$

$$= \left(\frac{C_n}{2}e^{j\theta_n}\right)e^{jn\omega_0 t} + \left(\frac{C_n}{2}e^{-j\theta_n}\right)e^{-jn\omega_0 t}$$

$$= D_n e^{jn\omega_0 t} + D_{-n}e^{-jn\omega_0 t} \tag{6.28}$$

where

$$D_n = \tfrac{1}{2}C_n e^{j\theta_n}$$

$$D_{-n} = \tfrac{1}{2}C_n e^{-j\theta_n} \tag{6.29}$$

It is clear that a sinusoid $C_n \cos\left(n\omega_0 t + \theta_n\right)$ can be expressed as sum of two exponentials $D_n e^{jn\omega_0 t}$ and $D_{-n}e^{-jn\omega_0 t}$.

The compact trigonometric Fourier series of a periodic signal $f(t)$ is given by

$$f(t) = C_0 + \sum_{n=1}^{\infty} C_n \cos\left(n\omega_0 t + \theta_n\right)$$

Use of Eq. (6.29) in the above equation yields

$$f(t) = D_0 + \sum_{n=1}^{\infty} D_n e^{jn\omega_0 t} + D_{-n}e^{-jn\omega_0 t}$$

$$= D_0 + \sum_{n=-\infty\,(n\neq 0)}^{\infty} D_n e^{jn\omega_0 t}$$

where $D_0 = C_0$ and D_n, D_{-n} are given by Eq. (6.29). If we allow n to take on value 0 in the above summation, then

$$f(t) = \sum_{n=-\infty}^{\infty} D_n e^{jn\omega_0 t} \tag{6.30}$$

This shows that a periodic signal can be expressed by an exponential form of Fourier series. The coefficients D_n of this series are directly related to C_n and θ_n, the trigonometric Fourier series coefficients, as shown in Eq. (6.29). From Eqs. (6.29) and (6.6), it follows that

$$D_n = \tfrac{1}{2}C_n e^{j\theta_n} = \frac{1}{2}[C_n \cos\theta_n + jC_n \sin\theta_n] = \tfrac{1}{2}(a_n - jb_n) \tag{6.31a}$$

$$D_{-n} = \tfrac{1}{2}C_n e^{-j\theta_n} = \frac{1}{2}[C_n \cos\theta_n - jC_n \sin\theta_n] = \tfrac{1}{2}(a_n + jb_n) \tag{6.31b}$$

Substitution of Eqs. (6.14b) and (6.14c) in Eq. (6.31a) yields†

$$
\begin{aligned}
D_n &= \frac{1}{T_0} \left[\int_{T_0} f(t) \cos n\omega_0 t \, dt - j \int_{T_0} f(t) \sin n\omega_0 t \, dt \right] \\
&= \frac{1}{T_0} \int_{T_0} f(t) \left[\cos n\omega_0 t - j \sin n\omega_0 t \right] dt \\
&= \frac{1}{T_0} \int_{T_0} f(t) e^{-jn\omega_0 t} \, dt
\end{aligned}
\tag{6.32}
$$

In the same way we obtain

$$
D_{-n} = \frac{1}{T_0} \int_{T_0} f(t) e^{jn\omega_0 t} \, dt
\tag{6.33}
$$

Observe that Eq. (6.33) is implied in Eq. (6.32) and can be obtained merely by changing the sign of n in Eq. (6.32). Also,

$$
D_0 = C_0 = a_0 = \frac{1}{T_0} \int_{T_0} f(t) \, dt
\tag{6.34}
$$

We can obtain D_0 from Eq. (6.32) by letting $n = 0$. Therefore Eq. (6.32) is valid for all values of n (positive, negative and zero).

In summary, a periodic signal of period T_0 can be expressed by an exponential Fourier series

$$
f(t) = \sum_{n=-\infty}^{\infty} D_n e^{jn\omega_0 t} \qquad \omega_0 = \frac{2\pi}{T_0}
\tag{6.35}
$$

where the coefficients D_n are complex, in general, and are given by

$$
D_n = \frac{1}{T_0} \int_{T_0} f(t) e^{-jn\omega_0 t} \, dt
\tag{6.36}
$$

---

†Equation (6.32) can also be obtained directly by multiplying both sides of Eq. (6.30) by $e^{-jm\omega_0 t}$ and then integrating over the interval T_0. In the derivation we need the identity

$$
\int_{T_0} e^{jn\omega_0 t} e^{-jm\omega_0 t} \, dt =
\begin{cases}
0 & m \neq n \\
T_0 & m = n
\end{cases}
$$

Remember also that $e^{\pm j2\pi k} = 1$ for any integral value of k.

Table 6.2

Fourier Series Representation of a Periodic Signal of Period T_0 ($\omega_0 = 2\pi/T_0$)

Series form	Coefficient computation	Conversion formulas
Trigonometric $$f(t) = a_0 + \sum_{n=1}^{\infty} a_n \cos n\omega_0 t + b_n \sin n\omega_0 t$$	$$a_0 = \frac{1}{T_0} \int_{T_0} f(t)\, dt$$ $$a_n = \frac{2}{T_0} \int_{T_0} f(t) \cos n\omega_0 t\, dt$$ $$b_n = \frac{2}{T_0} \int_{T_0} f(t) \sin n\omega_0 t\, dt$$	$$a_0 = C_0 = D_0$$ $$a_n - jb_n = C_n e^{j\theta_n} = 2D_n$$ $$a_n + jb_n = C_n e^{-j\theta_n} = 2D_{-n}$$
Compact Trigonometric $$f(t) = C_0 + \sum_{n=1}^{\infty} C_n \cos(n\omega_0 t + \theta_n)$$	$$C_0 = a_0$$ $$C_n = \sqrt{a_n^2 + b_n^2}$$ $$\theta_n = \tan^{-1}\left(\frac{-b_n}{a_n}\right)$$	$$C_0 = D_0$$ $$C_n = 2|D_n| \quad n \geq 1$$ $$\theta_n = \angle D_n$$
Exponential $$f(t) = \sum_{n=-\infty}^{\infty} D_n e^{jn\omega_0 t}$$	$$D_n = \frac{1}{T_0} \int_{T_0} f(t) e^{-jn\omega_0 t}\, dt$$	

Observe the compactness of expressions (6.35) and (6.36) and compare them to expressions corresponding to trigonometric Fourier series. These two equations demonstrate very clearly the principle virtue of exponential Fourier series. First, the form of the series is most compact. Second, the mathematical expression for deriving the coefficients of the series is also compact. It is much more convenient to handle the exponential series than the trigonometric one. For this reason we shall use exponential (rather than trigonometric) representation of signals in the rest of the book. Equations (6.32) and (6.33) show that for real $f(t)$, D_n and D_{-n} are conjugates. This fact also follows from Eqs. (6.31). Therefore, for a real periodic signal $f(t)$,

$$D_{-n} = D_n{}^* \tag{6.37}$$

and if

$$D_n = |D_n| e^{j\angle D_n} \tag{6.38}$$

then

$$D_{-n} = |D_n| e^{-j\angle D_n} \tag{6.39}$$

Therefore

$$|D_n| = |D_{-n}| \tag{6.40a}$$

$$\angle D_n = -\angle D_{-n} \tag{6.40b}$$

Note that $|D_n|$ are the amplitudes (magnitudes) and $\angle D_n$ are the angles of various exponential components. From Eqs. (6.40) it follows that the amplitude spectrum ($|D_n|$ vs. ω) is an even function of ω and the angle spectrum ($\angle D_n$ vs. ω) is an odd function of ω when $f(t)$ is a real signal.

■ **Example 6.5**
   Find the exponential Fourier series for the signal in Fig. 6.3a (Example 6.1).
   In this case $T_0 = \pi$, $\omega_0 = 2\pi/T_0 = 2$, and

$$f(t) = \sum_{n=-\infty}^{\infty} D_n e^{j2nt}$$

where

$$D_n = \frac{1}{T_0} \int_{T_0} f(t) e^{-j2nt}\, dt \tag{6.41}$$

$$= \frac{1}{\pi} \int_0^{\pi} e^{-t/2}\, e^{-j2nt}\, dt$$

$$= \frac{1}{\pi} \int_0^{\pi} e^{-(\frac{1}{2}+j2n)t}\, dt$$

$$= \frac{-1}{\pi\left(\frac{1}{2}+j2n\right)} e^{-(\frac{1}{2}+j2n)t} \bigg|_0^{\pi}$$

$$= \frac{0.504}{1+j4n} \tag{6.42}$$

and

$$f(t) = 0.504 \sum_{n=-\infty}^{\infty} \frac{1}{1+j4n} e^{j2nt} \tag{6.43a}$$

$$= 0.504 \left[1 + \tfrac{1}{1+j4} e^{j2t} + \tfrac{1}{1+j8} e^{j4t} + \tfrac{1}{1+j12} e^{j6t} + \cdots \right.$$
$$\left. + \tfrac{1}{1-j4} e^{-j2t} + \tfrac{1}{1-j8} e^{-j4t} + \tfrac{1}{1-j12} e^{-j6t} + \cdots \right] \tag{6.43b}$$

Observe that the coefficients D_n are complex. Moreover, D_n and D_{-n} are conjugates, as expected. ■

⊙ **Computer Example C6.3**

Using a computer, compute and plot the exponential Fourier spectra for the periodic signal $f(t)$ shown in Fig. 6.3a (Example 6.5). Use Eqs. (6.42) to compute D_n.

```
n=-10:1:10;
num=0.504;
den=1+j*4*n;
N=length(den);
for i=1:N
Dn(i)=num/den(i);
end
magDn=abs(Dn);
phaseDn=(180/pi)*imag(log(Dn))';
plot(n,magDn,'o'),grid,
xlabel('n'),ylabel('|Dn|'),
title('Plot of |Dn|'),pause,
plot(n,phaseDn,'o'),grid,
xlabel('n'),ylabel('< Dn'),
title('Plot of Angle Dn'),grid   ⊙
```

6.2-1   Exponential Fourier Spectra

In exponential spectra, we plot coefficients D_n as a function of ω. But since D_n is complex in general, we need two plots: the real and the imaginary parts of D_n or the amplitude (magnitude) and the angle of D_n. We prefer the latter because of its close connection to the amplitudes and phases of corresponding components of the trigonometric Fourier series. We therefore plot $|D_n|$ vs. ω and $\angle D_n$ vs. ω. This requires that the coefficients D_n be expressed in polar form as $|D_n| e^{j \angle D_n}$. For the series in Example 6.5 [Eq. (6.43b)], for instance,

$$D_0 = 0.504$$

$$D_1 = \frac{0.504}{1+j4} = 0.122 e^{-j75.96°} \implies |D_1| = 0.122, \ \angle D_1 = -75.96°$$

$$D_{-1} = \frac{0.504}{1-j4} = 0.122 e^{j75.96°} \implies |D_{-1}| = 0.122, \ \angle D_{-1} = 75.96°$$

and

$$D_2 = \frac{0.504}{1 + j8} = 0.0625 e^{-j82.87°} \implies |D_2| = 0.0625, \ \angle D_2 = -82.87°$$

$$D_{-2} = \frac{0.504}{1 - j8} = 0.0625 e^{j82.87°} \implies |D_{-2}| = 0.0625, \ \angle D_{-2} = 82.87°$$

and so on. Note that D_n and D_{-n} are conjugates, as expected [see Eqs. (6.40)].

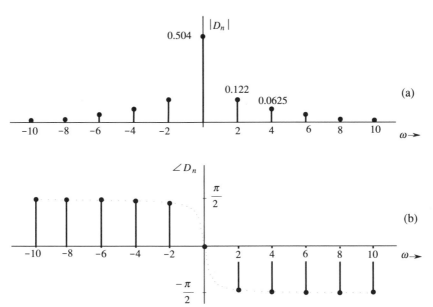

Fig. 6.10   Exponential Fourier spectra for the signal in Fig. 6.3a.

Figure 6.10 shows the frequency spectra (amplitude and angle) of the exponential Fourier series for the periodic signal $f(t)$ in Fig. 6.3a.

We notice some interesting features of these spectra. First, the spectra exist for positive as well as negative values of ω (the frequency). Second, the amplitude spectrum is an even function of ω and the angle spectrum is an odd function of ω. Finally, we see a close connection between these spectra and the spectra of the corresponding trigonometric Fourier series for $f(t)$ (Fig. 6.3).

The existence of the spectrum at negative frequencies is somewhat disturbing because by definition, the frequency (number of repetitions per second) is a positive quantity. How do we interpret a negative frequency? The cause of the confusion here is the use of the term "frequency" in a loose sense rather than in its conventional sense. We say that the frequency of $e^{j\omega_0 t}$ is ω_0. In this case, the frequency of $e^{-j\omega_0 t}$ is logically $-\omega_0$. In the conventional sense, however, both $e^{j\omega_0 t}$ and $e^{-j\omega_0 t}$ are signals of the same frequency ω_0 because

$$e^{\pm j\omega_0 t} = \cos \omega_0 t \pm j \sin \omega_0 t$$

Clearly, we are using the term "frequency" in a loose (exponential) sense when we say that the frequency of $e^{j\omega_0 t}$ is ω_0. A healthier way of looking at the situation is to say that **exponential spectra are a graphical representation of coefficients**

D_n **as a function of** ω**. Existence of the spectrum at** $\omega = -n\omega_0$ **is merely an indication of the fact that an exponential component** $e^{-jn\omega_0 t}$ **exists in the series**. Equation (6.28) shows that a sinusoid of frequency $n\omega_0$ can be expressed in terms of a pair of exponentials $e^{jn\omega_0 t}$ and $e^{-jn\omega_0 t}$. Moreover, for a real $f(t)$, the strengths of these pairs (D_n and D_{-n}) are complex conjugates in general. This results in the fact that $|D_n|$ is an even function and $\angle D_n$ is an odd function of ω.

Equation (6.29) shows the close connection between the trigonometric spectra (C_n and θ_n) with exponential spectra ($|D_n|$ and $\angle D_n$). It is clear that

$$D_0 = C_0$$

$$|D_n| = \frac{C_n}{2} \qquad n \neq 0$$

$$\angle D_n = \theta_n \qquad n > 0 \tag{6.44}$$

The dc components D_0 and C_0 are identical in both spectra. Moreover, the exponential amplitude spectrum $|D_n|$ is half of the trigonometric amplitude spectrum C_n for $n \geq 1$. The exponential angle spectrum $\angle D_n$ is identical to the trigonometric phase spectrum θ_n for $n \geq 0$. We can therefore produce the exponential spectra merely by inspection of trigonometric spectra, and vice versa. The following examples demonstrate this feature.

■ **Example 6.6**

The trigonometric Fourier spectra of a certain periodic signal $f(t)$ are shown in Fig. 6.11a. Through inspection of these spectra, sketch the corresponding exponential Fourier spectra and verify your results analytically.

The trigonometric spectral components exist at frequencies $0, 3, 6,$ and 9. The exponential spectral components exist at $0, 3, 6, 9$ and $-3, -6, -9$. Consider first the amplitude spectrum. The dc component remains unchanged; that is, $D_0 = C_0 = 16$. Now $|D_n|$ is an even function of ω and $|D_n| = |D_{-n}| = C_n/2$. Thus all the remaining spectrum $|D_n|$ for positive n is half the trigonometric amplitude spectrum C_n, and the spectrum $|D_n|$ for negative n is a reflection about the vertical axis of the spectrum for positive n, as shown in Fig. 6.11b.

The angle spectrum $\angle D_n$ is θ_n for positive n and is $-\theta_n$ for negative n, as shown in Fig. 6.11b. We shall now verify that both sets of spectra represent the same signal.

Signal $f(t)$ whose trigonometric spectra are shown in Fig. 6.11a, has four spectral components of frequencies $0, 3, 6,$ and 9. The dc component is 16. The amplitude and the phase of the component of frequency 3 are 12 and $-\frac{\pi}{4}$, respectively. Therefore this component can be expressed as $12\cos\left(3t - \frac{\pi}{4}\right)$. Proceeding in this manner, we can write the Fourier series for $f(t)$ as

$$f(t) = 16 + 12\cos\left(3t - \tfrac{\pi}{4}\right) + 8\cos\left(6t - \tfrac{\pi}{2}\right) + 4\cos\left(9t - \tfrac{\pi}{4}\right)$$

Consider now the exponential spectra in Fig. 6.11b. They contain components of frequencies 0 (dc), ± 3, ± 6, and ± 9. The dc component is $D_0 = 16$. The component e^{j3t} (frequency 3) has amplitude is 6 and angle is $-\frac{\pi}{4}$. Therefore this component strength is $6e^{-j\frac{\pi}{4}}$, and it can be expressed as $(6e^{-j\frac{\pi}{4}})e^{j3t}$. Similarly, the component of frequency -3 is $(6e^{j\frac{\pi}{4}})e^{-j3t}$. Proceeding in this manner, $\hat{f}(t)$, the signal corresponding to the spectra in Fig. 6.11b, is

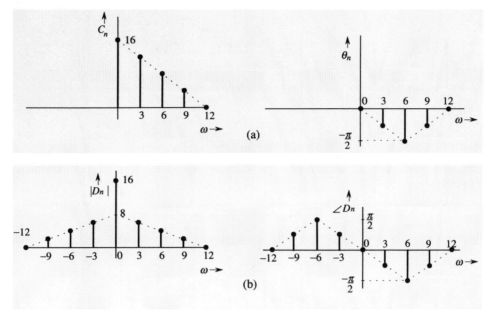

Fig. 6.11   Spectra for Example 6.6.

$$\hat{f}(t) = 16 + [6e^{-j\frac{\pi}{4}}e^{j3t} + 6e^{j\frac{\pi}{4}}e^{-j3t}] + [4e^{-j\frac{\pi}{2}}e^{j6t} + 4e^{j\frac{\pi}{2}}e^{-j6t}]$$

$$+ [2e^{-j\frac{\pi}{4}}e^{j9t} + 2e^{j\frac{\pi}{4}}e^{-j9t}]$$

$$= 16 + 6\left[e^{j(3t-\frac{\pi}{4})} + e^{-j(3t-\frac{\pi}{4})}\right] + 4\left[e^{j(6t-\frac{\pi}{2})} + e^{-j(6t-\frac{\pi}{2})}\right]$$

$$+ 2\left[e^{j(9t-\frac{\pi}{4})} + e^{-j(9t-\frac{\pi}{4})}\right]$$

$$= 16 + 12\cos\left(3t - \tfrac{\pi}{4}\right) + 8\cos\left(6t - \tfrac{\pi}{2}\right) + 4\cos\left(9t - \tfrac{\pi}{4}\right)$$

Clearly both sets of spectra represent the same periodic signal.   ■

Bandwidth of a Signal

The difference between the highest and the lowest frequencies of the spectral components of a signal is the **bandwidth** of the signal. The bandwidth of the signal whose exponential spectra are shown in Fig. 6.11b is 9 (in radians). The highest and lowest frequencies are 9 and 0 respectively. Note that the component of frequency 12 has zero amplitude and is nonexistent. Moreover, the lowest frequency is 0, not −9. Recall that the frequencies (in the conventional sense) of the spectral components at $\omega = -3, -6$, and -9 in reality are 3, 6, and 9.† The bandwidth can be more readily seen from the trigonometric spectra in Fig. 6.11a.

---

†Some authors *do* define bandwidth as the difference between the highest and the lowest (negative) frequency in the exponential spectrum. The bandwidth according to this definition is twice that defined here. In reality this definition defines not the signal bandwidth but the width of the exponential spectrum of the signal (*spectral width*).

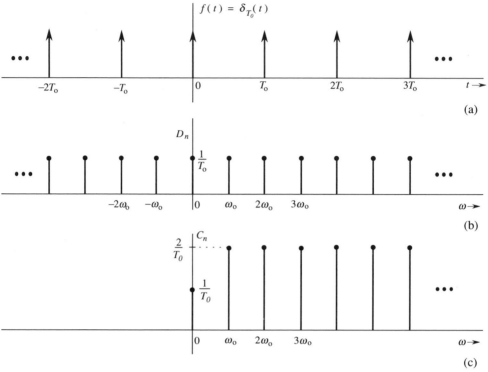

Fig. 6.12 Impulse train and its Fourier spectra.

■ **Example 6.7**

Find the exponential Fourier series and sketch the corresponding spectra for the impulse train $\delta_{T_0}(t)$ shown in Fig. 6.12a. From this result sketch the trigonometric spectrum and write the trigonometric Fourier series for $\delta_{T_0}(t)$.

The exponential Fourier series is given by

$$\delta_{T_0}(t) = \sum_{n=-\infty}^{\infty} D_n e^{jn\omega_0 t} \qquad \omega_0 = \frac{2\pi}{T_0} \tag{6.45}$$

where

$$D_n = \frac{1}{T_0} \int_{T_0} \delta_{T_0}(t) e^{-jn\omega_0 t}\, dt$$

Choosing the interval of integration $(\frac{-T_0}{2}, \frac{T_0}{2})$ and recognizing that over this interval $\delta_{T_0}(t) = \delta(t)$,

$$D_n = \frac{1}{T_0} \int_{-T_0/2}^{T_0/2} \delta(t) e^{-jn\omega_0 t}\, dt$$

In this integral the impulse is located at $t = 0$. From the sampling property (2.33a), the integral on the right-hand side is the value of $e^{-jn\omega_0 t}$ at $t = 0$ (where the impulse is located). Therefore

$$D_n = \frac{1}{T_0} \tag{6.46}$$

Substitution of this value in Eq. (6.45) yields the desired exponential Fourier series

$$\delta_{T_0}(t) = \frac{1}{T_0} \sum_{n=-\infty}^{\infty} e^{jn\omega_0 t} \qquad \omega_0 = \frac{2\pi}{T_0} \qquad (6.47)$$

Equation (6.46) shows that the exponential spectrum is uniform ($D_n = 1/T_0$) for all the frequencies, as shown in Fig. 6.12b. The spectrum, being real, requires only the amplitude plot. All phases are zero.

To sketch the trigonometric spectrum, we use Eq. (6.44) to obtain

$$C_0 = D_0 = \frac{1}{T_0}$$

$$C_n = 2|D_n| = \frac{2}{T_0} \qquad n = 1, 2, 3, \cdots$$

$$\theta_n = 0$$

Figure 6.12c shows the trigonometric Fourier spectrum. From this spectrum we can express $\delta_{T_0}(t)$ as

$$\delta_{T_0}(t) = \frac{1}{T_0}[1 + 2(\cos \omega_0 t + \cos 2\omega_0 t + \cos 3\omega_0 t + \cdots)] \qquad \omega_0 = \frac{2\pi}{T_0} \qquad (6.48)$$

■

△ **Exercise E6.4**

The exponential Fourier spectra of a certain periodic signal $f(t)$ are shown in Fig. 6.13. Determine and sketch the trigonometric Fourier spectra of $f(t)$ by inspection of Fig. 6.13. Now write the (compact) trigonometric Fourier series for $f(t)$.

Answer:
$$f(t) = 4 + 6\cos\left(3t - \frac{\pi}{6}\right) + 2\cos\left(6t - \frac{\pi}{4}\right) + 4\cos\left(9t - \frac{\pi}{2}\right)$$
$$= 4 + 6\cos\left(3t - \frac{\pi}{6}\right) + 2\cos\left(6t - \frac{\pi}{4}\right) + 4\sin 9t \qquad \triangledown$$

△ **Exercise E6.5**

Find the exponential Fourier series and sketch the corresponding Fourier spectrum D_n vs. ω for the full-wave rectified sine wave shown in Fig. 6.14.

Answer:
$$f(t) = \frac{2}{\pi} \sum_{n=-\infty}^{\infty} \frac{1}{1 - 4n^2} e^{j2nt} \qquad \triangledown$$

△ **Exercise E6.6**

Find the exponential Fourier series and sketch the corresponding Fourier spectra for the periodic signals shown in Figs. 6.7a and 6.7b.

Answer:

(a) $$f(t) = \frac{1}{3} + \frac{2}{\pi^2} \sum_{n=-\infty}^{\infty} \frac{(-1)^n}{n^2} e^{jn\pi t}$$

(b) $$f(t) = \frac{jA}{\pi} \sum_{n=-\infty \, (n\neq 0)}^{\infty} \frac{(-1)^n}{n} e^{jn\pi t} \qquad \triangledown$$

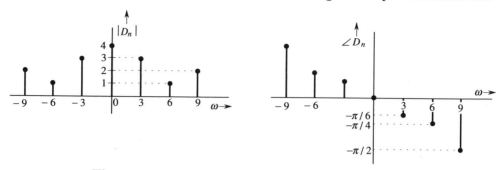

Fig. 6.13   Fourier spectra for the signal in Exercise E6.4.

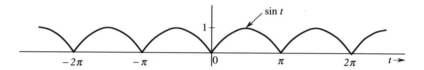

Fig. 6.14   A full-wave rectified sine wave in exercise E6.5.

6.2-2   Why Use Exponentials?

The exponential Fourier series is just another way of representing trigonometric Fourier series (or vice versa). The two forms carry identical information—no more, no less. The reasons for preferring the exponential form have been discussed: This form is very compact, and the expression for deriving the exponential coefficients is also very compact, as compared to those in the trigonometric series. Furthermore, as seen in Chapters 4 and 5, the system response to exponential signals is also much simpler (compact) than the system response to sinusoids. Moreover, mathematical manipulation and handling of exponential form prove much easier† than the trigonometric form in the area of signals as well as systems. For this reason in our future discussion we shall use exponential form exclusively.

A minor disadvantage of the exponential form is that it is less easy to visualize than the sinusoids. For intuitive and qualitative understanding, the sinusoids have the edge over exponentials. Fortunately, this difficulty can be overcome readily because of close connection between exponential and Fourier spectra. For the purpose of mathematical analysis we shall continue to use exponential signals and spectra, but to understand the physical situation intuitively or qualitatively we shall speak in terms of sinusoids and trigonometric spectra. Thus in future discussions, although all mathematical manipulation will be in terms of exponential spectra, we shall speak of exponential and sinusoids interchangeably when discussing intuitive and qualitative insights and the understanding of physical situations. This is an important point; readers should make extra effort in familiarizing themselves with the two forms of spectra, their relationships, and their convertibility.

---

†This is the direct result of the fact that exponentials have some interesting properties which simplify multiplication, division, differentiation, and integration of exponentials. For example, when exponentials are multiplied (divided), the exponents simply add (subtract). The derivative of e^{st} is se^{st} and the integral is e^{st}/s.

The discussion so far shows that a periodic signal has a dual personality—the time domain and frequency domain. It can be described by its waveform or by its Fourier spectra. In the next chapter we show that this is also true of nonperiodic signals. We shall also see that the time- and frequency-domain descriptions provide complementary insights into a signal. For in-depth perspective, we need to understand both of these identities. It is important for the reader to learn to think of a signal from both of these perspectives. In the next chapter, we shall see that systems also have this dual personality, which offers complementary insights into the system behavior.

6.3 AN ALTERNATIVE VIEW OF FOURIER REPRESENTATION: SIGNAL-VECTOR ANALOGY‡

We now consider a very general approach to signal representation with far-reaching consequences. There is a perfect analogy between signals and vectors; the analogy is so strong that the term "analogy" understates the reality. In fact, signals *are* vectors! A vector can be represented as a sum of its components in a variety of ways, depending upon the choice of coordinate system. A signal can also be represented as a sum of its components in a variety of ways one of which is the trigonometric (or exponential) Fourier series. Before discussing this subject, we shall describe the perfect vector-signal analogy.

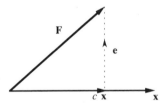

Fig. 6.15 Component (projection) of a vector along another vector.

6.3-1 Vector Comparison

A vector is specified by its magnitude and its direction. We shall denote all vectors by boldface type and their magnitudes by their absolute values. For example, $\mathbf{A}$ is a certain vector with magnitude $|\mathbf{A}|$. Consider two vectors $\mathbf{F}$ and $\mathbf{x}$, as shown in Fig. 6.15. Let the component of $\mathbf{F}$ along $\mathbf{x}$ be $c\mathbf{x}$. Geometrically the component of $\mathbf{F}$ along $\mathbf{x}$ is the projection of $\mathbf{F}$ on $\mathbf{x}$, and is obtained by drawing a perpendicular from the tip of $\mathbf{F}$ on the vector $\mathbf{x}$, as shown in Fig. 6.15. But what does this mean mathematically? The vector $\mathbf{F}$ can be expressed in terms of vector $\mathbf{x}$ as

$$\mathbf{F} = c\mathbf{x} + \mathbf{e} \tag{6.49}$$

‡This section closely follows the author's earlier work.[6] Derivation of Fourier series through signal-vector analogy provides an interesting insight into signal representation and other topics such as signal correlation, the Gibbs phenomenon, data truncation, window functions, and many more. However, if time constraints necessitate exclusion of the rest of this chapter, it may be omitted without causing discontinuity in the rest of the book.

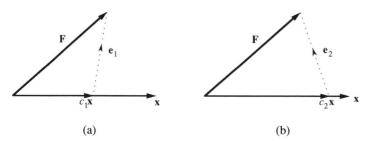

(a)   (b)

Fig. 6.16   Approximation of a vector in terms of another vector.

However, this is not the only way to express $\mathbf{F}$ in terms of $\mathbf{x}$. Figure 6.16 shows two of the infinite other possibilities. Thus, in Fig. 6.16a,

$$\mathbf{F} = c_1\mathbf{x} + \mathbf{e}_1 \tag{6.50a}$$

and in Fig. 6.16b

$$\mathbf{F} = c_2\mathbf{x} + \mathbf{e}_2 \tag{6.50b}$$

In each of these three representations [Eqs. (6.49), (6.50a), and (6.50b)] $\mathbf{F}$ is represented in terms of $\mathbf{x}$ plus another vector called the **error vector**. If we approximate $\mathbf{F}$ by $c\mathbf{x}$,

$$\mathbf{F} \simeq c\mathbf{x}$$

the error in the approximation is the vector $\mathbf{e}$. Similarly, the errors in approximations in Figs. 6.16a and 6.16b are $\mathbf{e}_1$ and $\mathbf{e}_2$. What is unique about the approximation in Fig. 6.15 is that the error vector is the smallest. We can now define mathematically the component of a vector $\mathbf{F}$ along vector $\mathbf{x}$ to be $c\mathbf{x}$ where c is chosen to minimize the length of the error vector $\mathbf{e} = \mathbf{F} - c\mathbf{x}$.

If $c\mathbf{x}$ is the component of $\mathbf{F}$ along $\mathbf{x}$ then the magnitude of c is an indication of similarity of the two vectors. If $c = 0$, then the vector $\mathbf{F}$ has no component along the vector $\mathbf{x}$. Geometrically this means that $\mathbf{F}$ and $\mathbf{x}$ are mutually perpendicular. Such vectors are called **orthogonal vectors**. Mathematically we define $\mathbf{F}$ and $\mathbf{x}$ to be orthogonal if $c = 0$. For convenience we define the dot product of two vectors $\mathbf{F}$ and $\mathbf{x}$ as

$$\mathbf{F} \cdot \mathbf{x} = |\mathbf{F}||\mathbf{x}| \cos \theta$$

where θ is the angle between vectors $\mathbf{F}$ and $\mathbf{x}$. Now, the length of the component of $\mathbf{F}$ along $\mathbf{x}$ is $|\mathbf{F}| \cos \theta$, but it is also $c|\mathbf{x}|$. Therefore

$$c|\mathbf{x}| = |\mathbf{F}| \cos \theta$$

and

$$c = \frac{|\mathbf{F}| \cos \theta}{|\mathbf{x}|}$$

$$= \frac{|\mathbf{F}||\mathbf{x}| \cos \theta}{|\mathbf{x}|^2}$$

$$= \frac{\mathbf{F} \cdot \mathbf{x}}{\mathbf{x} \cdot \mathbf{x}} \tag{6.51}$$

If **F** and **x** are orthogonal, $c = 0$. We therefore define **F** and **x** to be orthogonal if

$$\mathbf{F} \cdot \mathbf{x} = 0 \tag{6.52}$$

6.3-2   Signal Comparison

The concept of vector comparison and orthogonality can be extended to signals. Consider the problem of approximating a real signal $f(t)$ in terms of another real signal $x(t)$ over an interval $[t_1, t_2]$:

$$f(t) \simeq cx(t) \tag{6.53}$$

The error $e(t)$ in this approximation is

$$e(t) = f(t) - cx(t) \tag{6.54}$$

For the best approximation we must minimize the error (or some measure of the error) averaged over the interval $[t_1, t_2]$. Minimizing average error can be misleading because there could be large positive and negative errors that may cancel each other out in the process of averaging and give a false indication of zero error. A better measure is the **mean-squared error (MSE)**: that is, the value of $e^2(t)$ averaged over interval $[t_1, t_2]$. Because the mean-squared value is the square of the rms (root mean squared) value of a signal, it is a useful indication of the strength of a signal. Minimizing the MSE means minimizing the rms value of the error $e(t)$ over the interval $[t_1, t_2]$.

Now ϵ, the MSE of $e(t)$, is given by

$$\epsilon = \frac{1}{t_2 - t_1} \int_{t_1}^{t_2} e^2(t)\, dt$$

$$= \frac{1}{t_2 - t_1} \int_{t_1}^{t_2} [f(t) - cx(t)]^2\, dt$$

Note that the right-hand side is a definite integral with t as the dummy variable. Hence ϵ is a function of parameter c (not t) and ϵ is minimum for some choice of c. To minimize ϵ, a necessary condition is

$$\frac{d\epsilon}{dc} = 0 \tag{6.55}$$

or

$$\frac{d}{dc} \left[\int_{t_1}^{t_2} [f(t) - cx(t)]^2\, dt \right] = 0$$

Expanding the squared term, we obtain

$$\frac{d}{dc} \left[\int_{t_1}^{t_2} f^2(t)\, dt \right] - \frac{d}{dc} \left[2c \int_{t_1}^{t_2} f(t)x(t)\, dt \right] + \frac{d}{dc} \left[c^2 \int_{t_1}^{t_2} x^2(t)\, dt \right] = 0$$

From which we obtain

$$-2 \int_{t_1}^{t_2} f(t)x(t)\, dt + 2c \int_{t_1}^{t_2} x^2(t)\, dt = 0$$

and

$$c = \dfrac{\displaystyle\int_{t_1}^{t_2} f(t)x(t)\,dt}{\displaystyle\int_{t_1}^{t_2} x^2(t)\,dt} \tag{6.56}$$

We observe a remarkable similarity between the behavior of vectors and of signals, as indicated by Eqs. (6.51) and (6.56). It is evident from these two parallel expressions that **the integral of the product of two signals corresponds to the dot product of two vectors.**

To summarize our discussion, if a signal $f(t)$ is approximated by a signal $x(t)$ as

$$f(t) \simeq cx(t)$$

then the optimum value of c that minimizes the mean squared error (MSE) in this approximation is given by Eq. (6.56).

Taking our clue from vectors, we say that the component of signal $f(t)$ of the form $x(t)$ is $cx(t)$ where c is given by Eq. (6.56). Continuing with the analogy, we say that if the component of a signal $f(t)$ of the form $x(t)$ is zero (that is, $c = 0$), the signals $f(t)$ and $x(t)$ are orthogonal over the interval $[t_1, t_2]$. Therefore we define the real signals $f(t)$ and $x(t)$ to be orthogonal over the interval $[t_1, t_2]$ if†

$$\int_{t_1}^{t_2} f(t)x(t)\,dt = 0 \tag{6.57}$$

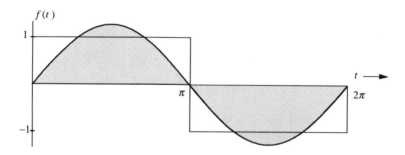

Fig. 6.17   Approximation of square signal in terms of a single sinusoid.

■ **Example 6.8**

For the square signal $f(t)$ defined by (Fig. 6.17)

$$f(t) = \begin{cases} 1 & 0 \le t \le \pi \\ -1 & \pi < t \le 2\pi \end{cases}$$

find the component in $f(t)$ of the form $\sin t$. In other words, approximate $f(t)$ in terms of $\sin t$:

$$f(t) \simeq c\sin t \qquad 0 \le t \le 2\pi$$

such that the MSE in this approximation is minimum.

---

†For complex signals the definition is modified as in Eq. (6.85).

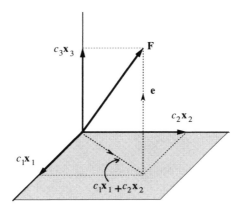

Fig. 6.18 Representation of a vector in three-dimensional space.

From Eq. (6.56), we find

$$c = \frac{\int_0^{2\pi} f(t) \sin t \, dt}{\int_0^{2\pi} \sin^2 t \, dt} = \frac{1}{\pi} \left[\int_0^{\pi} \sin t \, dt + \int_{\pi}^{2\pi} - \sin t \, dt \right] = \frac{4}{\pi}$$

Thus

$$f(t) \simeq \frac{4}{\pi} \sin t \tag{6.58}$$

represents the best approximation of $f(t)$ by a function $\sin t$, which will minimize the mean square error. This sinusoidal component of $f(t)$ is shown shaded in Fig. 6.17.

By analogy with vectors, we say that the square function $f(t)$ shown in Fig. 6.17 has a component of signal $\sin t$ and that the magnitude of this component is $4/\pi$. ∎

6.3-3 Orthogonal Vector Space

The analogy between vectors and signals may be extended further. Let us now investigate a three-dimensional Cartesian vector space described by three mutually orthogonal vectors $\mathbf{x}_1$, $\mathbf{x}_2$, and, $\mathbf{x}_3$, as shown in Fig. 6.18. First, we will seek to approximate a three-dimensional vector $\mathbf{F}$ in terms of two mutually orthogonal vectors $\mathbf{x}_1$ and $\mathbf{x}_2$:

$$\mathbf{F} \simeq c_1 \mathbf{x}_1 + c_2 \mathbf{x}_2$$

The error $\mathbf{e}$ in this approximation is

$$\mathbf{e} = \mathbf{F} - (c_1 \mathbf{x}_1 + c_2 \mathbf{x}_2)$$

or

$$\mathbf{F} = c_1 \mathbf{x}_1 + c_2 \mathbf{x}_2 + \mathbf{e}$$

As in the earlier geometrical argument, we see from Fig 6.18 that the length of $\mathbf{e}$ is minimum when $\mathbf{e}$ is perpendicular to the $\mathbf{x}_1 - \mathbf{x}_2$ plane. This means that $c_1 \mathbf{x}_1$ and $c_2 \mathbf{x}_2$ are the projections (components) of $\mathbf{F}$ on $\mathbf{x}_1$ and $\mathbf{x}_2$, respectively.

Now let us determine the best approximation to $\mathbf{F}$ in terms of all the three mutually orthogonal vectors $\mathbf{x}_1, \mathbf{x}_2$, and $\mathbf{x}_3$:

$$\mathbf{F} \simeq c_1 \mathbf{x}_1 + c_2 \mathbf{x}_2 + c_3 \mathbf{x}_3 \tag{6.59}$$

In this case, we find from Fig. 6.18 that a unique choice of c_1, c_2, and c_3 exists, for which (6.59) is no longer an approximation but an equality:

$$\mathbf{F} = c_1\mathbf{x}_1 + c_2\mathbf{x}_2 + c_3\mathbf{x}_3 \tag{6.60}$$

Once more, $c_1\mathbf{x}_1, c_2\mathbf{x}_2$, and $c_3\mathbf{x}_3$ are the projections (components) of $\mathbf{F}$ on $\mathbf{x}_1, \mathbf{x}_2$, and $\mathbf{x}_3$, respectively; that is,

$$c_i = \frac{\mathbf{F} \cdot \mathbf{x}_i}{\mathbf{x}_i \cdot \mathbf{x}_i} \qquad i = 1, 2, 3 \tag{6.61}$$

Note that the error in the approximation is zero when $\mathbf{F}$ is approximated in terms of three mutually orthogonal vectors: $\mathbf{x}_1, \mathbf{x}_2$, and $\mathbf{x}_3$. This is because $\mathbf{F}$ is a three-dimensional vector, and the vectors $\mathbf{x}_1, \mathbf{x}_2$, and $\mathbf{x}_3$ represent a *complete set* of orthogonal vectors in three-dimensional space. Completeness here means that it is impossible to find another vector $\mathbf{x}_4$ in this space, which is orthogonal to all three vectors $\mathbf{x}_1, \mathbf{x}_2$, and $\mathbf{x}_3$. Any vector in this space can then be represented (with zero error) in terms of these three vectors. Such vectors are known as *basis* vectors. If the set of vectors $\{\mathbf{x}_i\}$ is not complete, the error in the approximation will generally not be zero. Thus, in the three-dimensional case discussed above, it is not possible to represent a general vector $\mathbf{F}$ in terms of only two basis vectors without an error.

The choice of basis vectors is not unique. In fact, each set of basis vectors corresponds to a particular choice of coordinate system.

In this particular example, we had in mind ordinary physical space, which has only three dimensions, but these arguments can be generalized to N-dimensional space for an arbitrary $N \geq 1$. Any entity $\mathbf{F}$ that is specified by N ordered numbers (N-tuple) is an N-dimensional vector. If $\mathbf{x}_1, \mathbf{x}_2, \ldots, \mathbf{x}_N$ is a complete set of basis vectors for this space, then we can express $\mathbf{F}$ in terms of this basis set as

$$\mathbf{F} = c_1\mathbf{x}_1 + c_2\mathbf{x}_2 + c_3\mathbf{x}_3 + \cdots + c_N\mathbf{x}_N \tag{6.62}$$

All the vectors $\mathbf{x}_1, \mathbf{x}_2, \ldots, \mathbf{x}_N$ are mutually orthogonal, and the set must be complete if any general vector $\mathbf{F}$ is to be represented by Eq. (6.62). The condition of orthogonality implies that the dot product of any two vectors $\mathbf{x}_n$ and $\mathbf{x}_m$ must be zero, and the dot product of any vector with itself must be a constant:

$$\mathbf{x}_m \cdot \mathbf{x}_n = \begin{cases} 0 & m \neq n \\ K_n{}^2 & m = n \end{cases} \tag{6.63}$$

When $K_n{}^2 = 1$ for all n, the set is normalized, and it is called an **orthonormal** set.

The constants $c_1, c_2, \ldots, c_N$ in Eq. (6.62) can also be derived directly by taking the dot product of both sides of Eq. (6.62) with $\mathbf{x}_n$. This yields

$$\mathbf{F} \cdot \mathbf{x}_n = c_1\mathbf{x}_1 \cdot \mathbf{x}_n + c_2\mathbf{x}_2 \cdot \mathbf{x}_n + \cdots + c_n\mathbf{x}_n \cdot \mathbf{x}_n + \cdots + c_N\mathbf{x}_N \cdot \mathbf{x}_n \tag{6.64}$$

Use of Eq. (6.63) now yields

$$c_n = \frac{\mathbf{F} \cdot \mathbf{x}_n}{\mathbf{x}_n \cdot \mathbf{x}_n} = \frac{1}{K_n{}^2}\mathbf{F} \cdot \mathbf{x}_n \quad n = 1, 2, \ldots, N \tag{6.65}$$

Note that if $\{x_n\}$ were an orthonormal set, $K_n{}^2 = 1$, and

$$c_n = \mathbf{F} \cdot \mathbf{x}_n \tag{6.66}$$

6.3-4 Orthogonal Signal Space

We continue with our signal approximation problem using clues and insights developed for vector approximation. Consider the problem of approximating a signal $f(t)$ over the interval $[t_1, t_2]$ by a set of N real, mutually orthogonal signals $x_1(t)$, $x_2(t)$, ..., $x_N(t)$. As before, we define orthogonality of a real signal set over interval $[t_1, t_2]$ by

$$\int_{t_1}^{t_2} x_m(t)x_n(t)\, dt = \begin{cases} 0 & m \neq n \\ K_n{}^2 & m = n \end{cases} \tag{6.67}$$

If $K_n{}^2 = 1$ for all n, then the set is normalized and is called an **orthonormal set**. An orthogonal set can always be normalized by dividing $x_n(t)$ by K_n for all n.

Let us now approximate $f(t)$ in terms of the orthogonal signal set $\{x_i(t)\}$ over an interval $[t_1, t_2]$ as

$$f(t) \simeq c_1 x_1(t) + c_2 x_2(t) + \cdots + c_N x_N(t) \tag{6.68a}$$

$$= \sum_{n=1}^{N} c_n x_n(t) \tag{6.68b}$$

The error $e(t)$ in this approximation is

$$e(t) = f(t) - \sum_{n=1}^{N} c_n x_n(t) \tag{6.69}$$

and ϵ_N, the MSE, is

$$\epsilon_N = \frac{1}{t_2 - t_1} \int_{t_1}^{t_2} e^2(t)\, dt$$

$$= \frac{1}{t_2 - t_1} \int_{t_1}^{t_2} \left[f(t) - \sum_{n=1}^{N} c_n x_n(t) \right]^2 dt \tag{6.70}$$

We now use our criterion of best approximation to be that which minimizes ϵ_N, the mean-squared error. Since ϵ_N is a function of N parameters c_1, c_2, ..., c_N, to minimize ϵ_N, a necessary condition is

$$\frac{\partial \epsilon_N}{\partial c_i} = 0 \qquad i = 1, 2, \ldots, N$$

or

$$\frac{\partial}{\partial c_i} \int_{t_1}^{t_2} \left[f(t) - \sum_{n=1}^{N} c_n x_n(t) \right]^2 dt = 0 \tag{6.71}$$

When we expand the integrand, we find that all the cross-multiplication terms arising from the orthogonal signals are zero by virtue of orthogonality; that is, all terms of the form $\int x_m(t)x_n(t)\,dt$ with $m \neq n$ vanish. Similarly, the derivative with respect to c_i of all terms that do not contain c_i is zero. For each i, this leaves only two nonzero terms in Eq. (6.71):

$$\frac{\partial}{\partial c_i} \int_{t_1}^{t_2} \left[-2c_i f(t)x_i(t) + c_i^2 x_i^2(t) \right] dt = 0$$

or

$$-2 \int_{t_1}^{t_2} f(t)x_i(t)\,dt + 2c_i \int_{t_1}^{t_2} x_i^2(t)\,dt = 0 \qquad i = 1, 2, \dots, n$$

Therefore

$$c_i = \frac{\displaystyle\int_{t_1}^{t_2} f(t)x_i(t)\,dt}{\displaystyle\int_{t_1}^{t_2} x_i^2(t)\,dt} \tag{6.72a}$$

$$= \frac{1}{K_i^2} \int_{t_1}^{t_2} f(t)x_i(t)\,dt \quad i = 1, 2, \dots, N \tag{6.72b}$$

A comparison of Eq. (6.72) with Eq. (6.65) forcefully brings out the analogy of signals with vectors.

Equation (6.72) shows one interesting property of the coefficients of c_1, c_2, ..., c_N; the optimum value of any coefficient in the approximation (6.72a) is independent of the number of terms used in the approximation. For example, if we have used only one term $(N = 1)$ or two terms $(N = 2)$ or any number of terms, the optimum value of the coefficient c_1 would be the same [as given by Eq. (6.72)]. The advantage of this approximation of a signal $f(t)$ by a set of mutually orthogonal signals is that we can continue to add terms to the approximation without disturbing the previous terms. This property of **finality** of the values of the coefficients is very important from practical point of view.†

Mean-Squared Error (MSE)

When the coefficients c_i in the approximation (6.68) are chosen according to Eq. (6.72), the mean-squared error (MSE) in the approximation (6.68) is minimized.

---

†Contrast this situation with polynomial approximation of $f(t)$. Suppose we wish to approximate $f(t)$ by a polynomial in t such that the polynomial is equal to $f(t)$ at two points t_1 and t_2. This can be done by choosing a first order polynomial $a_0 + a_1 t$ with

$$f(t_1) = a_0 + a_1 t_1 \quad \text{and} \quad f(t_2) = a_0 + a_1 t_2$$

Solution of these equations yields the desired values of a_0 and a_1. For a three-point approximation, we must choose the polynomial $a_0 + a_1 t + a_2 t^2$ with

$$f(t_i) = a_0 + a_1 t_i + a_2 t_i^2 \quad i = 1, 2, \text{ and } 3$$

The approximation improves with larger number of points (higher-order polynomial), but the coefficients a_0, a_1, a_2, $\cdots$ do not have the finality property. As we increase the number of terms in the polynomial, we need to calculate the coefficients again.

This minimum value of ϵ_N is given by Eq. (6.70):

$$\epsilon_N = \frac{1}{t_2 - t_1} \int_{t_1}^{t_2} \left[f(t) - \sum_{n=1}^{N} c_n x_n(t) \right]^2 dt$$

$$= \frac{1}{t_2 - t_1} \left[\int_{t_1}^{t_2} f^2(t)\, dt + \sum_{n=1}^{N} c_n^2 \int_{t_1}^{t_2} x_n^2(t)\, dt - 2 \sum_{n=1}^{N} c_n \int_{t_1}^{t_2} f(t) x_n(t)\, dt \right]$$

Substitution of Eqs. (6.67) and (6.72) in this equation yields

$$\epsilon_N = \frac{1}{t_2 - t_1} \left[\int_{t_1}^{t_2} f^2(t)\, dt + \sum_{n=1}^{N} c_n^2 K_n^2 - 2 \sum_{n=1}^{N} c_n^2 K_n^2 \right]$$

$$= \frac{1}{t_2 - t_1} \left[\int_{t_1}^{t_2} f^2(t)\, dt - \sum_{n=1}^{N} c_n^2 K_n^2 \right] \tag{6.73}$$

Signal Representation by a Complete Orthogonal Set

So far we have approximated $f(t)$ only in terms of a set of mutually orthogonal signals. Is it possible to have exact representation of $f(t)$ rather than an approximation? For exact representation, ϵ_N, the MSE, must vanish. An examination of Eq. (6.73) shows that the MSE is a nonincreasing function of N, the number of terms. Furthermore, by definition the MSE can never be negative but may go to zero. Hence it is possible that $\epsilon \to 0$ as $N \to \infty$; the sum $\sum_n c_n K_n^2$ in Eq. (6.73) converges to $\int f^2(t)\, dt$. In this case

$$\int_{t_1}^{t_2} f^2(t)\, dt = \sum_{n=1}^{\infty} c_n^2 K_n^2 \tag{6.74}$$

and the approximation (6.68) becomes an equality

$$f(t) = c_1 x_1(t) + c_2 x_2(t) + \cdots + c_n x_n(t) + \cdots$$

$$= \sum_{n=1}^{\infty} c_n x_n(t) \tag{6.75}$$

with the coefficients c_n given by Eq. (6.72). The series on the right-hand side of Eq. (6.75) is called the **generalized Fourier series** of $f(t)$ with respect to the set $\{x_n(t)\}$. When the set $\{x_n(t)\}$ is such that $\epsilon_N \to 0$ as $N \to \infty$ for every member of some particular class† of function $f(t)$, we say that the set $\{x_n(t)\}$ is complete on $[t_1, t_2]$ for that class of $f(t)$, and the set $\{x_n(t)\}$ is called a set of **basis functions** or **basis signals**. Thus when the set $\{x_n(t)\}$ is complete, we have the equality (6.75). One subtle point that must be understood clearly is the meaning of equality in Eq. (6.75). **The equality here is not an equality in the ordinary sense, but in the sense that the mean-squared error (MSE) of the difference**

---

†In this book we shall consider only a class of functions which are square-integrable over the interval $[t_1, t_2]$.

between the two sides of Eq. (6.75) approaches zero. If the equality exists in the ordinary sense, the MSE of this difference is always zero, but the converse is not true. The MSE can approach zero even though $e(t)$, the difference between the two sides, is nonzero at some isolated instants. This is because even if $e(t)$ is nonzero at such instants, the area under $e^2(t)$ is still zero; thus the Fourier series on the right-hand side of Eq. (6.75) may differ from $f(t)$ at a finite number of points. In fact, when $f(t)$ has a jump discontinuity at $t = t_0$, the corresponding Fourier series at t_0 converges to the mean of $f(t_0{}^+)$ and $f(t_0{}^-)$.

Equation (6.74) is very important, and goes by the name of **Parseval's theorem**. Recall that the area under the squared value of a signal is analogous to the square of the length of a vector in the vector-signal analogy. In vector space we know that the square of the length of a vector is equal to the sum of the squares of the lengths of its orthogonal components. Parseval's theorem (6.74) is the statement of this fact as applied to signals. Parseval's theorem can also be interpreted in another way by expressing Eq. (6.74) as

$$\frac{1}{t_2 - t_1} \int_{t_1}^{t_2} f^2(t)\, dt = \frac{1}{t_2 - t_1} \left[\sum_{n=1}^{\infty} \int_{t_1}^{t_2} [c_n x_n(t)]^2 \, dt \right] \tag{6.76}$$

This equation states that if $f(t)$ is expressed in terms of a complete set of mutually orthogonal components $c_n x_n(t)$ over a certain interval, then the mean-squared value of $f(t)$ [left-hand side of Eq. (6.76)] is equal to the sum of the mean-squared values of all its orthogonal components [right-hand side of Eq. (6.76)].

■ **Example 6.9**

As an example, we shall consider again the square signal $f(t)$ in Fig. 6.17. In Example 6.8 this signal was approximated by a single sinusoid $\sin t$. Actually the set $\sin t$, $\sin 2t$, ..., $\sin nt$, $\cdots$ is orthogonal over the interval‡ $[0, 2\pi]$. The reader can verify this fact by showing that

$$\int_0^{2\pi} \sin mt \, \sin nt \, dt = \begin{cases} 0 & m \neq n \\ \pi & m = n \end{cases} \tag{6.77}$$

Let us approximate the square signal in Fig. 6.17, using this set, and see how the approximation improves with number of terms

$$f(t) \simeq c_1 \sin t + c_2 \sin 2t + \cdots + c_n \sin Nt$$

where

$$c_n = \frac{\int_0^{2\pi} f(t) \sin nt \, dt}{\int_0^{2\pi} \sin^2 nt \, dt}$$

$$= \frac{1}{\pi} \left[\int_0^{\pi} \sin nt \, dt + \int_{\pi}^{2\pi} -\sin nt \, dt \right]$$

$$= \begin{cases} \frac{4}{\pi n} & n \text{ odd} \\ 0 & n \text{ even} \end{cases} \tag{6.78}$$

---

‡In fact, this set is orthogonal over any interval of duration 2π.

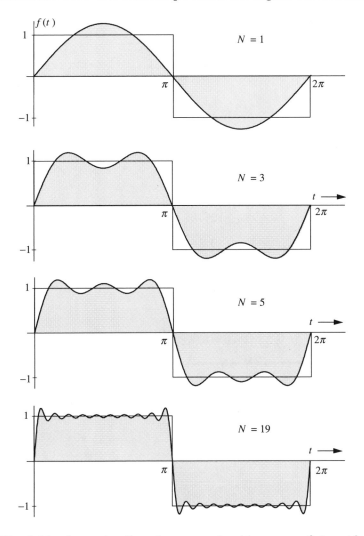

Fig. 6.19   Approximation of a square signal by a sum of sinusoids.

Therefore

$$f(t) \simeq \frac{4}{\pi} \left(\sin t + \frac{1}{3} \sin 3t + \frac{1}{5} \sin 5t + \cdots + \frac{1}{N} \sin Nt \right) \tag{6.79}$$

Note that coefficients of terms $\sin kt$ are zero for even values of k. Figure 6.19 shows the square function and the approximated function for $N = 1$, 3, 5, and 19 respectively. Observe how the approximation improves as N increases. The intriguing question is whether $f(t)$ can be represented exactly if we let $N \to \infty$. To answer this question, let us investigate ϵ_N, the MSE. From Eq. (6.73)

$$\epsilon_N = \frac{1}{2\pi} \left[\int_0^{2\pi} f^2(t)\,dt - \sum_{n=1}^{\infty} c_n{}^2 K_n{}^2 \right]$$

Note that

$$\int_0^{2\pi} f^2(t)\, dt = \int_0^{\pi} 1^2 \, dt + \int_{\pi}^{2\pi} -1^2 \, dt = 2\pi$$

$$c_n{}^2 = \begin{cases} \dfrac{16}{n^2\pi^2} & n \text{ odd} \\[6pt] 0 & n \text{ even} \end{cases}$$

and from Eq. (6.77)

$$K_n{}^2 = \pi$$

Therefore

$$\epsilon_N = \frac{1}{2\pi}\left[2\pi - \sum_{n=1,3,5,\cdots}^{N} \frac{16}{n^2\pi^2}\pi \right]$$

$$= 1 - \frac{8}{\pi^2} \sum_{n=1,3,5,\cdots}^{N} \frac{1}{n^2}$$

For a single-term approximation $(N = 1)$,

$$\epsilon_1 = 1 - \tfrac{8}{\pi^2} = 0.19$$

For a two-term approximation $(N = 3)$

$$\epsilon_3 = 1 - \tfrac{8}{\pi^2}\left(1 + \tfrac{1}{9}\right) = 0.1$$

For a three-term approximation $(N = 5)$

$$\epsilon_5 = 1 - \tfrac{8}{\pi^2}\left(1 + \tfrac{1}{9} + \tfrac{1}{25}\right) = 0.0669$$

For a four-term approximation $(N = 7)$

$$\epsilon_7 = 0.0504$$

Continuing in this way, we find that for a 50-term approximation $(N = 99)$

$$\epsilon_{99} = 0.00405$$

The reader can verify that

$$1 + \tfrac{1}{9} + \tfrac{1}{25} + \tfrac{1}{49} + \cdots + \cdots = \tfrac{\pi^2}{8}$$

and

$$\epsilon_\infty = 0$$

Clearly $f(t)$ can be represented by the infinite series

$$f(t) = \frac{4}{\pi}\left(\sin t + \frac{1}{3}\sin 3t + \frac{1}{5}\sin 5t + \cdots \right)$$

$$= \frac{4}{\pi} \sum_{n=1,3,5,\cdots}^{\infty} \frac{1}{n}\sin nt \tag{6.80}$$

The equality exists in the sense that the MSE of the difference between the two sides of Eq. (6.80) $\to 0$ as $N \to \infty$. ■

6.3-5   The Gibbs Phenomenon

Figure 6.19 shows that the series on the right-hand side of Eq. (6.80) exhibits oscillatory behavior at the points of discontinuity, with an overshoot of about 9% (of the amount of discontinuity) occurring near the points of discontinuities. Because the MSE approaches 0 as $N \to \infty$, we are tempted to think that the overshoot and oscillations would vanish as $N \to \infty$. Unfortunately this does not happen; the oscillations have frequency N, and as $N \to \infty$, the frequency of oscillation increases correspondingly. The overshoot remains at about 9% regardless of the value of N. This fact is readily seen from Fig. 6.19, which shows the plots for $N = 1$, 3, 5, and 19. This curious behavior appears to contradict the mathematical result that MSE $\to 0$ as $N \to \infty$. In fact, it puzzled many people at the turn of the century. Josiah Willard Gibbs gave a mathematical explanation of this behavior and therefore it came to be called the **Gibbs phenomenon**. We can reconcile the two conflicting notions by observing that the frequency of oscillation is N, so the width of the spike with 9% overshoot is $1/2N$. As we increase N, the number of terms in the series, the frequency of oscillation increases and the spike width $1/2N$ diminishes. As $N \to \infty$, the spike width $\to 0$, causing the MSE to approach zero. Therefore, as $N \to \infty$, the series on the right-hand side of Eq. (6.80) differs from $f(t)$ by about 9% at the immediate left and right of the points of discontinuity, and yet the MSE $\to 0$. We alluded to the cause of this problem on p. 461 where, it was explained that MSE can approach zero even if the series differs from $f(t)$ at some isolated instants.

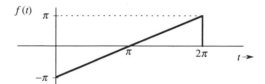

Fig. 6.20   The signal for Exercise E6.7.

$\triangle$ **Exercise E6.7**

Approximate the signal $f(t) = t - \pi$ (Fig. 6.20) over the interval $[0, 2\pi]$ in terms of the set of sinusoids $\{\sin nt\}$, $n = 0$, 1, 2, $\cdots$ used in Example 6.9. Find ϵ_N and the MSE. Show that $\epsilon_N \to 0$ as $N \to \infty$.

Answer:

$$f(t) \simeq -2 \sum_{n=1}^{N} \frac{1}{N} \sin nt \quad \text{and} \quad \epsilon_N = \frac{2}{3}\pi^3 - \sum_{n=1}^{N} \frac{4\pi}{n^2} \qquad \triangledown$$

A Historical Note on the Gibbs Phenomenon

Normally speaking, troublesome functions with strange behavior are invented by mathematicians, although we rarely see such oddities in practice. In the case of the Gibbs phenomenon, however, the tables were turned. A rather puzzling behavior was observed in such a mundane object as a mechanical wave synthesizer, and then well-known mathematicians of the day were dispatched on the scent of it to discover its hide-out.

Albert Michelson (of Michelson-Morley fame) was an intense, practical man who developed ingenious physical instruments of extraordinary precision, mostly in the field of optics. In 1898 he developed an instrument (the harmonic analyzer)

Albert Michelson (left) and **Willard J. Gibbs** (right).

which could compute the first 80 coefficients of the Fourier series of a signal $f(t)$ specified by any graphical description. The harmonic analyzer could also be used as a harmonic synthesizer, which could plot a function $f(t)$ generated by summing the first 80 harmonics (Fourier components) of arbitrary amplitudes and phases. This instrument therefore had the ability of self-checking its operation by analyzing a signal $f(t)$ and then adding the resulting 80 components to see whether the sum yielded a close approximation of $f(t)$.

Michelson found that the instrument checked very well with most of signals analyzed. However, when he tried a square wave, a curious behavior (discussed on p. 435) was observed. The sum of 80 components showed oscillatory behavior (ringing) with an overshoot of 9% in the vicinity of the points of discontinuity. Moreover, this behavior was a constant feature regardless of the number of terms added. Larger number of terms made the oscillations proportionately faster, but the overshoot remained 9% regardless of the terms added. This puzzling behavior caused Michelson to suspect some mechanical defect in his synthesizer. He wrote about his observation in a letter to *Nature* (December 1898). Josiah Willard Gibbs, an eminent mathematical physicist (inventor of vector analysis), and a professor at Yale, investigated and clarified this behavior for a sawtooth periodic signal in a letter to Nature.[7] Later, in 1906, Bôcher generalized the result for any function with discontinuity.[8] It was Bôcher who gave the name *Gibbs phenomenon* to this behavior. Gibbs showed that the peculiar behavior in the synthesis of a square wave was inherent in the behavior of the Fourier series because of nonuniform convergence at the points of discontinuity.

This, however, is not the end of the story. Both Bôcher and Gibbs were under the impression that this property had remained undiscovered until Gibbs's letter in 1899. It is now known that the Gibbs phenomenon had been observed (although not explained) in 1848 by Wilbraham of Trinity College, Cambridge, who saw quite clearly the behavior of the sum of components of the Fourier series in the sawtooth periodic signal investigated by Gibbs.[9] This work, however, remained unknown to most people, including Gibbs and Bôcher.

Representation of Complex Signals

So far we have restricted ourselves to real functions of t. To generalize the results to complex functions of t, consider again the problem of approximating a function $f(t)$ by another function $x(t)$ over an interval $[a, b]$:

$$f(t) \simeq cx(t) \tag{6.81}$$

where $f(t)$ and $x(t)$ now can be complex functions of t. In this case, both the coefficient c and the error

$$e(t) = f(t) - cx(t)$$

are complex (in general). For the best approximation, we choose c such that the mean-squared value of the error magnitude $|e(t)|$ is minimum. In other words, the optimum value of c is that value which minimizes the integral

$$I = \int_{t_1}^{t_2} |f(t) - cx(t)|^2 \, dt \tag{6.82}$$

Now let

$$\int_{t_1}^{t_2} x(t)x^*(t) \, dt = \int_{t_1}^{t_2} |x(t)|^2 \, dt = K^2 \tag{6.83}$$

where $x^*(t)$ denotes the complex conjugate of $x(t)$. Remembering that

$$|u + v|^2 = (u + v)(u^* + v^*) = |u|^2 + |v|^2 + u^*v + uv^*$$

we can, after some manipulation, express the integral I in Eq. (6.82) as

$$I = \int_{t_1}^{t_2} |f(t)|^2 \, dt - \left| \frac{1}{K} \int_{t_1}^{t_2} f(t)x^*(t) \, dt \right|^2 + \left| Kc - \frac{1}{K} \int_{t_1}^{t_2} f(t)x^*(t) \, dt \right|^2$$

Since the first two terms on the right-hand side are independent of c, it is clear that I is minimized by choosing c such that the third term is zero or

$$c = \frac{1}{K^2} \int_{t_1}^{t_2} f(t)x^*(t) \, dt \tag{6.84}$$

In light of the above result, we need to redefine orthogonality for the complex case as follows: Two complex functions $x_1(t)$ and $x_2(t)$ are orthogonal over an interval $[t_1, t_2]$ if

$$\int_{t_1}^{t_2} x_1(t)x_2^*(t) \, dt = 0 \qquad \text{or} \qquad \int_{t_1}^{t_2} x_1^*(t)x_2(t) \, dt = 0 \tag{6.85}$$

Either equality suffices. This is a general definition of orthogonality, which reduces to Eq. (6.57) when the functions are real.

Earlier results can now be generalized as follows: A set of functions $x_1(t)$, $x_2(t)$, ..., $x_N(t)$ is mutually orthogonal over the interval $[t_1, t_2]$ if

$$\int_a^b x_m(t) x_n^*(t)\, dt = \begin{cases} 0 & m \neq n \\ K_n{}^2 & m = n \end{cases} \tag{6.86}$$

If this set is complete, then a function $f(t)$ can be expressed as

$$f(t) = c_1 x_1(t) + c_2 x_2(t) + \cdots + c_i x_i(t) + \cdots \tag{6.87a}$$

where

$$c_n = K_n{}^2 \int_a^b f(t) x^*{}_n(t)\, dt \tag{6.87b}$$

6.3-6 Some Examples of Orthogonal Functions

There exists a large number of orthogonal function sets which can be used as basis functions for generalized Fourier series. Some well-known sets are trigonometric (sinusoid) functions, exponential functions, Walsh functions, Laguerre functions, Bessel functions, Legendre polynomials, Jacobi polynomials, Hermite polynomials, and Chebyshev polynomials. The functions that concern us most in this book are the trigonometric and the exponential function sets. To obtain a general idea, we discuss Legendere and trigonometric sets here.

Legendre Fourier Series

A set of Legendre polynomials $P_n(t)$ $(n = 0, 1, 2, 3, \cdots)$ forms a complete set of mutually orthogonal functions over an interval $(-1 \leq t \leq 1)$. These polynomials can be defined by the Rodrigues formula:

$$P_n(t) = \frac{1}{2^n n!} \frac{d^n}{dt^n} (t^2 - 1)^n, \qquad n = (0, 1, 2, \cdots)$$

It follows from this equation that

$$P_0(t) = 1, \qquad\qquad P_1(t) = t$$

$$P_2(t) = \left(\tfrac{3}{2} t^2 - \tfrac{1}{2} \right), \qquad P_3(t) = \left(\tfrac{5}{2} t^3 - \tfrac{3}{2} t \right)$$

and so on.

We may verify the orthogonality of these polynomials by showing that

$$\int_{-1}^1 P_m(t) P_n(t)\, dt = \begin{cases} 0 & m \neq n \\ \frac{2}{2m+1} & m = n \end{cases} \tag{6.88}$$

We can express a function $f(t)$ in terms of Legendre polynomials over an interval $(-1 \leq t \leq 1)$ as:

$$f(t) = c_0 P_0(t) + c_1 P_1(t) + \cdots + \cdots \tag{6.89}$$

where

$$c_r = \frac{\int_{-1}^1 f(t) P_r(t)\, dt}{\int_{-1}^1 P_r{}^2(t)\, dt}$$

$$= \frac{2r + 1}{2} \int_{-1}^1 f(t) P_r(t)\, dt \tag{6.90}$$

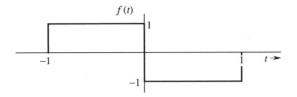

Fig. 6.21   The square signal for Example 6.10.

Note that although the series representation is valid over the interval $[-1, 1]$, it can be extended to any interval by the appropriate time-scaling (see Prob. 6.3-8).

■ **Example 6.10**

Let us consider the square signal shown in Fig. 6.21. This function can be represented by Legendre Fourier series:

$$f(t) = c_0 P_0(t) + c_1 P_1(t) + \cdots + c_r P_r(t) + \cdots$$

The coefficients c_0, c_1, c_2, ..., c_r etc. may be found from Eq. (6.90). We have

$$f(t) = \begin{cases} 1 & \cdots -1 \leq t \leq 0 \\ -1 & \cdots 0 \leq t \leq 1 \end{cases}$$

and

$$c_0 = \frac{1}{2} \int_{-1}^{1} f(t)\, dt = 0$$

$$c_1 = \frac{3}{2} \int_{-1}^{1} t f(t)\, dt$$

$$= \frac{3}{2} \left(\int_{-1}^{0} t\, dt - \int_{0}^{1} t\, dt \right) = -\frac{3}{2}$$

$$c_2 = \frac{5}{2} \int_{-1}^{1} f(t) \left(\tfrac{3}{2} t^2 - \tfrac{1}{2} \right)\, dt = 0$$

This result follows immediately from the fact that the integrand is an odd function of t. In fact, this is true of all c_r for even values of r, that is,

$$c_0 = c_2 = c_4 = c_6 = \cdots = 0$$

Also,

$$c_3 = \frac{7}{2} \int_{-1}^{1} f(t) \left(\tfrac{5}{2} t^3 - \tfrac{3}{2} t \right)\, dt$$

$$= \frac{7}{2} \left[\int_{-1}^{0} \left(\tfrac{5}{2} t^3 - \tfrac{3}{2} t \right)\, dt + \int_{0}^{1} - \left(\tfrac{5}{2} t^3 - \tfrac{3}{2} t \right)\, dt \right] = \frac{7}{8}$$

In a similar way, coefficients c_5, $c_7 \cdots$ etc., can be evaluated. We now have

$$f(t) = -\tfrac{3}{2} t + \tfrac{7}{8} \left(\tfrac{5}{2} t^3 - \tfrac{3}{2} t \right) + \cdots \tag{6.91}$$

■

Trigonometric Fourier Series

We have already proved [see Eq. (6.10)] that the trigonometric signal set

$$\{1, \cos \omega_0 t, \cos 2\omega_0 t, \ldots, \cos n\omega_0 t, \ldots; \tag{6.92}$$

$$\sin \omega_0 t, \sin 2\omega_0 t, \ldots, \sin n\omega_0 t, \ldots\}$$

is orthogonal over any interval of duration T_0, where $T_0 = 1/\mathcal{F}_0$ is the period of the sinusoid of frequency $\mathcal{F}_0$. This is a complete set.[10,11] Therefore we can express a signal $f(t)$ by a trigonometric Fourier series over any interval of duration T_0 sec. as

$$f(t) = a_0 + a_1 \cos \omega_0 t + a_2 \cos 2\omega_0 t + \cdots$$

$$+ b_1 \sin \omega_0 t + b_2 \sin 2\omega_0 t + \cdots$$

or

$$f(t) = a_0 + \sum_{n=1}^{\infty} a_n \cos n\omega_0 t + b_n \sin n\omega_0 t \quad t_1 \le t \le t_1 + T_0 \tag{6.93a}$$

where

$$\omega_0 = 2\pi \mathcal{F}_0 = \frac{2\pi}{T_0} \tag{6.93b}$$

Using Eq. (6.72), we can determine the Fourier coefficients a_0, a_n, and b_n. Thus

$$a_n = \frac{\int_{t_1}^{t_1+T_0} f(t) \cos n\omega_0 t\, dt}{\int_{t_1}^{t_1+T_0} \cos{}^2 n\omega_0 t\, dt} \tag{6.94}$$

The integral in the denominator of Eq. (6.94) has already been found to be $T_0/2$ when $n \ne 0$ [Eq. (6.10a) with $m = n$]. For $n = 0$, the denominator is T_0. Hence

$$a_0 = \frac{1}{T_0} \int_{t_1}^{t_1+T_0} f(t)\, dt \tag{6.95a}$$

and

$$a_n = \frac{2}{T_0} \int_{t_1}^{t_1+T_0} f(t) \cos n\omega_0 t\, dt \quad n = 1, 2, 3, \ldots \tag{6.95b}$$

Similarly, we find that

$$b_n = \frac{2}{T_0} \int_{t_1}^{t_1+T_0} f(t) \sin n\omega_0 t\, dt \quad n = 1, 2, 3, \ldots \tag{6.95c}$$

Note that the Fourier series in Eq. (6.79) of Example 6.9 is indeed a trigonometric Fourier series with $T_0 = 2\pi$ and $\omega_0 = 2\pi/T_0$. In this particular example, it is easy to verify from Eqs. (6.95a) and (6.95b) that $a_n = 0$ for all n, including $n = 0$. Hence the Fourier series in that example consisted only of sine terms.

Exponential Fourier Series

The set of exponentials $e^{jn\omega_0 t}$ $(n = 0, \pm 1, \pm 2, \ldots)$ is a complete set of orthogonal functions and therefore is a basis set in its own right capable of representing

$f(t)$. To verify that the exponentials are orthogonal over an interval $[t_1, t_1 + T_0]$, where $T_0 = 2\pi/\omega_0$, we use the orthogonality test for complex functions [Eq. (6.85)] with $x_n(t) = e^{jn\omega_0 t}$:

$$\int_{t_1}^{t_1+T_0} e^{jm\omega_0 t}(e^{jn\omega_0 t})^* \, dt = \int_{t_1}^{t_1+T_0} e^{j(m-n)\omega_0 t} \, dt = \begin{cases} 0 & m \neq n \\ T_0 & m = n \end{cases} \tag{6.96}$$

Clearly, this set of exponentials is orthogonal over any interval of duration $T_0 = 1/\mathcal{F}_0 = 2\pi/\omega_0$. An arbitrary signal $f(t)$ can now be expressed over an interval $[t_1, t_1 + T_0]$ as

$$f(t) = \sum_{n=-\infty}^{\infty} D_n e^{jn\omega_0 t} \qquad t_1 \leq t \leq t_1 + T_0 \tag{6.97a}$$

where [see Eq. (6.87)]

$$D_n = \frac{1}{T_0} \int_{t_1}^{t_1+T_0} f(t) e^{-jn\omega_0 t} \, dt \tag{6.97b}$$

6.4 SUMMARY

In this chapter we showed how a periodic signal can be represented as a sum of sinusoids or exponentials. If the frequency of a periodic signal is $\mathcal{F}_0$, then it can be expressed as a sum of a sinusoid of frequency $\mathcal{F}_0$ and its harmonics (trigonometric Fourier series). We can reconstruct the periodic signal from a knowledge of the amplitudes and phases of these sinusoidal components (amplitude and phase spectra).

A sinusoid can be expressed in terms of exponentials. Therefore the Fourier series of a periodic signal can also be expressed as a sum of exponentials (the exponential Fourier series). The exponential form of the Fourier series and the expressions for the series coefficients are more compact than those of the trigonometric Fourier series. Also, the response of LTIC systems to an exponential input is much simpler than that for a sinusoidal input. Moreover, the exponential form of representation lends itself better to mathematical manipulations than does the trigonometric form. For these reasons, the exponential form of the series is preferred in modern practice in the areas of signals and systems.

The plots of amplitudes and angles of various exponential components of the Fourier series as functions of the frequency are the exponential Fourier spectra (amplitude and angle spectra) of the signal. Because a sinusoid $\cos \omega_0 t$ can be represented as a sum of two exponentials, $e^{j\omega_0 t}$ and $e^{-j\omega_0 t}$, the frequencies in the exponential spectra range from $-\infty$ to ∞. By definition, frequency of a signal is always a positive quantity. Presence of a spectral component of a negative frequency $-n\omega_0$ merely indicates that the Fourier series contains terms of the form $e^{-jn\omega_0 t}$. The spectra of the trigonometric and exponential Fourier series are closely related, and one can be found by the inspection of the other.

In Sec. 6.3 we discuss a method of representing signals by generalized Fourier series, of which the trigonometric and exponential Fourier series are special cases.

Signals are vectors in every sense. Just as a vector can be represented as a sum of its components in a variety of ways, depending upon the choice of the coordinate system, a signal can be represented as a sum of its components in a variety of ways of which the trigonometric and exponential Fourier series are only two examples. Just as we have vector coordinate systems formed by mutually orthogonal vectors, we also have signal coordinate systems (basis signals) formed by mutually orthogonal signals. Any signal in this signal space can be represented as a sum of the basis signals. Each set of basis signals yields a particular Fourier series representation of the signal. The signal is equal to its Fourier series, not in the ordinary sense, but in the special sense that the mean squared value of the difference between the signal and its Fourier series approaches zero. This allows for the signal to differ from its Fourier series at some isolated points.

REFERENCES

1. Guillemin, E. A., *Theory of Linear Physical Systems*, Wiley, New York, 1963.

2. Bell, E. T., *Men of Mathematics*, Simon and Schuster, New York, 1937.

3. Durant, Will, and Ariel, *The Age of Napoleon*, History of Civilization, Part XI, Simon and Schuster, New York, 1975.

4. Calinger, R., 4th ed., *Classics of Mathematics*, Moore Publishing Co., Oak Park, Il. 1982.

5. Lanczos, C., *Discourse on Fourier Series*, Oliver Boyd Ltd., London, 1966.

6. Lathi, B. P., *Signals, Systems and Communication*, Wiley, New York, 1965.

7. Gibbs, W. J., *Nature*, vol. 59, p. 606, April 1899.

8. Bôcher, M., Annals of Mathematics, (2), vol. 7, 1906.

9. Carslaw, H. S., Bulletin, American Mathematical Society, vol. 31, pp. 420-424, Oct. 1925.

10. Walker P. L., *The Theory of Fourier Series and Integrals*, Wiley, New York, 1986.

11. Churchill, R. V., and J. W. Brown, *Fourier Series and Boundary Value Problems*, 3d ed., McGraw-Hill, New York, 1978.

PROBLEMS

6.1-1 For each of the periodic signals shown in Fig. P6.1-1, find the compact trigonometric Fourier series and sketch the amplitude and phase spectra. If either the sine or cosine terms are absent in the Fourier series, explain why.

6.1-2 If the two halves of one period of a periodic signal are of identical shape except that the one is the negative of the other, the periodic signal is said to have a **half-wave symmetry**. If a periodic signal $f(t)$ with a period T_0 satisfies the half-wave symmetry condition, then

$$f\left(t - \frac{T_0}{2}\right) = -f(t)$$

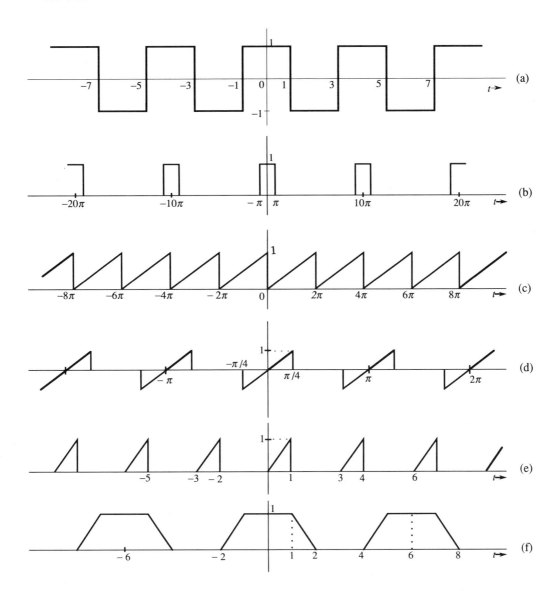

Fig. P6.1-1.

In this case, show that all the even-numbered harmonics vanish, and that the odd-numbered harmonic coefficients are given by

$$a_n = \frac{4}{T_0} \int_0^{T_0/2} f(t) \cos n\omega_0 t \, dt \quad \text{and} \quad b_n = \frac{4}{T_0} \int_0^{T_0/2} f(t) \sin n\omega_0 t \, dt$$

Using these results, find the Fourier series for the periodic signals in Fig. P6.1-2.

6.1-3 State with reasons whether the following signals are periodic or nonperiodic. For periodic signals, find the period and state which of the harmonics are present in the series.

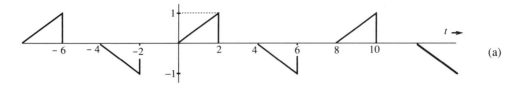

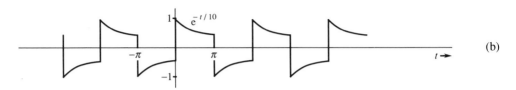

Fig. P6.1-2.

<table>
<tr><td>

(a) $3\sin t + 2\sin 3t$

(b) $2 + 5\sin 4t + 4\cos 7t$

(c) $2\sin 3t + 7\cos \pi t$

(d) $7\cos \pi t + 5\sin 2\pi t$

(e) $3\cos \sqrt{2}t + 5\cos 2t$

</td><td>

(f) $\sin \frac{5t}{2} + 3\cos \frac{6t}{5} + 3\sin \left(\frac{t}{7} + 30°\right)$

(g) $\sin 3t + \cos \frac{15}{4}t$

(h) $(3\sin 2t + \sin 5t)^2$

(i) $(5\sin 2t)^3$

</td></tr>
</table>

Hint for parts **h** and **i**: If a signal is periodic then any power of that signal is also periodic. To find the period use identities in Sec. B.10-6.

6.2-1 Derive Eq. (6.32) directly from Eq. (6.30), as explained in the footnote on P.000.

6.2-2 For each of the periodic signals in Fig. P6.1-1, find exponential Fourier series and sketch the corresponding spectra.

6.2-3 A periodic signal $f(t)$ is expressed by the following Fourier series:

$$f(t) = 3\cos t + \sin \left(5t - \frac{\pi}{6}\right) - 2\cos \left(8t - \frac{\pi}{3}\right)$$

(a) Sketch the amplitude and phase spectra for the trigonometric series.
(b) By inspection of spectra in part **a**, sketch the exponential Fourier series spectra.
(c) By inspection of spectra in part **b**, write the exponential Fourier series for $f(t)$.

6.2-4 The trigonometric Fourier series of a certain periodic signal is given by

$$f(t) = 3 + \sqrt{3}\cos 2t + \sin 2t + \sin 3t - \frac{1}{2}\cos \left(5t + \frac{\pi}{3}\right)$$

(a) Sketch the trigonometric Fourier spectra.
(b) By inspection of spectra in part **a**, sketch the exponential Fourier series spectra.
(c) By inspection of spectra in part **b**, write the exponential Fourier series for $f(t)$.

6.2-5 The exponential Fourier series of a certain function is given as

$$f(t) = (2 + j2)e^{-j3t} + j2e^{-jt} + 3 - j2e^{jt} + (2 - j2)e^{j3t}$$

(a) Sketch the exponential Fourier spectra.
(b) By inspection of the spectra in part **a**, sketch the trigonometric Fourier spectra for $f(t)$. Find the compact trigonometric Fourier series from these spectra.
(c) Find the signal bandwidth.

6.2-6 Figure P6.2-6 shows the trigonometric Fourier spectra of a periodic signal $f(t)$.
(a) By inspection of Fig. P6.2-6, find the trigonometric Fourier series representing $f(t)$.
(b) By inspection of Fig. P6.2-6, sketch the exponential Fourier spectra of $f(t)$.
(c) By inspection of the exponential Fourier spectra obtained in part **b**, find the exponential Fourier series for $f(t)$.
(d) Show that the series found in parts **a** and **c** are equivalent.

6.2-7 Figure P6.2-7 shows the exponential Fourier spectra of a periodic signal $f(t)$.

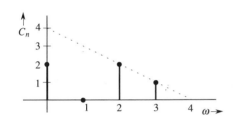 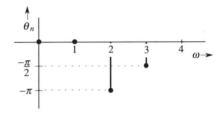

Fig. P6.2-6

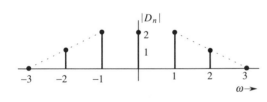 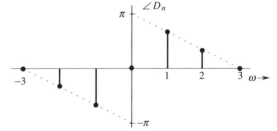

Fig. P6.2-7.

(a) By inspection of Fig. P6.2-7, find the exponential Fourier series representing $f(t)$.
(b) By inspection of Fig. P6.2-7, sketch the trigonometric Fourier spectra for $f(t)$.
(c) By inspection of the trigonometric Fourier spectra found in part **b**, find the trigonometric Fourier series for $f(t)$.
(d) Show that the series found in parts **a** and **c** are equivalent.

6.2-8 (a) Find the exponential Fourier series for the signal in Fig. P6.2-8a.
(b) Using the results in part (a), find the Fourier series for the signal $\hat{f}(t)$ in Fig. P6.2-8b, which is a time-shifted version of the signal $f(t)$.
(c) Using the results in part (a), find the Fourier series for the signal $\tilde{f}(t)$ in Fig. P6.2-8c, which is a time-scaled version of the signal $f(t)$.

6.2-9 If a periodic signal $f(t)$ is expressed as an exponential Fourier series

$$f(t) = \sum_{n=-\infty}^{\infty} D_n e^{jn\omega_0 t}$$

(a) Show that the exponential Fourier series for $\hat{f}(t) = f(t - T)$ is given by

$$\hat{f}(t) = \sum_{n=-\infty}^{\infty} \hat{D}_n e^{jn\omega_0 t}$$

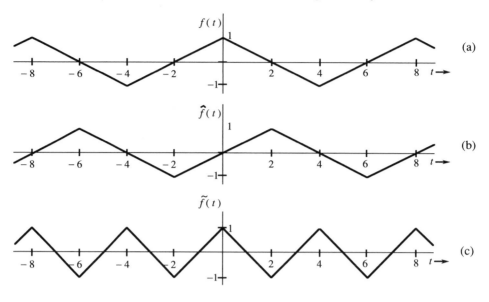

Fig. P6.2-8.

in which

$$|\hat{D}_n| = |D_n| \quad \text{and} \quad \angle\hat{D}_n = \angle D_n - n\omega_0 T$$

(b) Verify this result by finding the exponential Fourier series for the signals in Figs. P6.2-8a and P6.2-8b.

6.3-1 Let $x_1(t)$ and $x_2(t)$ be two signals orthogonal over an interval $[t_1, t_2]$. Consider a signal $f(t)$ where

$$f(t) = c_1 x_1(t) + c_2 x_2(t) \qquad t_1 \le t \le t_2$$

Represent this signal by a two-dimensional vector $\mathbf{F}(c_1, c_2)$.

(a) Determine the vector representation of the following six signals in the two-dimensional vector space

    (i) $f_1(t) = 2x_1(t) - x_2(t)$ **(iv)** $f_4(t) = x_1(t) + 2x_2(t)$

    (ii) $f_2(t) = -x_1(t) + 2x_2(t)$ **(v)** $f_5(t) = 2x_1(t) + x_2(t)$

    (iii) $f_3(t) = -x_2(t)$ **(vi)** $f_6(t) = 3x_1(t)$

(b) Point out pairs of mutually orthogonal vectors among these six vectors. Verify that the pairs of signal corresponding to these orthogonal vectors are also orthogonal.

6.3-2 A signal $f(t)$ is approximated in terms of a signal $x(t)$ over an interval $[t_1, t_2]$:

$$f(t) \simeq cx(t) \qquad t_1 \le t \le t_2$$

where c is chosen to minimize the mean-squared error.

(a) Show that $x(t)$ and the error $e(t) = f(t) - cx(t)$ are orthogonal over the interval $[t_1, t_2]$.

(b) Can you explain the result in terms of analogy with vectors?

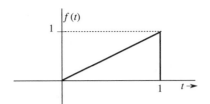

Fig. P6.3-3.

(c) Verify this result for the square signal $f(t)$ in Fig. 6.17 and its approximation in terms of signal $\sin t$.

6.3-3 Represent the signal $f(t)$ shown in Fig. P6.3-3 over the interval $[0, 1]$ by a trigonometric Fourier series of fundamental frequency $\omega_0 = 2\pi$. Compute the mean-squared error in the representation of $f(t)$ by only the first k terms of this series for $k = 1, 2, 3,$ and 4.

6.3-4 Represent $f(t) = t$ over the interval $[0, 1]$ by a trigonometric Fourier series which has
(a) $\omega_0 = 2\pi$ and only sine terms.
(b) $\omega_0 = \pi$ and only sine terms.
(c) $\omega_0 = \pi$ and only cosine terms.
You may use a dc term in these series if necessary.

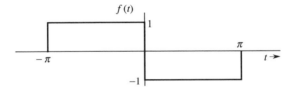

Fig. P6.3-5.

6.3-5 In Example 6.10, we represented the function in Fig. 6.21 by Legendre polynomials.
(a) Using the results in this example, represent the signal $g(t)$ in Fig. P6.3-5 by Legendre polynomials. Hint: Use appropriate time-scaling.
(b) Compute the mean-squared error for the one- and two-(nonzero)term approximations.

COMPUTER PROBLEMS

C6-1 Find the compact trigonometric Fourier series for the periodic signal $f(t)$ shown in Fig. 6.4a. Use a computer to plot the amplitude and phase spectra of $f(t)$ for 5, 10, 50 100 points.

C6-2 For the periodic signal in Fig. 6.3a, use a computer to observe how a sum of harmonics [the Fourier series in Eq. (6.19)] approaches $f(t)$ as the number of harmonics are increased.

7

Continuous-Time Signal Analysis: The Fourier Transform

In Chapter 6 we succeeded in representing periodic signals as a sum of (everlasting) sinusoids or exponentials. In this chapter we extend this spectral representation to nonperiodic signals.

7.1 NONPERIODIC SIGNAL REPRESENTATION BY FOURIER INTEGRAL

Applying a limiting process, we now show that nonperiodic signals can be expressed as a continuous sum (integral) of everlasting exponentials. To represent a nonperiodic signal $f(t)$ such as the one shown in Fig. 7.1a by everlasting exponential signals, let us construct a new periodic signal $f_{T_0}(t)$ formed by repeating the signal $f(t)$ every T_0 seconds, as shown in Fig. 7.1b. The period T_0 is made long enough to avoid overlap between the repeating pulses. The periodic signal $f_{T_0}(t)$ can be represented by an exponential Fourier series. If we let $T_0 \to \infty$, the pulses in the periodic signal repeat after an infinite interval, and therefore

$$\lim_{T_0 \to \infty} f_{T_0}(t) = f(t)$$

Thus the Fourier series representing $f_{T_0}(t)$ will also represent $f(t)$ in the limit $T_0 \to \infty$. The exponential Fourier series for $f_{T_0}(t)$ is given by

$$f_{T_0}(t) = \sum_{n=-\infty}^{\infty} D_n e^{jn\omega_0 t} \tag{7.1}$$

in which

$$D_n = \frac{1}{T_0} \int_{-T_0/2}^{T_0/2} f_{T_0}(t) e^{-jn\omega_0 t} \, dt \tag{7.2a}$$

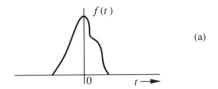

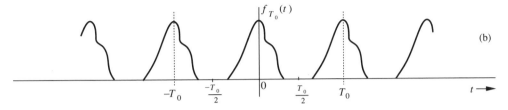

Fig. 7.1   Construction of a periodic signal by periodic extension of $f(t)$.

and

$$\omega_0 = \frac{2\pi}{T_0} \tag{7.2b}$$

Observe that integrating $f_{T_0}(t)$ over $(-\frac{T_0}{2}, \frac{T_0}{2})$ is the same as integrating $f(t)$ over $(-\infty, \infty)$. Therefore Eq. (7.2a) can be expressed as

$$D_n = \frac{1}{T_0} \int_{-\infty}^{\infty} f(t) e^{-jn\omega_0 t}\, dt \tag{7.2c}$$

It is interesting to see how the nature of the spectrum changes as T_0 increases. To understand this fascinating behavior, let us define $F(\omega)$, a continuous function of ω, as

$$F(\omega) = \int_{-\infty}^{\infty} f(t) e^{-j\omega t}\, dt \tag{7.3}$$

A glance at Eqs. (7.2c) and (7.3) shows that

$$D_n = \frac{1}{T_0} F(n\omega_0) \tag{7.4}$$

This shows that the Fourier coefficients D_n are $(1/T_0)$ times the samples of $F(\omega)$ taken every ω_0 rad/s, as shown in Fig. 7.2a.† Therefore, $(1/T_0)F(\omega)$ is the envelope for the coefficients D_n. We now let $T_0 \to \infty$ by doubling T_0 repeatedly. Doubling T_0 halves the fundamental frequency ω_0, so that there are now twice as many components (samples) in the spectrum. However, by doubling T_0, the envelope $(1/T_0)$ $F(\omega)$ is halved, as shown in Fig. 7.2b. If we continue this process of doubling T_0 repeatedly, the spectrum progressively becomes denser while its magnitude becomes smaller. Note, however, that the relative shape of the envelope remains the same [proportional to $F(\omega)$ in Eq. (7.3)]. In the limit as $T_0 \to \infty$, $\omega_0 \to 0$, and $D_n \to 0$. This means that the spectrum is so dense that the spectral components are spaced at zero (infinitesimal) interval. At the same time, the amplitude of each component

---

†For the sake of simplicity we assume D_n and therefore $F(\omega)$ in Fig. 7.2 to be real. The argument, however, is also valid for complex D_n [or $F(\omega)$].

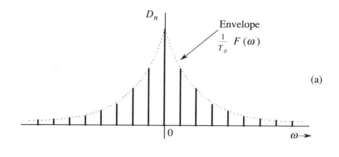

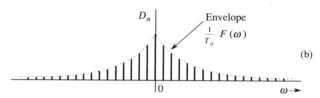

Fig. 7.2   Change in the Fourier spectrum when the period T_0 in Fig. 7.1 is doubled.

is zero (infinitesimal). We have **nothing of everything, yet we have something!** This sounds like *Alice in Wonderland*, but as we shall see, these are the classic characteristics of a very familiar phenomenon.

Substitution of Eq. (7.4) in Eq. (7.1) yields

$$f_{T_0}(t) = \sum_{n=-\infty}^{\infty} \frac{F(n\omega_0)}{T_0} e^{jn\omega_0 t} \tag{7.5}$$

As $T_0 \to \infty$, ω_0 becomes infinitesimal ($\omega_0 \to 0$). Because of this, we shall replace ω_0 by a more appropriate notation, $\Delta\omega$. In terms of this new notation, Eq. (7.2b) becomes

$$\Delta\omega = \frac{2\pi}{T_0}$$

and Eq. (7.5) becomes

$$f_{T_0}(t) = \sum_{n=-\infty}^{\infty} \left[\frac{F(n\Delta\omega)\Delta\omega}{2\pi} \right] e^{(jn\Delta\omega)t} \tag{7.6a}$$

Equation (7.6a) shows that $f_{T_0}(t)$ can be expressed as a sum of everlasting exponentials of frequencies $0, \pm\Delta\omega, \pm2\Delta\omega, \pm3\Delta\omega, \cdots$, etc. The amount of the component of frequency $n\Delta\omega$ is $[F(n\Delta\omega)\Delta\omega]/2\pi$. In the limit as $T_0 \to \infty, \Delta\omega \to 0$, and $f_{T_0}(t) \to f(t)$, therefore

$$f(t) = \lim_{T_0 \to \infty} f_{T_0}(t) = \lim_{\Delta\omega \to 0} \frac{1}{2\pi} \sum_{n=-\infty}^{\infty} F(n\Delta\omega)e^{(jn\Delta\omega)t}\Delta\omega \tag{7.6b}$$

The sum on the right-hand side of Eq. (7.6b) can be viewed as the area under the function $F(\omega)e^{j\omega t}$, as shown in Fig. 7.3. Therefore

$$f(t) = \frac{1}{2\pi} \int_{-\infty}^{\infty} F(\omega)e^{j\omega t}d\omega \tag{7.7}$$

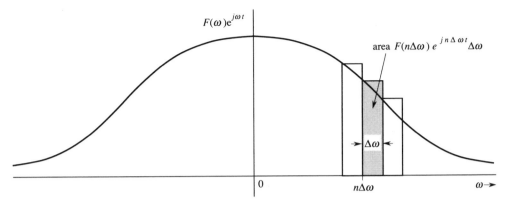

Fig. 7.3 The Fourier series becomes the Fourier integral in the limit as $T_0 \to \infty$.

The integral on the right hand side is called the **Fourier integral**. We have now succeeded in representing a nonperiodic signal $f(t)$ by a Fourier integral† (rather than a Fourier series). This integral is basically a Fourier series (in the limit) with fundamental frequency $\Delta\omega \to 0$, as seen from Eq. (7.6b). The amount of the exponential $e^{jn\Delta\omega t}$ is $F(n\Delta\omega)\Delta\omega/2\pi$. Thus the function $F(\omega)$ given by Eq. (7.3) acts as a spectral function.

We call $F(\omega)$ the direct Fourier transform of $f(t)$, and $f(t)$ the inverse Fourier transform of $F(\omega)$. The same information is conveyed by the statement that $f(t)$ and $F(\omega)$ are a Fourier transform pair. Symbolically, this is expressed as

$$F(\omega) = \mathcal{F}[f(t)] \qquad \text{and} \qquad f(t) = \mathcal{F}^{-1}[F(\omega)]$$

or

$$f(t) \Longleftrightarrow F(\omega)$$

To recapitulate,

$$\mathcal{F}[f(t)] = F(\omega) = \int_{-\infty}^{\infty} f(t)e^{-j\omega t}\, dt \tag{7.8a}$$

and

$$\mathcal{F}^{-1}[F(\omega)] = f(t) = \frac{1}{2\pi}\int_{-\infty}^{\infty} F(\omega)e^{j\omega t}\, d\omega \tag{7.8b}$$

It is helpful to keep in mind that the Fourier integral in Eq. (7.8b) is of the nature of a Fourier series with fundamental frequency $\Delta\omega$ approaching zero [Eq. (7.6b)]. Therefore most of the discussion and properties of Fourier series apply to the Fourier transform as well. We can plot the spectrum $F(\omega)$ as a function of ω. Since $F(\omega)$ is complex, we have both amplitude and angle (or phase) spectra:

$$F(\omega) = |F(\omega)|e^{j\angle F(\omega)} \tag{7.9}$$

in which $|F(\omega)|$ is the amplitude and $\angle F(\omega)$ is the angle (or phase) of $F(\omega)$. From Eq. (7.8a),

---

†This should not be considered as a rigorous proof of Eq. (7.7). The situation is not as simple as we have made it appear.[1]

$$F(-\omega) = \int_{-\infty}^{\infty} f(t)e^{j\omega t}\, dt$$

From this equation and Eq. (7.8a), it follows that if $f(t)$ is a real function of t, then $F(\omega)$ and $F(-\omega)$ are complex conjugates. Therefore

$$|F(-\omega)| = |F(\omega)| \tag{7.10a}$$

$$\angle F(-\omega) = -\angle F(\omega) \tag{7.10b}$$

Thus, for real $f(t)$, the amplitude spectrum $|F(\omega)|$ is an even function, and the phase spectrum $\angle F(\omega)$ is an odd function of ω. These results were derived earlier for the Fourier spectrum of a periodic signal [Eq. (6.40)] and should come as no surprise. **The transform $F(\omega)$ is the frequency-domain specification of $f(t)$.**

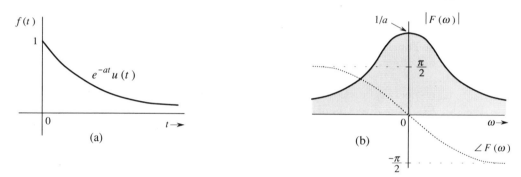

Fig. 7.4 $e^{-at}u(t)$ and its Fourier spectra.

■ **Example 7.1**
    Find the Fourier transform of $e^{-at}u(t)$.
    By definition [Eq. (7.8a)],

$$F(\omega) = \int_{-\infty}^{\infty} e^{-at}u(t)e^{-j\omega t}\, dt = \int_{0}^{\infty} e^{-(a+j\omega)t}\, dt = \left. \frac{-1}{a+j\omega}e^{-(a+j\omega)t}\right|_{0}^{\infty}$$

But $|e^{-j\omega t}| = 1$. Therefore, as $t \to \infty$, $e^{-(a+j\omega)t} = e^{-at}e^{-j\omega t} = 0$ if $a > 0$. Therefore

$$F(\omega) = \frac{1}{a+j\omega} \qquad a > 0 \tag{7.11a}$$

Expressing $a + j\omega$ in the polar form as $\sqrt{a^2 + \omega^2}\, e^{j\tan^{-1}(\frac{\omega}{a})}$, we obtain

$$F(\omega) = \frac{1}{\sqrt{a^2 + \omega^2}} e^{-j\tan^{-1}(\frac{\omega}{a})} \tag{7.11b}$$

Therefore

$$|F(\omega)| = \frac{1}{\sqrt{a^2 + \omega^2}} \qquad \text{and} \qquad \angle F(\omega) = -\tan^{-1}\left(\frac{\omega}{a}\right) \tag{7.12}$$

The amplitude spectrum $|F(\omega)|$ and the phase spectrum $\angle F(\omega)$ are shown in Fig. 7.4b. Observe that $|F(\omega)|$ is an even function of ω, and $\angle F(\omega)$ is an odd function of ω, as expected.   ■

⊙ **Computer Example C7.1**
Use a computer to find and plot the Fourier transform of $e^{-2t}u(t)$.

```
T=.01;
t=0:T:8;
y=exp(-2*t);
Y=fft(y,512);
magY=abs(Y)*T;
phaseY=(180/pi)*angle(Y);
f=(1/T)*(1:128)/256;
w=2*pi*f;
plot(w,magY(1:128)),grid, xlabel('w'),ylabel('|F(w)|'),pause,
plot(w,phaseY(1:128)),grid, xlabel('w'),ylabel('< F(w)'),pause,
disp('The FFT method used here has built-in aliasing, which has badly')
disp('distorted the phase spectrum.  For better accuracy, we need to ')
disp('use more points in FFT.')
disp('Press any key for plots of C7.1 by an alternate (simple) method.')
pause
w=-5*pi:pi/100:5*pi;w=w';w1=w.^2;
f1=sqrt(w1+4);f=f1.^(-1); fangle=-atan(w/2)*(180/pi);
plot(w,f),grid,xlabel('w'),ylabel('|F(w)|'),pause
plot(w,fangle),grid,xlabel('w'),ylabel('< H(w)')    ⊙
```

Existence of the Fourier Transform

In Example 7.1 we observed that the Fourier transform of $e^{-at}u(t)$ does not exist if $a < 0$ (growing exponential). Clearly, not all signals are Fourier-transformable. The existence of the Fourier transform is assured for any $f(t)$ satisfying the Dirichlet conditions mentioned on p. 421. The first of these conditions is†

$$\int_{-\infty}^{\infty} |f(t)|dt \; < \; \infty \tag{7.13}$$

Observe that because $|e^{-j\omega t}| = 1$

$$|F(\omega)| \le \int_{-\infty}^{\infty} |f(t)|dt$$

Therefore the existence of the Fourier transform is assured if condition (7.13) is satisfied.

This condition is sufficient but not necessary for the existence of the Fourier transform of a signal. For example, the signal $(\sin at)/t$ violates condition (7.13), but does have a Fourier transform. Any signal that can be generated in practice satisfies Dirichlet conditions and therefore has a Fourier transform. Thus the physical existence of a signal is a sufficient condition for the existence of its transform.

---

†The remaining Dirichlet conditions are: $f(t)$ may have only a finite number of maxima and minima and a finite number of discontinuities in every finite interval. When these conditions are satisfied, the Fourier integral (7.8b) converges to $f(t)$ at all points where $f(t)$ is continuous and converges to the average of the right-hand and left-hand limits of $f(t)$ at points where $f(t)$ is discontinuous.

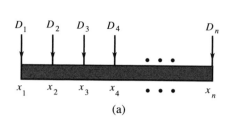

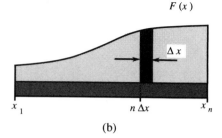

Fig. 7.5   Analogy for Fourier transform.

Linearity of the Fourier Transform

Fourier transform is linear; that is, if

$$f_1(t) \Longleftrightarrow F_1(\omega) \qquad \text{and} \qquad f_2(t) \Longleftrightarrow F_2(\omega)$$

then

$$a_1 f_1(t) + a_2 f_2(t) \Longleftrightarrow a_1 F_1(\omega) + a_2 F_2(\omega) \tag{7.14}$$

The proof is trivial and follows directly from the definition of the Fourier transform.

7.2   PHYSICAL APPRECIATION OF THE FOURIER TRANSFORM

In understanding any aspect of Fourier transform, we should remember that Fourier representation is a way of expressing a signal in terms of sinusoids (or exponentials) that start at $t = -\infty$ (everlasting). The Fourier spectrum of a signal indicates the relative amplitudes and phases of the sinusoids that are required to synthesize that signal. A periodic signal Fourier spectrum has finite amplitudes and exists at discrete frequencies (ω_0 and its multiples). Such a spectrum is easy to visualize, but the spectrum of a nonperiodic signal is not easy to visualize because it has a continuous spectrum. The continuous spectrum concept can be appreciated by considering an analogous, more tangible phenomenon. One familiar example of a continuous distribution is the loading of a beam. Consider a beam loaded with weights $D_1, D_2, D_3, \cdots, D_n$ units at the uniformly spaced points $x_1, x_2, \cdots, x_n$, as shown in Fig. 7.5a. The total load W_T on the beam is given by the sum of these loads at each of the n points:

$$W_T = \sum_{i=1}^{n} D_i$$

Consider now the case of a continuously loaded beam, as shown in Fig. 7.5b. In this case, although there appears to be a load at every point, the load at any one point is zero. This does not mean that there is no load on the beam. A meaningful measure of load in this situation is not the load at a point, but rather the loading density per unit length at that point. Let $F(x)$ be the loading density per unit length of beam. This means that the load over a beam length $\Delta x (\Delta x \to 0)$ at some point x is $F(x)\Delta x$. To find the total load on the beam, we divide the beam into segments of interval $\Delta x \, (\Delta x \to 0)$. The load over the nth such segment of length Δx is $[F(n\Delta x)] \, \Delta x$. The total load W_T is given by

$$W_T = \lim_{\Delta x \to 0} \sum_{x_1}^{x_n} F(n\Delta x)\,\Delta x$$

$$= \int_{x_1}^{x_n} F(x)\,dx$$

In the case of discrete loading (Fig. 7.5a), the load exists only at the n discrete points. At other points there is no load. On the other hand, in the continuously loaded case, the load exists at every point, but at any specific point x the load is zero. The load over a small interval Δx , however, is $[F(n\Delta x)]\,\Delta x$ (Fig. 7.5b). Thus, even though the load at a point x is zero, the relative load at that point is $F(x)$.

An exactly analogous situation exists in the case of a signal spectrum. When $f(t)$ is periodic, the spectrum is discrete, and $f(t)$ can be expressed as a sum of discrete exponentials with finite amplitudes:

$$f(t) = \sum_{n} D_n e^{jn\omega_0 t}$$

For a nonperiodic signal, the spectrum becomes continuous; that is, the spectrum exists for every value of ω, but the amplitude of each component in the spectrum is zero. The meaningful measure here is not the amplitude of a component of some frequency but the spectral density per unit bandwidth. From Eq. (7.6b) it is clear that $f(t)$ is synthesized by adding exponentials of the form $e^{jn\Delta\omega t}$, in which the contribution by any one exponential component is zero. But the contribution by exponentials in an infinitesimal band $\Delta\omega$ located at $\omega = n\Delta\omega$ is $\frac{1}{2\pi}F(n\Delta\omega)\Delta\omega$, and the addition of all these components yields $f(t)$ in the integral form:

$$f(t) = \lim_{\Delta\omega \to 0} \frac{1}{2\pi} \sum_{n=-\infty}^{\infty} F(n\Delta\omega)e^{(jn\Delta\omega)t}\Delta\omega = \frac{1}{2\pi}\int_{-\infty}^{\infty} F(\omega)e^{j\omega t}\,d\omega \qquad (7.15)$$

The contribution by components within the band $d\omega$ is $\frac{1}{2\pi}F(\omega)d\omega = F(\omega)d\mathcal{F}$, in which $d\mathcal{F}$ is the bandwidth in Hertz. Clearly $F(\omega)$ is the **spectral density** per unit bandwidth (in Hertz). This also means that even if the amplitude of any one component is zero, the relative amount of a component of frequency ω is $F(\omega)$. Although $F(\omega)$ is a spectral density, in practice it is customarily called the **spectrum** of $f(t)$ rather than the spectral density of $f(t)$. Deferring to this convention, we shall call $F(\omega)$ the Fourier spectrum (or Fourier transform) of $f(t)$.

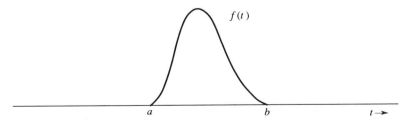

Fig. 7.6

An important point to remember here is that $f(t)$ is represented (or synthesized) by exponentials or sinusoids that are everlasting (not causal). This leads to a rather fascinating picture when we try to visualize the synthesis of a timelimited pulse signal $f(t)$ (Fig. 7.6) according to the sinusoidal components in its Fourier spectrum. The signal $f(t)$ exists only over an interval (a, b) and is zero outside this interval. The spectrum of $f(t)$ contains an infinite number of exponentials (or sinusoids) which start at $t = -\infty$ and continue forever. The amplitudes and phases of these components are such that they add up exactly to $f(t)$ over the finite interval (a, b) and add up to zero everywhere outside this interval. Juggling with such a perfect and delicate balance of amplitudes and phases of an infinite number of components boggles the human imagination. Yet the Fourier transform accomplishes it routinely, without much thinking on our part. Indeed, we become so involved in mathematical manipulations that we fail to notice this marvel.

7.2-1   Connection between the Fourier and the Laplace Transform

Equation (7.3) defines the Fourier transform as

$$F(\omega) = \int_{-\infty}^{\infty} f(t)e^{-j\omega t}dt \tag{7.16}$$

The general (bilateral) Laplace transform, as defined in Eq. (4.1), is given by

$$F(s) = \int_{-\infty}^{\infty} f(t)e^{-st}dt \tag{7.17}$$

These equations show that the Fourier transform is a special case of the Laplace transform in which $s = j\omega$. Note that since the integral in Eq. (7.17) is denoted by $F(s)$, the integral in Eq. (7.16) is $F(j\omega)$. From the viewpoint of unification of two transforms it would have been proper to use the notation $F(j\omega)$ for the Fourier transform. But because the notation $F(\omega)$ is more compact and more convenient for manipulation, we prefer it to the notation $F(j\omega)$. This fact should be kept in mind in future discussions whenever we connect the Fourier with the Laplace transform. In this book, both $F(\omega)$ and $F(j\omega)$ mean the same thing.

Although the two transforms developed independently, our discussion shows that the Fourier transform is a special case of Laplace transform in which the complex frequency $s = j\omega$; that is, when s is restricted to the imaginary axis of the complex frequency plane. Because of this restriction, exponentially growing signals do not have Fourier transform. Alternatively, the Laplace transform may be considered as a generalization of the Fourier transform in which the frequency is generalized from $j\omega$ to $s = \sigma + j\omega$. Since the Fourier transform is a special case of Laplace transform, we may use Fourier transform to analyze systems, provided that all the signals (input and output) are Fourier-transformable. Now, unstable systems generate exponentially growing signals which are not Fourier-transformable. Consequently the Fourier transform can be used to analyze only asymptotically stable systems and only if the inputs are Fourier-transformable. In addition, the Laplace transform of a signal is more compact than the Fourier transform. Moreover, the variable $j\omega$ is more cumbersome than the variable s. Clearly, the Laplace transform is superior to the Fourier transform for system analysis.

In signal analysis, on the other hand, Fourier transform is preferred over Laplace transform. The reason lies in the fact that both the Laplace and the Fourier transforms furnish spectral representation of a signal $f(t)$ in terms of exponential or sinusoidal components. The exponentials $e^{j\omega t}$ in the Fourier transform are ordinary sinusoids, whereas an exponential e^{st} in the Laplace transform can be expressed as

$$e^{st} = e^{(\sigma+j\omega)t} = e^{\sigma t}e^{j\omega t} = e^{\sigma t}(\cos \omega t + j \sin \omega t)$$

This shows that exponentials e^{st} are sinusoids with exponentially varying amplitudes. Therefore the Laplace transform of a signal represents the signal in terms of sinusoids with exponentially varying amplitudes. The Fourier transform with $s = j\omega$ provides the spectral representation of a signal $f(t)$ in terms of sinusoidal components of the form $\cos \omega t$. A better physical insight or understanding of a signal $f(t)$ is gained through its spectral representation by using sinusoidal components of the form $\cos \omega t$ rather than those of the form $e^{\sigma t} \cos \omega t$. For this reason the Fourier transform is preferred to the Laplace transform in signal analysis.

A word of caution is in order. Because the Fourier transform is the Laplace transform with $s = j\omega$, it should not be assumed automatically that the Fourier transform of a signal $f(t)$ is the Laplace transform of the signal with s replaced by $j\omega$. If a signal is absolutely integrable (that is, if $f(t)$ satisfies condition (7.13)), the Fourier transform of a signal is the same as its Laplace transform with s replaced by $j\omega$. This is readily verified for signal $e^{-at}u(t)$, whose Laplace and Fourier transforms are $1/(s+a)$ and $1/(j\omega+a)$, respectively. This is not generally true of signals which are not absolutely integrable. The Laplace transform of the unit-step function is

$$u(t) \Longleftrightarrow \frac{1}{s} \qquad \mathrm{Re}\, s > 0 \tag{7.18}$$

Later we show [Eq. (7.29)] that the Fourier transform of the unit step function is

$$u(t) \Longleftrightarrow \frac{1}{j\omega} + \pi\delta(\omega) \tag{7.19}$$

The reason for this discrepancy has something to do with the region of convergence. Both the Laplace and the Fourier transform synthesize $f(t)$, using everlasting exponentials of the form e^{st}. The frequency s can be anywhere in the complex plane for the Laplace transform but it must be restricted to the $j\omega$ axis in the case of Fourier transform. The unit step function is readily synthesized in the Laplace transform by spectrum $F(s) = 1/s$, in which the frequencies s are chosen in the RHP (the region of convergence for $u(t)$ is Re $s > 0$). In the Fourier transform, however, we are restricted to values of s only on the $j\omega$ axis. The function $u(t)$ can still be synthesized by frequencies along $j\omega$ axis but the spectrum is more complicated when we are free to choose the frequencies in the RHP. This is the reason for the difference between the Laplace and the Fourier transforms of functions which are not absolutely integrable.

The inverse Laplace transform can be derived from the inverse Fourier transform [Eq. (7.17)] by a change of variable from $j\omega$ to $s = \sigma + j\omega$. This yields $ds = jd\omega$. The limits of integration for ω from $-\infty$ to ∞ become from $\sigma - j\infty$ to $\sigma + j\infty$ for s. We select $\sigma = c$, such that the path of integration lies in the region of convergence. With these changes Eq. (7.17) becomes

$$f(t) = \frac{1}{2\pi j} \int_{c-j\infty}^{c+j\infty} F(s)e^{st}ds$$

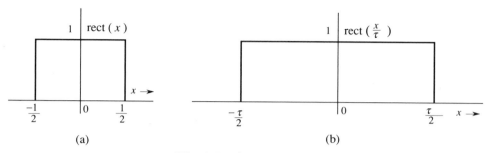

Fig. 7.7 A gate pulse.

7.3   TRANSFORMS OF SOME USEFUL FUNCTIONS

Before deriving the Fourier transform of any function, we introduce gate, triangle, and sinc functions.

The Unit Gate Function

We define a unit gate function rect (x) as a gate pulse of unit height and unit width, centered at the origin, as shown in Fig. 7.7a:

$$\text{rect}\,(x) = \begin{cases} 0 & |x| > \frac{1}{2} \\ \frac{1}{2} & |x| = \frac{1}{2} \\ 1 & |x| < \frac{1}{2} \end{cases} \tag{7.20}$$

The gate pulse in Fig. 7.7b is the unit gate pulse rect (x) expanded by a factor τ and therefore can be expressed as rect $\left(\frac{x}{\tau}\right)$ (see Sec. B.4-2). Observe that τ, the denominator of the argument of rect$\left(\frac{x}{\tau}\right)$, indicates the width of the pulse.

The Unit Triangle Function

We define a unit triangle function $\Delta(x)$ as a triangular pulse of unit height and unit width, centered at the origin, as shown in Fig. 7.8a:

$$\Delta(x) = \begin{cases} 0 & |x| > \frac{1}{2} \\ 1 - 2|x| & |x| < \frac{1}{2} \end{cases} \tag{7.21}$$

The pulse in Fig. 7.8b is $\Delta(\frac{x}{\tau})$. Observe that here, as for the gate pulse, the denominator τ of the argument of $\Delta(\frac{x}{\tau})$ indicates the pulse width.

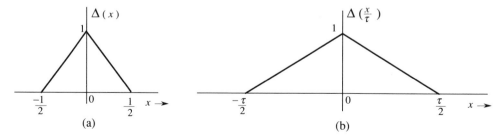

Fig. 7.8 A triangle pulse.

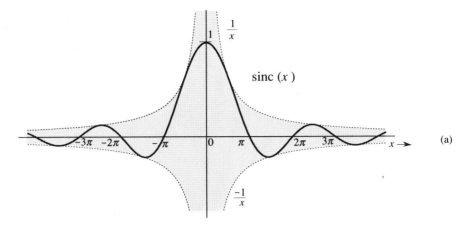

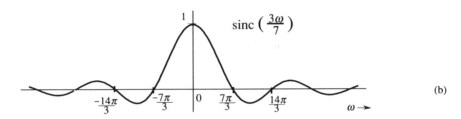

Fig. 7.9 A sinc pulse.

The Sinc Function

The function $\sin x / x$ is the "sine over argument" function denoted by $\mathrm{sinc}\,(x)$.† This function plays an important role in signal processing. It is also known as the *filtering or interpolating function*. We define

$$\mathrm{sinc}\,(x) = \frac{\sin x}{x} \tag{7.22}$$

Inspection of Eq. (7.22) shows that

1. $\mathrm{sinc}\,(x)$ is an even function of x.

2. $\mathrm{sinc}\,(x) = 0$ when $\sin x = 0$ except at $x = 0$, where it is indeterminate. This means that $\mathrm{sinc}\,x = 0$ for $x = \pm\pi, \pm2\pi, \pm3\pi, \cdots$

3. Using L'Hôpital's rule, we find $\mathrm{sinc}\,(0) = 1$.

4. $\mathrm{sinc}\,(x)$ is the product of an oscillating signal $\sin x$ (of period 2π) and a monotonically decreasing function $1/x$. Therefore $\mathrm{sinc}\,(x)$ exhibits sinusoidal oscillations of period 2π, with amplitude decreasing continuously as $1/x$.

Figure 7.9a shows $\mathrm{sinc}\,(x)$. Observe that $\mathrm{sinc}\,(x) = 0$ for values of x that are positive and negative integral multiples of π. Figure 7.9b shows $\mathrm{sinc}\left(\frac{3\omega}{7}\right)$. The

---

†$\mathrm{sinc}\,(x)$ is also denoted by $\mathrm{Sa}\,(x)$ in the literature. Some authors define $\mathrm{sinc}\,(x)$ as

$$\mathrm{sinc}\,(x) = \frac{\sin \pi x}{\pi x}$$

argument $\frac{3\omega}{7} = \pi$ when $\omega = \frac{7\pi}{3}$. Therefore the first zero of this function occurs at $\omega = \frac{7\pi}{3}$.

$\triangle$ **Exercise E7.1**

Sketch: **(a)** $\text{rect}\left(\frac{x}{8}\right)$ **(b)** $\Delta\left(\frac{\omega}{10}\right)$ **(c)** $\text{sinc}\left(\frac{3\pi\omega}{2}\right)$ **(d)** $\text{sinc}\,(t)\text{rect}\left(\frac{t}{4\pi}\right)$. $\triangledown$

■ **Example 7.2**

Find the Fourier transform of $f(t) = \text{rect}\left(\frac{t}{\tau}\right)$ (Fig. 7.10a).

$$F(\omega) = \int_{-\infty}^{\infty} \text{rect}\left(\frac{t}{\tau}\right) e^{-j\omega t}\,dt$$

Since $\text{rect}\left(\frac{t}{\tau}\right) = 1$ for $|t| < \frac{\tau}{2}$, and since it is zero for $|t| > \frac{\tau}{2}$,

$$F(\omega) = \int_{-\tau/2}^{\tau/2} e^{-j\omega t}\,dt$$

$$= -\frac{1}{j\omega}(e^{-j\omega\tau/2} - e^{j\omega\tau/2}) = \frac{2\sin\left(\frac{\omega\tau}{2}\right)}{\omega}$$

$$= \tau \frac{\sin\left(\frac{\omega\tau}{2}\right)}{\left(\frac{\omega\tau}{2}\right)} = \tau\,\text{sinc}\left(\frac{\omega\tau}{2}\right)$$

Therefore

$$\text{rect}\left(\frac{t}{\tau}\right) \Longleftrightarrow \tau\,\text{sinc}\left(\frac{\omega\tau}{2}\right) \tag{7.23}$$

Observe that $\text{sinc}\left(\frac{\omega\tau}{2}\right) = 0$ when $\frac{\omega\tau}{2} = \pm n\pi$; that is, when $\omega = \pm\frac{2n\pi}{\tau}, (n = 1, 2, 3, \cdots)$, as shown in Fig. 7.10b. ■

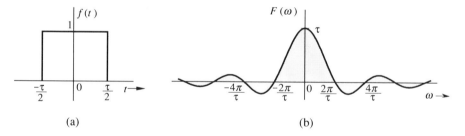

Fig. 7.10 A gate pulse and its Fourier spectrum.

Bandwidth of rect $\left(\frac{t}{\tau}\right)$

The spectrum $F(\omega)$ in Fig. 7.10 peaks at $\omega = 0$ and decays at higher frequencies. Therefore rect $\left(\frac{t}{\tau}\right)$ is a lowpass signal with most of the signal energy in lower frequency components. Strictly speaking, because the spectrum extends from 0 to ∞, the bandwidth is ∞. However, much of the spectrum is concentrated within the first lobe (from $\omega = 0$ to $\omega = \frac{2\pi}{\tau}$). Therefore a rough estimate of bandwidth† of a rectangular pulse of width τ seconds is $\frac{2\pi}{\tau}$ rad/s, or $\frac{1}{\tau}$ Hz. Note the reciprocal

---

†To compute bandwidth, we must consider the spectrum only for positive values of ω. See discussion on p. 445.

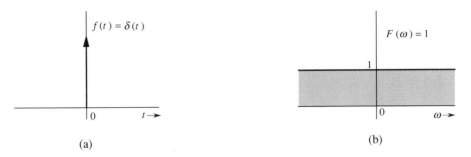

Fig. 7.11   Unit impulse and its Fourier spectrum.

relationship of the pulse width with its bandwidth. We shall observe later that this result is true in general.

■ **Example 7.3**

Find the Fourier transform of the unit impulse $\delta(t)$.

Using the sampling property of impulse [Eq. (2.33)], we obtain

$$\mathcal{F}[\delta(t)] = \int_{-\infty}^{\infty} \delta(t)e^{-j\omega t}dt = 1 \tag{7.24a}$$

or

$$\delta(t) \Longleftrightarrow 1 \tag{7.24b}$$

Figure 7.11 shows $\delta(t)$ and its spectrum.   ■

■ **Example 7.4**

Find the inverse Fourier transform of $\delta(\omega)$.

From Eq. (7.8b) and the sampling property of the impulse function,

$$\mathcal{F}^{-1}[\delta(\omega)] = \frac{1}{2\pi} \int_{-\infty}^{\infty} \delta(\omega)e^{j\omega t}\, d\omega = \frac{1}{2\pi}$$

Therefore

$$\frac{1}{2\pi} \Longleftrightarrow \delta(\omega) \tag{7.25a}$$

or

$$1 \Longleftrightarrow 2\pi\delta(\omega) \tag{7.25b}$$

This shows that the spectrum of a constant signal $f(t) = 1$ is an impulse $2\pi\delta(\omega)$, as shown in Fig. 7.12.

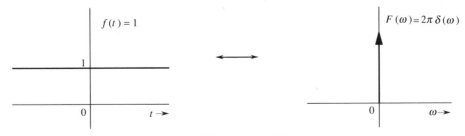

Fig. 7.12   A constant (dc) signal and its Fourier spectrum.

The result (7.25b) also could have been anticipated on qualitative grounds. Recall that the Fourier transform of $f(t)$ is a spectral representation of $f(t)$ in terms of everlasting exponential components of the form $e^{j\omega t}$. Now to represent a constant signal $f(t) = 1$, we need a single everlasting exponential‡ $e^{j\omega t}$ with $\omega = 0$. This results in a spectrum at single frequency $\omega = 0$. Another way of looking at the situation is that $f(t) = 1$ is a dc signal which has a single frequency $\omega = 0$ (dc).   ■

If an impulse at $\omega = 0$ is a spectrum of a dc signal, what does an impulse at $\omega = \omega_0$ represent? We shall answer this question in the next example.

■ **Example 7.5**

Find the inverse Fourier transform of $\delta(\omega - \omega_0)$.

Using the sampling property of the impulse function, we obtain

$$\mathcal{F}^{-1}[\delta(\omega - \omega_0)] = \frac{1}{2\pi} \int_{-\infty}^{\infty} \delta(\omega - \omega_0)e^{j\omega t}\, d\omega = \frac{1}{2\pi}e^{j\omega_0 t}$$

Therefore

$$\frac{1}{2\pi}e^{j\omega_0 t} \Longleftrightarrow \delta(\omega - \omega_0)$$

or

$$e^{j\omega_0 t} \Longleftrightarrow 2\pi\delta(\omega - \omega_0) \tag{7.26a}$$

This result shows that the spectrum of an everlasting exponential $e^{j\omega_0 t}$ is a single impulse at $\omega = \omega_0$. We reach the same conclusion by qualitative reasoning. To represent everlasting exponential $e^{j\omega_0 t}$, we need a single everlasting exponential $e^{j\omega t}$ with $\omega = \omega_0$. Therefore the spectrum consists of a single component at frequency $\omega = \omega_0$.

From Eq. (7.26a) it follows that

$$e^{-j\omega_0 t} \Longleftrightarrow 2\pi\delta(\omega + \omega_0) \tag{7.26b}$$

■

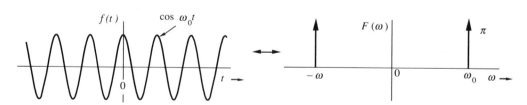

Fig. 7.13   A cosine signal and its Fourier spectrum.

---

‡The constant multiplier 2π in the spectrum $[F(\omega) = 2\pi\delta(\omega)]$ may be a bit puzzling. Since $1 = e^{j\omega t}$ with $\omega = 0$, it appears that the Fourier transform of $f(t) = 1$ should be an impulse of strength unity rather than 2π. Recall that in the Fourier transform $f(t)$ is synthesized not by exponentials of amplitude $F(n\Delta\omega)\Delta\omega$, but of amplitude $[F(n\Delta\omega)\Delta\omega]/2\pi$, as seen from Eq. (7.6b). Therefore $F(n\Delta\omega)\Delta\omega$ (the area of impulse) must be 2π to synthesize $f(t) = 1$. Had we used variable $\mathcal{F}$ (Hertz) instead of ω, the spectrum would have been a unit impulse.

■ Example 7.6

Find the Fourier transforms of everlasting sinusoid $\cos \omega_0 t$.

Recall the Euler formula [Eq. (B.24)]

$$\cos \omega_0 t = \frac{1}{2}(e^{j\omega_0 t} + e^{-j\omega_0 t})$$

Adding Eqs. (7.26a) and (7.26b), and using the above formula, we obtain

$$\cos \omega_0 t \Longleftrightarrow \pi[\delta(\omega + \omega_0) + \delta(\omega - \omega_0)] \tag{7.27}$$

The spectrum of $\cos \omega_0 t$ consists of two impulses at ω_0 and $-\omega_0$, as shown in Fig. 7.13. The result also follows from qualitative reasoning. An everlasting sinusoid $\cos \omega_0 t$ can be synthesized by two everlasting exponentials, $e^{j\omega_0 t}$ and $e^{-j\omega_0 t}$. Therefore the Fourier spectrum consists of only two components of frequencies ω_0 and $-\omega_0$. ■

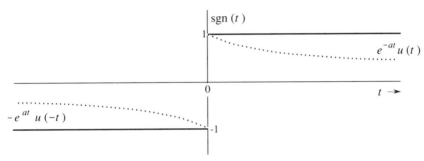

Fig. 7.14   A sign function.

■ Example 7.7

Find the Fourier transform of the sign function $\operatorname{sgn} t$ (pronounced signum t), shown in Fig. 7.14. Its value is $+1$ or -1 depending on whether t is positive or negative:

$$\operatorname{sgn} t = \begin{cases} 1 & t > 0 \\ -1 & t < 0 \end{cases} \tag{7.28}$$

The transform of $\operatorname{sgn} t$ can be obtained by considering $\operatorname{sgn} t$ as a sum of two exponentials, as shown in Fig. 7.14 in the limit as $a \to 0$:

$$\operatorname{sgn} t = \lim_{a \to 0} \left[e^{-at} u(t) - e^{at} u(-t) \right]$$

Therefore

$$\mathcal{F}[\operatorname{sgn} t] = \lim_{a \to 0} \left[\mathcal{F}[e^{-at} u(t)] - \mathcal{F}[e^{at} u(-t)] \right]$$

$$= \lim_{a \to 0} \left[\frac{1}{a + j\omega} - \frac{1}{a - j\omega} \right] \qquad \text{(see Pairs 1 and 2 in Table 7.1)}$$

$$= \lim_{a \to 0} \left[\frac{-2j\omega}{a^2 + \omega^2} \right] = \frac{2}{j\omega} \qquad ■$$

■ **Example 7.8**

Find the Fourier transform of the unit step function $u(t)$.

Observe that addition of a constant 1 in Fig. 7.12 to sgn t in Fig. 7.14 results in $2u(t)$. Therefore

$$u(t) = \tfrac{1}{2}[1 + \text{sgn } t]$$

Consequently

$$\mathcal{F}[u(t)] = \mathcal{F}[\tfrac{1}{2}] + \tfrac{1}{2}\mathcal{F}[\text{sgn } t]$$

$$= \pi\delta(\omega) + \frac{1}{j\omega} \tag{7.29}$$

Note that $u(t)$ is not a "true" dc signal because it is not constant over the interval $(-\infty$ to $\infty)$. To synthesize a "true" dc, we require only one everlasting exponential with $\omega = 0$ (impulse at $\omega = 0$). The signal $u(t)$ has a jump discontinuity at $t = 0$. It is impossible to synthesize such a signal with a single everlasting exponential $e^{j\omega t}$. To synthesize this signal from everlasting exponentials, we need, in addition to an impulse at $\omega = 0$, all frequency components, as indicated by the term $1/j\omega$ in Eq. (7.29). ■

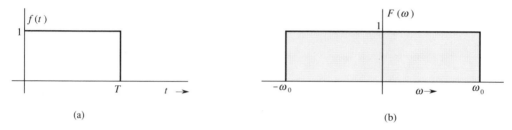

Fig. 7.15

△ **Exercise E7.2**

Show that the Fourier transform of a gate pulse $f(t)$ shown in Fig. 7.15a is $\frac{1}{j\omega}(1 - e^{-j\omega T})$ ▽

△ **Exercise E7.3**

Show that the inverse Fourier transform of $F(\omega)$ shown in Fig. 7.15b is $f(t) = \frac{\omega_0}{\pi}$ sinc $(\omega_0 t)$. Sketch $f(t)$. ▽.

△ **Exercise E7.4**

Show that $\cos(\omega_0 t + \theta) \Longleftrightarrow \pi[\delta(\omega + \omega_0)e^{-j\theta} + \delta(\omega - \omega_0)e^{j\theta}]$.

Hint: $\cos(\omega_0 t + \theta) = \tfrac{1}{2}[e^{j(\omega_0 t + \theta)} + e^{-j(\omega_0 t + \theta)}] = \tfrac{1}{2}[e^{j\theta}e^{j\omega_0 t} + e^{-j\theta}e^{-j\omega_0 t}]$ ▽

7.4 SOME PROPERTIES OF THE FOURIER TRANSFORM

We now study some of the important properties of the Fourier transform and their implications as well as their applications. Since the Fourier transform is a special case of the bilateral Laplace transform, the properties of the two transforms are similar. However, we studied the properties of unilateral (not bilateral) Laplace transform in Chapter 4. For this reason we shall find some differences (along with many similarities) between the properties of the Fourier and the (unilateral) Laplace transform. Before embarking on the study of the properties of the Fourier transform, it is important to point out a pervasive aspect of the Fourier transform: **the time-frequency duality**.

Table 7.1
A Short Table of Fourier Transforms

	$f(t)$	$F(\omega)$	
1	$e^{-at}u(t)$	$\dfrac{1}{a+j\omega}$	$a > 0$
2	$e^{at}u(-t)$	$\dfrac{1}{a-j\omega}$	$a > 0$
3	$e^{-a|t|}$	$\dfrac{2a}{a^2+\omega^2}$	$a > 0$
4	$te^{-at}u(t)$	$\dfrac{1}{(a+j\omega)^2}$	$a > 0$
5	$t^n e^{-at}u(t)$	$\dfrac{n!}{(a+j\omega)^{n+1}}$	$a > 0$
6	$\delta(t)$	1	
7	1	$2\pi\delta(\omega)$	
8	$e^{j\omega_0 t}$	$2\pi\delta(\omega-\omega_0)$	
9	$\cos\omega_0 t$	$\pi[\delta(\omega-\omega_0)+\delta(\omega+\omega_0)]$	
10	$\sin\omega_0 t$	$j\pi[\delta(\omega+\omega_0)-\delta(\omega-\omega_0)]$	
11	$u(t)$	$\pi\delta(\omega)+\dfrac{1}{j\omega}$	
12	$\operatorname{sgn} t$	$\dfrac{2}{j\omega}$	
13	$\cos\omega_0 t\, u(t)$	$\dfrac{\pi}{2}[\delta(\omega-\omega_0)+\delta(\omega+\omega_0)]+\dfrac{j\omega}{\omega_0^2-\omega^2}$	
14	$\sin\omega_0 t\, u(t)$	$\dfrac{\pi}{2j}[\delta(\omega-\omega_0)-\delta(\omega+\omega_0)]+\dfrac{\omega_0}{\omega_0^2-\omega^2}$	
15	$e^{-at}\sin\omega_0 t\, u(t)$	$\dfrac{\omega_0}{(a+j\omega)^2+\omega_0^2}$	$a > 0$
16	$e^{-at}\cos\omega_0 t\, u(t)$	$\dfrac{a+j\omega}{(a+j\omega)^2+\omega_0^2}$	$a > 0$
17	$\operatorname{rect}\left(\dfrac{t}{\tau}\right)$	$\tau\operatorname{sinc}\left(\dfrac{\omega\tau}{2}\right)$	
18	$\dfrac{W}{\pi}\operatorname{sinc}(Wt)$	$\operatorname{rect}\left(\dfrac{\omega}{2W}\right)$	
19	$\Delta\left(\dfrac{t}{\tau}\right)$	$\dfrac{\tau}{2}\operatorname{sinc}^2\left(\dfrac{\omega\tau}{4}\right)$	
20	$\dfrac{W}{2\pi}\operatorname{sinc}^2\left(\dfrac{Wt}{2}\right)$	$\Delta\left(\dfrac{\omega}{2W}\right)$	
21	$\displaystyle\sum_{n=-\infty}^{\infty}\delta(t-nT)$	$\omega_0\displaystyle\sum_{n=-\infty}^{\infty}\delta(\omega-n\omega_0)$	$\omega_0=\dfrac{2\pi}{T}$
22	$e^{-t^2/2\sigma^2}$	$\sigma\sqrt{2\pi}\,e^{-\sigma^2\omega^2/2}$	

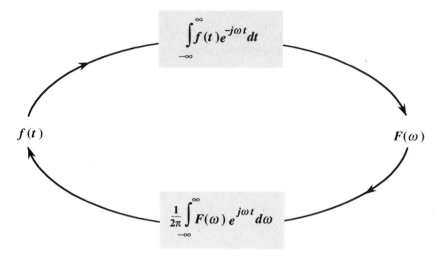

Fig. 7.16   A near symmetry between the direct and the inverse Fourier transforms.

7.4-1   Symmetry of the Direct and the Inverse Transform Operations: Time-Frequency Duality

Equations (7.8a) and (7.8b) show an interesting fact: the direct and the inverse transform operations are remarkably similar. These operations, which are required to go from $f(t)$ to $F(\omega)$ and then from $F(\omega)$ to $f(t)$, are shown graphically in Fig. 7.16. There are only two minor differences in these operations: the factor 2π appears only in the inverse operator, and the exponential index in the two operations have opposite signs. Otherwise the two operations are symmetrical.† This observation has far-reaching consequences in the study of Fourier transform. It is the basis of the so-called "duality" of time and frequency. **Duality principle may be compared with a photograph and its negative. A photograph can be obtained from its negative, and by using an identical procedure, the negative can be obtained from the photograph.** For any result or relationship between $f(t)$ and $F(\omega)$, there exists a dual result or relationship, obtained by interchanging the roles of $f(t)$ and $F(\omega)$ in the original result (along with some minor modifications arising because of the factor 2π and a sign change). For example, the time-shifting property, to be proved later, states that if $f(t) \iff F(\omega)$, then

$$f(t - t_0) \iff F(\omega)e^{-j\omega t_0} \tag{7.30a}$$

---

†Of the two differences, the former can be eliminated by change of variable from ω to $\mathcal{F}$ (in Hz). In this case

$$\omega = 2\pi\mathcal{F} \qquad \text{and} \qquad d\omega = 2\pi \, d\mathcal{F}$$

Therefore the direct and the inverse transforms are given by

$$F(2\pi\mathcal{F}) = \int_{-\infty}^{\infty} f(t)e^{-j2\pi\mathcal{F}t}dt \quad \text{and} \quad f(t) = \int_{-\infty}^{\infty} F(2\pi\mathcal{F})e^{j2\pi\mathcal{F}t}d\mathcal{F}$$

This leaves only one significant difference, that of sign change in the exponential index. Otherwise the two operations are symmetrical.

The dual of this property (the frequency-shifting property) states that

$$f(t)e^{j\omega_0 t} \iff F(\omega - \omega_0) \tag{7.30b}$$

Observe the role reversal of time and frequency in these two equations (with the minor difference of the sign change in the exponential index). The value of this principle lies in the fact that *whenever we derive any result, we can be sure that result has a dual.* This can give valuable insights about many unsuspected properties or results in signal processing.

Properties of Fourier transform are useful in deriving not only the direct and inverse transforms of many functions, but also in obtaining several valuable results in signal processing. The reader should not fail to observe the ever-present duality in this discussion. We begin with the so-called duality property, which is one of the consequences of the duality principle discussed above.

7.4-2   The Duality Property

This property states that if

$$f(t) \iff F(\omega)$$

then

$$F(t) \iff 2\pi f(-\omega) \tag{7.31}$$

Proof: from Eq. (7.8b)

$$f(t) = \frac{1}{2\pi} \int_{-\infty}^{\infty} F(x)e^{jxt}\, dx$$

Hence

$$2\pi f(-t) = \int_{-\infty}^{\infty} F(x)e^{-jxt}\, dx$$

Changing t to ω yields Eq. (7.31).

■ **Example 7.9**

In this example we shall apply the duality [Eq. (7.31)] to the pair in Fig. 7.17a. From Eq. (7.23) we have

$$\text{rect}\left(\frac{t}{\tau}\right) \iff \tau\, \text{sinc}\left(\frac{\omega\tau}{2}\right) \tag{7.32}$$

In this case

$$F(\omega) = \tau\, \text{sinc}\left(\frac{\omega\tau}{2}\right) \quad \text{and} \quad f(t) = \text{rect}\left(\frac{t}{\tau}\right)$$

Therefore

$$F(t) = \tau\, \text{sinc}\left(\frac{t\tau}{2}\right) \quad \text{and} \quad f(-\omega) = \text{rect}\left(\frac{-\omega}{\tau}\right) = \text{rect}\left(\frac{\omega}{\tau}\right)$$

Therefore, from Eq. (7.31)

$$\tau\, \text{sinc}\left(\frac{\tau t}{2}\right) \iff 2\pi\, \text{rect}\left(\frac{\omega}{\tau}\right) \tag{7.33}$$

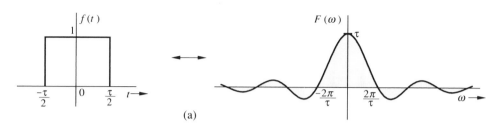

(a)

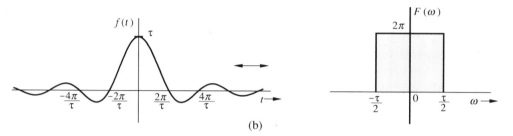

(b)

Fig. 7.17 Duality property of the Fourier transform.

Figure 7.17b shows this pair graphically. Observe the interchange of roles of t and ω (with a minor adjustment of factor 2π).

As an interesting exercise, by applying the duality property, the reader should generate a dual of every Pair in Table 7.1. ■

△ **Exercise E7.5**

Apply duality property to Pairs 1, 3, and 9 (Table 7.1) to show that

(a) $\frac{1}{jt+a} \Longleftrightarrow 2\pi e^{a\omega} u(-\omega)$ **(b)** $\frac{2a}{t^2+a^2} \Longleftrightarrow 2\pi e^{-a|\omega|}$

(c) $\delta(t+t_0) + \delta(t-t_0) \Longleftrightarrow 2\cos t_0\omega$ ▽

7.4-3   The Scaling Property

If

$$f(t) \Longleftrightarrow F(\omega)$$

then, for any real constant a,

$$f(at) \Longleftrightarrow \frac{1}{|a|} F\left(\frac{\omega}{a}\right) \tag{7.34}$$

Proof: For a positive real constant a,

$$\mathcal{F}[f(at)] = \int_{-\infty}^{\infty} f(at) e^{-j\omega t}\, dt$$

If we let $x = at$, then for $a > 0$

$$\mathcal{F}[f(at)] = \frac{1}{a} \int_{-\infty}^{\infty} f(x) e^{(-j\omega/a)x}\, dx = \frac{1}{a} F\left(\frac{\omega}{a}\right)$$

Similarly, it can be shown that if $a < 0$,

$$f(at) \Longleftrightarrow \frac{-1}{a} F\left(\frac{\omega}{a}\right)$$

Hence follows Eq. (7.34)

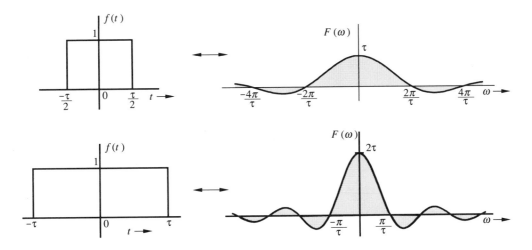

Fig. 7.18 Scaling property of the Fourier transform.

Significance of the Scaling Property

The function $f(at)$ represents the function $f(t)$ compressed in time by factor a (see Sec. B.4-2). Similarly, a function $F(\frac{\omega}{a})$ represents a function $F(\omega)$ expanded in frequency by the same factor a. **The scaling property states that time compression of a signal results in its spectral expansion, and time expansion of the signal results in its spectral compression.** Intuitively compression in time by a factor a means that the signal is varying rapidly by the same factor. To synthesize such a signal, the frequencies of its sinusoidal components must be increased by the factor a, implying that its frequency spectrum is expanded by the factor a. Similarly, a signal expanded in time varies more slowly; hence the frequencies of its components are lowered, implying that its frequency spectrum is compressed. For instance, the signal $\cos 2\omega_0 t$ is the same as the signal $\cos \omega_0 t$ time-compressed by a factor of 2. Clearly, the spectrum of the former (impulse at $\pm 2\omega_0$) is an expanded version of the spectrum of the latter (impulse at $\pm \omega_0$). The effect of this scaling is demonstrated in Fig. 7.18.

Reciprocity of Signal Duration and Its Bandwidth

The scaling property implies that if $f(t)$ is wider, its spectrum is narrower, and vice versa. Doubling the signal duration halves its bandwidth and vice versa. This suggests that the bandwidth of a signal is inversely proportional to the signal duration or width (in seconds). We have already verified this fact for the gate pulse, where we found that the bandwidth of a gate pulse of width τ seconds is $\frac{1}{\tau}$ Hz. More discussion of this interesting topic, can be found in the literature.[2]

■ Example 7.10

Show that

$$f(-t) \Longleftrightarrow F(-\omega) \tag{7.35}$$

Using this result and the fact that $e^{-at}u(t) \Longleftrightarrow \frac{1}{a+j\omega}$, find the Fourier transforms of $e^{at}u(-t)$ and $e^{-a|t|}$.

Equation (7.35) follows from Eq. (7.34) by letting $a = -1$. Application of this result to Pair

$$e^{-at}u(t) \Longleftrightarrow \frac{1}{a + j\omega}$$

yields

$$e^{at}u(-t) \Longleftrightarrow \frac{1}{a - j\omega}$$

Also

$$e^{-a|t|} = e^{-at}u(t) + e^{at}u(-t)$$

Therefore

$$e^{-a|t|} \Longleftrightarrow \frac{1}{a + j\omega} + \frac{1}{a - j\omega} = \frac{2a}{a^2 + \omega^2} \tag{7.36}$$

The signal $e^{-a|t|}$ and its spectrum are shown in Fig. 7.19.   ∎

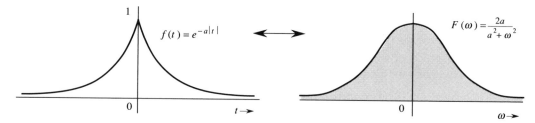

Fig. 7.19 $e^{-a|t|}$ and its Fourier spectrum.

7.4-4   The Time-Shifting Property

If

$$f(t) \Longleftrightarrow F(\omega)$$

then

$$f(t - t_0) \Longleftrightarrow F(\omega)e^{-j\omega t_0} \tag{7.37a}$$

Proof: By definition,

$$\mathcal{F}[f(t - t_0)] = \int_{-\infty}^{\infty} f(t - t_0)e^{-j\omega t}\, dt$$

Letting $t - t_0 = x$, we have

$$\mathcal{F}[f(t - t_0)] = \int_{-\infty}^{\infty} f(x)e^{-j\omega(x + t_0)}\, dx$$

$$= e^{-j\omega t_0} \int_{-\infty}^{\infty} f(x)e^{-j\omega x}\, dx$$

$$= F(\omega)e^{-j\omega t_0} \tag{7.37b}$$

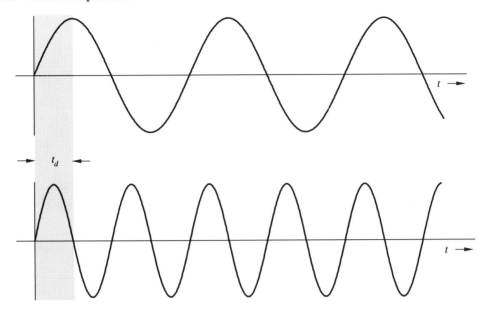

Fig. 7.20 Physical explanation of the time-shifting property.

This result shows that **delaying a signal by t_0 seconds does not change its amplitude spectrum. The phase spectrum, however, is changed by** $-\omega t_0$.

Physical Explanation of the Linear Phase

Time delay in a signal causes a linear phase shift in its spectrum. This result can also be derived by heuristic reasoning. Imagine $f(t)$ being synthesized by its Fourier components, which are sinusoids of certain amplitudes and phases. The delayed signal $f(t - t_0)$ can be synthesized by the same sinusoidal components, each delayed by t_0 seconds. The amplitudes of the components remain unchanged. Therefore amplitude spectrum of $f(t - t_0)$ is identical to that of $f(t)$. The time delay of t_0 in each sinusoid, however, does change the phase of each component. Now, a sinusoid $\cos \omega t$ delayed by t_0 is given by

$$\cos \omega(t - t_0) = \cos(\omega t - \omega t_0)$$

Therefore a time delay t_0 in a sinusoid of frequency ω manifests as a phase delay of ωt_0. This is a linear function of ω, meaning that higher-frequency components must undergo proportionately higher phase shifts to achieve the same time delay. This effect is shown in Fig. 7.20 with two sinusoids, the frequency of the lower sinusoid being twice that of the upper. The same time delay t_0 amounts to a phase shift of $\pi/2$ in the upper sinusoid and a phase shift of π in the lower sinusoid. This verifies the fact that **to achieve the same time delay, higher-frequency sinusoids must undergo proportionately higher phase shifts**. The phase shift required to achieve a time delay t_0 is $-\omega t_0$. In summary, a time delay in a signal $f(t)$ leaves its amplitude spectrum unchanged but causes a change of $-\omega t_0$ in its phase spectrum. This is precisely the result in Eq. (7.37b).

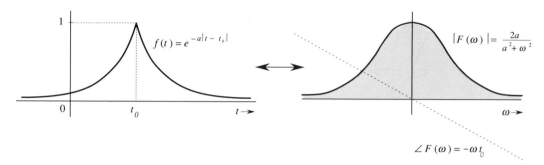

Fig. 7.21   Effect of time-shifting on the Fourier spectrum of a signal.

■ **Example 7.11**

Find the Fourier transform of $e^{-a|t-t_0|}$.

This function, shown in Fig. 7.21a, is a time-shifted version of $e^{-a|t|}$ (shown in Fig. 7.19a). From (7.36) and (7.37), we have

$$e^{-a|t-t_0|} \iff \frac{2a}{a^2 + \omega^2} e^{-j\omega t_0} \qquad (7.38)$$

The spectrum of $e^{-a|t-t_0|}$ (Fig. 7.21b) is the same as that of $e^{-a|t|}$ (Fig. 7.19b), except for an added phase shift of $-\omega t_0$.

Observe that the time delay t_0 causes a linear phase spectrum $-\omega t_0$. This example clearly demonstrates the effect of time shift.   ■

△ **Exercise E7.6**

Using Pair 18, and the time-shifting property, show that the Fourier transform of $\text{sinc} \, [\omega_0(t - T)]$ is $\frac{\pi}{\omega_0} \, \text{rect} \, (\frac{\omega}{2\omega_0}) e^{-j\omega T}$ ▽.

7.4-5   The Frequency-Shifting Property

If

$$f(t) \iff F(\omega)$$

then

$$f(t)e^{j\omega_0 t} \iff F(\omega - \omega_0) \qquad (7.39)$$

Proof: By definition,

$$\mathcal{F}[f(t)e^{j\omega_0 t}] = \int_{-\infty}^{\infty} f(t)e^{j\omega_0 t}e^{-j\omega t}dt = \int_{-\infty}^{\infty} f(t)e^{-j(\omega-\omega_0)t}dt = F(\omega - \omega_0)$$

This property states that **multiplication of a signal by a factor** $e^{j\omega_0 t}$ **shifts the spectrum of that signal by** $\omega = \omega_0$. **Note the duality between the time-shifting and the frequency-shifting properties.**

Changing ω_0 to $-\omega_0$ in Eq. (7.39) yields

$$f(t)e^{-j\omega_0 t} \iff F(\omega + \omega_0) \qquad (7.40)$$

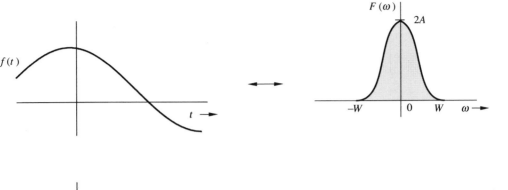

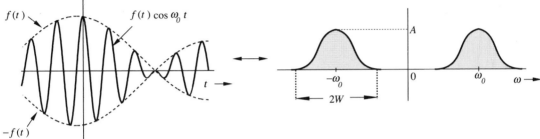

Fig. 7.22 Amplitude modulation of a signal causes spectral shifting.

Because $e^{j\omega_0 t}$ is not a real function that can be generated, frequency shifting in practice is achieved by multiplying $f(t)$ by a sinusoid. This can be seen from the fact that

$$f(t) \cos \omega_0 t = \tfrac{1}{2}[f(t)e^{j\omega_0 t} + f(t)e^{-j\omega_0 t}]$$

From Eq. (7.39) and (7.40), it follows that

$$f(t) \cos \omega_0 t \iff \tfrac{1}{2}[F(\omega - \omega_0) + F(\omega + \omega_0)] \qquad (7.41)$$

This shows that multiplication of a signal $f(t)$ by a sinusoid of frequency ω_0 shifts the spectrum $F(\omega)$ by $\pm\omega_0$. Multiplication of a sinusoid $\cos \omega_0 t$ by $f(t)$ amounts to modulating the sinusoid amplitude. This type of modulation is known as **amplitude modulation**. The sinusoid $\cos \omega_0 t$ is called the **carrier**, the signal $f(t)$ is the **modulating signal**, and the signal $f(t) \cos \omega_0 t$ is the **modulated signal**. More discussion of modulation and demodulation can be found in Reference 3.

To sketch a signal $f(t) \cos \omega_0 t$, we observe that

$$f(t) \cos \omega_0 t = \begin{cases} f(t) & \text{when } \cos \omega_0 t = 1 \\ -f(t) & \text{when } \cos \omega_0 t = -1 \end{cases}$$

Therefore $f(t) \cos \omega_0 t$ touches $f(t)$ when the sinusoid $\cos \omega_0 t$ is at its positive peaks and touches $-f(t)$ when $\cos \omega_0 t$ is at its negative peaks. This means that $f(t)$ and $-f(t)$ act as envelopes for the signal $f(t) \cos \omega_0 t$ (see Fig. 7.22). The signal $-f(t)$ is a mirror image of $f(t)$ about the horizontal axis. Figure 7.22 shows the signal $f(t)$, $f(t) \cos \omega_0 t$ and their spectra. Further explanation of sketching modulated signals is given in Section B.3.

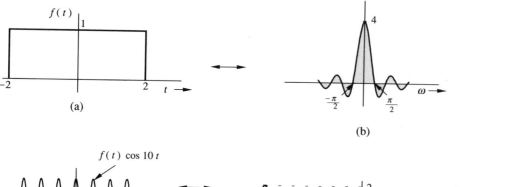

(a)

(b)

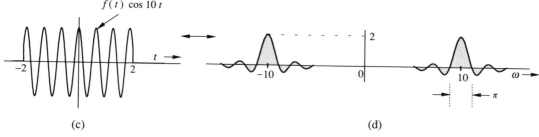

(c)   (d)

Fig. 7.23   An example of spectral shifting by amplitude modulation.

■ **Example 7.12**

Find and sketch the Fourier transform of the modulated signal $f(t) \cos 10t$ in which $f(t)$ is a gate pulse rect $\left(\frac{t}{4}\right)$ shown in Figure 7.23a.

From Pair 17 (Table 7.1) we find rect $\left(\frac{t}{4}\right) \Longleftrightarrow 4 \operatorname{sinc}(2\omega)$, shown in Fig. 7.23b. From (7.41) it follows that

$$f(t) \cos 10t \Longleftrightarrow \tfrac{1}{2}[F(\omega + 10) + F(\omega - 10)]$$

In this case $F(\omega) = 4 \operatorname{sinc}(2\omega)$. Therefore

$$f(t) \cos 10t \Longleftrightarrow 2 \operatorname{sinc}[2(\omega + 10)] + 2 \operatorname{sinc}[2(\omega - 10)]$$

The spectrum of $f(t) \cos 10t$ is obtained by shifting $F(\omega)$ in Fig. 7.23b to the left by 10 and also to the right by 10, and then multiplying it by half, as shown in Fig. 7.23d.
■

△ **Exercise E7.7**

Sketch signal $e^{-|t|} \cos 10t$. Find the Fourier transform of this signal and sketch its spectrum.
Answer: $F(\omega) = \frac{1}{(\omega - 10)^2 + 1} + \frac{1}{(\omega + 10)^2 + 1}$. The spectrum is that in Fig. 7.19, shifted to ± 10 and multiplied by half.   ▽

Application of Modulation

Modulation is used to shift signal spectra. Some of the situations where spectrum shifting is necessary are given below.

1. If several signals, each occupying the same frequency band, are transmitted simultaneously over the same transmission medium, they will all interfere; it will be impossible to separate or retrieve them at a receiver. For example, if all radio stations decide to broadcast audio signals simultaneously, the receiver will not be able to separate them. This problem is solved by using modulation, whereby each radio station is assigned a distinct carrier frequency. Each

station transmits a modulated signal, thus shifting the signal spectrum to its allocated band which is not occupied by any other station. A radio receiver can pick up any station by tuning to the band of the desired station. The receiver must now demodulate the received signal (undo the effect of modulation). Demodulation therefore consists of another spectral shift required to restore the signal to its original band. Note that both modulation and demodulation implement spectral shifting; consequently, demodulation operation is similar to modulation (see Prob. 7.4-8).

This method of transmitting several signals simultaneously over a channel by sharing its frequency band is known as **frequency-division multiplexing (FDM)**.

2. For effective radiation of power over a radio link, the antenna size must be of the order of the wavelength of the signal to be radiated. Audio signal frequencies are so low (wavelengths are so large) that impracticably large antennas will be required for radiation. Here, shifting the spectrum to a higher frequency (a smaller wavelength) by modulation solves the problem.

3. Amplification of very low-frequency signals requires impracticably large coupling capacitors between cascaded amplifier stages. In this case, the signal spectrum is shifted to a higher frequency where it can be easily amplified. The amplified signal is then demodulated. The so-called chopper amplifiers use this principle.

7.4-6 Convolution

The time-convolution property and its dual, the frequency-convolution property, state that if

$$f_1(t) \Longleftrightarrow F_1(\omega) \qquad \text{and} \qquad f_2(t) \Longleftrightarrow F_2(\omega)$$

then (**time-convolution**)

$$f_1(t) * f_2(t) \Longleftrightarrow F_1(\omega)F_2(\omega) \tag{7.42}$$

and (**frequency-convolution**)

$$f_1(t)f_2(t) \Longleftrightarrow \frac{1}{2\pi}F_1(\omega) * F_2(\omega) \tag{7.43}$$

Proof: By definition†

$$\mathcal{F}[f_1(t) * f_2(t)] = \int_{-\infty}^{\infty} e^{-j\omega t} \left[\int_{-\infty}^{\infty} f_1(\tau) f_2(t - \tau)d\tau \right] dt$$

$$= \int_{-\infty}^{\infty} f_1(\tau) \left[\int_{-\infty}^{\infty} e^{-j\omega t} f_2(t - \tau)dt \right] d\tau$$

The inner integral is the Fourier transform of $f_2(t - \tau)$, given by [time-shifting property in Eq. (7.37)] $F_2(\omega)e^{-j\omega\tau}$. Hence

$$\mathcal{F}[f_1(t)*f_2(t)] = \int_{-\infty}^{\infty} f_1(\tau)e^{-j\omega\tau} F_2(\omega)d\tau = F_2(\omega) \int_{-\infty}^{\infty} f_1(\tau)e^{-j\omega\tau} d\tau = F_1(\omega)F_2(\omega)$$

The Frequency convolution property (7.43) can be proved in exactly the same way by reversing the roles of $f(t)$ and $F(\omega)$.

■ **Example 7.13**

Using the time-convolution property, show that if

$$f(t) \Longleftrightarrow F(\omega)$$

then

$$\int_{-\infty}^{t} f(\tau)d\tau \Longleftrightarrow \frac{F(\omega)}{j\omega} + \pi F(0)\delta(\omega) \tag{7.44}$$

Because

$$u(t - \tau) = \begin{cases} 1 & \tau < t \\ 0 & \tau > t \end{cases}$$

it follows that

$$f(t) * u(t) = \int_{-\infty}^{\infty} f(\tau)u(t - \tau)d\tau = \int_{-\infty}^{t} f(\tau)\,d\tau$$

---

†Here we use Fubini's theorem, which states that the order of integration can be changed; that is,

$$\int_{-\infty}^{\infty} \int_{-\infty}^{\infty} g(x, y)\, dx\, dy = \int_{-\infty}^{\infty} dx \int_{-\infty}^{\infty} g(x, y)dy = \int_{-\infty}^{\infty} dy \int_{-\infty}^{\infty} g(x, y)\, dx$$

provided that any one of these double integrals is finite when $g(x, y)$ is replaced by $|g(x, y)|$.

Now from the time-convolution property [Eq. 7.42], it follows that

$$f(t) * u(t) = \int_{-\infty}^{t} f(\tau)\, d\tau \Longleftrightarrow F(\omega)\left[\frac{1}{j\omega} + \pi\delta(\omega)\right]$$

$$= \frac{F(\omega)}{j\omega} + \pi F(0)\delta(\omega) \quad \blacksquare$$

△ **Exercise E7.8**

   Using the time-convolution property, show that $f(t) * \delta(t) = f(t)$ ▽

△ **Exercise E7.9**

   Using the time-convolution property, show that

$$e^{-at}u(t) * e^{-bt}u(t) = \frac{1}{b-a}\left[e^{-at} - e^{-bt}\right]u(t)$$

Hint: Use property (7.42) to find the Fourier transform of $e^{-at}u(t) * e^{-bt}u(t)$. Then use partial fraction expansion to find the inverse Fourier transform.   ▽

7.4-7 Time Differentiation and Time Integration

   If

$$f(t) \Longleftrightarrow F(\omega)$$

then **(time differentiation)**†

$$\frac{df}{dt} \Longleftrightarrow j\omega F(\omega) \tag{7.45}$$

and **(time integration)**

$$\int_{-\infty}^{t} f(\tau)d\tau \Longleftrightarrow \frac{F(\omega)}{j\omega} + \pi F(0)\delta(\omega) \tag{7.46}$$

   Proof: Differentiation of both sides of Eq. (7.8b) yields

$$\frac{df}{dt} = \frac{1}{2\pi}\int_{-\infty}^{\infty} j\omega F(\omega)e^{j\omega t}\, d\omega$$

This shows that

$$\frac{df}{dt} \Longleftrightarrow j\omega F(\omega)$$

Repeated application of this property yields

$$\frac{d^n f}{dt^n} \Longleftrightarrow (j\omega)^n F(\omega) \tag{7.47}$$

   The time-integration property [Eq. (7.46)] already has been proved in Example 7.13. Note that from Eq. (7.8a), we have

$$F(0) = \int_{-\infty}^{\infty} f(t)\, dt$$

---

†Valid only if the transform of df/dt exists.

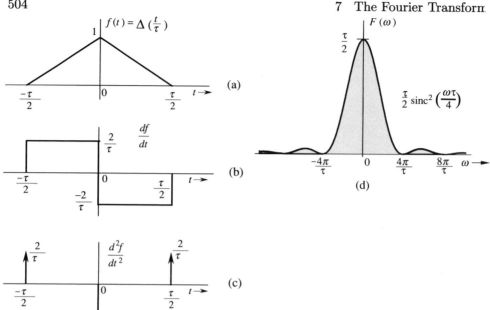

Fig. 7.24 Finding the Fourier transform of a piecewise-linear signal using the time-differentiation property.

Hence, if the area under $f(t)$ is zero, then $F(0) = 0$, and Eq. (7.46) becomes

$$\int_{-\infty}^{t} f(x)\,dx \iff \frac{F(\omega)}{j\omega} \tag{7.48}$$

These results also can be appreciated intuitively. Recall that $F(\omega)$ is the spectrum of exponential components of the form $e^{j\omega t}$ required to synthesize $f(t)$. Therefore the derivative of $f(t)$ can be synthesized by the derivatives of the exponential components of $f(t)$. Because the derivative of $e^{j\omega t}$ is $j\omega e^{j\omega t}$, the spectrum of df/dt is the spectrum of $f(t)$ multiplied by $j\omega$. A similar argument can be made for the spectrum of the integral of $f(t)$.

■ Example 7.14

Using the time-differentiation property, find the Fourier transform of a triangle pulse $\Delta(\frac{t}{\tau})$ shown in Fig. 7.24a.

To find the Fourier transform of this pulse we differentiate it successively, as shown in Figs. 7.24b and 7.24c. The second derivative consists of sequence of impulses, as shown in Fig. 7.24c. Recall that the derivative of a signal at a jump discontinuity is an impulse of strength equal to the amount of jump. The function df/dt has a positive jump of $2/\tau$ at $t = \pm\frac{\tau}{2}$, and a negative jump of $4/\tau$ at $t = 0$. Therefore

$$\frac{d^2 f}{dt^2} = \frac{2}{\tau}\delta(t + \tfrac{\tau}{2}) - \frac{4}{\tau}\delta(t) + \frac{2}{\tau}\delta(t - \tfrac{\tau}{2})$$

$$= \frac{2}{\tau}[\delta(t + \tfrac{\tau}{2}) - 2\delta(t) + \delta(t - \tfrac{\tau}{2})] \tag{7.49}$$

From the time-differentiation property (7.47)

$$\frac{d^2 f}{dt^2} \Longleftrightarrow (j\omega)^2 F(\omega) = -\omega^2 F(\omega) \tag{7.50a}$$

Also, from the time-shifting property (7.37)

$$\delta(t - t_0) \Longleftrightarrow e^{-j\omega t_0} \tag{7.50b}$$

Taking Fourier transform of Eq. (7.49) and using the results in Eqs. (7.50), we obtain

$$-\omega^2 F(\omega) = \frac{2}{\tau}[e^{j\frac{\omega\tau}{2}} - 2 + e^{-j\frac{\omega\tau}{2}}] = \frac{4}{\tau}\left(\cos\frac{\omega\tau}{2} - 1\right) = -\frac{8}{\tau}\sin^2\left(\frac{\omega\tau}{4}\right)$$

and

$$F(\omega) = \frac{8}{\omega^2\tau}\sin^2\left(\frac{\omega\tau}{4}\right) = \frac{\tau}{2}\left[\frac{\sin\left(\frac{\omega\tau}{4}\right)}{\frac{\omega\tau}{4}}\right]^2 = \frac{\tau}{2}\operatorname{sinc}^2\left(\frac{\omega\tau}{4}\right) \tag{7.51}$$

The spectrum $F(\omega)$ is shown in Fig. 7.24d. This procedure of finding the Fourier transform can be applied to any function made up of straight line segments. The second derivative of such a signal yields a sequence of impulses whose Fourier transform can be found by inspection. This example suggests a numerical method of finding the Fourier transform of an arbitrary signal $f(t)$ by approximating the signal by straight-line segments. ■

△ **Exercise E7.10**

Find the Fourier transform of rect $\left(\frac{t}{\tau}\right)$, using the time-differentiation property. ▽

Table 7.2

Fourier Transform Operations

Operation	$f(t)$	$F(\omega)$
Addition	$f_1(t) + f_2(t)$	$F_1(\omega) + F_2(\omega)$
Scalar multiplication	$kf(t)$	$kF(\omega)$
Duality	$F(t)$	$2\pi f(-\omega)$
Scaling	$f(at)$	$\frac{1}{|a|}F\left(\frac{\omega}{a}\right)$
Time shift	$f(t - t_0)$	$F(\omega)e^{-j\omega t_0}$
Frequency shift	$f(t)e^{j\omega_0 t}$	$F(\omega - \omega_0)$
Time convolution	$f_1(t) * f_2(t)$	$F_1(\omega)F_2(\omega)$
Frequency convolution	$f_1(t)f_2(t)$	$\frac{1}{2\pi}F_1(\omega) * F_2(\omega)$
Time differentiation	$\frac{d^n f}{dt^n}$	$(j\omega)^n F(\omega)$
Time integration	$\int_{-\infty}^{t} f(x)\,dx$	$\frac{F(\omega)}{j\omega} + \pi F(0)\delta(\omega)$

7.5   LTIC SYSTEM ANALYSIS BY FOURIER TRANSFORM

The time-differentiation property can also be used to find the zero-state response of LTIC systems. The procedure is identical to that used in deriving similar result for the Laplace transform in Sec. 4.3-1. Consider an LTIC system specified by a differential equation

$$Q(D)y(t) = P(D)f(t) \tag{7.52a}$$

Because of the time-differentiation property,

$$D^k y(t) = \frac{d^k y}{dt^k} \Longleftrightarrow (j\omega)^k Y(\omega) \quad \text{and} \quad D^k f(t) = \frac{d^k f}{dt^k} \Longleftrightarrow (j\omega)^k F(\omega) \tag{7.53}$$

The D operator is replaced by $j\omega$ in the corresponding Fourier transform. Therefore the Fourier transform of Eq. (7.52a) yields

$$Q(j\omega)Y(\omega) = P(j\omega)F(\omega)$$

and

$$Y(\omega) = H(\omega)F(\omega) \tag{7.54}$$

in which

$$H(\omega) = \frac{P(j\omega)}{Q(j\omega)} = H(s)|_{s=j\omega}$$

Note that

$$H(s)|_{s=j\omega} = H(j\omega)$$

Strictly speaking, we should use the notation $H(j\omega)$ rather than $H(\omega)$ in Eq. (7.54). However, for compactness and convenience we have chosen the notation $H(\omega)$ in this chapter. Yet it should be remembered throughout our discussion that $H(\omega)$ is the same as $H(j\omega)$ in Chapter 4. The result (7.54) is the same as the Laplace transform counterpart $Y(s) = H(s)F(s)$ with $s = j\omega$.

Equation (7.54) can be expressed as

$$\mathcal{F}[y(t)] = H(\omega)\mathcal{F}[f(t)]$$

If $f(t) = \delta(t)$, by definition $y(t) = h(t)$, and it follows that

$$\mathcal{F}[h(t)] = H(\omega)\mathcal{F}[\delta(t)] = H(\omega)$$

Therefore

$$h(t) \Longleftrightarrow H(\omega)$$

This shows that the transfer function $H(\omega)$ of a system is the Fourier transform[†] of its unit impulse response $h(t)$. We could also derive Eq. (7.54) from the fact that for an LTIC system

$$y(t) = h(t) * f(t)$$

[†]Assuming that the Fourier transform of $h(t)$ exists. This means that the system is asymptotically stable.

Application of the time-convolution property to this equation yields Eq. (7.54).

For reasons discussed earlier, Laplace transform is preferable to Fourier transform for linear systems analysis, in general. First, the Fourier transform can be used to analyze only asymptotically stable systems and that too only when the inputs are Fourier transformable. Exponentially growing signals, for example, are Laplace transformable but not Fourier transformable. Secondly the frequency variable s used in Laplace transforms is more compact and convenient than the variable $j\omega$ used in Fourier transform. Lastly, the Fourier transforms of some functions, such as step functions, sinusoids etc., contain impulses, which needlessly complicate the work. The Laplace transform of these functions is more compact and contains no impulses. However, the use of Fourier transforms may have a slight edge over the Laplace transform in analysis of stable noncausal systems with or without noncausal inputs that are Fourier transformable. These situations can be analyzed by bilateral Laplace transform with concomitant complexity. Use of Fourier transform may be preferable in these cases. We now present three examples which demonstrate strong and weak points of the use of Fourier transform in analysis of linear systems.

■ **Example 7.15**

Find the zero-state response of an LTIC system with transfer function

$$H(s) = \frac{1}{s+2}$$

and the input $f(t) = e^{-t}u(t)$.

In this case

$$F(\omega) = \frac{1}{j\omega + 1} \quad \text{and} \quad H(\omega) = H(s)|_{s=j\omega} = \frac{1}{j\omega + 2}$$

Therefore

$$Y(\omega) = H(\omega)F(\omega)$$
$$= \frac{1}{(j\omega + 2)(j\omega + 1)}$$
$$= \frac{1}{j\omega + 1} - \frac{1}{j\omega + 2}$$

and

$$y(t) = (e^{-t} - e^{-2t})u(t)$$

Here we observe that the procedure is identical to that of the Laplace transform method, with s replaced by $j\omega$. Use of $j\omega$ in place of s is inconvenient and clumsy. Clearly, the Laplace transform is preferable for this problem. ■

■ **Example 7.16**

For the system in Example 7.15, find the zero-state response to input $f(t) = u(t)$.

In this case

$$F(\omega) = \pi\delta(\omega) + \frac{1}{j\omega}$$

and

$$Y(\omega) = \frac{1}{j\omega + 2}\left[\pi\delta(\omega) + \frac{1}{j\omega}\right]$$

$$= \pi\frac{\delta(\omega)}{j\omega + 2} + \frac{1}{j\omega(j\omega + 2)}$$

$$= \frac{\pi}{2}\delta(\omega) + \frac{1}{j\omega(j\omega + 2)} \qquad \text{[recall that } f(x)\delta(x) = f(0)\delta(x)\text{]}$$

$$= \frac{\pi}{2}\delta(\omega) + \frac{\frac{1}{2}}{j\omega} - \frac{\frac{1}{2}}{j\omega + 2}$$

$$= \frac{1}{2}\left\{\left[\pi\delta(\omega) + \frac{1}{j\omega}\right] - \frac{1}{j\omega + 2}\right\}$$

and

$$y(t) = \frac{1}{2}(1 - e^{-2t})u(t)$$

In this situation, because of the presence of the impulse in the transform of $u(t)$, the Fourier transform is not only clumsier, but it needlessly complicates matters in comparison to the Laplace transform. ∎

■ Example 7.17

For the system in Example 7.15, find the zero-state response to input $f(t) = e^t u(-t)$. The input in this case is noncausal. From Pair 2 (Table 7.1), we obtain

$$F(\omega) = \frac{-1}{j\omega - 1}$$

and

$$Y(\omega) = H(\omega)F(\omega)$$

$$= \frac{-1}{(j\omega - 1)(j\omega + 2)}$$

$$= \frac{-\frac{1}{3}}{j\omega - 1} + \frac{\frac{1}{3}}{j\omega + 2}$$

and

$$y(t) = \frac{1}{3}[e^t u(-t) + e^{-2t}u(t)]$$

Because of the noncausal input, this problem cannot be solved by the (unilateral) Laplace transform; it can be solved only by the bilateral Laplace transform. However, since the system is stable and the input is Fourier-transformable, this problem can be solved by the Fourier transform. Similar comments apply to analysis of noncausal systems. ∎

△ **Exercise E7.11**

A first-order all pass filter transfer function is given by

$$H(s) = -\frac{s - a}{s + a}$$

Show that the zero-state response of this filter to input $e^{at}u(-t)$ is $e^{-at}u(t)$. Sketch the input and the output. ▽

7.6 SIGNAL DISTORTION DURING TRANSMISSION

For a system with transfer function $H(\omega)$, if $F(\omega)$ and $Y(\omega)$ are the spectra of the input and the output signals, respectively, then

$$Y(\omega) = F(\omega)H(\omega) \tag{7.55}$$

The transmission of the input signal $f(t)$ through the system changes it into the output signal $y(t)$. Equation (7.55) shows the nature of this change or modification. Here $F(\omega)$ and $Y(\omega)$ are the spectra of the input and the output, respectively, and the transfer function $H(\omega)$ is the spectral response of the system. The output spectrum is given by the input spectrum multiplied by the spectral response of the system. Equation (7.55) clearly brings out the spectral shaping (or modification) of the signal by the system. Equation (7.55) can be expressed in polar form as

$$|Y(\omega)|e^{j\angle Y(\omega)} = |F(\omega)||H(\omega)|e^{j[\angle F(\omega)+\angle H(j\omega)]}$$

Therefore

$$|Y(\omega)| = |F(\omega)|\,|H(\omega)| \tag{7.56a}$$

$$\angle Y(\omega) = \angle F(\omega) + \angle H(\omega) \tag{7.56b}$$

During the transmission, the input signal amplitude spectrum $|F(\omega)|$ is changed to $|F(\omega)|\,|H(\omega)|$ in which $|H(\omega)|$ is the amplitude response of the system. Similarly, the input signal phase spectrum $\angle F(\omega)$ is changed to $\angle F(\omega) + \angle H(\omega)$, in which $\angle H(\omega)$ is the phase response of the system. An input signal spectral component of frequency ω is modified in amplitude by a factor $|H(\omega)|$ and is shifted in phase by an angle $\angle H(\omega)$. Some frequency components may be boosted in amplitude, while others may be attenuated. The relative phases of the various components also change. In general, the output waveform will be different from the input waveform.

In many applications we deliberately want to change the waveform. Yet there are also several applications, such as signal amplification or message signal transmission over a communication channel, where the output waveform must be a replica of the input waveform. In such cases we need to minimize the distortion caused by the amplifier or the communication channel. It is therefore of practical interest to determine the characteristics of a system which allows a signal to pass without distortion (**distortionless transmission**).

Transmission is said to be distortionless if the input and the output have identical waveshapes within a multiplicative constant. A delayed output that retains the input waveform is also considered distortionless. Thus, in distortionless transmission, the input $f(t)$ and the output $y(t)$ satisfy the condition

$$y(t) = kf(t - t_d) \tag{7.57}$$

The Fourier transform of this equation yields

$$Y(\omega) = kF(\omega)e^{-j\omega t_d}$$

But

$$Y(\omega) = F(\omega)H(\omega)$$

Therefore

$$H(\omega) = k\,e^{-j\omega t_d}$$

This is the transfer function required for distortionless transmission. From this equation it follows that

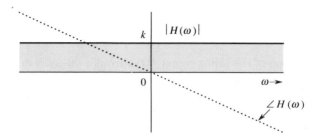

Fig. 7.25 LTIC system Frequency response for distortionless transmission.

$$|H(\omega)| = k \qquad (7.58a)$$

$$\angle H(\omega) = -\omega t_d \qquad (7.58b)$$

This shows that for distortionless transmission, the amplitude response $|H(\omega)|$ must be a constant, and the phase response $\angle H(\omega)$ must be a linear function of ω with slope of $-t_d$, where t_d is the delay of the output with respect to input (Fig. 7.25).

It is instructive to derive the conditions for distortionless transmission heuristically. Once again, imagine $f(t)$ to be composed of various sinusoids (its spectral components), which are being passed through the distortionless system. The output signal is the input signal multiplied by k and delayed by t_d. To synthesize such a signal we need exactly the same components as those of $f(t)$, with each component multiplied by k and delayed by t_d. This means that the system transfer function $H(\omega)$ should be such that each sinusoidal component suffers the same attenuation k and each component undergoes the same time delay of t_d seconds. The first condition requires that

$$|H(\omega)| = k$$

We have seen, in our discussion on p. 497, that a constant time delay t_d for all frequency components requires a linear phase delay ωt_d. To achieve the same time delay, higher frequencies must undergo proportionately higher phase delays, as demonstrated in Fig. 7.20. Therefore

$$\angle H(\omega) = -\omega t_d$$

The time delay resulting from the signal transmission through a system is the negative of the slope of the system phase response $\angle H(\omega)$; that is,

$$t_d(\omega) = -\frac{d}{d\omega}\angle H(\omega) \qquad (7.59)$$

If the slope of $\angle H(\omega)$ is constant (that is, if $\angle H(\omega)$ is linear with ω), all the components are delayed by the same time interval t_d. But if the slope is not constant, t_d, the time delay varies with frequency. This means that different frequency components undergo different amounts of time delay, and consequently the output waveform will not be a replica of the input waveform. A good way of judging phase linearity is to plot t_d as a function of frequency. For a distortionless system, t_d should be constant over the band of interest.

It is often thought (erroneously) that flatness of amplitude response $|H(\omega)|$ alone can guarantee signal quality. A system may have a flat amplitude response and yet distort a signal beyond recognition if the phase response is not linear (t_d not constant).

The Nature of Distortion in Audio and Video Signals

Generally speaking, a human ear can readily perceive amplitude distortion, although it is relatively insensitive to phase distortion. For the phase distortion to become noticeable, the variation in delay [variation in the slope of $\angle H(\omega)$] should be comparable to the signal duration (or the physically perceptible duration, in case the signal itself is long). In the case of audio signals, each spoken syllable can be considered as an individual signal. The average duration of a spoken syllable is of the order of magnitude from 0.01 to 0.1 seconds. The audio systems may have nonlinear phases, yet no noticeable signal distortion results because in practical audio systems, maximum variation in the slope of $\angle H(\omega)$ is only a small fraction of a millisecond. This is the real reason behind the statement that "the human ear is relatively insensitive to phase distortion."[4] As a result, the audio equipment manufacturers make available only the amplitude response characteristic of their systems.

For video signals, on the other hand, the situation is exactly the opposite. The human eye is sensitive to phase distortion but is relatively insensitive to amplitude distortion. The effect of amplitude distortion in television signals manifests as partial destruction of the relative half-tone values of the resulting picture, which is not readily perceived by the human eye. The phase distortion (nonlinear phase), on the other hand, causes different time delays in different picture elements. This results in a smeared picture, which is readily perceived by the human eye. Phase distortion is also very important in digital communication systems because the nonlinear phase characteristic of a channel causes pulse dispersion (spreading out), which in turn causes pulses to interfere with neighboring pulses. This interference can cause an error in the pulse amplitude at the receiver: a binary **1** may read as **0**, and vice versa.

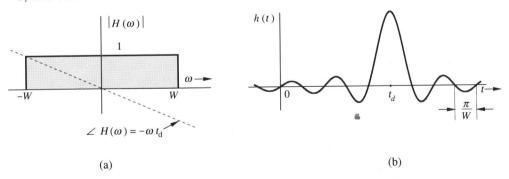

(a)   (b)

Fig. 7.26 Ideal lowpass filter frequency response and its impulse response.

7.7 IDEAL AND PRACTICAL FILTERS

Ideal filters allow distortionless transmission of a certain band of frequencies and suppress all the remaining frequencies. The ideal lowpass filter (Fig. 7.26), for example, allows all components below $\omega = W$ rad/s to pass without distortion and

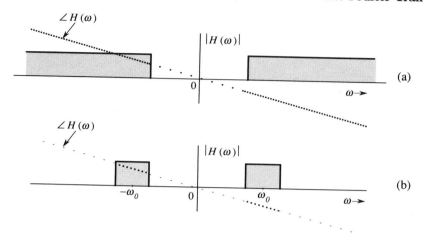

Fig. 7.27 Ideal highpass and bandpass filter frequency response.

suppresses all components above $\omega = W$. Figure 7.27 shows ideal highpass and bandpass filter characteristics.

The ideal lowpass filter in Fig. 7.26a has a linear phase of slope $-t_d$, which results in a time delay of t_d seconds of all its input components of frequencies below W rad/s. Therefore, if the input is a signal $f(t)$ bandlimited to W rad/s, the output $y(t)$ is $f(t)$ delayed by t_d, that is,

$$y(t) = f(t - t_d) \tag{7.60}$$

The signal $f(t)$ is transmitted by this system without distortion but with time delay t_d. For this filter $|H(\omega)| = \text{rect}\left(\frac{\omega}{2W}\right)$, and $\angle H(\omega) = e^{-j\omega t_d}$, so that $H(\omega) = \text{rect}\left(\frac{\omega}{2W}\right)e^{-j\omega t_d}$. The unit impulse response $h(t)$ of this filter is found from Pair 18 (Table 7.1) and the time-shifting property:

$$h(t) = \mathcal{F}^{-1}\left[\text{rect}\left(\frac{\omega}{2W}\right)e^{-j\omega t_d}\right]$$
$$= \frac{W}{\pi}\,\text{sinc}\left[W(t - t_d)\right]$$

Recall that $h(t)$ is the system response to impulse input $\delta(t)$, which is applied at $t = 0$. Figure 7.26b shows a curious fact: the response $h(t)$ begins even before the input is applied (at $t = 0$). Clearly, the filter is noncausal and therefore physically unrealizable. Similarly, one can show that other ideal filters (such as the ideal highpass or ideal bandpass filters shown in Fig. 7.27) are also physically unrealizable.

For a physically realizable system, $h(t)$ must be causal; that is,

$$h(t) = 0 \qquad \text{for } t < 0$$

In the frequency domain, this condition is equivalent to the well-known **Paley-Wiener criterion** which states that the necessary and sufficient condition for amplitude response $|H(\omega)|$ to be realizable is†

$$\int_{-\infty}^{\infty} \frac{|\ln|H(\omega)||}{1 + \omega^2}\,d\omega < \infty \tag{7.61}$$

---

†$|H(\omega)|$ is assumed to be square-integrable, that is,

If $H(\omega)$ does not satisfy this condition, it is unrealizable. Note that if $|H(\omega)| = 0$ over any finite band, $|\ln|H(\omega)|| = \infty$ over that band, and the condition (7.61) is violated. If, however, $H(\omega) = 0$ at a single frequency (or a set of discrete frequencies), the integral in (7.61) may still be finite even though the integrand is infinite. Therefore, for a physically realizable system, $H(\omega)$ may be zero at some discrete frequencies, but it cannot be zero over any finite band. According to this criterion, ideal filter characteristics (Figs. 7.26, 7.27) are clearly unrealizable.

Fig. 7.28   Approximate realization of an ideal lowpass filter by truncating its impulse response.

The impulse response $h(t)$ in Fig. 7.26 is not realizable. One practical approach to filter design is to cut off the tail of $h(t)$ for $t < 0$. The resulting causal impulse response $\widehat{h}(t)$, where

$$\widehat{h}(t) = h(t)u(t)$$

is physically realizable because it is causal (Fig. 7.28). If t_d is sufficiently large, $\widehat{h}(t)$ will be a close approximation of $h(t)$, and the resulting filter $\widehat{H}(\omega)$ will be a good approximation of an ideal filter. This close realization of the ideal filter is achieved because of the increased value of time-delay t_d. This means that the price of close realization is higher delay in the output; this is often true of noncausal systems. Of course, theoretically a delay $t_d = \infty$ is needed to realize the ideal characteristics. But a glance at Fig. 7.26b shows that a delay t_d of three or four times $\frac{\pi}{W}$ will make $\widehat{h}(t)$ a reasonably close version of $h(t - t_d)$. For instance, an audio filter is required to handle frequencies of up to 20 kHz. In this case a t_d of about 10^{-4} (0.1 ms) would be a reasonable choice.

In practice, we can realize a variety of filter characteristics to approach ideal characteristics. Practical (realizable) filter characteristics are gradual, without jump discontinuities in amplitude response. We have encountered some filter families (Butterworth and Chebyshev) in Sec. 4.7-1. Figure 4.40 shows the amplitude response of lowpass Butterworth filters. Further discussion of practical filters can be found in Reference 5.

△ **Exercise E7.12**

Show that a filter with Gaussian transfer function $H(\omega) = e^{-\alpha\omega^2}$ is unrealizable. Do this in two ways: first by showing that its impulse response is noncausal, and then by showing that $|H(\omega)|$ violates the Paley-Wiener criterion.

Hint: Use Pair 22 in Table 7.1   ▽

$$\int_{-\infty}^{\infty} |H(\omega)|^2 \, d\omega \text{ is finite}$$

Note that the Paley-Wiener criterion is a criterion for the realizability of the amplitude response $|H(\omega)|$.

7.8   THINKING IN TIME DOMAIN AND FREQUENCY DOMAIN: A TWO-DIMENSIONAL VIEW OF SIGNALS AND SYSTEMS

Both signals and systems have dual personalities; the time domain and the frequency domain. For a deeper perspective, we should examine and understand both these identities because they offer complementary insights.

For example, an exponential signal can be specified by its time domain description such as $e^{-2t}u(t)$ or by its Fourier transform (its frequency domain description) $\frac{1}{j\omega+2}$. The time-domain description depicts the waveform of a signal. The frequency-domain description portrays its spectral composition (relative amplitudes of its sinusoidal (or exponential) components and their phases). For the signal e^{-2t}, for instance, the time-domain description portrays the exponentially decaying signal with a time constant 0.5. The frequency-domain description characterizes it as a lowpass signal, which can be synthesized by sinusoids with amplitudes decaying with frequency roughly as $1/\omega$.

An LTIC system can also be described or specified in the time domain by its impulse response $h(t)$ or in the frequency domain by its transfer function $H(\omega)$. In Sec. 2.6 we studied intuitive insights in the system behavior offered by the impulse response, which consists of characteristic modes of the system. By purely qualitative reasoning, we saw that the system responds well to signals that are similar to the characteristic modes and responds poorly to signals which are very different from those modes. We also saw that the shape of the impulse response $h(t)$ determines how fast the system is (response time or time constant of a system), and how it affects the pulse spreading and thereby determines the rate of pulse transmission.

The transfer function $H(\omega)$ specifies the frequency response, that is, the system response to sinusoidal input of various frequencies. This is precisely the filtering characteristic of the system. In Sec. 4.7-2 we saw how the filtering characteristics can be determined by a mere inspection of the poles and zeros of the system transfer function.

Experienced electrical engineers instinctively think in both (the time and frequency) domains whenever possible. When they look at a signal, they consider its waveform, the signal width (duration), and how fast the waveform decays. This is basically a time-domain perspective. They also think of the signal in terms of its frequency spectrum, that is, in terms of its sinusoidal components and their relative amplitudes and phases. This is the frequency-domain perspective. When they think of a system, they think of its impulse response $h(t)$. The width of $h(t)$ indicates the time constant (response time); that is, how fast the system is capable of responding to an input, and how much dispersion (spreading) it will cause. This is the time-domain perspective. From the frequency-domain perspective, these engineers view the systems as a filter, which selectively transmits certain frequency components and suppresses the others [frequency response $H(\omega)$]. Knowing the input signal spectrum and the frequency response of the system, they have a mental image of the output signal spectrum. This is precisely expressed by $Y(\omega) = F(\omega)H(\omega)$.

We can analyze LTI systems by time-domain techniques or by frequency-domain techniques. Then why learn both? The reason is that the two domains offer complementary insights into system behavior. Some aspects are easily grasped in one domain; other aspects may be easier to see in the other domain. Both the time-domain and the frequency-domain methods are as essential for the study of signals and systems as two eyes are essential to a human being for correct visual perception of reality. A person can see with either eye, but for proper perception of three dimensional-reality, both eyes are essential.

It is important to keep the two domains separate, and not to mix the entities in the two domains. If we are using the frequency domain to determine the system response, we must deal with all signals in terms of their spectra (Fourier transforms) and all systems in terms of their transfer functions. For example, to determine the system response $y(t)$ to an input $f(t)$, we must first convert the input signal into its frequency domain description $F(\omega)$. The system description also must be in the frequency-domain, that is, the transfer function $H(\omega)$. The output signal spectrum $Y(\omega) = F(\omega)H(\omega)$. Thus the result (output) is also in the frequency domain. To determine the final answer $y(t)$, we must take the inverse transform of $Y(\omega)$.

7.9 SIGNAL ENERGY

We define E_f, the energy of a real signal $f(t)$, as the energy dissipated when a voltage $f(t)$ is applied across (or if a current $f(t)$ is passed through) a $1\ \Omega$ resistor. In either case the voltage across the resistor is $f(t)$, and the current through it is also $f(t)$. The instantaneous power is $f^2(t)$ and the energy dissipated is

$$E_f = \int_{-\infty}^{\infty} f^2(t)\, dt \tag{7.62}$$

The concept of energy of a signal is meaningful only if the area under $f^2(t)$ is finite (that is, if its energy is finite). Signals that satisfy this condition are called **energy signals**. Every signal encountered in practice has finite energy and is therefore an energy signal. A signal such as $\cos \omega_0 t$ is not an energy signal because it has infinite energy. However, this signal cannot be generated in practice because it starts at $t = -\infty$ and continues forever.

Signal energy can also be related to the signal spectrum $F(\omega)$ by substituting Eq. (7.7) in Eq. (7.62).

$$E_f = \int_{-\infty}^{\infty} f(t) \left[\frac{1}{2\pi} \int_{-\infty}^{\infty} F(\omega) e^{j\omega t}\, d\omega \right] dt$$

Interchanging the order of integration yields

$$E_f = \frac{1}{2\pi} \int_{-\infty}^{\infty} F(\omega) \left[\int_{-\infty}^{\infty} f(t) e^{j\omega t}\, dt \right] d\omega$$

$$= \frac{1}{2\pi} \int_{-\infty}^{\infty} F(\omega) F(-\omega)\, d\omega$$

For a real signal, $F(\omega)$ and $F(-\omega)$ are conjugates [see Eqs. (7.10)]. Therefore

$$E_f = \frac{1}{2\pi} \int_{-\infty}^{\infty} |F(\omega)|^2\, d\omega \tag{7.63}$$

Consequently, for a real signal $f(t)$†

$$E_f = \int_{-\infty}^{\infty} f^2(t)\,dt = \frac{1}{2\pi} \int_{-\infty}^{\infty} |F(\omega)|^2\,d\omega \tag{7.64}$$

This is the statement of the well-known **Parseval's theorem**. A similar result was obtained for a periodic signal and its Fourier series on p. 458. This result allows us to determine the signal energy from either the time-domain specification $f(t)$ or the frequency-domain specification $F(\omega)$ of the same signal.

Equation (7.63) can be interpreted to mean that the energy of a signal $f(t)$ results from energies contributed by all the spectral components of the signal $f(t)$. The total signal energy is the area under $|F(\omega)^2|$ (divided by 2π). If we consider a small band $\Delta\omega$ ($\Delta\omega \to 0$), as shown in Fig. 7.29, the energy ΔE_f of the spectral components in this band is the area under $|F(\omega)|^2$ under this band (divided by 2π):

$$\Delta E_f = \frac{1}{2\pi}|F(\omega)|^2\,\Delta\omega = |F(\omega)|^2\,\Delta\mathcal{F} \qquad \frac{\Delta\omega}{2\pi} = \Delta\mathcal{F} \text{ Hz} \tag{7.65}$$

Therefore the energy contributed by the components in this band of $\Delta\mathcal{F}$ (in Hertz) is $|F(\omega)|^2\Delta\mathcal{F}$. The total signal energy is the sum of energies of all such bands and is given by the area under $|F(\omega)|^2$ as in Eq. (7.63). Therefore $|F(\omega)|^2$ is the **energy spectral density** (per unit bandwidth in Hertz).

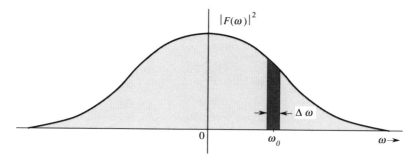

Fig. 7.29   Interpretation of Energy spectral density of a signal.

For real signals, $F(\omega)$ and $F(-\omega)$ are conjugates, and $|F(\omega)|^2$ is an even function of ω because

$$|F(\omega)|^2 = F(\omega)F^*(\omega) = F(\omega)F(-\omega)$$

Therefore Eq. (7.63) can be expressed as†

$$E_f = \frac{1}{\pi} \int_0^{\infty} |F(\omega)|^2\,d\omega \tag{7.66}$$

---

†For a complex signal $f(t)$, the corresponding relationship is

$$E_f = \int_{-\infty}^{\infty} |f(t)|^2\,dt = \frac{1}{2\pi} \int_{-\infty}^{\infty} |F(\omega)|^2\,d\omega$$

†In Eq. (7.66) it is assumed that $F(\omega)$ does not contain an impulse at $\omega = 0$. If such an impulse exists, it should be integrated separately with a multiplying factor of $1/2\pi$ rather than $1/\pi$.

The signal energy E_f, which results from contributions from all the frequency components from $\omega = 0$ to ∞, is given by ($1/\pi$ times) the area under $|F(\omega)|^2$ from $\omega = 0$ to ∞. It follows that the energy contributed by spectral components of frequencies between ω_1 and ω_2 is

$$\Delta E_f = \frac{1}{\pi} \int_{\omega_1}^{\omega_2} |F(\omega)|^2 \, d\omega \qquad (7.67)$$

■ **Example 7.18**

Find the energy of signal $f(t) = e^{-at} u(t)$. Determine the frequency W (rad/s) such that the energy contributed by the spectral components of all the frequencies below W is 95% of the signal energy E_f.

By definition

$$E_f = \int_{-\infty}^{\infty} f^2(t) dt = \int_0^{\infty} e^{-2at} dt = \frac{1}{2a}$$

We can verify this result by Parseval's theorem. For this signal

$$F(\omega) = \frac{1}{j\omega + a}$$

and

$$E_f = \frac{1}{\pi} \int_0^{\infty} |F(\omega)|^2 d\omega = \frac{1}{\pi} \int_0^{\infty} \frac{1}{\omega^2 + a^2} \, d\omega = \frac{1}{\pi a} \tan^{-1} \frac{\omega}{a} \Big|_0^{\infty} = \frac{1}{2a}$$

The band $\omega = 0$ to $\omega = W$ contains 95% of the signal energy, that is, $0.95/2a$. Therefore from Eq. (7.67) with $\omega_1 = 0$ and $\omega_2 = W$ obtain

$$\frac{0.95}{2a} = \frac{1}{\pi} \int_0^W \frac{d\omega}{\omega^2 + a^2}$$

$$= \frac{1}{\pi a} \tan^{-1} \frac{\omega}{a} \Big|_0^W = \frac{1}{\pi a} \tan^{-1} \frac{W}{a}$$

or

$$\frac{0.95\pi}{2} = \tan^{-1} \frac{W}{a} \implies W = 12.706a \text{ rad/s}$$

This means that the spectral components of $f(t)$ in the band from 0 (dc) to 12.706 rad/s (2.02 Hz) contribute 95% of the total signal energy; all the remaining spectral components (in the band from 12.706 rad/s to ∞) contribute only 5% of the signal energy. ■

△ **Exercise E7.13**

Show that the energy of the signal

$$f(t) = \frac{2a}{t^2 + a^2}$$

is $\frac{2\pi}{a}$. Verify this result by using Parseval's theorem.

Hint: Using Pair 3 and the duality property, show that $\frac{2a}{t^2+a^2} \Longleftrightarrow 2\pi e^{-a|\omega|}$ ▽

The Essential Bandwidth of a Signal

Spectra of most of the signals extend to infinity. However, because the energy of any practical signal is finite, the signal spectrum must approach 0 as $\omega \to \infty$. Most of the signal energy is contained within a certain band B Hz, and the energy content of the components of frequencies greater than B is negligible. We can therefore suppress the signal spectrum beyond B Hz with little effect on the signal shape and energy. The bandwidth B is called the **essential bandwidth** of the signal. The criterion for selecting B depends on the error tolerance in a particular application. We may, for example, select B to be that band which contains 95% of the signal energy.† This figure may be higher or lower than 95% depending on the precision needed. Using such a criterion, we can determine the essential bandwidth of a signal. Suppression of all the spectral components of $f(t)$ beyond the essential bandwidth results in a signal $\hat{f}(t)$ which is a close approximation of $f(t)$. If we use the 95% criterion for the essential bandwidth, the energy of the error (the difference) $f(t) - \hat{f}(t)$ is 5% of E_f.

7.10 DATA TRUNCATION: WINDOW FUNCTIONS

We often need to truncate data in diverse situations from numerical computations to filter design. For example, if we need to compute numerically the Fourier transform of some signal, say $e^{-t}u(t)$, on a computer, we will have to truncate the signal $e^{-t}u(t)$ beyond a sufficiently large value of t (typically five time constants and above). Similarly, if we wish to design a lowpass filter, we may want to realize its impulse response $h(t)$, which is noncausal, and which approaches zero asymptotically as $t \to \pm\infty$. For a practical design, we may want to truncate $h(t)$ beyond a sufficiently large t where $h(t)$ becomes small and negligible. In signal sampling, to eliminate aliasing, we need to truncate the signal spectrum beyond the half sampling frequency $\omega_s/2$, using an anti-aliasing filter. Again, we may want to synthesize a periodic signal by adding the first n harmonics and truncating all the higher harmonics. These examples show that data truncation can occur in both time and frequency domain. On the surface, truncation appears to be a simple problem of cutting off the data at a point where it is deemed to be sufficiently small. Unfortunately, this is not the case. Simple truncation can cause some unsuspected problems. By simple truncation we mean giving unit weight to all the data up to some cutoff point and zero weight to data beyond that point.

Window Functions

Truncation operation may be regarded as multiplying a signal of infinite width by a window function of finite width. Simple truncation amounts to using a **rectangular window** $w_R(t)$ (Fig. 7.30a) in which we assign unit weight to all the data within the window width ($|t| < \frac{T}{2}$), and assign zero weight to all the data lying outside the window ($|t| > \frac{T}{2}$). It is also possible to use a window in which the weight assigned to the data within the window may not be constant. In a **triangular window** $w_T(t)$, for example, the weight assigned to data decreases linearly

---

†Essential bandwidth for a lowpass signal may also be defined as a frequency at which the value of the amplitude spectrum is a small fraction (about 5 to 10%) of its peak value. In Example 7.18, the peak of $|F(\omega)|$ is $1/a$, and it occurs at $\omega = 0$.

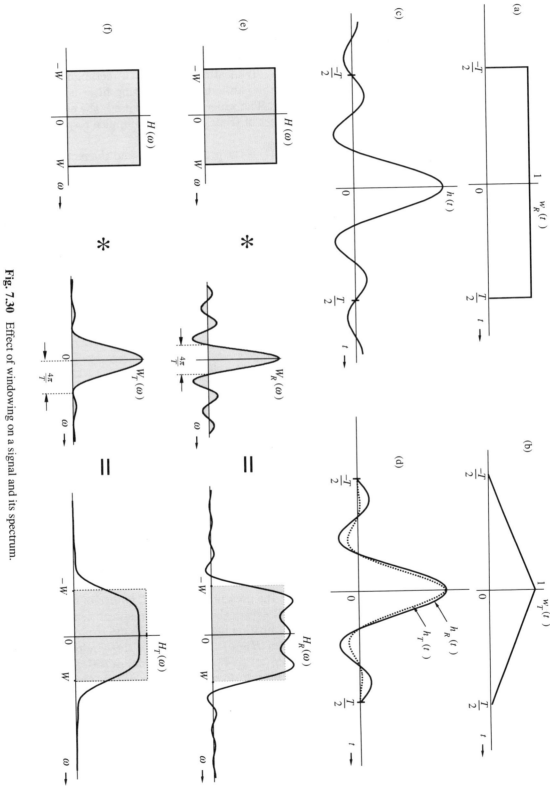

Fig. 7.30 Effect of windowing on a signal and its spectrum.

over the window width (Fig. 7.30b). We shall explain the effects of truncation with an example.

Consider the problem of designing a lowpass filter of bandwidth W radians/s. For this filter the impulse response is $\frac{W}{\pi}\text{sinc}(Wt)$ (Fig. 7.30c). This impulse response, being noncausal, is unrealizable. Truncation of $h(t)$ by a rectangular window $w_R(t)$ (Fig. 7.30a) makes it realizable, although the resulting filter is now an approximation to the desired ideal filter. The wider the window used, the closer the approximation.† We shall study the effect of truncation of the impulse response on the resulting filter transfer function.

The truncated impulse response $h_R(t)$ is the product of $h(t)$ and the rectangular window function $w_R(t)$, as shown in Fig. 7.30d.

$$h_R(t) = h(t)w_R(t)$$

Therefore, from the frequency-convolution property [Eq. (7.43)],

$$H_R(\omega) = \frac{1}{2\pi}H(\omega) * W_R(\omega)$$

in which $H(\omega)$ and $W_R(\omega)$ are the Fourier transforms of $h(t)$ and $w_R(t)$ respectively, given by

$$H(\omega) = \text{rect}\left(\frac{\omega}{2W}\right) \qquad \text{and} \qquad W_R(\omega) = T\,\text{sinc}\left(\frac{\omega T}{2}\right)$$

Therefore

$$H_R(\omega) = \frac{T}{2\pi}\text{rect}\left(\frac{\omega}{2W}\right) * \text{sinc}\left(\frac{\omega T}{2}\right)$$

Figure 7.30e shows the process of convolution of $H(\omega)$ and $W_R(\omega)$ with the resulting spectrum $H_R(\omega)$. Truncation of $h(t)$ causes two distinct changes in its spectrum because of the Gibbs phenomenon:

1. The spectrum $H_R(\omega)$ shows a **spectral spreading** at the edges, and instead of a sudden switch there is a gradual transition from the passband to the stopband.

2. Although $H(\omega)$ is bandlimited, $H_R(\omega)$ is not. It exhibits an oscillating gain in the stopband, which decays slowly with frequency (as $1/\omega$). This effect is called the **spectral leakage**.

The first effect (spectral spreading), which causes the loss of spectral resolution, can be explained by the width property of the convolution. Because $H_R(\omega)$ is the convolution of $H(\omega)$ with $W_R(\omega)$, the width of $H_R(\omega)$ is the sum of the width of $H(\omega)$ and the width of $W_R(\omega)$, which in the present case is $4\pi/T$ (width of the main lobe). Therefore truncation (windowing) causes the spectrum to spread (smear) by $\frac{2\pi}{T}$ on either side of the spectrum. The transition band (from passband to stopband), which is zero for $H(\omega)$, increases to $\frac{2\pi}{T}$ for $H_R(\omega)$. The narrower the window $w_R(t)$ (smaller T), the greater the width of $W_R(\omega)$ (reciprocity of signal duration and its bandwidth). Therefore the **narrower the window $w_R(t)$ (less data), the larger the spectral spreading**, as expected. Conversely, a wider window (more data) results in less spectral spreading, and $H_R(\omega)$ is a better approximation of $H(\omega)$.

---

†In addition to truncation, we also need to delay the truncated function by $\frac{T}{2}$ in order to render it causal. However, the time delay only adds a linear phase to the spectrum without changing the amplitude spectrum. For this reason we shall ignore the delay in order to simplify our discussion.

The second effect (spectral leakage) can be explained by the fact that $W_R(\omega)$, the sinc spectrum of the rectangular window, is an oscillatory function with a poor high frequency convergence, so that it decays rather slowly with frequency (as $1/\omega$). Therefore $H_R(\omega)$ will acquire this undesirable oscillatory and slowly decaying characteristic in the stopband, which is not present in the $H(\omega)$.

Incidentally, this discussion also explains the Gibbs phenomenon. Truncation by a rectangular window involves convolution with a sinc spectrum, which is oscillatory with amplitude decaying as $1/\omega$ regardless of the window width. This causes the Gibbs phenomenon.

Leakage Reduction Using Tapered Windows

The reason for the undesirable slowly decaying oscillatory behavior lies in the fact that the rectangular window function has a jump discontinuity (sudden change). Therefore its spectrum $W_R(\omega)$ decays slowly as $1/\omega$ (see asymptotic rate of spectral decay in Sec. 6.1-4), indicating the presence of large amounts of high frequency components. Generally speaking, the spectrum of a smoother window function decays faster with frequency (see asymptotic rate of spectral decay in Sec. 6.1-4). This suggests that a tapered (smoother) window may have better leakage reduction properties than a rectangular window. For example, the triangular window of width T shown in Fig. 7.30b has no jump discontinuities, and its spectrum $W_T(\omega) = \frac{T}{2}\text{sinc}^2\left(\frac{\omega T}{4}\right)$ decays as $1/\omega^2$. Figure 7.30f shows the spectrum $H_T(\omega)$ resulting from the convolution of $H(\omega)$ with $W_T(\omega)$. Observe that $H_T(\omega)$ decays faster (as $1/\omega^2$) in the stopband. However, for a given T, the width of $W_T(\omega)$ (triangular window) is twice that of the $W_R(\omega)$ (rectangular window), as seen in Figs. 7.30e and 7.30f. This results in higher spectral spreading for the triangular window. The transition band doubles. In general, smoother (tapered) windows have better leakage behavior, but this is achieved at the cost of higher spectral spreading (smearing) at the edges. In other words, a smoother window achieves a better stopband behavior at the cost of the sharpness of the filter characteristics at the cutoff frequency (higher transition bandwidth). Of all the windows of a given width, the rectangular window has the least spectral spreading (smearing) at the edges.† If we try to reduce spectral leakage by using a smoother window, the spectral spreading increases. Fortunately, as shown earlier, spectral spreading can be reduced by increasing the window width. Therefore we can achieve a given combination of spectral spread (transition bandwidth) and leakage characteristics by choosing a suitable tapered window function of a sufficiently longer width T.

So far we have discussed the effect of signal truncation (truncation in time domain) on the signal spectrum. Because of time-frequency duality, the effect of spectral truncation (truncation in frequency domain) on the signal shape is similar.

There are several well-known tapered-window functions such as Hanning, Hamming, Blackman, and Kaiser, which truncate the data gradually and have better properties than the triangular window.[6,7] Figure 7.31 shows two well-known tapered-window functions, the Von Hann (or Hanning) window $w_{\text{HAN}}(x)$ and the Hamming window $w_{\text{HAM}}(x)$. We have intentionally used the independent variable

---

†A tapered window yields higher spectral spreading because the effective width of a tapered window is smaller than that of the rectangular window (see Sec. 2.6-2 for the definition of effective width). Therefore, from the reciprocity of the signal width and its spectral width, it follows that the width of the rectangular window spectrum is smaller than that of a tapered window.

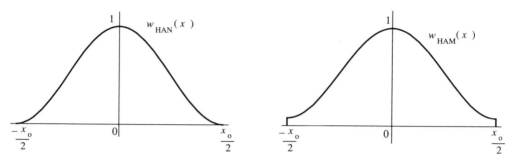

Fig. 7.31 Hanning and Hamming windows.

x because windowing can be performed in time domain as well as in frequency domain; so x could be t or ω depending on the application.

$$w_{\text{HAN}}(x) = \frac{1}{2}\left[1 + \cos\left(\frac{2\pi x}{x_0}\right)\right] \qquad |x| < \frac{x_0}{2} \qquad (7.68a)$$

$$w_{\text{HAM}}(x) = 0.54 + 0.46\cos\left(\frac{2\pi x}{x_0}\right) \qquad |x| < \frac{x_0}{2} \qquad (7.68b)$$

⊙ **Computer Example C7.2**

There are several well known windowing functions such as Hanning, Hamming, Blackman, Kaiser which truncate data smoothly and have desirable effects on data weighting. This example illustrates Hamming and Hanning (Von Hann) windowing functions.

Hamming and Hanning Windows

```
% These commands generate an N-point Hamming and Hanning windows.
% n is the number of points desired for these windows.
n = 100;
w = .54 - .46*cos(2*pi*(0:n-1)'/(n-1));
plot(w),grid,
xlabel('x'),ylabel('wHAM'),
title('Hamming Window Function'),pause
w=.5*(1-cos(2*pi*(0:n-1)'/(n-1)));
plot(w),grid,
xlabel('x'),ylabel('wHAN'),
title('Hanning (Von Hann) Window Function')    ⊙
```

7.11 SUMMARY

In Chapter 6 we represented periodic signals as a sum of (everlasting) sinusoids or exponentials (Fourier series). In this chapter we extended this result to nonperiodic signals, which are represented by the Fourier integral (instead of the Fourier series). A nonperiodic signal $f(t)$ may be regarded as a periodic signal with period $T_0 \to \infty$, so that the Fourier integral is basically a Fourier series with fundamental frequency approaching zero. Therefore, for nonperiodic signals, the Fourier spectra are continuous. This means that a signal is represented as a sum of sinusoids (or exponentials) of all frequencies over a continuous frequency interval. The Fourier transform $F(\omega)$, therefore, is the spectral density (per unit bandwidth in Hz).

The Fourier transform is a special case of Laplace transform with the complex variable s restricted to imaginary axis; that is, with $s = j\omega$. Alternatively, the Laplace transform may be viewed as generalization of the Fourier transform with the frequency $j\omega$ generalized to $s = \sigma + j\omega$. Laplace transform is superior to Fourier transform for the analysis of LTIC systems, whereas Fourier transform is preferred for better insights into signals and signal analysis.

An ever-present aspect of the Fourier transform is the duality between time and frequency, which also implies duality between signal $f(t)$ and its transform $F(\omega)$. This arises because of near-symmetrical equations for direct and inverse Fourier transforms. The duality principle has far-reaching consequences and yields many valuable insights into signal analysis. The scaling property of the Fourier transform leads to the conclusion that the signal bandwidth is inversely proportional to signal duration (signal width). Time shifting of a signal does not change its amplitude spectrum, but adds a linear phase spectrum. Multiplication of a signal by an exponential $e^{j\omega_0 t}$ results in shifting the spectrum to the right by ω_0. In practice, spectral shifting is achieved by multiplying a signal with a sinusoid such as $\cos \omega_0 t$ (rather than exponential $e^{j\omega_0 t}$). This process is known as amplitude modulation. Multiplication of two signals results in convolution of their spectra, whereas convolution of two signals results in multiplication of their spectra.

For distortionless transmission of a signal through an LTIC system, the amplitude response $|H(\omega)|$ of the system must be constant, and the phase response $\angle H(\omega)$ should be a linear function of ω over a band of interest. Ideal filters, which allow distortionless transmission of a certain band of frequencies and suppress all the remaining frequencies, are physically unrealizable (noncausal). In fact, it is impossible to build a physical system with zero gain $[H(\omega) = 0]$ over a finite band of frequencies. Such systems (which include ideal filters) can be realized only with infinite time delay in the response.

The energy of a signal $f(t)$ is defined as the area under $f^2(t)$. The signal energy can also be expressed as $1/2\pi$ times the area under $|F(\omega)^2|$ (Parseval's theorem). The signal energy is a measure of the signal strength. The energy contributed by spectral components within a band $\Delta\mathcal{F}$ (in Hz) is given by $|F(\omega)|^2 \Delta\mathcal{F}$. Therefore $|F(\omega)|^2$ is the energy spectral density per unit bandwidth (in Hz).

In practical operation, we often need to truncate data. Truncating data is like viewing it through a window, which permits a view of only certain portions of the data and hiding (suppressing) the remainder. Abrupt truncation of data amounts to a rectangular window, which assigns a unit weight to data seen from the window, and assigns zero weight to the remaining data. Tapered windows, on the other hand, reduce the weight gradually from 1 to 0. Data truncation can cause some unsuspected problems. For example, in computation of the Fourier transform, windowing (data truncation) causes spectral spreading (spectral smearing), characteristic of the window function used. A rectangular window results in the least spreading, but it does so at the cost of oscillatory spectral leakage outside the signal band which decays slowly as $1/\omega$. Compared to rectangular window, tapered windows in general have larger spectral spreading (smearing), but the spectral leakage decays faster with frequency.

REFERENCES

1. Churchill, R.V., and J.W. Brown, *Fourier Series and Boundary Value Problems, 3d ed.*, McGraw-Hill, New York, 1978.

2. Bracewell, R.N., *Fourier Transform, and Its Applications, revised 2nd Ed.*, McGraw-Hill, New York, 1986.

3. Lathi, B.P., *Modern Digital and Analog Communication Systems, 2nd ed.*, Holt Rinehart and Winston, New York, 1989.

4. Guillemin, E.A., *Theory of Linear Physical Systems*, Wiley, New York, 1963.

5. Van Valkenberg, M.E., *Analog Filter Design*, Holt, Rinehart and Winston, New York, 1982.

6. Hamming, R.W., *Digital Filters, 2nd Ed.*, Prentice-Hall, Englewood Cliffs, N.J. 1983.

7. Harris, F.J., "On the Use of Windows for Harmonic Analysis with the Discrete Fourier Transform", *Proc. IEEE*, vol. 66, No. 1, January 1978, pp 51-83.

PROBLEMS

7.1-1 Show that if $f(t)$ is an even function of t, then

$$F(\omega) = 2 \int_0^\infty f(t) \cos \omega t \, dt$$

and if $f(t)$ is an odd function of t, then

$$F(\omega) = -2j \int_0^\infty f(t) \sin \omega t \, dt$$

Hence, prove that if $f(t)$ is a real and even function of t, then $F(\omega)$ is a real and even function of ω. In addition if $f(t)$ is a real and odd function of t, then $F(\omega)$ is an imaginary and odd function of ω.

7.1-2 Show that for a real $f(t)$, Eq. (7.7) can be expressed as

$$f(t) = \frac{1}{\pi} \int_0^\infty |F(\omega)| \cos[\omega t + \angle F(\omega)] \, d\omega$$

This is the trigonometric form of the Fourier integral. Compare this with the compact trigonometric Fourier series.

7.1-3 A signal $f(t)$ can be expressed as the sum of even and odd components (see Sec. B.6-2):

$$f(t) = f_e(t) + f_o(t)$$

(a) If $f(t) \Longleftrightarrow F(\omega)$, show that for real $f(t)$,

$$f_e(t) \Longleftrightarrow \text{Re}[F(\omega)] \qquad \text{and} \qquad f_o(t) \Longleftrightarrow j \, \text{Im}[F(\omega)]$$

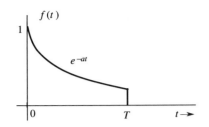

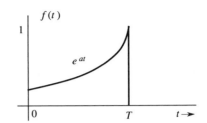

Fig. P7.1-4

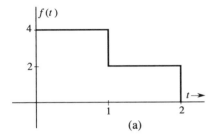

(a)

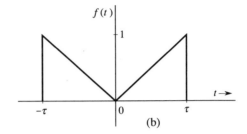

(b)

Fig. P7.1-5

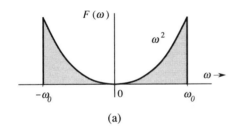

(a)

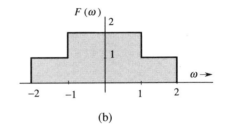

(b)

Fig. P7.1-6

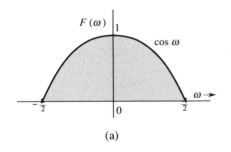

(a)

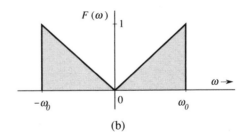

(b)

Fig. P7.1-7

(b) Verify these results by finding the Fourier transforms of the even and odd components of the following signals: **(i)** $u(t)$ **(ii)** $e^{-at}u(t)$.

7.1-4 From definition (7.8a), find the Fourier transforms of the signals $f(t)$ shown in Fig. P7.1-4.

7.1-5 From definition (7.8a), find the Fourier transforms of signals shown in Fig. P7.1-5.

7.1-6 From definition (7.8b), find the inverse Fourier transforms of the spectra shown in Fig. P7.1-6

7.1-7   From definition (7.8b), find the inverse Fourier transforms of spectra shown in Fig. P7.1-7.

7.3-1   Sketch the following functions:
  (a) $\text{rect}\left(\frac{t}{2}\right)$ (b) $\Delta\left(\frac{3\omega}{100}\right)$ (c) $\text{rect}\left(\frac{t-10}{8}\right)$ (d) $\text{sinc}\left(\frac{\pi\omega}{5}\right)$ (e) $\text{sinc}\left(\frac{\omega-10\pi}{5}\right)$
  (f) $\text{sinc}\left(\frac{t}{5}\right)\text{rect}\left(\frac{t}{10\pi}\right)$.
  Hint: $f\left(\frac{x-a}{b}\right)$ is $f\left(\frac{x}{b}\right)$ right-shifted by a.

7.3-2   From definition (7.8a), show that the Fourier transform of $\text{rect}\,(t-5)$ is $\text{sinc}\left(\frac{\omega}{2}\right)e^{-j5\omega}$.

7.3-3   From definition (7.8b), show that the inverse Fourier transform of $\text{rect}\left(\frac{\omega-10}{2\pi}\right)$ is $\text{sinc}\,(\pi t)\,e^{j10t}$.
  Hint: $\text{rect}\left(\frac{\omega-10}{2\pi}\right)$ is a gate pulse of width 2π that is centered at $\omega = 10$.

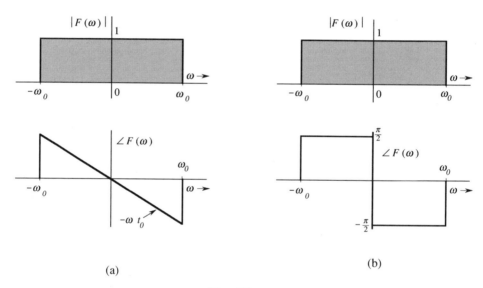

(a)   (b)

Fig. P7.3-4

7.3-4   Find the inverse Fourier transform of $F(\omega)$ for the spectra shown in Fig. P7.3-4.
  Hint: $F(\omega) = |F(\omega)|e^{j\angle F(\omega)}$. For part a, $F(\omega) = 1e^{-j\omega t_0}$ $|\omega| \leq \omega_0$. For part b,

$$F(\omega) = \begin{cases} 1e^{-j\pi/2} = -j & 0 < \omega \leq \omega_0 \\ 1e^{j\pi/2} = j & 0 > \omega \geq -\omega_0 \end{cases}$$

7.4-1   Apply the duality property to the appropriate Pair in Table 7.1 to show that
  (a) $\frac{1}{2}[\delta(t) + \frac{j}{\pi t}] \Longleftrightarrow u(\omega)$ (b) $\frac{1}{t} \Longleftrightarrow -j\pi\,\text{sgn}(\omega)$
  (c) $\delta(t+T) + \delta(t-T) \Longleftrightarrow 2\cos T\omega$ (d) $\delta(t+T) - \delta(t-T) \Longleftrightarrow 2j\sin T\omega$.
  Hint: $f(-t) \Longleftrightarrow F(-\omega)$ and $\delta(t) = \delta(-t)$.

7.4-2   The Fourier transform of the triangular pulse $f(t)$ in Fig. P7.4-2a is given to be

$$F(\omega) = \frac{1}{\omega^2}(e^{j\omega} - j\omega e^{j\omega} - 1)$$

  Using this information, and the time-shifting and time-scaling properties, find the Fourier transforms of signals shown in Figs. P7.4-2b, c, d, e, and f.

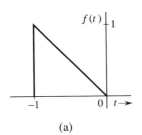

(a)

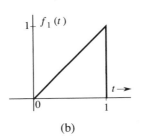

(b)

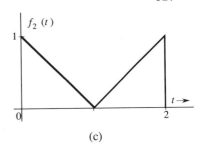

(c)

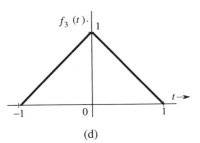

(d)

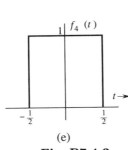

(e)

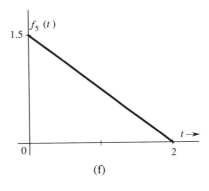

(f)

Fig. P7.4-2

Hint: See Sec. B.4 for explanation of various signal operations. Time inversion in $f(t)$ results in the pulse $f_1(t)$ in Fig. b; consequently $f_1(t) = f(-t)$. The pulse in Fig. c can be expressed as $f(t - T) + f_1(t - T)$ (the sum of $f(t)$ and $f_1(t)$ both delayed by T. The pulses in Figs. (d) and (e) both can be expressed as $f(t - T) + f_1(t + T)$ (sum of $f(t)$ delayed by T and $f_1(t)$ advanced by T) for some suitable choice of T. The pulse in Fig. f can be obtained by time-expanding $f(t)$ by a factor of 2 and then delaying the resulting pulse by two seconds (or by first delaying $f(t)$ by one second and then time-expanding by a factor of 2).

7.4-3 Using only the time-shifting property and Table 7.1, find the Fourier transforms of signals shown in Fig. P7.4-3.

Hint: The signal in Fig. a is a sum of two shifted gate pulses. The signal in Fig. b is $\sin t \, [u(t) - u(t - \pi)] = \sin t \, u(t) - \sin t \, u(t - \pi) = \sin t \, u(t) + \sin (t - \pi) \, u(t - \pi)$. The reader should verify that addition of the two sinusoids above indeed results in the pulse in Fig. b. In the same way we can express the signal in figs. c as $\cos t u(t) + \sin (t - \frac{\pi}{2}) u(t - \frac{\pi}{2})$ (verify this by sketching these signals). The signal in Fig. d is $e^{-at}[u(t) - u(t - T)] = e^{-at} u(t) - e^{-aT} e^{-a(t-T)} u(t - T)$.

7.4-4 Using the time-shifting property, show that if $f(t) \iff F(\omega)$, then

$$f(t + T) + f(t - T) \iff 2F(\omega) \cos T\omega$$

This is the dual of Eq. (7.41). Using this result and Pairs 17 and 19 in Table 7.1, find the Fourier transforms of signals shown in Fig. P7.4-4.

7.4-5 Prove the following results, which are duals of each other:

$$f(t) \sin \omega_0 t \iff \tfrac{1}{2j}[F(\omega - \omega_0) - F(\omega + \omega_0)]$$

$$\tfrac{1}{2j}[f(t + T) - f(t - T)] \iff F(\omega) \sin T\omega$$

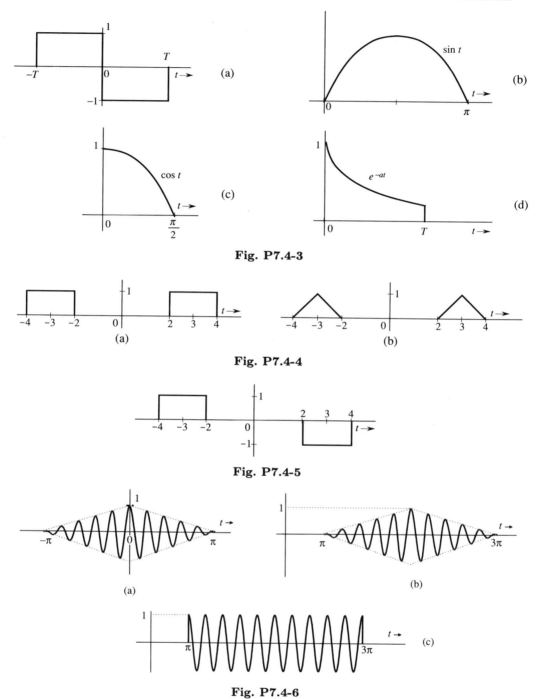

Fig. P7.4-3

Fig. P7.4-4

Fig. P7.4-5

Fig. P7.4-6

Using the latter result and Table 7.1, find the Fourier transform of the signal in Fig. P7.4-5.

7.4-6 The signals in Fig. P7.4-6 are modulated signals with carrier $\cos 10t$. Find the Fourier transforms of these signals using appropriate properties of the Fourier transform and

Table 7.1. Sketch the amplitude and phase spectra for parts **a** and **b**.
Hint: These functions can be expressed in the form $g(t) \cos \omega_0 t$.

7.4-7 Using the frequency-shifting property and Table 7.1, find the inverse Fourier transform of spectra shown in Fig. P7.4-7.

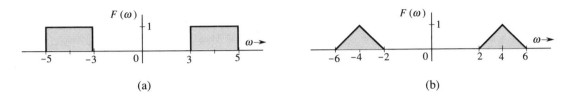

$$(a) \qquad\qquad (b)$$

Fig. P7.4-7

7.4-8 The process of recovering a signal $f(t)$ from the modulated signal $f(t) \cos \omega_0 t$ is called **demodulation**. Show that the signal $f(t) \cos \omega_0 t$ can be demodulated by multiplying it with $2 \cos \omega_0 t$ and passing the product through a lowpass filter of bandwidth W rps [the bandwidth of $f(t)$]. Assume $W < \omega_0$.
Hint: $2 \cos^2 \omega_0 t = 1 + \cos 2\omega_0 t$. Recognize that the spectrum of $f(t) \cos 2\omega_0 t$ is centered at $2\omega_0$ and will be suppressed by a lowpass filter of bandwidth W rps.

7.4-9 Using the time-convolution property, prove Pairs 2, 4, 13 and 14 in Table 2.1 (assume $\lambda < 0$ in Pair 2, λ_1 and $\lambda_2 < 0$ in Pair 4, $\lambda_1 < 0$ and $\lambda_2 > 0$ in Pair 13, and λ_1 and $\lambda_2 > 0$ in Pair 14).

7.4-10 A signal $f(t)$ is bandlimited to B Hz. Show that the signal $f^n(t)$ is bandlimited to nB Hz. Hint: $f^2(t) \Longleftrightarrow [F(\omega) * F(\omega)]/2\pi$, and so on. Use the width property of convolution).

7.4-11 Find the Fourier transform of the signal in Fig. 7.4-3a by three different methods:
(a) By direct integration using the definition (7.8a).
(b) Using only Pair 17 Table 7.1 and the time-shifting property.
(c) Using the time-differentiation and the time-shifting properties, along with the fact that $\delta(t) \Longleftrightarrow 1$.
Hint: $1 - \cos 2x = 2 \sin^2 x$.

7.4-12 **(a)** Prove the frequency differentiation property (dual of the time differentiation):

$$-jtf(t) \Longleftrightarrow \frac{d}{d\omega} F(\omega)$$

(b) Using this property and Pair 1 (Table 7.1), determine the Fourier transform of $te^{-at}u(t)$.

7.5-1 Find the (zero-state) response of an LTIC system described by the equation

$$(D^2 + 3D + 2)y(t) = (D + 3)f(t)$$

if the input $f(t)$ is **(a)**$e^{-3t}u(t)$ **(b)** $e^{-4t}u(t)$ **(c)** $e^{-2t}u(t)$ **(d)** $e^t u(-t)$ **(e)** $u(t)$.

7.5-2 A noncausal LTIC system is specified by the transfer function

$$H(\omega) = \frac{-(j\omega + 2)}{(j\omega - 2)(j\omega + 3)}$$

Find the (zero-state) response of this system if the input $f(t)$ is
(a) $e^{-t}u(t)$ **(b)** $e^t u(-t)$ **(c)** $e^{2t}u(-t)$ **(d)** $e^{-3t}u(t)$.

7.5-3 For an LTIC system with transfer function

$$H(s) = \frac{1}{s+1}$$

find the (zero-state) response if the input $f(t)$ is **(a)** $e^{-2t}u(t)$ **(b)** $e^{-t}u(t)$
(c) $e^t u(-t)$ **(d)** $u(t)$.

7.7-1 Consider a filter with the transfer function

$$H(\omega) = e^{-(k\omega^2 + j\omega t_0)}$$

Show that this filter is physically unrealizable by using the time-domain criterion
[noncausal $h(t)$] and frequency-domain (Paley-Wiener) criterion. Can this filter be
made approximately realizable by choosing a sufficiently large t_0? Use your own
(reasonable) criterion of approximate realizability to determine t_0.
Hint: Use Pair 22 in Table 7.1.

7.7-2 Show that a filter with transfer function

$$H(\omega) = \frac{2(10^5)}{\omega^2 + 10^{10}} e^{-j\omega t_0}$$

is unrealizable. Can this filter be made approximately realizable by choosing a sufficiently large t_0? Use your own (reasonable) criterion of approximate realizability to
determine t_0.
Hint: Show that the impulse response is noncausal.

7.9-1 Show that the energy of a Gaussian pulse

$$f(t) = \frac{1}{\sigma\sqrt{2\pi}} e^{-\frac{t^2}{2\sigma^2}}$$

is $\frac{1}{2\sigma\sqrt{\pi}}$. Verify this result by deriving the energy E_f from $F(\omega)$ using the Parseval's
theorem.
Hint: See Pair 22 in Table 7.1. Use the fact that

$$\int_{-\infty}^{\infty} e^{-x^2/2}\, dx = \sqrt{2\pi}$$

7.9-2 Show that

$$\int_{-\infty}^{\infty} \text{sinc}^2\,(kx)\, dx = \frac{\pi}{k}$$

Hint: Recognize that the integral is the energy of $f(t) = \text{sinc}\,(kt)$. Find this energy
by using the Parseval's theorem.

7.9-3 Generalize Parseval's theorem to show that for real, Fourier transformable signals
$f_1(t)$ and $f_2(t)$

$$\int_{-\infty}^{\infty} f_1(t)f_2(t)\, dt = \frac{1}{2\pi}\int_{-\infty}^{\infty} F_1(-\omega)F_2(\omega)\, d\omega = \frac{1}{2\pi}\int_{-\infty}^{\infty} F_1(\omega)F_2(-\omega)\, d\omega$$

7.9-4 For the signal

$$f(t) = \frac{2a}{t^2 + a^2}$$

determine the essential bandwidth B (Hz) of $f(t)$ such that the energy contained in the spectral components of $f(t)$ of frequencies below B Hz is 99% of the signal energy E_f.

7.9-5 **(a)** If $f_1(t)$, $f_2(t)$ are both real, Fourier-transformable signals, and if

$$g(t) = f_1(t) + f_2(t)$$

then show that in general

$$E_g \neq E_{f_1} + E_{f_2}$$

in which E_g, E_{f_1}, and E_{f_2} are the energies of $g(t), f_1(t)$, and $f_2(t)$, respectively.
(b) Show that $E_g = E_{f_1} + E_{f_2}$ if $f_1(t)$ and $f_2(t)$ are orthogonal; that is, if

$$\int_{-\infty}^{\infty} f_1(t) f_2(t)\, dt = 0$$

Moreover, E_z, the energy of the signal $z(t) = f_1(t) - f_2(t)$ is also $E_{f_1} + E_{f_2}$ if $f_1(t)$ and $f_2(t)$ are orthogonal.
(c) If

$$w(t) = a_1 f_1(t) \pm a_2 f_2(t)$$

then for orthogonal $f_1(t)$ and $f_2(t)$

$$E_w = a_1^2 E_{f_1} + a_2^2 E_{f_2}$$

COMPUTER PROBLEMS

C7-1 Use a computer to find and plot the Fourier transform of $[e^{-t} + e^{-3t}]u(t)$.

C7-2 There are several well known windowing functions such as Hanning, Hamming, Blackman, Kaiser which truncate data smoothly and have desirable effects on data weighting. Write a computer example illustrating a Blackman windowing function given by

$$w_B(x) = 0.42 + 0.5 \cos\left(\frac{2\pi x}{x_0}\right) + 0.08 \cos\left(\frac{4\pi x}{x_0}\right) \qquad |x| < \frac{x_0}{2}$$

Sampling

In earlier chapters we saw how a continuous-time signal can be processed by processing its samples through a discrete-time system. It is important to maintain the signal sampling rate sufficiently high so that the original signal can be reconstructed from these samples without error (or with an error within a given tolerance). The necessary quantitative framework for this is provided by the sampling theorem derived in the following section.

8.1 THE SAMPLING THEOREM

We now show that a real signal whose spectrum is bandlimited to B Hz [$F(\omega) = 0$ for $|\omega| > 2\pi B$] can be reconstructed exactly (without any error) from its samples taken uniformly at a rate $R > 2B$ samples per second. In other words, the minimum sampling frequency is $\mathcal{F}_s = 2B$ Hz.‡

To prove the sampling theorem, consider a signal $f(t)$ (Fig. 8.1a) whose spectrum is bandlimited to B Hz (Fig. 8.1b).‡ For convenience spectra are shown as functions of ω as well as of $\mathcal{F}$ (Hz). Sampling $f(t)$ at a rate of $\mathcal{F}_s$ Hz ($\mathcal{F}_s$ samples per second) can be accomplished by multiplying $f(t)$ by an impulse train $\delta_T(t)$(Fig. 8.1c), consisting of unit impulses repeating periodically every T seconds where $T = 1/\mathcal{F}_s$. This results in the sampled signal $\overline{f}(t)$ shown in Fig. 8.1d. The sampled signal consists of impulses spaced every T seconds (the sampling interval); the nth impulse, located at $t = nT$, has a strength $f(nT)$, the value of $f(t)$ at $t = nT$.

$$\overline{f}(t) = f(t)\delta_T(t) = \sum_n f(nT)\delta(t - nT) \qquad (8.1)$$

---

‡The theorem stated here (and proved subsequently) applies to lowpass signals. A bandpass signal whose spectrum exists over a frequency band $\mathcal{F}_c - \frac{B}{2} < |\mathcal{F}| < \mathcal{F}_c + \frac{B}{2}$ has a bandwidth B Hz. Such a signal is uniquely determined by $2B$ samples per second. In general, the sampling scheme is a bit more complex in this case. It uses two interlaced sampling trains, each at a rate of B samples per second (known as second-order sampling). See, for example, Reference 1 or 2.

‡The spectrum $F(\omega)$ in Fig. 8.1b is shown as real, for convenience. However, our arguments are valid for complex $F(\omega)$ as well.

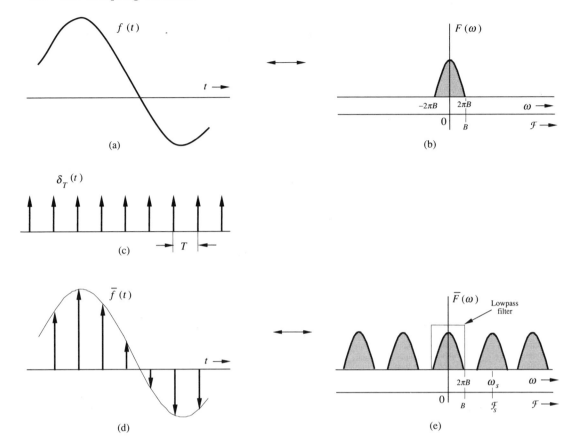

Fig. 8.1 Sampled signal and its Fourier spectrum.

Because the impulse train $\delta_T(t)$ is a periodic signal of period T, it can be expressed as a Fourier series. The trigonometric form of the Fourier series, already found in Example 6.7 [Eq. (6.48)], is

$$\delta_T(t) = \frac{1}{T}[1 + 2\cos \omega_s t + 2\cos 2\omega_s t + 2\cos 3\omega_s t + \cdots] \qquad \omega_s = \frac{2\pi}{T} = 2\pi\mathcal{F}_s$$

Therefore

$$\overline{f}(t) = f(t)\delta_T(t) = \frac{1}{T}[f(t) + 2f(t)\cos \omega_s t + 2f(t)\cos 2\omega_s t + 2f(t)\cos 3\omega_s t + \cdots]$$

To find $\overline{F}(\omega)$, the Fourier transform of $\overline{f}(t)$, we take the Fourier transform of the right-hand side of the above equation, term by term. The transform of the first term in the brackets is $F(\omega)$. The transform of the second term $2f(t)\cos \omega_s t$ is $F(\omega-\omega_s)+F(\omega+\omega_s)$ [see Eq. (7.41)]. This represents spectrum $F(\omega)$ shifted to ω_s and $-\omega_s$. Similarly, the transform of the third term $2f(t)\cos 2\omega_s t$ is $F(\omega - 2\omega_s) + F(\omega + 2\omega_s)$, which represents the spectrum $F(\omega)$ shifted to $2\omega_s$ and $-2\omega_s$, and so on to infinity. This means that the spectrum $\overline{F}(\omega)$ consists of $F(\omega)$ repeating periodically with period $\omega_s = \frac{2\pi}{T}$ radians/s, or $\mathcal{F}_s = \frac{1}{T}$ Hz, as shown in Fig. 8.1e.

Therefore

$$\overline{F}(\omega) = \frac{1}{T} \sum_{n=-\infty}^{\infty} F(\omega - n\omega_s) \tag{8.2}$$

If we are to reconstruct $f(t)$ from $\overline{f}(t)$, we should be able to recover $F(\omega)$ from $\overline{F}(\omega)$. This is possible if there is no overlap between successive cycles of $\overline{F}(\omega)$. Figure 8.1e shows that this requires

$$\mathcal{F}_s > 2B \tag{8.3}$$

Also, the sampling interval $T = 1/\mathcal{F}_s$. Therefore

$$T < \frac{1}{2B} \tag{8.4}$$

Thus, as long as the sampling frequency $\mathcal{F}_s$ is greater than twice the signal bandwidth B (in Hz), $\overline{F}(\omega)$ will consist of nonoverlapping repetitions of $F(\omega)$. When this is true, Fig. 8.1e shows that $f(t)$ can be recovered from its samples $\overline{f}(t)$ by passing the sampled signal $\overline{f}(t)$ through an ideal lowpass filter of bandwidth B Hz. The minimum sampling rate $\mathcal{F}_s = 2B$ required to recover $f(t)$ from its samples $\overline{f}(t)$ is called the **Nyquist rate** for $f(t)$, and the corresponding sampling interval $T = 1/2B$ is called the **Nyquist interval** for $f(t)$.†

8.1-1   Practical Difficulties in Signal Reconstruction

If a signal is sampled at the Nyquist rate $\mathcal{F}_s = 2B$, the spectrum $\overline{F}(\omega)$ consists of repetitions of $F(\omega)$ without any gap between successive cycles as shown in Fig. 8.2a. To recover $f(t)$ from $\overline{f}(t)$, we need to pass the sampled signal $\overline{f}(t)$ through an ideal lowpass filter, shown dotted in Fig. 8.2a. As seen in Sec. 7.7, such a filter is unrealizable; it can be closely approximated only with infinite time delay in the response. This means that we can recover the signal $f(t)$ from its samples with infinite time delay. A practical solution to this problem is to sample the signal at a rate higher than the Nyquist rate ($\mathcal{F}_s > 2B$ or $\omega_s > 4\pi B$). This yields $\overline{F}(\omega)$, consisting of repetition of $F(\omega)$ with a finite band gap between successive cycles, as shown in Fig. 8.2b. We can now recover $F(\omega)$ from $\overline{F}(\omega)$ using a lowpass filter with a gradual cutoff characteristic, shown dotted in Fig. 8.2b. But even in this case, the filter gain is required to be zero beyond the first cycle of $F(\omega)$ (see Fig. 8.2b). By the Paley-Wiener criterion, it is impossible to realize even this filter. The only advantage in this case is that the required filter can be closely approximated with a smaller time delay. This means that it is impossible in practice to recover a bandlimited signal $f(t)$ exactly from its samples even if the sampling rate is higher than the Nyquist rate. However, as the sampling rate increases, the recovered signal approaches the desired signal.

Aliasing

There is another fundamental practical difficulty in reconstructing a signal from its samples. The sampling theorem was proved on the assumption that signal $f(t)$

---

†We have proved that the sampling rate $R > 2B$. However, if the spectrum $F(\omega)$ has no impulse (or its derivatives) at the highest frequency B, the signal can be recovered from its samples taken at a rate $R = 2B$ Hz (Nyquist samples). In case $F(\omega)$ contains an impulse at the highest frequency B, the rate R must be greater than $2B$ Hz. Such is the case when $f(t) = \sin 2\pi Bt$. This signal is bandlimited to B Hz, but all of its samples are zero when taken at a rate $\mathcal{F}_s = 2B$ (starting at $t = 0$), and $f(t)$ cannot be recovered from its Nyquist samples.

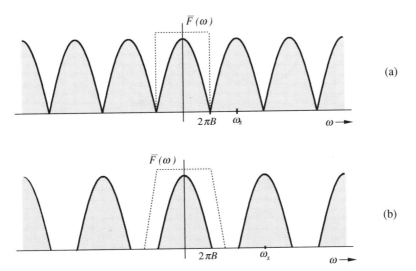

Fig. 8.2   Spectra of a signal sampled at (a) the Nyquist rate (b) above the Nyquist rate.

is bandlimited. **All practical signals are timelimited**, that is, they are of finite duration or width. It can be shown (see Prob. 8.1-3) that a signal cannot be timelimited and bandlimited simultaneously. If a signal is timelimited, it cannot be bandlimited and vice versa (but it can be simultaneously nontimelimited and nonbandlimited). This means that all practical signals, which are timelimited, are non-bandlimited; they have infinite bandwidth, and the spectrum $\overline{F}(\omega)$ consists of overlapping cycles of $F(\omega)$ repeating every $\mathcal{F}_s$ Hz (the sampling frequency), as shown in Fig. 8.3. Because of the overlapping tails, $\overline{F}(\omega)$ no longer has complete information about $F(\omega)$, and it is no longer possible to recover $f(t)$ from the sampled signal $\overline{f}(t)$. If the sampled signal is passed through the ideal lowpass filter, the output is not $F(\omega)$ but a version of $F(\omega)$ distorted as a result of two separate causes:

1. The loss of the tail of $F(\omega)$ beyond $|\mathcal{F}| > \mathcal{F}_s/2$ Hz;

2. The reappearance of this tail inverted or folded onto the spectrum at the cutoff or *folding frequency* $\frac{\mathcal{F}_s}{2}$ Hz. For instance, a component of frequency $\frac{\mathcal{F}_s}{2} + \mathcal{F}_x$ shows up as or "impersonates" a component of frequency $\frac{\mathcal{F}_s}{2} - \mathcal{F}_x$ in the reconstructed signal. This tail inversion, known as **spectral folding** or **aliasing**, is shown shaded in Fig. 8.3. The folding frequency is half the sampling frequency. For the physical explanation of how sampling permits a higher frequency sinusoid to masquerade as a lower frequency sinusoid, see Sec. 5.5-2 (Fig. 5.11).†

Elimination of Aliasing: The Anti-Aliasing Filter

For a bandlimited signal there is no aliasing if the signal is sampled at a rate above its Nyquist rate. When a signal is not bandlimited, aliasing results regardless of the value of the sampling rate. Therefore aliasing can be eliminated by bandlimiting a signal before sampling. If the essential bandwidth of a signal $f(t)$ is B Hz,

---

†Observe that the sampling frequency $\omega_s = 2\pi\mathcal{F}_s = \frac{2\pi}{T}$, and the folding frequency is $\frac{\omega_s}{2} = \frac{\pi}{T}$ rad/s. Therefore all components with frequency $\omega > \frac{\pi}{T}$ will be mistaken for frequencies below $\frac{\pi}{T}$. We derived the same result in Sec. 5.5-1.

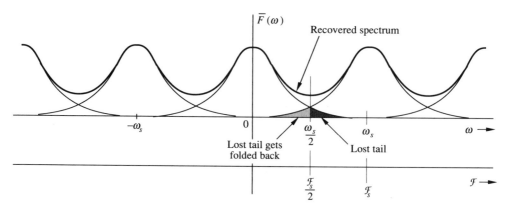

Fig. 8.3 Aliasing effect.

the spectral components beyond B Hz contain a negligible fraction of the signal energy, and bandlimiting the signal to B Hz will cause only a small distortion (caused by loss of the spectral tail beyond B Hz). The resulting signal $\hat{f}(t)$ is bandlimited to B Hz, and aliasing can be avoided by sampling it at a rate no less than $2B$ Hz. Bandlimiting is achieved by passing the signal $f(t)$ through an ideal lowpass filter of bandwidth B Hz. This filter is called the **anti-aliasing filter**. For the signal $e^{-at}u(t)$, for instance, we found the essential bandwidth (using the 95% criterion) in Example 7.18 to be 12.706 rad/s or $2.02a$ Hz. The anti-aliasing filter in this case is an ideal lowpass filter of bandwidth $2.02a$ Hz. This filter is used to bandlimit the signal e^{-at} to $2.02a$ Hz by suppressing all the spectral components above $2.02a$ Hz. The Nyquist rate of the resulting signal $\hat{f}(t)$ is $4.04a$ Hz. If we reconstruct a signal from these samples, the resulting signal $\hat{f}(t)$ will be a close approximation of the original signal $f(t)$. The energy of the error signal $f(t) - \hat{f}(t)$ will be 5% (or whatever criterion is selected) of the signal energy E_f.

The anti-aliasing filter, being an ideal filter, is unrealizable. In practice we use a steep cutoff filter, which leaves a sharply attenuated residual spectrum beyond the essential frequency B.

Practical Sampling

In proving sampling theorem, we assumed ideal samples obtained by multiplying a signal $f(t)$ by an impulse train which is physically nonexistent. In practice we multiply a signal $f(t)$ by a train of pulses of finite width, shown in Fig. 8.4b. The sampled signal is shown in Fig. 8.4c. We wonder whether it is possible to recover or reconstruct $f(t)$ from the sampled signal $\overline{f}(t)$ in Fig. 8.4c. Surprisingly, the answer is positive, provided that the sampling rate is not below the Nyquist rate. The signal $f(t)$ can be recovered by lowpass filtering $\overline{f}(t)$ as if it were sampled by impulse train.

The plausibility of this result can be seen from the fact that to reconstruct $f(t)$, we need the knowledge of the Nyquist sample values. This information is available or built in the sampled signal $\overline{f}(t)$ in Fig. 8.4c because the kth sampled pulse strength is $f(kT)$. To prove the result analytically, we observe that the sampling pulse train $p_T(t)$ shown in Fig. 8.4b can be expressed by a Fourier series

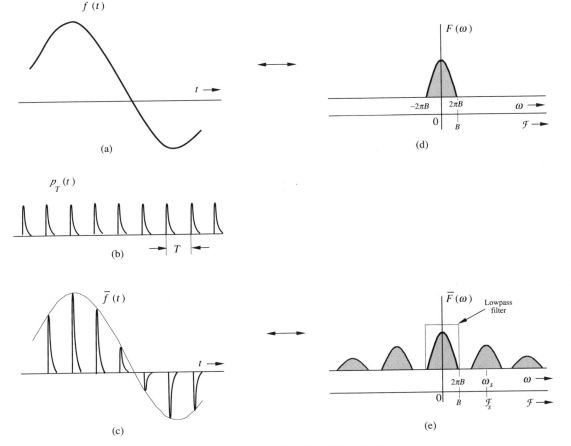

Fig. 8.4 Sampled signal and its Fourier spectrum.

$$p_T(t) = C_0 + \sum_{n=1}^{\infty} C_n \cos(n\omega_s t + \theta_n) \qquad \omega_s = \frac{2\pi}{T}$$

and

$$\overline{f}(t) = f(t)p_T(t) = f(t)\left[C_0 + \sum_{n=1}^{\infty} C_n \cos(n\omega_s t + \theta_n)\right]$$

$$= C_0 f(t) + \sum_{n=1}^{\infty} C_n f(t) \cos(n\omega_s t + \theta_n)$$

The sampled signal $\overline{f}(t)$ consists of $C_0 f(t)$, $C_1 f(t)\cos(\omega_s t + \theta_1)$, $C_2 f(t)\cos(2\omega_s t + \theta_2)$, $\cdots$. Note that the first term $C_0 f(t)$ is the desired signal and all the other terms are modulated signals with spectra centered at $\pm\omega_s, \pm 2\omega_s, \pm\omega_s, \cdots$, as shown in Fig. 8.4e. Clearly the signal $f(t)$ can be recovered by lowpass filtering of $\overline{f}(t)$ provided that $\omega_s > 4\pi B$ (or $\mathcal{F}_s > 2B$).

△ **Exercise E8.1**
    Find the Nyquist rate and the Nyquist interval for the signals (a) sinc $(100\pi t)$ and
(b) sinc $(100\pi t)+$ sinc $(50\pi t)$.

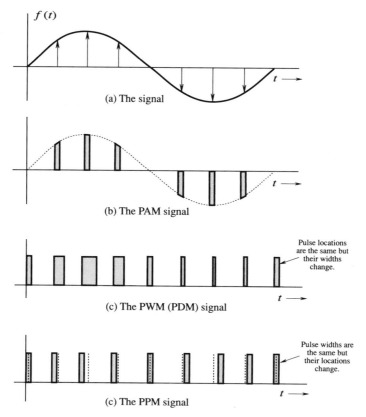

Fig. 8.5 Pulse modulated signals.

Answer: The Nyquist interval is 0.01 sec. and the Nyquist sampling rate is 100 Hz for both the signals. ▽

8.1-2 Some Applications of the Sampling Theorem

The sampling theorem is very important in signal analysis, processing, and transmission because it allows us to replace a continuous-time signal by a discrete sequence of numbers. Processing a continuous-time signal is therefore equivalent to processing a discrete sequence of numbers. This leads us directly into the area of digital filtering (discrete-time systems) introduced in Chapters 3 and 5. In the field of communication, the transmission of a continuous-time message reduces to the transmission of a sequence of numbers. This opens doors to many new techniques of communicating continuous-time signals by pulse trains. The continuous-time signal $f(t)$ is sampled, and sample values are used to modify certain parameters of a periodic pulse train. We may vary the amplitudes (Fig. 8.5b), widths (Fig. 8.5c), or positions (Fig. 8.5d) of the pulses in proportion to the sample values of the signal $f(t)$. Accordingly, we have **pulse-amplitude modulation** (PAM), **pulse-width modulation** (PWM), or **pulse position modulation** (PPM). The most important form of pulse modulation today is the **pulse code modulation** (PCM), discussed below. In all these cases, instead of transmitting $f(t)$, we transmit the corresponding pulse-modulated signal. At the receiver, we read the information of the pulse-modulated signal and reconstruct the analog signal $f(t)$.

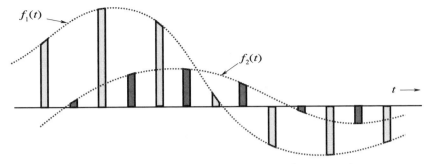

Fig. 8.6   Time-division multiplexing of two signals.

One advantage of using pulse modulation is that it permits simultaneous transmission of several signals on a time-sharing basis **(time-division multiplexing, or TDM)**. Because a pulse-modulated signal occupies only a part of the channel time, we can transmit several pulse-modulated signals on the same channel by interweaving them. Figure 8.6 shows the time-division multiplexing of two PAM signals. We can multiplex several signals, depending upon the channel available.

Pulse Code Modulation (PCM)

Pulse code modulation is the most useful and widely used of all the pulse modulations mentioned earlier. Basically, PCM is a method of converting an analog signal into a digital signal (A/D conversion). An **analog** signal is characterized by the fact that its amplitude can take on any value over a continuous range; this means that it can take on an infinite number of values. On the other hand, a **digital** signal amplitude can take on only a finite number of values. An analog signal can be converted into a digital signal by means of sampling and **quantizing** (rounding off). Sampling an analog signal alone will not yield a digital signal because a sampled analog signal can still take on any value in a continuous range. It is digitized by rounding off its value to one of the closest permissible numbers (or **quantized levels**), as shown in Fig. 8.7. The amplitudes of the analog signal $f(t)$ lie in the range $(-V, V)$, which is partitioned into L subintervals, each of magnitude $\Delta v = 2V/L$. Next, each sample amplitude is approximated by the midpoint value of the subinterval in which the sample falls (see Fig. 8.7a). This means that each sample is approximated to one of the L numbers. Thus the signal is digitized, with quantized samples taking on any one of the L values. Such a signal is known as an **L-ary digital signal**.

From practical viewpoint a binary digital signal (a signal that can take on only two values) is very desirable because of its simplicity, economy, and ease of engineering. We can convert an L-ary signal into a binary signal by using pulse coding. Figure 8.7b shows such a coding for the case of $L = 16$. Here the 16 decimal digits are expressed in terms of binary numbers consisting of four binary digits each. Each of the 16 numbers to be transmitted is assigned one binary code of four digits. Now each binary **0** is encoded by a negative pulse, and each binary **1** is encoded by a positive pulse. Thus each sample is now transmitted by a group of four binary pulses (pulse code). The resulting signal is a binary signal.

Audio signal bandwidth is about 15 kHz, but subjective tests show that signal articulation (intelligibility) is not affected if all the components above 3400 Hz

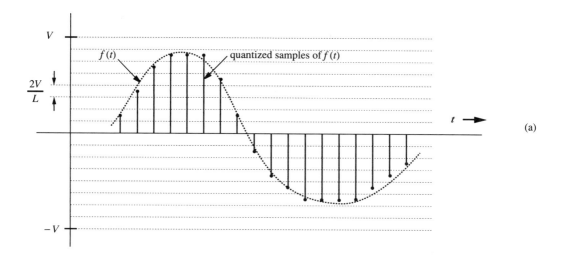

(a)

Digit	Binary equivalent	Pulse code waveform
0	0000	
1	0001	
2	0010	
3	0011	
4	0100	
5	0101	
6	0110	
7	0111	
8	1000	
9	1001	
10	1010	
11	1011	
12	1100	
13	1101	
14	1110	
15	1111	

(b)

Fig. 8.7 Analog-to-digital (A/D) conversion of a signal.

are suppressed.[3]† Since, in telephone communications, the objective is intelligibility rather than high fidelity, the components above 3400 Hz are eliminated by a lowpass filter. The resulting signal is then sampled at a rate of 8000 samples/s (8 kHz). This rate is intentionally kept higher than the Nyquist sampling rate of 6.8 kHz to avoid unrealizable filters, as required for signal reconstruction. Each sample is finally quantized into 256 levels ($L = 256$), which requires a group of eight binary pulses to encode each sample ($2^8 = 256$). Thus a telephone signal requires $8 \times 8000 = 64000$ binary pulses/second.

The compact disc (CD) is a recent application of PCM. This is a high-fidelity situation requiring the audio signal bandwidth to be 15 kHz. Although the Nyquist sampling rate is only 30 kHz, an actual sampling rate of 44.1 kHz is used for the reason mentioned earlier. The signal is quantized into a rather large number of levels ($L = 65536$) to reduce quantizing error. The binary-coded samples are now recorded on the CD.

Digital communication has several advantages over analog communication. The central problem in communication systems is that as the message signal travels along the channel (transmission path), it grows progressively weaker, while the channel noise and the signal distortion, being cumulative, become progressively stronger. Ultimately the signal, overwhelmed by noise and distortion, is mutilated. Amplification is of little help because it enhances the signal and the noise in the same proportion. Consequently the distance over which an analog message can be transmitted is limited by the transmitted power.

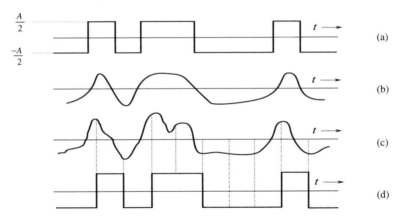

Fig. 8.8 Digital signal: (a) transmitted (b) received distorted signal (without noise) (c) received distorted signal (with noise) (d) regenerated signal at the receiver.

Digital communication is more rugged than analog communication because it can withstand channel noise and distortion much better as long as the noise and the distortion are within limits. The digital (binary) message in Fig. 8.8a is distorted by the channel, as shown in Fig. 8.8b. Yet if the distortion is within a limit, we can recover the data without error because we need only to make a simple binary decision as to whether the received pulse is positive or negative. Figure 8.8c shows the same data with channel distortion and noise. Here again the data can be recovered correctly as long as the distortion and the noise are within limits. Such is

†Components below 300 Hz may also be suppressed without affecting the articulation.

not the case with analog messages. Any distortion or noise, no matter how small, will distort the received signal.

The greatest advantage of digital communication over analog communication, however, is the viability of regenerative repeaters in the former. If a transmission path is long enough, the channel distortion and noise will accumulate sufficiently to overwhelm even a digital signal. The trick is to set up repeater stations along the transmission path at distances short enough to be able to detect signal pulses before the noise and distortion have a chance to accumulate sufficiently. At each repeater station the pulses are detected, and new, clean pulses are transmitted to the next repeater station, which, repeats the process. If the noise and distortion are within limits (which is possible because of the closely spaced repeaters), pulses can be detected correctly.† This way the messages can be transmitted over longer distances with greater reliability. In contrast, analog messages cannot be cleaned up periodically, and the transmission is therefore less reliable. The only error appearing in PCM comes from the quantizing. This error can be reduced as much as desired by increasing the number of quantizing levels, the price of which is paid in an increased bandwidth of the transmission medium (channel). Another important advantage of digital communication is that error correction can be incorporated into the code.[3]

8.2 DUAL OF THE SAMPLING THEOREM: SPECTRAL SAMPLING

As in other cases, the sampling theorem has dual. In Sec. 8.1, we discussed the time-sampling theorem where we showed that a signal bandlimited to B Hz can be reconstructed from the signal samples taken at a rate $R > 2B$ samples/second. Note that the signal spectrum exists over the frequency range $-B$ to B Hz. Therefore $2B$ is the spectral width (not the bandwidth, which is B) of the signal. This means that a signal $f(t)$ can be reconstructed from samples taken at a rate $R >$ the spectral width (in Hz) of the signal.

The dual of time-sampling theorem is the frequency-sampling theorem. This applies to timelimited signals, which are duals of bandlimited signals. We now prove that the spectrum $F(\omega)$ of a signal time-limited to τ seconds can be reconstructed from the samples of $F(\omega)$ taken at a rate $R' > \tau$ (the signal width) samples per Hertz.

Figure 8.9a shows a signal $f(t)$ that is time-limited to τ seconds along with its Fourier transform $F(\omega)$. Although $F(\omega)$ is complex in general, it is adequate for our line of reasoning to show it as a real function.

$$F(\omega) = \int_{-\infty}^{\infty} f(t)e^{-j\omega t}\,dt = \int_{0}^{\tau} f(t)e^{-j\omega t}\,dt \qquad (8.5)$$

We now construct $f_{T_0}(t)$, a periodic signal formed by repeating $f(t)$ every T_0 seconds $(T_0 > \tau)$, as shown in Fig. 8.9b. This periodic signal can be expressed by the exponential Fourier series:

$$f_{T_0}(t) = \sum_{n=-\infty}^{\infty} D_n e^{jn\omega_0 t} \qquad \omega_0 = \frac{2\pi}{T_0} \qquad (8.6a)$$

---

†The error in pulse detection can be made negligible.

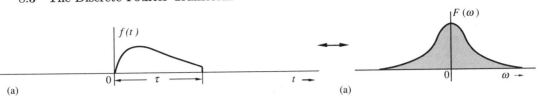

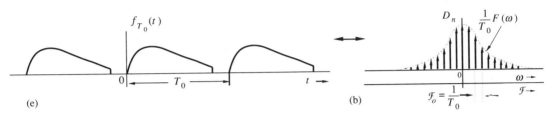

Fig. 8.9   Periodic repetition of a signal amounts to sampling its spectrum.

in which (assuming $\tau < T_0$)

$$D_n = \frac{1}{T_0} \int_0^{T_0} f(t) e^{-jn\omega_0 t}\, dt = \frac{1}{T_0} \int_0^{\tau} f(t) e^{-jn\omega_0 t}\, dt \qquad (8.6b)$$

From Eq. (8.5) it follows that

$$D_n = \frac{1}{T_0} F(n\omega_0) \qquad (8.7)$$

This rather surprising result states that the coefficients of the Fourier series for $F_{T_0}(t)$ are $(1/T_0)$ times the sample values of the spectrum $F(\omega)$ taken at intervals of ω_0. This shows that the spectrum of the periodic signal $f_{T_0}(t)$ is the sampled spectrum $F(\omega)$, as shown in Fig. 8.9b. Now as long as $\tau < T_0$, the successive cycles of $f(t)$ appearing in $f_{T_0}(t)$ do not overlap, so that $f(t)$ can be recovered from $f_{T_0}(t)$. This means indirectly that $F(\omega)$ can be reconstructed from its samples. These samples are separated by the fundamental frequency $\mathcal{F}_0 = 1/T_0$ Hz of the periodic signal $f_{T_0}(t)$. The condition for recovery is that $T_0 > \tau$. This means that

$$\mathcal{F}_0 < \frac{1}{\tau}\,\text{Hz} \qquad (8.8)$$

Therefore, to be able to reconstruct the spectrum $F(\omega)$ from the samples of $F(\omega)$, the samples should be taken at frequency intervals not greater than $\mathcal{F}_0 = 1/\tau$ Hz. If R' is the sampling rate (samples/Hz), then

$$R' > \frac{1}{\mathcal{F}_0} = \tau\ \text{samples/Hz} \qquad (8.9)$$

8.3   NUMERICAL COMPUTATION OF THE FOURIER TRANS-FORM: THE DISCRETE FOURIER TRANSFORM (DFT)

Numerical computation of the Fourier transform of $f(t)$ requires sample values of $f(t)$ because a digital computer can work only with discrete data (sequence of numbers). Moreover, a computer can compute $F(\omega)$ only at some discrete values of ω [samples of $F(\omega)$]. We therefore need to relate the samples of $F(\omega)$ to samples of $f(t)$. This can be done from the results of the two sampling theorems developed in Secs. 8.1 and 8.2.

We begin with a timelimited signal $f(t)$ (Fig. 8.10a) and its spectrum $F(\omega)$ (Fig. 8.10b). Since $f(t)$ is timelimited, $F(\omega)$ is non-bandlimited. For convenience we shall show all spectra as functions of frequency variable $\mathcal{F}$ (in Hertz) rather than ω. According to the sampling theorem, the spectrum $\overline{F}(\omega)$ of the sampled signal $\overline{f}(t)$ consists of $F(\omega)$ repeating every $\mathcal{F}_s$ Hz where $\mathcal{F}_s = 1/T$. This is shown in Fig. 8.10c and 8.10d.† In the next step, the sampled signal in Fig. 8.10c is repeated periodically every T_0 seconds, as shown in Fig. 8.10e. According to spectral sampling theorem, such an operation results in sampling the spectrum at a rate of T_0 samples per Hz. This means that the samples are spaced $\mathcal{F}_o = 1/T_0$ Hz, as shown in Fig. 8.10f.

The above discussion shows that, when a signal $f(t)$ is sampled and then periodically repeated, the corresponding spectrum is also sampled and periodically repeated. Our goal is to relate the samples of $f(t)$ to the samples of $F(\omega)$.

Number of Samples

One interesting observation in Figs. 8.10e and 8.10f is that N_0, the number of samples of the signal in Fig. 8.10e in one period T_0 is identical to N_0', the number of samples of the spectrum in Fig. 8.10f in one period $\mathcal{F}_s$. This is because

$$N_0 = \frac{T_0}{T} \quad \text{and} \quad N_0' = \frac{\mathcal{F}_s}{\mathcal{F}_o} \tag{8.10a}$$

But because

$$\mathcal{F}_s = \frac{1}{T} \quad \text{and} \quad \mathcal{F}_o = \frac{1}{T_0} \tag{8.10b}$$

$$N_0 = \frac{T_0}{T} = \frac{\mathcal{F}_s}{\mathcal{F}_o} = N_0' \tag{8.10c}$$

Aliasing and Leakage in Numerical Computation

Figure 8.10f shows the presence of aliasing in the samples of the spectrum $F(\omega)$. This aliasing error can be reduced as much as desired by increasing the sampling frequency $\mathcal{F}_s$ (decreasing the sampling interval $T = \frac{1}{\mathcal{F}_s}$). The aliasing can never be eliminated for timelimited $f(t)$, however, because its spectrum $F(\omega)$ is non-bandlimited. Had we started out with a signal having a bandlimited spectrum $F(\omega)$, there would be no aliasing in the spectrum in Fig. 8.10f. Unfortunately such a signal is non-timelimited and its repetition (in Fig. 8.10e) would result in signal overlapping (aliasing in time domain). In this case we shall have to contend with errors in signal samples. In other words, in computing the direct or inverse Fourier transform numerically, we can reduce the error as much as we wish, but it can never be eliminated. This is true of numerical computation of direct and inverse Fourier transforms regardless of the method used. For example, if we determine the Fourier transform by direct integration numerically, using Eq. (7.8a), there will be an error because the interval of integration Δt can never be made zero. Similar remarks apply to numerical computation of the inverse transform. Therefore we

---

†There is a multiplying constant $1/T$ for the spectrum in Fig. 8.10d [see Eq. (8.2)], but this is irrelevant to our discussion here.

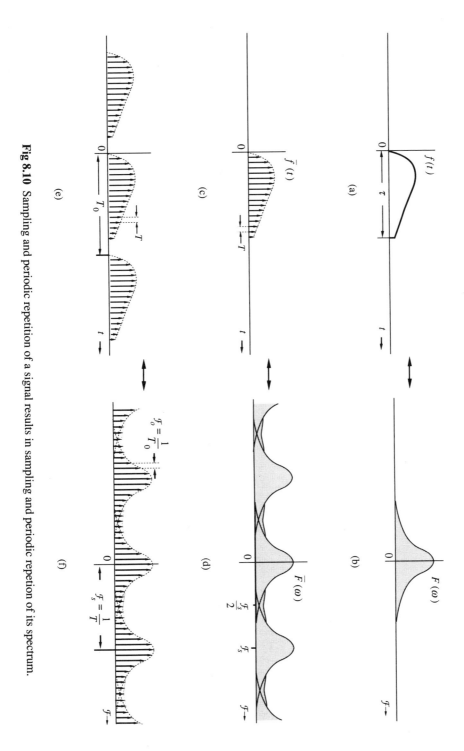

Fig 8.10 Sampling and periodic repetition of a signal results in sampling and periodic repetion of its spectrum.

should always keep in mind the nature of this error in our results. In our discussion (Fig. 8.10) we assumed $f(t)$ to be a timelimited signal. If $f(t)$ is not timelimited it will have to be timelimited because numerical computations can work only with finite data. Further, this data truncation causes error because of spectral spreading (smearing) and leakage, as discussed in Sec. 7.10. The leakage also causes aliasing. Leakage can be reduced by using a tapered window for signal truncation. But this increases spectral spreading or smearing. The spectral spreading can be reduced by increasing the window width (i.e. more data). This increases T_0, and reduces $\mathcal{F}_0$ (increases spectral resolution).

Picket Fence Effect

The numerical computation method yields only the uniform sample values of $F(\omega)$ [or $f(t)$]. This is like viewing a signal and its spectrum through a "picket-fence". The major peaks of $F(\omega)$ [or $f(t)$] could lie between two samples and may remain hidden, giving a wrong picture of the reality. Such misleading results can be avoided by using a sufficiently large N_0, the number of samples, which increases resolution.

Derivation of the Discrete Fourier transform (DFT)

If $f(kT)$ and $F(r\omega_0)$ are the kth and rth samples of $f(t)$ and $F(\omega)$, respectively, then we define new variables f_k and F_r as

$$f_k = Tf(kT)$$

$$= \frac{T_0}{N_0}f(kT) \tag{8.11a}$$

and

$$F_r = F(r\omega_0) \tag{8.11b}$$

in which

$$\omega_0 = 2\pi\mathcal{F}_o = \frac{2\pi}{T_0} \tag{8.11c}$$

We shall now show that f_k and F_r are related by the following equations:

$$F_r = \sum_{k=0}^{N_0-1} f_k e^{-jr\Omega_0 k} \tag{8.12a}$$

$$f_k = \frac{1}{N_0}\sum_{r=0}^{N_0-1} F_r e^{jr\Omega_0 k} \qquad \Omega_0 = \frac{2\pi}{N_0} \tag{8.12b}$$

These equations define the direct and the inverse **discrete Fourier transforms (DFT)**, with F_r the direct discrete Fourier transform (DFT) of f_k, and f_k the inverse discrete Fourier transform (IDFT) of F_r. The notation

$$f_k \Longleftrightarrow F_r$$

is also used to indicate that f_k and F_r are a DFT pair. Remember that f_k is T_0/N_0 times the kth sample of $f(t)$ and F_r is the rth sample of $F(\omega)$. Knowing the sample values of $f(t)$, we can compute the sample values of $F(\omega)$—and vice versa—using the DFT. Note, however, that f_k is a function of k ($k = 0, 1, 2, \cdots, N_0 - 1$) rather than of t and that F_r is a function of r ($r = 0, 1, 2, \cdots, N_0 - 1$) rather than of ω. Moreover, both f_k and F_r are periodic sequences with a period N_0 (Figs. 8.10e and 8.10f). Such sequences are called N_0-**periodic sequences**. The proof of the DFT relationships in Eqs. (8.12) follows directly from the results of the sampling theorem. The sampled signal $\overline{f}(t)$ (Fig. 8.10c) can be expressed as

$$\overline{f}(t) = \sum_{k=0}^{N_0-1} f(kT)\delta(t - kT) \tag{8.13}$$

Since $\delta(t - kT) \Longleftrightarrow e^{-jk\omega T}$, the Fourier transform of Eq. (8.13) yields

$$\overline{F}(\omega) = \sum_{k=0}^{N_0-1} f(kT)e^{-jk\omega T} \tag{8.14}$$

But from Fig. 8.1e [or Eq. (8.2)], it is clear that $\overline{F}(\omega)$, the Fourier transform of $\overline{f}(t)$, is $\frac{F(\omega)}{T}$ for $|\omega| \le \frac{\omega_s}{2}$, assuming negligible aliasing. Hence

$$F(\omega) = T\overline{F}(\omega) = T\sum_{k=0}^{N_0-1} f(kT)e^{-jk\omega T} \qquad |\omega| \le \frac{\omega_s}{2}$$

and

$$F_r = F(r\omega_0) = T\sum_{k=0}^{N_0-1} f(kT)e^{-jkr\omega_0 T} \tag{8.15}$$

If we let $\omega_0 T = \Omega_0$, then from Eqs. (8.10a) and (8.10b)

$$\Omega_0 = \omega_0 T = 2\pi \mathcal{F}_o T = \frac{2\pi}{N_0} \tag{8.16}$$

Also, from Eq. (8.11a),

$$Tf(kT) = f_k$$

Therefore Eq. (8.15) becomes

$$F_r = \sum_{k=0}^{N_0-1} f_k e^{-jr\Omega_0 k} \qquad \Omega_0 = \frac{2\pi}{N_0} \tag{8.17}$$

The inverse transform relationship (8.12b), could be derived by using a similar procedure with the roles of t and ω reversed, but here we shall use a more direct proof. To prove the inverse relation in Eq. (8.12b), we multiply both sides of Eq. (8.17) by $e^{jm\Omega_0 r}$ and sum over r:

$$\sum_{r=0}^{N_0-1} F_r e^{jm\Omega_0 r} = \sum_{r=0}^{N_0-1} \left[\sum_{k=0}^{N_0-1} f_k e^{-jr\Omega_0 k} \right] e^{jm\Omega_0 r}$$

By Interchanging the order of summation on the right hand side

$$\sum_{r=0}^{N_0-1} F_r e^{jm\Omega_0 r} = \sum_{k=0}^{N_0-1} f_k \left[\sum_{r=0}^{N_0-1} e^{j(m-k)\Omega_0 r} \right]$$

It is shown in Sec. 9.1-1 [see Eq. (9.8) on p. 569] that the inner sum on the right hand side is zero for $k \neq m$, and the sum is N_0 when $k = m$. Therefore the outer sum will have only one nonzero term when $k = m$, and it is $N_0 f_k = N_0 f_m$. Therefore

$$f_m = \frac{1}{N_0} \sum_{r=0}^{N_0-1} F_r e^{jm\Omega_0 r} \qquad \Omega_0 = \frac{2\pi}{N_0} \tag{8.18}$$

Because F_r is N_0-periodic, we need to determine the values of F_r over any one period. It is customary to determine F_r over the range $(0, N_0 - 1)$ rather than over the range $(-\frac{N_0}{2}, \frac{N_0}{2} - 1)$.‡

Choice of T, T_0, and N_0

In DFT computation, we first need to select suitable values for N_0, T, and T_0. For this purpose we should first decide B, the essential bandwidth of the signal. The sampling frequency $\mathcal{F}_s$ must be at least $2B$; that is,

$$\frac{\mathcal{F}_s}{2} \geq B \tag{8.19}$$

Moreover, the sampling interval $T = \frac{1}{\mathcal{F}_s}$ [Eq. (8.10b)], and

$$T \leq \frac{1}{2B} \tag{8.20}$$

Once we pick B, we can choose T according to Eq. (8.20)). Also,

$$\mathcal{F}_o = \frac{1}{T_0} \tag{8.21}$$

in which $\mathcal{F}_o$ is the **frequency resolution** [separation between samples of $F(\omega)$]. Hence, if $\mathcal{F}_o$ is given, we can pick T_0 according to Eq. (8.21). Knowing T_0 and T, we determine N_0 from

$$N_0 = \frac{T_0}{T} \tag{8.22}$$

Points of Discontinuity

If $f(t)$ has a jump discontinuity at a sampling point, the sample value should be taken as the average of the values on the two sides of the discontinuity because the Fourier representation at a point of discontinuity converges to the average value.

■ **Example 8.1**

Find the DFT of a gate function $f(t) = 8 \,\mathrm{rect}\,(t)$.

The gate function and its Fourier transform are shown in Figs. 8.11a and 8.11b respectively. To determine the value of the sampling interval T, we must first decide upon the essential bandwidth B. From Fig. 8.11, we see that $F(\omega)$ is quite small beyond $\omega = 8\pi$ or $f = 4$ Hz. Hence the essential bandwidth $B = 4$ Hz is a reasonable choice. This gives the sampling interval $T = \frac{1}{2B} = \frac{1}{8}$. Looking again at the spectrum in Fig. 8.11b, we see that a choice of frequency resolution $\mathcal{F}_o = \frac{1}{4}$ Hz is reasonable. This will give four samples in each lobe of $F(\omega)$. In this case $T_0 = \frac{1}{\mathcal{F}_o} = 4$ seconds. and $N_0 = \frac{T_0}{T} = 32$. The

---

‡The DFT relationships represent a transform in their own right and they are exact. If, however, we identify f_k and F_r as the samples of a signal $f(t)$ and of its Fourier transform $F(\omega)$, respectively, then the DFT relationships are approximations because of the aliasing and leakage effects.

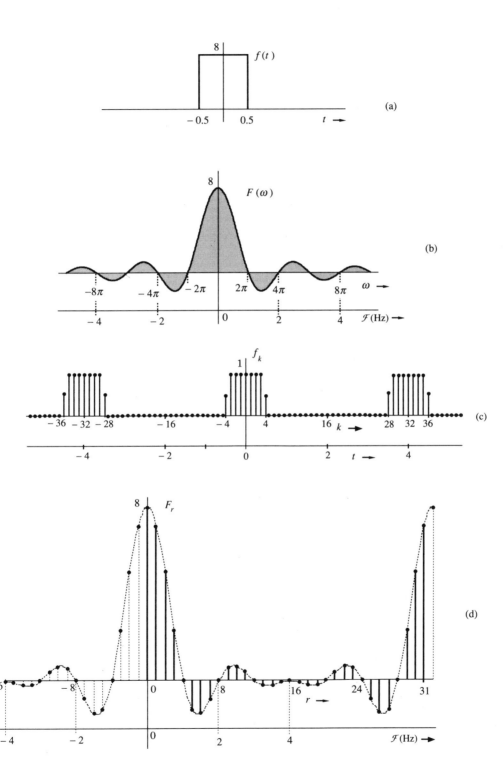

Fig. 8.11 Discrete Fourier transform of a gate pulse.

duration of $f(t)$ is only 1 second. We must repeat it every 4 seconds ($T_0 = 4$), as shown in Fig. 8.11c, and take samples every $\frac{1}{8}$ second. This gives us 32 samples ($N_0 = 32$). Also,

$$f_k = \frac{T_0}{N_0} f(kT)$$

$$= \frac{1}{8} f(kT)$$

Since $f(t) = 8\,\text{rect}\,(t)$, the values of f_k are 1, 0, or 0.5 (at the points of discontinuity), as shown in Fig. 8.11c. In this figure, f_k is shown as a function of t as well as k, for convenience.

In the derivation of the DFT, we assumed that $f(t)$ begins at $t = 0$ (Fig. 8.10a), and then took N_0 samples over the interval $(0, T_0)$. In the present case, however, $f(t)$ begins at $-\frac{1}{2}$. This difficulty is easily resolved when we realize that the DFT found by this procedure is actually the DFT of f_k repeating periodically every T_0 seconds. From Fig. 8.11c, it is clear that repeating the segment of f_k over the interval from -2 to 2 seconds periodically is identical to repeating the segment of f_k over the interval from 0 to 4 seconds. Hence the DFT of the samples taken from -2 to 2 seconds is the same as that of the samples taken from 0 to 4 seconds. Therefore, regardless of where $f(t)$ starts, we can always take the samples of $f(t)$ and its periodic extension over the interval 0 to T_0. In the present example, the 32 sample values are

$$f_k = \begin{cases} 1 & 0 \le k \le 3 \quad \text{and} \quad 29 \le k \le 31 \\ 0 & 5 \le k \le 27 \\ 0.5 & k = 4, 28 \end{cases}$$

We also have $N_0 = 32$ and $\Omega_0 = 2\pi/32 = \pi/16$. Therefore [see Eq. (8.12a)]

$$F_r = \sum_{k=0}^{31} f_k e^{-jr\frac{\pi}{16}k}$$

We can also compute F_r numerically on a computer using an FFT algorithm, as discussed in Sec. 8.3-2. Figure 8.11d shows the plot of F_r.

The samples F_r are separated by $\mathcal{F}_o = \frac{1}{T_0}$ Hz. In this case $T_0 = 4$, so the frequency resolution $\mathcal{F}_o$ is $\frac{1}{4}$ Hz, as desired. The folding frequency $\frac{\mathcal{F}_s}{2} = B = 4$ Hz corresponds to $r = \frac{N_0}{2} = 16$. Because F_r is N_0-periodic ($N_0 = 32$), the values of F_r for $r = -16$ to $n = -1$ are the same as those for $r = 16$ to $n = 31$. ■

⊙ **Computer Example C8.1**
Do Example 8.1 using a computer.

```
% This routine calculates and plots the DFT for f(t) = 8 rect (t).
% This section generates fk.
   for k=1:32    % Create the discrete time index k.
     if k<=4
       fk(k)=1;
     end
     if k==5
       fk(k)=0.5;
   end
     if k>=6,
```

```
        if k<=28
          fk(k)=0;
        end
      end
    if k==29
      fk(k)=0.5;
    end
    if k>=30,
        if k<=32
          fk(k)=1;
        end
      end
    end
k=1:32; plot(k,fk,'o'),grid,
grid,xlabel('k'),ylabel('fk'),
title('Samples of f(t)'),pause
% This section calculates the DFT for f(t) using FFT algorithm.
Fr=fft(fk);
plot(k,fft(fk),'x'),grid,
xlabel('r'),ylabel('Fr'),
title('Plot of Fr'),   ⊙.
```

Zero padding

Recall that the DFT F_r is like observing the spectrum $F(\omega)$ through a "picket-fence". If the frequency sampling interval $\mathcal{F}_o$ is not sufficiently small, we could miss out on some significant details, and obtain a misleading picture. In Example 8.1, for instance, had we inadvertently chosen the frequency resolution $\mathcal{F}_o = 1$ Hz, the FFT would have given values of $F(\omega)$ at frequencies $0, 1, 2, \cdots$, Hz or $0, 2\pi, 4\pi, \cdots$, rad/s. Figure 8.11b shows that these are all zero except for the value at dc. This would give us $F_r = 0$ for all $r \neq 0$ and $F_r = 1$ for $r = 0$. This does not mean that FFT has failed; it simply means we require a higher resolution (smaller $\mathcal{F}_o$).

Because $\mathcal{F}_o = 1/T_0$, to obtain higher resolution (smaller $\mathcal{F}_o$), we need to increase the value of T_0, the period of repetition for $f(t)$. This increases N_0, the number of samples of $f(t)$ by adding dummy samples with a value of 0. In Example 8.1, we selected $T_0 = 4$ s so that $\mathcal{F}_o = 0.25$ Hz. Use of $T_0 = 4$ amounts to adding dummy samples of zero value for $k = 5$ through 27 (Fig. 8.11c). This addition of dummy samples is also known as **zero padding**. It is important to remember, however, that increasing resolution (reducing $\mathcal{F}_o$) by zero padding does not increase the accuracy of the result. It merely allows us to peek behind the "picket-fence" more often (more samples). The accuracy of observation can be increased only by reducing aliasing, which requires reduction in the sampling interval T, which also increases N_0 ($N_0 = T_0/T$). Thus, to reduce aliasing error or to improve resolution, we need to increase the number of signal samples N_0.

In the example above, we knew the Fourier transform $F(\omega)$ beforehand and hence could make intelligent choices for $\mathcal{F}_s$ (or B) and $\mathcal{F}_o$. In practice, we generally do not know $F(\omega)$ beforehand. In fact, that is exactly what we are trying to determine. In such a case, we must make intelligent guesses for $\mathcal{F}_s$ and $\mathcal{F}_o$ from circumstantial evidence. We should then continue reducing the value of T and recomputing the transform until the result stabilizes within the desired number of significant digits.

■ **Example 8.2**

A signal $f(t)$ has a duration of 2 ms and an essential bandwidth of 10 kHz. It is desired to have a frequency resolution of 100 Hz in the DFT ($\mathcal{F}_o = 100$). Determine N_0.

The effective signal duration T_0 is given by

$$T_0 = \frac{1}{\mathcal{F}_o} = \frac{1}{100} = 10 \text{ ms}$$

Since the signal duration is only 2 ms, we need zero padding over 8 ms. Also, $B = 10,000$. Hence $\mathcal{F}_s = 2B = 20,000$, and $T = 1/\mathcal{F}_s = 50\,\mu s$. Further,

$$N_0 = \frac{\mathcal{F}_s}{\mathcal{F}_o} = \frac{20,000}{100} = 200$$

In the FFT algorithm used to compute DFT, it proves convenient (although not necessary) to select N_0 as a power of 2, that is, $N_0 = 2^n$ (n, integer). Let us choose $N_0 = 256$. Increasing N_0 from 200 to 256 can be used to reduce aliasing error (by reducing T), to improve resolution (by increasing T_0), or a combination of both.

(i) Reducing aliasing error

We maintain the same T_0; i.e., the same $\mathcal{F}_o = 100$. This yields

$$\mathcal{F}_s = N_0 \mathcal{F}_o = 256 \times 100 = 25600$$

and

$$T = \frac{1}{\mathcal{F}_s} = 39\,\mu s$$

Thus increasing N_0 from 200 to 256 permits us to reduce the sampling interval T from $50\,\mu s$ to $39\,\mu s$ while maintaining the same frequency resolution ($\mathcal{F}_o = 100$).

(ii) Improving resolution

Here we maintain the same T; i.e., $T = 50\,\mu s$. This gives

$$T_0 = N_0 T = 256(50 \times 10^{-6}) = 12.8 \text{ ms}$$

and

$$\mathcal{F}_o = \frac{1}{T_0} = 78.125 \text{ Hz}$$

Thus increasing N_0 from 200 to 256 can improve the frequency resolution from 100 to 78.125 Hz while maintaining the same aliasing error ($T = 50\,\mu s$).

(iii) Combination of these two options

We could choose $T = 45\,\mu s$ and $T_0 = 11.5$ ms so that $\mathcal{F}_o = 86.96$ Hz. ■

8.3-1 Some Properties of DFT

The discrete Fourier transform is basically the Fourier transform of a sampled signal repeated periodically. Hence the properties derived earlier for the Fourier transform apply to the DFT as well. For this reason we shall merely list some of the important properties of the DFT without proof.

1. Linearity

If $f_k \iff F_r$ and $g_k \iff G_r$, then

$$a_1 f_k + a_2 g_k \iff a_1 F_r + a_2 G_r \tag{8.23}$$

2. Time Shifting (Cyclic Shifting)

$$f_{k-n} \iff F_r e^{-jr\Omega_0 n} \tag{8.24}$$

3. Frequency Shifting

$$f_k e^{jk\Omega_0 m} \iff F_{r-m} \tag{8.25}$$

4. Cyclic Convolution

$$f_k * g_k \iff F_r G_r \tag{8.26a}$$

and

$$f_k g_k \iff \frac{1}{N_0} F_r * G_r \tag{8.26b}$$

For two N_0-periodic sequences f_k and g_k, the cyclic convolution is defined by

$$f_k * g_k = \sum_{n=0}^{N_0-1} f_n g_{k-n} = \sum_{n=0}^{N_0-1} g_n f_{k-n} \tag{8.27}$$

For nonperiodic sequences, the convolution can be visualized in terms of two sequences, in which one sequence is fixed and the other sequence is inverted and moved past the fixed sequence, one digit at a time (Fig. 2.23). If the two sequences are N_0-periodic, the same configuration will repeat after N_0 shifts of the sequence. Clearly the convolution $f_k * g_k$ becomes N_0-periodic (cyclic), and it can be conveniently visualized as shown in Fig. 8.12 for the case of $N_0 = 4$. The inner clockwise sequence f_k is fixed, and the outer counterclockwise sequence g_k is rotated clockwise one unit at a time. We multiply the overlapping numbers and add. For example, the value of $f_k * g_k$ at $k = 0$ (Fig. 8.12) is

$$f_0 g_0 + f_1 g_3 + f_2 g_2 + f_3 g_1$$

and the value of $f_k * g_k$ at $k = 1$ is (Fig. 8.12)

$$f_0 g_1 + f_1 g_0 + f_2 g_3 + f_3 g_2$$

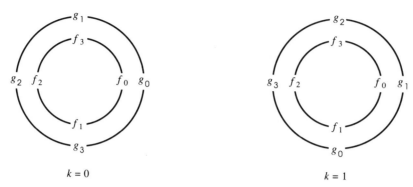

$$k = 0 \qquad\qquad\qquad\qquad\qquad k = 1$$

Fig. 8.12   Graphical picture of cyclic convolution.

Applications of DFT

The DFT is useful not only in the computation of direct and inverse Fourier transforms, but also in other applications such as convolution, correlation, and filtering. Use of the efficient FFT algorithm discussed in Sec. 8.3-2 makes it particularly appealing.

Convolution

Let $f(t)$ and $g(t)$ be the two signals to be convolved. In general, these signals may have different time duration. To convolve them by using their samples, they must be sampled at the same rate (not below the Nyquist rate of either signal). If f_k $(0 \le k \le N_1 - 1)$ and g_k $(0 \le k \le N_2 - 1)$ are the corresponding discrete sequences representing these samples, and if

then†
$$c(t) = f(t) * g(t)$$
$$c_k = f_k * g_k$$

in which c_k is the discrete sequence corresponding to $c(t)$; that is, $c_k = Tc(kT)$.

Because of the width property of the convolution, c_k exists for $0 \le k \le N_1 + N_2 - 1$. To be able to use the DFT technique of cyclic convolution, we must make sure that the cyclic convolution will yield the same result as the linear convolution. This means that the resulting signal of the cyclic convolution must have the same length $(N_1 + N_2 - 1)$ as that of the signal resulting from linear convolution. This can be brought about by adding $N_2 - 1$ dummy samples of zero value to f_k and $N_1 - 1$ dummy samples of zero value to g_k (zero padding). This changes the length of both f_k and g_k to be $N_1 + N_2 - 1$. We can now use the FFT to find the convolution $f_k * g_k$ in three steps, as follows:

1. Find the DFTs F_r and G_r corresponding to f_k and g_k.
2. Multiply F_r by G_r.
3. Find the IDFT of $F_r G_r$. In Chapter 2 we computed $f(t) * h(t)$ numerically for signals, shown in Fig. 2.24a and 2.24b. The reader is encouraged to perform this convolution using the DFT (or FFT).

---

†This follows from the fact that $f_k = Tf(kT), g_k = Tg(kT)$, and $c_k = Tc(kT)$. From Eq. (2.79), $c(kT) = Tf(kT) * g(kT)$. Therefore $c_k = f_k * g_k$.

Filtering

We generally think of filtering in terms of some hardware-oriented solution (namely, building a circuit with RLC components and op amps). However, filtering also has a software-oriented solution [a computer algorithm that yields the filtered output $y(t)$ for a given input $f(t)$]. This can be conveniently accomplished by using the DFT. If $f(t)$ is the signal to be filtered, then F_r, the DFT of f_k, is found. The spectrum F_r is then shaped (filtered) as desired by multiplying F_r by H_r, where H_r are the samples of the filter transfer function $H(\omega)$ [$H_r = H(r\omega_0)$]. Finally, we take the IDFT of $F_r H_r$ to obtain the filtered output y_k [$y_k = \frac{T_0}{N_0} y(kT)$].

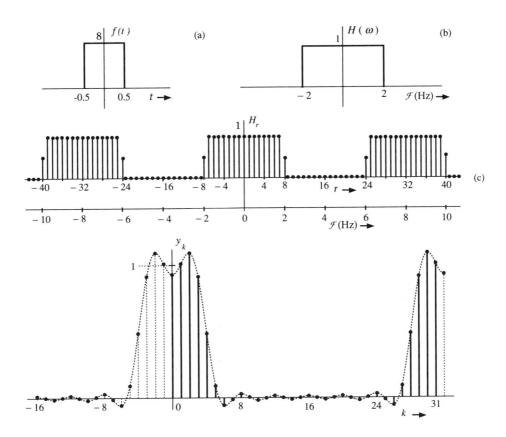

Fig. 8.13 Filtering $f(t)$ through $H(\omega)$.

Consider, for example, the problem of filtering the gate function $f(t)$ in Fig. 8.13a through an ideal lowpass filter $H(\omega)$, as shown in Fig. 8.13b. We have already found the 32 point DFT of $f(t)$ (see Fig. 8.11d). Next we multiply F_r by H_r. To compute H_r, we remember that in computing the 32 point DFT of $f(t)$, we have used $\mathcal{F}_o = \frac{1}{4}$. Because F_r is 32-periodic, H_r must also be 32-periodic with samples separated by $\frac{1}{4}$ Hz. This means that H_r must be repeated every 8 Hz or 16π rad/sec (see Fig. 8.13c). This gives the 32 samples of H_r over $(0 \le \omega \le 16\pi)$ as follows:

No.	fk	Fr	Hr	FrHr	yk
0	1.0	8.000 E0	1.0	8.000 E0	9.285 E-1
1	1.0	7.179 E0	1.0	7.179 E0	1.009 E0
2	1.0	5.027 E0	1.0	5.027 E0	1.090 E0
3	1.0	2.331 E0	1.0	2.331 E0	9.123 E-1
4	0.5	0.000 E0	1.0	0.000 E0	4.847 E-1
5	0.0	-1.323 E0	1.0	-1.323 E0	8.884 E-2
6	0.0	-1.497 E0	1.0	-1.497 E0	-5.698 E-2
7	0.0	-8.616 E-1	1.0	-8.616 E-1	-1.383 E-2
8	0.0	0.000 E0	0.5	0.000 E0	2.933 E-2
9	0.0	5.803 E-1	0.0	0.000 E0	4.837 E-3
10	0.0	6.682 E-1	0.0	0.000 E0	-1.966 E-2
11	0.0	3.778 E-1	0.0	0.000 E0	-2.156 E-3
12	0.0	0.000 E0	0.0	0.000 E0	1.534 E-2
13	0.0	-2.145 E-1	0.0	0.000 E0	9.828 E-4
14	0.0	-1.989 E-1	0.0	0.000 E0	-1.338 E-2
15	0.0	-6.964 E-2	0.0	0.000 E0	-2.876 E-4
16	0.0	0.000 E0	0.0	0.000 E0	1.280 E-2
17	0.0	-6.964 E-2	0.0	0.000 E0	-2.876 E-4
18	0.0	-1.989 E-1	0.0	0.000 E0	-1.338 E-2
19	0.0	-2.145 E-1	0.0	0.000 E0	9.828 E-4
20	0.0	0.000 E0	0.0	0.000 E0	1.534 E-2
21	0.0	3.778 E-1	0.0	0.000 E0	-2.156 E-3
22	0.0	6.682 E-1	0.0	0.000 E0	-1.966 E-2
23	0.0	5.803 E-1	0.0	0.000 E0	4.837 E-3
24	0.0	0.000 E0	0.5	0.000 E0	2.933 E-2
25	0.0	-8.616 E-1	1.0	-8.616 E-1	-1.383 E-2
26	0.0	-1.497 E0	1.0	-1.497 E0	-5.698 E-2
27	0.0	-1.323 E0	1.0	-1.323 E0	8.884 E-2
28	0.5	0.000 E0	1.0	0.000 E0	4.847 E-1
29	1.0	2.331 E0	1.0	2.331 E0	9.123 E-1
30	1.0	5.027 E0	1.0	5.027 E0	1.090 E0
31	1.0	7.179 E0	1.0	7.179 E0	1.009 E0

Table 8.1

$$H_r = \begin{cases} 1 & 0 \leq r \leq 7 \quad \text{and} \quad 25 \leq r \leq 31 \\ 0 & 9 \leq r \leq 23 \\ 0.5 & r = 8, 24 \end{cases}$$

We multiply F_r by H_r and take the IDFT. The resulting output signal is shown in Fig. 8.13d. Table 8.1 shows a printout of f_k, F_r, H_r, Y_r, and y_k.

⊙ **Computer Example C8.2**

Filtering can be done on a computer in software, much like it can be done in hardware (e.g. using op-amps, resistors, and capacitors). However, in software it is often much easier to implement elaborate filters, as we only need to program their equations. We shall consider filtering a square pulse in Example 8.1 through an ideal lowpass filter of transfer function $H(\omega)$ shown in Fig. 8.13b. The DFT of the filter output is Yr=FrHr, where Fr is the DFT of $f(t)$ (already computed in Example C8.1 and stored in a file 'C81.m') and Hr are the samples of the ideal lowpass filter transfer function $H(\omega)$. Since Fr is already computed and stored, we would import it by a command 'C81.m'.

```
C81; % import Fr, the DFT of f(k) from the previous Example C8.1.
for r=1:32,    % Create the discrete time index r.
    if r<=8
      Hr(r)=1;
    end
    if r==9
      Hr(r)=0.5;
  end
    if r>=10,
      if k<=24
        Hr(r)=0;
```

```
            end
         end
         if r==25
            Hr(r)=0.5;
         end
         if r>=26,
            if r<=32
               Hr(r)=1;
            end
         end
      end
   pause
   yk=ifft(Fr.*Hr); % The output samples yk are the inverse DFT of FrHr.
   plot(k,ifft(Fr.*Hr),'o'),grid,
   xlabel('r'),ylabel('Yr'),
   title('Plot of y[k]'),pause,
   disp('Table of k, fk, and Fr is shown next.  Press a key')
   pause
   [k' fk' Fr']
   disp('Table of Hr, FrHr and yk is shown next.  Press a key')
   pause
   [Hr' (Fr.*Hr)' yk']  ⊙
```

8.3-2   The Fast Fourier Transform (FFT)

The number of computations required in performing the DFT was dramatically reduced by an algorithm developed by Tukey and Cooley in 1965.[4] This algorithm, known as the **Fast Fourier transform** (FFT), reduces the number of computations from something on the order of N_0^2 to $N_0 \log N_0$. To compute one sample F_r from Eq. (8.12a), we require N_0 complex multiplications and $N_0 - 1$ complex additions. To compute N_0 such values (F_r for $r = 0, 1, \cdots, N_0 - 1$), we require a total of N_0^2 complex multiplications and $N_0(N_0 - 1)$ complex additions. For a large N_0, this can be prohibitively time-consuming, even for a very high-speed computer.

Although there are many variations of the original Tukey-Cooley algorithm, these can be grouped into two basic types: **decimation-in-time** and **decimation-in-frequency**. The algorithm is simplified if we choose N_0 to be a power of 2, although this is not essential. To simplify the notation, let

$$W_{N_0} = e^{-(j2\pi/N_0)} = e^{-j\Omega_0} \tag{8.28}$$

so that

$$F_r = \sum_{k=0}^{N_0-1} f_k W_{N_0}^{kr} \qquad 0 \le r \le N_0 - 1 \tag{8.29a}$$

and

$$f_k = \frac{1}{N_0} \sum_{r=0}^{N_0-1} F_r W_{N_0}^{-kr} \qquad 0 \le k \le N_0 - 1 \tag{8.29b}$$

The Decimation-in-Time Algorithm

Here we divide the N_0-point data sequence f_k into two $(\frac{N_0}{2})$-point sequences consisting of even- and odd-numbered samples respectively, as follows:

$$\underbrace{f_0, f_2, f_4, \cdots, f_{N_0-2}}_{\text{sequence } g_k}, \underbrace{f_1, f_3, f_5, \cdots, f_{N_0-1}}_{\text{sequence } h_k}$$

Then from Eq. (8.29a),

$$F_r = \sum_{k=0}^{\frac{N_0}{2}-1} f_{2k} W_{N_0}^{2kr} + \sum_{k=0}^{\frac{N_0}{2}-1} f_{2k+1} W_{N_0}^{(2k+1)r} \tag{8.30}$$

Also, since

$$W_{\frac{N_0}{2}} = W_{N_0}^2 \tag{8.31}$$

we have

$$F_r = \sum_{k=0}^{\frac{N_0}{2}-1} f_{2k} W_{\frac{N_0}{2}}^{kr} + W_{N_0}^r \sum_{k=0}^{\frac{N_0}{2}-1} f_{2k+1} W_{\frac{N_0}{2}}^{kr}$$

$$= G_r + W_{N_0}^r H_r \qquad 0 \le r \le N_0 - 1 \tag{8.32}$$

in which G_r and H_r are the $(\frac{N_0}{2})$-point DFTs of the even- and odd-numbered sequences, g_k and h_k, respectively. Also, G_r and H_r, being the $(\frac{N_0}{2})$-point DFTs, are $(\frac{N_0}{2})$-periodic. Hence

$$G_{r+(\frac{N_0}{2})} = G_r$$

$$H_{r+(\frac{N_0}{2})} = H_r \tag{8.33}$$

Moreover,

$$W_{N_0}^{r+(\frac{N_0}{2})} = W_{N_0}^{\frac{N_0}{2}} W_{N_0}^r = e^{-j\pi} W_{N_0}^r = -W_{N_0}^r \tag{8.34}$$

From Eqs. (8.32), (8.33), and (8.34), we obtain

$$F_{r+(\frac{N_0}{2})} = G_r - W_{N_0}^r H_r \tag{8.35}$$

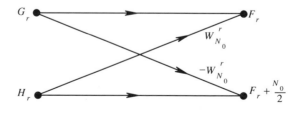

Fig. 8.14 Butterfly.

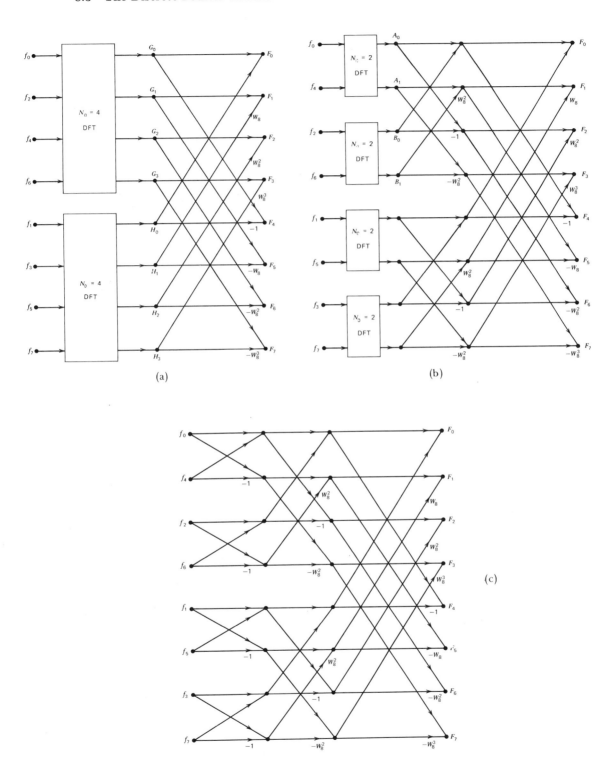

Fig. 8.15 Successive steps in 8-point FFT.

This property can be used to reduce the number of computations. We can, for example, compute F_r using Eq. (8.32) for the first $(\frac{N_0}{2})$ points $(0 \leq n \leq \frac{N_0}{2} - 1)$ and using Eq. (8.35) for the last $\frac{N_0}{2}$ points:

$$F_r = G_r + W_{N_0}^r H_r \qquad 0 \leq r \leq \frac{N_0}{2} - 1 \tag{8.36a}$$

$$F_{r+(\frac{N_0}{2})} = G_r - W_{N_0}^r H_r \qquad 0 \leq r \leq \frac{N_0}{2} - 1 \tag{8.36b}$$

Thus an N_0-point DFT can be computed by combining the two $(\frac{N_0}{2})$-point DFTs, as in Eq. (8.36). These equations can be represented conveniently by the **signal flow** graph shown in Fig. 8.14. This structure is known as a **butterfly**. Figure 8.15a shows the implementation of Eqs. (8.33) for the case of $N_0 = 8$.

The next step is to compute the $(\frac{N_0}{2})$-point DFTs G_r and H_r. We repeat the same procedure by dividing g_k and h_k into two $(\frac{N_0}{4})$-point sequences corresponding to the even- and odd-numbered samples. Then we continue this process until we reach the one-point DFT. Figures 8.15a, 8.15b, and 8.15c show these steps for the case of $N_0 = 8$. Figure 8.15c shows that the two-point DFTs require no multiplication.

To count the number of computations required in the first step, assume that G_r and H_r are known. From Eqs. (8.36), it is clear that to compute all the N_0 points of the F_r, we require N_0 complex additions and $\frac{N_0}{2}$ complex multiplications† (corresponding to $W_{N_0}^r H_r$).

In the second step, to compute the $(\frac{N_0}{2})$-point DFT G_r from the $(\frac{N_0}{4})$-point DFT, we require $\frac{N_0}{2}$ complex additions and $\frac{N_0}{4}$ complex multiplications. We require an equal number of computations for H_r. Hence, in the second step, there are N_0 complex additions and $\frac{N_0}{2}$ complex multiplications. The number of computations required remains the same in each step. Since a total of $\log_2 N_0$ steps is needed to arrive at a one-point DFT, we require, conservatively, a total of $N_0 \log_2 N_0$ complex additions and $(\frac{N_0}{2}) \log_2 N_0$ complex multiplications, to compute the N_0-point DFT.

The procedure for obtaining IDFT is identical to that used to obtain the DFT except that $W_{N_0} = e^{j(2\pi/N_0)}$ instead of $e^{-j(2\pi/N_0)}$ (in addition to the multiplier $1/N_0$). Another FFT algorithm, the **decimation-in-frequency** algorithm, is similar to the decimation-in-time algorithm. The only difference is that instead of dividing f_k into two sequences of even- and odd-numbered samples, we divide f_k into two sequences formed by the first $\frac{N_0}{2}$ and the last $\frac{N_0}{2}$ digits, proceeding in the same way until a single-point DFT is reached in $\log_2 N_0$ steps. The total number of computations in this algorithm is the same as that in the decimation-in-time algorithm.

8.4   SUMMARY

A signal bandlimited to B Hz can be reconstructed exactly from its samples if the sampling rate $R > 2B$ Hz (the sampling theorem). Such a reconstruction,

---

†Actually, $\frac{N_0}{2}$ is a conservative figure because some multiplications corresponding to the cases in which $W_{N_0}^r = 1, j$, etc., are eliminated.

although possible theoretically, poses practical problems such as the need for filters with zero gain over a band (or bands) of frequencies. Such filters are unrealizable, or realizable with infinite time delay. Therefore, in practice, there is always an error in reconstructing a signal from its samples. Moreover, practical signals are not bandlimited, which causes an additional error (aliasing error) in signal reconstruction from its samples. Aliasing error can be eliminated by bandlimiting a signal to its effective bandwidth.

The sampling theorem is very important in signal analysis, processing, and transmission because it allows us to replace a continuous-time signal by a discrete sequence of numbers. Processing a continuous-time signal is therefore equivalent to processing a discrete sequence of numbers. This leads us directly into the area of digital filtering (discrete-time systems). In the field of communication, the transmission of a continuous-time message reduces to the transmission of a sequence of numbers. This opens doors to many new techniques of communicating continuous-time signals by pulse trains.

The dual of the sampling theorem states that for a signal timelimited to τ seconds, its spectrum $F(\omega)$ can be reconstructed from the samples of $F(\omega)$ taken at a uniform intervals not greater than $1/\tau$ Hz. In other words the spectrum should be sampled at a rate not less than τ samples/Hz.

To compute the direct or inverse Fourier transform numerically, we need relationship between the samples of $f(t)$ and $F(\omega)$. The sampling theorem and its dual provide such a quantitative relationship in the form of discrete Fourier transform (DFT). The DFT computations are greatly facilitated by fast Fourier transform (FFT) algorithm, which reduces the number of computations from something of the order of N_0^2 to $N_0 \log N_0$.

REFERENCES

1. Linden, D.A., "A Discussion of Sampling Theorem," *Proc. IRE*, vol. 47, pp 1219-1226, July 1959.

2. Bracewell, R.N., *The Fourier Transform and Its Applications, 2nd revised ed.*, McGraw-Hill, New York, 1986.

3. Bennett, W.R., *Introduction to Signal Transmission*, McGraw-Hill, New York, 1970.

4. Cooley, J.W., and J.W., Tukey, "An Algorithm for the Machine Calculation of Complex Fourier Series", *Mathematics of Computation*, Vol. 19, pp 297-301, April 1965.

PROBLEMS

8.1-1 Figure P8.1-1 shows Fourier spectra of signals $f_1(t)$ and $f_2(t)$. Determine the Nyquist sampling rates for signals $f_1(t)$, $f_2(t)$, $f_1^2(t)$, $f_2^3(t)$, and $f_1(t)f_2(t)$.
Hint: Use the frequency convolution and the width property of the convolution.

8.1-2 Determine the Nyquist sampling rate and the Nyquist sampling interval for the signals
(a) $\text{sinc}^2(100\pi t)$ **(b)** $\text{sinc}(100\pi t) + 3\,\text{sinc}^2(60\pi t)$ **(c)** $\text{sinc}(50\pi t)\text{sinc}(100\pi t)$.

8.1-3 Prove that a signal cannot be simultaneously time-limited and bandlimited.

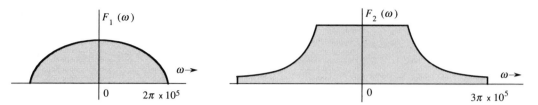

Fig. P8.1-1

Hint: Show that contrary assumption leads to contradiction. Assume a signal simultaneously timelimited and bandlimited so that $F(\omega) = 0$ for $|\omega| > 2\pi B$. In this case $F(\omega) = F(\omega)\text{rect}\left(\frac{\omega}{4\pi B'}\right)$ for $B' > B$. This means that $f(t)$ is equal to $f(t) * 2B'\text{sinc}(2\pi B't)$. The latter cannot be timelimited because the sinc function tail extends to infinity.

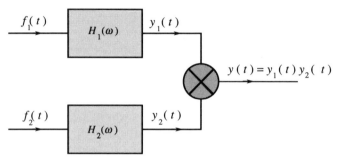

Fig. P8.1-4

8.1-4 Signals $f_1(t) = 10^4\text{rect}(10^4 t)$ and $f_2(t) = \delta(t)$ are applied at the inputs of ideal lowpass filters $H_1(\omega) = \text{rect}\left(\frac{\omega}{40,000\pi}\right)$ and $H_2(\omega) = \text{rect}\left(\frac{\omega}{20,000\pi}\right)$ (Fig. P8.1-4). The outputs $y_1(t)$ and $y_2(t)$ of these filters are multiplied to obtain the signal $y(t) = y_1(t)y_2(t)$.
(a) Sketch $F_1(\omega)$ and $F_2(\omega)$.
(b) Sketch $H_1(\omega)$ and $H_2(\omega)$.
(c) Sketch $Y_1(\omega)$ and $Y_2(\omega)$.
(d) Find the Nyquist sampling rate of $y_1(t), y_2(t)$, and $y(t)$.
Hint for part **(d)**: Use the convolution property and the width property of convolution to determine the bandwidth of $y_1(t)y_2(t)$. See Prob. 8.1-1.

8.1-5 For the signal $e^{-at}u(t)$, determine the bandwidth of an anti-aliasing filter if the essential bandwidth of the signal contains 99% of the signal energy.

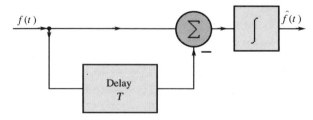

Fig. P8.1-6

8.1-6 A zero-order hold circuit (Fig. P8.1-6) is often used to reconstruct a signal $f(t)$ from its samples.
(a) Find the unit impulse response of this circuit.
(b) Find the transfer function $H(\omega)$, and sketch $|H(\omega)|$.
(c) Show that when a sampled signal $\bar{f}(t)$ is applied at the input of this circuit, the output is a staircase approximation of $f(t)$. The sampling interval is T.

8.1-7 A compact disc (CD) records audio signals digitally by using PCM. Assume the audio signal bandwidth to be 15 kHz.
(a) What is the Nyquist sampling rate?
(b) If the Nyquist samples are quantized into 65536 levels ($L = 65536$) and then binary-coded, determine the number of binary digits required to encode a sample.
(c) Determine the number of binary digits/second (bits/second) required to encode the audio signal.
(d) For practical reasons discussed in the text, signals are sampled at a rate well above the Nyquist rate. Practical CDs use 44100 samples/second. If $L = 65536$, determine the number of pulses/second required to encode the signal.

8.1-8 A TV signal (video and audio) has a bandwidth of 4.5 MHz. This signal is sampled, quantized and binary-coded to obtain a PCM (pulse code modulated) signal.
(a) Determine the sampling rate if the signal is to be sampled at a rate 20% above the Nyquist rate.
(b) If the samples are quantized into 1024 levels, determine the number of binary pulses required to encode each sample.
(c) Determine the binary pulse rate (bits/second) of the binary coded signal.

8.3-1 Starting from Eqs. (8.12a) and (8.12b), prove the linearity [(Eq. 8.23)], time-shifting [Eq. (8.24)], frequency-shifting [Eq. (8.25)], and cyclic convolution [Eq. (8.26)] properties of the DFT.

8.3-2 For a signal $f(t)$ that is timelimited to 10 ms and has an essential bandwidth of 10 kHz, determine N_0, the number of signal samples necessary to compute a power of 2-FFT with a frequency resolution $\mathcal{F}_o$ of at least 50 Hz. Explain if any zero padding is necessary.

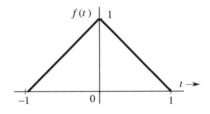

Fig. P8.3-3

8.3-3 To compute the DFT corresponding to the signal $f(t)$ in Fig. P8.3-3, write the sequence f_k (for $k = 0$ to $N_0 - 1$) if the frequency resolution $\mathcal{F}_o$ must be at least 0.25 Hz and if the folding frequency [essential bandwidth of $f(t)$] must be at least 3 Hz. Do not compute the DFT; just write the appropriate sequence f_k.

8.3-4 Choose appropriate values for N_0 and T and compute the FFT of the signal $e^{-t}u(t)$. Hint: The choice of N_0 and T is not unique; it will depend on your assumptions. What is important here is the reasoning that you use to arrive at your answer.

8.3-5 Repeat Problem 8.3-4 for the signal

$$f(t) = \frac{2}{t^2 + 1}$$

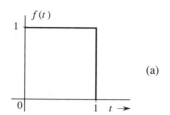

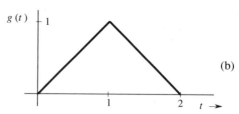

(a)

(b)

Fig. P8.3-6

Hint: $\frac{2}{t^2+1} \Longleftrightarrow 2\pi e^{-|\omega|}$

8.3-6 For the functions $f(t)$ and $g(t)$ shown in Fig. P8.3-6, write the appropriate sequences f_k and g_k necessary for the computation of the fast convolution of $f(t)$ and $g(t)$. Use $T = \frac{1}{8}$.

COMPUTER PROBLEMS

C8-1 Write a routine to band pass filter the triangular wave in Example 6.2 (Fig. 6.4a). Show the frequency and time domain plots of your input and output, and show the frequency domain plot of your filter.

Analysis Of
Discrete-Time Signals

In this chapter we analyze discrete-time signals. Our approach is parallel to that used for continuous-time signals. We first represent a periodic $f[k]$ as a Fourier series formed by a discrete-time exponential (or sinusoid) and its harmonics. Later we extend this representation to nonperiodic signals $f[k]$ by considering $f[k]$ as a limiting case of a periodic signal with the period approaching infinity.

9.1 DISCRETE-TIME PERIODIC SIGNALS

A discrete-time signal $f[k]$ is said to be **periodic** if

$$f[k] = f[k + N_0] \tag{9.1}$$

for some positive integer N_0. The smallest value of N_0 that satisfies Eq. (9.1) is the **period** of $f[k]$. A periodic signal of period N_0 is also said to be a N_0-periodic signal. Figure 9.1 shows an example of a periodic signal of period 6. Observe that each period contains 6 samples (or values). If we consider the first cycle to start at $k = 0$, the last sample (or value) in this cycle is at $k = N_0 - 1 = 5$ (not at $k = N_0 = 6$). Note also that by definition, a periodic signal must begin at $k = -\infty$ (everlasting signal) for the reasons discussed in Sec. 6.1.

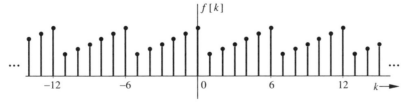

Fig. 9.1 Discrete-time periodic signal.

565

9.1-1   Representation of Periodic Signals: Discrete-Time Fourier Series (DTFS)

A continuous-time sinusoid $\cos \omega t$ is a periodic signal regardless of the value of ω. Such is not the case for the discrete-time sinusoid $\cos \Omega k$. A sinusoid $\cos \Omega k$ is periodic only if $\frac{\Omega}{2\pi}$ is a rational number. This can be proved by observing that if this sinusoid is N_0-periodic, then

$$\cos \Omega(k + N_0) = \cos \Omega k$$

This is possible only if

$$\Omega N_0 = 2\pi m, \quad m, \text{ an integer} \tag{9.2}$$

This means that

$$\frac{\Omega}{2\pi} = \frac{m}{N_0} \text{ (a rational number)}$$

A sinusoid $\cos \frac{2\pi}{N_0} k$ (or an exponential $e^{j\frac{2\pi}{N_0}k}$) is an N_0-periodic signal because

$$\cos \frac{2\pi}{N_0}(k + N_0) = \cos \left(\frac{2\pi}{N_0}k + 2\pi\right) = \cos \frac{2\pi}{N_0}k$$

As in the case of continuous-time signals, an N_0-periodic signal $f[k]$ can be represented by a discrete-time Fourier series with fundamental frequency $\Omega_0 = \frac{2\pi}{N_0}$ and its harmonics. We could start with the trigonometric Fourier series and then proceed to the exponential Fourier series, which is preferable because of its compactness and ease of mathematical manipulations. By now, however, we have gained sufficient understanding of the connection between the trigonometric and the exponential functions to be able to dispense with the trigonometric form and tackle the exponential form directly.

The exponential Fourier series consists of harmonics e^{j0k}, $e^{\pm j\Omega_0 k}$, $e^{\pm j2\Omega_0 k}$, $\cdots$, $e^{\pm jn\Omega_0 k}$, $\cdots$, and so on. There would be infinite harmonics, except for the fact shown in Sec. 5.5-1 that discrete-time exponentials whose frequencies are separated by 2π (or integral multiples of 2π) are identical because

$$e^{j(\Omega \pm 2\pi)k} = e^{j\Omega k}e^{\pm 2\pi k} = e^{j\Omega k} \tag{9.3}$$

The consequence of this result is that the rth harmonic is identical to the $(r + N_0)$th harmonic. To show this, let g_n denote the nth harmonic $e^{jn\Omega_0 k}$. Then

$$g_{r+N_0} = e^{j(r+N_0)\Omega_0 k} = e^{j(r\Omega_0 k + 2\pi k)} = e^{jr\Omega_0 k} = g_r \tag{9.4}$$

and

$$g_r = g_{r+N_0} = g_{r+2N_0} = \cdots = g_{r+mN_0} \quad m, \text{ integer} \tag{9.5}$$

Thus the first harmonic is identical to $(N_0 + 1)$st harmonic, the second harmonic is identical to the $(N_0 + 2)$nd harmonic, and so on. This means that there are only N_0 independent harmonics. We may choose these N_0 independent harmonics $e^{jr\Omega_0 k}$ over $0 \le r \le N_0 - 1$, or $-1 \le r \le N_0 - 2$, or $1 \le r \le N_0$, or any other suitable choice for that matter. Let us take the first choice ($0 \le r \le N_0 - 1$). This corresponds to exponentials $e^{jr\Omega_0 k}$ for $r = 0, 1, 2, \ldots, N_0 - 1$. The Fourier series consists of only these N_0 harmonics, and can be expressed as

$$f[k] = \sum_{r=0}^{N_0-1} \mathcal{D}_r e^{jr\Omega_0 k} \qquad \Omega_0 = \frac{2\pi}{N_0} \tag{9.6}$$

We now show that the coefficients $\mathcal{D}_r$ are given by

$$\mathcal{D}_r = \frac{1}{N_0} \sum_{k=0}^{N_0-1} f[k] e^{-jr\Omega_0 k} \tag{9.7}$$

To derive this result we use the following equation, which will be proved on p. 569 in order to avoid interruption of our discussion:

$$\sum_{k=0}^{N_0-1} e^{jm\Omega_0 k} = \begin{cases} N_0 & m = 0, \pm N_0, \pm 2N_0, \cdots \\ 0 & \text{otherwise} \end{cases} \tag{9.8}$$

To compute coefficients $\mathcal{D}_r$ in the Fourier series (9.6), we multiply both sides of (9.6) by $e^{-jm\Omega_0 k}$ and sum over k from $k = 0$ to $(N_0 - 1)$.

$$\sum_{k=0}^{N_0-1} f[k] e^{-jm\Omega_0 k} = \sum_{k=0}^{N_0-1} \sum_{r=0}^{N_0-1} \mathcal{D}_r e^{j(r-m)\Omega_0 k} \tag{9.9}$$

The right-hand sum, after interchanging the order of summation, results in

$$\sum_{r=0}^{N_0-1} \mathcal{D}_r \left[\sum_{k=0}^{N_0-1} e^{j(r-m)\Omega_0 k} \right]$$

The inner sum, according to Eq. (9.8), is zero for all values of $r \neq m$. The sum is nonzero and has a value $\mathcal{D}_r N_0 = \mathcal{D}_m N_0$ only when $r = m$. The outside sum has only one term $\mathcal{D}_m N_0$ (corresponding to $r = m$), and therefore the right-hand side is equal to $\mathcal{D}_m N_0$. Therefore

$$\sum_{k=0}^{N_0-1} f[k] e^{-jm\Omega_0 k} = \mathcal{D}_m N_0$$

and

$$\mathcal{D}_m = \frac{1}{N_0} \sum_{k=0}^{N_0-1} f[k] e^{-jm\Omega_0 k}$$

We now have a discrete-time Fourier series representation of an N_0-periodic signal $f[k]$ as

$$f[k] = \sum_{r=0}^{N_0-1} \mathcal{D}_r e^{jr\Omega_0 k} \tag{9.10}$$

where

$$\mathcal{D}_r = \frac{1}{N_0} \sum_{k=0}^{N_0-1} f[k] e^{-jr\Omega_0 k} \qquad \Omega_0 = \frac{2\pi}{N_0} \tag{9.11}$$

The Fourier series consists of N_0 components

$$\mathcal{D}_0, \ \mathcal{D}_1 e^{j\Omega_0 k}, \ \mathcal{D}_2 e^{j2\Omega_0 k}, \ \cdots, \ \mathcal{D}_{N_0-1} e^{j(N_0-1)\Omega_0 k}$$

The frequencies of these components are $0, \Omega_0, 2\Omega_0, \ldots, (N_0 - 1)\Omega_0$ where $\Omega_0 = 2\pi/N_0$. In general, the Fourier coefficients $\mathcal{D}_r$ are complex, and they can be represented as

$$\mathcal{D}_r = |\mathcal{D}_r| e^{j\angle \mathcal{D}_r} \tag{9.12}$$

where $|\mathcal{D}_r|$ is the amplitude spectrum and $\angle \mathcal{D}_r$ is the angle spectrum.

The results are very similar to the representation of a continuous-time periodic signal by exponential Fourier series except that, generally, the continuous-time signal spectrum bandwidth is infinite, and consists of infinite number of exponential components. The spectrum of the discrete-time periodic signal, on the other hand, is bandlimited and has at most N_0 components.

The set of Eqs. (9.10) and (9.11) defining the DTFS is identical (except for a scaling factor N_0) to the set of Eqs. (8.12) defining the DFT (discrete Fourier transform). Therefore the FFT algorithm discussed in Sec. 8.3-2 can be used for numerical computations of DTFS.

Periodic Extension of the Fourier Spectrum

Note that if $\phi[r]$ is N_0-periodic, then

$$\sum_{r=0}^{N_0-1} \phi[r] = \sum_{r=<N_0>} \phi[r] \tag{9.13}$$

where $r =< N_0 >$ indicates summation over any N_0 successive values of r. This follows because for an N_0-periodic signal $\phi[r]$, the same values repeat at every N_0 interval. Thus the sum over any successive N_0 points is the same. Now if $f[k]$ is N_0-periodic, $f[k]e^{-jr\Omega_0 k}$ is also N_0-periodic because $e^{-jr\Omega_0 k}$ is N_0-periodic. Therefore from Eq. (9.7) it follows that $\mathcal{D}_r$ is also N_0-periodic, as is $\mathcal{D}_r e^{jr\Omega_0 k}$. Now, by using the property (9.13), Eqs. (9.6) and (9.7) can be expressed as

$$f[k] = \sum_{r=<N_0>} \mathcal{D}_r e^{jr\Omega_0 k} \tag{9.14}$$

and

$$\mathcal{D}_r = \frac{1}{N_0} \sum_{k=<N_0>} f[k]e^{-jr\Omega_0 k} \tag{9.15}$$

If we plot $\mathcal{D}_r$ for all values of r (rather than only $0 \leq r \leq N_0 - 1$), then the spectrum $\mathcal{D}_r$ is N_0-periodic. Moreover, Eq. (9.14) shows that $f[k]$ can be synthesized by not only the N_0 exponentials corresponding to $0 \leq r \leq N_0 - 1$, but by any successive N_0 exponentials in this spectrum, starting at any value of r (positive or negative). For this reason it is customary to show the spectrum $\mathcal{D}_r$ for all values of r (not only over the interval $0 \leq r \leq N_0 - 1$). Yet we must remember that to synthesize $f[k]$ from this spectrum, we need to add only N_0 successive components.

The frequency of the exponential $e^{jr\Omega_0 k}$ is $r\Omega_0$. Therefore the frequency range corresponding to $0 \leq r \leq N_0 - 1$ is $0 \leq \Omega \leq (N_0 - 1)\Omega_0$. But $N_0\Omega_0 = 2\pi$, so that on frequency scale Ω, $\mathcal{D}_r$ repeats every 2π interval. These ideas will be clarified by the following examples.

Proof of Eq. (9.8)

Recall that $\Omega_0 N_0 = 2\pi$ and

$$e^{jm\Omega_0 k} = 1 \qquad m = 0, \pm N_0, \pm 2N_0, \cdots$$

Therefore

$$\sum_{k=0}^{N_0-1} e^{jm\Omega_0 k} = \sum_{k=0}^{N_0-1} 1 = N_0 \quad m = 0, \pm N_0, \pm 2N_0, \cdots$$

To compute the sum for other values of m, we note that the sum on the left-hand side of Eq. (9.8) is a geometric progression with common ratio $\alpha = e^{jm\Omega_0}$. Therefore (see Sec. B.10-4)

$$\sum_{k=0}^{N_0-1} e^{jm\Omega_0 k} = \sum_{k=0}^{N_0-1} \alpha^k = \frac{\alpha^{N_0} - 1}{\alpha - 1} \qquad \alpha = e^{jm\Omega_0}$$

$$= \frac{e^{jm\Omega_0 k N_0} - 1}{e^{jm\Omega_0 k} - 1}$$

$$= 0 \qquad (e^{jm\Omega_0 k N_0} = e^{j2\pi mk} = 1)$$

■ **Example 9.1**

Find the discrete-time Fourier series (DTFS) for $f[k] = \cos 0.1\pi k$ (Fig. 9.2a). Sketch the amplitude and phase spectra.

In this case the sinusoid $\cos 0.1\pi k$ is periodic because $\frac{\Omega}{2\pi} = \frac{1}{20}$ is a rational number and the period N_0 is [see Eq. (9.2)].

$$N_0 = m\left(\frac{2\pi}{\Omega}\right) = m\left(\frac{2\pi}{0.1\pi}\right) = 20m$$

The smallest value of m that makes $20m$ an integer is $m = 1$. Therefore the period $N_0 = 20$, so that $\Omega_0 = \frac{2\pi}{N_0} = 0.1\pi$, and

$$f[k] = \sum_{r=0}^{19} \mathcal{D}_r e^{j0.1\pi r k}$$

where, from Eq. (9.11),

$$\mathcal{D}_r = \frac{1}{20} \sum_{k=0}^{19} \cos 0.1\pi k\, e^{-j0.1\pi rk}$$

$$= \frac{1}{40} \sum_{k=0}^{19} \left(e^{j0.1\pi k} + e^{-j0.1\pi k}\right) e^{-j0.1\pi rk}$$

$$= \frac{1}{40} \left[\sum_{k=0}^{19} e^{j0.1\pi k(1-r)} + \sum_{k=0}^{19} e^{-j0.1\pi k(1+r)}\right]$$

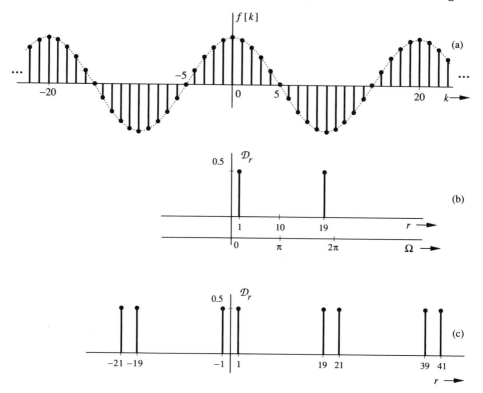

Fig. 9.2   Discrete-time sinusoid $\cos 0.1\pi k$ and its Fourier spectrum.

From Eq. (9.8) it follows that the first summation on right-hand side is zero for all values of r except $r = 1$, when it is equal to $N_0 = 20$. Similarly, the second summation is zero for all values of r except $r = 19$, when it is equal to $N_0 = 20$. Therefore

$$\mathcal{D}_r = \mathcal{D}_{19} = \frac{1}{2}$$

and all other coefficients are zero. Therefore

$$f[k] = \cos 0.1\pi k = \frac{1}{2}\left(e^{j0.1\pi k} + e^{j1.9\pi k}\right) \qquad (9.16a)$$

But

$$e^{j1.9\pi k} = e^{-j2\pi k}\, e^{j1.9\pi k} = e^{-j0.1\pi k}$$

and

$$f[k] = \cos 0.1\pi k = \frac{1}{2}\left(e^{j0.1\pi k} + e^{-j0.1\pi k}\right) \qquad (9.16b)$$

Observe that the result (9.16b) is a trigonometric identity, and could have been found immediately without the formality of finding the Fourier coefficients. The Fourier series is a way of expressing a periodic signal $f[k]$ in terms of exponentials of the form $e^{jr\Omega_0 k}$ and its harmonics. The result in Eqs. (9.16b) is merely a statement of the (obvious) fact that $\cos 0.1\pi k$ can be expressed as a sum of two exponentials $e^{j0.1\pi k}$ and $e^{-j0.1\pi k}$.

In this case $\mathcal{D}_r$ are real, so there is only an amplitude spectrum $|\mathcal{D}_r| = \mathcal{D}_r$. Figure 9.2b shows the sketch of $\mathcal{D}_r$ for the interval $(0 \leq r < 20)$. According to Eq. (9.16a), there are only two components corresponding to $r = 1$ and 19. The rth component $\mathcal{D}_r$ is the

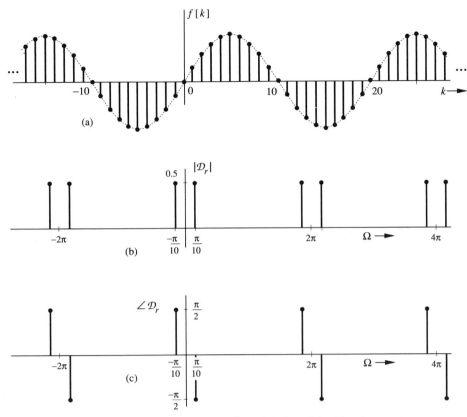

Fig. 9.3   Discrete-time sinusoid $\sin 0.1\pi k$ and its Fourier spectra.

amplitude of the frequency $r\Omega_0 = 0.1r\pi$. Therefore the frequency interval corresponding to $0 \leq r < 20$ is $0 \leq \Omega < 2\pi$, as shown in Fig. 9.2b. Although the spectrum in the interval $0 \leq r < 20$ (or $0 \leq \Omega < 2\pi$) is adequate to specify the frequency-domain description (Fourier series), for the reasons explained on p. 568, we repeat this spectrum periodically with period $N_0 = 20$ (or $\Omega = 2\pi$), as shown in Fig. 9.2c. Because of the periodicity property of the discrete-time exponentials $e^{jr\Omega_0 k}$, the Fourier series components can be selected in any interval of width $N_0 = 20$ (or $\Omega = 2\pi$). For example, if we select an interval $-\pi \leq \Omega < \pi$ (or $-10 \leq r < 10$), we obtain the Fourier series in Eq. (9.16b). ■

■ **Example 9.2**
Find the discrete-time Fourier series (DTFS) for $f[k] = \sin 0.1\pi k$ shown in Fig. 9.3a. In this case we shall find the result directly by using the trigonometric identity

$$\sin 0.1\pi k = \tfrac{1}{2j}\left(e^{j0.1\pi k} - e^{-j0.1\pi k}\right) \tag{9.17}$$

Here the fundamental frequency $\Omega_0 = 0.1\pi$, and there are only two nonzero components:

$$\mathcal{D}_1 = \tfrac{1}{2j} = \tfrac{1}{2}e^{-j\pi/2}$$

$$\mathcal{D}_{-1} = -\tfrac{1}{2j} = \tfrac{1}{2}e^{j\pi/2}$$

Therefore

$$|\mathcal{D}_1| = |\mathcal{D}_{-1}| = \tfrac{1}{2}$$

$$\angle \mathcal{D}_1 = -\frac{\pi}{2} \quad \text{and} \quad \angle \mathcal{D}_{-1} = \frac{\pi}{2}$$

Figures 9.3b and 9.3c show the amplitude and phase plots over the interval $-\pi \le \Omega < \pi$ as well as their periodic extension every 2π interval. Had we used the systematic way of finding the Fourier series, as in Example 9.1, we would have obtained only two nonzero coefficients, $\mathcal{D}_1 = \frac{1}{2j}$ and $\mathcal{D}_{19} = -\frac{1}{2j}$. Periodic repetition of these components would be identical to the spectra in Figs. 9.3b and 8.3c. ■

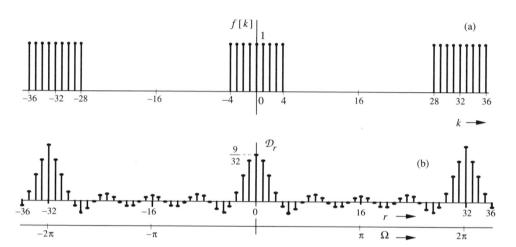

Fig. 9.4 Periodic sampled gate pulse and its Fourier spectrum.

■ **Example 9.3**

Find the discrete-time Fourier series (DTFS) for the periodic sampled gate function shown in Fig. 9.4a.

In this case $N_0 = 32$ and $\Omega_0 = \frac{2\pi}{32} = \frac{\pi}{16}$. Therefore

$$f[k] = \sum_{r=<32>} \mathcal{D}_r e^{jr\frac{\pi}{16}k} \tag{9.18}$$

where

$$\mathcal{D}_r = \frac{1}{32} \sum_{k=<32>} f[k] e^{-jr\frac{\pi}{16}k} \tag{9.19}$$

For our convenience we shall choose the interval $-16 \le k \le 15$ for the summation (9.19), although any other interval of the same width (32 points) would give the same result.

$$\mathcal{D}_r = \frac{1}{32} \sum_{k=-16}^{15} f[k] e^{-jr\frac{\pi}{16}k}$$

Now $f[k] = 1$ for $-4 \le k \le 4$ and is zero for all other values of k. Therefore

$$\mathcal{D}_r = \frac{1}{32} \sum_{k=-4}^{4} e^{-jr\frac{\pi}{16}k}$$

This is a geometric progression with a common ratio $e^{-j\frac{\pi}{16}r}$. Therefore [see Sec. (B.10-4)]

$$\mathcal{D}_r = \frac{1}{32} \left[\frac{e^{-j\frac{5\pi r}{16}} - e^{j\frac{4\pi r}{16}}}{e^{-j\frac{\pi r}{16}} - 1} \right]$$

$$= \left(\frac{1}{32} \right) \frac{e^{-j\frac{0.5\pi r}{16}} \left[e^{-j\frac{4.5\pi r}{16}} - e^{j\frac{4.5\pi r}{16}} \right]}{e^{-j\frac{0.5\pi r}{16}} \left[e^{-j\frac{0.5\pi r}{16}} - e^{j\frac{0.5\pi r}{16}} \right]}$$

$$= \left(\frac{1}{32} \right) \frac{\sin\left(\frac{4.5\pi r}{16} \right)}{\sin\left(\frac{0.5\pi r}{16} \right)}$$

$$= \left(\frac{1}{32} \right) \frac{\sin(4.5r\Omega_0)}{\sin(0.5r\Omega_0)} \qquad \Omega_0 = \frac{\pi}{16} \qquad\qquad (9.20)$$

This spectrum (with its periodic extension) is shown in Fig. 9.4b. In this example we used the same equations as those in DFT in Example 8.1, with a minor difference. In the present example, the values of $f[k]$ at discontinuity (at $k = 4$ and -4) are taken as 1, whereas in Example 8.1 these values are 0.5. This is the reason for the difference in spectra in Fig. 9.4b and Fig. 8.11d. The correct spectrum for the discrete-time signal in Fig. 9.4a is that in Fig. 9.4b. For the continuous-time rectangular pulse signal in Fig. 8.11a, on the other hand, the spectrum in Fig. 8.11d is more accurate than that in Fig. 9.4b. ∎

⊙ Computer Example C9.1

Use a computer to find and plot the discrete-time Fourier series of the periodic sampled gate function shown in Fig. 9.4a (Example 9.3). Use Eq. (9.19) to compute $\mathcal{D}_r$.

```
fk=[zeros(1,12) ones(1,9) zeros(1,11)];
k=-16:15; plot(k,fk,'o'),grid,
xlabel('k'),ylabel('f[k]'),
title('Plot of f[k]'),pause
r=-16:15;  r=r';  omega=pi/16;
    for m=1:length(k)
    Dr=(1/32)*(fk(m)*exp(-j*r*omega*k(m)));
    DRR=[DRR Dr];
    end
    [p,q]=size(DRR); DRTT=0;
    for s=1:p
    DRT=DRR(:,s)+DRTT;
    DRTT=DRT;
    end
plot(r,DRTT,'o'),grid,
xlabel('r'),ylabel('Dr'),
title('Plot of Dr'),pause,
plot(r,abs(DRTT),'o'),grid,
xlabel('r'),ylabel('|Dr|'),
title('|Dr|'),pause
plot(r,angle(DRTT),'o'),grid,
xlabel('r'),ylabel('< Dr'),
title('Angle Dr')   ⊙.
```

△ **Exercise E9.1**

Find the DTFS for

$$f[k] = 4\cos 0.2\pi k + 6\sin 0.5\pi k$$

over the interval $0 \le r \le 19$.

Answer:

$$f[k] = 2e^{j0.2\pi k} + (3e^{-j\pi/2})e^{j0.5\pi k} + (3e^{j\pi/2})e^{j1.5\pi k} + 2e^{j1.8\pi k}$$

Hint: The frequencies of the two sinusoids are 0.2π and 0.5π, which are integral multiples of $\Omega_0 = 0.1\pi$. Therefore $N_0 = 20$.   ▽

9.2 NONPERIODIC SIGNALS: THE DISCRETE-TIME FOURIER TRANSFORM (DTFT)

In this section we extend the results in Sec. 9.1 to represent a nonperiodic discrete-time signal $f[k]$ as a sum of discrete-time exponentials. The approach is parallel to that used for continuous-time signals. By repeating a nonperiodic signal $f[k]$ periodically at intervals of N_0, we generate an N_0-periodic signal $f_{N_0}[k]$ which then can be represented by a discrete-time Fourier series. If we let $N_0 \to \infty$, $f_{N_0}[k] \to f[k]$. Therefore the Fourier series, in the limit $N_0 \to \infty$, represents $f[k]$.

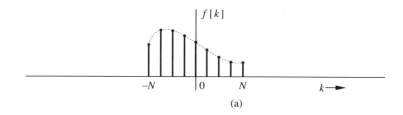

(a)

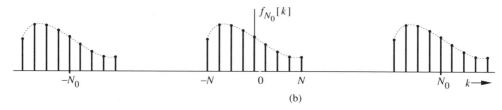

(b)

Fig. 9.5   Generation of a periodic signal by periodic extension of a signal $f[k]$.

Figure 9.5a shows a hypothetical nonperiodic signal $f[k]$. Figure 9.5b shows a periodic signal $f_{N_0}[k]$ obtained by repeating $f[k]$ periodically at every N_0 interval. The periodic signal $f_{N_0}[k]$ can be expressed by a discrete-time Fourier series as

$$f_{N_0}[k] = \sum_{r=\langle N_0\rangle} \mathcal{D}_r e^{jr\Omega_0 k} \qquad \Omega_0 = \frac{2\pi}{N_0} \tag{9.21}$$

where

$$\mathcal{D}_r = \frac{1}{N_0}\sum_{k=-\infty}^{\infty} f[k]e^{-jr\Omega_0 k} \tag{9.22}$$

Observe that the limits in the summation on the right-hand side are taken from $-\infty$ to ∞ rather than from $-\frac{N_0}{2}$ to $\left(\frac{N_0}{2}\right) - 1$ because $f[k] = 0$ for $|k| > \frac{N_0}{2}$.

As $N_0 \to \infty$, $f_{N_0}[k] \to f[k]$, but now, as seen from Eq. (9.22), the Fourier coefficients $\mathcal{D}_r \to 0$. Let us define a function $F(\Omega)$ as

$$F(\Omega) = \sum_{k=-\infty}^{k=\infty} f[k]e^{-j\Omega k} \qquad (9.23)$$

From this definition and Eq. (9.22), we have

$$\mathcal{D}_r = \frac{1}{N_0}F(r\Omega_0) \qquad (9.24)$$

and from Eq. (9.23)

$$F(r\Omega_0) = \sum_{k=-\infty}^{\infty} f[k]e^{-jr\Omega_0 k} \qquad (9.25)$$

Using Eq. (9.24), we can express Eq. (9.21) as

$$f_{N_0}[k] = \frac{1}{N_0} \sum_{r=<N_0>} F(r\Omega_0)e^{jr\Omega_0 k} \qquad (9.26a)$$

$$= \sum_{r=<N_0>} F(r\Omega_0)e^{jr\Omega_0 k}\frac{\Omega_0}{2\pi} \qquad (9.26b)$$

In the limit as $N_0 \to \infty$, $\Omega_0 \to 0$ and $f_{N_0}[k] \to f[k]$. Therefore

$$f[k] = \lim_{\Omega_0 \to 0} \frac{1}{2\pi} \sum_{r=<N_0>} F(r\Omega_0)e^{jr\Omega_0 k}\Omega_0 \qquad (9.27)$$

As Ω_0, the fundamental frequency is reduced, the spectrum becomes denser, but in compensation the coefficients $\mathcal{D}_r$ become smaller [see Eq. (9.24)]. In the limit as $\Omega_0 \to 0$, the spectrum becomes continuous and the coefficients $\mathcal{D}_r \to 0$, as in the continuous-time case. Because Ω_0 is an infinitesimal ($\Omega_0 \to 0$), it would be appropriate to replace it with an infinitesimal notation $\triangle\Omega$:

$$\triangle\Omega = \frac{2\pi}{N_0} \qquad (9.28)$$

Equations (9.27) and (9.24) can be expressed as

$$f[k] = \lim_{\triangle\Omega \to 0} \frac{1}{2\pi} \sum_{r=<N_0>} F(r\triangle\Omega)e^{jr\triangle\Omega k}\triangle\Omega \qquad (9.29)$$

where

$$F(r\triangle\Omega) = \sum_{k=-\infty}^{\infty} f[k]e^{-jr\triangle\Omega k} \qquad (9.30)$$

The range $r =< N_0 >$ implies the interval $N_0\triangle\Omega = 2\pi$ for the frequency variable Ω in Eq. (9.29). The right hand side of Eq. (9.29), by definition, is an integral and therefore

$$f[k] = \frac{1}{2\pi} \int_{2\pi} F(\Omega)e^{jk\Omega}\, d\Omega \tag{9.31}$$

where $\int_{2\pi}$ indicates integration over any continuous interval of 2π. The spectrum $F(\Omega)$ is given by [Eq. (9.23)]

$$F(\Omega) = \sum_{k=-\infty}^{\infty} f[k]e^{-j\Omega k} \tag{9.32}$$

Equation (9.31) shows the representation of $f[k]$ in terms of everlasting exponentials $e^{j\Omega k}$. Observe that

$$F(\Omega + 2\pi) = \sum_{k=-\infty}^{\infty} f[k]e^{-j(\Omega+2\pi)k}$$

$$= \sum_{k=-\infty}^{\infty} f[k]e^{-j\Omega k}e^{-j2\pi k}$$

$$= F(\Omega) \tag{9.33}$$

Clearly, the spectrum $F(\Omega)$ is a continuous and periodic function of Ω with a period 2π. The function $F(\Omega)$ is the direct discrete-time Fourier transform (DTFT) of $f[k]$, and $f[k]$ is the inverse discrete-time Fourier transform (IDTFT) of $F(\Omega)$. This can be conveniently expressed symbolically as

$$f[k] \Longleftrightarrow F(\Omega) \tag{9.34}$$

In Sec. 9.5 we discuss the relationship between the DTFT, the Fourier transform, and the z-transform. Basically, the DTFT is the Fourier transform of a sampled (continuous-time) signal. In Sec. 8.1 we saw that sampling a signal results in periodic repetition of its spectrum. This explains the periodicity of $F(\Omega)$.

Linearity of the DTFT
From Eq. (9.32) it follows that if

$$f_1[k] \Longleftrightarrow F_1(\Omega) \qquad \text{and} \qquad f_2[k] \Longleftrightarrow F_2(\Omega)$$

then

$$a_1 f_1[k] + a_2 f_2[k] \Longleftrightarrow a_1 F_1(\Omega) + a_2 F_2(\Omega) \tag{9.35}$$

Existence of the DTFT
From Eq. (9.32) it follows that the existence of $F(\Omega)$ is guaranteed if $f[k]$ is absolutely summable; that is,

$$\sum_{k=-\infty}^{\infty} |f[k]| < \infty \tag{9.36}$$

This is because the magnitude of $e^{-jk\Omega}$ is unity.

Amplitude and Phase Spectra

If $f[k]$ is a real function of k, then from Eq. (9.32), it follows that

$$F(-\Omega) = F^*(\Omega) \tag{9.37}$$

so that

$$|F(\Omega)| = |F(-\Omega)| \tag{9.38a}$$

$$\angle F(\Omega) = -\angle F(-\Omega) \tag{9.38b}$$

Therefore the amplitude spectrum $|F(\Omega)|$ is an even function of Ω and the phase spectrum $\angle F(\Omega)$ is an odd function of Ω for real $f[k]$. This behavior is identical to that of continuous-time signal spectra.

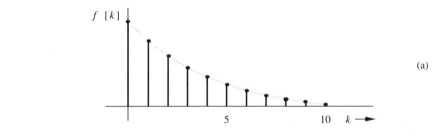
(a)

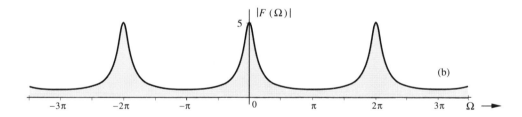
(b)

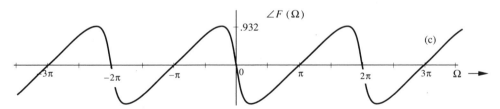
(c)

Fig. 9.6   Discrete-time exponential $a^k u[k]$ and its Fourier spectra.

■ **Example 9.4**

Find the DTFT of $f[k] = a^k u[k]$.

$$F(\Omega) = \sum_{k=0}^{\infty} a^k e^{-j\Omega k}$$

$$= \sum_{k=0}^{\infty} (ae^{-j\Omega})^k$$

This is a geometric progression with a common ratio $ae^{-j\Omega}$. Therefore [see Sec. (B.10-4)]

$$F(\Omega) = \frac{1}{1 - ae^{-j\Omega}}$$

provided that $|ae^{-j\Omega}| < 1$. But because $|e^{-j\Omega}| = 1$, this condition implies $|a| < 1$. Therefore

$$F(\Omega) = \frac{1}{1 - ae^{-j\Omega}} \qquad |a| < 1 \tag{9.39}$$

If $|a| > 1$, $F(\Omega)$ does not converge. This result is in conformity with condition (9.36). From Eq. (9.39)

$$F(\Omega) = \frac{1}{1 - a\cos\Omega + ja\sin\Omega} \tag{9.40}$$

so that

$$|F(\Omega)| = \frac{1}{\sqrt{(1 - a\cos\Omega)^2 + (a\sin\Omega)^2}} \tag{9.41a}$$

$$= \frac{1}{\sqrt{1 + a^2 - 2a\cos\Omega}}$$

$$\angle F(\Omega) = -\tan^{-1}\left[\frac{a\sin\Omega}{1 - a\cos\Omega}\right] \tag{9.41b}$$

Figure 9.6 shows $f[k] = a^k u[k]$ and its spectra for $a = 0.8$. Observe that $|F(\Omega)|$ is an even function of Ω and $\angle F(\Omega)$ is an odd function of Ω, as expected. These spectra are identical to those in Example 5.9 (Fig. 5.9). The reason for this interesting behavior is discussed in Sec. 9.2-1. ■

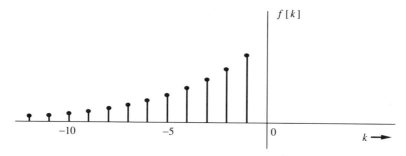

Fig. 9.7   Exponential $a^k u[-(k+1)]$.

■ **Example 9.5**
   Find the DTFT of $a^k u[-(k+1)]$ shown in Fig. 9.7.

$$F(\Omega) = \sum_{k=-\infty}^{\infty} a^k[-(k+1)]e^{-j\Omega k}$$

$$= \sum_{k=-\infty}^{-1} (ae^{-j\Omega})^k = \sum_{m=1}^{\infty} (\frac{1}{a}e^{j\Omega})^m$$

$$= \frac{1}{a}e^{j\Omega} + \left(\frac{1}{a}e^{j\Omega}\right)^2 + \left(\frac{1}{a}e^{j\Omega}\right)^3 + \cdots$$

This is a geometric progression with a common ratio $e^{j\Omega}/a$. Therefore, from Sec. B.10-4,

$$F(\Omega) = \frac{1}{ae^{-j\Omega} - 1} \qquad |a| > 1 \tag{9.42}$$

$$= \frac{1}{(a\cos\Omega - 1) - ja\sin\Omega}$$

Therefore

$$|F(\Omega)| = \frac{1}{\sqrt{1 + a^2 - 2a\cos\Omega}}$$

$$\angle F(\Omega) = \tan^{-1}\left[\frac{a\sin\Omega}{a\cos\Omega - 1}\right] \qquad \blacksquare$$

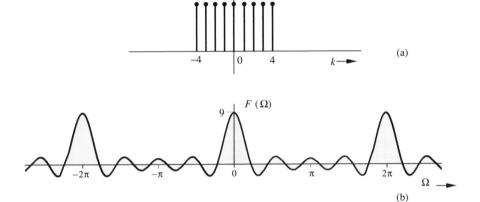

Fig. 9.8 Discrete-time gate pulse and its Fourier spectrum.

■ **Example 9.6**

Find the DTFT of the discrete-time rectangular pulse shown in Fig. 9.8a.

$$F(\Omega) = \sum_{k=-\infty}^{\infty} f[k]e^{-j\Omega k}$$

$$= \sum_{k=-4}^{4} (e^{-j\Omega})^k$$

This is a geometric progression with a common ratio $e^{-j\Omega}$ and [see Sec. B.10-4]

$$F(\Omega) = \frac{e^{-j5\Omega} - e^{j4\Omega}}{e^{-j\Omega} - 1}$$

$$= \frac{e^{-j\Omega/2}(e^{-j4.5\Omega} - e^{j4.5\Omega})}{e^{-j\Omega/2}(e^{-j\Omega/2} - e^{j\Omega/2})}$$

$$= \frac{\sin(4.5\Omega)}{\sin(0.5\Omega)} \tag{9.43}$$

Figure 9.8b shows the spectrum $F(\Omega)$.

Note that the spectrum $\mathcal{D}_r$ in Fig. 9.4b [Eq. (9.20)] is a sampled version of $F(\Omega)$ in Fig. 9.8b [Eq. (9.43)]:

$$\mathcal{D}_r = \frac{1}{32}F(r\Omega_0) \qquad \Omega_0 = \frac{\pi}{16}$$

Therefore $F(\Omega)$ in Fig. 9.8b is the envelope of $\mathcal{D}_r$ (within a multiplicative constant 32) in Fig. 9.4b. The reason for this lies in the fact that the signal $f[k]$ in Fig. 9.4a is formed by periodically repeating the signal $f[k]$ in Fig. 9.8a. We showed in Sec. 8.2 that periodic repetition of a signal causes the spectrum to be sampled (see Fig. 8.9). The proof of this result is given again in Sec. 9.5. ■

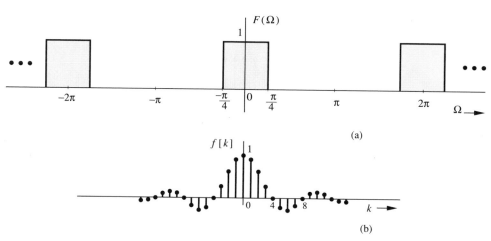

Fig. 9.9 Inverse Discrete-time Fourier transform of a periodic gate spectrum.

■ **Example 9.7**

Find the inverse DTFT of the rectangular pulse spectrum $F(\Omega) = \text{rect}\left(\frac{\Omega}{2\Omega_c}\right)$ with $\Omega_c = \frac{\pi}{4}$ and repeating at the intervals of 2π, as shown in Fig. 9.9a.

From Eq. (9.31)

$$f[k] = \frac{1}{2\pi}\int_{-\pi}^{\pi} F(\Omega)e^{jk\Omega}\,d\Omega$$

$$= \frac{1}{2\pi}\int_{-\Omega_c}^{\Omega_c} e^{jk\Omega}\,d\Omega$$

$$= \frac{1}{j2\pi k}e^{jk\Omega}\Big|_{-\Omega_c}^{\Omega_c}$$

$$= \frac{\sin(\Omega_c k)}{\pi k}$$

$$= \frac{\Omega_c}{\pi}\,\text{sinc}\,(\Omega_c k) \tag{9.44}$$

The signal $f[k]$ is shown in Fig. 9.9b (for the case $\Omega_c = \pi/4$). ■

⊙ **Computer Example C9.2**

Using a computer, find the inverse DTFT of the rectangular pulse spectrum in Fig. 9.9a (Example 9.7).

```
% This routine calculates the inverse DTFT.
Fk=0;          % Initialize the intermediate variable f[k].
deltaOmega=0.1;
    for k=-16:16
        for r=-8:deltaOmega:8
          fk=exp(j*r*deltaOmega*k)*deltaOmega;
          Fk=Fk+fk;
        end
          Fk=(1/(2*pi))*Fk;
          FK(k+17)=Fk;
          Fk=0;
    end
k=-16:16;
plot(k,FK,'o'),grid,
title('Plot of IDFT (f[k])'),   ⊙
```

△ **Exercise E9.2**

Find and sketch the amplitude and phase spectra of the DTFT of the signal $f[k] = a^{|k|}$ with $|a| < 1$.

Answer:

$$F(\Omega) = \frac{1 - a^2}{1 - 2a\cos\Omega + a^2}$$

Hint:

$$f[k] = \begin{cases} a^k & k \geq 0 \\ a^{-k} & k \leq -1 \end{cases} \qquad \triangledown$$

9.2-1 DTFT of a signal is its $\mathcal{Z}$-Transform with $z = e^{j\Omega}$

The (bilateral) z-transform of a signal $f[k]$ is given by

$$F[z] = \sum_{k=-\infty}^{\infty} f[k]z^{-k}$$

From Eq. (9.32), it is clear that

$$F(\Omega) = F[z]\big|_{z=e^{j\Omega}}$$
$$= F[e^{j\Omega}]$$

Therefore the DTFT of a signal is its z-transform with $z = e^{j\Omega}$ (z is restricted to the unit circle in the z-plane). This is readily verified from Example 9.4. The z-transform of $a^k u[k]$ is (Table 5.1, Pair 7)

$$F[z] = \frac{z}{z - a} \qquad |z| > 1$$

If we let $z = e^{j\Omega}$ in $F[z]$, we obtain $F(\Omega)$, the DTFT of $a^k u[k]$:

$$F(\Omega) = F[z]\big|_{z=e^{j\Omega}} = \frac{e^{j\Omega}}{e^{j\Omega} - a} = \frac{1}{1 - ae^{-j\Omega}}$$

The relationship of DTFT to z-transform is parallel to that of the Fourier transform to the Laplace transform.

9.3   PROPERTIES OF THE DTFT

Because the DTFT is basically a z-transform, its properties are similar to those of the (bilateral) z-transform. The linearity property [Eq. (9.35)] has been already discussed. Other useful properties of the DTFT are as follows:

Time-Shifting and Frequency-Shifting Properties
If

$$f[k] \Longleftrightarrow F(\Omega)$$

then

and

$$f[k - k_0] \Longleftrightarrow F(\Omega)e^{-jk_0\Omega} \qquad (9.45)$$

$$f[k]e^{jk\Omega_0} \Longleftrightarrow F(\Omega - \Omega_0) \qquad (9.46)$$

These properties can be proved by direct substitution in equations defining the inverse and the direct DTFT.

Time and Frequency Inversion

$$f[-k] \Longleftrightarrow F(-\Omega) \qquad (9.47)$$

This property can be proved directly by finding the DTFT of $f[-k]$, using Eq. (9.32).

Multiplication by k: Frequency Differentiation

$$kf[k] \Longleftrightarrow j\frac{dF(\Omega)}{d\Omega} \qquad (9.48)$$

The result follows immediately by differentiating both sides of Eq. (9.32) with respect to Ω.

Time and Frequency Convolution Properties
If

$$f_1[k] \Longleftrightarrow F_1(\Omega) \quad \text{and} \quad f_2[k] \Longleftrightarrow F_2(\Omega)$$

then

and

$$f_1[k] * f_2[k] \Longleftrightarrow F_1(\Omega)F_2(\Omega) \qquad (9.49a)$$

$$f_1[k]f_2[k] \Longleftrightarrow \frac{1}{2\pi}F_1(\Omega) * F_2(\Omega) \qquad (9.49b)$$

where

and

$$f_1[k] * f_2[k] = \sum_{m=-\infty}^{\infty} f_1[m]f_2[k - m]$$

$$F_1(\Omega) * F_2(\Omega) = \int_{2\pi} F_1(u)F_2(\Omega - u)\, du$$

The proof of time convolution is identical to that of the time-convolution property in Chapter 5, with z replaced by $e^{j\Omega}$. The proof of the frequency-convolution property is similar to that of the frequency-convolution property of the Fourier transform in Chapter 7.

9.4 LTID SYSTEM ANALYSIS BY DTFT

Consider a linear time-invariant discrete-time system (LTID) with the unit impulse response $h[k]$ and the transfer function $H[z]$. We shall find the (zero-state) system response $y[k]$ for the input $f[k]$. Because

$$y[k] = f[k] * h[k] \tag{9.50}$$

from Eq. (9.49) it follows that

$$Y(\Omega) = F(\Omega)H(\Omega) \tag{9.51}$$

where $F(\Omega)$, $Y(\Omega)$, and $H(\Omega)$ are DTFTs of $f[k]$, $y[k]$, and $h[k]$, respectively. Recall that DTFT is the z-transform with $z = e^{j\Omega}$. Therefore Eq. (9.51) is the same as

$$Y[z] = F[z]H[z] \tag{9.52}$$

with $z = e^{j\Omega}$. This shows that finding a linear system response using DTFT is the same as finding the response using the z-transform method with $z = e^{j\Omega}$. This is a special case of z-transform, with z restricted to the unit circle in the z-plane ($z = e^{j\Omega}$). This development is parallel to restricting s to the imaginary axis ($s = j\omega$) in continuous-time systems. Clearly, the z-transform method is more general than the DTFT method, in regard to finding the system response. The DTFT method is restricted only to those input signals for which the DTFT exists. Moreover this method can be used only if the system is asymptotically stable because unstable or even marginally stable systems can produce an output for which the DTFT does not exist. For these reasons DTFT should be avoided, if possible, in system analysis. The main use of DTFT is in signal analysis, where we are interested in studying the sinusoidal spectral components of a discrete-time signal. The situation here is exactly analogous to that of the Fourier transform in the use of continuous-time signals and systems.

There are some situations in system analysis where the DTFT method may be simpler than the z-transform method. These are the cases where either the system or the input or both are noncausal (assuming that the system is stable and the inputs are DTF-transformable). These situations require the use of bilateral z-transform and the use of DTFT may therefore be simpler. This discussion is parallel to that in Sec. 7.5, where we discussed the use of Laplace and Fourier transforms in continuous-time system analysis. The following examples will clarify these observations.

■ Example 9.8
For a system with unit impulse response $h[k] = (0.5)^k u[k]$, determine the zero-state response $y[k]$ for the input $f[k] = (0.8)^k u[k]$.

In this case, from Eq. (9.51)

$$Y(\Omega) = F(\Omega)H(\Omega)$$

From the results in Eq. (9.39), we obtain

$$F(\Omega) = \frac{1}{1 - 0.8e^{-j\Omega}} = \frac{e^{j\Omega}}{e^{j\Omega} - 0.8} \tag{9.53}$$

$$H(\Omega) = \frac{1}{1 - 0.5e^{-j\Omega}} = \frac{e^{j\Omega}}{e^{j\Omega} - 0.5} \tag{9.54}$$

Therefore

$$Y(\Omega) = \frac{e^{j2\Omega}}{(e^{j\Omega} - 0.5)(e^{j\Omega} - 0.8)}$$

and

$$\frac{Y(\Omega)}{e^{j\Omega}} = \frac{e^{j\Omega}}{(e^{j\Omega} - 0.5)(e^{j\Omega} - 0.8)}$$

$$= \frac{-\frac{5}{3}}{e^{j\Omega} - 0.5} + \frac{\frac{8}{3}}{e^{j\Omega} - 0.8}$$

so that

$$Y(\Omega) = -\left(\frac{5}{3}\right)\frac{e^{j\Omega}}{e^{j\Omega} - 0.5} + \left(\frac{8}{3}\right)\frac{e^{j\Omega}}{e^{j\Omega} - 0.8}$$

$$= -\left(\frac{5}{3}\right)\frac{1}{1 - 0.5e^{-j\Omega}} + \left(\frac{8}{3}\right)\frac{1}{1 - 0.8e^{-j\Omega}}$$

The inverse DTFT of this equation yields

$$y[k] = \left[-\tfrac{5}{3}(0.5)^k + \tfrac{8}{3}(0.8)^k\right]u[k] \tag{9.55}$$

The procedure here is identical to that of z-transform method with z replaced by $e^{j\Omega}$. This is rather clumsy and inconvenient. Clearly, the z-transform method, which has more compact notation, is preferable here. ∎

■ **Example 9.9**

Repeat Example 9.8 if the input

$$f[k] = (1.5)^{-k}u[-(k+1)]$$

From Eq. (9.42)

$$F(\Omega) = \frac{1}{1.5e^{-j\Omega} - 1}$$

$$= \frac{-e^{j\Omega}}{e^{j\Omega} - 1.5}$$

and

$$Y(\Omega) = F(\Omega)H(\Omega)$$

$$= \frac{-e^{j2\Omega}}{(e^{j\Omega} - 1.5)(e^{j\Omega} - 0.5)}$$

Therefore

$$\frac{Y(\Omega)}{e^{j\Omega}} = \frac{-e^{j\Omega}}{(e^{j\Omega} - 1.5)(e^{j\Omega} - 0.5)}$$

$$= \frac{-1.5}{e^{j\Omega} - 1.5} + \frac{0.5}{e^{j\Omega} - 0.5}$$

and

$$Y(\Omega) = 1.5\frac{-e^{j\Omega}}{e^{j\Omega} - 1.5} + 0.5\frac{e^{j\Omega}}{e^{j\Omega} - 0.5}$$

Note that in the first term on the right-hand side $|a| > 1$, and in the second term $|a| < 1$. Hence the inverse DTFTs of the first and the second terms are obtained from Eqs. (9.42) and (9.39), respectively.

$$y[k] = 1.5(1.5)^k u[-(k+1)] + 0.5(0.5)^k u[k]$$

This problem can be solved by using bilateral z-transform. The DTFT, though a little clumsy, avoids the consideration of regions of convergence that is inherent in the bilateral z-transform. ■

The Frequency Response of an LTID

If $H[z]$ is the transfer function of an LTID, its response (to input $e^{j\Omega k}$) is given by $H[e^{j\Omega}]e^{j\Omega k}$ (see Sec. 5.5). The frequency response of the system, by definition, is $H[e^{j\Omega}]$. The amplitude and the phase response are $|H[e^{j\Omega}]|$, and $\angle H[e^{j\Omega}]$, respectively. But

$$H(\Omega) = H[z]|_{z=e^{j\Omega}} = H[e^{j\Omega}]$$

Therefore $H(\Omega)$, the DTFT of $h[k]$, is the frequency response of the system. This means that $|H(\Omega)|$ and $\angle H(\Omega)$ are the amplitude and the phase response of the system, which explains why the spectra in Fig. 9.6 are identical to those in Fig. 5.9. Also

$$Y(\Omega) = F(\Omega)\,H(\Omega)$$

This shows that the frequency spectrum of the output signal is the product of the frequency spectrum of the input signal and the frequency response of the system. From the above equation we can also show that $H(\Omega)$ is the DTFT of $h[k]$.

9.5 RELATIONSHIPS AMONG VARIOUS TRANSFORMS

We showed earlier that the DTFT is the z-transform with z restricted to the unit circle in the complex plane ($z = e^{j\Omega}$). We now show that the DTFS, the DTFT, and Fourier transform of a continuous-time signal are closely related. We also know that the Fourier, the Laplace, and the z-transform are related. **Therefore all the series and transforms discussed in this book are basically a family.**

Relationship between DTFS and DTFT

Let us consider a finite-duration (timelimited) signal $f[k]$ of width N_0. We construct a periodic signal $f_{N_0}[k]$ by repeating $f[k]$ periodically at every N_0 interval. Now

$$f_{N_0}[k] = \sum_{r=<N_0>} \mathcal{D}_r e^{jr\Omega_0 k} \quad \Omega_0 = \frac{2\pi}{N_0} \tag{9.56}$$

where

$$\mathcal{D}_r = \frac{1}{N_0} \sum_k f[k]e^{-jr\Omega_0 k} \tag{9.57}$$

If $F(\Omega)$ is the DTFT of $f[k]$, then

$$f[k] = \frac{1}{2\pi} \int_{2\pi} F(\Omega)e^{jk\Omega}\, d\Omega \tag{9.58}$$

where

$$F(\Omega) = \sum_k f[k]e^{-jk\Omega} \tag{9.59}$$

consequently

$$F(r\Omega_0) = \sum_k f[k]e^{-jr\Omega_0 k} \tag{9.60}$$

From Eqs. (9.57) and (9.60) it follows that

$$F(r\Omega_0) = N_0 \mathcal{D}_r \tag{9.61}$$

This shows that the samples of $F(\Omega)$ (the DTFT of $f[k]$) taken at intervals of Ω_0 are $N_0 \mathcal{D}_r$, the DTFS coefficients (multiplied by N_0) of $f_{N_0}[k]$. This result is readily verified from Examples 9.3 and 9.6 (Figs. 9.4 and 9.8). Note the parallel of this situation with that of continuous-time signals [see Fig. 8.9 or Eq. (8.7)]. Observe also that a periodic signal has a discrete spectrum, whereas a nonperiodic signal has a continuous spectrum.

This discussion shows that the samples of $F(\Omega)$ are equal to DTFS coefficients $\mathcal{D}_r$ (within a multiplicative constant N_0), which in turn can be computed by using the DFT (or the FFT algorithm). Therefore the DTFT and the inverse DTFT can be computed by using the DFT or the FFT, as discussed in Sec. 8.3.

The DTFT and the Fourier Transform (FT)

The DTFT of a signal $f[k]$ is defined as

$$F(\Omega) = \sum_{k=-\infty}^{\infty} f[k]e^{-jk\Omega} \tag{9.62}$$

Figures 9.10a and 9.10b show a hypothetical signal $f[k]$ and its DTFT $F(\Omega)$. Recall that the Fourier transform of a continuous-time impulse $\delta(t-k)$ is $e^{-jk\omega}$. A glance at the right hand side of Eq. (9.62) shows that $F(\Omega)$ is, in fact, the Fourier transform of a sequence of impulses at unit interval, with $f[k]$ as the strength of the kth impulse (at $t = k$). This impulse train $\overline{f}(t)$ can be expressed as

$$\overline{f}(t) = \sum_{k=-\infty}^{\infty} f[k]\delta(t - k)$$

the Fourier transform of the above equation yields

$$\overline{F}(\omega) = \sum_{k=-\infty}^{\infty} f[k]e^{-jk\omega} \tag{9.63}$$

Comparison of (9.63) with (9.62) shows that

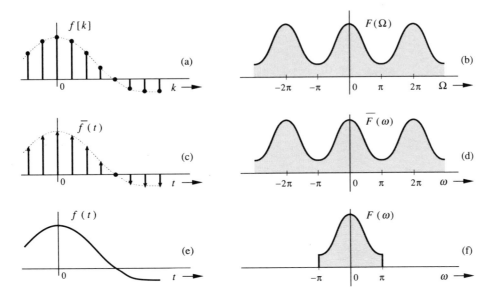

Fig. 9.10 Connection between the DTFT and Fourier transform.

$$\overline{F}(\omega) = F(\Omega)|_{\Omega=\omega} \qquad (9.64a)$$

$$= F(\omega) \qquad (9.64b)$$

This interesting result brings out the direct connection between the spectra of a sampled continuous-time signal (Fig. 9.10c) and those of the corresponding discrete-time signal $f[k]$ in Fig. 9.10a.†

If $\overline{f}(t)$ is the signal $f(t)$ sampled at the Nyquist rate, then from Sec. 8.1 we know (see Fig. 8.1e) that

$$F(\omega) = \begin{cases} \overline{F}(\omega) & |\omega| \le \pi \\ 0 & |\omega| > \pi \end{cases}$$

Observe that the Nyquist interval in the present case is $T = 1$ and the bandwidth of $f(t)$ is $1/2T = 0.5$ Hz or π rad/s. Figure 9.10e shows the continuous-time signal $f(t)$ constructed from $f[k]$ as its Nyquist samples. The Fourier transform of $f(t)$ is

---

†If we consider the impulses in the sequence at an interval of T seconds (rather than at the unit interval),

$$\overline{f}(t) = \sum_{k=-\infty}^{\infty} f[k]\delta(t - kT)$$

and

$$\overline{F}(\omega) = \sum_{k=-\infty}^{\infty} f[k]e^{-j\omega kT} = F(\Omega)|_{\Omega=\omega T} = F(\omega T)$$

Therefore $F(\Omega)$ is equal to $\overline{F}(\omega)$ with a scaled frequency variable $\Omega = \omega T$. The period 2π of $F(\Omega)$ corresponds to a period of $2\pi/T$ on the ω scale.

bandlimited to $\omega = \pi$ and is identical to $F(\Omega)$ truncated beyond $|\Omega| = \pi$, as shown in Fig. 9.10f. Therefore if $F(\Omega)$ is the DTFT of $f[k]$, then $F(\omega)$ truncated beyond $|\omega| \geq \pi$ is the Fourier transform of a continuous time signal $f(t)$ whose Nyquist samples are $f[k]$. Note that the bandwidth of $f(t)$ is 0.5 Hz and its Nyquist interval is unity.

In Example 9.7, if we truncate $F(\Omega)$ beyond $|\Omega| > \pi$, the resulting spectrum is $F(\omega)$

$$F(\omega) = \mathrm{rect}\left(\frac{\omega}{2\omega_0}\right) \tag{9.65}$$

and

$$f(t) = \frac{\omega_0}{\pi}\mathrm{sinc}(\omega_0 t) \tag{9.66}$$

The Nyquist samples of $f(t)$ (at unit time interval) are

$$f[k] = \frac{\omega_0}{\pi}\,\mathrm{sinc}\,(\omega_0 k) \tag{9.67}$$

9.6 DERIVATION OF THE Z-TRANSFORM PAIR

We now derive the z-transform pair from the DTFT pair. For this purpose it is convenient to use the notation $F(j\Omega)$ for the DTFT. Using this notation, the DTFT pair becomes

$$F(j\Omega) = \sum_{k=-\infty}^{\infty} f[k]\,e^{-j\Omega k} \tag{9.68}$$

and

$$f[k] = \frac{1}{2\pi}\int_{-\pi}^{\pi} F(j\Omega)\,e^{j\Omega k}\,d\Omega \tag{9.69}$$

Consider now the DTFT of $f[k]\,e^{-\sigma k}$ (σ real):

$$\mathrm{DTFT}\,[f[k]\,e^{-\sigma k}] = \sum_{k=-\infty}^{\infty} f[k]\,e^{-\sigma k}\,e^{-j\Omega k} \tag{9.70}$$

$$= \sum_{k=-\infty}^{\infty} f[k]\,e^{-(\sigma+j\Omega)k} \tag{9.71}$$

It follows from Eq. (9.68) that the above sum is $F(\sigma + j\Omega)$. Thus

$$\mathrm{DTFT}\,[f[k]\,e^{-\sigma k}] = \sum_{k=-\infty}^{\infty} f[k]\,e^{-(\sigma+j\Omega)k} = F(\sigma + j\Omega) \tag{9.72}$$

The inverse DTFT of the above equation yields

$$f[k]\,e^{-\sigma k} = \frac{1}{2\pi}\int_{-\pi}^{\pi} F(\sigma + j\Omega)\,e^{j\Omega k}\,d\Omega \tag{9.73}$$

Therefore

$$f[k] = \frac{1}{2\pi} \int_{-\pi}^{\pi} F(\sigma + j\Omega) \, e^{(\sigma+j\Omega)k} \, d\Omega \tag{9.74}$$

Let us define a new variable z as

$$z = e^{\sigma+j\Omega} \quad \text{so that} \quad \ln z = \sigma + j\Omega \quad \text{and} \quad \frac{1}{z} \, dz = j \, d\Omega \tag{9.75}$$

Observe that $z = r \, e^{j\Omega}$, where $r = e^{\sigma}$. Thus, z lies on a circle of radius r, and as Ω varies from $-\pi$ to π, z circumambulates along this circle, completing exactly one rotation in counterclockwise direction. Change to variable z and use of Eqs. (9.75) in Eq. (9.74) yields

$$f[k] = \frac{1}{2\pi j} \oint F(\ln z) \, z^{k-1} \, dz \tag{9.76a}$$

and From Eq. (9.72), we obtain

$$F(\ln z) = \sum_{k=-\infty}^{\infty} f[k] \, z^{-k} \tag{9.76b}$$

where the integral $\oint$ indicates a contour integral around a circle of radius r in counterclockwise direction.

The above two equations are the desired extensions. They are, however, in a clumsy form. For the sake of convenience, we make a notational change by noting that $F(\ln z)$ is a function of z; we denote it by a simpler notation $F[z]$. Thus Eqs. (9.76) become

$$f[k] = \frac{1}{2\pi j} \oint F[z] \, z^{k-1} \, dz \tag{9.77}$$

and

$$F[z] = \sum_{k=-\infty}^{\infty} f[k] \, z^{-k} \tag{9.78}$$

This is the (bilateral) z-transform pair. Equation (9.77) expresses $f[k]$ as a continuous sum of exponentials of the form $z^k = e^{(\sigma+j\Omega)k} = r^k \, e^{j\Omega k}$. Thus by selecting a proper value for r (or σ), we can make the exponential grow (or decay) at any exponential rate we desire. Note that the DTFT is just a special case of the z-transform and can be obtained by letting $z = e^{j\Omega}$ in Eqs. (9.77) and (9.78).

9.7 SUMMARY

Analysis of discrete-time signals is discussed in this chapter. Our approach is parallel to that used for continuous-time signals. We first represent a periodic $f[k]$ as a Fourier series formed by a discrete-time exponential and its harmonics. Later we extend this representation to a nonperiodic signal $f[k]$ by considering $f[k]$ as a limiting case of a periodic signal with period approaching infinity. Periodic signals are represented by discrete-time Fourier series (DTFS); nonperiodic signals are represented by the discrete-time Fourier transform (DTFT). The development, although similar to that of continuous-time signals, also shows some significant differences. The basic difference in the two cases arises because a continuous-time exponential $e^{j\omega t}$ has a unique waveform for every value of ω in the range $-\infty$ to ∞. On the other hand, a discrete-time exponential $e^{j\Omega k}$ has a unique waveform only for values of Ω in a continuous interval of 2π. Therefore, if Ω_0 is the fundamental frequency, then at most $\frac{2\pi}{\Omega_0}$ number of exponentials in the Fourier series are independent. Consequently the discrete-time exponential Fourier series has only $N_0 = \frac{2\pi}{\Omega_0}$ terms.

The discrete-time Fourier transform (DTFT) of a nonperiodic signal is a continuous function of Ω and is periodic with period 2π. We can analyze linear time-invariant discrete-time (LTID) systems using DTFT if the input signals are DTF-transformable and if the system is causal and stable. In general, z-transform is superior to DTFT for analysis of LTID systems, whereas DTFT is preferred for better insights in signals and signal analysis. The relationship of DTFT to z-transform is similar to that of the Fourier transform to the Laplace transform.

If $H(\Omega)$ is the DTFT of the system's impulse response $h[k]$, then $|H(\Omega)|$ is the amplitude response, and $\angle H(\Omega)$ is the phase response of the system. Moreover, if $F(\Omega)$ and $Y(\Omega)$ are the DTFTs of the input $f[k]$ and the corresponding output $y[k]$, then $Y(\Omega) = H(\Omega)F(\Omega)$. Therefore the output spectrum is the product of the input spectrum and the system's frequency response.

In Sec. 9.5 we showed that DTFS, DTFT, z-transform, Fourier series, Fourier transform, and Laplace transform are all related and form a family. In fact, we can derive all the other members of this family from the Fourier series.

PROBLEMS

9.1-1 Express the following exponentials in the form $e^{j(\Omega k + \theta)}$ where $0 \le \Omega < 2\pi$

    (a) $e^{j(8.2\pi k + \theta)}$ **(b)** $e^{j4\pi k}$ **(c)** $e^{-1.95k}$

    (d) $e^{-j10.7\pi k}$

9.1-2 Find the discrete-time Fourier series (DTFS) and sketch their spectra for the following periodic signal:

$$f[k] = 4\cos 2.4\pi k + 2\sin 3.2\pi k$$

9.1-3 Repeat Prob. 9.1-2 if

$$f[k] = \cos 2.2\pi k \cos 3.3\pi k$$

9.1-4 Repeat Prob. 9.1-2 if $f[k] = 2\cos 3.2\pi(k-3)$.

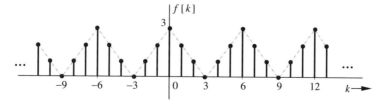

Fig. P9.1-5

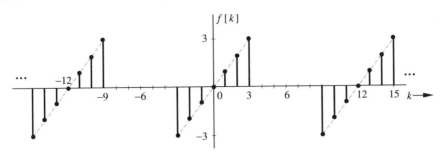

Fig. P9.1-6

9.1-5 Find the discrete-time Fourier series for the $f[k]$ shown in Fig. P9.1-5.

9.1-6 Repeat Prob. 9.1-5 for the $f[k]$ shown in Fig. P9.1-6.

9.2-1 For the following signals, find the DTFT directly, using the definition in Eq. (9.32).

    (a) $f[k] = \delta[k]$ **(b)** $\delta[k - k_0]$ **(c)** $a^k u[k - 1]$ $|a| < 1$

    (d) $f[k] = a^k u[k + 1]$ $|a| < 1$.

In each case, sketch the signals and their amplitude spectra. Sketch phase spectra for parts **(a)** and **(b)** only.

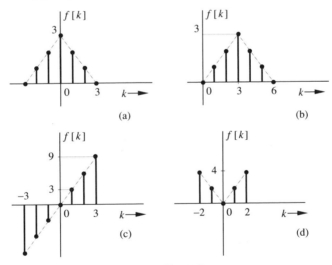

Fig. P9.2-2

9.2-2 Find the DTFT for the signals shown in Fig. P9.2-2.

9.2-3 Find the inverse DTFT for the spectra shown in Fig. P9.2-3.

9.3-1 Using the DTFT method, find the zero-state response $y[k]$ of a causal system with frequency response

$$H(\Omega) = \frac{e^{j\Omega} + 0.32}{e^{j2\Omega} + e^{j\Omega} + 0.16}$$

and the input

$$f[k] = (-0.5)^k u[k]$$

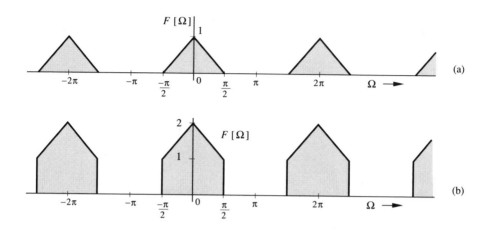

Fig. P9.2-3

9.3-2   Repeat Prob. 9.3-1 if

$$H(\Omega) = \frac{e^{j\Omega} - 0.5}{(e^{j\Omega} + 0.5)(e^{j\Omega} - 1)}$$

and

$$f[k] = 3^{-(k+1)}u[k]$$

9.3-3   Repeat Prob. 9.3-1 if

$$H(\Omega) = \frac{e^{j\Omega}}{e^{j\Omega} - 0.5}$$

and

$$f[k] = 0.8^k u[k] + 2(2)^k u[-(k+1)]$$

9.4-1   Using the time-shifting property and the results in Examples 9.4 and 9.6, find the DTFT of **(a)** $a^k\{u[k] - u[k-9]\}$ **(b)** $u[k] - u[k-9]$.

9.4-2   Using appropriate properties and the result in Example 9.4, find the DTFT of **(a)** $(k+1)a^k u[k]$ $(|a| < 1)$ **(b)** $a^k \cos \Omega_0 k\, u[k]$.

9.5-1   Find the DTFT of the following signals, using the fact that DTFT of a signal is its z-transform with $z = e^{j\Omega}$ when the region of convergence for $F[z]$ includes the unit circle. Use Table 5.1 freely.

(a) $a^k \cos \Omega_0 k\, u[k]$ $|a| < 1$

(b) $a^k \sin \Omega_0 k\, u[k]$ $|a| < 1$

(c) $a^k u[k-1]$ $|a| < 1$ Hint: $a^k = a^{-1}a^{k-1}$

(d) $a^k u[k+1]$ $|a| < 1$ Hint: $a^k u[k+1] = a^k u[k] + a^{-1}\delta[k+1]$

COMPUTER PROBLEMS

C9-1   Use a computer to find and plot the discrete-time Fourier series of a periodic sampled gate function in Fig. 9.4a with $N_0 = 64$.

C9-2   Using a computer, find the inverse DTFT of the rectangular pulse spectrum in Fig. 9.9 (Example 9.7) with $\omega_0 = \frac{\pi}{6}$, and the period 2π.

V

STATE-SPACE
ANALYSIS

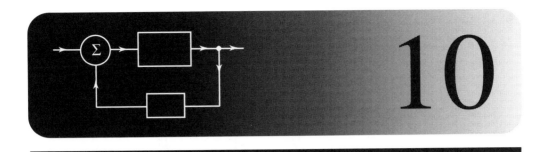

State-Space Analysis

Before starting this chapter, the reader should review the introduction to state variables in Sec. 1.6. This chapter requires some understanding of matrix algebra. Section B.9 is a self-contained treatment of matrix algebra, which should be more than adequate for the purposes of this chapter.

10.1 INTRODUCTION

From the discussion in Chapter 1, we know that to determine a system's response(s) at any instant t, we need to know the system's inputs during its entire past, from $-\infty$ to t. If the inputs are known only for $t > t_0$, we can still determine the system output(s) for any $t > t_0$, provided we know certain initial conditions in the system at $t = t_0$. These initial conditions collectively are called the **initial state** of the system (at $t = t_0$).

The state of a system at any instant t_0 is the smallest set of numbers $x_1(t_0)$, $x_2(t_0)$, ... , $x_n(t_0)$ which is sufficient to determine the behavior of the system for all time $t > t_0$ when the input(s) to the system are known for $t > t_0$. The variables $x_1, x_2, \ldots, x_n$ are known as **state variables.**

The initial conditions of a system can be specified in many different ways. Consequently the system state can also be specified in many different ways. This means that state variables are not unique. The concept of a system state is very important. We know that an output $y(t)$ at any instant $t > t_0$ can be determined from the initial state $\{x(t_0)\}$ and a knowledge of the input $f(t)$ during the interval (t_0, t). Therefore the output $y(t_0)$ (at $t = t_0$) is determined from the initial state $\{x(t_0)\}$ and the input $f(t)$ during the interval (t_0, t_0). The latter is $f(t_0)$. Hence the output at any instant is determined completely from a knowledge of the system state and the input at that instant. This result is also valid for multiple-input, multiple-output (MIMO) systems, where every possible system output at any instant t is determined completely from a knowledge of the system state and the input(s) at the instant t. These ideas should become clear from the following example of an *RLC* circuit.

■ **Example 10.1**

Find the state-space description of the RLC circuit shown in Fig. 10.1. Verify that all possible system outputs at some instant t can be determined from a knowledge of the system state and the input at that instant t.

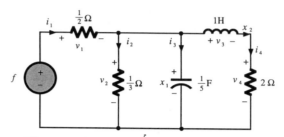

Fig. 10.1 RLC network for Example 10.1.

It is known that inductor currents and capacitor voltages in an RLC circuit can be used as one possible choice of state variables. For this reason we shall choose x_1 (the capacitor voltage) and x_2 (the inductor current) as our state variables.

The node equation at the intermediate node is

$$i_3 = i_1 - i_2 - x_2$$

but $i_3 = 0.2\dot{x}_1$, $i_1 = 2(f - x_1)$, $i_2 = 3x_1$. Hence

or

$$0.2\dot{x}_1 = 2(f - x_1) - 3x_1 - x_2$$

$$\dot{x}_1 = -25x_1 - 5x_2 + 10f$$

This is the first state equation. To obtain the second state equation, we sum the voltages in the extreme right loop formed by C, L, and the 2 Ω resistor so that they are equal to zero:

or

$$-x_1 + \dot{x}_2 + 2x_2 = 0$$

$$\dot{x}_2 = x_1 - 2x_2$$

Thus the two state equations are

$$\dot{x}_1 = -25x_1 - 5x_2 + 10f \tag{10.1a}$$

$$\dot{x}_2 = x_1 - 2x_2 \tag{10.1b}$$

Every possible output can now be expressed in terms of x_1, x_2, and f. From Fig. 10.1, we have

$$v_1 = f - x_1$$

$$i_1 = 2(f - x_1)$$

$$v_2 = x_1$$

$$i_2 = 3x_1$$

$$i_3 = i_1 - i_2 - x_2 = 2(f - x_1) - 3x_1 - x_2 = -5x_1 - x_2 + 2f$$

$$i_4 = x_2$$

$$v_4 = 2i_4 = 2x_2$$

$$v_3 = x_1 - v_4 = x_1 - 2x_2 \tag{10.2}$$

This set of equations is known as the **output equation** of the system. It is clear from this set that every possible output at some instant t can be determined from a knowledge of $x_1(t)$, $x_2(t)$, and $f(t)$, the system state and the input at the instant t. Once we solve the state equations (10.1) to obtain $x_1(t)$ and $x_2(t)$, we can determine every possible output for any given input $f(t)$. ∎

If we already have a system equation in the form of an nth-order differential equation, we can convert it into a state equation as follows. Consider the system equation

$$\frac{d^n y}{dt^n} + a_{n-1}\frac{d^{n-1}y}{dt^{n-1}} + \cdots + a_1\frac{dy}{dt} + a_0 y = f(t) \tag{10.3}$$

One possible set of initial conditions is $y(0)$, $\dot{y}(0)$, ... , $y^{(n-1)}(0)$. Let us define y, $\dot{y}$, $\ddot{y}$, ... , $y^{(n-1)}$ as the state variables and, for convenience, let us rename the n state variables as x_1, x_2, ... , x_n:

$$x_1 = y$$
$$x_2 = \dot{y}$$
$$x_3 = \ddot{y}$$
$$\vdots$$
$$x_n = y^{(n-1)} \tag{10.4}$$

From Eq. (10.4), we have

$$\dot{x}_1 = x_2$$
$$\dot{x}_2 = x_3$$
$$\vdots$$
$$\dot{x}_{n-1} = x_n$$

and, from Eq. (10.3),

$$\dot{x}_n = -a_{n-1}x_n - a_{n-2}x_{n-1} - \cdots - a_1 x_2 - a_0 x_1 + f \tag{10.5a}$$

These n simultaneous first-order differential equations are the state equations of the system. The output equation is

$$y = x_1 \tag{10.5b}$$

For continuous-time systems, the state equations are n simultaneous first-order differential equations in n state variables x_1, x_2, ... , x_n of the form

$$\dot{x}_i = g_i(x_1, x_2, \ldots, x_n, f_1, f_2, \ldots, f_j) \qquad i = 1, 2, \ldots, n$$

where f_1, f_2, ... , f_n are the j system inputs. For a linear system, these equations reduce to a simpler linear form:

$$\dot{x}_k = a_{k1}x_1 + a_{k2}x_2 + \cdots + a_{kn}x_n + b_{k1}f_1 + b_{k2}f_2 + \cdots + b_{kj}f_j \qquad k = 1, 2, \ldots, n \tag{10.6a}$$

and the output equations are of the form

$$y_m = c_{m1}x_1 + c_{m2}x_2 + \cdots + c_{mn}x_n + d_{m1}f_1 + d_{m2}f_2 + \cdots + d_{mj}f_j \qquad m = 1, 2, \ldots, k$$
$$\tag{10.6b}$$

The set of Equations (10.6a) and (10.6b) is called a **dynamical** equation. When it is used to describe a system, it is called the **dynamical-equation description** or **state-variable** description of the system. The n simultaneous first-order state equations are also known as the **normal-form** equations.

These equations can be written more conveniently in matrix form:

$$
\underbrace{\begin{bmatrix} \dot{x}_1 \\ \dot{x}_2 \\ \cdots \\ \dot{x}_n \end{bmatrix}}_{\dot{\mathbf{x}}} = \underbrace{\begin{bmatrix} a_{11} & a_{12} & \cdots & a_{1n} \\ a_{21} & a_{22} & \cdots & a_{2n} \\ \cdots & \cdots & \cdots & \cdots \\ a_{n1} & a_{n2} & \cdots & a_{nn} \end{bmatrix}}_{\mathbf{A}} \underbrace{\begin{bmatrix} x_1 \\ x_2 \\ \cdots \\ x_n \end{bmatrix}}_{\mathbf{x}} + \underbrace{\begin{bmatrix} b_{11} & b_{12} & \cdots & b_{1j} \\ b_{21} & b_{22} & \cdots & b_{2j} \\ \cdots & \cdots & \cdots & \cdots \\ b_{n1} & b_{n2} & \cdots & b_{nj} \end{bmatrix}}_{\mathbf{B}} \underbrace{\begin{bmatrix} f_1 \\ f_2 \\ \cdots \\ f_j \end{bmatrix}}_{\mathbf{f}} \tag{10.7a}
$$

and

$$
\underbrace{\begin{bmatrix} y_1 \\ y_2 \\ \cdots \\ y_k \end{bmatrix}}_{\mathbf{y}} = \underbrace{\begin{bmatrix} c_{11} & c_{12} & \cdots & c_{1n} \\ c_{21} & c_{22} & \cdots & c_{2n} \\ \cdots & \cdots & \cdots & \cdots \\ c_{k1} & c_{k2} & \cdots & c_{kn} \end{bmatrix}}_{\mathbf{C}} \underbrace{\begin{bmatrix} x_1 \\ x_2 \\ \cdots \\ x_n \end{bmatrix}}_{\mathbf{x}} + \underbrace{\begin{bmatrix} d_{11} & d_{12} & \cdots & d_{1j} \\ d_{21} & d_{22} & \cdots & d_{2j} \\ \cdots & \cdots & \cdots & \cdots \\ d_{k1} & d_{k2} & \cdots & d_{kj} \end{bmatrix}}_{\mathbf{D}} \underbrace{\begin{bmatrix} f_1 \\ f_2 \\ \cdots \\ f_j \end{bmatrix}}_{\mathbf{f}} \tag{10.7b}
$$

or

$$\dot{\mathbf{x}} = \mathbf{A}\mathbf{x} + \mathbf{B}\mathbf{f} \tag{10.8a}$$

$$\mathbf{y} = \mathbf{C}\mathbf{x} + \mathbf{D}\mathbf{f} \tag{10.8b}$$

Equation (10.8a) is the state equation and Eq. (10.8b) is the output equation. The vectors $\mathbf{x}$, $\mathbf{y}$, and $\mathbf{f}$ are the state vector, the output vector, and the input vector, respectively.

For discrete-time systems, the state equations are n simultaneous first-order difference equations. Discrete-time systems are discussed in Sec. 10.6.

10.2   A SYSTEMATIC PROCEDURE FOR DETERMINING STATE EQUATIONS

We shall discuss here a systematic procedure for determining the state-space description of linear time-invariant systems. In particular, we shall consider two types of systems: (1) *RLC* networks and (2) systems specified by block diagrams or *n*th-order transfer functions.

10.2-1 Electrical Circuits

The method used in Example 10.1 can be used for most of the simple cases. The steps are as follows:

1. Choose all independent capacitor voltages and inductor currents to be the state variables.

2. Choose a set of loop currents; express the state variables and their first derivatives in terms of these loop currents.

3. Write the loop equations and eliminate all variables other than state variables (and their first derivatives) from the equations derived in Steps 2 and 3.

■ **Example 10.2**

Write the state equations for the network shown in Fig. 10.2.

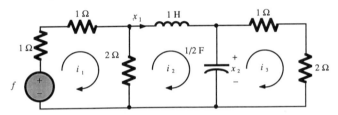

Fig. 10.2 *RLC* network for Example 10.2.

Step 1. There is one inductor and one capacitor in the network. Therefore we shall choose the inductor current x_1 and the capacitor voltage x_2 as the state variables.

Step 2. The relationship between the loop currents and the state variables can be written by inspection:

$$x_1 = i_2 \tag{10.9a}$$

$$\tfrac{1}{2}\dot{x}_2 = i_2 - i_3 \tag{10.9b}$$

Step 3. The loop equations are

$$4i_1 - 2i_2 = f \tag{10.10a}$$

$$2(i_2 - i_1) + \dot{x}_1 + x_2 = 0 \tag{10.10b}$$

$$-x_2 + 3i_3 = 0 \tag{10.10c}$$

Now we eliminate i_1, i_2, and i_3 from Eqs. (10.9) and (10.10) as follows. From Eq. (10.10b), we have

$$\dot{x}_1 = 2(i_1 - i_2) - x_2$$

We can eliminate i_1 and i_2 from this equation by using Eqs. (10.9a) and (10.10a) to obtain

$$\dot{x}_1 = -x_1 - x_2 + \tfrac{1}{2}f$$

The substitution of Eqs. (10.9a) and (10.10c) in Eq. (10.9b) yields

$$\dot{x}_2 = 2x_1 - \tfrac{2}{3}x_2$$

These are the desired state equations. We can express them in matrix form as

$$\begin{bmatrix} \dot{x}_1 \\ \dot{x}_2 \end{bmatrix} = \begin{bmatrix} -1 & -1 \\ 2 & -\frac{2}{3} \end{bmatrix} \begin{bmatrix} x_1 \\ x_2 \end{bmatrix} + \begin{bmatrix} \frac{1}{2} \\ 0 \end{bmatrix} f \qquad (10.11)$$

The derivation of state equations from loop equations is facilitated considerably by choosing loops in such a way that only one loop current passes through each of the inductors or capacitors. ∎

An Alternative procedure

We can also determine the state equations by the following procedure:

1. Choose all independent capacitor voltages and inductor currents to be the state variables.

2. Replace each capacitor by a fictitious voltage source equal to the capacitor voltage, and replace each inductor by a fictitious current source equal to the inductor current. This will transform the *RLC* network into a network consisting only of resistors, current sources and voltage sources.

3. Find the current through each capacitor and equate it to $C\dot{x}_i$, where x_i is the capacitor voltage. Similarly, find the voltage across each inductor and equate it to $L\dot{x}_j$, where x_j is the inductor current.

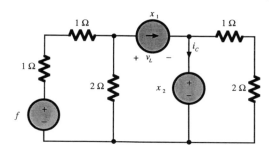

Fig. 10.3 Equivalent circuit of the network in Fig. 10.2.

∎ Example 10.3

Using the above procedure, write the state equations for the network in Fig. 10.2.

In the network in Fig. 10.2, we replace the inductor by a current source of current x_1 and the capacitor by a voltage source of voltage x_2, as shown in Fig. 10.3. The resulting network consists of five resistors, two voltage sources, and one current source. We can find the voltage v_L across the inductor and the current i_c through the capacitor by using the principle of superposition. This can be done by inspection. For example, v_L has three components due to three sources. To compute the component due to f, we assume that $x_1 = 0$ (open circuit) and $x_2 = 0$ (short circuit). Under these conditions all of the network to the right of the $2\,\Omega$ resistor is opened, and the component of v_L due to f is the voltage across the $2\,\Omega$ resistor. This is clearly $\frac{1}{2}f$. Similarly, to find the component of v_L due to x_1, we short f and x_2. The source x_1 sees an equivalent resistor of $1\,\Omega$ across it, and hence $v_L = -x_1$. Continuing the process, we find that the component of v_L due to x_2 is $-x_2$. Hence

$$v_L = \dot{x}_1 = \frac{1}{2}f - x_1 - x_2 \qquad (10.12a)$$

Using the same procedure, we find

$$v_c = \frac{1}{2}\dot{x}_2 = x_1 - \frac{1}{3}x_2 \tag{10.12b}$$

These equations are identical to the state equations (10.11) obtained earlier.† ∎

10.2-2 State Equations From Transfer Function

It is relatively easy to determine the state equations of a system specified by its transfer function. Consider, for example, a first-order system with the transfer function

$$H(s) = \frac{1}{s+a} \tag{10.13}$$

The system realization is shown in Fig. 10.4. The integrator output serves as a natural state variable since, in practical realization, initial conditions are placed on the integrator output. From Fig. 10.4, we have

$$\dot{x} = -ax + f \tag{10.14a}$$

$$y = x \tag{10.14b}$$

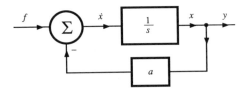

Fig. 10.4

In Sec. 4.6 we saw that a given transfer function can be realized in several ways. Consequently we should be able to obtain different state-space descriptions of the same system by using different realizations. This will be clarified by the following example.

∎ **Example 10.4**

Determine the state-space description of a system specified by the transfer function

$$H(s) = \frac{2s+10}{s^3 + 8s^2 + 19s + 12} \tag{10.15a}$$

$$= \left(\frac{2}{s+1}\right)\left(\frac{s+5}{s+3}\right)\left(\frac{1}{s+4}\right) \tag{10.15b}$$

$$= \frac{\frac{4}{3}}{s+1} - \frac{2}{s+3} + \frac{\frac{2}{3}}{s+4} \tag{10.15c}$$

---

†This procedure requires modification if the system contains all capacitor-voltage source tie sets or all-inductor-current source cut sets. In the case of all capacitor-voltage source tie sets, all capacitor voltages cannot be independent. One capacitor voltage can be expressed in terms of the remaining capacitor voltages and the voltage source(s) in that tie set. Consequently, one of the capacitor voltages should not be used as a state variable, and that capacitor should not be replaced by a voltage source. Similarly, in all inductor-current source tie sets, one inductor should not be replaced by a current source. If there are all-capacitor tie sets or all-inductor cut sets only, no further complications occur. In all-capacitor-voltage source tie sets and/or all-inductor-current source cut sets, we have additional difficulties in that the terms involving derivatives of the input may occur. This problem can be solved by redefining the state variables. The final state variables will not be capacitor voltages and inductor currents.

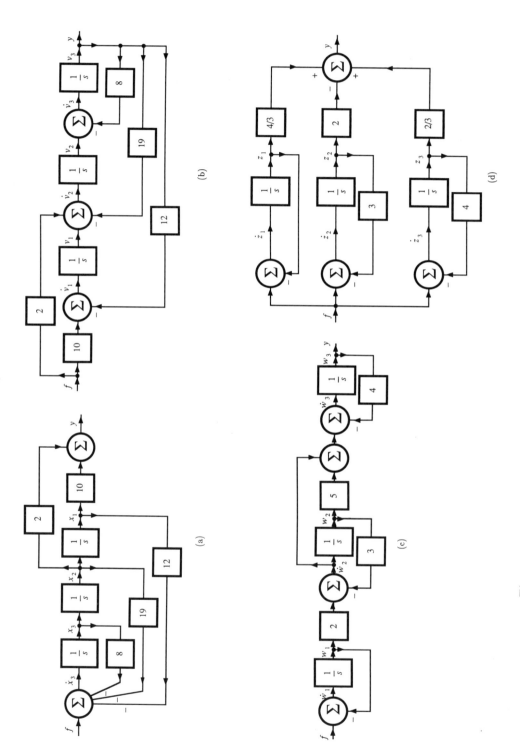

Fig. 10.5 Canonical, cascade, and parallel realizations of the system in Example 10.4.

Using the procedure developed in Sec. 4.6, we shall realize $H(s)$ in Eq. (10.15) with four different realizations: (i) the controller canonical form [Eq. (10.15a)], (ii) the observer canonical form [Eq. (10.15a)], (iii) cascade realization [Eq. (10.15b)] and (iv) parallel realization [Eq. (10.15c)]. These realizations are shown in Figs. 10.5a, 10.5b, 10.5c, and 10.5d, respectively. As mentioned earlier, the output of each integrator serves as a natural state variable.

1. Canonical Forms

Here we shall realize the system using the first (controller) canonical form discussed in Sec. 4.6-1. If we choose the state variables to be the three integrator outputs x_1, x_2, and x_3, then, from Fig. 10.5a,

$$\dot{x}_1 = x_2$$

$$\dot{x}_2 = x_3 \tag{10.16a}$$

$$\dot{x}_3 = -12x_1 - 19x_2 - 8x_3 + f$$

Also, the output y is given by

$$y = 10x_1 + 2x_2 \tag{10.16b}$$

Equations (10.16a) are the state equations, and Eq. (10.16b) is the output equation. In matrix form we have

$$\begin{bmatrix} \dot{x}_1 \\ \dot{x}_2 \\ \dot{x}_3 \end{bmatrix} = \underbrace{\begin{bmatrix} 0 & 1 & 0 \\ 0 & 0 & 1 \\ -12 & -19 & -8 \end{bmatrix}}_{\mathbf{A}} \begin{bmatrix} x_1 \\ x_2 \\ x_3 \end{bmatrix} + \underbrace{\begin{bmatrix} 0 \\ 0 \\ 1 \end{bmatrix}}_{\mathbf{B}} f \tag{10.17a}$$

and

$$\mathbf{y} = \underbrace{\begin{bmatrix} 10 & 2 & 0 \end{bmatrix}}_{\mathbf{C}} \begin{bmatrix} x_1 \\ x_2 \\ x_3 \end{bmatrix} \tag{10.17b}$$

We can also realize $H(s)$ by using the second (observer) canonical form (discussed in Appendix 4.1), as shown in Fig. 10.5b. If we label the output of the three integrators from left to right as the state variables v_1, v_2, and v_3, then, from Fig. 10.5b,

$$\dot{v}_1 = -12v_3 + 10f$$

$$\dot{v}_2 = v_1 - 19v_3 + 2f \tag{10.18a}$$

$$\dot{v}_3 = v_2 - 8v_3$$

and the output y is given by

$$y = v_3 \tag{10.18b}$$

Hence

$$\begin{bmatrix} \dot{v}_1 \\ \dot{v}_2 \\ \dot{v}_3 \end{bmatrix} = \underbrace{\begin{bmatrix} 0 & 0 & -12 \\ 1 & 0 & -19 \\ 0 & 1 & -8 \end{bmatrix}}_{\hat{\mathbf{A}}} \begin{bmatrix} v_1 \\ v_2 \\ v_3 \end{bmatrix} + \underbrace{\begin{bmatrix} 10 \\ 2 \\ 0 \end{bmatrix}}_{\hat{\mathbf{B}}} f \tag{10.19a}$$

and

$$
\mathbf{y} = \underbrace{\begin{bmatrix} 0 & 0 & 1 \end{bmatrix}}_{\hat{\mathbf{C}}} \begin{bmatrix} v_1 \\ v_2 \\ v_3 \end{bmatrix} \tag{10.19b}
$$

Observe closely the relationship between the state-space descriptions of $H(s)$ that use the controller canonical form [Eqs. (10.17)] and those using the observer canonical form [Eqs. (10.19)]. The **A** matrices in these two cases are the transpose of one another; also, the **B** of one is the transpose of **C** in the other, and vice versa. Hence

$$
(\mathbf{A})^T = \hat{\mathbf{A}}
$$

$$
(\mathbf{B})^T = \hat{\mathbf{C}} \tag{10.20}
$$

$$
(\mathbf{C})^T = \hat{\mathbf{B}}
$$

This is no coincidence. This duality relation is generally true.[1]

2. Series Realization

The three integrator outputs w_1, w_2, and w_3 in Fig. 10.5c are the state variables. The state equations are

$$
\dot{w}_1 = -w_1 + f \tag{10.21a}
$$

$$
\dot{w}_2 = 2w_1 - 3w_2 \tag{10.21b}
$$

$$
\dot{w}_3 = 5w_2 + \dot{w}_2 - 4w_3 \tag{10.21c}
$$

and the output equation is

$$
y = w_3
$$

The elimination of $\dot{w}_2$ from Eq. (10.21c) by using Eq. (10.21b) converts these equations into the desired state form:

$$
\begin{bmatrix} \dot{w}_1 \\ \dot{w}_2 \\ \dot{w}_3 \end{bmatrix} = \begin{bmatrix} -1 & 0 & 0 \\ 2 & -3 & 0 \\ 2 & 2 & -4 \end{bmatrix} \begin{bmatrix} w_1 \\ w_2 \\ w_3 \end{bmatrix} + \begin{bmatrix} 1 \\ 0 \\ 0 \end{bmatrix} f \tag{10.22a}
$$

and

$$
y = \begin{bmatrix} 0 & 0 & 1 \end{bmatrix} \begin{bmatrix} w_1 \\ w_2 \\ w_3 \end{bmatrix} \tag{10.22b}
$$

3. Parallel Realization (Diagonal Representation)

The three integrator outputs z_1, z_2, and z_3 in Fig. 10.5d are the state variables. The state equations are

$$\dot{z}_1 = -z_1 + f$$

$$\dot{z}_2 = -3z_2 + f$$

$$\dot{z}_3 = -4z_3 + f \tag{10.23a}$$

and the output equation is

$$y = \tfrac{4}{3}z_1 - 2z_2 + \tfrac{2}{3}z_3 \tag{10.23b}$$

Therefore the equations in the matrix form are

$$\begin{bmatrix} \dot{z}_1 \\ \dot{z}_2 \\ \dot{z}_n \end{bmatrix} = \begin{bmatrix} -1 & 0 & 0 \\ 0 & -3 & 0 \\ 0 & 0 & -4 \end{bmatrix} \begin{bmatrix} z_1 \\ z_2 \\ z_3 \end{bmatrix} + \begin{bmatrix} 1 \\ 1 \\ 1 \end{bmatrix} f \tag{10.24a}$$

$$y = \begin{bmatrix} \tfrac{4}{3} & -2 & \tfrac{2}{3} \end{bmatrix} \begin{bmatrix} z_1 \\ z_2 \\ z_3 \end{bmatrix} \tag{10.24b}$$

■

It is clear that a system has several state-space descriptions. Notable among these are the canonical-form variables and the diagonalized variables (in the parallel realization). State equations in these forms can be written immediately by inspection of the transfer function. Consider the general nth-order transfer function

$$H(s) = \frac{b_m s^m + b_{m-1} s^{m-1} + \cdots + b_1 s + b_0}{s^n + a_{n-1} s^{n-1} + \cdots + a_1 s + a_0} \tag{10.25a}$$

$$= \frac{b_m s^m + b_{m-1} s^{m-1} + \cdots + b_1 s + b_0}{(s - \lambda_1)(s - \lambda_2) \cdots (s - \lambda_n)}$$

$$= \frac{k_1}{s - \lambda_1} + \frac{k_2}{s - \lambda_2} + \cdots + \frac{k_n}{s - \lambda_n} \tag{10.25b}$$

Figures 10.6a and 10.6b show the realizations of $H(s)$, using the controller canonical form [Eq. (10.25a)] and the parallel form [Eq. (10.25b)], respectively.

The n integrator outputs x_1, x_2, $\ldots$, x_n in Fig. 10.6a are the state variables. It is clear that

$$\dot{x}_1 = x_2$$

$$\dot{x}_2 = x_3$$

$$\cdots \cdots \cdots \tag{10.26a}$$

$$\dot{x}_{n-1} = x_n$$

$$\dot{x}_n = -a_{n-1} x_n - a_{n-2} x_{n-1} - \cdots - a_1 x_2 - a_0 x_1 + f$$

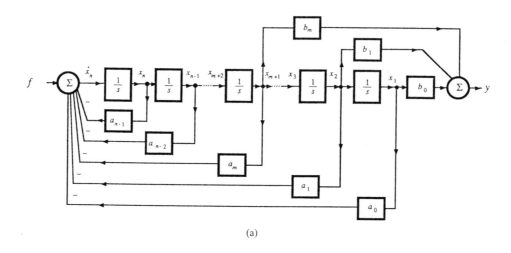

(a)

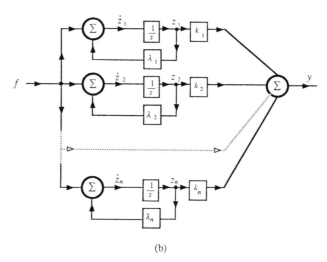

(b)

Fig. 10.6 Controller canonical and parallel realizations for an nth order LTIC system.

and output y is

$$y = b_0 x_1 + b_1 x_2 + \cdots + b_m x_{m+1} \tag{10.26b}$$

or

$$\begin{bmatrix} \dot{x}_1 \\ \dot{x}_2 \\ \cdots \\ \dot{x}_{n-1} \\ \dot{x}_n \end{bmatrix} = \begin{bmatrix} 0 & 1 & 0 & \cdots & 0 & 0 \\ 0 & 0 & 1 & \cdots & 0 & 0 \\ \cdots\cdots\cdots\cdots\cdots\cdots\cdots\cdots\cdots\cdots\cdots\cdots \\ 0 & 0 & 0 & \cdots & 0 & 1 \\ -a_0 & -a_1 & -a_2 & \cdots & -a_{n-2} & -a_{n-1} \end{bmatrix} \begin{bmatrix} x_1 \\ x_2 \\ \cdots \\ x_{n-1} \\ x_n \end{bmatrix} + \begin{bmatrix} 0 \\ 0 \\ \cdots \\ 0 \\ 1 \end{bmatrix} f$$

$$\tag{10.27a}$$

and

$$
[\, b_0 \quad b_1 \quad \cdots \quad b_m \quad 0 \quad \cdots \quad 0\,]
\begin{bmatrix} x_1 \\ x_2 \\ \vdots \\ x_n \end{bmatrix}
\tag{10.27b}
$$

Observe that these equations (state equations and output equation) can be written immediately by inspection of $H(s)$.

The n integrator outputs $z_1, z_2, \ldots, z_n$ in Fig. 10.6b are the state variables. It is clear that

$$
\dot{z}_1 = \lambda_1 z_1 + f
$$

$$
\dot{z}_2 = \lambda_2 z_2 + f
$$

$$
\cdots\cdots\cdots\cdots\cdots
\tag{10.28a}
$$

$$
\dot{z}_n = \lambda_n z_n + f
$$

and

$$
y = k_1 z_1 + k_2 z_2 + \cdots + k_n z_n
\tag{10.28b}
$$

or

$$
\begin{bmatrix} \dot{z}_1 \\ \dot{z}_2 \\ \cdots \\ \dot{z}_{n-1} \\ \dot{z}_n \end{bmatrix}
=
\begin{bmatrix}
\lambda_1 & 0 & \cdots & 0 & 0 \\
0 & \lambda_2 & \cdots & 0 & 0 \\
\cdots & \cdots & \cdots & \cdots & \cdots \\
0 & 0 & \cdots & \lambda_{n-1} & 0 \\
0 & 0 & \cdots & 0 & \lambda_n
\end{bmatrix}
\begin{bmatrix} z_1 \\ z_2 \\ \cdots \\ z_{n-1} \\ z_n \end{bmatrix}
+
\begin{bmatrix} 1 \\ 1 \\ \cdots \\ 1 \\ 1 \end{bmatrix} f
\tag{10.29a}
$$

and

$$
y = [\, k_1 \quad k_2 \quad \cdots \quad k_{n-1} \quad k_n\,]
\begin{bmatrix} z_1 \\ z_2 \\ \vdots \\ z_{n-1} \\ z_n \end{bmatrix}
\tag{10.29b}
$$

The state equation (10.29a) and the output equation (10.29b) can be written immediately by inspection of the transfer function $H(s)$ in Eq. (10.25b). Observe that the diagonalized form of the state matrix [Eq. (10.29a)] has the transfer function poles as its diagonal elements. The presence of repeated poles in $H(s)$ will modify the procedure slightly. The handling of these cases is discussed in Sec. 4.6.

It is clear from the above discussion that a state-space description is not unique. For any realization of $H(s)$ using integrators, scalar multipliers and adders, a corresponding state-space description exists. Since there are many possible realizations of $H(s)$, there are many possible state-space descriptions.

⊙ **Computer Example C10.1**

Use a computer to convert the following transfer function to its equivalent controller form state equation.

$$H(s) = \frac{s+1}{s^2 + 2s + 3}$$

```
% Enter coefficient of the numerator polynomial in descending order
num=[1 1];
% Enter coefficient of the denominator polynomial in descending order.
den=[1 2 3];
% Convert the transfer function to state-space form.
% Determine the size of the numerator (num) and denominator (den).
[M,N]=size(num);    [m,n]=size(den);
num=[zeros(M,n-N) num]; num=num./den(1); den=den./den(1);
A=[-den(2:n); eye(n-2,n-1)];   B=eye(n-1,1);
C=num(:,2:n)-num(:,1)*den(2:n); D=num(:,1);
% If the denominator is a constant scalar, then use the special case.
if n==1
A=0;   B=0;   C=0; D=num./den;
end
% Display the state space representation.
A,B,C,D,
disp('For a simpler method to Example C10.1, hit any key'),pause
```
Alternate (simpler) Method

```
% Enter coefficients of the numerator polynomial in descending order.
num=[1 1];
% Enter coefficients of the denominator polynomial in descending order.
den=[1 2 3];
% Convert the transfer function to state space form.
[A,B,C,D]=tf2ss(num,den);
% Display the state-space representation.
A,B,C,D   ⊙
```

⊙ **Computer Example C10.2**

Use a computer to convert the following transfer function to its equivalent observer form state equation.

$$H(s) = \frac{s+1}{s^2 + 2s + 3}$$

```
% Enter coefficients of the numerator polynomial in descending order.
num=[1 1];
% Enter coefficients of the denominator polynomial in descending order.
den=[1 2 3];
% Convert transfer function to observer state space form.
% Determine the size of the numerator (num) and denominator (den).
[M,N]=size(num); [m,n]=size(den);
num=[zeros(M,n-N) num]; num=num./den(1);   den=den./den(1);
```

```
A=[-den(2:n); eye(n-2,n-1)]'; B=(num(:,2:n)-num(:,1)*den(2:n))';
C=(eye(n-1,1))';  D=num(:,1);
% If the denominator is a constant scalar, then use this special case.
if n==1
A=0; B=0; C=0; D=num./den;
end
% Display the observer state-space representation.
A,B,C,D,
disp('For a simpler method to Example C10.2.  Hit any key'),pause
```

Alternate (simpler) Method

```
% Enter Coefficients of the numerator polynomial in descending order.
num=[1 1];
% Enter coefficients of the denominator polynomial in descending order.
den=[1 2 3];
% Convert the transfer function to state space form.
[a,b,c,d]=tf2ss(num,den);
% Converts to observer form.
A=a';    B=c';    C=b';    D=d;
% Displays the observer state-space representation.
A,B,C,D   ⊙
```

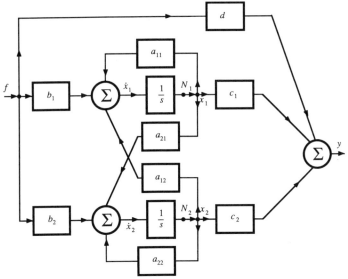

Fig. 10.7   Realization of a second-order system.

Realization

Consider a system with a single input f, a single output y, and two state variables, x_1 and x_2. The system equations are

$$\dot{x}_1 = a_{11}x_1 + a_{12}x_2 + b_1 f$$

and

$$\dot{x}_2 = a_{21}x_1 + a_{22}x_2 + b_2 f \qquad (10.30a)$$

$$y = c_1 x_1 + c_2 x_2 + df \qquad (10.30b)$$

Figure 10.7 shows the block diagram of the realized system. The initial conditions $x_1(0)$ and $x_2(0)$ should be applied at N_1 and N_2. This procedure can be easily extended to general multiple-input, multiple-output systems with n state variables.

10.3   SOLUTION OF STATE EQUATIONS

The state equations of a linear system are n simultaneous linear differential equations of the first order. We studied the techniques of solving simultaneous linear differential equations in Chapters 2 and 4. The same techniques can be applied to state equations without any modification. However, it is more convenient to carry out the solution in the framework of matrix notation.

These equations can be solved in both the time domain and frequency domain (Laplace transform). The latter requires fewer new concepts and is therefore easier to deal with than the time-domain solution. For this reason, we shall first consider the Laplace transform solution.

10.3-1   Laplace Transform Solution of State Equations

The kth state equation [Eq. (10.6a)] is of the form

$$\dot{x}_k = a_{k1}x_1 + a_{k2}x_2 + \cdots + a_{kn}x_n + b_{k1}f_1 + b_{k2}f_2 + \cdots + b_{kj}f_j \qquad (10.31a)$$

We shall take the Laplace transform of this equation. Let

$$x_k(t) \Longleftrightarrow X_k(s)$$

so that

$$\dot{x}_k(t) \Longleftrightarrow sX_k(s) - x_k(0)$$

Also, let

$$f_i(t) \Longleftrightarrow F_i(s)$$

The Laplace transform of Eq. (10.31a) yields

$$sX_k(s) - x_k(0) = a_{k1}X_1(s) + a_{k2}X_2(s) + \cdots + a_{kn}X_n(s) + b_{k1}F_1(s)$$
$$+ b_{k2}F_2(s) + \cdots + b_{kj}F_j(s) \qquad (10.31b)$$

Taking the Laplace transforms of all n state equations, we obtain

$$s\underbrace{\begin{bmatrix} X_1(s) \\ X_2(s) \\ \cdots \\ X_n(s) \end{bmatrix}}_{\mathbf{X}(s)} - \underbrace{\begin{bmatrix} x_1(0) \\ x_2(0) \\ \cdots \\ x_n(0) \end{bmatrix}}_{\mathbf{x}(0)} = \underbrace{\begin{bmatrix} a_{11} & a_{12} & \cdots & a_{1n} \\ a_{21} & a_{22} & \cdots & a_{2n} \\ \cdots\cdots\cdots\cdots\cdots\cdots\cdots \\ a_{n1} & a_{n2} & \cdots & a_{nn} \end{bmatrix}}_{\mathbf{A}} \underbrace{\begin{bmatrix} X_1(s) \\ X_2(s) \\ \cdots \\ X_n(s) \end{bmatrix}}_{\mathbf{X}(s)}$$

$$+ \underbrace{\begin{bmatrix} b_{11} & b_{12} & \cdots & b_{1j} \\ b_{21} & b_{22} & \cdots & b_{2j} \\ \cdots\cdots\cdots\cdots\cdots\cdots \\ b_{n1} & b_{n2} & \cdots & b_{nj} \end{bmatrix}}_{\mathbf{B}} \underbrace{\begin{bmatrix} F_1(s) \\ F_2(s) \\ \cdots \\ F_j(s) \end{bmatrix}}_{\mathbf{F}(s)} \qquad (10.32a)$$

Defining the vectors as shown above, we have

$$s\mathbf{X}(s) - \mathbf{x}(0) = \mathbf{A}\mathbf{X}(s) + \mathbf{B}\mathbf{F}(s)$$

or

$$s\mathbf{X}(s) - \mathbf{A}\mathbf{X}(s) = \mathbf{x}(0) + \mathbf{B}\mathbf{F}(s)$$

and

$$(s\mathbf{I} - \mathbf{A})\mathbf{X}(s) = \mathbf{x}(0) + \mathbf{B}\mathbf{F}(s) \qquad (10.32b)$$

where $\mathbf{I}$ is the $n \times n$ identity matrix. From Eq. 10.32b, we have

$$\mathbf{X}(s) = (s\mathbf{I} - \mathbf{A})^{-1}[\mathbf{x}(0) + \mathbf{B}\mathbf{F}(s)] \qquad (10.33a)$$

$$= \mathbf{\Phi}(s)[\mathbf{x}(0) + \mathbf{B}\mathbf{F}(s)] \qquad (10.33b)$$

where

$$\mathbf{\Phi}(s) = (s\mathbf{I} - \mathbf{A})^{-1} \qquad (10.34)$$

Thus, from Eq. (10.33b),

$$\mathbf{X}(s) = \mathbf{\Phi}(s)\mathbf{x}(0) + \mathbf{\Phi}(s)\mathbf{B}\mathbf{F}(s) \qquad (10.35a)$$

and

$$\mathbf{x}(t) = \underbrace{\mathcal{L}^{-1}[\mathbf{\Phi}(s)]\mathbf{x}(0)}_{\text{zero-input component}} + \underbrace{\mathcal{L}^{-1}[\mathbf{\Phi}(s)\mathbf{B}\mathbf{F}(s)]}_{\text{zero-state component}} \qquad (10.35b)$$

Equation (10.35b) gives the desired solution. Observe the two components of the solution. The first component yields $x(t)$ when the input $f(t) = 0$. Hence this is the zero-input component. In a similar manner, we see that the second component is the zero-state component.

■ Example 10.5

Find the state vector $\mathbf{x}(t)$ for the system whose state equation is given by

$$\dot{\mathbf{x}} = \mathbf{A}\mathbf{x} + \mathbf{B}f$$

where

$$\mathbf{A} = \begin{bmatrix} -12 & \frac{2}{3} \\ -36 & -1 \end{bmatrix} \qquad \mathbf{B} = \begin{bmatrix} \frac{1}{3} \\ 1 \end{bmatrix} \qquad \mathbf{f}(t) = u(t)$$

and the initial conditions are: $x_1(0) = 2$, $x_2(0) = 1$.

From Eq. (10.33b), we have

$$\mathbf{X}(s) = \mathbf{\Phi}(s)[\mathbf{x}(0) + \mathbf{B}\mathbf{F}(s)]$$

Let us first find $\mathbf{\Phi}(s)$. We have

and

$$(s\mathbf{I} - \mathbf{A}) = s \begin{bmatrix} 1 & 0 \\ 0 & 1 \end{bmatrix} - \begin{bmatrix} -12 & \frac{2}{3} \\ -36 & -1 \end{bmatrix} = \begin{bmatrix} s+12 & -\frac{2}{3} \\ 36 & s+1 \end{bmatrix}$$

$$\mathbf{\Phi}(s) = (s\mathbf{I} - \mathbf{A})^{-1} = \begin{bmatrix} \frac{s+1}{(s+4)(s+9)} & \frac{2/3}{(s+4)(s+9)} \\ \frac{-36}{(s+4)(s+9)} & \frac{s+12}{(s+4)(s+9)} \end{bmatrix} \qquad (10.36a)$$

Now $\mathbf{x}(0)$ is given as

$$\mathbf{x}(0) = \begin{bmatrix} 2 \\ 1 \end{bmatrix}$$

Also, $F(s) = \frac{1}{s}$, and

$$\mathbf{B}F(s) = \begin{bmatrix} \frac{1}{3} \\ 1 \end{bmatrix} \frac{1}{s} = \begin{bmatrix} \frac{1}{3s} \\ \frac{1}{s} \end{bmatrix}$$

Therefore

$$\mathbf{x}(0) + \mathbf{B}F(s) = \begin{bmatrix} 2 + \frac{1}{3s} \\ 1 + \frac{1}{s} \end{bmatrix} = \begin{bmatrix} \frac{6s+1}{3s} \\ \frac{s+1}{s} \end{bmatrix}$$

and

$$\mathbf{X}(s) = \mathbf{\Phi}(s)[\mathbf{x}(0) + \mathbf{B}F(s)]$$

$$= \begin{bmatrix} \frac{s+1}{(s+4)(s+9)} & \frac{\frac{2}{3}}{(s+4)(s+9)} \\ \frac{-36}{(s+4)(s+9)} & \frac{s+12}{(s+4)(s+9)} \end{bmatrix} \begin{bmatrix} \frac{6s+1}{3s} \\ \frac{s+1}{s} \end{bmatrix}$$

$$= \begin{bmatrix} \frac{2s^2+3s+1}{s(s+4)(s+9)} \\ \frac{s-59}{(s+4)(s+9)} \end{bmatrix}$$

$$= \begin{bmatrix} \frac{\frac{1}{36}}{s} - \frac{\frac{21}{20}}{s+4} + \frac{\frac{136}{45}}{s+9} \\ \frac{-\frac{63}{5}}{s+4} + \frac{\frac{68}{5}}{s+9} \end{bmatrix}$$

The inverse Laplace transform of this equation yields

$$\begin{bmatrix} x_1(t) \\ x_2(t) \end{bmatrix} = \begin{bmatrix} \left(\frac{1}{36} - \frac{21}{20}e^{-4t} + \frac{136}{45}e^{-9t} \right)u(t) \\ \left(-\frac{63}{5}e^{-4t} + \frac{68}{5}e^{-9t} \right)u(t) \end{bmatrix} \qquad (10.36b)$$

■

The Output

The output equation is given by

$$\mathbf{y} = \mathbf{Cx} + \mathbf{Df}$$

and

$$\mathbf{Y}(s) = \mathbf{CX}(s) + \mathbf{DF}(s)$$

The substitution of Eq. (10.33b) into this equation yields

$$\mathbf{Y}(s) = \mathbf{C}\{\mathbf{\Phi}(s)[\mathbf{x}(0) + \mathbf{BF}(s)]\} + \mathbf{DF}(s)$$

$$= \underbrace{\mathbf{C}\mathbf{\Phi}(s)\mathbf{x}(0)}_{\text{zero-input response}} + \underbrace{[\mathbf{C}\mathbf{\Phi}(s)\mathbf{B} + \mathbf{D}]\mathbf{F}(s)}_{\text{zero-state response}} \qquad (10.37)$$

The zero-state response (that is, the response $\mathbf{Y}(s)$ when $\mathbf{x}(0)=0$), is given by

$$\mathbf{Y}(s) = [\mathbf{C}\mathbf{\Phi}(s)\mathbf{B} + \mathbf{D}]\mathbf{F}(s) \qquad (10.38a)$$

Note that the transfer function of a system is defined under the zero-state condition [see Eq. (4.40)]. The matrix $\mathbf{C}\mathbf{\Phi}(s)\mathbf{B} + \mathbf{D}$ is the **transfer function matrix** $\mathbf{H}(s)$ of the system, which relates the responses y_1, y_2, ... , y_k to the inputs f_1, f_2, ... , f_j:

$$\mathbf{H}(s) = \mathbf{C}\mathbf{\Phi}(s)\mathbf{B} + \mathbf{D} \qquad (10.38b)$$

and the zero-state response is

$$\mathbf{Y}(s) = \mathbf{H}(s)\mathbf{F}(s) \qquad (10.39)$$

The matrix $\mathbf{H}(s)$ is a $k \times j$ matrix (k is the number of outputs and j being the number of inputs). The ijth element $H_{ij}(s)$ of $H(s)$ is the transfer function that relates the output $y_i(t)$ to the input $f_j(t)$.

■ **Example 10.6**

Let us consider a system with a state equation

$$\begin{bmatrix} \dot{x}_1 \\ \dot{x}_2 \end{bmatrix} = \begin{bmatrix} 0 & 1 \\ -2 & -3 \end{bmatrix} \begin{bmatrix} x_1 \\ x_2 \end{bmatrix} + \begin{bmatrix} 1 & 0 \\ 1 & 1 \end{bmatrix} \begin{bmatrix} f_1 \\ f_2 \end{bmatrix} \qquad (10.40a)$$

and an output equation

$$\begin{bmatrix} y_1 \\ y_2 \\ y_3 \end{bmatrix} = \begin{bmatrix} 1 & 0 \\ 1 & 1 \\ 0 & 2 \end{bmatrix} \begin{bmatrix} x_1 \\ x_2 \end{bmatrix} + \begin{bmatrix} 0 & 0 \\ 1 & 0 \\ 0 & 1 \end{bmatrix} \begin{bmatrix} f_1 \\ f_2 \end{bmatrix} \qquad (10.40b)$$

In this case,

$$\mathbf{A} = \begin{bmatrix} 0 & 1 \\ -2 & -3 \end{bmatrix} \quad \mathbf{B} = \begin{bmatrix} 1 & 0 \\ 1 & 1 \end{bmatrix} \quad \mathbf{C} = \begin{bmatrix} 1 & 0 \\ 1 & 1 \\ 0 & 2 \end{bmatrix} \quad \mathbf{D} = \begin{bmatrix} 0 & 0 \\ 1 & 0 \\ 0 & 1 \end{bmatrix} \qquad (10.40c)$$

and

$$\mathbf{\Phi}(s) = (s\mathbf{I} - \mathbf{A})^{-1} = \begin{bmatrix} s & -1 \\ 2 & s+3 \end{bmatrix}^{-1} = \begin{bmatrix} \frac{s+3}{(s+1)(s+2)} & \frac{1}{(s+1)(s+2)} \\ \frac{-2}{(s+1)(s+2)} & \frac{s}{(s+1)(s+2)} \end{bmatrix} \qquad (10.41)$$

Hence the transfer function matrix $\mathbf{H}(s)$ is given by

$$\mathbf{H}(s) = \mathbf{C\Phi}(s)\mathbf{B} + \mathbf{D}$$

$$
= \begin{bmatrix} 1 & 0 \\ 1 & 1 \\ 0 & 2 \end{bmatrix}
\begin{bmatrix} \frac{s+3}{(s+1)(s+2)} & \frac{1}{(s+1)(s+2)} \\ \frac{-2}{(s+1)(s+2)} & \frac{s}{(s+1)(s+2)} \end{bmatrix}
\begin{bmatrix} 1 & 0 \\ 1 & 1 \end{bmatrix}
+ \begin{bmatrix} 0 & 0 \\ 1 & 0 \\ 0 & 1 \end{bmatrix}
$$

$$
= \begin{bmatrix} \frac{s+4}{(s+1)(s+2)} & \frac{1}{(s+1)(s+2)} \\ \frac{s+4}{s+2} & \frac{1}{s+2} \\ \frac{2(s-2)}{(s+1)(s+2)} & \frac{s^2+5s+2}{(s+1)(s+2)} \end{bmatrix}
\tag{10.42}
$$

and the zero-state response is

$$\mathbf{Y}(s) = \mathbf{H}(s)\mathbf{F}(s)$$

Remember that the ijth element of the transfer function matrix in Eq. (10.42) represents the transfer function that relates the output $y_i(t)$ to the input $f_j(t)$. Thus the transfer function that relates the output y_3 to the input f_2 is $H_{32}(s)$, where

$$H_{32}(s) = \frac{s^2 + 5s + 2}{(s+1)(s+2)} \qquad \blacksquare$$

Characteristic Roots (Eigenvalues) of a Matrix

It is interesting to observe that the denominator of every transfer function in Eq. (10.42) is $(s+1)(s+2)$ with the exception of $H_{21}(s)$ and $H_{22}(s)$, where the cancellation of the factor $(s+1)$ occurs. This is no coincidence. It can be seen easily that the denominator of every element of $\mathbf{\Phi}(s)$ is $|s\mathbf{I} - \mathbf{A}|$ because $\mathbf{\Phi}(s) = (s\mathbf{I} - \mathbf{A})^{-1}$, and the inverse of a matrix has its determinant in the denominator. Since $\mathbf{C}$, $\mathbf{B}$, and $\mathbf{D}$ are matrices with constant elements, it can be seen from Eq. (10.38b) that the denominator of $\mathbf{\Phi}(s)$ will also be the denominator of $\mathbf{H}(s)$. Hence the denominator of every element of $\mathbf{H}(s)$ is $|s\mathbf{I} - \mathbf{A}|$, except for the possible cancellation of the common factors mentioned earlier. In other words, the poles of all transfer functions of the system are also the zeros of the polynomial $|s\mathbf{I} - \mathbf{A}|$. **Therefore the zeros of the polynomial $|s\mathbf{I} - \mathbf{A}|$ are the characteristic roots of the system.** Hence the characteristic roots of the system are the roots of the equation

$$|s\mathbf{I} - \mathbf{A}| = 0 \tag{10.43a}$$

Since $|s\mathbf{I} - \mathbf{A}|$ is an nth-order polynomial in s with n zeros $\lambda_1, \lambda_2, \ldots, \lambda_n$, we can write Eq. (10.43a) as

$$
\begin{aligned}
|s\mathbf{I} - \mathbf{A}| &= s^n + a_{n-1}s^{n+1} + \cdots + a_1 s + a_0 \\
&= (s - \lambda_1)(s - \lambda_2) \cdots (s - \lambda_n) = 0
\end{aligned}
\tag{10.43b}
$$

For the system in Example 10.6,

$$|s\mathbf{I} - \mathbf{A}| = \begin{vmatrix} s & 0 \\ 0 & s \end{vmatrix} - \begin{vmatrix} 0 & 1 \\ -2 & -3 \end{vmatrix}$$

$$= \begin{vmatrix} s & -1 \\ 2 & s+3 \end{vmatrix}$$

$$= s^2 + 3s + 2 \tag{10.44a}$$

$$= (s+1)(s+2) \tag{10.44b}$$

Hence

$$\lambda_1 = -1 \quad \text{and} \quad \lambda_2 = -2$$

Equation (10.43) is known as the **characteristic equation of the matrix A**, and λ_1, λ_2, ... , λ_n are the characteristic roots of **A**. The term **eigenvalue**, meaning "characteristic value" in German, is also commonly used in the literature. Thus we have shown that the characteristic roots of a system are the eigenvalues (characteristic values) of the matrix **A**.

At this point, the reader will recall that if λ_1, λ_2, ... , λ_n are the poles of the transfer function, then the zero-input response is of the form

$$y_0(t) = c_1 e^{\lambda_1 t} + c_2 e^{\lambda_2 t} + \cdots + c_n e^{\lambda_n t} \tag{10.45}$$

This fact is also obvious from Eq. (10.38). The denominator of every element of the zero-input response matrix $\mathbf{C\Phi}(s)\mathbf{x}(0)$ is $|s\mathbf{I} - \mathbf{A}| = (s - \lambda_1)(s - \lambda_2)\cdots(s - \lambda_n)$. Therefore the partial-fraction expansion and the subsequent inverse Laplace transform will yield a zero-input component of the form in Eq. (10.45).

10.3-2 Time-Domain Solution of State Equations

The state equation is

$$\dot{\mathbf{x}} = \mathbf{A}\mathbf{x} + \mathbf{B}\mathbf{f} \tag{10.46}$$

We now show that the solution of the vector differential Equation (10.46) is

$$\mathbf{x}(t) = e^{\mathbf{A}t}\mathbf{x}(0) + \int_0^t e^{\mathbf{A}(t-\tau)}\mathbf{B}\mathbf{f}(\tau)\,d\tau \tag{10.47}$$

Before proceeding further, we must define the exponential of the matrix appearing in Eq. (10.47). An exponential of a matrix is defined by an infinite series identical to that used in defining an exponential of a scalar. We shall define

$$e^{\mathbf{A}t} = \mathbf{I} + \mathbf{A}t + \frac{\mathbf{A}^2 t^2}{2!} + \frac{\mathbf{A}^3 t^3}{3!} + \cdots + \frac{\mathbf{A}^n t^n}{n!} + \cdots + \tag{10.48a}$$

$$= \sum_{k=0}^{\infty} \frac{\mathbf{A}^k t^k}{k!} \tag{10.48b}$$

Thus if

$$\mathbf{A} = \begin{bmatrix} 0 & 1 \\ 2 & 1 \end{bmatrix}$$

then

$$\mathbf{A}t = \begin{bmatrix} 0 & 1 \\ 2 & 1 \end{bmatrix} t = \begin{bmatrix} 0 & t \\ 2t & t \end{bmatrix} \tag{10.49}$$

and

$$\frac{\mathbf{A}^2 t^2}{2!} = \begin{bmatrix} 0 & 1 \\ 2 & 1 \end{bmatrix} \begin{bmatrix} 0 & 1 \\ 2 & 1 \end{bmatrix} \frac{t^2}{2} = \begin{bmatrix} 2 & 1 \\ 2 & 3 \end{bmatrix} \frac{t^2}{2} = \begin{bmatrix} t^2 & \frac{t^2}{2} \\ t^2 & \frac{3t^2}{2} \end{bmatrix} \tag{10.50}$$

and so on.

It can be shown that the infinite series in Eq. (10.48) is absolutely and uniformly convergent for all values of t. Consequently it can be differentiated or integrated term by term. Thus, to find $(d/dt)e^{\mathbf{A}t}$, we differentiate the series on the right-hand side of Eq. (10.48a) term by term:

$$\frac{d}{dt}e^{\mathbf{A}t} = \mathbf{A} + \mathbf{A}^2 t + \frac{\mathbf{A}^3 t^2}{2!} + \frac{\mathbf{A}^4 t^3}{3!} + \cdots \tag{10.51a}$$

$$= \mathbf{A}\left[\mathbf{I} + \mathbf{A}t + \frac{\mathbf{A}^2 t^2}{2!} + \frac{\mathbf{A}^3 t^3}{3!} + \cdots\right]$$

$$= \mathbf{A}e^{\mathbf{A}t} \tag{10.51b}$$

Note that the infinite series on the right-hand side of Eq. (10.51a) also may be expressed as

$$\frac{d}{dt}e^{\mathbf{A}t} = \left[\mathbf{I} + \mathbf{A}t + \frac{\mathbf{A}^2 t^2}{2!} + \frac{\mathbf{A}^3 t^3}{3!} + \cdots + \cdots\right] \mathbf{A}$$

$$= e^{\mathbf{A}t}\mathbf{A}$$

Hence

$$\frac{d}{dt}e^{\mathbf{A}t} = \mathbf{A}e^{\mathbf{A}t} = e^{\mathbf{A}t}\mathbf{A} \tag{10.52}$$

Also note that from the definition (10.48a), it follows that

$$e^{\mathbf{0}} = \mathbf{I} \tag{10.53a}$$

where

$$\mathbf{I} = \begin{bmatrix} 1 & 0 \\ 0 & 1 \end{bmatrix}$$

If we premultiply or postmultiply the infinite series for $e^{\mathbf{A}t}$ [Eq. (10.48a)] by an infinite series for $e^{-\mathbf{A}t}$, we find that

$$(e^{-\mathbf{A}t})(e^{\mathbf{A}t}) = (e^{\mathbf{A}t})(e^{-\mathbf{A}t}) = \mathbf{I} \tag{10.53b}$$

It is shown in Sec. B.9-3 that

$$\frac{d}{dt}(\mathbf{PQ}) = \frac{d\mathbf{P}}{dt}\mathbf{Q} + \mathbf{P}\frac{d\mathbf{Q}}{dt}$$

Using this relationship, we observe that

$$\frac{d}{dt}[e^{-\mathbf{A}t}\mathbf{x}] = \left(\frac{d}{dt}e^{-\mathbf{A}t}\right)\mathbf{x} + e^{-\mathbf{A}t}\dot{\mathbf{x}}$$

$$= -e^{-\mathbf{A}t}\mathbf{A}\mathbf{x} + e^{-\mathbf{A}t}\dot{\mathbf{x}} \tag{10.54}$$

We now premultiply both sides of Eq. (10.46) by $e^{-\mathbf{A}t}$ to yield

$$e^{-\mathbf{A}t}\dot{\mathbf{x}} = e^{-\mathbf{A}t}\mathbf{A}\mathbf{x} + e^{-\mathbf{A}t}\mathbf{B}\mathbf{f} \tag{10.55a}$$

or

$$e^{-\mathbf{A}t}\dot{\mathbf{x}} - e^{-\mathbf{A}t}\mathbf{A}\mathbf{x} = e^{-\mathbf{A}t}\mathbf{B}\mathbf{f} \tag{10.55b}$$

A glance at Eq. (10.54) shows that the left-hand side of Eq. (10.55b) is $\frac{d}{dt}[e^{-\mathbf{A}t}]$. Hence

$$\frac{d}{dt}[e^{-\mathbf{A}t}] = e^{-\mathbf{A}t}\mathbf{B}\mathbf{f}$$

The integration of both sides of this equation from 0 to t yields

$$e^{-\mathbf{A}t}\mathbf{x}\Big|_0^t = \int_0^t e^{-\mathbf{A}\tau}\mathbf{B}\mathbf{f}(\tau)\,d\tau \tag{10.56a}$$

or

$$e^{-\mathbf{A}t}\mathbf{x}(t) - \mathbf{x}(0) = \int_0^t e^{-\mathbf{A}\tau}\mathbf{B}\mathbf{f}(\tau)\,d\tau \tag{10.56b}$$

Hence

$$e^{-\mathbf{A}t}\mathbf{x} = \mathbf{x}(0) + \int_0^t e^{-\mathbf{A}\tau}\mathbf{B}\mathbf{f}(\tau)\,d\tau \tag{10.56c}$$

Premultiplying Eq. (10.56c) by $e^{\mathbf{A}t}$ and using Eq. (10.53b), we have

$$\mathbf{x}(t) = \underbrace{e^{\mathbf{A}t}\mathbf{x}(0)}_{\text{zero-input component}} + \underbrace{\int_0^t e^{\mathbf{A}(t-\tau)}\mathbf{B}\mathbf{f}(\tau)\,d\tau}_{\text{zero-state component}} \tag{10.57a}$$

This is the desired solution. The first term on the right-hand side represents $x(t)$ when the input $f(t) = 0$. Hence it is the zero-input component. The second term, by a similar argument, is seen to be the zero-state component.

The results of Eq. (10.57a) can be expressed more conveniently in terms of the matrix convolution. We can define the convolution of two matrices in a manner similar to the multiplication of two matrices, except that the multiplication of two elements is replaced by their convolution. For example,

$$\begin{bmatrix} f_1 & f_2 \\ f_3 & f_4 \end{bmatrix} * \begin{bmatrix} g_1 & g_2 \\ g_3 & g_4 \end{bmatrix} = \begin{bmatrix} (f_1 * g_1 + f_2 * g_3) & (f_1 * g_2 + f_2 * g_4) \\ (f_3 * g_1 + f_4 * g_3) & (f_3 * g_2 + f_4 * g_4) \end{bmatrix}$$

Using this definition of matrix convolution, we can express Eq. (10.57a) as

$$\mathbf{x}(t) = e^{\mathbf{A}t}\mathbf{x}(0) + e^{\mathbf{A}t} * \mathbf{B}\mathbf{f}(t) \tag{10.57b}$$

Note that the limits of the convolution integral [Eq. (10.57a)] are from 0 to t. Hence all the elements of $e^{\mathbf{A}t}$ in the convolution term of Eq. (10.57b) are implicitly assumed to be multiplied by $u(t)$.

The result of Eq. (10.57) can be easily generalized for any initial value of t. It is left as an exercise for the reader to show that the solution of the state equation can be expressed as

$$\mathbf{x}(t) = e^{\mathbf{A}(t-t_0)}\mathbf{x}(t_0) + \int_{t_0}^{t} e^{\mathbf{A}(t-\tau)}\mathbf{B}\mathbf{f}(\tau)\, d\tau \tag{10.58}$$

Determining $e^{\mathbf{A}t}$

The exponential $e^{\mathbf{A}t}$ required in Eq. (10.57) can be computed from the definition in Eq. (10.51a). Unfortunately, this is an infinite series, and its computation can be quite laborious. Moreover, we may not be able to recognize the closed-form expression for the answer. There are several efficient methods of determining $e^{\mathbf{A}t}$ in closed form. It is shown in Sec. B.9-5 that for an $n \times n$ matrix $\mathbf{A}$,

$$e^{\mathbf{A}t} = \beta_0 \mathbf{I} + \beta_1 \mathbf{A} + \beta_2 \mathbf{A}^2 + \cdots + \beta_{n-1}\mathbf{A}^{n-1} \tag{10.59a}$$

where

$$\begin{bmatrix} \beta_0 \\ \beta_1 \\ \cdots \\ \beta_{n-1} \end{bmatrix} = \begin{bmatrix} 1 & \lambda_1 & \lambda_1^2 & \cdots & \lambda_1^{n-1} \\ 1 & \lambda_2 & \lambda_2^2 & \cdots & \lambda_2^{n-1} \\ \cdots\cdots\cdots\cdots\cdots\cdots\cdots \\ 1 & \lambda_n & \lambda_n^2 & \cdots & \lambda_n^{n-1} \end{bmatrix}^{-1} \begin{bmatrix} e^{\lambda_1 t} \\ e^{\lambda_2 t} \\ \cdots \\ e^{\lambda_n t} \end{bmatrix}$$

and $\lambda_1, \lambda_2, \ldots, \lambda_n$ are the n characteristic values (eigenvalues) of $\mathbf{A}$.

We can also determine $e^{\mathbf{A}t}$ by comparing Eqs. (10.57a) and (10.35b). It is clear that

$$e^{\mathbf{A}t} = \mathcal{L}^{-1}[\mathbf{\Phi}(s)] \tag{10.59b}$$

$$= \mathcal{L}^{-1}[(s\mathbf{I} - \mathbf{A})^{-1}] \tag{10.59c}$$

Thus $e^{\mathbf{A}t}$ and $\mathbf{\Phi}(s)$ are a Laplace transform pair. To be consistent with Laplace transform notation, $e^{\mathbf{A}t}$ is often denoted by $\boldsymbol{\phi}(t)$, **the state transition matrix** (STM):

$$e^{\mathbf{A}t} = \boldsymbol{\phi}(t)$$

■ **Example 10.7**

Find the solution to the problem in Example 10.5 using the time-domain method. For this case, the characteristic roots are given by

$$|s\mathbf{I} - \mathbf{A}| = \begin{vmatrix} s + 12 & -\frac{2}{3} \\ 36 & s + 1 \end{vmatrix} = s^2 + 13s + 36 = (s+4)(s+9) = 0$$

The roots are $\lambda_1 = -4$ and $\lambda_2 = -9$, so

$$\begin{bmatrix} \beta_0 \\ \beta_1 \end{bmatrix} = \begin{bmatrix} 1 & -4 \\ 1 & -9 \end{bmatrix}^{-1} \begin{bmatrix} e^{-4t} \\ e^{-9t} \end{bmatrix} = \frac{1}{5} \begin{bmatrix} 9e^{-4t} - 4e^{-9t} \\ e^{-4t} - e^{-9t} \end{bmatrix}$$

and

$$e^{\mathbf{A}t} = \beta_0 \mathbf{I} + \beta_1 \mathbf{A}$$

$$= \left(\frac{9}{5}e^{-4t} - \frac{4}{5}e^{-9t} \right) \begin{bmatrix} 1 & 0 \\ 0 & 1 \end{bmatrix} + \left(\frac{1}{5}e^{-4t} - \frac{1}{5}e^{-9t} \right) \begin{bmatrix} -12 & \frac{2}{3} \\ -36 & -1 \end{bmatrix}$$

$$= \begin{bmatrix} \left(\frac{-3}{5}e^{-4t} + \frac{8}{5}e^{-9t} \right) & \frac{2}{15}(e^{-4t} - e^{-9t}) \\ \frac{36}{5}(-e^{-4t} + e^{-9t}) & \left(\frac{8}{5}e^{-4t} - \frac{3}{5}e^{-9t} \right) \end{bmatrix} \quad (10.60)$$

The zero-input component is given by [see Eq. (10.57a)]

$$e^{\mathbf{A}t}\mathbf{x}(0) = \begin{bmatrix} \left(-\frac{3}{5}e^{-4t} + \frac{8}{5}e^{-9t} \right) & \frac{2}{15}(e^{-4t} - e^{-9t}) \\ \frac{36}{5}(-e^{-4t} + e^{-9t}) & \left(\frac{8}{5}e^{-4t} - \frac{3}{5}e^{-9t} \right) \end{bmatrix} \begin{bmatrix} 2 \\ 1 \end{bmatrix}$$

$$= \begin{bmatrix} \left(\frac{-16}{15}e^{-4t} + \frac{46}{15}e^{-9t} \right) u(t) \\ \left(\frac{-64}{5}e^{-4t} + \frac{69}{5}e^{-9t} \right) u(t) \end{bmatrix} \quad (10.61a)$$

Note the presence of $u(t)$ in Eq. (10.61a), indicating that the response begins at $t = 0$.
The zero-state component is $e^{\mathbf{A}t} * \mathbf{B}f$ [see Eq. (10.57b)], where

$$\mathbf{B}f = \begin{bmatrix} \frac{1}{3} \\ 1 \end{bmatrix} u(t) = \begin{bmatrix} \frac{1}{3}u(t) \\ u(t) \end{bmatrix}$$

and

$$e^{\mathbf{A}t} * \mathbf{B}f(t) = \begin{bmatrix} (\frac{-3}{5}e^{-4t} + \frac{8}{5}e^{-9t})u(t) & \frac{2}{15}(e^{-4t} - e^{-9t})u(t) \\ \frac{36}{5}(-e^{-4t} + e^{-9t}u(t)) & (\frac{8}{5}e^{-4t} - \frac{3}{5}e^{-9t})u(t) \end{bmatrix} * \begin{bmatrix} \frac{1}{3}u(t) \\ u(t) \end{bmatrix}$$

Note again the presence of the term $u(t)$ in every element of $e^{\mathbf{A}t}$. This is the case because
the limits of the convolution integral run from 0 to t [Eq. (10.56)]. Thus

$$e^{\mathbf{A}t} * \mathbf{B}f(t) = \begin{bmatrix} \left(-\frac{3}{5}e^{-4t} + \frac{8}{5}e^{-9t} \right) u(t) * \frac{1}{3}u(t) & \frac{2}{15}(e^{-4t} - e^{-9t})u(t) * u(t) \\ \frac{36}{5}(-e^{-4t} + e^{-9t})u(t) * \frac{1}{3}u(t) & \left(\frac{8}{5}e^{-4t} - \frac{3}{5}e^{-9t} \right) u(t) * u(t) \end{bmatrix}$$

$$= \begin{bmatrix} -\frac{1}{15}e^{-4t}u(t) * u(t) + \frac{2}{5}e^{-9t}u(t) * u(t) \\ -\frac{4}{5}e^{-4t}u(t) * u(t) + \frac{9}{5}e^{-9t}u(t) * u(t) \end{bmatrix}$$

The above convolution integrals are now found from the convolution table (Table 2.2). This yields

$$
e^{\mathbf{A}t} * \mathbf{B}f(t) = \begin{bmatrix} -\frac{1}{60}(1 - e^{-4t})u(t) + \frac{2}{45}(1 - e^{-9t})u(t) \\ -\frac{1}{5}(1 - e^{-4t})u(t) + \frac{1}{5}(1 - e^{-9t})u(t) \end{bmatrix}
$$

$$
= \begin{bmatrix} (\frac{1}{36} + \frac{1}{60}e^{-4t} - \frac{2}{45}e^{-9t})u(t) \\ \frac{1}{5}(e^{-4t} - e^{-9t})u(t) \end{bmatrix} \tag{10.61b}
$$

The sum of the two components [Eq. (10.61a) and Eq. (10.61b)] now gives the desired solution for $\mathbf{x}(t)$:

$$
\mathbf{x}(t) = \begin{bmatrix} x_1(t) \\ x_2(t) \end{bmatrix} = \begin{bmatrix} \left(\frac{1}{36} - \frac{21}{20}e^{-4t} + \frac{136}{45}e^{-9t} \right)u(t) \\ \left(\frac{-63}{5}e^{-4t} + \frac{68}{5}e^{-9t} \right)u(t) \end{bmatrix} \tag{10.61c}
$$

This is consistent with the solution obtained by using the frequency-domain method [see Eq. (10.36b)]. Once the state variables x_1 and x_2 are found for $t \geq 0$, all the remaining variables are determined from the output equation. ■

The Output

The output equation is given by

$$
\mathbf{y}(t) = \mathbf{C}\mathbf{x}(t) + \mathbf{D}f(t)
$$

The substitution of the solution for $\mathbf{x}$ [Eq. (10.57)] in this equation yields

$$
\mathbf{y}(t) = \mathbf{C}[e^{\mathbf{A}t}\mathbf{x}(0) + e^{\mathbf{A}t} * \mathbf{B}f(t)] + \mathbf{D}f(t) \tag{10.62a}
$$

Since the elements of $\mathbf{B}$ are constants,

$$
e^{\mathbf{A}t} * \mathbf{B}f(t) = e^{\mathbf{A}t}\mathbf{B} * f(t)
$$

Using this result, Eq. (10.62a) becomes

$$
\mathbf{y}(t) = \mathbf{C}[e^{\mathbf{A}t}\mathbf{x}(0) + e^{\mathbf{A}t}\mathbf{B} * f(t)] + \mathbf{D}f(t) \tag{10.62b}
$$

Now recall that the convolution of $f(t)$ with the unit impulse $\delta(t)$ yields $f(t)$. Let us define a $j \times j$ diagonal matrix $\boldsymbol{\delta}(t)$ such that all its diagonal terms are unit impulse functions. It is then obvious that

$$
\boldsymbol{\delta}(t) * \mathbf{f}(t) = \mathbf{f}(t)
$$

and Eq. (10.62b) can be expressed as

$$
\mathbf{y}(t) = \mathbf{C}[e^{\mathbf{A}t}\mathbf{x}(0) + e^{\mathbf{A}t}\mathbf{B} * \mathbf{f}(t)] + \mathbf{D}\boldsymbol{\delta}(t) * \mathbf{f}(t) \tag{10.63a}
$$

$$
= \mathbf{C}e^{\mathbf{A}t}\mathbf{x}(0) + [\mathbf{C}e^{\mathbf{A}t}\mathbf{B} + \mathbf{D}\boldsymbol{\delta}(t)] * \mathbf{f}(t) \tag{10.63b}
$$

By Using the notation $\boldsymbol{\phi}(t)$ for $e^{\mathbf{A}t}$, Eq. (10.63b) may be expressed as

$$\mathbf{y}(t) = \underbrace{\mathbf{C}\boldsymbol{\phi}(t)\mathbf{x}(0)}_{\text{zero-input response}} + \underbrace{[\mathbf{C}\boldsymbol{\phi}(t)\mathbf{B} + \mathbf{D}\boldsymbol{\delta}(t)] * \mathbf{f}(t)}_{\text{zero-state response}} \qquad (10.63c)$$

The zero-state response, that is, the response when $\mathbf{x}(0) = \mathbf{0}$, is

$$\mathbf{y}(t) = [\mathbf{C}\boldsymbol{\phi}(t)\mathbf{B} + \mathbf{D}\boldsymbol{\delta}(t)] * \mathbf{f}(t) \qquad (10.64a)$$

$$= \mathbf{h}(t) * \mathbf{f}(t) \qquad (10.64b)$$

where

$$\mathbf{h}(t) = \mathbf{C}\boldsymbol{\phi}(t)\mathbf{B} + \mathbf{D}\boldsymbol{\delta}(t) \qquad (10.65)$$

The matrix $\mathbf{h}(t)$ is a $k \times j$ matrix known as the **impulse response matrix**. The reason for this designation is obvious. The ijth element of $\mathbf{h}(t)$ is $h_{ij}(t)$, which represents the zero-state response y_i when the input $f_j(t) = \delta(t)$ and when all other inputs (and all the initial conditions) are zero. It can also be seen from Eq. (10.39) and (10.64b) that

$$\mathcal{L}[\mathbf{h}(t)] = \mathbf{H}(s)$$

■ **Example 10.8**

For the system described by Eqs. (10.40a) and (10.40b), determine $e^{\mathbf{A}t}$ using Eq. (10.59b):

$$\boldsymbol{\phi}(t) = e^{\mathbf{A}t} = \mathcal{L}^{-1}\boldsymbol{\Phi}(s)$$

This problem was solved earlier with frequency-domain techniques. From Eq. (10.41), we have

$$\boldsymbol{\phi}(t) = \mathcal{L}^{-1}\begin{bmatrix} \frac{s+3}{(s+1)(s+2)} & \frac{1}{(s+1)(s+2)} \\ \frac{-2}{(s+1)(s+2)} & \frac{s}{(s+1)(s+2)} \end{bmatrix}$$

$$= \mathcal{L}^{-1}\begin{bmatrix} \frac{2}{s+1} - \frac{1}{s+2} & \frac{1}{s+1} - \frac{1}{s+2} \\ \frac{-2}{s+1} + \frac{2}{s+2} & \frac{-1}{s+1} + \frac{2}{s+2} \end{bmatrix}$$

$$= \begin{bmatrix} 2e^{-t} - e^{-2t} & e^{-t} - e^{-2t} \\ -2e^{-t} + 2e^{-2t} & -e^{-t} + 2e^{-2t} \end{bmatrix}$$

The same result is obtained in Sec. B.9-5 by using Eq. (10.59a) [see Eq. (B.101)].

Also, $\boldsymbol{\delta}(t)$ is a diagonal $j \times j$ or 2×2 matrix:

$$\boldsymbol{\delta}(t) = \begin{bmatrix} \delta(t) & 0 \\ 0 & \delta(t) \end{bmatrix}$$

Substituting the matrices $\boldsymbol{\phi}(t)$, $\boldsymbol{\delta}(t)$, $\mathbf{C}$, $\mathbf{D}$, and $\mathbf{B}$ [Eq. (10.40c)] into Eq. (10.65), we have

$$\mathbf{h}(t) = \begin{bmatrix} 1 & 0 \\ 1 & 1 \\ 0 & 2 \end{bmatrix} \begin{bmatrix} 2e^{-t} - e^{-2t} & e^{-t} - e^{-2t} \\ -2e^{-t} + 2e^{-2t} & -e^{-t+2} + e^{-2t} \end{bmatrix} \begin{bmatrix} 1 & 0 \\ 1 & 1 \end{bmatrix} + \begin{bmatrix} 0 & 0 \\ 1 & 0 \\ 0 & 1 \end{bmatrix} \begin{bmatrix} \delta(t) & 0 \\ 0 & \delta(t) \end{bmatrix}$$

$$= \begin{bmatrix} 3e^{-t} - 2e^{-2t} & e^{-t} - e^{-2t} \\ \delta(t) + 2e^{-2t} & e^{-2t} \\ -6e^{-t} + 8e^{-2t} & \delta(t) - 2e^{-2t} + 4e^{-2t} \end{bmatrix} \qquad (10.66)$$

The reader can verify that the transfer-function matrix $\mathbf{H}(s)$ in Eq. (10.42) is the Laplace transform of the unit-impulse response matrix $\mathbf{h}(t)$ in Eq. (10.66). ∎

10.4 LINEAR TRANSFORMATION OF STATE VECTORS

In Sec. 10.1 we saw that the state of a system can be specified in several ways. The sets of all possible state variables must be related—in other words, if we are given one set of state variables, we should be able to relate it to any other set. We are particularly interested in a linear type of relationship.

Let $x_1, x_2, \ldots, x_n$ and $w_1, w_2, \ldots, w_n$ be two different sets of state variables specifying the same system. Let these sets be related by linear equations as

$$w_1 = p_{11}x_1 + p_{12}x_2 + \cdots + p_{1n}x_n$$

$$w_2 = p_{21}x_1 + p_{22}x_2 + \cdots + p_{2n}x_n$$

$$\cdots\cdots\cdots\cdots\cdots\cdots\cdots\cdots\cdots\cdots\cdots\cdots\cdots \tag{10.67a}$$

$$w_n = p_{n1}x_1 + p_{n2}x_2 + \cdots + p_{nn}x_n$$

or

$$\underbrace{\begin{bmatrix} w_1 \\ w_2 \\ \cdots \\ w_n \end{bmatrix}}_{\mathbf{w}} = \underbrace{\begin{bmatrix} p_{11} & p_{12} & \cdots & p_{1n} \\ p_{21} & p_{22} & \cdots & p_{2n} \\ \cdots\cdots\cdots\cdots\cdots \\ p_{n1} & p_{n2} & \cdots & p_{nn} \end{bmatrix}}_{\mathbf{P}} \underbrace{\begin{bmatrix} x_1 \\ x_2 \\ \cdots \\ x_n \end{bmatrix}}_{\mathbf{x}} \tag{10.67b}$$

Defining the vector $\mathbf{w}$ and matrix $\mathbf{P}$ as shown above, we can write Eq. (10.67b) as

$$\mathbf{w} = \mathbf{Px} \tag{10.67c}$$

and

$$\mathbf{x} = \mathbf{P}^{-1}\mathbf{w} \tag{10.67d}$$

Thus the state vector $\mathbf{x}$ is transformed into another state vector $\mathbf{w}$ through the linear transformation in Eq. (10.67c).

If we know $\mathbf{w}$, we can determine $\mathbf{x}$ from Eq. (10.67d), provided that $\mathbf{P}^{-1}$ exists. This is equivalent to saying that $\mathbf{P}$ is a nonsingular matrix† ($|\mathbf{P}| \neq 0$). Thus, if $\mathbf{P}$ is a nonsingular matrix, the vector $\mathbf{w}$ defined by Eq. (10.67c) is also a state vector. Consider the state equation of a system:

$$\dot{\mathbf{x}} = \mathbf{Ax} + \mathbf{Bf} \tag{10.68a}$$

If

$$\mathbf{w} = \mathbf{Px} \tag{10.68b}$$

then

$$\mathbf{x} = \mathbf{P}^{-1}\mathbf{w}$$

---

†This condition is equivalent to saying that all n equations in Eq. (10.67a) are linearly independent, that is, none of the n equations can be expressed as a linear combination of the remaining equations.

and
$$\dot{\mathbf{x}} = \mathbf{P}^{-1}\dot{\mathbf{w}}$$

Hence the state equation (10.68a) now becomes
$$\mathbf{P}^{-1}\dot{\mathbf{w}} = \mathbf{A}\mathbf{P}^{-1}\mathbf{w} + \mathbf{B}\mathbf{f}$$
or

$$\dot{\mathbf{w}} = \mathbf{P}\mathbf{A}\mathbf{P}^{-1}\mathbf{w} + \mathbf{P}\mathbf{B}\mathbf{f} \qquad (10.68c)$$

$$= \hat{\mathbf{A}}\mathbf{w} + \hat{\mathbf{B}}\mathbf{f} \qquad (10.68d)$$

where
$$\hat{\mathbf{A}} = \mathbf{P}\mathbf{A}\mathbf{P}^{-1} \qquad (10.69a)$$
and
$$\hat{\mathbf{B}} = \mathbf{P}\mathbf{B} \qquad (10.69b)$$

Equation (10.68d) is a state equation for the same system, but now it is expressed in terms of the state vector $\mathbf{w}$.

The output equation is also modified. Let the original output equation be

$$\mathbf{y} = \mathbf{C}\mathbf{x} + \mathbf{D}\mathbf{f}$$

In terms of the new state variable $\mathbf{w}$, this becomes

$$\mathbf{y} = \mathbf{C}(\mathbf{P}^{-1}\mathbf{w}) + \mathbf{D}\mathbf{f}$$

$$= \hat{\mathbf{C}}\mathbf{w} + \mathbf{D}\mathbf{f}$$

where
$$\hat{\mathbf{C}} = \mathbf{C}\mathbf{P}^{-1} \qquad (10.69c)$$

■ **Example 10.9**

The state equations of a certain system are given by

$$\begin{bmatrix} \dot{x}_1 \\ \dot{x}_2 \end{bmatrix} = \begin{bmatrix} 0 & 1 \\ -2 & -3 \end{bmatrix} \begin{bmatrix} x_1 \\ x_2 \end{bmatrix} + \begin{bmatrix} 1 \\ 2 \end{bmatrix} f(t) \qquad (10.70a)$$

Find the state equations for this system when the new state variables w_1 and w_2 are

$$w_1 = x_1 + x_2$$

$$w_2 = x_1 - x_2$$

or

$$\begin{bmatrix} w_1 \\ w_2 \end{bmatrix} = \begin{bmatrix} 1 & 1 \\ 1 & -1 \end{bmatrix} \begin{bmatrix} x_1 \\ x_2 \end{bmatrix} \qquad (10.70b)$$

From Eq. (10.70b), the state equation for the state variable $\mathbf{w}$ is given by

$$\dot{\mathbf{w}} = \hat{\mathbf{A}}\mathbf{w} + \hat{\mathbf{B}}\mathbf{f}$$

where [see Eq. (10.69)]

$$\hat{\mathbf{A}} = \mathbf{PAP^{-1}} = \begin{bmatrix} 1 & 1 \\ 1 & -1 \end{bmatrix} \begin{bmatrix} 0 & 1 \\ -2 & -3 \end{bmatrix} \begin{bmatrix} 1 & 1 \\ 1 & -1 \end{bmatrix}^{-1}$$

$$= \begin{bmatrix} 1 & 1 \\ 1 & -1 \end{bmatrix} \begin{bmatrix} 0 & 1 \\ -2 & -3 \end{bmatrix} \begin{bmatrix} \frac{1}{2} & \frac{1}{2} \\ \frac{1}{2} & -\frac{1}{2} \end{bmatrix}$$

$$= \begin{bmatrix} -2 & 0 \\ 3 & -1 \end{bmatrix}$$

and

$$\hat{\mathbf{B}} = \mathbf{PB} = \begin{bmatrix} 1 & 1 \\ 1 & -1 \end{bmatrix} \begin{bmatrix} 1 \\ 2 \end{bmatrix} = \begin{bmatrix} 3 \\ -1 \end{bmatrix}$$

Therefore

$$\begin{bmatrix} \dot{w}_1 \\ \dot{w}_2 \end{bmatrix} = \begin{bmatrix} -2 & 0 \\ 3 & -1 \end{bmatrix} \begin{bmatrix} w_1 \\ w_2 \end{bmatrix} + \begin{bmatrix} 3 \\ -1 \end{bmatrix} f(t)$$

This is the desired state equation for the state vector $\mathbf{w}$. The solution of this equation requires a knowledge of the initial state $\mathbf{w}(0)$. This can be obtained from the given initial state $\mathbf{x}(0)$ by using Eq. (10.70b). ∎

⊙ **Computer Example C10.3**
The state equations for a certain system are given by

$$\begin{bmatrix} \dot{x}_1 \\ \dot{x}_2 \end{bmatrix} = \begin{bmatrix} 0 & 1 \\ -2 & -3 \end{bmatrix} \begin{bmatrix} x_1 \\ x_2 \end{bmatrix} + \begin{bmatrix} 1 \\ 2 \end{bmatrix} f(t)$$

Use a computer to find the state equations for this system when the new state variables $w1$ and $w2$ are

$$w_1 = x_1 + x_2$$
$$w_2 = x_1 - x_2$$

```
% Ahat and Bhat need to be calculated using the transformation matrix P.
% Enter the matrix A.
A=[0 1;-2 -3];
% Enter the vector B.
B=[1;2];
% Enter the transformation matrix P.
P=[1 1;1 -1];
% All necessary information has been input.
% Calculate Ahat for the transformed state-space representation.
Ahat=P*A*inv(P);
% Calculate Bhat for the transformed state-space representation.
Bhat=P*B;
% Display Ahat and Bhat for the new representation.
Ahat,Bhat    ⊙
```

Invariance of Eigenvalues

We have seen (Sec. 10.3) that the poles of all possible transfer functions of a system are the eigenvalues of the matrix $\mathbf{A}$. If we transform a state vector from $\mathbf{x}$ to $\mathbf{w}$, the variables $w_1, w_2, \ldots, w_n$ are linear combinations of $x_1, x_2, \ldots, x_n$ and therefore may be considered as outputs. Hence the poles of the transfer functions relating $w_1, w_2, \ldots, w_n$ to the various inputs must also be the eigenvalues of matrix $\mathbf{A}$. On the other hand, the system is also specified by Eq. (10.68d). This means that the poles of the transfer functions must be the eigenvalues of $\hat{\mathbf{A}}$. Therefore the eigenvalues of matrix $\mathbf{A}$ remain unchanged for the linear transformation of variables represented by Eq. (10.67), and the eigenvalues of matrix $\mathbf{A}$ and matrix $\hat{\mathbf{A}}(\hat{\mathbf{A}} = \mathbf{P}\mathbf{A}\mathbf{P}^{-1})$ are identical, implying that the characteristic equations of $\mathbf{A}$ and $\hat{\mathbf{A}}$ are also identical. This result also can be proved alternately as follows.

Consider the matrix $\mathbf{P}(s\mathbf{I} - \mathbf{A})\mathbf{P}^{-1}$. We have

$$\mathbf{P}(s\mathbf{I} - \mathbf{A})\mathbf{P}^{-1} = \mathbf{P}s\mathbf{I}\mathbf{P}^{-1} - \mathbf{P}\mathbf{A}\mathbf{P}^{-1} = s\mathbf{P}\mathbf{I}\mathbf{P}^{-1} - \hat{\mathbf{A}} = s\mathbf{I} - \hat{\mathbf{A}}$$

Taking the determinants of both sides, we obtain

$$|\mathbf{P}||s\mathbf{I} - \mathbf{A}||\mathbf{P}^{-1}| = |s\mathbf{I} - \hat{\mathbf{A}}|$$

The determinants $|\mathbf{P}|$ and $|\mathbf{P}^{-1}|$ are reciprocals of each other. Hence

$$|s\mathbf{I} - \mathbf{A}| = |s\mathbf{I} - \hat{\mathbf{A}}| \tag{10.71}$$

This is the desired result. We have shown that the characteristic equations of $\mathbf{A}$ and $\hat{\mathbf{A}}$ are identical. Hence the eigenvalues of $\mathbf{A}$ and $\hat{\mathbf{A}}$ are identical.

In Example 10.9, matrix $\mathbf{A}$ is given as

$$\mathbf{A} = \begin{bmatrix} 0 & 1 \\ -2 & -3 \end{bmatrix}$$

The characteristic equation is

$$|s\mathbf{I} - \mathbf{A}| = \begin{vmatrix} s & -1 \\ 2 & s+3 \end{vmatrix} = s^2 + 3s + 2 = 0$$

Also

$$\hat{\mathbf{A}} = \begin{bmatrix} -2 & 0 \\ 3 & -1 \end{bmatrix}$$

and

$$|s\mathbf{I} - \hat{\mathbf{A}}| = \begin{bmatrix} s+2 & 0 \\ -3 & s+1 \end{bmatrix} = s^2 + 3s + 2 = 0$$

This verifies that the characteristic equations of $\mathbf{A}$ and $\hat{\mathbf{A}}$ are identical.

⊙ **Computer Example C10.4**

Using a computer, show that the eigenvalues of the system in Computer Example 10.3 are invariant under linear transformations.

```
% Enter the matrix A.
A=[0 1;-2 -3];
% Enter the matrix Ahat.
Ahat=[-2 0;3 -1];
% Calculate the eigenvalues of A.
eA=eig(A)
% Calculate the eigenvalues of Ahat.
eAhat=eig(Ahat)    ⊙
```

Note: You may wish to try other matrices, and transformations to convince yourself.

10.4-1   Diagonalization of Matrix A

For several reasons, it is desirable to make matrix $\mathbf{A}$ diagonal. If $\mathbf{A}$ is not diagonal, we can transform the state variables such that the resulting matrix $\hat{\mathbf{A}}$ is diagonal.† One can show that for any diagonal matrix $\mathbf{A}$, the diagonal elements of this matrix must necessarily be $\lambda_1, \lambda_2, \ldots, \lambda_n$ (the eigenvalues) of the matrix. Consider the diagonal matrix $\mathbf{A}$:

$$\mathbf{A} = \begin{bmatrix} a_1 & 0 & 0 & 0 \\ 0 & a_2 & 0 & 0 \\ \multicolumn{4}{c}{\cdots\cdots\cdots\cdots} \\ 0 & 0 & 0 & a_n \end{bmatrix}$$

The characteristic equation is given by

$$|s\mathbf{I} - \mathbf{A}| = \begin{bmatrix} (s-a_1) & 0 & 0 & \cdots & 0 \\ 0 & (s-a_2) & 0 & \cdots & 0 \\ \multicolumn{5}{c}{\cdots\cdots\cdots\cdots\cdots\cdots\cdots\cdots} \\ 0 & 0 & 0 & \cdots & (s-a_n) \end{bmatrix} = 0$$

or

$$(s-a_1)(s-a_2)\cdots(s-a_n) = 0$$

Hence the eigenvalues of $\mathbf{A}$ are $a_1, a_2, \ldots, a_n$. The nonzero (diagonal) elements of a diagonal matrix are therefore its eigenvalues $\lambda_1, \lambda_2, \ldots, \lambda_n$. We shall denote the diagonal matrix by a special symbol, Λ:

---

†In this discussion we assume distinct eigenvalues. If the eigenvalues are not distinct, we can reduce the matrix to a modified diagonalized (Jordan) form.

$$\Lambda = \begin{bmatrix} \lambda_1 & 0 & 0 & \cdots & 0 \\ 0 & \lambda_2 & 0 & \cdots & 0 \\ \cdots\cdots\cdots\cdots\cdots\cdots \\ 0 & 0 & 0 & \cdots & \lambda_n \end{bmatrix} \qquad (10.72)$$

Let us now consider the transformation of the state vector A such that the resulting matrix $\hat{\mathbf{A}}$ is a diagonal matrix Λ.

Consider the system

$$\dot{\mathbf{x}} = \mathbf{Ax} + \mathbf{Bf}$$

We shall assume that $\lambda_1, \lambda_2, \ldots, \lambda_n$, the eigenvalues of $\mathbf{A}$, are distinct (no repeated roots). Let us transform the state vector $\mathbf{x}$ into the new state vector $\mathbf{z}$, using the transformation

$$\mathbf{z} = \mathbf{Px} \qquad (10.73a)$$

Then, after the development of Eq. (10.68c), we have

$$\dot{\mathbf{z}} = \mathbf{PAP}^{-1}\mathbf{z} + \mathbf{PBf} \qquad (10.73b)$$

We desire the transformation to be such that $\mathbf{PAP}^{-1}$ is a diagonal matrix Λ given by Eq. (10.72), or

$$\dot{\mathbf{z}} = \Lambda\mathbf{z} + \hat{\mathbf{B}}\mathbf{f} \qquad (10.73c)$$

Hence

$$\Lambda = \mathbf{PAP}^{-1} \qquad (10.74a)$$

or

$$\Lambda\mathbf{P} = \mathbf{PA} \qquad (10.74b)$$

We know Λ and $\mathbf{A}$. Equation (10.74b) therefore can be solved to determine $\mathbf{P}$.

■ **Example 10.10**

Find the diagonalized form of the state equation for the system in Example 10.9.

In this case,

$$\mathbf{A} = \begin{bmatrix} 0 & 1 \\ -2 & -3 \end{bmatrix}$$

We found $\lambda_1 = -1$ and $\lambda_2 = -2$. Hence

$$\Lambda = \begin{bmatrix} -1 & 0 \\ 0 & -2 \end{bmatrix}$$

and Eq. (10.74b) becomes

$$\begin{bmatrix} -1 & 0 \\ 0 & -2 \end{bmatrix} \begin{bmatrix} p_{11} & p_{12} \\ p_{21} & p_{22} \end{bmatrix} = \begin{bmatrix} p_{11} & p_{12} \\ p_{21} & p_{22} \end{bmatrix} \begin{bmatrix} 0 & 1 \\ -2 & -3 \end{bmatrix}$$

Equating the four elements on two sides, we obtain

$$-p_{11} = -2p_{12} \tag{10.75a}$$

$$-p_{12} = p_{11} - 3p_{12} \tag{10.75b}$$

$$-2p_{21} = -2p_{22} \tag{10.75c}$$

$$-2p_{22} = p_{21} - 3p_{22} \tag{10.75d}$$

The reader will immediately recognize that Eqs. (10.75a) and (10.75b) are identical. Similarly, Eqs. (10.75c) and (10.75d) are identical. Hence two equations may be discarded, leaving us with only two equations [Eqs. (10.75a) and (10.75c)] and four unknowns. This means there is no unique solution. There is, in fact, an infinite number of solutions. We can assign any value to p_{11} and p_{21} to yield one possible solution.† If $p_{11} = k_1$ and $p_{21} = k_2$, then from Eqs. (10.75a) and (10.75c) we have $p_{12} = k_1/2$ and $p_{22} = k_2$:

$$\mathbf{P} = \begin{bmatrix} k_1 & \frac{k_1}{2} \\ k_2 & k_2 \end{bmatrix} \tag{10.75e}$$

We may assign any values to k_1 and k_2. For convenience, let $k_1 = 2$ and $k_2 = 1$. This yields

$$\mathbf{P} = \begin{bmatrix} 2 & 1 \\ 1 & 1 \end{bmatrix} \tag{10.75f}$$

The transformed variables [Eq. (10.73a)] are

$$\begin{bmatrix} z_1 \\ z_2 \end{bmatrix} = \begin{bmatrix} 2 & 1 \\ 1 & 1 \end{bmatrix} \begin{bmatrix} x_1 \\ x_2 \end{bmatrix} = \begin{bmatrix} 2x_1 + x_2 \\ x_1 + x_2 \end{bmatrix} \tag{10.76}$$

Thus the new state variables z_1 and z_2 are related to x_1 and x_2 by Eq. (10.76). The system equation with $\mathbf{z}$ as the state vector is given by [see Eq. (10.73c)]

$$\dot{\mathbf{z}} = \Lambda \mathbf{z} + \hat{\mathbf{B}} \mathbf{f}$$

where

$$\hat{\mathbf{B}} = \mathbf{PB} = \begin{bmatrix} 2 & 1 \\ 1 & 1 \end{bmatrix} \begin{bmatrix} 1 \\ 2 \end{bmatrix} = \begin{bmatrix} 4 \\ 3 \end{bmatrix}$$

Hence

$$\begin{bmatrix} \dot{z}_1 \\ \dot{z}_2 \end{bmatrix} = \begin{bmatrix} -1 & 0 \\ 0 & -2 \end{bmatrix} \begin{bmatrix} z_1 \\ z_2 \end{bmatrix} + \begin{bmatrix} 4 \\ 3 \end{bmatrix} f \tag{10.77a}$$

or

$$\dot{z}_1 = -z_1 + 4f$$

$$\dot{z}_2 = -2z_2 + 3f \tag{10.77b}$$

---

†If, however, we want the state equations in diagonalized form, as in Eq. (10.29a), where all the elements of $\hat{\mathbf{B}}$ matrix are unity, there is a unique solution. This is because the equation $\hat{\mathbf{B}} = \mathbf{PB}$, where all the elements of $\hat{\mathbf{B}}$ are unity, imposes additional constraints. In the present example, this will yield $p_{11} = \frac{1}{2}$, $p_{12} = \frac{1}{4}$, $p_{21} = \frac{1}{3}$, and $p_{22} = \frac{1}{3}$. The relationship between $\mathbf{z}$ and $\mathbf{x}$ is then

$$z_1 = \tfrac{1}{2}x_1 + \tfrac{1}{4}x_2 \quad \text{and} \quad z_2 = \tfrac{1}{3}x_1 + \tfrac{1}{3}x_2$$

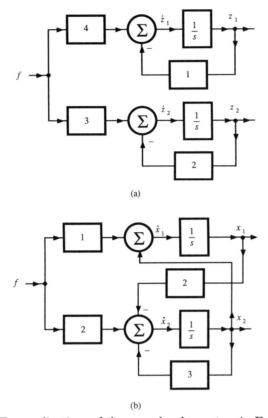

(a)

(b)

Fig. 10.8 Two realizations of the second-order system in Example 10.10.

Note the distinctive nature of these state equations. Each state equation involves only one variable and therefore can be solved by itself. A general state equation has the derivative of one state variable equal to a linear combination of all state variables. Such is not the case with the diagonalized matrix Λ. Each state variable z_i is so chosen that it is uncoupled from the rest of the variables; hence a system with n eigenvalues is split into n decoupled systems, each with an equation of the form

$$\dot{z}_i = \lambda_i z_i + \text{(input terms)}$$

This fact also can be readily seen from Fig. 10.8a, which is a realization of the system represented by Eq. (10.77). In contrast, consider the original state equations [see Eq. 10.70a)]

$$\dot{x}_1 = x_2 + f(t)$$

$$\dot{x}_2 = -2x_1 - 3x_2 + 2f(t)$$

A realization for these equations is shown in Fig. 10.8b. It can be seen from Fig. 10.8a that the states z_1 and z_2 are decoupled, whereas the states x_1 and x_2 (Fig. 10.8b) are coupled. It should be remembered that Figs. 10.8a and 10.8b are simulations of the same system.† ∎

––––––––––––––––––––––––––––––

†Here we only have a simulated state equation; the outputs are not shown. The outputs are linear

⊙ **Computer Example C10.5**
Use a computer to find the diagonalized form of the following state equation:

$$\begin{bmatrix} \dot{x}_1 \\ \dot{x}_2 \end{bmatrix} = \begin{bmatrix} 0 & 1 \\ -2 & -3 \end{bmatrix} \begin{bmatrix} x_1 \\ x_2 \end{bmatrix} + \begin{bmatrix} 1 \\ 2 \end{bmatrix} f(t)$$

```
% Enter the state space system.
A=[0 1;-2 -3];
B=[1;2];
% Calculate the eigenvalues and eigenvectors of A.
[P,lambda]=eig(A)
% Calculate Ahat for the transformed state-space representation.
Ahat=inv(P)*A*P;
% Calculate Bhat for the transformed state space representation.
Bhat=inv(P)*B;
% Display Ahat and Bhat for the diagonal representation.
Ahat,Bhat
disp('Strike a key to see more.....')
pause
% To get the same answer for Ahat as in Example 10.10 you need to
% make the norm of each basis vector unity.
P=[P(1,:)/P(1,2);P(2,:)/P(2,2)]
Ahat=P'*A*inv(P');      % Calculates Ahat.
Bhat=P'*B;              % Calculates Bhat.
% To verify the answer, display Ahat, Bhat.
% Display Ahat and Bhat for the diagonal representation.
Ahat,Bhat    ⊙
```

10.5   CONTROLLABILITY AND OBSERVABILITY

Consider a diagonalized state-space description of a system:

$$\dot{\mathbf{z}} = \Lambda\mathbf{z} + \hat{\mathbf{B}}\mathbf{f} \tag{10.78a}$$

and

$$\mathbf{Y} = \hat{\mathbf{C}}\mathbf{z} + \mathbf{D}\mathbf{f} \tag{10.78b}$$

We shall assume that all n eigenvalues λ_1, λ_2, ... , λ_n are distinct. The state equations (10.78a) are of the form

$$\dot{z}_m = \lambda_m z_m + \hat{b}_{m1} f_1 + \hat{b}_{m2} f_2 + \cdots + \hat{b}_{mj} f_j \qquad m = 1, 2, \ldots n$$

If $\hat{b}_{m1}, \hat{b}_{m2}, \ldots, \hat{b}_{mj}$ (the mth row in matrix $\hat{\mathbf{B}}$) are all zero, then

---

combinations of state variables (and inputs). Hence the output equation can be easily incorporated into these diagrams (see Fig. 10.7).

$$\dot{z}_m = \lambda_m z_m$$

and the variable z_m is uncontrollable because z_m is not connected to any of the inputs. Moreover, z_m is decoupled from all the remaining $(n-1)$ state variables because of the diagonalized nature of the variables. Hence there is no direct or indirect coupling of z_m with any of the inputs, and the system is uncontrollable. On the other hand, if at least one element in the mth row of $\hat{\mathbf{B}}$ is nonzero, z_m is coupled to at least one input and is therefore controllable. **Thus a system with a diagonalized state [Eqs. (10.78)] is completely controllable if and only if the matrix $\hat{\mathbf{B}}$ has no row of zero elements.**

The outputs [Eq.(10.78b)] are of the form

$$y_i = \hat{c}_{i1} z_1 + \hat{c}_{i2} z_2 + \cdots + \hat{c}_{in} z_n + \sum_{m=1}^{j} d_{im} f_m$$

If $\hat{c}_{im} = 0$, then the state z_m will not appear in the expression for y_i. Since all the states are decoupled because of the diagonalized nature of the equations, the state z_m cannot be observed directly or indirectly (through other states) at the output y_i. Hence the mth mode $e^{\lambda_m t}$ will not be observed at the output y_i. If $\hat{c}_{1m}, \hat{c}_{2m}, \ldots$, $\hat{c}_{km}$ (the mth column in matrix $\hat{\mathbf{C}}$) are all zero, the state z_m will not be observable at any of the k outputs, and the state z_m is unobservable. On the other hand, if at least one element in the mth column of $\hat{\mathbf{C}}$ is nonzero, z_m is observable at least at one output. **Thus a system with diagonalized equations of the form in Eqs. (10.78) is completely observable if and only if the matrix $\hat{\mathbf{C}}$ has no column of zero elements.** In the above discussion, we assumed distinct eigenvalues; for repeated eigenvalues, the modified criteria can be found in the literature.[1, 2]

If the state-space description is not in diagonalized form, it may be converted into diagonalized form using the procedure in Example 10.10. It is also possible to test for controllability and observability even if the state-space description is in undiagonalized form[1, 2].†

■ **Example 10.11**

Investigate the controllability and observability of the systems in Figs. 10.9a and 10.9b.

In both cases, the state variables are identified as the two integrator outputs, x_1 and x_2. The state equations for the system in Fig. 10.9a are

$$\dot{x}_1 = x_1 + f_1$$
$$\dot{x}_2 = x_1 - x_2 \tag{10.79}$$

and

$$y = x_1 - 2x_2$$

Hence

$$\mathbf{A} = \begin{bmatrix} 1 & 0 \\ 1 & -1 \end{bmatrix}, \quad \mathbf{B} = \begin{bmatrix} 1 \\ 0 \end{bmatrix}, \quad \mathbf{C} = \begin{bmatrix} 1 & -2 \end{bmatrix}, \quad \mathbf{D} = 0$$

---

†It can be shown that a system is completely controllable if and only if the $n \times nj$ composite matrix $[\mathbf{B}, \mathbf{AB}, \mathbf{A}^2\mathbf{B}, \ldots, \mathbf{A}^{n-1}\mathbf{B}]$ has a rank n. Similarly a system is completely observable if and only if the $n \times nk$ composite matrix $[\mathbf{C}', \mathbf{A}'\mathbf{C}', \mathbf{A}'^2\mathbf{C}', \ldots, \mathbf{A}'^{n-1}\mathbf{C}']$ has a rank n.

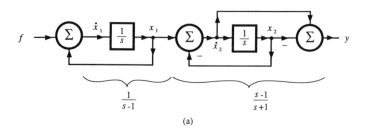

(a)

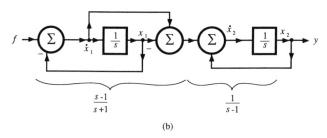

(b)

Fig. 10.9 Systems for Example 10.11.

$$|s\mathbf{I} - \mathbf{A}| = \begin{vmatrix} s - 1 & 0 \\ -1 & s + 1 \end{vmatrix} = (s-1)(s+1)$$

Therefore

$$\lambda_1 = 1 \qquad \text{and} \qquad \lambda_2 = -1$$

and

$$\Lambda = \begin{bmatrix} 1 & 0 \\ 0 & -1 \end{bmatrix} \tag{10.80}$$

We shall now use the procedure in Sec. 10.4-1 to diagonalize this system. From Eq. (10.74b), we have

$$\begin{bmatrix} 1 & 0 \\ 0 & -1 \end{bmatrix} \begin{bmatrix} p_{11} & p_{12} \\ p_{21} & p_{22} \end{bmatrix} = \begin{bmatrix} p_{11} & p_{12} \\ p_{21} & p_{22} \end{bmatrix} \begin{bmatrix} 1 & 0 \\ 1 & -1 \end{bmatrix}$$

The solution of this equation yields

$$p_{12} = 0 \qquad \text{and} \qquad -2p_{21} = p_{22}$$

Choosing $p_{11} = 1$ and $p_{21} = 1$, we have

$$p = \begin{bmatrix} 1 & 0 \\ 1 & -2 \end{bmatrix}$$

and

$$\hat{\mathbf{B}} = \mathbf{PB} = \begin{bmatrix} 1 & 0 \\ 1 & -2 \end{bmatrix} \begin{bmatrix} 1 \\ 0 \end{bmatrix} = \begin{bmatrix} 1 \\ 1 \end{bmatrix} \tag{10.81a}$$

All the rows of $\hat{\mathbf{B}}$ are nonzero. Hence the system is controllable. Also,

$$\mathbf{Y} = \mathbf{Cx}$$
$$= \mathbf{CP}^{-1}\mathbf{z}$$
$$= \hat{\mathbf{C}}\mathbf{z} \tag{10.81b}$$

and

$$\hat{\mathbf{C}} = \mathbf{C}\mathbf{P}^{-1} = [1 \quad -2]\begin{bmatrix} 1 & 0 \\ 1 & -2 \end{bmatrix}^{-1} = [1 \quad -2]\begin{bmatrix} 1 & 0 \\ \frac{1}{2} & -\frac{1}{2} \end{bmatrix} = [0 \quad 1] \qquad (10.81c)$$

The first column of $\hat{\mathbf{C}}$ is zero. Hence the mode z_1 (corresponding to $\lambda_1 = 1$) is unobservable. The system is therefore controllable but not observable. We come to the same conclusion by realizing the system with the state variables z_1 and z_2, whose state equations are

$$\dot{\mathbf{z}} = \Lambda\mathbf{z} + \hat{\mathbf{B}}f$$

$$y = \hat{\mathbf{C}}\mathbf{z}$$

From Eqs. (10.80) and (10.81), we have

$$\dot{z}_1 = z_1 + f$$

$$\dot{z}_2 = -z_2 + f$$

and

$$y = z_2$$

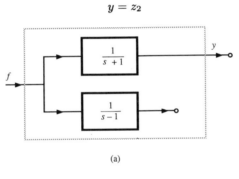

(a)

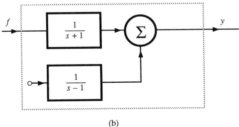

(b)

Fig. 10.10   Equivalent of the systems in Fig. 10.9.

Figure 10.10a shows a realization of these equations. It is clear that each of the two modes is controllable, but the first mode (corresponding to $\lambda = 1$) is not observable at the output.

The state equations for the system in Fig. 10.9b are

$$\dot{x}_1 = -x_1 + f$$

$$\dot{x}_2 = -2x_1 + x_2 + f \qquad (10.82)$$

and

$$y = x_2$$

Hence

$$\mathbf{A} = \begin{bmatrix} -1 & 0 \\ -2 & 1 \end{bmatrix}, \quad \mathbf{B} = \begin{bmatrix} 1 \\ 1 \end{bmatrix}, \quad \mathbf{C} = [0 \quad 1], \quad \mathbf{D} = 0$$

$$|s\mathbf{I} - \mathbf{A}| = \begin{vmatrix} s+1 & 0 \\ -1 & s-1 \end{vmatrix} = (s+1)(s-1)$$

so that $\lambda_1 = -1$, $\lambda_2 = 1$, and

$$\Lambda = \begin{bmatrix} -1 & 0 \\ 1 & 1 \end{bmatrix} \tag{10.83}$$

Diagonalizing the matrix, we have

$$\begin{bmatrix} 1 & 0 \\ 0 & -1 \end{bmatrix} \begin{bmatrix} p_{11} & p_{12} \\ p_{21} & p_{22} \end{bmatrix} = \begin{bmatrix} p_{11} & p_{12} \\ p_{21} & p_{22} \end{bmatrix} \begin{bmatrix} -1 & 0 \\ -2 & 1 \end{bmatrix}$$

The solution of this equation yields $p_{11} = -p_{12}$ and $p_{22} = 0$. Choosing $p_{11} = -1$ and $p_{21} = 1$, we obtain

$$\mathbf{P} = \begin{bmatrix} -1 & 1 \\ 1 & 0 \end{bmatrix}$$

and

$$\hat{\mathbf{B}} = \mathbf{PB} = \begin{bmatrix} -1 & 1 \\ 1 & 0 \end{bmatrix} \begin{bmatrix} 1 \\ 1 \end{bmatrix} = \begin{bmatrix} 0 \\ 1 \end{bmatrix} \tag{10.84a}$$

$$\hat{\mathbf{C}} = \mathbf{CP}^{-1} = [0 \quad 1] \begin{bmatrix} 0 & 1 \\ 1 & 1 \end{bmatrix} = [1 \quad 1] \tag{10.84b}$$

The first row of $\hat{\mathbf{B}}$ is zero. Hence the second mode (corresponding to $\lambda_1 = 1$) is not controllable. However, since none of the columns of $\hat{\mathbf{C}}$ vanish, both modes are observable at the output. Hence the system is observable but not controllable.

We come to the same conclusion by realizing the system with the state variables z_1 and z_2. The two state equations are

$$\dot{\mathbf{z}} = \Lambda \mathbf{z} + \hat{\mathbf{B}} f$$

$$y = \hat{\mathbf{C}} \mathbf{z}$$

From Eqs. (10.83) and (10.84), we have

$$\dot{z}_1 = z_1$$

$$\dot{z}_2 = -z_2 + f$$

and thus

$$y = z_1 + z_2 \tag{10.85}$$

Figure 10.10b shows a realization of these equations. It is clear that each of the two modes is observable at the output, but the mode corresponding to $\lambda_1 = 1$ is not controllable. ∎

⊙ **Computer Example C10.6**

Use a computer to investigate the controllability and observability of the following state equation:

$$\begin{bmatrix} \dot{x}_1 \\ \dot{x}_2 \end{bmatrix} = \begin{bmatrix} 1 & 0 \\ 1 & -1 \end{bmatrix} \begin{bmatrix} x_1 \\ x_2 \end{bmatrix} + \begin{bmatrix} 1 \\ 0 \end{bmatrix} f(t)$$

```
% Enter the state-space system.
A=[1 0;1 -1];
B=[1;0];
C=[1 -2];
% Calculate the eigenvalues and eigenvectors of A.
[P,lamda]=eig(A)
% Calculate Ahat for the transformed state-space representation.
Ahat=inv(P)*A*P;
% Calculate Bhat for the transformed state-space representation.
Bhat=inv(P)*B;
% Display Ahat and Bhat for the diagonal representation.
Ahat
disp('Strike a key to see more.....'),pause
Bhat
disp(' All the rows of Bhat are nonzero; the system is controllable')
% Determine whether the system is observable.
Chat=C*P;
% Display Chat for the diagonal representation.
Chat
disp('2nd column of Chat is zero; the mode for lamda=1 unobservable')
```
⊙

10.5-1   Inadequacy of Transfer Function Description of a System

Example 10.11 demonstrates the inadequacy of the transfer function to describe an LTI system in general. The systems in Figs. 10.9a and 10.9b both have the same transfer function

$$H(s) = \frac{1}{s+1}$$

Yet the two systems are very different. Their true nature is brought out in Figs. 10.10a and 10.10b, respectively. Both the systems are unstable, but their transfer function $H(s) = \frac{1}{s+1}$ does not give any hint of it. The system in Fig. 10.9a appears stable from the external terminals, but it is internally unstable. The system in Fig. 10.9b, on the other hand, will show instability at the external terminals, but its transfer function $H(s) = \frac{1}{s+1}$ is silent about it. The system in Fig. 10.9a is controllable but not observable, whereas the system in Fig. 10.9b is observable but not controllable.

The transfer function description of a system looks at a system only from the input and output terminals. Consequently, it can describe only the part of the system which is coupled to the input and the output terminals. Figures 10.10a and 10.10b show that in both cases only a part of the system that has a transfer function

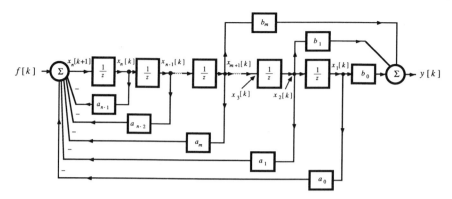

Fig. 10.11   Controller canonical realization of an nth-order discrete-time system.

$H(s) = \frac{1}{s+1}$ is coupled to the input and the output terminals. This is the reason why both systems have the same transfer function $H(s) = \frac{1}{s+1}$.

The state variable description Eqs. 10.79 and 10.82, on the other hand contains all the information about these systems to describe them completely. This is because the state variable description is an internal description, not the external description obtained from the system behavior at external terminals.

Mathematically, the reason the transfer function fails to describe these systems completely is the fact that their transfer function has a common factor $s - 1$ in the numerator and denominator; this common factor is canceled out with a consequent loss of the information about these systems. Such a situation occurs when a system is uncontrollable or/and unobservable. If a system is both controllable and observable (which is the case with most of the practical systems) the transfer function describes the system completely. In such a case the internal and external descriptions are equivalent.

10.6   STATE-SPACE ANALYSIS OF DISCRETE-TIME SYSTEMS

We have shown that an nth-order differential equation can be expressed in terms of n first-order differential equations. In the following analogous procedure, we show that an nth-order difference equation can be expressed in terms of n first-order difference equations.

Consider the z-transfer function

$$H[z] = \frac{b_m z^m + b_{m-1} z^{m-1} + \cdots + b_1 z + b_0}{z^n + a_{n-1} z^{n-1} + \cdots + a_1 z + a_0} \tag{10.86a}$$

The input $f[k]$ and the output $y[k]$ of this system are related by the difference equation

$$(E^n + a_{n-1} E^{n-1} + \cdots + a_1 E + a_0) y[k] =$$
$$(b_m E^m + b_{m-1} E^{m-1} + \cdots + b_1 E + b_0) f[k] \tag{10.86b}$$

The controller canonical realization of this equation is shown in Fig. 10.11. Signals appearing at the outputs of n delay elements are denoted by $x_1[k], x_2[k], \ldots, x_n[k]$. The input of the first delay is $x_n[k+1]$. We can now write n equations, one at the input of each delay:

$$x_1[k+1] = x_2[k]$$

$$x_2[k+1] = x_3[k]$$

$$\cdots \cdots \cdots \cdots \cdots \qquad (10.87)$$

$$x_{n-1}[k+1] = x_n[k]$$

$$x_n[k+1] = -a_0 x_1[k] - a_1 x_2[k] - \cdots - a_{n-1}x_n[k] + f[k]$$

and

$$y(k) = b_0 x_1(k) + b_1 x_2(k) + \cdots + b_m x_{m+1}(k) \qquad (10.88)$$

Equations (10.87) are n first-order difference equations in n variables $x_1(k)$, $x_2(k)$, $\ldots$, $x_n(k)$. These variables should immediately be recognized as state variables since the specification of the initial values of these variables in Fig. 10.11 will uniquely determine the response $y[k]$ for a given $f[k]$. Thus Eqs. (10.87) represent the state equations, and Eq. (10.88) is the output equation. In matrix form we can write these equations as

$$\underbrace{\begin{bmatrix} x_1[k+1] \\ x_2[k+1] \\ \cdots \\ x_{n-1}[k+1] \\ x_n[k+1] \end{bmatrix}}_{\mathbf{x}[k+1]} = \underbrace{\begin{bmatrix} 0 & 1 & 0 & \cdots & 0 & 0 \\ 0 & 0 & 1 & \cdots & 0 & 0 \\ \cdot & \cdot & \cdot & \cdots & \cdot & \cdot \\ 0 & 0 & 0 & \cdots & 0 & 1 \\ -a_0 & -a_1 & -a_2 & \cdots & -a_{n-2} & -a_{n-1} \end{bmatrix}}_{\mathbf{A}} \underbrace{\begin{bmatrix} x_1[k] \\ x_2[k] \\ \cdots \\ x_{n-1}[k] \\ x_n[k] \end{bmatrix}}_{\mathbf{x}[k]} + \underbrace{\begin{bmatrix} 0 \\ 0 \\ \cdots \\ 0 \\ 1 \end{bmatrix}}_{\mathbf{B}} f[k]$$

$$(10.89\text{a})$$

and

$$\mathbf{y}[k] = \underbrace{\begin{bmatrix} b_0 & b_1 & \cdots & b_m \end{bmatrix}}_{\mathbf{C}} \begin{bmatrix} x_1[k] \\ x_2[k] \\ \vdots \\ x_{m+1}[k] \end{bmatrix} \qquad (10.89\text{b})$$

In general,

$$\mathbf{x}[k+1] = \mathbf{A}\mathbf{x}[k] + \mathbf{B}f[k] \qquad (10.90\text{a})$$

$$\mathbf{y}[k] = \mathbf{C}\mathbf{x}[k] + \mathbf{D}f[k] \qquad (10.90\text{b})$$

Here we have represented a discrete-time system with state equations in controller canonical form. There are several other possible representations, as discussed in Sec. 10.2. We may, for example, realize the system by using a series, parallel, or observer canonical form. In all cases, the output of each delay element qualifies as a state variable. We then write the equation at the input of each delay element. The n equations thus obtained are the n state equations.

10.6-1 Solution in State-Space

Consider the state equation

$$\mathbf{x}[k+1] = \mathbf{A}\mathbf{x}[k] + \mathbf{B}\mathbf{f}[k] \tag{10.91}$$

From this equation it follows that

$$\mathbf{x}[k] = \mathbf{A}\mathbf{x}[k-1] + \mathbf{B}\mathbf{f}[k-1] \tag{10.92a}$$

and

$$\mathbf{x}[k-1] = \mathbf{A}\mathbf{x}[k-2] + \mathbf{B}\mathbf{f}[k-2] \tag{10.92b}$$

$$\mathbf{x}[k-2] = \mathbf{A}\mathbf{x}[k-3] + \mathbf{B}\mathbf{f}[k-3] \tag{10.92c}$$

$$\cdots\cdots\cdots\cdots\cdots\cdots\cdots\cdots\cdots$$

$$\mathbf{x}[1] = \mathbf{A}\mathbf{x}[0] + \mathbf{B}\mathbf{f}[0]$$

Substituting Eq. (10.92b) in Eq. (10.92a), we obtain

$$\mathbf{x}[k] = \mathbf{A}^2\mathbf{x}[k-2] + \mathbf{A}\mathbf{B}\mathbf{f}[k-2] + \mathbf{B}\mathbf{f}[k-2]$$

Substituting Eq. (10.92c) in this equation, we obtain

$$\mathbf{x}[k] = \mathbf{A}^3\mathbf{x}[k-3] + \mathbf{A}^2\mathbf{B}\mathbf{f}[k-3] + \mathbf{A}\mathbf{B}\mathbf{f}[k-2] + \mathbf{B}\mathbf{f}[k-1]$$

Continuing in this way, we obtain

$$\mathbf{x}[k] = \mathbf{A}^k\mathbf{x}[0] + \mathbf{A}^{k-1}\mathbf{B}\mathbf{f}[0] + \mathbf{A}^{k-2}\mathbf{B}\mathbf{f}[1] + \cdots + \mathbf{B}\mathbf{f}[k-1]$$

$$= \mathbf{A}^k\mathbf{x}[0] + \sum_{j=0}^{k-1} \mathbf{A}^{k-1-j}\mathbf{B}\mathbf{f}[j] \tag{10.93a}$$

The upper limit on the summation in Eq. (10.93a) is nonnegative. Hence $k \geq 1$, and the summation is recognized as the convolution sum:

$$\mathbf{A}^{k-1}u[k-1] * \mathbf{B}\mathbf{f}[k]$$

Hence

$$\mathbf{x}[k] = \underbrace{\mathbf{A}^k\mathbf{x}[0]}_{\text{zero-input}} + \underbrace{\mathbf{A}^{k-1}u[k-1] * \mathbf{B}\mathbf{f}[k]}_{\text{zero-state}} \tag{10.93b}$$

and

$$\mathbf{y}[k] = \mathbf{C}\mathbf{x} + \mathbf{D}\mathbf{f}$$

$$= \mathbf{C}\mathbf{A}^k\mathbf{x}[0] + \sum_{j=0}^{k-1} \mathbf{C}\mathbf{A}^{k-1-j}\mathbf{B}\mathbf{f}[j] + \mathbf{D}\mathbf{f} \tag{10.94a}$$

$$= \mathbf{C}\mathbf{A}^k\mathbf{x}[0] + \mathbf{C}\mathbf{A}^{k-1}u[k-1] * \mathbf{B}\mathbf{f}[k] + \mathbf{D}\mathbf{f} \tag{10.94b}$$

In Sec. B.9-5, we showed that

$$\mathbf{A}^k = \beta_0 \mathbf{I} + \beta_1 \mathbf{A} + \beta_2 \mathbf{A}^2 + \cdots + \beta_{n-1} \mathbf{A}^{n-1} \tag{10.95a}$$

where (assuming n distinct eigenvalues of $\mathbf{A}$)

$$\begin{bmatrix} \beta_0 \\ \beta_1 \\ \cdots \\ \beta_{n-1} \end{bmatrix} = \begin{bmatrix} 1 & \lambda_1 & \lambda_1^2 & \cdots & \lambda_1^{n-1} \\ 1 & \lambda_2 & \lambda_2^2 & \cdots & \lambda_2^{n-1} \\ \cdots\cdots\cdots\cdots\cdots \\ 1 & \lambda_n & \lambda_n^2 & \cdots & \lambda_n^{n-1} \end{bmatrix}^{-1} \begin{bmatrix} \lambda_1^k \\ \lambda_2^k \\ \cdots \\ \lambda_n^k \end{bmatrix} \tag{10.95b}$$

and $\lambda_1, \lambda_2, \ldots, \lambda_n$ are the n eigenvalues of $\mathbf{A}$.

We can also determine $\mathbf{A}^k$ from the z-transform formula, which will be derived later in Eq. (10.102):

$$\mathbf{A}^k = \mathcal{Z}^{-1}[(\mathbf{I} - z^{-1}\mathbf{A})^{-1}] \tag{10.95c}$$

■ **Example 10.12**

Give a state-space description of the system in Fig. 10.12. Find the output $y[k]$ if the input $f[k] = u[k]$ and the initial conditions are $x_1[0] = 2$ and $x_2[0] = 3$.

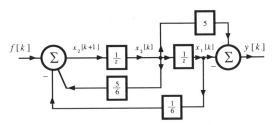

Fig. 10.12   System for Example 10.12.

The state equations are [see Eq. (10.89)]

$$\begin{bmatrix} x_1[k+1] \\ x_2[k+1] \end{bmatrix} = \begin{bmatrix} 0 & 1 \\ -\frac{1}{6} & \frac{5}{6} \end{bmatrix} \begin{bmatrix} x_1[k] \\ x_2[k] \end{bmatrix} + \begin{bmatrix} 0 \\ 1 \end{bmatrix} f \tag{10.96a}$$

and

$$y[k] = [\,-1 \quad 5\,] \begin{bmatrix} x_1[k] \\ x_2[k] \end{bmatrix} \tag{10.96b}$$

To find the solution [Eq. (10.94)], we must first determine $\mathbf{A}^k$. The characteristic equation of $\mathbf{A}$ is

$$|\lambda\mathbf{I} - \mathbf{A}| = \begin{vmatrix} \lambda & -1 \\ \frac{1}{6} & \lambda - \frac{5}{6} \end{vmatrix} = \lambda^2 - \frac{5}{6}\lambda + \frac{1}{6} = \left(\lambda - \frac{1}{3}\right)\left(\lambda - \frac{1}{2}\right) = 0$$

Hence $\lambda_1 = \frac{1}{3}$ and $\lambda_2 = \frac{1}{2}$ are the eigenvalues of $\mathbf{A}$ and [see Eq. (10.95)]

$$\mathbf{A}^k = \beta_0 \mathbf{I} + \beta_1 \mathbf{A}$$

where [see Eq. (B.99b)]

$$\begin{bmatrix} \beta_0 \\ \beta_1 \end{bmatrix} = \begin{bmatrix} 1 & \frac{1}{3} \\ 1 & \frac{1}{2} \end{bmatrix}^{-1} \begin{bmatrix} (\frac{1}{3})^k \\ (\frac{1}{2})^k \end{bmatrix} = \begin{bmatrix} 3 & -2 \\ -6 & 6 \end{bmatrix} \begin{bmatrix} 3^{-k} \\ 2^{-k} \end{bmatrix} = \begin{bmatrix} 3(3^{-k}) - 2(2^{-k}) \\ -6(3^{-k}) + 6(2^{-k}) \end{bmatrix}$$

and

$$\mathbf{A}^k = [3(3^{-k}) - 2(2^{-k})] \begin{bmatrix} 1 & 0 \\ 0 & 1 \end{bmatrix} + [-6(3^{-k}) + 6(2^{-k})] \begin{bmatrix} 0 & 1 \\ -\frac{1}{6} & \frac{5}{6} \end{bmatrix}$$

$$= \begin{bmatrix} 3(3^{-k}) - 2(2^{-k}) & -6(3^{-k}) + 6(2^{-k}) \\ 3^{-k} - 2^{-k} & -2(3^{-k}) + 3(2^{-k}) \end{bmatrix} \qquad (10.97)$$

We can now determine the state vector $\mathbf{x}[k]$ from Eq. (10.93b). Since we are interested in the output $y[k]$, we shall use Eq. (10.94b) directly. Note that

$$\mathbf{C}\mathbf{A}^k = [-1 \quad 5]\mathbf{A}^k = [2(3^{-k}) - 3(2^{-k}) \quad -4(3^{-k}) + 9(2^{-k})] \qquad (10.98)$$

and the zero-input response is $\mathbf{C}\mathbf{A}^k\mathbf{x}[0]$, with

$$\mathbf{x}[0] = \begin{bmatrix} 2 \\ 3 \end{bmatrix}$$

Hence the zero-input response is

$$\mathbf{C}\mathbf{A}^k\mathbf{x}[0] = -8(3^{-k}) + 21(2^{-k}) \qquad (10.99a)$$

The zero-state component is given by the convolution sum of $\mathbf{C}\mathbf{A}^{k-1}u[k-1]$ and $\mathbf{B}f[k]$. Using the shifting property of convolution sum [Eq. (3.60)], we can obtain this by finding the convolution sum of $\mathbf{C}\mathbf{A}^k u[k]$ and $\mathbf{B}f[k]$ and then replacing k by $k-1$ in the result. We use this procedure because the convolution sums are listed in Table 5.2 for functions of the type $f[k]u[k]$ rather than $f[k]u[k-1]$.

$$\mathbf{C}\mathbf{A}^k u[k] * \mathbf{B}f[k] = [2(3^{-k}) - 3(2^{-k}) \quad -4(3^{-k}) + 9(2^{-k})] * \begin{bmatrix} 0 \\ u[k] \end{bmatrix}$$

$$= -4(3^{-k}) * u[k] + 9(2^{-k}) * u[k]$$

Using Table 5.2 (Pair 2a), we obtain

$$\mathbf{C}\mathbf{A}^k u[k] * \mathbf{B}f[k] = -4 \left[\frac{1 - 3^{-(k+1)}}{1 - \frac{1}{3}} \right] u[k] + 9 \left[\frac{1 - 2^{-[k+1]}}{1 - \frac{1}{2}} \right] u[k]$$

$$= [12 + 6(3^{-(k+1)}) - 18(2^{-(k+1)})]u[k]$$

Now the desired (zero-state) response is obtained by replacing k by $k-1$. Hence

$$\mathbf{C}\mathbf{A}^k u[k] * \mathbf{B}f[k-1] = [12 + 6(3^{-k}) - 18(2^{-k})]u[k-1] \qquad (10.99b)$$

It follows that

$$y[k] = [-8(3^{-k}) + 21(2^{-k})u[k] + [12 + 6(3^{-k}) - 18(2^{-k})]u[k-1] \qquad (10.100a)$$

This is the correct answer. We can simplify this answer by observing that $12 + 6(3^{-k}) - 18(2^{-k}) = 0$ for $k = 0$. Hence $u[k-1]$ may be replaced by $u[k]$ in Eq. (10.99b), and

$$y[k] = [12 - 2(3^{-k}) + 3(2^{-k})]u[k] \qquad (10.100b)$$

∎

10.6-2   The $\mathcal{Z}$-Transform Solution

The z-transform of Eq. (10.91) is given by

$$z\mathbf{X}[z] - z\mathbf{x}[0] = \mathbf{A}\mathbf{X}[z] + \mathbf{B}\mathbf{F}[z]$$

Therefore

$$(z\mathbf{I} - \mathbf{A})\mathbf{X}[z]) = z\mathbf{x}[0] + \mathbf{B}\mathbf{F}[z]$$

and

$$\begin{aligned}
\mathbf{X}[z] &= (z\mathbf{I} - \mathbf{A})^{-1}z\mathbf{x}[0] + (z\mathbf{I} - \mathbf{A})^{-1}\mathbf{B}\mathbf{F}[z] \\
&= (\mathbf{I} - z^{-1}\mathbf{A})^{-1}\mathbf{x}[0] + (z\mathbf{I} - \mathbf{A})^{-1}\mathbf{B}\mathbf{F}[z] \qquad (10.101a)
\end{aligned}$$

Hence

$$\mathbf{x}[k] = \underbrace{\mathcal{Z}^{-1}[(\mathbf{I} - z^{-1}\mathbf{A})^{-1}]\mathbf{x}[0]}_{\text{zero-input component}} + \underbrace{\mathcal{Z}^{-1}[(z\mathbf{I} - \mathbf{A})^{-1}\mathbf{B}\mathbf{F}[z]]}_{\text{zero-state component}} \qquad (10.101b)$$

A comparison of Eq. (10.101b) with Eq. (10.93b) shows that

$$\mathbf{A}^k = \mathcal{Z}^{-1}[(\mathbf{I} - z^{-1}\mathbf{A})^{-1}] \qquad (10.102)$$

The output equation is given by

$$\begin{aligned}
\mathbf{Y}[z] &= \mathbf{C}\mathbf{X}[z] + \mathbf{D}\mathbf{F}[z] \\
&= \mathbf{C}[(\mathbf{I} - z^{-1}\mathbf{A})^{-1}\mathbf{x}[0] + (z\mathbf{I} - \mathbf{A})^{-1}\mathbf{B}\mathbf{F}[z]] + \mathbf{D}\mathbf{F}[z] \\
&= \mathbf{C}(\mathbf{I} - z^{-1}\mathbf{A})^{-1}\mathbf{x}[0] + [\mathbf{C}(z\mathbf{I} - \mathbf{A})^{-1}\mathbf{B} + \mathbf{D}]\mathbf{F}[z] \qquad (10.103a) \\
&= \underbrace{\mathbf{C}(\mathbf{I} - z^{-1}\mathbf{A})^{-1}\mathbf{x}[0]}_{\text{zero-input response}} + \underbrace{\mathbf{H}[z]\mathbf{F}[z]}_{\text{zero-state response}}
\end{aligned}$$

where

$$\mathbf{H}[z] = \mathbf{C}(\mathbf{I} - z^{-1}\mathbf{A})^{-1}\mathbf{B} + \mathbf{D} \qquad (10.103b)$$

Note that $\mathbf{H}[z]$ is the transfer function matrix of the system, and $H_{ij}[z]$, the ijth element of $\mathbf{H}[z]$, is the transfer function relating the output $y_i(k)$ to the input $f_j(k)$. If we define $\mathbf{h}[k]$ as

$$\mathbf{h}[k] = \mathcal{Z}^{-1}[\mathbf{H}[z]]$$

then $\mathbf{h}[k]$ represents the unit impulse function response matrix of the system. Thus $h_{ij}[k]$, the ijth element of $\mathbf{h}(k)$, represents the zero-state response $y_i(k)$ when the input $f_j(k) = \delta(k)$ and all other inputs are zero.

■ **Example 10.13**

Using the z-transform, find the response $y[k]$ for the system in Example 10.12. From Eq. (10.103a)

$$\mathbf{Y}[z] = [-1 \quad 5] \begin{bmatrix} 1 & -\frac{1}{z} \\ \frac{1}{6z} & 1 - \frac{5}{6z} \end{bmatrix}^{-1} \begin{bmatrix} 2 \\ 3 \end{bmatrix} + [-1 \quad 5] \begin{bmatrix} z & -1 \\ \frac{1}{6} & z - \frac{5}{6} \end{bmatrix}^{-1} \begin{bmatrix} 0 \\ \frac{z}{z-1} \end{bmatrix}$$

$$= [-1 \quad 5] \begin{bmatrix} \frac{z(6z-5)}{6z^2-5z+1} & \frac{6z}{6z^2-5z+1} \\ \frac{-z}{6z^2-5z+1} & \frac{6z^2}{6z^2-5z+1} \end{bmatrix} \begin{bmatrix} 2 \\ 3 \end{bmatrix} + [-1 \quad 5] \begin{bmatrix} \frac{z}{(z-1)(z^2-\frac{5}{6}z+\frac{1}{6})} \\ \frac{z^2}{(z-1)(z^2-\frac{5}{6}z+\frac{1}{6})} \end{bmatrix}$$

$$= \frac{13z^2 - 3z}{z^2 - \frac{5}{6}z + \frac{1}{6}} + \frac{(5z-1)z}{(z-1)(z^2 - \frac{5}{6}z + \frac{1}{6})}$$

$$= \frac{-8z}{z - \frac{1}{3}} + \frac{21z}{z - \frac{1}{2}} + \frac{12z}{z-1} + \frac{12z}{z-1} + \frac{6z}{z-\frac{1}{3}} - \frac{18z}{z-\frac{1}{2}}$$

Therefore

$$y[k] = \underbrace{[-8(3^{-k}) + 21(2^{-k})}_{y_x[k]} + \underbrace{12 + 6(3^{-k}) - 18(2^{-k})]u[k]}_{y_f[k]}$$ ■

Linear Transformation, Controllability, And Observability

The procedure for linear transformation is parallel to that in the continuous-time case (Sec. 10.4). If $\mathbf{w}$ is the transformed-state vector given by

$$\mathbf{w} = \mathbf{Px}$$

then

and

$$\mathbf{w}[k+1] = \mathbf{PAP}^{-1}\mathbf{w}[k] + \mathbf{PBf}$$

$$\mathbf{y}[k] = (\mathbf{CP}^{-1})\mathbf{w} + \mathbf{Df}$$

Controllability and observability may be investigated by diagonalizing the matrix.

10.7 SUMMARY

An nth-order system can be described in terms of n key variables—the state variables of the system. The state variables are not unique, but can be selected in a variety of ways. Every possible system output can be expressed as a linear combination of the state variables and the inputs. Therefore the state variables describe the entire system, not merely the relationship between certain input(s) and output(s). For this reason the state variable description is an internal description of the system. Such a description is therefore the most general system description, and it contains the information of the external descriptions, such as the impulse response and the transfer function. State-variable description can also be extended to time-varying parameter systems and nonlinear systems. An external system description may not describe a system completely.

The state equations of a system can be written directly from the knowledge of the system structure, from the system equations, or from the block diagram representation of the system. State equations consist of a set of n first-order differential equations and can be solved by time-domain or frequency-domain (transform) methods. Because a set of state variables is not unique, we can have a variety of state-space descriptions of the same system. It is possible to transform one given set of state variables into another by a linear transformation. Using such a transformation, we can see clearly which of the system states are controllable and which are observable.

REFERENCES

1. Kailath, Thomas, *Linear Systems*, Prentice-Hall, Englewood Cliffs, N.J., 1980.

2. Zadeh, L., and C. Desoer, *Linear System Theory*, McGraw-Hill, New York, 1963.

PROBLEMS

10.1-1 Convert each of the following second-order differential equations into a set of two first-order differential equations (state equations). State which of the sets represent nonlinear equations.

(a) $\ddot{y} + 10\dot{y} + 2y = f$
(b) $\ddot{y} + 2e^y \dot{y} + \log y = f$
(c) $\ddot{y} + \phi_1(y)\dot{y} + \phi_2(y)y = f$

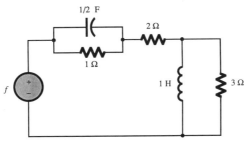

Fig. P10.2-1

10.2-1 Write the state equations for the RLC network in Fig. P10.2-1.

10.2-2 Write the state and output equations for the network in Fig. P10.2-2.

10.2-3 Write the state and output equations for the network in Fig. P10.2-3.

10.2-4 Write the state and output equations for the electrical network in Fig. P10.2-4.

10.2-5 Write the state and output equations for the network in Fig. P10.2-5.

10.2-6 Write the state and output equations of the network shown in Fig. P10.2-6.

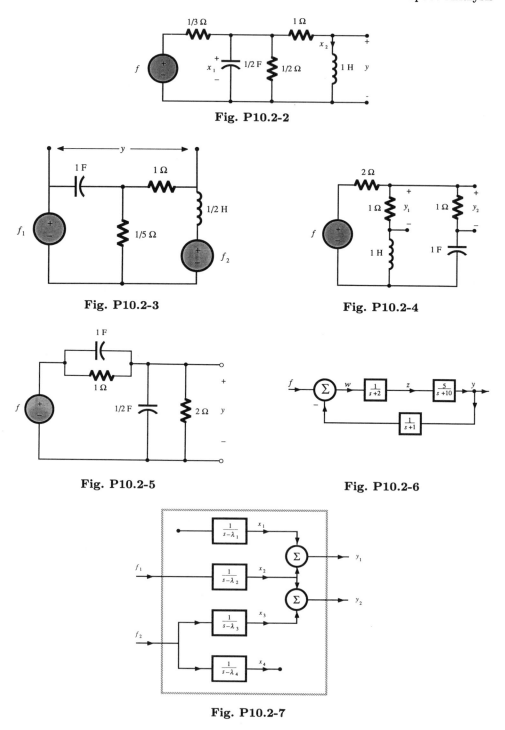

Fig. P10.2-2

Fig. P10.2-3

Fig. P10.2-4

Fig. P10.2-5

Fig. P10.2-6

Fig. P10.2-7

10.2-7 Write the state and output equations of the system shown in Fig. P10.2-7.

10.2-8 Write the different sets of state equations (two canonical, series, and parallel forms) and the output equation for a system having a transfer function

$$H(s) = \frac{3s + 10}{s^2 + 7s + 12}$$

10.2-9 Repeat Problem 10.2-8 if

(a) $H(s) = \dfrac{4s}{(s+1)(s+2)^2}$ (b) $H(s) = \dfrac{s^3 + 7s^2 + 12s}{(s+1)^3(s+2)}$

10.3-1 Find the state vector $\mathbf{x}(t)$ using the Laplace transform method if

$$\dot{\mathbf{x}} = \mathbf{A}\mathbf{x} + \mathbf{B}\mathbf{f}$$

where

$$\mathbf{A} = \begin{bmatrix} 0 & 2 \\ -1 & -3 \end{bmatrix} \quad \text{and} \quad \mathbf{B} = \begin{bmatrix} 0 \\ 1 \end{bmatrix} \quad \mathbf{x}(0) = \begin{bmatrix} 2 \\ 1 \end{bmatrix} \quad \text{and} \quad f(t) = 0$$

10.3-2 Repeat Prob. 10.3-1 if

$$\mathbf{A} = \begin{bmatrix} -5 & -6 \\ 1 & 0 \end{bmatrix} \quad \mathbf{B} = \begin{bmatrix} 1 \\ 0 \end{bmatrix} \quad \mathbf{x}(0) = \begin{bmatrix} 5 \\ 4 \end{bmatrix} \quad \text{and} \quad f(t) = \sin 100t$$

10.3-3 Repeat Prob. 10.3-1 if

$$\mathbf{A} = \begin{bmatrix} -2 & 0 \\ 1 & -1 \end{bmatrix} \quad \mathbf{B} = \begin{bmatrix} 1 \\ 0 \end{bmatrix} \quad \mathbf{x}(0) = \begin{bmatrix} 0 \\ -1 \end{bmatrix} \quad \text{and} \quad f(t) = u(t)$$

10.3-4 Repeat Prob. 10.3-1 if

$$\dot{\mathbf{x}} = \mathbf{A}\mathbf{x} + \mathbf{B}\mathbf{f}$$

where

$$\mathbf{A} = \begin{bmatrix} -1 & 1 \\ 0 & -2 \end{bmatrix} \quad \mathbf{B} = \begin{bmatrix} 1 & 1 \\ 0 & 1 \end{bmatrix} \quad \mathbf{f} = \begin{bmatrix} u(t) \\ \delta(t) \end{bmatrix} \quad \mathbf{x}(0) = \begin{bmatrix} 1 \\ 2 \end{bmatrix}$$

10.3-5 Using the Laplace transform method, find the response y if

$$\dot{\mathbf{x}} = \mathbf{A}\mathbf{x} + \mathbf{B}f(t)$$

$$y = \mathbf{C}\mathbf{x} + \mathbf{D}f(t)$$

where

$$\mathbf{A} = \begin{bmatrix} -3 & 1 \\ -2 & 0 \end{bmatrix} \quad \mathbf{B} = \begin{bmatrix} 1 \\ 0 \end{bmatrix} \quad \mathbf{C} = \begin{bmatrix} 0 & 1 \end{bmatrix} \quad \mathbf{D} = 0$$

and

$$f(t) = u(t) \qquad \mathbf{x}(0) = \begin{bmatrix} 2 \\ 0 \end{bmatrix}$$

10.3-6 Repeat Prob. 10.3-5 if

$$\mathbf{A} = \begin{bmatrix} -1 & 1 \\ -1 & -1 \end{bmatrix} \quad \mathbf{B} = \begin{bmatrix} 0 \\ 1 \end{bmatrix} \quad \mathbf{C} = \begin{bmatrix} 1 & 1 \end{bmatrix} \quad \mathbf{D} = 1$$

$$f(t) = u(t) \qquad \mathbf{x}(0) = \begin{bmatrix} 2 \\ 1 \end{bmatrix}$$

10.3-7   The transfer function $H(s)$ in Prob. 10.2-8 is realized as a cascade of $H_1(s)$ followed by $H_2(s)$ where

$$H_1(s) = \frac{1}{s+3} \quad \text{and} \quad H_2(s) = \frac{3s+10}{s+4}$$

Let the outputs of these subsystems be state variables x_1 and x_2, respectively. Write the state equations and the output equation for this system and verify that $H(s) = \mathbf{C}\boldsymbol{\phi}(s)\mathbf{B} + \mathbf{D}$.

10.3-8   Find the transfer-function matrix $\mathbf{H}(s)$ for the system in Prob. 10.3-5.

10.3-9   Find the transfer-function matrix $\mathbf{H}(s)$ for the system in Prob. 10.3-6.

10.3-10   Find the transfer-function matrix $\mathbf{H}(s)$ for the system

$$\dot{\mathbf{x}} = \mathbf{A}\mathbf{x} + \mathbf{B}\mathbf{f}$$

$$y = \mathbf{C}\mathbf{x} + \mathbf{D}\mathbf{f}$$

where

$$\mathbf{A} = \begin{bmatrix} 0 & 1 \\ -1 & -2 \end{bmatrix} \quad \mathbf{B} = \begin{bmatrix} 0 & 1 \\ 1 & 0 \end{bmatrix} \quad \mathbf{f} = \begin{bmatrix} f_1(t) \\ f_2(t) \end{bmatrix}$$

$$\mathbf{C} = \begin{bmatrix} 1 & 2 \\ 4 & 1 \\ 1 & 1 \end{bmatrix} \quad \mathbf{D} = \begin{bmatrix} 0 & 0 \\ 0 & 0 \\ 1 & 0 \end{bmatrix}$$

10.3-11   Repeat Prob. 10.3-1, using the time-domain method.

10.3-12   Repeat Prob. 10.3-2, using the time-domain method.

10.3-13   Repeat Prob. 10.3-3, using the time-domain method.

10.3-14   Repeat Prob. 10.3-4, using the time-domain method.

10.3-15   Repeat Prob. 10.3-5, using the time-domain method.

10.3-16   Repeat Prob. 10.3-6, using the time-domain method.

10.3-17   Find the unit impulse response matrix $\mathbf{h}(t)$ for the system in Prob. 10.3-7, using Eq. (10.65).

10.3-18   Find the unit impulse response matrix $\mathbf{h}(t)$ for the system in Prob. 10.3-6.

10.3-19   Find the unit impulse response matrix $\mathbf{h}(t)$ for the system in Prob. 10.3-10.

10.4-1   The state equations of a certain system are given as

$$\dot{x}_1 = x_2 + 2f$$

$$\dot{x}_2 = -x_1 - x_2 + f$$

Define a new state vector $\mathbf{w}$ such that

$$w_1 = x_2$$

$$w_2 = x_2 - x_1$$

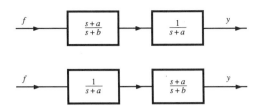

Fig. P10.5-1

Find the state equations of the system with **w** as the state vector. Determine the characteristic roots (eigenvalues) of the matrix **A** in the original and the transformed state equations.

10.4-2 The state equations of a certain system are

$$\dot{x}_1 = x_2$$

$$\dot{x}_2 = -2x_1 - 3x_2 + 2f$$

(a) Determine a new state vector **w** (in terms of vector **x**) such that the resulting state equations are in diagonalized form.

(b) If the output **y** is given by

$$\mathbf{y} = \mathbf{Cx} + \mathbf{Df}$$

where

$$\mathbf{C} = \begin{bmatrix} 1 & 1 \\ -1 & 2 \end{bmatrix} \quad \text{and} \quad \mathbf{D} = 0$$

determine the output **y** in terms of the new state vector **w**.

10.4-3 Given a system

$$\dot{\mathbf{x}} = \begin{bmatrix} 0 & 1 & 0 \\ 0 & 0 & 1 \\ 0 & -2 & -3 \end{bmatrix} \mathbf{x} + \begin{bmatrix} 0 \\ 0 \\ 1 \end{bmatrix} f$$

Determine a new state vector **w** for this system such that the state equations are diagonalized.

10.4-4 The state equations of a certain system are given in diagonalized form as

$$\dot{\mathbf{x}} = \begin{bmatrix} -1 & 0 & 0 \\ 0 & -3 & 0 \\ 0 & 0 & -2 \end{bmatrix} \mathbf{x} + \begin{bmatrix} 1 \\ 1 \\ 1 \end{bmatrix} f$$

The output equation is given by

$$y = \begin{bmatrix} 1 & 3 & 1 \end{bmatrix} \mathbf{x}$$

Determine the output y if

$$\mathbf{x}(0) = \begin{bmatrix} 1 \\ 2 \\ 1 \end{bmatrix} \quad \text{and} \quad f(t) = u(t)$$

10.5-1   Write the state equations for the systems shown in Fig. P10.5-1. Determine a new state vector **w** such that the resulting state equations are in diagonalized form. Write the output **y** in terms of **w**. Determine in each case whether the system is controllable and observable.

10.6-1   An LTI discrete-time system is specified by:

$$\mathbf{A} = \begin{bmatrix} 2 & 0 \\ 1 & 1 \end{bmatrix} \quad \mathbf{B} = \begin{bmatrix} 0 \\ 1 \end{bmatrix} \quad \mathbf{C} = \begin{bmatrix} 0 & 1 \end{bmatrix} \quad \text{and} \quad \mathbf{D} = [1]$$

and

$$\mathbf{x}(0) = \begin{bmatrix} 2 \\ 1 \end{bmatrix} \quad \text{and} \quad f[k] = u[k]$$

(a) Find the output $y[k]$, using the time-domain method.
(b) Find the output $y[k]$, using the frequency-domain method.

10.6-2   An LTI discrete-time system is specified by the difference equation

$$y[k+2] + y[k+1] + 0.16y[k] = f[k+1] + 0.32f[k]$$

(a) Show the two canonical, series and parallel realizations of this system.
(b) Write the state and the output equations from these realizations, using the output of each delay element as a state variable.

10.6-3   Repeat Problem 10.6-1 if

$$y[k+2] + y[k+1] - 6y[k] = 2f[k+2] + f[k+1]$$

COMPUTER PROBLEMS

C10-1   Use a computer to convert the following transfer functions to their equivalent state equation.

$$\textbf{(a)} \ \frac{s^2 + 1}{s^3 + 2s^2 + 3s + 6} \quad \textbf{(b)} \ \frac{s + 2}{s^4 + 4s^2 + 7s + 9} \quad \textbf{(c)} \ \frac{s + 2}{4s^2 + 7s + 9}$$

C10-2   Use a computer to convert the transfer functions in problem C10-1 to their equivalent observer form state equation.

C10-3   Redo computer Example C10.3 if the state equations for the system are

$$\begin{bmatrix} \dot{x}_1 \\ \dot{x}_2 \end{bmatrix} = \begin{bmatrix} 1 & 0 \\ -3 & -2 \end{bmatrix} \begin{bmatrix} x_1 \\ x_2 \end{bmatrix} + \begin{bmatrix} 1 \\ 0 \end{bmatrix} f(t)$$

C10-4   Using a computer, show that the eigenvalues of the system in Computer problem C10-3 are invariant under linear transformations.

C10-5   Using a computer, find the diagonalized form of the following state equation:

$$\begin{bmatrix} \dot{x}_1 \\ \dot{x}_2 \end{bmatrix} = \begin{bmatrix} 1 & 0 \\ -3 & -2 \end{bmatrix} \begin{bmatrix} x_1 \\ x_2 \end{bmatrix} + \begin{bmatrix} 1 \\ 0 \end{bmatrix} f(t)$$

C10-6   Use a computer to investigate the controllability and observability of the state equation in the computer problem C10-5.

Supplementary Reading

The following books are listed for readers desiring alternate treatment of topics covered in this book.

Cadzow, J. A., and H.F. Van Landingham, *Signals and Systems*, Prentice-Hall, Englewood Cliffs, N.J., 1985.

Chen, C.T., *Systems and Signal Analysis*, Saunders College Publishing, New York, 1989.

Gabel, R.A., and R.A. Roberts, *Signals and Linear Systems,* 3rd ed., Wiley, New York, 1987.

Glisson, T.H., *Introduction to System Analysis*, McGraw-Hill, New York, 1985.

Houts, R. C., *Signal Analysis in Linear Systems*, Saunders College Publishing, New York, 1991.

Jackson, L.B., *Signals, Systems, and Transforms*, Addison Wesley, Boston, Mass, 1991.

Kamen, E., *Introduction to Signals and System*, Macmillan, New York, 1987.

Kwakernaak, H., and R. Sivan, *Modern Signals and Systems*, Prentice-Hall, Englewood Cliffs, N.J., 1991.

Mayhan, Robert J., *Discrete-time and Continuous-time Linear Systems*, Addison Wesley, Reading, Mass, 1984.

McGillem, C.D., and G.R. Cooper, *Continuous and Discrete Signal and System Analysis*, 3rd ed., Holt, Rinehart and Winston, New York, 1991.

Neff, H.P., *Continuous and Discrete Linear systems*, Harper and Row, New York, 1984.

Oppenheim, A.V., and A.S. Willsky, *Signals and Systems*, Prentice-Hall, Englewood Cliffs, N.J., 1983.

Papoulis, A., *Circuits and Systems*, Holt, Rinehart, and Winston, New York, 1980.

Poularikas, A., and S. Seely, *Signals and Systems*, PWS-Kent, Boston, 1985.

1985.

Siebert, W.M., *Circuits, Signals, and Systems,* The MIT Press/McGraw-Hill, New York, 1986.

Soliman, S., and M. Srinath, *Continuous and Discrete Signal and Systems,* Prentice-Hall, New York, 1990.

Ziemer, R.E., W.H. Tranter, and D.R. Fannin, *Signals and Systems,* Macmillan, New York, 1983.

Index

651